Advances in Nucleic Acid Therapeutics

Titles in the Series:
1: Metabolism, Pharmacokinetics and Toxicity of Functional Groups
2: Emerging Drugs and Targets for Alzheimer's Disease; Volume 1
3: Emerging Drugs and Targets for Alzheimer's Disease; Volume 2
4: Accounts in Drug Discovery
5: New Frontiers in Chemical Biology
6: Animal Models for Neurodegenerative Disease
7: Neurodegeneration
8: G Protein-coupled Receptors
9: Pharmaceutical Process Development
10: Extracellular and Intracellular Signaling
11: New Synthetic Technologies in Medicinal Chemistry
12: New Horizons in Predictive Toxicology
13: Drug Design Strategies: Quantitative Approaches
14: Neglected Diseases and Drug Discovery
15: Biomedical Imaging
16: Pharmaceutical Salts and Cocrystals
17: Polyamine Drug Discovery
18: Proteinases as Drug Targets
19: Kinase Drug Discovery
20: Drug Design Strategies: Computational Techniques and Applications

21: Designing Multi-target Drugs
22: Nanostructured Biomaterials for Overcoming Biological Barriers
23: Physico-chemical and Computational Approaches to Drug Discovery
24: Biomarkers for Traumatic Brain Injury
25: Drug Discovery from Natural Products
26: Anti-inflammatory Drug Discovery
27: New Therapeutic Strategies for Type 2 Diabetes: Small Molecules
28: Drug Discovery for Psychiatric Disorders
29: Organic Chemistry of Drug Degradation
30: Computational Approaches to Nuclear Receptors
31: Traditional Chinese Medicine
32: Successful Strategies for the Discovery of Antiviral Drugs
33: Comprehensive Biomarker Discovery and Validation for Clinical Application
34: Emerging Drugs and Targets for Parkinson's Disease
35: Pain Therapeutics; Current and Future Treatment Paradigms
36: Biotherapeutics: Recent Developments using Chemical and Molecular Biology
37: Inhibitors of Molecular Chaperones as Therapeutic Agents
38: Orphan Drugs and Rare Diseases
39: Ion Channel Drug Discovery
40: Macrocycles in Drug Discovery
41: Human-based Systems for Translational Research
42: Venoms to Drugs: Venom as a Source for the Development of Human Therapeutics
43: Carbohydrates in Drug Design and Discovery
44: Drug Discovery for Schizophrenia
45: Cardiovascular and Metabolic Disease: Scientific Discoveries and New Therapies
46: Green Chemistry Strategies for Drug Discovery
47: Fragment-based Drug Discovery
48: Epigenetics for Drug Discovery
49: New Horizons in Predictive Drug Metabolism and Pharmacokinetics
50: Privileged Scaffolds in Medicinal Chemistry: Design, Synthesis, Evaluation
51: Nanomedicines: Design, Delivery and Detection
52: Synthetic Methods in Drug Discovery: Volume 1

53: Synthetic Methods in Drug Discovery: Volume 2
54: Drug Transporters: Role and Importance in ADME and Drug Development
55: Drug Transporters: Recent Advances and Emerging Technologies
56: Allosterism in Drug Discovery
57: Anti-aging Drugs: From Basic Research to Clinical Practice
58: Antibiotic Drug Discovery: New Targets and Molecular Entities
59: Peptide-based Drug Discovery: Challenges and New Therapeutics
60: Drug Discovery for Leishmaniasis
61: Biophysical Techniques in Drug Discovery
62: Acute Brain Impairment Through Stroke: Drug Discovery and Translational Research
63: Theranostics and Image Guided Drug Delivery
64: Pharmaceutical Formulation: The Science and Technology of Dosage Forms
65: Small-molecule Transcription Factor Inhibitors in Oncology
66: Therapies for Retinal Degeneration: Targeting Common Processes
67: Kinase Drug Discovery: Modern Approaches
68: Advances in Nucleic Acid Therapeutics

How to obtain future titles on publication:
A standing order plan is available for this series. A standing order will bring delivery of each new volume immediately on publication.

For further information please contact:
Book Sales Department, Royal Society of Chemistry, Thomas Graham House, Science Park, Milton Road, Cambridge, CB4 0WF, UK
Telephone: +44 (0)1223 420066, Fax: +44 (0)1223 420247,
Email: booksales@rsc.org
Visit our website at www.rsc.org/books

Advances in Nucleic Acid Therapeutics

Edited by

Sudhir Agrawal
Arnay Sciences LLC, Shrewsbury, MA, USA
Email: sagrawal@arnaysciences.com

and

Michael J. Gait
MRC Laboratory of Molecular Biology, Cambridge, UK
Email: mgait@mrc-lmb.cam.ac.uk

ROYAL SOCIETY OF CHEMISTRY

Drug Discovery Series No. 68

Print ISBN: 978-1-78801-209-6
PDF ISBN: 978-1-78801-571-4
EPUB ISBN: 978-1-78801-732-9
Print ISSN: 2041-3203
Electronic ISSN: 2041-3211

A catalogue record for this book is available from the British Library

The Royal Society of Chemistry is a charity, registered in England and Wales, Number 207890, and a company incorporated in England by Royal Charter (Registered No. RC000524), registered office: Burlington House, Piccadilly, London W1J 0BA, UK, Telephone: +44 (0) 20 7437 8656.

For further information see our web site at www.rsc.org

Printed in the United Kingdom by CPI Group (UK) Ltd, Croydon, CR0 4YY, UK

Foreword

The initial foundation for using either DNA or RNA (oligonucleotides) as therapeutic drugs was formulated by Zamecnik and Stephenson in a classic paper published several years ago (P. C. Zamecnik and M. L. Stephenson, *Proc. Natl. Acad. Sci. U. S. A.*, 1978, **75**, 280). Many of us considered that this concept would prove to be a new, refreshing approach for providing revolutionary drugs useful in the treatment of nondruggable diseases. Over time these expectations have been realized as several oligonucleotide drugs (Macugen, Fomivirsen, Mipomersen, Eteplirsen, Nusinersen, Inotersen, and Patisiran) are currently available for treating a diverse group of diseases. There are also a large number of additional oligonucleotides in various stages of drug development.

The path towards identifying drugs in the nucleic acids therapeutic arena has involved many unexpected revelations. Initially the focus was strictly on using antisense oligonucleotides. Over time, new research has opened possibilities for oligonucleotide drugs in such diverse fields as interfering RNA, microRNA, noncoding RNA, splicing modulation of RNA transcripts, targeting toxic repeats in RNA and DNA, investigating RNA and DNA aptamers and ribozymes for treating various disease states, and formulating synthetic agonists of Toll-like receptors. In this book the editors, through various chapters, provide a broad and complete perspective on the history of these fields. As each chapter unfolds, the reader discovers the logic behind why various DNA and RNA analogues were developed, how they were applied in clinical studies, and the limitations and advantages of these analogues. Also clearly presented are details on the unexpected side effects with some of these being very serious, such as the Toll receptor problem, the retention and targeting of oligonucleotides in tissues, and the variation of clinical studies with animal models.

Drug Discovery Series No. 68
Advances in Nucleic Acid Therapeutics
Edited by Sudhir Agrawal and Michael J. Gait

Published by the Royal Society of Chemistry, www.rsc.org

Drug development in the nucleic acids field continues at an increasing rate. Moreover so do the challenges that are important for successfully identifying an oligonucleotide therapeutic drug. This book provides an excellent road map for navigating all that has come before and also outlines for the reader a multitude of possible directions on how to proceed with further research. The editors have assembled an excellent set of authors who are experts in the science as presented in various chapters. This would be expected as both Michael Gait and Sudhir Agrawal are among the most highly respected and experienced research scientists in this field. Thus it is not at all surprising that the science is both current and focused on important concepts. I can enthusiastically recommend this book as both a reference and also as a guide for further research in the nucleic acids therapeutics arena.

Marv Caruthers
University of Colorado

Preface

Nucleic acids therapeutics are now recognized as the third major drug discovery platform, in addition to small molecules and proteins. In the past three decades, tremendous progress has been made towards the realization of the potential of nucleic acids therapeutics for the treatment of a broad range of diseases. In addition, multiple mechanisms of actions have been elucidated. Recently, several nucleic acid drugs have been approved for clinical use.

Chemical modifications of the three components of nucleic acids – heterocyclic bases, five-membered sugars, and internucleotide linkages – as well as the nucleotide sequences themselves are the key drivers for the creation of nucleic acid drugs. Rational combinations of these have provided drug-like properties. Further advances in the chemistry of nucleic acids and additional insights into their mechanisms of action have expanded their applications to include antisense targeting of mRNA, microRNA, non-coding RNA and splicing modulation, ribozymes, RNA interference (RNAi) and short interfering RNA (siRNA), gene editing, aptamers, and the modulation of immune responses. During the past ten years, since the excellent publication of Jens Kurreck's book (*Therapeutic Oligonucleotides*), progress in this field has been so rapid and broad that we felt it was appropriate now to document the key developments in the field in the form of a new book.

In Chapter 1 we provide a brief history of the development of nucleotide analogues, early experience in the use of modified antisense oligonucleotides (ONs) from preclinical studies to human trials, as well as the importance of nucleotide sequence and its implications in interaction with innate immune receptors. The next three chapters provide updates on applications of antisense technology. In Chapter 2, David Corey and Zhongtian Liu discuss various mechanisms of action of antisense ONs. In Chapter 3, Eric Swayze and Punit Seth describe the medicinal chemistry of RNase H-activating antisense

Drug Discovery Series No. 68
Advances in Nucleic Acid Therapeutics
Edited by Sudhir Agrawal and Michael J. Gait

Published by the Royal Society of Chemistry, www.rsc.org

ONs and in Chapter 4, Richard Geary, Brenda Baker and Brett Monia provide an update on and experience of the application of antisense ONs in clinical development.

During the development of antisense technology, it was realized that subcutaneous delivery of antisense ONs led to activation of host immune responses. Initially, this was thought to be a side effect but soon the discovery of the family of Toll-like receptors (TLRs) led to an understanding of immune activation triggered by receptor-mediated engagement. Tremendous progress has been made in translating these observations into a novel therapeutic platform. In Chapter 13, Shin-Ichiroh Saitoh and Kensuke Miyake provide a detailed background on immune receptors that are known to recognize nucleic acid sequence, patterns, and motifs. In Chapters 5 and 14, one of us (S.A.) and Ekambar Kandimalla have discussed the chemistry of novel nucleic acid compounds and how they modulate receptor-mediated immune responses, along with their therapeutic applications, including clinical proof of concept trials.

A more recent therapeutic application of antisense involves splicing modulation to affect the translation of the targeted pre-mRNA. In Chapter 6, Elena Daoutsali and Annemieke Aartsma-Rus provide an up to date survey on this subject through a variety of examples. Similarly, applications of antisense have been expanded to targeting toxic RNA repeats (Chapter 7 by Derick Wansink and colleagues), microRNA (Chapter 8 by Anna Malinowksi and Jonathan Hall) and long non-coding RNA (Chapter 9 by Claes Wahlestedt and colleagues).

In parallel, significant advances have also been made in RNAi technology for therapeutics. In Chapter 10, Anastasia Khvorova and colleagues discuss in detail the challenges of delivery of RNAi-based therapeutics and how these obstacles have been addressed. In Chapter 11, Muthiah Manoharan and Kallanthottathil Rajeev describe the clinical development of siRNA candidates targeted to liver. In Chapter 12 Jiehua Zhou and John Rossi describe the application of RNAi for treatment of HIV infection. Therapeutic application of ribozyme technology had shown early promise, but has now been found to have significant limitations. In Chapter 18, Darko Balke and Sabine Müller describe novel ribozyme constructs in the search for potential therapeutic applications.

In the past few years, we have also seen explosive growth in the development of gene editing using nucleic acids towards therapeutics. In Chapter 17, Carine Giovannangeli and colleagues provide details on this subject. Synthetic nucleic acids have been studied as aptamers to target proteins and other cellular targets and their clinical evaluation is reviewed in Chapter 15 by Paloma Giangrande and colleagues and in Chapter 16 by Gerald Zon. Through understanding the various mechanisms of actions of nucleic acids, extensive experience has been gained on their safety and pharmacokinetics, both in preclinical and in clinical use. In Chapter 20, Cathaline den Besten and Patrik Andersson discuss this topic in detail and provide their analysis.

To maintain successes in the field, significant advances have also been made in manufacturing and quality control, discussed in Chapter 19 by Yogesh Sanghvi.

We are immensely grateful to all these co-authors for their outstanding contributions that provide a detailed story of their respective subjects along with a current bibliography. We are also grateful to the editorial team at the Royal Society of Chemistry (Katie Morrey and Drew Gwilliams, Rowan Frame and Robin Driscoll) for their timely publishing and encouragement and members of the Royal Society of Chemistry staff for their assistance, as well as the many members of the Oligonucleotide Therapeutic Society, the premier professional society in the field, who have contributed to this book, which we hope will become a manual for the state of the art.

In this relatively young field of nucleic acid therapeutics, the use of their sequences to target drugs very precisely in cells and *in vivo* and the development of nucleic acids chemistry have been paramount and resulted in a substantial broadening of their applications. Rapidly developed and newly approved drugs are now available for the treatment of some rare diseases and other more prevalent diseases are surely following. Despite some setbacks, the list of RNA targets and approved drugs is expanding quickly. We are excited at the future prospects for this field.

Sudhir Agrawal and Michael J. Gait

Contents

Chapter 1 **History and Development of Nucleotide Analogues in Nucleic Acids Drugs** 1
Sudhir Agrawal and Michael J. Gait

1.1 Introduction 1
1.2 The Antisense Concept 2
1.3 Developments in Oligonucleotide Synthesis 2
1.4 Choices for Antisense Oligodeoxynucleotide Modifications 3
1.4.1 Backbone Modifications 3
1.4.2 Heterocyclic Bases 9
1.4.3 Sugar Modifications 10
1.5 Gapmers Using Combinations of Modified Oligodeoxy and/or Oligoribonucleotides 12
1.6 Antisense Conjugates 13
1.7 The Role of Innate Immune Receptors in Nucleic Acid Therapeutics 14
1.8 Future Directions 15
Acknowledgements 16
References 17

Chapter 2 **Mechanisms of Antisense Oligonucleotides** 22
Zhongtian Liu and David R. Corey

2.1 Introduction 22
2.2 RNase H and ASO Action 23
2.3 ASOs and Regulation of Splicing 25

Drug Discovery Series No. 68
Advances in Nucleic Acid Therapeutics
Edited by Sudhir Agrawal and Michael J. Gait

Published by the Royal Society of Chemistry, www.rsc.org

2.4 ASOs and Activation of Frataxin, a Case Study for an Emerging Mechanism 27
2.5 Summary 29
Acknowledgements 29
References 29

Chapter 3 The Medicinal Chemistry of RNase H-activating Antisense Oligonucleotides 32
Punit P. Seth and Eric E. Swayze

3.1 Introduction to Gapmers 32
3.2 Human RNase H1 33
3.2.1 Biochemistry of Human RNase H1 33
3.2.2 Structural Biology of Human RNase H1 34
3.3 Structure–Activity Relationships of Gap Modifications 36
3.3.1 Phosphorothioate (PS) DNA 38
3.3.2 Chiral PS DNA 39
3.3.3 Methyl Phosphonates, Phosphoramidates, Phosphotriesters and Boranophosphonate DNA 40
3.3.4 2′-Fluoro Arabino Nucleic Acids (FANA) 40
3.3.5 DNA-like Modifications 41
3.4 SAR of Wing Modifications 41
3.4.1 2′-*O*-Methoxyethyl RNA (MOE) 42
3.4.2 Locked Nucleic Acids (LNA) and Constrained Ethyl 2′-4′-Bridged Nucleic Acids (cEt) 42
3.4.3 α-L-LNA and Related Analogs 43
3.4.4 F-HNA and F-CeNA 43
3.4.5 Tricyclo DNA (tcDNA) 44
3.4.6 Phosphorodiamidate Linked Morpholinos (PMOs) 44
3.5 Design of Gapmer ASOs 45
3.5.1 MOE Gapmer ASOs 45
3.5.2 LNA Gapmers 46
3.5.3 cEt Gapmers 47
3.5.4 ASO Gapmer Duplexes 47
3.6 Control of Protein Binding 47
3.6.1 Interaction of Gapmer ASOs with Plasma Proteins 48
3.6.2 Interaction of Gapmer-ASOs with Cell-surface Proteins 48
3.6.3 Targeting Cell-surface Proteins for Cell-specific Delivery of Gapmers 49

3.6.4 Avoiding Interactions with TLR Receptors to Avoid Immune-stimulatory Toxicities 50
3.6.5 Optimizing Intracellular Distribution 51
3.7 Conclusions 51
References 52

Chapter 4 Antisense Technology: Liver Targeting and Beyond for Drug Discovery 62
Richard S. Geary, Brenda F. Baker and Brett P. Monia

4.1 Introduction 62
4.2 Liver Targeting 64
4.2.1 The Beginning 64
4.2.2 Broadening the Liver-targeting Pipeline 66
4.3 Innovations in Liver Targeting 67
4.4 Beyond the Liver 70
4.5 Conclusions 72
References 73

Chapter 5 Oligonucleotide-based Toll-like Receptor Antagonists and Therapeutic Applications 80
Ekambar R. Kandimalla and Sudhir Agrawal

5.1 Introduction 80
5.2 Oligonucleotide-based TLR Antagonists 82
5.2.1 Structure–Activity Relationship Studies 83
5.2.2 TLR Antagonists 83
5.2.3 Inhibitory Activity of TLR Antagonists 85
5.2.4 Inhibitory Activity of Clinical Candidates 85
5.3 Studies of TLR Antagonists in Disease Models 87
5.3.1 Psoriasis 87
5.3.2 Systemic Lupus Erythematosus (SLE) 88
5.3.3 Rheumatoid Arthritis (RA) 89
5.3.4 Duchenne Muscular Dystrophy (DMD) 90
5.3.5 MyD88 L265P-positive B Cell Lymphoma (BCL) 90
5.3.6 Restenosis and Atherosclerosis 91
5.3.7 Inflammatory Bowel Disease (IBD) 92
5.3.8 HIV-1 92
5.4 Clinical Development of Lead TLR Antagonist Candidates 93
5.4.1 Moderate-to-severe Plaque Psoriasis 93
5.4.2 Waldenström's Macroglobulinemia 94
5.5 Conclusions 96
Acknowledgements 96
References 96

Chapter 6 Splicing Modulation for Therapeutics **103**
Elena Daoutsali and Annemieke Aartsma-Rus

6.1 Introduction 103
6.1.1 RNA Splicing 103
6.1.2 Constitutive Splicing 106
6.1.3 Alternative Splicing 106
6.1.4 Cryptic Splicing 108
6.2 Therapeutic Exon Skipping Options 108
6.2.1 Antisense Oligonucleotides (ASOs) and Chemical Modifications 108
6.2.2 Restoring Cryptic Splicing 109
6.2.3 Reading Frame Restoration 111
6.2.4 Exon Inclusion 116
6.2.5 Generating Less Protein or Non-toxic Protein 118
6.3 Future Perspectives, Towards Additional Approved Splice-modulating ASOs 120
6.3.1 The Challenge of Personalized Medicine Development 120
6.3.2 ASO Delivery 121
6.3.3 Future Perspective 121
References 122

Chapter 7 Targeting Toxic Repeats **126**
M. Leontien van der Bent, Remco T. P. van Cruchten and Derick G. Wansink

7.1 Introduction 126
7.2 Expanded Repetitive Sequences and Human Disease 127
7.2.1 Repeat Instability 127
7.2.2 Molecular Mechanisms of Disease 128
7.3 Why are Expanded Repeats so Special as Therapeutic Targets? 130
7.3.1 Structures Formed by Expanded Repeats 131
7.3.2 Structures of Repeat-Expanded Transcripts 133
7.3.3 Structural Implications for Therapeutic Targeting 134
7.4 Therapeutics to Target Expanded Repeats 135
7.4.1 Therapeutic Strategies 135
7.4.2 Small Molecules Targeting Repeat Structures 136

7.4.3 Antisense Strategies 138
7.4.4 CRISPR/Cas9, TALEN, ZFN and Other Protein Effectors 141
7.5 Challenges for Multisystemic Repeat Diseases 142
Abbreviations 143
Acknowledgements 144
References 144

Chapter 8 Research and Development of Oligonucleotides Targeting MicroRNAs (miRNAs) 151
A. L. Malinowska and J. Hall

8.1 Introduction: MicroRNA Biogenesis and Functions 151
8.2 miRNAs as Targets for Drugs 154
8.3 AntimiR Oligonucleotides as Drugs 154
8.3.1 The Development of AntimiR Medicinal Chemistry 154
8.3.2 New Insights into Mechanisms of Oligonucleotide-based miRNA Targeting 162
8.3.3 New Chemistries and Alternative Approaches for Targeting miRNAs 165
8.4 AntimiRs in Clinical Trials 167
8.4.1 AntimiRs Targeting miR-122 for Treatment of HCV 167
8.4.2 AntimiR-targeting of Other miRNAs in Clinical Studies 169
8.5 Conclusions 170
Acknowledgement 171
References 171

Chapter 9 Oligonucleotide Targeting of Long Non-coding RNAs 181
I. Blokhin, O. Khorkova, J. Hsiao and C. Wahlestedt

9.1 Introduction 181
9.2 History of lncRNAs 182
9.3 Biology and Functions of lncRNA 185
9.3.1 lncRNAs as Regulators of Transcription 186
9.3.2 lncRNA as Regulators of Post-transcriptional Processing 187
9.3.3 lncRNA as Regulators of Translation 189
9.4 Classification of lncRNA 190
9.4.1 Functional Classification 191
9.4.2 Genomic Classification 191

9.5 Targeting of Long Non-coding RNA by Oligonucleotides 192
9.5.1 Antisense Oligonucleotides 192
9.5.2 siRNAs 192
9.5.3 CRISPR and Other Approaches 193
9.6 Therapeutic Applications 194
9.6.1 Neurology and Psychiatry 194
9.6.2 Oncology 198
9.6.3 Cardiology 201
9.6.4 Gastroenterology 202
9.7 Perspectives 202
References 203

Chapter 10 Conjugate-mediated Delivery of RNAi-based Therapeutics: Enhancing Pharmacokinetics–Pharmacodynamics Relationships of Medicinal Oligonucleotides 206
Bruno M. D. C. Godinho, Andrew H. Coles and Anastasia Khvorova

10.1 Introduction 206
10.2 Chemical Stabilization as a Prerequisite for Conjugate-mediated Delivery of siRNAs: Effects on Clearance, Distribution and Sustained Gene Silencing 207
10.3 Modulating Biodistribution of Therapeutic Oligonucleotides Using Conjugated Modalities: Targeted *versus* Broad Functional Delivery 213
10.3.1 Broad Functional Delivery of Conjugated siRNAs 213
10.3.2 Targeted Delivery of Conjugated siRNAs 216
10.4 Productive Delivery of Therapeutic Oligonucleotides: Overcoming the Endosomal Barrier 218
10.5 The Effects of the Route of Administration: Local *versus* Systemic Delivery 219
10.5.1 Local Delivery of Conjugated siRNAs 219
10.5.2 Systemic Delivery of Conjugated siRNAs 221
10.6 Enhancing PK Properties of Conjugated siRNAs: Reducing Clearance Kinetics and Accelerating Target Tissue Uptake 223
10.7 Conjugation Chemistry for RNAi-based Therapeutics: Future Perspectives 224
Acknowledgements 226
References 226

Chapter 11 Liver-targeted RNAi Therapeutics: Principles and Applications 233
Kallanthottathil G. Rajeev and Muthiah Manoharan

11.1 Introduction 233
11.2 The Role of Chemistry 234
11.3 Liver-specific Delivery of siRNA 237
11.3.1 Ionizable Lipid Nanoparticles (iLNPs) 238
11.3.2 Lipid-conjugated siRNA provided Proof of Concept for RNAi Therapeutics 239
11.3.3 Discovery of GalNAc Conjugates 240
11.4 Clinical Candidates 240
11.4.1 ONPATTRO™ (Patisiran) 240
11.4.2 Inclisiran 254
11.4.3 Givosiran 255
11.4.4 Fitusiran 256
11.4.5 TTRsc02 256
11.4.6 Revusiran 256
11.5 Conclusions and Outlook 257
References 257

Chapter 12 Advances and Challenges of RNAi-Based Anti-HIV Therapeutics 266
J. Zhou and J. J. Rossi

12.1 Introduction 266
12.2 Potential Targets for Anti-HIV-1 RNAi Therapeutics 268
12.2.1 Targeting the HIV-1 Viral Genome 268
12.2.2 Targeting Host Factors 270
12.3 Challenges in Obtaining Effective Anti-HIV-1 RNAi Activity 272
12.3.1 Key Barriers to *In Vivo* RNAi Efficacy 272
12.3.2 The Need for Rational Design and for Chemical Modifications 274
12.3.3 The Need for Combinatorial RNAi 275
12.4 Recent Progress and Clinical Development of Anti-HIV-1 RNAi 276
12.4.1 *In Vivo* Delivery of Anti-HIV-1 RNAi Effectors 276
12.4.2 *Ex vivo* Delivery of Anti-HIV-1 shRNAs 281
12.4.3 Clinical Development of Anti-HIV-1 RNAi 282
12.5 Conclusions and Perspective 283
Conflict of Interest Declaration 284
Acknowledgements 285
References 285

Chapter 13 Nucleic Acid Innate Immune Receptors **292**
Shin-Ichiroh Saitoh and Kensuke Miyake

13.1 Introduction 292
13.2 Toll-like Receptors 293
13.2.1 TLR3 Recognizes dsRNA 293
13.2.2 TLR7 and TLR8 Recognize ssRNA and Guanosine or Uridine 293
13.2.3 TLR9 Recognizes CpG-DNA 294
13.2.4 Chaperones Regulate the Maturation of NA-sensing TLRs 295
13.2.5 Unc93B1 Regulates the Balance of TLR7 and TLR9 Responses 296
13.2.6 Proteolytic Cleavage of NA-sensing TLRs is Essential for Their Function 296
13.2.7 Trafficking of TLR7 and TLR9 is Essential for Type I Interferon Production in pDCs 297
13.3 Nucleic Acids Sensing in the Cytoplasm 298
13.3.1 Cytosolic DNA Sensors Recognize dsDNA 298
13.3.2 RIG-I and MDA5 Recognize dsRNA and Activate MAVS to Induce Immune Responses 299
13.4 Conclusions 299
References 300

Chapter 14 Synthetic Agonists of Toll-like Receptors and Therapeutic Applications **306**
Sudhir Agrawal and Ekambar R. Kandimalla

14.1 Introduction 306
14.1.1 RIG-I-like Receptors 307
14.1.2 AIM2-like Receptors (ALRs) 308
14.1.3 NOD-like Receptors, NLRP3 Inflammasome 308
14.1.4 Cyclic GMP–AMP Synthase (cGAS) and the STING Pathway 309
14.1.5 Toll-like Receptors (TLR) 309
14.2 Agonists of TLR3 310
14.2.1 Synthetic Agonists of TLR3 310
14.3 Agonists of TLR 7 and TLR 8 311
14.3.1 Synthetic Agonists of TLR7 and TLR8 312
14.3.2 Preclinical Studies of Agonists of TLR7 and TLR8 314
14.4 Agonists of TLR9 315
14.4.1 Synthetic Agonists of TLR9 316
14.5 Therapeutic Applications of Synthetic Agonists of TLR9 322
14.5.1 Preclinical Studies in Cancer 322

14.5.2 Treatment for Asthma and Allergies 324
14.5.3 Use as Vaccine Adjuvants 325
14.6 Clinical Development of Synthetic Agonists of TLR9 325
14.6.1 Clinical Trials in Hepatitis C Patients 325
14.6.2 Clinical Trials in Cancer 326
14.7 Conclusions 327
References 328

Chapter 15 Prostate-specific Membrane Antigen (PSMA) Aptamers for Prostate Cancer Imaging and Therapy 339
Shambhavi Shubham, Li-Hsien Lin, Ofonime Udofot, Sven Krupse and Paloma H. Giangrande

15.1 Introduction 339
15.1.1 Aptamers and SELEX 339
15.1.2 DNA *vs.* RNA Aptamers 342
15.2 Prostate Specific Membrane Antigen (PSMA) Aptamers 343
15.2.1 PSMA 343
15.2.2 DNA and RNA PSMA Aptamers 345
15.3 PSMA Aptamers Applications 348
15.3.1 Imaging and Diagnostic Applications 348
15.3.2 PSMA Aptamers as Therapeutic Inhibitors 351
15.3.3 Targeted Delivery Applications 351
15.4 Conclusions 359
15.5 Future Perspectives 359
Acknowledgements 361
References 361

Chapter 16 Aptamers and Clinical Applications 367
G. Zon

16.1 Introduction 367
16.2 Recent Preclinical Studies of Aptamer Drugs 369
16.2.1 Aptamer Structures 369
16.2.2 Non-ocular Diseases 376
16.3 Recent Studies of Aptamer-based Targeting of Drugs 381
16.4 Clinical Studies of Aptamer Drugs Registered in ClinicalTrials.gov 384
16.4.1 Completed Clinical Studies 385
16.4.2 Terminated or Withdrawn Studies 390
16.4.3 Active Studies 391
16.5 Conclusions and Prospects 392
References 394

Chapter 17 CRISPR-based Technologies for Genome Engineering: Properties, Current Improvements and Applications in Medicine 400
Sylvain Geny, Elaheh Sadat Hosseini, Jean-Paul Concordet and Carine Giovannangeli

17.1 Introduction 400
17.2 Sequence-specific CRISPR Nucleases and Improved Variants 401
17.2.1 Cas 9 from *Streptococcus pyogenes*, Orthologues and Variants 401
17.2.2 Expanding Targeted Functions with "CRISPR Fusions" 404
17.2.3 Specificity of CRISPR Systems 405
17.3 Genome Editing Mechanisms and Current Improvements 408
17.3.1 Principles: Genome Editing Takes Place During Repair of DSB Breaks Induced by Cas9 408
17.3.2 Improvement of HDR-based CRISPR Strategies for Programmable Genome Modification 410
17.4 Epigenome Editing and Base Editing with the CRISPR Systems 414
17.4.1 Modulation of Transcription, CRISPRa and CRISPRi Systems 414
17.4.2 Modulation of Chromatin Status 416
17.4.3 Base Editors 417
17.5 Applications in Medicine and Challenges 419
17.5.1 Applications in Biomedical Research 419
17.5.2 The Delivery Challenge 420
17.5.3 The Genome Editing Precision Challenge 422
17.5.4 Clinical Trials Based on Genome Editing 422
17.6 Conclusions 423
References 424

Chapter 18 Therapeutic Potential of Ribozymes 434
Darko Balke and Sabine Müller

18.1 Introduction 434
18.2 *Trans*-cleaving Ribozymes 435
18.3 Ribozyme-mediated Genetic Repair 439
18.3.1 Spliceosome-mediated RNA *Trans*-splicing (SMaRT) 440
18.3.2 Group I Intron Ribozyme-mediated *Trans*-splicing 441

18.3.3 Twin Ribozyme-mediated RNA Repair 443
18.3.4 Correction of Genetic Disorders by Retro-homing Group II Introns 445
18.4 Conclusions 445
References 447

Chapter 19 Large-scale Automated Synthesis of Therapeutic Oligonucleotides: A Status Update 453
Yogesh S. Sanghvi

19.1 Introduction 453
19.2 Chemical Modifications in Clinical and Commercial Products 454
19.2.1 First-generation Backbone Modifications – the Phosphorothioate (PS) Internucleotide Linkage 454
19.2.2 Second-generation Sugar Modifications 455
19.3 The Oligonucleotide Manufacturing Process 457
19.3.1 Use of an Automated Synthesizer 457
19.3.2 Starting Materials 458
19.3.3 Reagent-related Impurities 462
19.3.4 The Four-step Synthesis Cycle 464
19.3.5 Cleavage and Deprotection 465
19.3.6 The Purification Process 466
19.3.7 The Lyophilization Process 468
19.4 Analytical Protocols 468
19.5 Synthesis Yield and Product Purity 470
19.6 Conclusions and Future Outlook 470
References 471

Chapter 20 Preclinical and Clinical Drug-metabolism, Pharmacokinetics and Safety of Therapeutic Oligonucleotides 474
Patrik Andersson and Cathaline den Besten

20.1 Introduction 474
20.2 Oligonucleotide Chemistries and Mode of Action (MOA) 475
20.2.1 Chemistry and Design Considerations of Therapeutic ONDs 475
20.2.2 Delivery Approaches 478
20.3 Distribution, Metabolism and Pharmacokinetics (DMPK) 479
20.3.1 DMPK Properties of ONDs 479
20.3.2 Delivery Strategies 485
20.4 Class Profile of Toxicity 488

20.5 Hybridization-dependent Toxicities 490
20.5.1 On-target Safety and Exaggerated Pharmacology 490
20.5.2 Off-target Pharmacology 491
20.6 Hybridization-independent Toxicities 493
20.6.1 Effects Related to Transient Protein Binding 494
20.6.2 Immune-mediated Effects 496
20.6.3 Toxicity in High-exposure Organs 505
20.7 Getting to the Clinic 511
20.7.1 Regulatory Considerations for Preclinical Development 511
20.7.2 Precision Medicine and Opportunity for Accelerated Timelines 515
20.8 Conclusions 515
Acknowledgements 516
References 517

Subject Index **532**

CHAPTER 1

History and Development of Nucleotide Analogues in Nucleic Acids Drugs

SUDHIR AGRAWAL*[a] AND MICHAEL J. GAIT*[b]

[a]Arnay Sciences LLC, Shrewsbury, MA 01545, USA; [b]Medical Research Council, Laboratory of Molecular Biology, Francis Crick Avenue, Cambridge CB2 0QH, UK
*E-mail: mgait@mrc-lmb.cam.ac.uk, sagrawal@arnaysciences.com

1.1 Introduction

Forty years ago, Zamecnik and Stephenson proposed the therapeutic use of antisense oligonucleotides on the basis of their finding that Rous sarcoma virus (RSV) replication could be inhibited by a synthetic oligonucleotide complimentary to the RSV genome.[1] This concept opened up a new approach to drug discovery, namely an oligonucleotide binding sequence-specifically *via* Watson–Crick base-pairing to a complementary target RNA.

Since then, continuous progress has been made towards realizing the potential of this novel scientific approach and this has led recently to the approval of five antisense drugs. While the underlying concept of antisense is very simple, a rigorous understanding of the chemistry of nucleic acids had to be developed for its use in humans. In this chapter we discuss the history of this chemistry of oligonucleotides in antisense and the lessons

Drug Discovery Series No. 68
Advances in Nucleic Acid Therapeutics
Edited by Sudhir Agrawal and Michael J. Gait

Published by the Royal Society of Chemistry, www.rsc.org

learned from preclinical studies and clinical trials that have guided the development in conferring drug-like properties.

In parallel to the development of antisense (see Chapters 2–4), the application of synthetic oligonucleotides as therapeutic agents has evolved into broad applications involving multiple modalities. These applications include ribozymes (see Chapter 18), small interfering RNA (siRNA, see Chapters 10, 11 and 12), microRNA (see Chapter 8), aptamers (see Chapters 15 and 16), non-coding RNA (see Chapter 9), splicing modulation (Chapter 6), targeting toxic repeats (Chapter 7), gene editing (Chapter 17), and immune modulations (see Chapters 5, 13 and 14).

The common feature of these applications is that drug candidates are composed of natural nucleosides or nucleoside analogues linked *via* phosphodiester or modified linkages.

1.2 The Antisense Concept

In 1976, RSV was the only purified virus for which a sufficient quantity was available for potential sequencing. Maxam and Gilbert sequenced this RNA virus and noted that both ends of the linear viral genome bore the same primary sequence and were in the same polarity. It occurred to Zamecnik that the new piece of DNA synthesized by reverse transcription at the 5′-end of this retrovirus might circularize and hybridize with the 3′-end. Thus he considered the possibility of inhibiting viral replication by adding a piece of synthetic DNA to the replication system to block the circularization step by hybridizing specifically with the 3′-end of the viral RNA in a competitive way.

This experiment led to startling observations, including the inhibition of new virus particles and the prevention of transformation of chick fibroblasts into sarcoma cells. In a cell-free system, translation of the Rous sarcoma viral message was also dramatically impaired. These observations were the first to show proof of the antisense concept.[1,2]

Not much further progress was made in the field up to 1985, primarily for three reasons. First, there was still widespread disbelief that oligonucleotides could enter eukaryotic cells. Second, there was very little DNA (or RNA) genomic sequence available for targeting by antisense, and third, efficient automated methodologies to synthesize oligonucleotides in sufficient quantities were only just beginning to become established.

1.3 Developments in Oligonucleotide Synthesis

Although the principle of solid-phase oligonucleotide synthesis was first introduced by Letsinger and Mahadevan in 1965,[3] development of more efficient methods of oligodeoxynucleotide (ODN) synthesis on solid support took place from 1975 in the Gait laboratory by the phosphodiester chemistry[4] and from 1979 by the phosphotriester method in the Itakura laboratory[5] and the Gait laboratory.[6,7] These methods were superseded by the outstanding phosphoramidite chemistry of Caruthers and colleagues,[8] which was automated by Applied Biosystems and other companies. This transformed the ability of non-chemists to obtain ODNs for biological purposes.

1.4 Choices for Antisense Oligodeoxynucleotide Modifications

In the mid-1980s the discovery of human immunodeficiency virus 1 (HIV-1) and the availability of its RNA sequence led Zamecnik to employ an antisense approach in attempts to inhibit HIV replication.[2] Antisense ODNs with phosphodiester linkages (PO-ASO, Figure 1.1) complementary to various regions of HIV-1 mRNA were synthesized using automated synthesizers. In HIV-1-infected cells, these PO-ASOs inhibited HIV-1 replication and suppressed expression of HIV-1 related markers.[9] Such experiments were carried out using primary human cells and cellular uptake of ODNs was not a limiting factor. In these studies a control PO-ASO showed minimal inhibition of HIV-1 replication, providing evidence of sequence-specific antisense activity. These studies re-established the potential application of antisense as a therapeutic approach.

1.4.1 Backbone Modifications

It was realized that the use of PO-ASOs would have limited therapeutic application, since these ASOs would be degraded rapidly in biological fluids. Soon, the focus of research shifted into identifying novel analogues of oligonucleotides that would have increased stability against nucleases and maintain the sequence-specific hybridization for use in antisense studies. The mechanisms of ASO activities (RNase H and steric blocking) are reviewed in Chapter 2. The characteristics required for a good ASO oligonucleotide are summarized in Figure 1.2.

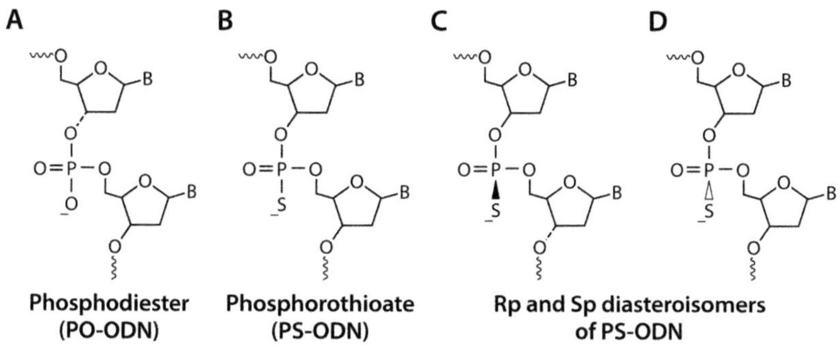

Figure 1.1 Chemical structures of DNA and DNA analogues (A) phosphodiester oligodeoxynucleotide (PO-ODN), (B) phosphorothioate (PS-ODN), (C) *R*p PS-ODN, (D) *S*p PS-ODN.

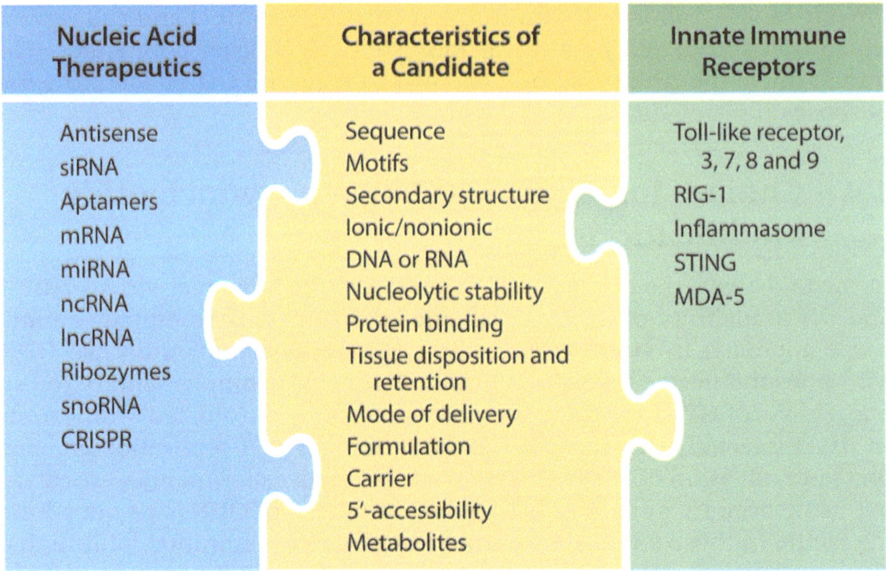

Figure 1.2 Nucleic acid-based therapeutics include the use of synthetic oligonucleotides as drug candidates. These candidates have various characteristics including DNA or RNA of varying sequence composition, single- or double-stranded, formulated, conjugated or complexed with lipid carriers *etc.* While the intended target of these candidates is largely RNA there are innate immune receptors including TLR 3, 7, 8 and 9, RIG-1, STING and inflammasomes. These receptors are known to recognize patterns of nucleic acids and activate appropriate immune responses. Modulation of these receptors has shown broad therapeutic applications. In design of nucleic acid-based therapeutics and their intended mechanism of action due consideration is needed to avoid overlapping mechanisms.

1.4.1.1 Phosphorothioates

The ASO therapeutics field took a major step forward in the mid 1980s through the chemical synthesis of phosphorothioate (PS) ODNs,[10] based on the pioneering work of Eckstein. In these analogues a simple sulfur atom replaces an oxygen atom (Figure 1.1). PS-linked ODNs are much more resistant to nuclease degradation than phosphodiesters and thus cellular activities were found to be much higher. However, mixed diastereomeric PS-ODNs, accessible readily by automated synthesis, have lower binding affinity to target RNA as compared with PO-ODN. Synthetic methodologies were optimized to synthesize milligram quantities of PS-ODNs for use in experiments as ASOs.

Studies with PS-ASO targeted to various regions of HIV-1 mRNA as well as non-complementary analogs including homo-oligomers, were conducted in HIV-1-infected cell-based assays and showed potent dose-dependent inhibition of viral replication and antiviral activity.[11,12] The antiviral activity was related to the base composition of the analogues, and longer PS oligonucleotides were more effective than shorter ones. These studies also established that in primary human cells, cellular uptake of PS-ASO was efficient and that no carrier was required. Furthermore, PS-ASOs also showed very potent and durable inhibition of HIV-1 replication in chronically HIV-1-infected cells.[13–15]

On the basis of the promise of these results, antisense technology gained the attention of the broader scientific community and PS-ASOs became the choice as first-generation ASOs. PS-ODNs thus quickly became the primary choice for therapeutics development by newly formed biotechnology companies, such as Gilead Sciences, Isis Pharmaceuticals (now Ionis), Hybridon (now Idera Pharmaceuticals) and several others. Over the next few years, hundreds of reports appeared in the literature on the use of PS-ASOs targeting various viruses,[16–18] oncogenes,[19,20] kinases[19,21] and other targets.[22] Soon it was realized that cellular uptake in transformed cells in culture was poor and lipid-based formulations were needed for efficient uptake and antisense activity. Also it was noted that the duplex of a PS-ASO with a target RNA elicited RNase H activity,[14,15] thereby allowing PS-ASOs to cleave RNA strands and thus to inhibit translation. This was more efficient than by using a steric blocking mechanism.[23,24] However the efficiency of RNase H cleavage of RNA by a PS-ASO was poor as compared with a PO-ASO.[14,15]

The first pharmacokinetic and tissue disposition study of a systemically delivered PS-ASO occurred in mice and showed that the plasma half-life was very short and that there was a broad tissue distribution.[25] The highest concentrations of the administered PS-ASO were observed in the liver and kidneys, with the lowest concentrations in the brain. Delivered PS-ASO remained stable in tissues for days and was excreted primarily in urine in degraded form. Degradation was primarily from the 3′-end, and modifications of the 3′-end increased the stability further.[26,27] The binding of PS-ASO to serum proteins played a major role in plasma half-life and tissue disposition and was affected by the presence of secondary structures.[28,29]

Studies with multiple PS-ASOs of varying sequence composition and length were conducted in animal models of viral diseases and cancer.[30,31] In these studies PS-ASOs exerted very potent activity. However it soon become clear that in some cases the control PS-ASOs employed also had some activity, leading to questions on what a good control would be for PS-ASO and/or if a PS-ASO had off-target activity.[32] In one study a PS-ASO targeted to human papillomavirus inhibited papillomavirus-induced growth of implanted human foreskin in a mouse xenograft model. However, it also showed activity in a lethal mouse cytomegalovirus (CMV) model, in which the PS-ASO was not expected to have antisense activity.[33] Detailed studies in immune-compromised mice led to an understanding that the antiviral activity of PS-ASO was largely due to immune activation of the host.

The impact of immune-stimulatory properties of PS-ASOs also became evident during the non-clinical safety evaluations of multiple candidates in support of investigational supporting studies. In mice and rats, repeated systemic administration of PS-ASO candidates caused general inflammation, splenomegaly, thrombocytopenia, elevation of transaminases and histological changes in multiple organs.[34,35] In addition, in primates, bolus administration of the first PS-ASO (GEM91) led to severe hemodynamic changes, which subsequently were found to be due to activation of the alternative complement pathway.[36] Complement activation thus became recognized as a second off-target effect of PS-ASOs. The alternative pathway complement activation cascade was attributed to the poly-anionic nature of PS-ASOs and largely found to be a plasma concentration threshold effect. As such, it could be mitigated by subcutaneous administration or by slow intravenous infusion to keep a low plasma concentration. Based on this observation, the US Food and Drug Administration (FDA) implemented guidelines on dosing of PS-ASOs and recommended use of non-human primates as the non-rodent species for non-clinical safety studies.[37] Meanwhile, early development programs with PS-ASOs continued to proliferate.

Over 25 PS-ASO drug candidates targeted to viral RNA, oncogenes and cellular targets had advanced to human trials.[38] Routes of administration included intravitreal, intravenous infusion or subcutaneous.[39–41] In humans the plasma half-life and excretion of PS-ASOs was similar to that which was observed in pre-clinical models.[42,43] Clinical development of most of the PS-ASO drug candidates were discontinued, either due to lack of activity or a poor therapeutic index.[44] Experience with a specific PS-ASO drug candidate, GEM91, in humans was very informative. Subcutaneous administration of GEM91 caused flu-like symptoms, swelling of the draining lymph nodes, prolongation of activated partial thromboplastin time (aPTT) and more importantly, rather than suppressing HIV-1, it increased HIV-1 RNA levels in blood.[45] Importantly, administration of the same dose by the intravenous route had minimal effects on these parameters. This was puzzling at that time, but much later it became clear that PS-ASO containing the unmethylated CpG motif were activating the immune responses by binding to Toll-like receptor 9 (TLR9), an innate immune receptor present in immune cells that recognizes DNA containing CpG dinucleotide motifs (see Chapter 14). Treatment with many other PS-ASO drug candidates also had shown flu-like symptoms and injection-site reactions. It is important to note that most of the PS-ASO drug candidates that had been advanced to human studies contained a CpG motif.[46] For example, the PS-ASO Fomivirsen, targeted to CMV and delivered intravitreally, had been approved but is no longer marketed. The mechanism of action of Fomivirsen has been questioned.

Collectively, the development of PS-ASOs from preclinical to clinical studies has provided very important insights into the properties of PS-ASOs.[47–49] These could be classified as a class effect, including an affinity to target RNA, stability towards nucleases, serum protein binding and poly-anionic-related side effects, such as complement activation and prolongation of aPTT. The immunostimulatory effects of PS-ASOs have been shown to be sequence-dependent and associated with the presence of CpG motifs.[50]

Much debated was whether some of the issues observed with PS-ASOs could be due to the presence of a mixture of diastereoisomers in the synthetic PS-ASOs. A sulfur substitution for a non-bridged oxygen on phosphorus results in both *R*p and *S*p diastereoisomers leading to 2^{19} isomers in a 20-mer (Figure 1.1). It was well known, based on the pioneering work of Stec and colleagues, that polymerases and nucleases interact and exert action on PS-ODN in a diasterioisomeric-selective fashion.[51] To obtain large amounts of stereo-enriched PS-ODN, nucleoside bicyclic oxazaphospholidium as synthons were introduced and used for synthesis of various configurations of stereo-enriched PS-ASOs, including all *R*p, all *S*p, or segments containing *S*p–*R*p–*S*p or *R*p–*S*p–*R*p.[52] As expected, binding affinity to RNA was higher for PS-ASO containing *R*p linkages followed by *S*p–*R*p–*S*p, equal to *R*p–*S*p–*R*p, more than stereo-random and *S*p isomers. Stability towards nucleases was in the reverse order where *S*p > *S*p–*R*p–*S*p > stereo-random > *R*p. RNase H activation was preferred by *R*p isomers followed by *R*p–*S*p–*R*p > *S*p isomers.[52] In cell culture studies, stereo-enriched PS-ASOs targeted to mouse double minute 2 homolog (*mdm2*) of different configurations showed very similar antisense activity compared with synthetic stereo-random PS-ASOs.[52] In an independent follow-up study stereo-enriched PS-ASOs targeted to scavenger receptor class B type 1 (SR-B1) showed similar antisense activity to stereo-random PS-ASOs.[53] However, in a recent study, stereo-pure PS-ASOs were shown to have improved activity in cell culture and in *in vivo* studies.[54] It is likely that the sequence composition of ASO and the particular placement of stereo-specific linkages at specific positions have an effect on the antisense activity.

1.4.1.2 Non-anionic Internucleotide Linkages

The most extensively studied non-anionic oligonucleotide analogues are those containing alkylphosphotriester[55,56] methylphosphonate[57] and phosphoramidate linkages[58] (Figure 1.3). These internucleotide linkages consist of a mixture of distereoisomers. Ts'o, Miller and colleagues had carried out extensive studies with methylphosphonate oligodeoxynucleotides (MP-ODNs) and had shown that MP-ODNs are stable under physiological pH, resistant to nucleolytic degradation and had a lower binding affinity to target RNA compared with PO-ODN.[57] In earlier studies, short MP-ODN had been employed as antisense in a cell-free system. To evaluate MP-ODNs as antisense, solid-phase synthetic methodology using methylphosphonamidites as synthons was developed[57,59] to obtain milligram quantities of longer length MP-ODN. MP-ASO of 15- to 20-mers targeted to HIV-1 showed dose-dependent inhibition of HIV-1 replication, longer MP-ASOs were more active and the effective concentration ranged from 30 to 150 micromolar.[59] This could have been due to a lower affinity to target RNA and a lack of RNase

H activation. Such ASOs failed to activate RNase H and inhibited translation by blocking the ribosome machinery, which is referred to as 'translation arrest'. In addition, it was noted that MP-ASOs of longer length had poor solubility under physiological conditions. *In vivo* delivery and disposition of MP-ASOs were limited due to their non-ionic nature. No MP-ASO candidates were advanced to human trials.

The anti-cancer agent GRN163 is a 13-mer ASO containing all-phosphoramidate internucleotide linkages, that inhibits telomerase[58] and did make it to a clinical trial, but did not receive regulatory approval as an antisense drug. Studies with an ASO containing phosphoramidate linkages (PN-ASO) targeted to HIV-1 had also been carried out in cell-based assays and showed similar results to those observed with MP-ASOs, and no further studies were conducted.[11]

1.4.1.2.1 Phosphorodiamidate Morpholino Oligonucleotodes. Phosphorodiamidate morpholino oligonucleotides (PMO) are charge-neutral oligonucleotides in which bases are attached to a morpholino ring, linked through a phosphorodiamidate group (Figure 1.3).[60] Replacement of anionic phosphates by uncharged phosphorodiamidate groups eliminates ionization in the physiological pH range. PMO-ASO inhibits translation by 'steric block' (see Chapter 2) since a duplex of PMO and RNA is not a substrate for RNase H[61](see Chapter 2). Cellular uptake of PMO is not very efficient and requires very high doses for *in vivo* delivery. There have been reports in which PMO ASOs have been evaluated against Ebola, Marburg and Chikungunya viruses, for example.[62] One of the PMO-ASO candidates designed to

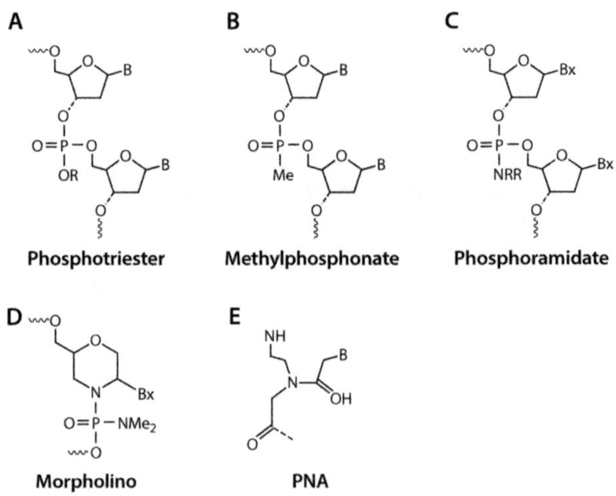

Figure 1.3 Chemical structures of non-anionic internucleotide linkages (A) phosphotriester, (B) methylphosphonate, (C) phosphoramidate, (D) phosphorodiamidate morpholino (PMO), (E) peptide nucleic acids (PNA).

induce alternative splicing of mutant dystrophin, eteplirsen, has shown an acceptable safety profile and restoration of dystrophin in Duchenne muscular dystrophy (DMD) patients receiving 50 mg kg^{-1} dose weekly and recently has been approved for clinical use (see Chapter 6). The chemistries used in exon-skipping activities have been reviewed.[63] More specifically, conjugation of cell-penetrating peptides to PMOs has allowed for significant improvements in delivery and increased exon-skipping activity in *in vivo* models in muscles as well as in heart.[64]

The major use of PMO has been in research to study the role of specific mRNA transcripts in embryos. Several studies have been conducted in which PMO-ASOs have been injected into eggs, embryos of zebrafish, *Xenopus* and sea urchins to elucidate gene function.[65]

1.4.1.2.2 Peptide Nucleic Acids. Peptide nucleic acids (PNA) are charge-neutral synthetic mimics of DNA or RNA containing *N*-2 aminoethylglycine units linked by amide bonds as repeating units in place of a sugar–phosphate backbone (Figure 1.3).[66] PNA provided two key properties for use as antisense, a significant increase in affinity to target RNA, due to the absence of electrostatic repulsion, and resistance to degradation by nucleases or proteases. Since the duplex of PNA with RNA does not activate RNase H, PNA-ASO would act by 'translation arrest'. PNA-ASOs have been broadly studied as anti-cancer,[67,68] anti-viral[69,70] and anti-bacterial agents[71,72] as well as inhibitors of micro RNAs.[73] However, once again very high doses are needed in *in vivo* applications, due to poor cellular uptake, unfavorable pharmacokinetics and the need for efficient delivery vehicles. No PNA-ASO agents have been advanced to clinical development.

1.4.2 Heterocyclic Bases

One of the approaches pursued to improve the affinity of ASO to target RNA was to employ modified heterocyclic bases. There were a number of studies carried out with the incorporation of heterocyclic modification that provided an additional hydrogen bond to its complementary base pair to improve binding affinity. These included modifications in the pyrimidine moiety, at positions C-2, C-4, C-5 and C-6. A large number of these were incorporated in ASOs. However, the results observed were not exciting. Incorporation of modified purines resulted in lowering of the binding affinity of antisense. In a detailed study, various heterocyclic bases including C-5 propynyl, 5-methyl cytosine (5-MeC), phenoxazine and G-clamp were compared for antisense activity and these showed that G-clamp had potent dose-dependent antisense activity.[74] However, these modified ASOs containing G-clamps were found to be highly toxic in *in vivo* studies. Out of hundreds of heterocyclic bases that have been studied in ASOs, only a few are currently being employed. The most used modified nucleoside in ASO is 5-methyl-2′-deoxycytidine (5-MedC).[75]

1.4.3 Sugar Modifications

1.4.3.1 *Oligoribonucleotides*

One of the key properties of ASOs is a strong affinity to target RNA, and it is well known that the affinity of the duplex of RNA–RNA is stronger than the duplex of DNA–RNA. Based on these criteria, the use of synthetic oligoribonucleotides (ORN) in antisense has been studied (Figure 1.4). To provide nucleolytic stability, phosphorothioate ORNs (PS-ORN-ASO) were synthesized. Evaluations of PS-ORN-ASOs showed an increased affinity to target RNA.[76] However they had lower potency as compared with the corresponding deoxynucleotide PS-ASOs, probably due to reduced nucleolytic stability and the lack of RNase H activation by the RNA–RNA duplex.

1.4.3.2 *2′-O-Alkyl Sugars*

To achieve further increases in the affinity of ASO to target RNA and an increase in nucleolytic stability, the use of 2′-*O*-methylribonucleosides[77] was evaluated. These are naturally occurring nucleosides, and, importantly, phosphoramidites of 2′-*O*-methylribonucleosides had become available commercially. Studies carried out with 2′-*O*-methyloligoribonucleotide phosphorothioates (2′-OMe PS-ASO), showed nucleolytic stability as compared with PS-ORN and higher affinity to target RNA.[78] However, they also showed less antisense activity compared with PS-ASOs, indicating that RNase H activation was a key parameter for antisense.[23,24] Since then, studies have been conducted with many additional 2′-*O*-alkyl analogs and the results have been similar. These studies led to the identification of 2′-*O*-methoxyethylribonucleosides (2′-MOE),[79] which are now used widely in ASO studies (see Chapters 2–4).

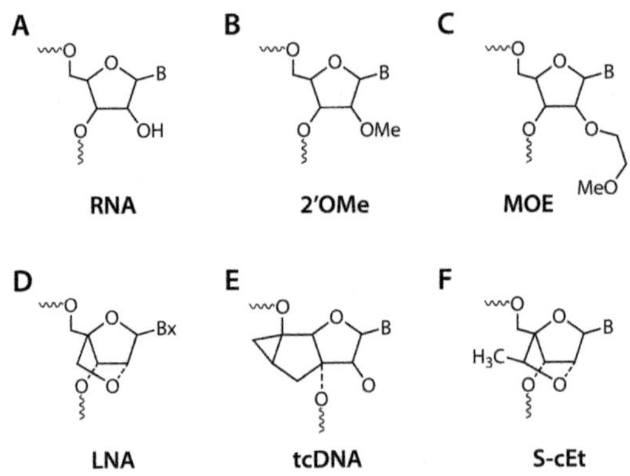

Figure 1.4 Chemical structures of RNA and RNA analogues (A) oligoribonucleotide (RNA), (B) 2′-*O*-methylribonucleotide (2′OMe), (C) 2′-*O*-methoxyethylribonucleotide (2′MOE), (D) Locked nucleic acid (LNA), (E) tricyclo-DNA (tcDNA), (F) constrained ethyl (cET).

Additional studies with PS-ORN and 2′-OMe PS-ASO provided very good insights into the nature of PS linkages in RNA and DNA. As discussed above, PS-ASOs had shown dose-dependent activation of complement and prolongation of aPTT and strong binding to serum proteins. These activities were believed to be due to the poly-anionic nature of the phosphorothioate backbone. Studies with PS-ORN and 2′-OMe-PS-ASO showed significantly reduced effects on these parameters, indicating that the poly-anionic nature of the PS backbone in PS-ORN and PS 2′-OMe had different characteristics.[80] This insight was helpful in designing gapmer antisense (see Chapters 2–4).

Lack of RNase H activation and increased affinity to target RNA also provided the opportunity for the application of 2′-*O*-alkyl modified PS-RNA for steric-blocking applications, such as splice switching[81,82] (see Chapter 6) and microRNA inhibition (see Chapter 8). Clinical development of 2′-*O*-alkyl ASOs targeting splice sites have yielded mixed results. Drisapersen, a 2′-OMe PS-RNA, failed to show clinical benefits in patients with DMD and also caused significant adverse events. In contrast in the last two years, an 18-mer 2′-MOE phosphorothioate (2′-MOE PS-ASO; nusinersen) that redirects the splicing of the SMN-2 gene has been approved for the treatment of spinal muscular atrophy (see Chapter 6). It is important to note that the safety issues observed with the use of drisapersen include thrombocytopenia and renal toxicity. This could be largely due to repeated administration of 2′-OMe PS-ASO, which is very stable to nucleolytic degradation. Drisapersen is retained and accumulated in tissues after repeated administration due to its long half-life. Nusinersen, a 2′-MOE PS-ASO, is also very stable to nucleolytic degradation, but it is administered intrathecally at lower doses and less frequently, which minimizes the effects of tissue accumulation.

1.4.3.3 *Bridged Nucleic Acids*

Reducing the conformational flexibility of nucleotides has been shown to increase their binding affinity to complementary RNA. These are collectively known as bridged nucleic acids (BNA). Locked nucleic acids (LNA) (Figure 1.4) link the 2′-oxygen and 4′-carbon of ribose and have shown a significant increase in affinity to target RNA.[83] All LNA oligomers of octamers or longer tend to aggregate, thereby limiting their therapeutic utility. Therefore they are generally used as mixmers with 2′-deoxy nucleotides or 2′OMe nucleotides (see Chapters 6 and 8). Other bridged nucleic acids which have been studied include a methylated analog of LNA, known as ‘constrained Ethyl’ (cET)[84] (see Chapter 3), tricyclo-DNA (tcDNA)[85] (Figure 1.4) and 2′-*O*,4′-C-ethylene linked nucleic acid (ENA)[86] (not shown). These bridged nucleic acids have also shown very strong affinity to target RNA and increased nucleolytic stability, but they are not substrates for RNase H. These modifications are being employed in ASOs mostly for splicing modulation and gapmer approaches. Encouraging results have been reported with ENA, tcDNA or combinations of 2′-OMe and ENA in preclinical models of DMD. Increased affinity has allowed for the use of shorter ASOs. In a recent study an ASO, DS-5141 containing segments of 2′-OMe PS-RNA and ENA has shown promising activity in an *mdx* mouse model and a phase 1/2 clinical trial in DMD has been initiated in Japan.[87]

1.5 Gapmers Using Combinations of Modified Oligodeoxy and/or Oligoribonucleotides

Based on the experience with PS-ASOs and other modified ASOs, including MP-ASOs and 2′-OMe PS-ASO's, it became clear that each of these modifications carry specific desirable properties. However, each of them lack one or more of the required characteristics for antisense (Figure 1.2). For example, desirable properties of PS-ASOs include activation of RNase H, whereas the characteristics of MP-ASOs only include nucleolytic stability and a non-anionic backbone. For 2′-OMe PS-ASO characteristics include increased nucleolytic stability, increased affinity to target RNA and reduced poly-anionic characteristics and immune-stimulatory effects. This led to a concept that an optimal ASO could be designed in which segments of PS-DNA and MP-ASO or 2′-OMe PS-RNA could be appropriately placed. For example, a PS-DNA segment could be placed in the middle flanked by segments of 2′-OMe PS-RNA at both the 3′- and 5′-ends[23,88] (see Chapter 2). Studies have also been carried out in which a segment of 2′-OMe PS-RNA has been placed in the middle and segments of PS-DNA are placed at both the 3′- and 5′-ends.

Early studies were carried out with ASOs containing segments of PS-ASO and MP-ASO, which showed that RNase H-mediated excision of the target could be directed to specific sites.[14] However, these ASOs had lower affinity to target RNA. ASOs containing segments of PS-ASO and 2′-OMe PS-ASO had many of the desired characteristics, including high affinity to target RNA, increased nucleolytic stability, reduction in poly-anionic characteristic and immune-stimulatory activity and activation of RNase H.[23,80,88] Over the years these types of ASO constructs have been referred to as mixed backbone-oligos, hybrids or, more recently, as gapmers. In these studies 2′-MOE nucleotides[84] have shown promising results and have been widely employed in gapmer ASOs (Chapters 2–4).

Gapmer ASOs of varying sequence and composition, length and configuration have been studied for hundreds of RNA targets in cell culture and in animal models and have yielded very potent antisense activity. Pharmacokinetic studies in mice and primates have shown very similar plasma half-life and tissue disposition as PS-ASO, with significantly increased *in vivo* stability, resulting in longer persistence in tissues of the administered gapmer ASO.[89,90] This increased *in vivo* stability of gapmer ASO should have allowed for less frequent dosing in preclinical and clinical studies. However, this has not yet been achieved. This indicates that gapmer ASO retained in tissues is not bio-available for antisense activity. Gapmer ASOs also showed reduced inflammatory responses[91] and improved safety parameters as compared with PS-ASO (discussed in detail in Chapter 20). The increased nucleolytic stability of gapmer ASO has also allowed for potential oral delivery.[28,92] Clinical experience with gapmers is described in Chapters 3 and 4. Unfortunately, the list of discontinuations of clinical development of gapmer ASO drug candidates has also been growing, largely either due to lack of efficacy or to safety signals that limit the therapeutic index.

Incorporation of LNA into gapmer ASOs has proved disappointing and the clinical development of a number of drug candidates have been discontinued due to safety signals. Incorporation of cEt in gapmer ASO has shown improvements in potency in preclinical studies and the clinical development of gapmer ASO drug candidates is in progress (see Chapter 4).

It is important to consider whether safety signals with gapmer ASOs (discussed in Chapter 20) are due to specific sequence and motifs and/or extensive tissue build-up. It was postulated that increased nucleolytic stability and tissue retention of gapmer ASO may result in less frequent dosing to achieve therapeutic benefit. However, this has not been the case, and in a majority of the studies, gapmer ASOs are being injected on a weekly regimen. It is evident from pre-clinical studies in mice and primates that after repeated injections significant amounts of administered gapmer ASO build up in certain tissues.[93] What are the fates of gapmer ASOs and their degradation products in tissues? Do they interact with local innate immune receptors? Does the accumulated gapmer ASO with very high affinity to RNA, and in some cases use of shorter ASO sequences, result in excessive activation of RNase H by binding to unrelated RNA targets? Insights into these questions will further provide guidance on improving the design of gapmer ASOs to improve therapeutic index.

1.6 Antisense Conjugates

The results of administration of PS-ASOs and gapmer ASOs have shown short plasma half-lives, wide deposition in multiple organs and elimination primarily *via* urinary excretion (Chapter 20). Serum protein binding is very important for both of these ASO types. ASOs of a non-ionic nature, like MP and PMO, do not bind to serum proteins, have a very short plasma half-life, are not distributed to many organs and are rapidly eliminated by urinary excretion. A number of chemical conjugates, including cholesterol,[94] lipids,[95] peptides,[96] antibodies[97] and sugars,[98] have been studied for targeted *in vivo* delivery. The formulation of ASOs with lipids led to preferential disposition to liver, but was also associated with immune activation. In the last few years, use of *N*-acetylgalactosamine (GalNAc), a ligand for asialoglycoprotein receptor expressed on hepatocytes has been widely studied (see Chapters 4 and 11). Conjugation of GalNAc has led to increased liver uptake of gapmer ASOs, resulting in improved efficacy and allowing use of lower doses. It is to be noted that a significant fraction of conjugated gapmer ASOs are also present in kidneys. Emerging clinical data with GalNAc gapmer ASO conjugate targeted to ApoCIII has been encouraging and long-term exposure data will guide further development. At the same time, it is important to note that clinical development of two GalNAc ASO conjugates, RG101 and RG125, has been discontinued due to clinical or pre-clinical safety signals.

Cell-penetrating peptide conjugates of PMO ASOs have shown enhanced uptake in muscles[99] and, in some cases, heart muscles.[64,100] There is now a significant effort to find specific receptor-mediated targeting into other tissue types.

1.7 The Role of Innate Immune Receptors in Nucleic Acid Therapeutics

As discussed above and in Chapter 14, the first indication of immune stimulatory activity of a PS-ASO in humans was observed in 1993–1994. It was not until the year 2000, when TLR9 was discovered, that this provided the basis for the observed immune activation. Since then, tremendous progress has been made in the discovery and the role of the family of innate immune receptors. Many of these receptors, including Toll-like-receptors (TLR) 3, 7, 8 and 9, RIG-1, Inflammasome and STING, are activated upon recognition of patterns of foreign nucleic acids of viral and bacterial origin and, under certain conditions, also with endogenous nucleic acids (see Chapters 5, 13 and 14).

Under normal conditions, activation of these receptors leads to a defense mechanism. Under pathological conditions, unintended activation of these receptors could lead to autoimmune inflammatory disorders. In brief, TLR9 is a receptor for bacterial DNA containing an unmethylated CpG motif, and upon recognition leads to the initiation of a signaling cascade resulting in activation of nuclear factor kappa B and induction of T helper 1 (Th1)-type cytokines. Many of the PS-ASOs advanced to clinical development contained a CpG motif and had shown immune-stimulatory properties, thereby leading to the questioning of the intended mechanism of action (see Chapter 14). To mitigate immune-stimulatory activity of PS-ASOs or gapmer ASOs containing CpG motifs, substitution of C in the CpG motif by 5-MedC has been employed. As has been reported recently, the presence of 5-MedC in the CpG motif is indeed not immune-stimulatory, but acts as an antagonist of TLR9,[101] so caution should be exercised. Such has been a recent case with a PS-ASO targeting small and mothers against decapentaplegic homolog 7 (SMAD-7), in which CpG motifs were substituted by 5′-MedC in dCpG motifs and which failed in a phase 3 trial.[102] These types of ASOs can act as antagonists to TLR9. Similarly gapmer ASOs, depending on the sequence composition and position of 2′-*O*-alkyl substitutions, could act as antagonists of TLR7, 8 and 9 and interfere with the intended mechanism of action (see Chapter 5). Nucleic acids that either activate or inhibit innate immune receptors have shown very broad therapeutic applications (see Chapters 5 and 14).

It is important to give due considerations to innate immune receptors when selecting ASO sequences and modifications to be employed (Figure 1.2). In addition, the role of these innate immune receptors with respect to accumulated gapmer ASOs and their metabolites in tissues has yet to be elucidated. Does this interaction lead to local signaling cascades and thereby inflammation and safety signals? In a preliminary study, an oligonucleotide which had shown inflammatory signals and histopathological changes in C57 black 6 mice, showed significantly reduced effects in TLR7 and 9 double-knockout mice (Agrawal, unpublished data).

1.8 Future Directions

Tremendous progress has been made in the last 40 years. The field of nucleic acid-based therapeutics has evolved to become a major drug-discovery platform. In parallel to continued developments in the antisense field (Chapters 2–4, 6 and 7), there has been tremendous progress made in several related fields. Understanding of the role of RNA in biology has led to identifications of miRNAs (Chapter 8), long non-coding RNAs (Chapter 9), small nucleolar RNA (snoRNA) *etc.* and their interplay in gene regulation, thereby expanding the pool of potential targets for antisense. New mechanisms of gene regulation have also been uncovered, including the harnessing of ribozymes (Chapter 18), siRNA (Chapters 10–12) and CRISPR (Chapter 17). Furthermore, the discovery of innate immune receptors for nucleic acids has helped in providing answers to many questions in the field of nucleic acid therapeutics and has resulted in a new therapeutic platform (Chapters 5, 13 and 14). In addition, aptamers have been developed as therapeutics (Chapters 15 and 16).

To date, a number of nucleic acid-based drugs have been approved. These include an aptamer, Macugen, and five antisense oligonucleotides, fomivirsen, mipomersen, eteplirsen, nusinersen and inotersen. A siRNA drug candidate, patisiran, has been approved recently. Clinical developments of a number of other ASO and siRNA candidates are in progress. This indicates that the potential of antisense and other nucleic acid-based therapeutics is being realized. This has required development of large-scale synthesis techniques (Chapter 19).

However challenges remain. Out of the over 70 ASO drug candidates that have been advanced to clinical development, the list of ASO drug candidates being discontinued also continues to grow. There could be number of reasons for discontinuations, such as an unvalidated gene target, ASO sequence composition and motifs, level of knockdown achieved not being sufficient for therapy, potency of the ASO, off-target activity and, most importantly, therapeutic-index-limiting treatment (Chapter 20).

Since 2001, the focus has been on the development of gapmer ASOs. Gapmer ASOs provide optimal properties for antisense, both in cell-based assays as well as *in vivo* models and have yielded positive phase 3 results. However, treatment has also been associated with safety signals. There could be many reasons for this, including sequence composition, off-target activity, nucleotide motifs, or even secondary structures, or interactions with innate immune receptors. A common feature in all of these gapmer ASOs is increased *in vivo* stability, resulting in accumulation and retention in many organs. The retained gapmer ASO and its metabolites are not functionally active, since repeated dosing is needed for sustained target RNA or protein knockdown.

The focus is shifting towards further improvements in the potency of gapmer ASOs by incorporation of novel nucleotide analogues (*e.g.* LNA, cEt and others). Improved affinity of these new gapmer ASOs has allowed for the use of shorter gapmer ASOs (Chapters 3 and 4). Studies carried out with a few of these gapmer ASO candidates have shown increased potency, resulting in the use of reduced dose and dosing schedule *in vivo* models. However, without addressing the *in vivo* stability issues repeated dosing would still lead to tissue retention and accumulation.

Throughout the studies with gapmer ASOs, it has been postulated that their increased potency is also related to the increased nucleolytic stability. A recent study provided a new insight into nucleolytic stability and its effects on antisense potency.[103,104] A comparative study of a 19-mer PS-ASO, 19-mer PS-ASOs linked together *via* their 3′-ends, or *via* their 5′-ends, or *via* a normal 3′–5′ linkage, in cell culture and an *in vivo* model, showed that 5′–5′ linked PS-ASO had more potent antisense activity than the other three PS-ASOs in cell-based assays and in the *in vivo* model. If nucleolytic stability is a key factor in exerting the antisense activity, a PS-ASO with 3′–3′ linkage, with the highest nucleolytic stability, would have been the most active as antisense. Furthermore a 5′–5′ linked PS-ASO showed more potent activity than a gapmer ASO. This indicates that nucleolytic stability of ASO may not be as important a factor as previously thought, but more important is how the duplex of ASO and RNA engages RNase H in excision of the target RNA. Increased antisense activity of the 5′–5′ linked PS-ASO was associated with excision of the target RNA in the center of the ASO, as compared with other ASOs, in which excision products were at the 5′-end. This insight provides further direction on the design of optimal ASOs.

One of the key emerging issues in the broad applicability of ASOs is how to address safety signals. This perhaps may be achieved by improving the potency of the ASOs, avoiding tissue build-up and by minimizing interactions with innate immune receptors. Applications of novel delivery vehicles could be of equal importance. There is more work to be done. With the proof-of-concept of antisense in humans established, it is not too distant to expect the application of ASOs for the treatment of genetic disorders, a true approach to precision medicine.

Acknowledgements

S.A. is indebted to my co-author and co-editor, Mike Gait for his mentorship. While working with him, I learned about aspects of nucleic acid chemistry, but more importantly, how to approach research to seek answers. My journey in the antisense field started in the lab of the late Paul Zamecnik, known as the 'father of antisense'. He was a great teacher, a friend, and together we started the journey of Hybridon, now Idera Pharmaceuticals. I am also grateful to all of my colleagues and collaborators whose names appear in the references cited.

References

1. P. C. Zamecnik and M. L. Stephenson, *Proc. Natl. Acad. Sci. U. S. A.*, 1978, **75**, 280.
2. P. C. Zamecnik, *Methods Mol. Med.*, 1996, **1**, 1.
3. R. L. Letsinger and V. Mahadevan, *J. Am. Chem. Soc.*, 1965, **87**, 3526.
4. M. J. Gait and R. C. Sheppard, *J. Am. Chem. Soc.*, 1976, **98**, 8514.
5. K. Miyoshi and K. Itakura, *Tetrahedron Lett.*, 1979, **20**, 3635.
6. M. J. Gait, M. Singh, R. C. Sheppard, M. Edge, A. R. Greene, G. R. Heathcliffe, T. C. Atkinson, C. R. Newton and A. F. Markham, *Nucleic Acids Res.*, 1980, **8**, 1080.
7. B. S. Sproat and M. J. Gait, in *Oligonucleotide Synthesis: A Practical Approach*, ed. M. J. Gait, IRL Press, Oxford, 1984, pp. 83–114.
8. S. L. Beaucage and M. H. Caruthers, *Tetrahedron Lett.*, 1981, **22**, 1859.
9. P. C. Zamecnik, J. Goodchild, Y. Taguchi and P. S. Sarin, *Proc. Natl. Acad. Sci. U. S. A.*, 1986, **83**, 4143.
10. W. J. Stec, G. Zon, W. Egan and B. Stec, *J. Am. Chem. Soc.*, 1984, **106**, 6077.
11. S. Agrawal, J. Goodchild, M. P. Civeira, A. H. Thornton, P. S. Sarin and P. C. Zamecnik, *Proc. Natl. Acad. Sci. U. S. A.*, 1988, **85**, 7079.
12. M. Matsukura, K. Shinozuka, G. Zon, H. Mitsuya, M. Reitz, J. Cohen and S. Broder, *Proc. Natl. Acad. Sci. U. S. A.*, 1987, **84**, 7706.
13. S. Agrawal, T. Ikeuchi, D. Sun, P. S. Sarin, A. Konopka, J. Maizel and P. C. Zamecnik, *Proc. Natl. Acad. Sci. U. S. A.*, 1989, **86**, 7790.
14. S. Agrawal, S. H. Mayrand, P. C. Zamecnik and T. Pederson, *Proc. Natl. Acad. Sci. U. S. A.*, 1990, **87**, 1401.
15. R. Y. Walder and J. A. Walder, *Proc. Natl. Acad. Sci. U. S. A.*, 1988, **85**, 5011–5015.
16. L. M. Cowsert, M. C. Fox, G. Zon and C. K. Mirabelli, *Antimicrob. Agents Chemother.*, 1993, **37**, 171.
17. M. Flores-Aguilar, W. R. Freeman, C. A. Wiley, P. Gangan, D. Munguia, M. Tatebayashi, C. Vuong and G. Besen, *J. Infect. Dis.*, 1997, **175**, 1308.
18. J. M. Leiter, S. Agrawal, P. Palese and P. C. Zamecnik, *Proc. Natl. Acad. Sci. U. S. A.*, 1990, **87**, 3430.
19. B. P. Monia, J. F. Johnston, T. Geiger, M. Muller and D. Fabbro, *Nat. Med.*, 1996, **2**, 668.
20. M. Z. Ratajczak, J. A. Kant, S. M. Luger, N. Hijiya, J. Zhang, G. Zon and A. M. Gewirtz, *Proc. Natl. Acad. Sci. U. S. A.*, 1992, **89**, 11823.
21. N. M. Dean and R. McKay, *Proc. Natl. Acad. Sci. U. S. A.*, 1994, **91**, 11762.

22. R. H. Barker Jr., V. Metelev, E. Rapaport and P. Zamecnik, *Proc. Natl. Acad. Sci. U. S. A.*, 1996, **93**, 514.
23. S. Agrawal, *Trends Biotechnol.*, 1992, **10**, 152.
24. B. P. Monia, E. A. Lesnik, C. Gonzalez, W. F. Lima, D. McGee, C. J. Guinosso, A. M. Kawasaki, P. D. Cook and S. M. Freier, *J. Biol. Chem.*, 1993, **268**, 14514.
25. S. Agrawal, J. Temsamani and J. Y. Tang, *Proc. Natl. Acad. Sci. U. S. A.*, 1991, **88**, 7595.
26. J. Temsamani, A. Roskey, C. Chaix and S. Agrawal, *Antisense Nucleic Acid Drug Dev.*, 1997, **7**, 159.
27. J. Temsamani, J. Y. Tang, A. Padmapriya, M. Kubert and S. Agrawal, *Antisense Res. Dev.*, 1993, **3**, 277.
28. S. Agrawal, J. Temsamani, W. Galbraith and J. Tang, *Clin. Pharmacokinet.*, 1995, **28**, 7.
29. S. Agrawal, X. Zhang, Q. Cai, E. R. Kandimalla, A. Manning, Z. Jiang, T. Marcel and R. Zhang, *J. Drug Targeting*, 1998, **5**, 303.
30. N. Dean, R. McKay, L. Miraglia, R. Howard, S. Cooper, J. Giddings, P. Nicklin, L. Meister, R. Ziel, T. Geiger, M. Muller and D. Fabbro, *Cancer Res.*, 1996, **56**, 3499.
31. K. Moriya, M. Matsukura, K. Kurokawa and K. Koike, *Biochem. Biophys. Res. Commun.*, 1996, **218**, 217.
32. T. Gura, *Science*, 1995, **270**, 575.
33. E. J. Lewis, S. Agrawal, J. Bishop, J. Chadwick, N. D. Cristensen, S. Cuthill, P. Dunford, A. K. Field, J. Francis, V. Gibson, A. K. Greenham, F. Kelly, R. Kilkushie, J. W. Kreider, J. S. Mills, M. Mulqueen, N. A. Roberts, P. Roberts and D. E. Szymkowski, *Antiviral Res.*, 2000, **48**, 187.
34. S. Agrawal, Q. Zhao, Z. Jiang, C. Oliver, H. Giles, J. Heath and D. Serota, *Antisense Nucleic Acid Drug Dev.*, 1997, **7**, 575.
35. A. A. Levin, *Biochim. Biophys. Acta*, 1999, **1489**, 69.
36. W. M. Galbraith, W. C. Hobson, P. C. Giclas, P. J. Schechter and S. Agrawal, *Antisense Res. Dev.*, 1994, **4**, 201.
37. L. E. Black, J. G. Farrelly, J. A. Cavagnaro, C. H. Ahn, J. J. DeGeorge, A. S. Taylor, A. F. DeFelice and A. Jordan, *Antisense Res. Dev.*, 1994, **4**, 299.
38. S. Agrawal and E. R. Kandimalla, *Curr. Cancer Drug Targets*, 2001, **1**, 197.
39. E. Bayever, P. L. Iversen, M. R. Bishop, J. G. Sharp, H. K. Tewary, M. A. Arneson, S. J. Pirruccello, R. W. Ruddon, A. Kessinger and G. Zon, *et al.*, *Antisense Res. Dev.*, 1993, **3**, 383.
40. M. D. de Smet, C. J. Meenken and G. J. van den Horn, *Ocul. Immunol. Inflammation*, 1999, **7**, 189.
41. J. Nemunaitis, J. T. Holmlund, M. Kraynak, D. Richards, J. Bruce, N. Ognoskie, T. J. Kwoh, R. Geary, A. Dorr, D. Von Hoff and S. G. Eckhardt, *J. Clin. Oncol.*, 1999, **17**, 3586.
42. J. M. Grindel, T. J. Musick, Z. Jiang, A. Roskey and S. Agrawal, *Antisense Nucleic Acid Drug Dev.*, 1998, **8**, 43.

43. D. Sereni, R. Tubiana, C. Lascoux, C. Katlama, O. Taulera, A. Bourque, A. Cohen, B. Dvorchik, R. R. Martin, C. Tournerie, A. Gouyette and P. J. Schechter, *J. Clin. Pharmacol.*, 1999, **39**, 47.
44. S. Agrawal, *Trends Biotechnol.*, 1996, **14**, 376.
45. S. Agrawal and E. R. Kandimalla, *Nat. Biotechnol.*, 2004, **22**, 1533.
46. S. Agrawal and R. R. Martin, *J. Immunol.*, 2003, **171**, 1621; author reply 1621.
47. S. Agrawal, *Biochim. Biophys. Acta*, 1999, **1489**, 53.
48. S. Agrawal and E. R. Kandimalla, *Mol. Med. Today*, 2000, **6**, 72.
49. S. Agrawal and Q. Zhao, *Curr. Opin. Chem. Biol.*, 1998, **2**, 519.
50. H. Hemmi, O. Takeuchi, T. Kawai, T. Kaisho, S. Sato, H. Sanjo, M. Matsumoto, K. Hoshino, H. Wagner, K. Takeda and S. Akira, *Nature*, 2000, **408**, 740.
51. M. R. Kurpiewski, M. Koziolkiewicz, A. Wilk, W. J. Stec and L. Jen-Jacobson, *Biochemistry*, 1996, **35**, 8846.
52. D. Yu, E. R. Kandimalla, A. Roskey, Q. Zhao, L. Chen, J. Chen and S. Agrawal, *Bioorg. Med. Chem.*, 2000, **8**, 275.
53. W. B. Wan, M. T. Migawa, G. Vasquez, H. M. Murray, J. G. Nichols, H. Gaus, A. Berdeja, S. Lee, C. E. Hart, W. F. Lima, E. E. Swayze and P. P. Seth, *Nucleic Acids Res.*, 2014, **42**, 13456.
54. N. Iwamoto, D. C. D. Butler, N. Svrzikapa, S. Mohapatra, I. Zlatev, D. W. Y. Sah, Meena, S. M. Standley, G. Lu, L. H. Apponi, M. Frank-Kamenetsky, J. J. Zhang, C. Vargeese and G. L. Verdine, *Nat. Biotechnol.*, 2017, **35**, 845.
55. P. M. Miller, S. Chandrasegaran, D. L. Dow, S. M. Pulford and L. S. Kan, *Biochemistry*, 1982, **21**, 5468.
56. P. S. Miller, K. N. Fang, N. S. Kondo and P. O. P. Ts'o, *J. Am. Chem. Soc.*, 1971, **93**, 6657.
57. P. S. Miller, C. H. Agris, A. Murakami, P. M. Reddy, S. A. Spitz and P. O. P. Ts'o, *Nucleic Acids Res.*, 1983, **11**, 6225.
58. A. Asai, Y. Oshima, Y. Yamamoto, T. Uochi, H. Kusaka, S. Akinaga, Y. Yamashita, K. Pongracz, R. Pruzan, E. Wunder, M. Piatyszek, S. Li, A. C. Chin, C. B. Harley and S. Gryaznov, *Cancer Res.*, 2003, **63**, 3931.
59. P. S. Sarin, S. Agrawal, M. P. Civeira, J. Goodchild, T. Ikeuchi and P. C. Zamecnik, *Proc. Natl. Acad. Sci. U. S. A.*, 1988, **85**, 7448.
60. J. Summerton, *US Pat.* 5185444, 1993.
61. J. Summerton, *Biochim. Biophys. Acta*, 1999, **1489**, 141.
62. K. L. Warfield, D. L. Swenson, G. G. Olinger, D. K. Nichols, W. D. Pratt, R. Blouch, D. A. Stein, M. J. Aman, P. L. Iversen and S. Bavari, *PLoS Pathog.*, 2006, **2**, e1.
63. P. Järver, L. O'Donovan and M. J. Gait, *Nucleic Acids Ther.*, 2014, **24**, 37.
64. C. Betts, A. F. Saleh, A. A. Arzumanov, S. M. Hammond, C. Godfrey, T. Coursindel, M. J. Gait and M. J. A. Wood, *Mol. Ther. Nucleic Acids*, 2012, **1**, e38.
65. S. C. Materna, *Methods Mol. Biol.*, 2017, **1565**, 87.

66. M. Egholm, O. Buchardt, P. E. Nielsen and R. H. Berg, *J. Am. Chem. Soc.*, 1992, **114**, 1895.
67. V. Rapozzi, B. E. Burm, S. Cogioi, G. A. van der Marel, J. H. van Boom, F. Quadrifoglio and L. E. Xodo, *Nucleic Acids Res.*, 2002, **30**, 3712.
68. R. Villa, M. Folini, S. Lualdi, S. Veronese, M. G. Daidone and N. Zaffaroni, *FEBS Lett.*, 2000, **473**, 241.
69. B. Chaubey, S. Tripathi, S. Ganguly, D. Harris, R. A. Casale and V. N. Pandey, *Virology*, 2005, **331**, 418.
70. B. Chaubey, S. Tripathi and V. N. Pandey, *Oligonucleotides*, 2008, **18**, 9.
71. L. Good, S. K. Awasthi, R. Dryselius, O. Larsson and P. E. Nielsen, *Nat. Biotechnol.*, 2001, **19**, 360.
72. L. Good and P. E. Nielsen, *Proc. Natl. Acad. Sci. U. S. A.*, 1998, **95**, 2073.
73. A. G. Torres, M. M. Fabani, E. Vigorito, D. Williams, N. Al-Obaidi, F. Wojcechowski, R. H. E. Hudosn, O. Seitz and M. J. Gait, *Nucleic Acids Res.*, 2012, **40**, 2152.
74. W. M. Flanagan, J. J. Wolf, P. Olson, D. Grant, K. Y. Lin, R. W. Wagner and M. D. Matteucci, *Proc. Natl. Acad. Sci. U. S. A.*, 1999, **96**, 3513.
75. S. Henry, K. Stecker, D. Brooks, D. Monteith, B. Conklin and C. F. Bennett, *J. Pharmacol. Exp. Ther.*, 2000, **292**, 468.
76. E. R. Kandimalla, D. R. Shaw and S. Agrawal, *Bioorg. Med. Chem. Lett.*, 1998, **8**, 2103.
77. H. Inoue, Y. Hayase, S. Iwai and E. Ohtsuka, *Nucleic Acids Symp. Ser.*, 1987, 221.
78. D. Yu, R. P. Iyer, D. R. Shaw, J. Lisziewicz, Y. Li, Z. Jiang, A. Roskey and S. Agrawal, *Bioorg. Med. Chem.*, 1996, **4**, 1685.
79. K.-H. Altmann, M.-O. Bévierre, A. de Mesmaeker and H. E. Moser, *Bioorg. Med. Chem. Lett.*, 1995, **5**, 431.
80. D. R. Shaw, P. K. Rustagi, E. R. Kandimalla, A. N. Manning, Z. Jiang and S. Agrawal, *Biochem. Pharmacol.*, 1997, **53**, 1123.
81. H. Sierakowska, M. J. Sambade, S. Agrawal and R. Kole, *Proc. Natl. Acad. Sci. U. S. A.*, 1996, **93**, 12840.
82. S. D. Wilton, F. Lloyd, K. Carville, S. Fletcher, K. Honeyman, S. Agrawal and R. Kole, *Neuromuscular Disord.*, 1999, **9**, 330.
83. J. Wengel, M. Petersen, K. E. Nielsen, G. A. Jensen, A. E. Håkansson, R. Kumar, M. D. Sørensen, V. K. Rajwanshi, T. Bryld and J. P. Jacobsen, *Nucleosides, Nucleotides Nucleic Acids*, 2001, 389.
84. P. P. Seth, A. Siwkowski, C. R. Allerson, G. Vasquez, S. Lee, T. P. Prakash, G. Kinberger, M. T. Migawa, H. Gaus, B. Bhat and E. E. Swayze, *Nucleic Acids Symp. Ser.*, 2008, 553.
85. D. Renneberg and C. J. Leumann, *J. Am. Chem. Soc.*, 2002, **124**, 5993.
86. K. Morita, C. Hasegawa, M. Kaneko, S. Tsutsumi, J. Sone, T. Ishikawa, T. Imanishi and M. Koizumi, *Nucleic Acids Res. Suppl.*, 2001, 241.
87. T. Lee, H. Awano, M. Yagi, M. Matsumoto, N. Watanabe, R. Goda, M. Koizumi, Y. Takeshima and M. Matsuo, *Genes*, 2017, **8**, 67.
88. V. Metelev and S. Agrawal, *US Pat.* 5652355, 1992.
89. R. Zhang, R. P. Iyer, D. Yu, W. Tan, X. Zhang, Z. Lu, H. Zhao and S. Agrawal, *J. Pharmacol. Exp. Ther.*, 1996, **278**, 971.

90. R. Zhang, Z. Lu, H. Zhao, X. Zhang, R. B. Diasio, I. Habus, Z. Jiang, R. P. Iyer, D. Yu and S. Agrawal, *Biochem. Pharmacol.*, 1995, **50**, 545.
91. S. Agrawal and Q. Zhao, *Antisense Nucleic Acid Drug Dev.*, 1998, **8**, 135.
92. H. Wang, Q. Cai, X. Zeng, D. Yu, S. Agrawal and R. Zhang, *Proc. Natl. Acad. Sci. U. S. A.*, 1999, **96**, 13989.
93. L. Shen, J. A. Engelhardt, G. Hung, J. Yee, R. Kikkawa, J. Matson, B. Tayefeh, T. Machemer, P. C. Giclas and S. P. Henry, *Nucleic Acid Ther.*, 2016, **26**, 236.
94. R. L. Letsinger, G. Zhang, D. K. Sun, T. Ikeuchi and P. S. Sarin, *Proc. Natl. Acad. Sci. U. S. A.*, 1989, **86**, 6553.
95. G. Shen, H. Fang, Y. Song, A. A. Bielska, Z. Wang and J. S. Taylor, *Bioconjugate Chem.*, 2009, **20**, 1729.
96. M. Lemaitre, B. Bayard and B. Lebleu, *Proc. Natl. Acad. Sci. U. S. A.*, 1987, **84**, 648.
97. K. Zhang, L. Hao, S. J. Hurst and C. A. Mirkin, *J. Am. Chem. Soc.*, 2012, **134**, 16488.
98. C. Sun, K. Ridderstrale, A. S. Hoglund, L. G. Larsson and C. Jansson, *Plant J.*, 2007, **52**, 1192.
99. H. M. Moulton, M. H. Nelson, S. A. Hatlevig, M. T. Reddy and P. L. Iversen, *Bioconjugate Chem.*, 2004, **15**, 290.
100. H. Yin, H. M. Moulton, Y. Seow, C. Boyd, J. Boutilier, P. Iversen and M. J. A. Wood, *Hum. Mol. Genet.*, 2008, **17**, 3909.
101. D. Yu, D. Wang, F. G. Zhu, L. Bhagat, M. Dai, E. R. Kandimalla and S. Agrawal, *J. Med. Chem.*, 2009, **52**, 5108.
102. I. Marafini and G. Monteleone, *Ann. Med.*, 2018, **50**, 361–370.
103. S. Agrawal, R. Improgo and W. Jiang, Patent application WO/2017/205384, 2017.
104. L. Bhagat, M. R. Putta, D. Wang, D. Yu, T. Lan, W. Jiang, Z. Sun, H. Wang, J. X. Tang, N. La Monica, E. R. Kandimalla and S. Agrawal, *J. Med. Chem.*, 2011, **54**, 3027.

CHAPTER 2

Mechanisms of Antisense Oligonucleotides

ZHONGTIAN LIU AND DAVID R. COREY*

Departments of Pharmacology & Biochemistry, University of Texas Southwestern Medical Center, 6001 Forest Park Road, Dallas, Texas 75390-9041, USA
*E-mail: david.corey@utsouthwestern.edu

2.1 Introduction

DNA makes RNA makes protein – the central dogma of biology – immediately suggests that agents that affect the synthesis of RNA or proteins would be a general class of agents for regulating gene expression and drug development.[1,2] Antisense oligonucleotides (ASOs) can be designed and chemically synthesized in order to bind any complementary nucleic acids sequence and provide the means to achieve recognition and to manipulate expression. The long-standing question has been whether this promise can be translated into reality, and an understanding of their mechanisms of action is one of the critical elements necessary to answer this question.

In theory, ASOs might recognize either DNA or RNA within cells. Chromosomal DNA is base-paired and thus prevents easy access by synthetic oligonucleotides. While it is possible for synthetic oligonucleotides to displace one strand and bind directly to chromosomal sequences or to recognize sites

Drug Discovery Series No. 68
Advances in Nucleic Acid Therapeutics
Edited by Sudhir Agrawal and Michael J. Gait

Published by the Royal Society of Chemistry, www.rsc.org

through triple helix formation, efficient recognition within cells is often difficult to achieve. Most studies, therefore, have focused on recognition of messenger RNA (mRNA).

mRNA can also be base-paired, but base-pairing is partial, thus making it is more accessible to binding by oligonucleotides than DNA. "More accessible", however, does not mean completely accessible. Some sequences will be more accessible than others and some oligonucleotides will therefore be much more functional than others. The correspondence between sequence target and function can be difficult to predict and ASOs designed to bind sequences close to one another can vary significantly in their activities.

ASOs are designed to target the sense strand of RNA (*i.e.* the strand coding for protein). In this chapter, we review the mechanisms that allow single-stranded ASOs to control gene expression. The focus is on ASOs that bind mature mRNA and repress translation as well as on ASOs that bind pre-mRNA to affect alternative splicing. Chapter 11 describes the mechanism of RNA interference (RNAi), which also employs a synthetic antisense strand (usually as one strand of a duplex) to control gene expression and which provides a useful counterpoint to the single-stranded ASO mechanisms described in this chapter.

2.2 RNase H and ASO Action

ASOs can recognize mRNA by Watson–Crick base-pairing. Once a target disease gene is identified, it is a simple matter to design a complementary sequence that will have the potential to bind the mRNA. However, recognition does not necessarily lead to control of gene expression. For example, an ASO that merely binds to an mRNA might be displaced by the ribosome and have no effect on protein expression. If the goal for therapy is to reduce the production of a protein, one challenge is to develop ASOs that will both bind and reduce RNA levels.

RNase H is a cellular enzyme that recognizes RNA–DNA hybrids.[3] Its normal cellular role is to degrade the RNA strand formed during lagging synthesis. While this physiological role takes place in the nucleus, RNase H is also found in the cytoplasm and can be active there.[4–6] Because RNase H can degrade RNA, agents that direct RNase H to specific sequences within RNA have the potential to reduce target mRNA levels. To take advantage of RNase H cleavage, the ASO should contain at least eight consecutive DNA bases so that a RNA–DNA hybrid will be formed upon binding, recruit RNase H, and promote cleavage of mRNA targets. Such cleavage is advantageous for controlling gene expression because one binding event is adequate to permanently destroy the RNA target. The ASO remains intact and it can move on to other mRNA targets, thus amplifying its potency.

However, the chemical and biochemical properties of DNA are not adequate for drug development. While DNA can bind RNA, DNA does not bind to RNA with optimal affinity. One option for improving the strength of binding is to make oligonucleotides longer, but synthesizing longer ASOs raises the cost, increases the likelihood of a harmful immune response, and creates a larger surface capable of making off-target interactions. Another drawback of relying solely on DNA is that it is a labile molecule that would be rapidly degraded by nucleases *in vivo*.

The drawbacks of DNA can be mitigated by introducing chemical modifications to enhance the affinity of binding and stabilize the ASO in order to block digestion by nucleases.[7,8] The dilemma for researchers designing ASOs was how to develop chemical modifications in order to achieve better binding whilst not interfering with the ASO's ability to direct cleavage of the target RNA by RNase H. The solution to this problem was the development of "gapmer" oligonucleotides consisting of an approximately eight base long DNA "gap" sandwiched between wings containing modified, usually RNA bases[9] (Figure 2.1A). The use of RNA-like sequences in the flanking regions, rather than DNA, is because RNA–RNA duplexes are more stable than DNA–RNA duplexes.

The DNA gap is designed to be long enough to support RNase H cleavage of the target mRNA. The modified flanking regions can be of various lengths and employ modifications that increase hybridization affinity. The increased binding affinity allows the overall length of the ASO to be reduced. Typical modifications include nucleotides with altered 2′ positions, such as 2′-fluoro, 2′-methoxyethyl (2′-MOE), or 2′-*O*-methyl[10] (see Chapter 1). Chemical

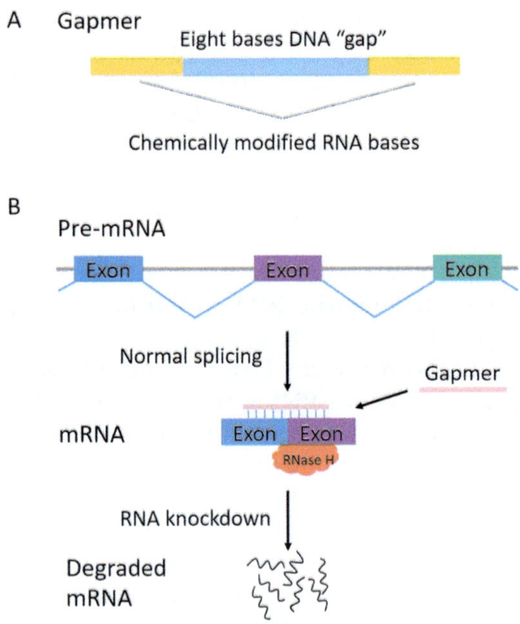

Figure 2.1 Gapmer ASOs. (A) Generic gapmer structure. Yellow: chemically modified RNA bases. Blue: Eight bases DNA "gap". (B) The gapmer binds to mRNA, recruits RNase H, and directs cleavage of the target mRNA.

modifications that constrain the ribose, like locked nucleic acid (LNA), can increase binding affinity even more[11–14] and can be used to tailor gapmers that combine RNase H activity with minimal length.[15] Phosphorothioate (PS) linkages stabilize ASOs against digestion by nucleases and improve binding to serum proteins.

Chemical modification allows gapmers to achieve all goals for recognition – tight binding, selective binding, and direction of RNase H for cleavage at a specific site (Figure 2.1B). The success of gapmers, however, is not absolute and not all gapmers are active. Some gapmers may have little or no activity, possibly because their complementary sequences within mRNA are not accessible. Other gapmers may produce RNase H-dependent toxic effects when added to cells through cleavage of non-target RNA sequences.[16–18] Therapeutic development programs typically involve screening dozens or hundreds of compounds to find the most potent ones for focused optimization, and gapmer activities can vary widely. While academic laboratories generally lack the resources to screen dozens of gapmers, at least six to ten should be screened to maximize the likelihood that a useful compound will be found.

2.3 ASOs and Regulation of Splicing

Pre-mRNA is produced during transcription in the nuclei of mammalian cells and consists of introns and exons that are spliced together to produce mature mRNA.[19] Splicing reactions occur at conserved sequences, consisting of the 5′ splice site at the 5′ end of an intron and the 3′ splice site at the 3′ end of an intron. Spliceosomes assemble at splice-sites, excise the introns, and link exons[20–22] (Figure 2.2A). Splicing is not uniform and exons from the same mRNA can be coupled in multiple ways to create different protein isoforms.[23] The functions of these protein isoforms may vary and some isoforms may have value for therapy. ASOs that redirect protein production towards therapeutic isoforms have the potential to be drugs by increasing a beneficial biological function that was previously missing or deficient.

Alteration of splicing is an important goal for oligonucleotide therapeutics. Most current therapeutic drugs are small molecules or antibodies. However, some disease targets are "undruggable" by these approaches (*i.e.* it has not proved possible to design a small molecule or antibody to bind and inhibit such targets). In addition, a small molecule or an antibody designed to bind a protein is unlikely to be able to achieve an increase in that protein's function. An ASO that affects splicing, however, might be able to increase production of a therapeutic isoform of a protein and thus achieve an outcome that is beyond the capability of more established approaches to drug development. Therefore, anti-splicing ASOs have the potential to fill unmet medical needs.

Increased production of a protein by anti-splicing ASOs also contrasts with the effects of gapmer ASOs that depend on cleavage by RNase H to reduce

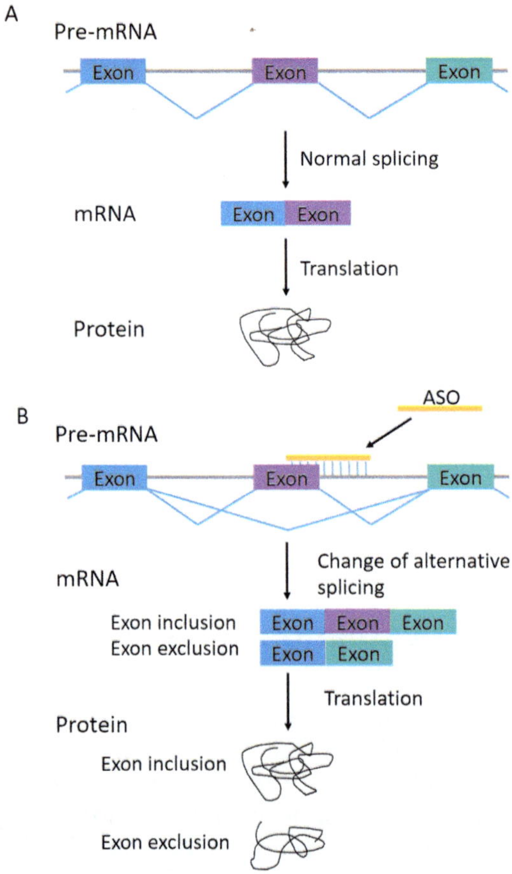

Figure 2.2 Mechanisms of alternative splicing of ASOs. (A) Pre-mRNA is spliced to yield mature mRNA, which is then translated into protein. (B) Regulation of alternative splicing by ASOs targeting regions close to intron–exon junctions. Outcomes can include ASO-mediated inclusion or exclusions of exons.

both RNA and protein levels. The generic advantage of anti-splicing ASOs *versus* gapmer ASOs is that they permit protein upregulation, an outcome that is difficult with other drug development strategies. The generic disadvantage is that the number of diseases that benefit from upregulation of a protein isoform and which are amenable to ASO-directed alternative splicing are much more limited than the large number of diseases that might benefit from the reduction of protein expression by an ASO.

Unlike gapmer ASOs, ASOs that modulate splicing should not induce cleavage of the target transcript. Instead, the ASOs use a "steric blocking" mechanism to prevent the binding of splicing factors and preserve the target transcript intact (Figure 2.2B). Depending on the gene and the location of the target sequence for the ASO, exon inclusion or exon exclusion can be achieved. The first step in development of an ASO to affect splicing is to use knowledge of gene splicing and biology to derive a hypothesis as to how redirecting alternative splicing might have a favorable functional outcome. The second step is to design candidate ASOs to bind complementary sequences at or near to key sequences within pre-mRNA to block the recognition of spliceosomes or splicing enhancement or inhibitory factors and thus modify alternative splicing.[24,25] It is likely that several ASOs will need to be tested to identify one or more that achieves adequate efficacy.

As described above for gapmer ASOs, chemical modifications also play an important role in the design of ASOs intended to modulate splicing. Chemical modifications are necessary to stabilize ASOs against degradation by nucleases, to increase the strength of hybridization to target, and to improve bio-distribution to target tissues *in vivo*. Common modifications are the same as those used for gapmers but the details of ASO design differ. PS backbone modifications increase ASOs' stability and the binding to the proteins in the plasma. Modifications at 2′ sugar positions of ASOs stabilize hybridization and increase *in vivo* potency. Commonly used 2′ modifications include 2′-*O*-methyl, 2′-*O*-methoxyethyl, and 2′-fluoro[10] (see Chapter 1). Phosphorodiamidate morpholino oligomers (PMOs) are another type of modification used to alter splicing. PMOs have a six-membered morpholino ring instead of the five-membered ribose ring and a phosphorodiamidate modified backbone instead of a PS backbone to increase stability *in vivo*.[26] Unlike gapmers, anti-splicing ASOs are chemically modified throughout the oligonucleotide because there is no need to preserve a central DNA region to act as a substrate for RNase H.

Therapeutic ASOs that alter splicing include compounds designed to treat Duchenne muscular dystrophy (DMD)[27–29] and spinal muscular atrophy (SMA).[30,31] In 2016, two of these ASOs were approved by the Food and Drug Administration (FDA), nusinersen (SMA) and eteplirsen (DMD).[32–34] These drugs are discussed in Chapter 6 and demonstrate the power of splice modulation as a therapeutic strategy. A similar steric blocking mechanism is used in the case of targeting of toxic RNA repeats (see below and Chapter 7).

2.4 ASOs and Activation of Frataxin, a Case Study for an Emerging Mechanism

Friedreich's ataxia (FA) is an inherited neuromuscular degenerative disease that is due to inadequate expression of the protein frataxin (FXN).[35,36] The disorder is caused by an expanded AAG trinucleotide repeat within intronic FXN mRNA. This expansion does not change the mRNA coding region or result in a mutant protein. The leading hypothesis to explain how reduced protein expression can arise from a mutant intron is that the expanded repeat binds chromosomal DNA through R-loop formation, induces histone modifications, and reduces gene transcription[37–39] (Figure 2.3).

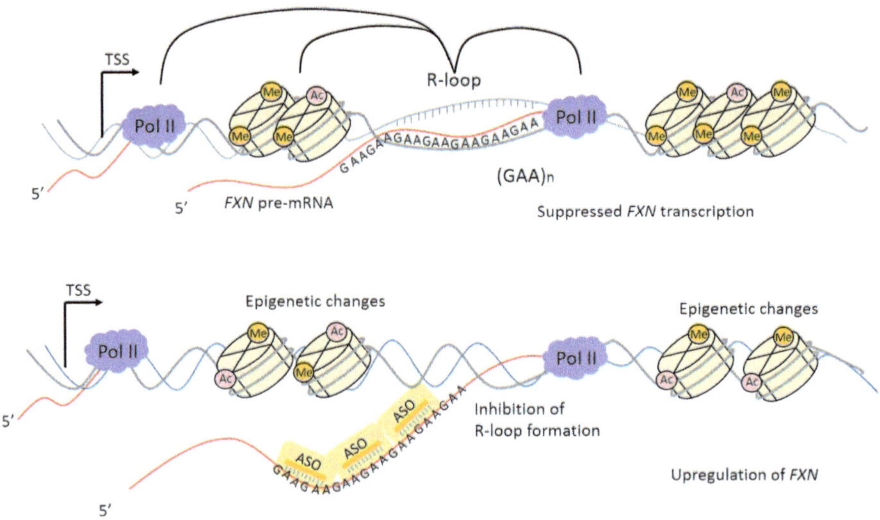

Figure 2.3 Activation of frataxin (*FXN*) expression using GAA-repeat-targeting ASOs. ASOs bind to the expanded repeat within intron 1 of *FXN* mRNA, prevent it from forming an R-loop with chromosomal DNA, and activate expression of *FXN*.

We reasoned that synthetic oligonucleotides designed to be complementary to the expanded repeat would block the expanded RNA repeat sequence and prevent it from binding to DNA. The resulting ASO–RNA hybrid would prevent R-loop formation and permit normal expression of *FXN* (Figure 2.3). This mechanism is unlike the gapmer mechanism (Figure 2.1) because it does not involve degradation of RNA. Whilst the mechanism does not alter splicing (Figure 2.2), it does involve steric blocking of the RNA target.

We designed ASOs that contain several LNA nucleotides. LNA nucleotides contain a 2′–4′ ribose linkage that "locks" the ribose in a confirmation that is ideal for hybridization (see Chapter 1) and was chosen as a starting point for design because it confers a substantial increase in binding affinity to target sequences. We introduced LNA ASOs complementary to the AAG repeat into fibroblast cells derived from patients with FA and observed activation of both *FXN* mRNA and protein expression to levels found in wild-type cells.[37] We then introduced anti-AAG LNA ASOs into cells with varying numbers of repeats and observed activation in all cell lines.[40]

To further explore ASO-mediated activation of *FXN* expression, we designed single-stranded ASOs containing other modified nucleotides including 2′-deoxy-2′-fluoro-beta-D-arabinonucleic acid (2′-F-ANA), 2′-MOE and phosphorothioate (PS) modifications[40] (see Chapter 1). ASOs containing 2′-F-ANA modification have enhanced resistance to nucleases and increased thermal stability.[41–43] We observed 1.5-fold to 3-fold activation of the *FXN* mRNA level by 2′-F-ANA, 2′-MOE, and 2′-fluoro modified ASOs. The results of these studies indicate that a diverse array of chemically modified ASOs can take advantage of a novel mechanism to activate *FXN* expression. More broadly, our findings confirm that ASOs are versatile molecules that can be adapted to control gene expression by emerging mechanisms.

2.5 Summary

ASOs are powerful tools for controlling gene expression. The two most widely used mechanisms include (1) recruitment of RNase H to reduce levels of a target transcript and (2) modulation of splicing to increase production of a therapeutic splice variant. However, RNA is involved in many mechanisms for controlling normal cellular process and disease. As we learn more about how RNA functions inside cells, emerging biological mechanisms will provide opportunities for exploiting ASOs to regulate gene expression and for developing ASOs for therapy.

Acknowledgements

This work was supported by the National Institutes of Health (GM106151 to DRC) and the Robert Welch Foundation (I-1244 to DRC). DRC holds the Rusty Kelley Professorship in Medical Science.

References

1. S. T. Crooke, *Nucleic Acid Ther.*, 2017, **27**, 70.
2. X. Shen and D. R. Corey, *Nucleic Acids Res.*, 2018, **46**, 1584.
3. H. Wu, W. F. Lima, H. Zhang, A. Fan, H. Sun and S. T. Crooke, *J. Biol. Chem.*, 2004, **279**, 17181.
4. T. A. Vickers and S. T. Crooke, *Nucleic Acids Res.*, 2015, **43**, 8955.
5. D. Castanotto, M. Lin, C. Kowolik, L. Wang, X. Q. Ren, H. S. Soifer, T. Koch, B. R. Hansen, H. Oerum and B. Armstrong, *et al.*, *Nucleic Acids Res.*, 2015, **43**, 9350.
6. K. A. Lennox and M. A. Behlke, *Nucleic Acids Res.*, 2016, **44**, 863.
7. G. F. Deleavey and M. J. Damha, *Chem. Biol.*, 2012, **19**, 937.
8. V. K. Sharma and J. K. Watts, *Future Med. Chem.*, 2015, **7**, 2221.
9. B. P. Monia, E. A. Lesnik, C. Gonzalez, W. F. Lima, D. McGee, C. J. Guinosso, A. M. Kawasaki, P. D. Cook and S. M. Freier, *J. Biol. Chem.*, 1993, **268**, 14514.
10. M. Manoharan, *Biochim. Biophys. Acta*, 1999, **1489**, 117.
11. A. A. Koshkin, S. K. Singh, P. Nielsen, V. K. Rajwanshi, R. Kumar, M. Meldgaard, C. E. Olsen and J. Wengel, *Tetrahedron*, 1998, **54**, 3607.
12. S. Obika, D. Nanbu, Y. Hari, J.-i. Andoh, K.-i. Morio, T. Doi and T. Imanishi, *Tetrahedron Lett.*, 1998, **39**, 5401.
13. D. A. Braasch and D. R. Corey, *Chem. Biol.*, 2001, **8**, 1.

14. P. P. Seth, G. Vasquez, C. A. Allerson, A. Berdeja, H. Gaus, G. A. Kinberger, T. P. Prakash, M. T. Migawa, B. Bhat and E. E. Swayze, *J. Org. Chem.*, 2010, **75**, 1569.
15. E. M. Straarup, N. Fisker, M. Hedtjarn, M. W. Lindholm, C. Rosenbohm, V. Aarup, H. F. Hansen, H. Orum, J. B. Hansen and T. Koch, *Nucleic Acids Res.*, 2010, **38**, 7100.
16. S. A. Burel, C. E. Hart, P. Cauntay, J. Hsiao, T. Machemer, M. Katz, A. Watt, H. H. Bui, H. Younis and M. Sabripour, *et al.*, *Nucleic Acids Res.*, 2016, **44**, 2093.
17. L. J. Kielpinski, P. H. Hagedorn, M. Lindow and J. Vinther, *Nucleic Acids Res.*, 2017, **45**, 12932.
18. P. J. Kamola, J. D. Kitson, G. Turner, K. Maratou, S. Eriksson, A. Panjwani, L. C. Warnock, G. A. Douillard Guilloux, K. Moores and E. L. Koppe, *et al.*, *Nucleic Acids Res.*, 2015, **43**, 8638.
19. D. L. Bentley, *Nat. Rev. Genet.*, 2014, **15**, 163.
20. D. L. Black, *Annu. Rev. Biochem.*, 2003, **72**, 291.
21. J. Bauman, N. Jearawiriyapaisarn and R. Kole, *Oligonucleotides*, 2009, **19**, 1.
22. N. A. Faustino and T. A. Cooper, *Genes Dev.*, 2003, **17**, 419.
23. E. S. Lander, L. M. Linton, B. Birren, C. Nusbaum, M. C. Zody, J. Baldwin, K. Devon, K. Dewar, M. Doyle and W. FitzHugh, *et al.*, *Nature*, 2001, **409**, 860.
24. Z. Dominski and R. Kole, *Proc. Natl. Acad. Sci. U. S. A.*, 1993, **90**, 8673.
25. M. A. Havens and M. L. Hastings, *Nucleic Acids Res.*, 2016, **44**, 6549.
26. J. Summerton, *Biochim. Biophys. Acta*, 1999, **1489**, 141.
27. R. Kole and A. M. Krieg, *Adv. Drug Delivery Rev.*, 2015, **87**, 104.
28. N. Goemans, M. Tulinius, A. K. Kroksmark, R. Wilson, M. van den Hauwe and G. Campion, *Neuromuscular Disord.*, 2017, **27**, 203.
29. J. R. Mendell, N. Goemans, L. P. Lowes, L. N. Alfano, K. Berry, J. Shao, E. M. Kaye and E. Mercuri, *Ann. Neurol.*, 2016, **79**, 257.
30. M. Hache, K. J. Swoboda, N. Sethna, A. Farrow-Gillespie, A. Khandji, S. Xia and K. M. Bishop, *J. Child Neurol.*, 2016, **31**, 899.
31. M. Scoto, R. S. Finkel, E. Mercuri and F. Muntoni, *Gene Ther.*, 2017, **24**, 514.
32. D. R. Corey, *Nat. Neurosci.*, 2017, **20**, 497.
33. A. Aartsma-Rus, *Nucleic Acid Ther.*, 2017, **27**, 67.
34. R. S. Finkel, E. Mercuri, B. T. Darras, A. M. Connolly, N. L. Kuntz, J. Kirschner, C. A. Chiriboga, K. Saito, L. Servais and E. Tizzano, *et al.*, *N. Engl. J. Med.*, 2017, **377**, 1723.
35. M. Pandolfo, *J. Neurol.*, 2009, **256**(Suppl 1), 3.
36. A. Collins, *Neurol. Clin.*, 2013, **31**, 1095.
37. L. Li, M. Matsui and D. R. Corey, *Nat. Commun.*, 2016, **7**, 10606.
38. M. Groh, M. M. Lufino, R. Wade-Martins and N. Gromak, *PLoS Genet.*, 2014, **10**, e1004318.
39. M. Groh, L. M. Silva and N. Gromak, *Biochem. Soc. Trans.*, 2014, **42**, 1123.

40. L. Li, X. Shen, Z. Liu, M. Norrbom, T. P. Prakash, D. O'Reilly, V. K. Sharma, M. J. Damha, J. K. Watts and F. Rigo, *et al.*, *Nucleic Acid Ther.*, 2018, **28**(1), 23–33.
41. A. Kalota, L. Karabon, C. R. Swider, E. Viazovkina, M. Elzagheid, M. J. Damha and A. M. Gewirtz, *Nucleic Acids Res.*, 2006, **34**, 451.
42. N. Souleimanian, G. F. Deleavey, H. Soifer, S. Wang, K. Tiemann, M. J. Damha and C. A. Stein, *Mol. Ther. Nucleic Acids*, 2012, **1**, e43.
43. M. M. Mangos, K. L. Min, E. Viazovkina, A. Galarneau, M. I. Elzagheid, M. A. Parniak and M. J. Damha, *J. Am. Chem. Soc.*, 2003, **125**, 654.

CHAPTER 3

The Medicinal Chemistry of RNase H-activating Antisense Oligonucleotides

PUNIT P. SETH AND ERIC E. SWAYZE*

Ionis Pharmaceuticals, 2855 Gazelle Court Carlsbad, CA 92010, USA
*E-mail: eswayze@ionisph.com

3.1 Introduction to Gapmers

Antisense oligonucleotides (ASOs) bind their cognate RNA in cells by Watson–Crick base-pairing and modulate their intermediary metabolism to produce a pharmacological effect.[1] ASOs can function through a variety of antisense mechanisms which can be broadly classified into those which promote degradation of the RNA by recruiting an endonuclease, or which are occupancy-based and modify RNA processing.[2] The chemistry of the ASO fundamentally determines which antisense mechanism it can harness after binding to its target RNA.[3] For example, duplexes formed between DNA ASOs and RNA are substrates for RNase H1, which selectively cleaves the RNA-strand of a DNA–RNA heteroduplex (see also Chapter 2 for mechanistic information).

Zamecnik and Stephenson showed that a 13-mer phosphodiester DNA oligonucleotide complementary to a region of the Rous sarcoma virus could inhibit viral replication.[4] However, unmodified phosphodiester

Drug Discovery Series No. 68
Advances in Nucleic Acid Therapeutics
Edited by Sudhir Agrawal and Michael J. Gait

Published by the Royal Society of Chemistry, www.rsc.org

DNA oligonucleotides lacked sufficient nuclease stability[5] which triggered a search for ASO designs that possessed improved nuclease stability and the ability to support RNase H1 cleavage.[6,7] DNA oligonucleotides modified with a phosphorothioate (PS) backbone modification fulfilled these requirements and were effective inhibitors of gene expression in cells and animals. This design enabled the approval of fomivirsen (Vitravene), which was locally administered *via* intravitreal injection.[8] While the PS DNA design was able to support a 1–3 day half-life in tissues, the resulting systemically administered drugs required frequent high doses.[9] Furthermore, PS reduces RNA-binding affinity [change in the melting temperature (ΔT_m) −0.5 °C per modification] of PS DNA ASOs.[10] These limitations conspired to limit the efficacy of PS DNA drugs in clinical trials. To address these limitations, the 'gapmer' design was introduced where a central region of PS DNA is flanked on either end with 2′-modified RNA (see Chapter 2).[11–13] The central gap-region of PS DNA permitted RNase H1 mediated cleavage of complementary RNA while the 2′-modifications enhanced affinity for complementary RNA and also protected the PS DNA gap region from exonuclease-mediated metabolism.[14]

In this review, we focus on the properties an RNase H-activating gapmer ASO must have to function effectively in animals, and on medicinal chemistry strategies to achieve these properties. The biochemistry and structural requirements for activating RNase H1 will be summarized, along with chemical modifications that can effect activation of RNase H1 when an ASO is bound to target RNA. The key modifications available to the medicinal chemist to engineer desired properties of the ASO will be briefly reviewed, as well as design strategies to achieve optimal activity in animal systems.

3.2 Human RNase H1

RNase H1 is ubiquitously expressed in all cells and selectively cleaves the RNA strand of a DNA–RNA heteroduplex.[15,16] RNase H1 is involved in the processing of R-loops[17] and Okazaki fragments[18] to maintain genome integrity.[19,20] RNase H1 is the critical enzyme implicated in the pharmacology of DNA-based gapmer antisense drugs which reduce RNA expression.[21] This has been supported by enzymology, reduction and overexpression of both H1 and H2[21] and conclusively demonstrated in conditional hepatocyte-specific RNase H1-knockout mice, where removal of RNase H1 eliminated ASO activity.[19] By far the great majority of ASOs that function to decrease target RNA expression require recruitment and activation of RNase H1 for their activity.

3.2.1 Biochemistry of Human RNase H1

The mammalian RNase H1 enzyme is comprised of three domains: catalytic, linker and hybrid-binding domains (HBD).[16] Lima *et al.* have reported extensive biochemical characterization of human RNase H1 to identify the structural requirements for binding and catalysis.[15] The full-length human RNase H1 prefers Mg^{+2} for catalysis and is inhibited by Mn^{2+}.[22] The enzyme is inhibited by *N*-ethylmaleimide (NEM), which alkylates a conserved cysteine residue (Cys148) adjacent to the catalytic site.[22] The enzyme is only active under reducing conditions, since Cys148 can form a disulfide bond with Cys147, which inactivates the enzyme but does not alter binding affinity for the duplex.[23] The enzyme uses two lysine residues and one tryptophan in the HBD to position the enzyme at the 5′-RNA–3′-DNA end of the heteroduplex.[24] Mutation of these key amino acids reduces binding affinity and the positional specificity of RNA cleavage. The enzyme cleaves the RNA strand 7–8 nucleotides from the 5′-RNA–3′-DNA end of the heteroduplex[25] and is sensitive to the positioning of 2′-modified nucleotides in the DNA strand of the heteroduplex.[26]

Wu and colleagues estimated the binding affinity of recombinant human RNase H1 for RNA–RNA (79 nM), RNA–2′-Fluoro (2′F)-RNA (76 nM), DNA–2′F-RNA (99 nM) and DNA–DNA (3608 nM) duplexes using competition inhibition assays. The enzyme also binds single-stranded (ss) DNA (1506 nM) and RNA (1400 nM) with higher affinity than DNA–DNA duplexes but the highest affinity was observed for phosphorothioate (PS)-modified ss DNA (36 nM). Interestingly, ss PS DNA also acts as a competitive inhibitor of human RNase H1 at higher concentrations.[27] Gaidamakov and colleagues showed that the full-length mouse RNase H1 bound DNA–RNA with high affinity (40 nM).[28] In contrast, the *Escherichia coli* enzyme, which is comprised of only the catalytic domain, bound DNA–RNA duplexes with lower affinity (<600 nM). The results of Wu and colleagues and Gaidamakov and colleagues together highlight the contributions of the HBD for enhancing the binding affinity of the mammalian enzyme for DNA–RNA heteroduplexes. Nowotny *et al.* confirmed these observations and reported dissociation constants of the isolated HBD for DNA–RNA (<200 nM), RNA–RNA (4900 nM) and DNA–DNA (23 000 nM) duplexes.[29]

3.2.2 Structural Biology of Human RNase H1

X-ray crystal structures of the catalytic domain and HBD provided insights into how the enzyme recognizes DNA–RNA heteroduplexes.[29–31] The catalytic domain of human RNase H1 makes several key contacts with the DNA and RNA strands of the heteroduplex (Figure 3.1). The majority of the contacts occur along the minor groove of the heteroduplex. The RNA strand is in the expected A-form (C3′-endo) conformation. The enzyme makes contact with two 2′-hydroxyl groups on either side of the scissile phosphate. The side chain of Gln186 and the backbone amides of Cys148, Ser150 and Asn151 and Met212 participate in these contacts.

The side-chains of amino acids Asn151, Asn182 and Gln183 make several key H-bonding interactions with the nucleobases in the major groove.[31] The enzyme also makes a key contact with a phosphodiester linkage of the

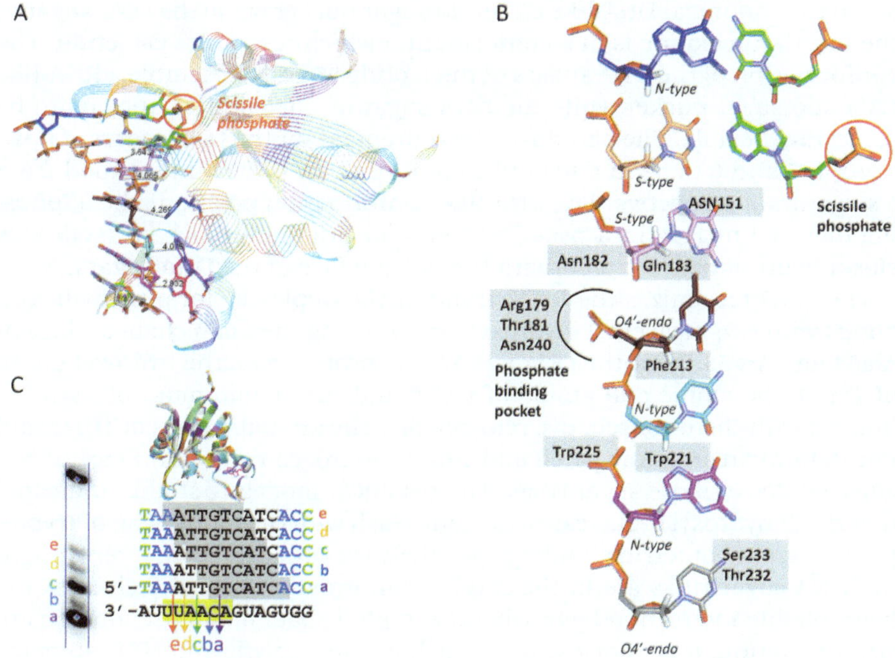

Figure 3.1 Interactions of the catalytic domain of RNase H1 with a DNA–RNA heteroduplex showing (A) the seven-nucleotide footprint on the DNA strand (only the RNA nucleotides (green) flanking the scissile phosphate shown for clarity), (B) conformation of individual DNA nucleotides and the important sites of contact, and (C) the catalytic domain, which has several distinct overlapping footprints on the heteroduplex for each cleavage site on the RNA strand.

DNA strand within the phosphate-binding pocket formed by amino acids Arg179, Thr181 and Asn240. Another key site of interaction is within the hydrophobic DNA-binding channel formed by the aromatic amino acids Phe213, Trp221 and Trp225. The aromatic side chains of these amino acids make close hydrophobic contacts with the bottom face of the DNA sugars. This provides specificity for DNA over RNA and the most common nucleoside modifications, since the 2′-hydroxyl groups in RNA and/or substitutions on the alpha face of the sugar disrupt the hydrophobic contacts. Lastly, Ser 233 and Thr 232 make weak contacts with the non-Watson–Crick face of the nucleobase at the last position of the footprint. The above contacts represent the seven-nucleotide footprint of the catalytic domain on the DNA–RNA heteroduplex.

The enzyme produces numerous structural distortions in the DNA strand which confer specificity for DNA over RNA.[31] For example, the DNA sugar of the first nucleotide of the footprint for a given cleavage site is in a RNA-like C3′-endo conformation. The DNA sugars of the second and third nucleotides are in the canonical DNA-like C2′-endo sugar pucker while the DNA sugar of the fourth nucleotide is in a conformation which resembles O4′-endo. The conformation of the DNA sugars of nucleotides 5 and 6 resemble a RNA-like C3′-endo sugar pucker while the DNA sugar of nucleotide 7 appears to be somewhat flexible. The paradoxical requirement for the DNA sugars at positions 4, 5 and 6 to be in a RNA-like conformation is best understood from a structural perspective. The RNA-like conformation positions the 3′-phosphodiester linkages in a pseudo-equatorial orientation, which facilitates closer hydrophobic contacts with the bottom face of the DNA sugars.

The HBD recognizes the RNA strand of the duplex by forming hydrogen bonds with two adjacent 2′-hydroxyl groups using the main chain amides of Ala52 and Arg55.[29] For the DNA strand, the protein uses the hydroxyl group of Tyr29, the amine and amide of Lys60 and the guanidinium of Arg57 to interact with three consecutive phosphates. The aromatic rings of Trp43 and Phe58 form an aromatic patch and closely approach the bottom face of two adjacent deoxyribose sugar rings. The results of modeling studies indicated that the 2′-hydroxyl group causes a steric clash with the indole ring of tryptophan, thereby conferring binding specificity for DNA over RNA. Interestingly, the DNA sugar rings are in the *N*-type conformation, which facilitates the hydrophobic interaction by moving the 3′-phosphate into a pseudo-equatorial orientation. It is not clear precisely how the catalytic and HBD domains interact with the gapmer ASO–RNA duplex, but several models have been proposed.[16,26]

3.3 Structure–Activity Relationships of Gap Modifications

The catalytic domain of RNase H1 has a distinct seven-nucleotide footprint on the heteroduplex for every cleavage site on the RNA.[24] Since RNase H1 cleaves the RNA strand at multiple sites, it has several overlapping seven-nucleotide footprints on the heteroduplex.[32] As a result, every nucleotide within the footprint has to have the ability to adopt several distinct conformations to support RNA cleavage at different sites along the heteroduplex. While PO DNA has the ability to fulfil these requirements, it does not have sufficient nuclease stability to be used in ASO drugs. The nuclease lability of PO DNA necessitated the use of chemical modifications that could enhance metabolic stability but also be compatible with the biochemistry of RNase H1 to support RNA cleavage. In the subsequent paragraphs, we describe chemical modifications that have been used to stabilize the DNA gap-region against nuclease degradation (see Figures 3.2–3.4). In addition, we also describe recent work where non-DNA modifications have been used to modulate RNase H1 cleavage to enhance specificity of gapmer designs for allele-selective gene repression.

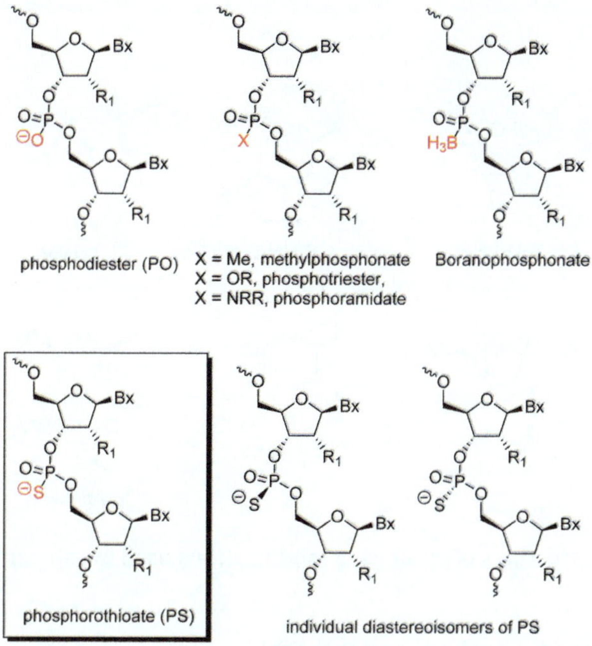

Figure 3.2 Structures of backbone modifications, including the phosphorothioate (PS) backbone and its individual diastereoisomers.

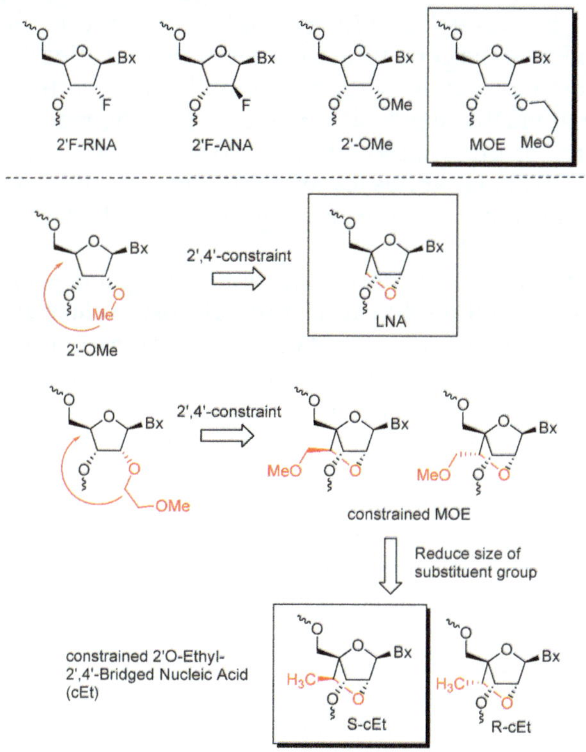

Figure 3.3 Structures of chemical modifications used in the wings of gapmer ASOs and the design of LNA, cMOE and cEt modifications.

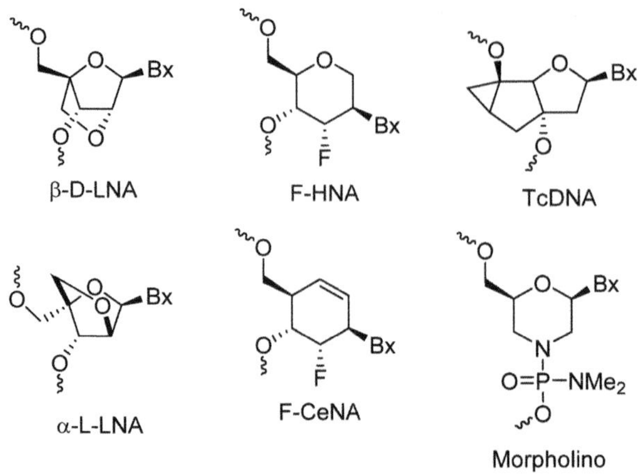

Figure 3.4 Structures of other wing modifications used in gapmer therapeutics.

3.3.1 Phosphorothioate (PS) DNA

The PS linkage, where one of the non-bridging oxygen atoms of natural phosphodiesters has been replaced by a sulfur atom, represents one of the oldest and most widely used modifications in nucleic acid therapeutics.[33] Interestingly, PS DNA has been found in certain bacteria, presumably to enhance nuclease stability of the genome at the sites of modification.[34] The PS linkage can be synthesized by treating the trivalent phosphite intermediate formed during automated DNA synthesis[35] with a sulfurizing reagent such as Beaucage's reagent[36] or phenylacetyl disulfide (PADS) (see Chapter 2).[37] Contributions by several laboratories towards the synthesis of PS-oligonucleotides helped to establish antisense technology in the early days of nucleic acid-based therapeutics.[38–40]

Since sulfur is larger and more hydrophobic than oxygen, and the negative charge in the PS modification rests almost exclusively on the sulfur atom, it is not surprising that the PS modification can impart dramatically different properties on ASOs. Each PS substitution reduces affinity for complementary nucleic acids (ΔT_m −0.5 °C per modification),[41] and although small, the loss in affinity can become significant when using fully PS modified 16- or 20-mer ASOs. PS modification dramatically enhances nuclease resistance, making unformulated PS DNA ASOs sufficiently stable for use in animals. This nuclease resistance is modest for exonucleases, but quite significant for endonucleases, and PS DNA-containing gapmer ASOs have multiple-week half-lives in animals and humans that correlate to long durations of effect.[42,43]

In addition to enhancing nuclease stability, the PS modification dramatically increases affinity of the oligonucleotide for proteins[44] and confers significant pharmacokinetic benefits for use in animals.[45] The

PS modification promotes transient association of the oligonucleotide drug with plasma proteins, which facilitates distribution to peripheral tissues.[46] The oligonucleotides are thought to partition onto cell-surface proteins, which facilitates cellular entry without the need for specialized delivery vehicles. While the molecular origins for the enhanced affinity of PS ASOs for proteins are not yet fully understood, they can be measured using biophysical techniques such as bioluminescence resonance energy transfer (BRET),[47] surface plasmon resonance[48] and fluorescence polarization.[49]

3.3.2 Chiral PS DNA

The PS linkage is chiral at phosphorus and a 20-mer oligonucleotide is a mixture of 2^{19} individual stereoisomers.[50] An NMR structure of a DNA–RNA duplex with *R*p or *S*p linkages showed that the *R*p linkage adopts similar backbone conformational geometry to the natural PO linkage.[51] In contrast, the *S*p linkage changes orientation of backbone torsion angles away from the canonical geometries observed in natural nucleic acid duplexes. The individual *R*p and *S*p stereoisomers also have distinct biophysical and biochemical properties. The *R*p isomer has similar RNA-binding properties but is only marginally more stable to most exonucleases as compared with PO oligonucleotides.[52] The *S*p isomer displays slightly reduced RNA-binding but is significantly more stable to most nucleases as compared with *R*p and PO oligonucleotides. Human RNase H1 exhibits slight preference for the *R*p isomer in the DNA strand but will cleave RNA in *S*p DNA–RNA heteroduplexes with slightly lower efficiency. As a result, pure *R*p DNA gapmer ASOs show similar potency to stereorandom PS ASOs in cell culture but greatly reduced activity in animals, where nuclease stability is critical for activity.[53] All *S*p DNA gapmer ASOs show reduced activity by virtue of reduced RNA-binding and RNase H1 efficiency, but are significantly more stable in animals.

The differential requirements for RNA-binding, nuclease stability and RNase H1 efficiency indicate that oligonucleotides with mixtures of *R*p and *S*p linkages in the DNA-gap display optimal properties for therapeutic applications.[53] Recent work has supported this, where ASOs with *S*p-DNA in the gap region interspersed with strategically placed *R*p-DNA linkages showed increased duration of pharmacological effects in mice.[54] Interestingly, neither potency nor tissue accumulation was improved, indicating that increased stability was not responsible for the improved duration of action. The ability of *R*p and *S*p linkages to modulate RNase H1 cleavage preferences has been utilized to confer allele selectivity by targeting SNPs in the mRNA.[55] This is achieved by placing a single *R*p linkage in the 3′-region of an otherwise *S*p gap, thereby favoring the phosphate binding pocket to form at the *R*p site and directing cleavage to near the mismatch site on the non-targeted allele. Controlling the stereochemistry of the PS linkage in the wings of gapmer ASOs showed no discernible benefits.[56] However, significantly more work with additional sequences will have to be done to draw any conclusions regarding the generality of these effects.

3.3.3 Methyl Phosphonates, Phosphoramidates, Phosphotriesters and Boranophosphonate DNA

The analogous isosteric substitutions of oxygen by carbon, nitrogen, alkoxy or boron yield alkyl phosphonate, phosphoramidate, phosphotriester or boranophosphonate modifications respectively (Figure 3.2).[3] While all these modifications also enhance nuclease stability and have been known for almost as long as the PS modification, they have found limited utility in the context of gapmer technology. Some of this may be related to synthetic challenges associated with incorporating these linkages into oligonucleotides using standard automated chemistry. Several of these modifications also ablate RNase H1 activity near the site of incorporation, which limits their use within the gap region. RNase H1 makes several key electrostatic contacts with the DNA backbone, and neutralizing the charge, such as with methylphosphonate, phosphoramidates and phosphotriesters, can interfere with these important interactions. Interestingly, boranophosphonate DNA displays a negative charge on the non-bridging oxygen atom and was shown to support RNase H1 cleavage in the context of polythymidylate oligonucleotides.[57]

3.3.4 2′-Fluoro Arabino Nucleic Acids (FANA)

2′-Fluoro-arabino nucleic acid (FANA) represents one of a handful of chemical modifications which can support RNase H1 activity near the site of incorporation (Figure 3.3).[58] Wilds and Damha have reported extensively on the biophysical and biochemical evaluation of FANA-modified oligonucleotides.[59] Incorporation of FANA enhances RNA-affinity (+1 °C per modification) and nuclease stability as compared with PS DNA.[60] Elegant work by the Damha and Gonzalez labs showed that the 2′-F atom in FANA can form pseudo C–H···F–C hydrogen-bonds with the nucleobase of the 3′-adjacent nucleotide.[61] Furthermore, the electronegative fluorine polarizes the vicinal hydrogen, which can now participate in duplex stabilizing C–H···O interactions with O4′ and O5′ of the 3′-adjacent nucleotide.[62]

Early work showed that FANA can enhance RNase H1 catalysis when incorporated in the DNA strand in the vicinity of the cleavage site on the RNA.[63] While complete substitution of FANA for DNA does not support RNase H1 activity, incorporation of PS FANA in the wings or as blockmers alternating with PS DNA nucleotides results in RNase H1-dependent antisense activity in cells upon transfection.[64] The FANA ASOs were shown to be more potent and have a longer duration of effect in cells as compared with the PS DNA ASO. PS FANA–DNA ASOs were also shown to be taken up by cells without the use of transfection methods,[65] a property shared by all gapmer ASOs by virtue of the ability of the PS backbone to adhere to cell-surface proteins.

3.3.5 DNA-like Modifications

Site-specific replacement of DNA in the gap-region has been investigated for improving the allele selectivity of gapmers. ASOs targeting SNPs associated with the CAG expansion in mutant *huntingtin* transcripts have been investigated as potential treatments for Huntington's disease (see also Chapter 2).[66] The mismatch created in the ASO–RNA heteroduplex by targeting a SNP in the mRNA shifts the RNase H1 cleavage preference towards less-preferred cleavage sites.[67] Insertion of non-DNA chemical modifications at strategic locations in the gap-region of the ASO ablates these less-preferred cleavage sites and increases allele selectivity. Østergaard *et al.* showed that 2-thio-deoxythymidine (2S-dT) in the DNA gap-region enhances allele selectivity by shifting RNase H1 cleavage by two nucleotides towards the 5′-DNA–3′-RNA end of the heteroduplex.[68] A structural model indicated that the increased size and hydrophobicity of the sulfur atom disrupts key contacts with asparagine and glutamine residues in the catalytic domain of RNase H1. Similarly, replacing PS DNA at specific locations in the gap with two isomers of α,β-constrained nucleic acid (CNA) also increased allele selectivity in a configuration-dependent manner.[69]

Recently, Østergaard and colleagues reported the effect of replacing each nucleotide position in the DNA gap with fluorinated modifications, such as FANA, 2′-fluoro RNA (FRNA), Fluoro Hexitol Nucleic Acid (FHNA), fluoro cyclohexenyl nucleic acid (F-CeNA) and 2′-fluoro *N*-methanocarba nucleic acid (F-NMC), on allele selectivity.[32] Replacing DNA in the middle of the gap with RNA-like modifications (FRNA and FHNA) reduced activity, while DNA-like modifications (FANA and F-CeNA) were better tolerated in the gap. A comprehensive analysis of the RNase H1 cleavage patterns indicated that the non-DNA modifications modulate cleavage preference by their ability, or lack thereof, to adopt preferred conformations along the seven-nucleotide footprint of the catalytic domain of RNase H1.

3.4 SAR of Wing Modifications

Gapmer ASOs have a central gap-region of 7–12 PS DNA nucleotides flanked on both ends with modified nucleotides, which are often referred to as 'wings'.[12] The wing region is typically two to five nucleotides in length and is comprised of modifications, which can protect the DNA-gap from exonuclease digestion and also enhance affinity for complementary RNA. Early gapmer designs utilized 2′-OMe, 2′-*O*-alkyl and 2′-FRNA modifications in the wings. While 2′-FRNA modifications can enhance metabolic stability, this is generally not sufficient for *in vivo* applications. FRNA gapmers can also show cellular toxicities resulting from non-specific interactions with nuclear proteins polypyrimidine tract binding protein-associated splicing factor (PSF) and nuclear RNA-binding protein 54-kDa (p54nrb).[70] 2′-*O*-alkyl modifications with larger alkyl groups, such as propyl, increase nuclease stability but reduce affinity for RNA.

In contrast, 2′-OMe nucleotides have increased metabolic stability relative to 2′-FRNA, but slightly reduced metabolic stability and RNA affinity as compared with the MOE modification.

The most advanced gapmer designs in clinical use utilize 2′-modifications like MOE, but more recent design iterations use higher-affinity modifications, such as LNA and cEt. Unlike the gap-region, which displays strong preference for PS DNA, the wing-region is surprisingly tolerant of structural diversity, indicating that this region is not extensively contacted by RNase H1. As a result, a very large number of nucleotide modifications have been evaluated in the wing regions of gapmer ASOs.[3] These include a plethora of 2′-modifications, bicyclic analogs, such as LNA, analogs of hexitol nucleic acids (HNA) and tricyclic analogs, such as tcDNA. In the following sections, we focus primarily on modifications that have been tested in animal models.

3.4.1 2′-*O*-Methoxyethyl RNA (MOE)

MOE represents a 2′-modification where an ethylene glycol moiety is appended to the 2′-hydroxyl group in RNA (Figure 3.3).[71] The sugar ring in MOE has the same conformational preference as RNA but restricted rotation around the MOE side-chain increases rigidity and boosts affinity for RNA (ΔT_m +1–2 °C per modification).[72] The increased steric bulk of the methoxyethyl group provides resistance to nuclease-mediated digestion, and the gauche orientation of the ethylene glycol traps water molecules in the vicinity of the phosphate backbone. This rigidifies the structure and contributes to the affinity improvement, as well as enhancing hydration, which reduces some of the non-specific protein interactions observed with PS DNA ASOs.[27] Multiple MOE-containing ASOs have entered clinical development in the past decade and two ASOs, mipomersen[73] and nusinersen,[74] were recently approved by the Food and Drug Administration (FDA) for the treatment of familial hypercholesterolemia and spinal muscular atrophy (SMA), respectively. The pharmacological properties of MOE gapmer ASOs are discussed in further detail in the ASO design section.

3.4.2 Locked Nucleic Acids (LNA) and Constrained Ethyl 2′-4′-Bridged Nucleic Acids (cEt)

LNA represents a conformationally restricted analog of 2′-OMe RNA where the methyl group has been constrained to the 4′-position of the furanose sugar ring (Figure 3.3).[75,76] This locks the sugar ring in LNA in the RNA-like C3′-endo conformation and provides unprecedented increases in binding affinity for complementary nucleic acids (ΔT_m +5 °C per modification). Structural studies revealed that LNA helices are compressed and comprise approximately 14 base pairs for every helical turn, indicating that the improved RNA-affinity is likely to be a result of enhanced base-stacking.[77] LNA oligonucleotides have been investigated extensively for nucleic acid-based diagnostic and for therapeutic applications.[78]

Conformational restriction of the methoxyethyl side-chain in MOE to the 4′-position yielded the constrained MOE modification.[79] Tethering of the MOE side-chain resulted in the formation of diastereoisomers, which were termed *R*- and *S*-constrained MOE depending on the configuration of the newly created chiral center. Shortening the size of the substituent group in cMOE provided the constrained ethyl (cEt) modification.[80] Both cMOE and cEt modifications showed similar enhancements in affinity for RNA to LNA, but a greatly enhanced nuclease resistance profile relative to MOE and LNA. A structural model indicated that the appended substituent groups in cMOE and cEt displace a key magnesium ion in the active site of the nuclease, accounting for the enhanced nuclease stability of these modifications.[81] The pharmacological properties of LNA and cEt gapmers are discussed in greater detail in the ASO design section.

3.4.3 α-L-LNA and Related Analogs

α–L-LNA represents the alpha-anomer of enantio-LNA and was first described by members of the Wengel lab during a screen of all eight stereoisomers of LNA (Figure 3.4).[82] α-L-LNA displays similar binding affinity for complementary RNA as does LNA.[83] Structural studies indicated that α-L-LNA modified nucleic acids form duplexes that resemble the H-form helical geometry recognized by RNase H1.[84] However, α-L-LNA nucleotides ablate RNA cleavage by RNase H1 near the site of incorporation in the DNA strand. α-L-LNA gapmers showed excellent activity for reducing gene expression in cells and in mice but showed a tendency to cause immuno-stimulation as evidenced by increased spleen and liver weights in mice.[85] These changes could be mitigated by using 5′-methyl, but not 6′- or 3′-methyl, analogs of α-L-LNA.

3.4.4 F-HNA and F-CeNA

Hexitol and cyclohexenyl nucleic acids represent nucleic acid analogs where the five-membered furanose ring in DNA and RNA is replaced by six-membered rings (Figure 3.4).[86] In HNA, the furanose ring is replaced by a hexitol ring system, which mimics the C3′-endo conformation in RNA. Appending fluorine in the 'ribo' configuration to the hexitol scaffold provides F-HNA (ΔT_m +3 °C per modification), which further enhances the RNA-binding property of HNA.[87] HNA-modified and F-HNA-modified gapmer ASOs were evaluated in mice, where the F-HNA ASO showed twofold improved potency as compared with the HNA ASO.[88] F-HNA ASOs showed similar RNA-binding properties and potency in mice to those of sequence- and design-matched LNA ASOs. However, F-HNA ASOs demonstrated an improved toxicity profile relative to LNA gapmers. The improved potency was attributed to the fluorine substitution, which presumably modulates the protein-binding and intracellular trafficking of F-HNA ASOs.

The cyclohexenyl ring in CeNA is more flexible than the hexitol ring in HNA and has been shown to adopt DNA-like or RNA-like conformations in nucleic acid duplexes depending on the sequence context of the incorporation.[89] The increased flexibility of the cyclohexenyl ring reduces the ability of CeNA to enhance duplex stability relative to the HNA modification. Appending fluorine in the ‘ribo’ configuration enhances RNA-binding affinity modestly relative to CeNA modified oligonucleotides.[90] Interestingly, a F-CeNA ASO showed similar RNA-binding but improved potency as compared with a sequence- and design-matched MOE gapmer ASO. Once again, the improved potency was attributed to fluorine substitution and its ability to modulate ASO protein binding and trafficking properties.

3.4.5 Tricyclo DNA (tcDNA)

tcDNA is a unique nucleic acid modification that restricts conformation flexibility around the C4′–C5′-exocyclic bond in DNA (Figure 3.4).[91] Structural studies of tcDNA-modified duplexes revealed that the cyclopropyl ring fusion results in compensatory changes in backbone torsion angles α and β which increases affinity for complementary nucleic acids.[92] Uniformly modified tcDNA ASOs have shown interesting properties for modulating splicing in mice.[93] A study of tcDNA gapmers where the length and tcDNA content were varied provided interesting insights into the therapeutic profile of this modification.[94] The 18- and 20-mer tcDNA gapmers showed similar RNA-binding and activity in the liver to the 20-mer MOE gapmer. The tcDNA gapmer showed improved activity in the muscle, indicating that the increased hydrophobicity of the tcDNA scaffold may improve delivery to this tissue. Interestingly, the 14-mer and 20-mer tcDNA ASOs showed similar cell and tissue accumulation, which highlight the ability of tcDNA to increase cellular uptake.[94]

3.4.6 Phosphorodiamidate Linked Morpholinos (PMOs)

PMO ASOs are charge-neutral nucleic acid analogs where the furanose ring of DNA is replaced by a six-membered morpholino ring and the phosphodiester backbone is replaced by a phosphorodiamidate linkage (Figure 3.4).[95] Uniform PMO ASOs have been primarily used for splice-correction and one PMO ASO, eteplirsen, was recently approved by the FDA for the treatment of Duchenne muscular dystrophy (DMD) (see Chapter 6).[96] Uniform PMO ASOs do not activate RNase H1 and the evaluation of PMO–DNA gapmers was thwarted, since the chemistry used for the preparation of PMOs is incompatible with the automated synthesis of DNA oligonucleotides. The Caruthers lab has recently addressed this issue and synthesized DNA–PMO gapmers and mixmers using boranophosphonate intermediates.[97] The PMO gapmers were reported to have similar RNA-binding properties to DNA oligonucleotides but biological data on these interesting gapmer ASOs has not yet been reported.

3.5 Design of Gapmer ASOs

Unmodified DNA ASOs did not possess sufficient metabolic stability or pharmacokinetic properties in animals and were largely replaced by PS DNA ASOs.[98,99] The first decade and a half of studies of antisense technology was devoted to understanding the pharmacological properties of first-generation PS DNA ASOs. This work resulted in the realization that PS DNA ASOs did not possess sufficient drug-like attributes to be used as effective systemically administered therapeutics. To address these limitations, gapmer designs were introduced where some of the DNA modifications in the flanks of the ASO were replaced by RNA-affinity and nuclease stability-enhancing 2′-modified nucleotides (Figure 3.5).

3.5.1 MOE Gapmer ASOs

The PS MOE gapmer is currently the most advanced and widely investigated oligonucleotide-based design in clinical use. Over 50 MOE gapmers targeting genes expressed in multiple tissues and spanning several diverse therapeutic areas have advanced into clinical development. Mipomersen, a second-generation ASO targeting apoB100 mRNA, was the first systemically delivered ASO approved by the FDA for the treatment of familial hypercholesteremia.[73]

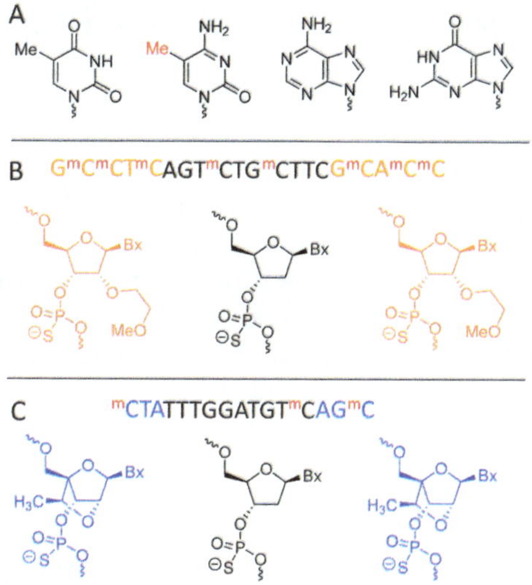

Figure 3.5 Design of gapmer ASOs (A) structures of nucleobases (B) Sequence and design of mipomersen, a representative MOE gapmer (C) Sequence and design of IONIS-STAT3-2.5$_{Rx}$ (AZD9150), a representative cEt gapmer. T = thymine, ^{m}C = 5-methylcytosine, A = adenine, G = guanine; Black indicates DNA; orange indicates 2′-MOE nucleosides; blue indicates cEt nucleosides; internucleoside linkages are all phosphorothioate.

Two additional MOE gapmers, inotersen[100] and volanosorsen,[101] completed successful phase 3 studies in 2017 for the treatment of transthyretin (TTR) amyloidosis and hypertriglyceridemia respectively. While these drugs target the liver, MOE gapmer ASOs can effect target reduction broadly when given systemically, although the required dose varies by tissue.[102] MOE gapmers are also active in the CNS if delivered locally.[103] A MOE gapmer targeting huntingtin mRNA recently completed a successful phase 1/2 trial where dose-dependent target reduction was observed.[104] The pharmacological, pharmacokinetic and toxicological properties of MOE gapmers have been extensively reviewed and are discussed in greater details in Chapter 4 of this book.

MOE gapmers show improved affinity for RNA and enhanced stability in tissues.[14] This enhances potency and duration of pharmacological effect, which effectively translates to reduced doses in clinical settings relative to first-generation PS DNA ASOs.[105] The increased hydration imparted by the MOE modification also mitigated some of the non-specific toxicities arising from interactions of the PS backbone with cellular proteins. The therapeutic profile of MOE gapmers has been further improved by cell-specific targeting strategies, which have provided 30-fold enhancement in potency for hepatocyte targets in clinical practice.[106] The use of cell-targeting strategies has also enabled the use of mixed-backbone designs, where some of the PS modifications in the MOE wings can be replaced by phosphodiester linkages.[107] These designs have the potential to reduce non-specific toxicities observed with some gapmer ASOs, without compromising cell-uptake and pharmacokinetic properties.

3.5.2 LNA Gapmers

LNA provides unprecedented increases in affinity for complementary RNA.[75,108] The use of high-affinity modifications allowed reduction in overall length of the ASO from 20-mers down to 14-, 16- and even 12-mers.[109] Somewhat counterintuitively, these shorter ASO designs exhibited improved potency in animal models despite lower overall tissue accumulation.[94] Interestingly, a bell-shaped relationship between affinity and *in vivo* activity with LNA gapmers has been reported, indicating that increasing RNA-affinity beyond a certain threshold will not increase, and may even decrease, activity.[110,111] LNA gapmer ASOs also showed an increased propensity to cause hepatotoxicity[112] which can be mitigated by small changes in sequence[113] or structure.[114,115] LNA ASOs do not appear to be genotoxic.[116] Recent research has implicated certain sequence motifs,[117] genes with long introns,[118] very long pre-mRNA transcripts,[119] endosomal pathways[120] and non-specific protein binding[121] as potential mechanisms for toxicity, and established *in vitro*[122,123] and *in vivo*[124] assays as screening tools to assess toxicity. While LNA has been broadly assimilated in the field of gapmer ASOs, comparatively few ASOs have progressed beyond early clinical studies.

3.5.3 cEt Gapmers

There are currently multiple cEt gapmer ASOs in clinical development. They exhibit 5–10-fold enhancement in potency relative to MOE gapmers for reducing gene expression in the liver,[114] and importantly show significant improvements in potency in tissues outside the liver.[94] As a result, the initial clinical development paths for cEt gapmers have focused on oncology indications, where they have shown encouraging data in early and mid-stage trials.[125–127] cEt gapmers have also shown enhanced potency in tissues such as muscle. A cEt ASO targeting dystrophia myotonica protein kinase (DMPK) mRNA was advanced into clinical development for the treatment of myotonic dystrophy.[128] cEt gapmers show 60-fold increased potency relative to MOE gapmers when combined with cell-specific targeting strategies for targeting genes expressed in liver hepatocytes.[129] The cEt modification has also been combined with MOE in the wings and evaluated as mixed gapmer designs. Combining these modifications can help titrate ASO affinity for its target RNA for specialized applications, such as allele-selective silencing,[130] and for modulating ASO tolerability.[131]

3.5.4 ASO Gapmer Duplexes

Gapmer ASOs have been almost exclusively evaluated as single-stranded designs in the literature. The single-stranded nature makes the nucleic acid more flexible and allows interactions of the PS backbone with plasma, cell-surface and intra-cellular proteins, which facilitates tissue distribution and cell uptake. Yokota and colleagues recently evaluated ASO duplexes where a PS LNA or cEt gapmer was duplexed with a PO RNA strand minimally protected with PS OMe nucleotides in the flanks.[132] The heteroduplex was further conjugated with tocopherol, which facilitated interactions of the ASO–RNA duplex with lipoproteins in plasma. In contrast, single-stranded PS gapmer ASOs are mostly associated with albumin. The tocopherol ASO duplexes showed surprisingly improved potency in the liver, which indicated that the change in plasma protein binding profile may alter tissue distribution and cellular uptake.

3.6 Control of Protein Binding

2′-Modifications can improve RNA-binding and nuclease stability of gapmer ASOs, but the ability to enhance interactions with proteins is a particularly advantageous property of the PS modification. The protein binding properties of PS ASOs allows them to be injected subcutaneously without the need for specialized formulation or delivery vehicles to aid distribution and/or uptake into tissues. Association with proteins in plasma reduces filtration in the kidney and allows the ASO to circulate and distribute to peripheral tissues. Association with cell-surface proteins promotes cellular uptake as these protein–ASO complexes get internalized into cells by endocytic processes.

Once inside the endosome, PS ASOs can interact with specific proteins, which appear to facilitate their release from these compartments. Upon release in the cytosol, PS ASOs associate with intracellular proteins, which facilitate trafficking to nuclear compartments and may even assist with hybridization to the target RNA. Thus, controlling these protein-binding events could aid in the design of the next generation of ASO therapeutics with tailored properties to suit the therapeutic indication.

3.6.1 Interaction of Gapmer ASOs with Plasma Proteins

PS ASOs bind to plasma proteins, such as albumin, which facilitates their distribution to peripheral tissues from the site of injection.[133] ASOs distribute across the entire body, except across the blood–brain barrier (BBB), with tissues such as kidney, liver, spleen, lymph nodes and bone marrow showing the highest accumulation. 16–20-mer PS ASOs associate with albumin with binding constants in the 10–50 μM range but interact with other less abundant plasma proteins with greater affinity. Gapmer ASOs showed two-fold enhancement in potency in α-2-macroglobin-knockout mice indicating that association with proteins such as α-2-macroglobin may steer the ASO towards less productive cellular or tissue compartments.[134] The reversible association of gapmer ASOs with plasma proteins allows for the ASO to partition onto cell-surface proteins, which promotes entry into cells and tissues.

3.6.2 Interaction of Gapmer-ASOs with Cell-surface Proteins

The identity of cell-surface proteins responsible for cellular uptake of PS gapmers has been an active area of research and debate for more than two decades.[135] Gewirtz and colleagues reported 200 000 binding sites for PS ASOs on the surface of K562 cells.[136] Stein and colleagues reported that PS ASOs interact with heparin-binding proteins and displace growth factors from the extracellular matrix.[137] Some reports implicated scavenger receptor A for the uptake of PS ASOs in Kupffer cells in the liver, on the basis of partial inhibition of cellular uptake with known scavenger receptor ligands.[138]

Miller *et al.* recently showed that the Stabilin receptors are responsible for uptake of PS gapmers in sinusoidal endothelial cells of the liver (LSECs) and in the spleen.[139] The Stabilin receptors are highly expressed in the spleen, LSECs, lymph nodes and bone-marrow,[140] which are also sites of high uptake for PS ASOs.[133] The physiological role of the Stabilin receptors is to clear negatively charged carbohydrate polymers of the extracellular matrix[141] suggesting a common pharmacophore with PS ASOs. HEK-293 cells expressing the Stabilin receptors showed 4–6-fold increased uptake of PS gapmers and a 10–20-fold enhancement of antisense activity when delivered without transfection.[139] Direct binding experiments with the purified ecto-domain of Stabilin-2 showed that PS gapmers interact with the receptor with an apparent dissociation constant (K_D) of approximately 140 nM. Binding was dependent on length, PS content and charge on the oligonucleotide. Stabilin-2-knockout mice showed reduced antisense activity and significantly reduced uptake of PS ASOs in the liver and the spleen.

While the Stabilins are responsible for the uptake of PS gapmers in LSECs, the asialoglycoprotein receptor (ASGR) was shown to be partially involved in the uptake of unconjugated gapmer ASOs into hepatocytes.[142] Gapmer ASOs showed increased uptake and antisense activity in HEK-293 cells overexpressing ASGR1, indicating that this receptor is capable of binding and internalizing PS-modified oligonucleotides.[143] Direct binding experiments with the ASGR demonstrated that the negatively charged PS-modification enhances interaction of certain ASO designs with the receptor in a length- and charge-dependent manner.[144] Furthermore, gapmer-ASOs showed a threefold reduced activity in ASGR-knockout mice or in mice where the ASGR1 was knocked out prior to administration of the active ASO.[145] These experiments demonstrated that the ASGR1 is involved in the uptake of gapmer-ASOs into liver hepatocytes but also indicated that cell-surface proteins other than ASGR1 are likely to be involved in the uptake of ASOs into this cell type. It is conceivable that different cell types have their own repertoire of cell-surface proteins that can interact with and internalize ASOs and this area continues to be under active investigation.[135]

3.6.3 Targeting Cell-surface Proteins for Cell-specific Delivery of Gapmers

While gapmer ASOs distribute to most tissues and cell-types after systemic administration, they preferentially accumulate in specific cell-types and tissues. For example, while the liver is a major site for accumulation of gapmer-ASOs, almost 70–95% of the drug delivered to the liver is taken up by non-parenchymal cells.[146] However, a majority of the genes of therapeutic interest are expressed in hepatocytes.[147] Furthermore, gapmer-ASOs appear to be intrinsically more active in hepatocytes as compared with non-parenchymal cells.[142] Thus, the ability to shift ASO distribution to cell-types where the genes of therapeutic interest are expressed has the potential to provide significant enhancements in potency.[148] This vision was recently realized by the delivery of ASOs to hepatocytes by actively targeting the ASGR using *N*-acetylgalactosamine (GalNAc)–ASO conjugates.[129] Both MOE and cEt gapmer designs showed almost tenfold enhancement in potency in mice, and up to 30-fold enhancement in potency in humans relative to unconjugated ASOs were reported.[106]

The ASGR is almost exclusively and abundantly (~500 000 copies per cell) expressed in hepatocytes in all mammals.[149] The physiological role of the receptor is to maintain homeostasis of glycoproteins terminating in sialic acid α-2,6-galactose in the plasma.[150] The functional receptor is expressed as a hetero-oligomer of two subunits in varying ratios (ASGR1 : ASGR2, 2–5 : 1).[151] ASGR1 possesses the carbohydrate recognition domain for interaction with ligands and the cytoplasmic adaptor motif to bind to clathrin adaptors.[152] As a result, the receptor is typically located in coated pits on the basolateral face of hepatocytes and internalizes cargo by clathrin-mediated endocytosis.[153] After internalization, the cargo is shuttled to endosomal compartments where it is released upon acidification, and the free receptor recycles back to the cell surface.[154] The ASGR binds galactose and GalNAc-terminated ligands with high affinity and the requirements for optimal receptor binding have been described previously.[155]

The SAR of GalNac conjugates of gapmer ASOs was recently reported.[107] Single-stranded gapmer ASOs with three or two GalNAc sugars showed maximal binding to the ASGR, while double-stranded ASOs showed slightly reduced binding.[144] This indicated that the single-stranded nature of the oligonucleotide facilitates interactions of the GalNAc sugar with the receptor. Interestingly, gapmer ASOs with even a single GalNAc sugar showed significant ASGR binding, which could be further modulated by the length, charge, PS content and single-strandedness of the oligonucleotide.[144,156] Evaluation of GalNAc–ASO conjugates in engineered cells expressing ASGR1 or ASGR2 individually, showed that only the ASGR1 subunit can mediate effective uptake of GalNAc–ASO conjugates into cells.[143] Up to 50-fold enhancements in potency with GalNAc–ASO conjugates, relative to unconjugated gapmer ASOs, were observed in these cells. Interestingly, GalNAc conjugation was able to promote internalization of phosphodiester oligonucleotides, which are otherwise not internalized into cells without the aid of lipid transfection agents.[143] This observation highlights the importance of adherence to cell-membrane proteins for the effective uptake of nucleic acid-based therapeutics.

3.6.4 Avoiding Interactions with TLR Receptors to Avoid Immune-stimulatory Toxicities

During early work with PS DNA ASOs, it was observed that some PS DNA oligonucleotides were immunostimulatory and caused B-cell proliferation, which was sequence-dependent and not related to their antisense effects.[157] Upon analysis of several hundred sequences, Krieg *et al.* noticed that DNA oligonucleotides with CpG motifs produced immunostimulatory effects in the context of certain sequence motifs.[158] CpG motifs in bacterial DNA are typically not methylated at the 5-position of the cytosine nucleobase, whereas this position is methylated in mammalian DNA to differentiate it from bacterial DNA. Thus, unmethylated CpG motifs in some ASOs were causing B-cell proliferation by activation of TLR9 independent of their antisense effects.[159] The immune-stimulatory effects of these DNA oligonucleotides could be suppressed simply by introducing a methyl group on the cytosine nucleobases.[160] While immune stimulation can be a desirable property for oligonucleotides being used as vaccine adjuvants,[161] gapmer ASOs are almost always methylated at the 5-position of cytidine to suppress immune stimulation.[160]

3.6.5 Optimizing Intracellular Distribution

PS ASOs interact with a host of cellular proteins.[44] However, the systematic study of intracellular proteins based on their ability to modulate ASO behavior was only recently accomplished.[135] Liang *et al.* reported a comprehensive catalogue of intracellular proteins which can interact with and/or modulate ASO activity and localization within cellular compartments.[162] Many of the identified proteins like nucleolin and La have RNA-recognition motifs (RRMs) to interact with nucleic acids. However, some proteins like heat shock protein 90 (HSP90) are chaperone proteins that lack any known nucleic acid binding domains but prefer binding gapmer ASOs with more hydrophobic 2′-modifications.[163] Shen *et al.* showed that gapmer ASOs can displace nuclear paraspeckle assembly transcript 1 (NEAT-1) RNA and form paraspeckle-like structures in the nucleus.[164] Furthermore, 2′-FRNA gapmer ASOs can cause degradation of P54nrb and PSF proteins *via* proteasome-mediated degradation, which can lead to cellular toxicities.[70] Nucleic acid binding proteins can also affect sub-cellular distribution of gapmer ASOs, causing them to become localized to cytoplasmic ribonucleoprotein granules, such as stress granules, and those seeded by the aggregation of specific ASO-binding proteins, such as fused in sarcoma (FUS) and PSF.[165] Certain proteins like Annexin-2 colocalize with ASOs in late endosomes and enhance ASO activity.[166] Similarly, Rap GTP-binding protein 5c (Rab5c) interacts with early endosome antigen 1 (EEA1) to facilitate PS ASO transport from the early to the late endosomes and Rab7a works in conjunction with lysobisphosphatidic acid (LBPA) to facilitate endosomal release of gapmer ASOs in Stabilin-2-expressing HEK-293 cells.[167] LBPA was also shown to be important for release of ASOs from late-endosomal compartments in A431 cells.[168] The interactions of gapmer ASOs with intracellular proteins represents an emerging field with exciting implications for design of future generations of ASO drugs based on control of intracellular protein recognition events.

3.7 Conclusions

Gapmer ASOs represent the most widely investigated nucleic acid-based therapeutic platform in clinical use today. Early work on gapmers focused on increasing metabolic stability by modification of the nucleic acid backbone. This was followed by efforts to identify suitable 2′-modifications to enhance affinity for complementary RNA and stability against exonuclease digestion. This body of work resulted in the MOE-gapmer design (generation 2) and mipomersen, a fully PS modified 5–10–5 MOE gapmer with 5-Me substitution on cytosine nucleobases, represents the first systemically administered antisense drug approved by the FDA. Further research led to the identification of conformationally restricted nucleotide analogs, like LNA and cEt (generation 2.5), which enhanced nuclease stability and affinity for RNA beyond MOE. Gapmers modified with LNA or cEt nucleotides show increased potency in the liver and in extrahepatic tissues, thus expanding the range of tissues available on which to practice antisense pharmacology. The mechanisms by which LNA and cEt ASOs increase potency appear to be a combination of improved binding affinity for the target RNA and alterations in intracellular protein binding and trafficking. The use of cell-targeting strategies, such as ASGR-mediated delivery to hepatocytes, have further increased ASO potency and proved to be transformative in clinical use. The successful journey of the gapmer, from the initial observations of Zamecnik to clinical success with multiple ASOs today, reflects the contributions of a generation of scientists from diverse fields, including nucleotide and nucleic acid synthesis, biochemistry, pharmacology, immunology, pharmacokinetics, toxicology and medicine. Lessons learnt during this journey laid the foundation for the field of nucleic acid therapeutics and will help shape the future of exciting new areas of research within the still emerging field of nucleic acid therapeutics.

References

1. S. T. Crooke, *Annu. Rev. Pharmacol. Toxicol.*, 1992, **32**, 329–376.
2. C. F. Bennett and E. E. Swayze, *Annu. Rev. Pharmacol. Toxicol.*, 2010, **50**, 259–293.
3. W. B. Wan and P. P. Seth, *J. Med. Chem.*, 2016, **59**, 9645–9667.
4. P. C. Zamecnik and M. L. Stephenson, *Proc. Natl. Acad. Sci. U. S. A.*, 1978, **75**, 280–284.
5. C. A. Stein and J. S. Cohen, *Cancer Res.*, 1988, **48**, 2659–2668.
6. K. J. Atabekov, L. G. Tyulkina, O. V. Karpova, V. G. Metelev, N. P. Rodionova, Z. A. Shabarova and J. G. Atabekov, *FEBS Lett.*, 1988, **232**, 96–98.
7. P. J. Furdon, Z. Dominski and R. Kole, *Nucleic Acids Res.*, 1989, **17**, 9193–9204.
8. V. S. Group, *Am. J. Ophthalmol.*, 2002, **133**, 475–483.
9. R. Z. Yu, J. Q. Su, J. S. Grundy, R. S. Geary, K. L. Sewell, A. Dorr and A. A. Levin, *Antisense Nucleic Acid Drug Dev.*, 2003, **13**, 57–66.
10. C. A. Stein, C. Subasinghe, K. Shinozuka and J. S. Cohen, *Nucleic Acids Res.*, 1988, **16**, 3209–3221.
11. H. Inoue, Y. Hayase, S. Iwai and E. Ohtsuka, *FEBS Lett.*, 1987, **215**, 327–330.
12. B. P. Monia, E. A. Lesnik, C. Gonzalez, W. F. Lima, D. McGee, C. J. Guinosso, A. M. Kawasaki, P. D. Cook and S. M. Freier, *J. Biol. Chem.*, 1993, **268**, 14514–14522.
13. V. Metelev, J. Lisziewicz and S. Agrawal, *Bioorg. Med. Chem. Lett.*, 1994, **4**, 2929–2934.

14. K.-H. Altmann, N. M. Dean, D. Fabbro, S. M. Freier, T. Geiger, R. Hanera, D. A. Hiisken, P. Martina, B. P. B. Monia, M. Miiller, F. Natt, P. Nicklind, J. Phillipsd, U. Pieles, H. Sasmor and H. E. Moser, *CHIMIA Int. J. Chem.*, 1996, **50**, 168–176.
15. W. Lima, H. Wu and S. T. Crooke, in *Antisense Drug Technology—Principles, Strategies, and Applications*, ed. S. T. Crooke, CRC Press, Boca Raton, 2nd edn, 2008, pp. 47–74.
16. S. M. Cerritelli and R. J. Crouch, *FEBS J.*, 2009, **276**, 1494–1505.
17. J. D. Amon and D. Koshland, *Elife*, 2016, **5**, e20533.
18. L. Balakrishnan and R. A. Bambara, *Cold Spring Harbor Perspect. Biol.*, 2013, **5**, a010173.
19. W. F. Lima, H. M. Murray, S. S. Damle, C. E. Hart, G. Hung, C. L. De Hoyos, X.-H. Liang and S. T. Crooke, *Nucleic Acids Res.*, 2016, **44**, 5299–5312.
20. W. Shen, H. Sun, C. L. De Hoyos, J. K. Bailey, X.-H. Liang and S. T. Crooke, *Nucleic Acids Res.*, 2017, **45**, 10672–10692.
21. H. Wu, W. F. Lima, H. Zhang, A. Fan, H. Sun and S. T. Crooke, *J. Biol. Chem.*, 2004, **279**, 17181–17189.
22. H. Wu, W. F. Lima and S. T. Crooke, *J. Biol. Chem.*, 1999, **274**, 28270–28278.
23. W. F. Lima, H. Wu, J. G. Nichols, S. M. Manalili, J. J. Drader, S. A. Hofstadler and S. T. Crooke, *J. Biol. Chem.*, 2003, **278**, 14906–14912.
24. W. F. Lima, H. Wu, J. G. Nichols, T. P. Prakash, V. Ravikumar and S. T. Crooke, *J. Biol. Chem.*, 2003, **278**, 49860–49867.
25. W. F. Lima, J. B. Rose, J. G. Nichols, H. Wu, M. T. Migawa, T. K. Wyrzykiewicz, A. M. Siwkowski and S. T. Crooke, *Mol. Pharmacol.*, 2007, **71**, 83–91.
26. W. F. Lima, J. B. Rose, J. G. Nichols, H. Wu, M. T. Migawa, T. K. Wyrzykiewicz, G. Vasquez, E. E. Swayze and S. T. Crooke, *Mol. Pharmacol.*, 2007, **71**, 73–82.
27. W. Y. Gao, F. S. Han, C. Storm, W. Egan and Y. C. Cheng, *Mol. Pharmacol.*, 1992, **41**, 223–229.
28. S. A. Gaidamakov, I. I. Gorshkova, P. Schuck, P. J. Steinbach, H. Yamada, R. J. Crouch and S. M. Cerritelli, *Nucleic Acids Res.*, 2005, **33**, 2166–2175.
29. M. Nowotny, S. M. Cerritelli, R. Ghirlando, S. A. Gaidamakov, R. J. Crouch and W. Yang, *EMBO J.*, 2008, **27**, 1172–1181.
30. M. Nowotny, S. A. Gaidamakov, R. J. Crouch and W. Yang, *Cell*, 2005, **121**, 1005–1016.
31. M. Nowotny, S. A. Gaidamakov, R. Ghirlando, S. M. Cerritelli, R. J. Crouch and W. Yang, *Mol. Cell*, 2007, **28**, 264–276.
32. M. E. Østergaard, J. Nichols, T. A. Dwight, W. Lima, M. E. Jung, E. E. Swayze and P. P. Seth, *Mol. Ther. Nucleic Acids*, 2017, **7**, 20–30.
33. F. Eckstein, *Nucleic Acid Ther.*, 2014, **24**, 374–387.
34. L. Wang, S. Chen, T. Xu, K. Taghizadeh, J. S. Wishnok, X. Zhou, D. You, Z. Deng and P. C. Dedon, *Nat. Chem. Biol.*, 2007, **3**, 709–710.
35. S. L. Beaucage and M. H. Caruthers, *Tetrahedron Lett.*, 1981, **22**, 1859–1862.

36. R. P. Iyer, W. Egan, J. B. Regan and S. L. Beaucage, *J. Am. Chem. Soc.*, 1990, **112**, 1253–1254.
37. Z. S. Cheruvallath, R. L. Carty, M. N. Moore, D. C. Capaldi, A. H. Krotz, P. D. Wheeler, B. J. Turney, S. R. Craig, H. J. Gaus, A. N. Scozzari, D. L. Cole and V. T. Ravikumar, *Org. Process Res. Dev.*, 2000, **4**, 199–204.
38. W. J. Stec and G. Zon, *Tetrahedron Lett.*, 1984, **25**, 5279–5282.
39. M. H. Caruthers, *Acc. Chem. Res.*, 1991, **24**, 278–284.
40. S. L. Beaucage and R. P. Iyer, *Tetrahedron*, 1993, **49**, 1925–1963.
41. S. M. Freier, W. F. Lima, Y. S. Sanghvi, T. Vickers, M. Zounes, P. D. Cook and D. J. Ecker, in *Gene Regulation: Biology of Antisense RNA and DNA*, ed. R. P. Erickson and J. G. Izant, Raven Press, New York, 1992, pp. 95–107.
42. R. Z. Yu, T.-W. Kim, A. Hong, T. A. Watanabe, H. J. Gaus and R. S. Geary, *Drug Metab. Dispos.*, 2007, **35**, 460–468.
43. J. J. P. Kastelein, M. K. Wedel, B. F. Baker, J. Su, J. D. Bradley, R. Z. Yu, E. Chuang, M. J. Graham and R. M. Crooke, *Circulation*, 2006, **114**, 1729–1735.
44. D. A. Brown, S. H. Kang, S. M. Gryaznov, L. DeDionisio, O. Heidenreich, S. Sullivan, X. Xu and M. I. Nerenberg, *J. Biol. Chem.*, 1994, **269**, 26801–26805.
45. R. S. Geary, R. Z. Yu, J. M. Leeds, M. V. Templin, T. A. Watanabe, S. P. Henry and A. A. Levin, in *Antisense Drug Technology: Principles, Strategies, and Applications*, ed. S. T. Crooke, Marcel Dekker, New York, 2001, pp. 119–154.
46. R. S. Geary, R. Z. Yu and A. A. Levin, *Curr. Opin. Invest. Drugs*, 2001, **2**, 562–573.
47. T. A. Vickers and S. T. Crooke, *PLoS One*, 2016, **11**, e0161930.
48. V. G. Vaidyanathan, L. Xu and B. P. Cho, *Protocol Exchange*, 2013, DOI: 10.1038/protex.2013.054.
49. J. M. Pagano, C. C. Clingman and S. P. Ryder, *RNA*, 2011, **17**, 14–20.
50. W. J. Stec, in *Antisense Research and Applications*, ed. S. T. Crooke and B. L. Lebleu, CRC Press, Boca Raton, FL, 1993, pp. 251–271.
51. M. Tonelli, N. B. Ulyanov, T. M. Billeci, B. Karwowski, P. Guga, W. J. Stec and T. L. James, *Biophys. J.*, 2003, **85**, 2525–2538.
52. M. Koziolkiewicz, A. Krakowiak, M. Kwinkowski, M. Boczkowska and W. J. Stec, *Nucleic Acids Res.*, 1995, **23**, 5000–5005.
53. W. B. Wan, M. T. Migawa, G. Vasquez, H. M. Murray, J. G. Nichols, H. Gaus, A. Berdeja, S. Lee, C. E. Hart, W. F. Lima, E. E. Swayze and P. P. Seth, *Nucleic Acids Res.*, 2014, **42**, 13456–13468.
54. N. Iwamoto, D. C. D. Butler, N. Svrzikapa, S. Mohapatra, I. Zlatev, D. W. Y. Sah, Meena, S. M. Standley, G. Lu, L. H. Apponi, M. Frank-Kamenetsky, J. J. Zhang, C. Vargeese and G. L. Verdine, *Nat. Biotechnol.*, 2017, **35**, 845.
55. C. Vargeese, Meena, N. Svrzikapa, S. Mohapatra, C. J. Francis, G. L. Verdine and A. Sokolovska (to Wave Life Sciences), WO2017015555A1, 2017.

56. M. Li, H. L. Lightfoot, F. Halloy, A. L. Malinowska, C. Berk, A. Behera, D. Schumperli and J. Hall, *Chem. Commun.*, 2017, **53**, 541–544.
57. V. K. Rait and B. R. Shaw, *Antisense Nucleic Acid Drug Dev.*, 1999, **9**, 53–60.
58. M. J. Damha, C. J. Wilds, A. Noronha, I. Brukner, G. Borkow, D. Arion and M. A. Parniak, *J. Am. Chem. Soc.*, 1998, **120**, 12976–12977.
59. C. J. Wilds and M. J. Damha, *Nucleic Acids Res.*, 2000, **28**, 3625–3635.
60. C.-N. Lok, E. Viazovkina, K.-L. Min, E. Nagy, C. J. Wilds, M. J. Damha and M. A. Parniak, *Biochemistry*, 2002, **41**, 3457–3467.
61. M. Y. Anzahaee, J. K. Watts, N. R. Alla, A. W. Nicholson and M. J. Damha, *J. Am. Chem. Soc.*, 2011, **133**, 728–731.
62. N. Martin-Pintado, G. F. Deleavey, G. Portella, R. Campos-Olivas, M. Orozco, M. J. Damha and C. González, *Angew. Chem., Int. Ed. Engl.*, 2013, **52**, 12065–12068.
63. W. F. Lima, J. G. Nichols, H. Wu, T. P. Prakash, M. T. Migawa, T. K. Wyrzykiewicz, B. Bhat and S. T. Crooke, *J. Biol. Chem.*, 2004, **279**, 36317–36326.
64. A. Kalota, L. Karabon, C. R. Swider, E. Viazovkina, M. Elzagheid, M. J. Damha and A. M. Gewirtz, *Nucleic Acids Res.*, 2006, **34**, 451–461.
65. N. Souleimanian, G. F. Deleavey, H. Soifer, S. Wang, K. Tiemann, M. J. Damha and C. A. Stein, *Mol. Ther. Nucleic Acids*, 2012, **1**, e43.
66. J. B. Carroll, S. C. Warby, A. L. Southwell, C. N. Doty, S. Greenlee, N. Skotte, G. Hung, C. F. Bennett, S. M. Freier and M. R. Hayden, *Mol. Ther.*, 2011, **19**, 2178–2185.
67. M. E. Ostergaard, A. L. Southwell, H. Kordasiewicz, A. T. Watt, N. H. Skotte, C. N. Doty, K. Vaid, E. B. Villanueva, E. E. Swayze, C. F. Bennett, M. R. Hayden and P. P. Seth, *Nucleic Acids Res.*, 2013, **41**, 9634–9650.
68. M. E. Østergaard, P. Kumar, J. Nichols, A. Watt, P. K. Sharma, P. Nielsen and P. P. Seth, *Nucleic Acid Ther.*, 2015, **25**, 266–274.
69. M. E. Ostergaard, B. Gerland, J. M. Escudier, E. E. Swayze and P. P. Seth, *ACS Chem. Biol.*, 2014, **9**, 1975–1979.
70. W. Shen, X.-H. Liang, H. Sun and S. T. Crooke, *Nucleic Acids Res.*, 2015, **43**, 4569–4578.
71. P. Martin, *Helv. Chim. Acta*, 1995, **78**, 486–504.
72. M. Teplova, G. Minasov, V. Tereshko, G. B. Inamati, P. D. Cook, M. Manoharan and M. Egli, *Nat. Struct. Biol.*, 1999, **6**, 535–539.
73. S. T. Crooke and R. S. Geary, *Br. J. Clin. Pharmacol.*, 2013, **76**, 269–276.
74. R. S. Finkel, E. Mercuri, B. T. Darras, A. M. Connolly, N. L. Kuntz, J. Kirschner, C. A. Chiriboga, K. Saito, L. Servais, E. Tizzano, H. Topaloglu, M. Tulinius, J. Montes, A. M. Glanzman, K. Bishop, Z. J. Zhong, S. Gheuens, C. F. Bennett, E. Schneider, W. Farwell and D. C. De Vivo, *N. Engl. J. Med.*, 2017, **377**, 1723–1732.
75. J. Wengel, *Acc. Chem. Res.*, 1999, **32**, 301–310.
76. T. Imanishi and S. Obika, *Chem. Commun.*, 2002, 1653–1659.
77. A. Eichert, K. Behling, C. Betzel, V. A. Erdmann, J. P. Furste and C. Forster, *Nucleic Acids Res.*, 2010, **38**, 6729–6736.
78. R. N. Veedu and J. Wengel, *Chem. Biodiversity*, 2010, **7**, 536–542.

79. P. P. Seth, A. Siwkowski, C. R. Allerson, G. Vasquez, S. Lee, T. P. Prakash, G. Kinberger, M. T. Migawa, H. Gaus, B. Bhat and E. E. Swayze, *Nucleic Acids Symp. Ser.*, 2008, 553–554.
80. P. P. Seth, G. Vasquez, C. A. Allerson, A. Berdeja, H. Gaus, G. A. Kinberger, T. P. Prakash, M. T. Migawa, B. Bhat and E. E. Swayze, *J. Org. Chem.*, 2010, **75**, 1569–1581.
81. P. S. Pallan, C. R. Allerson, A. Berdeja, P. P. Seth, E. E. Swayze, T. P. Prakash and M. Egli, *Chem. Commun.*, 2012, **48**, 8195–8197.
82. V. K. Rajwanshi, A. E. Hakansson, M. D. Sorensen, S. Pitsch, S. K. Singh, R. Kumar, P. Nielsen and J. Wengel, *Angew. Chem., Int. Ed. Engl.*, 2000, **39**, 1656–1659.
83. M. D. Sorensen, L. Kvaerno, T. Bryld, A. E. Hakansson, B. Verbeure, G. Gaubert, P. Herdewijn and J. Wengel, *J. Am. Chem. Soc.*, 2002, **124**, 2164–2176.
84. J. T. Nielsen, P. C. Stein and M. Petersen, *Nucleic Acids Res.*, 2003, **31**, 5858–5867.
85. P. P. Seth, A. Jazayeri, J. Yu, C. R. Allerson, B. Bhat and E. E. Swayze, *Mol. Ther. Nucleic Acids*, 2012, **1**, e47.
86. P. Herdewijn, *Chem. Biodiversity*, 2010, **7**, 1–59.
87. M. E. Ostergaard, T. Dwight, A. Berdeja, E. E. Swayze, M. E. Jung and P. P. Seth, *J. Org. Chem.*, 2014, **79**, 8877–8881.
88. M. Egli, P. S. Pallan, C. R. Allerson, T. P. Prakash, A. Berdeja, J. Yu, S. Lee, A. Watt, H. Gaus, B. Bhat, E. E. Swayze and P. P. Seth, *J. Am. Chem. Soc.*, 2011, **133**, 16642–16649.
89. K. Nauwelaerts, M. Fisher, M. Froeyen, E. Lescrinier, A. Van Aerschot, D. Xu, R. DeLong, H. Kang, R. L. Juliano and P. Herdewijn, *J. Am. Chem. Soc.*, 2007, **129**, 9340–9348.
90. P. P. Seth, J. Yu, A. Jazayeri, P. S. Pallan, C. R. Allerson, M. E. Ostergaard, F. Liu, P. Herdewijn, M. Egli and E. E. Swayze, *J. Org. Chem.*, 2012, **77**, 5074–5085.
91. R. Steffens and C. J. Leumann, *J. Am. Chem. Soc.*, 1999, **121**, 3249–3255.
92. P. S. Pallan, D. Ittig, A. Heroux, Z. Wawrzak, C. J. Leumann and M. Egli, *Chem. Commun.*, 2008, 883–885.
93. A. Goyenvalle, G. Griffith, A. Babbs, S. E. Andaloussi, K. Ezzat, A. Avril, B. Dugovic, R. Chaussenot, A. Ferry, T. Voit, H. Amthor, C. Buhr, S. Schurch, M. J. A. Wood, K. E. Davies, C. Vaillend, C. Leumann and L. Garcia, *Nat. Med.*, 2015, **21**, 270–275.
94. S. Murray, D. Ittig, E. Koller, A. Berdeja, A. Chappell, T. P. Prakash, M. Norrbom, E. E. Swayze, C. J. Leumann and P. P. Seth, *Nucleic Acids Res.*, 2012, **40**, 6135–6143.
95. J. Summerton and D. Weller, *Antisense Nucleic Acid Drug Dev.*, 1997, **7**, 187–195.
96. J. R. Mendell, L. R. Rodino-Klapac, Z. Sahenk, K. Roush, L. Bird, L. P. Lowes, L. Alfano, A. M. Gomez, S. Lewis, J. Kota, V. Malik, K. Shontz, C. M. Walker, K. M. Flanigan, M. Corridore, J. R. Kean, H. D. Allen,

C. Shilling, K. R. Melia, P. Sazani, J. B. Saoud, E. M. Kaye and the Eteplirsen Study Group, *Ann. Neurol.*, 2013, **74**, 637–647.

97. S. Paul and M. H. Caruthers, *J. Am. Chem. Soc.*, 2016, **138**, 15663–15672.
98. M. Matsukura, K. Shinozuka, G. Zon, H. Mitsuya, M. Reitz, J. S. Cohen and S. Broder, *Proc. Natl. Acad. Sci. U. S. A.*, 1987, **84**, 7706–7710.
99. S. Agrawal, J. Goodchild, M. P. Civeira, A. H. Thornton, P. S. Sarin and P. C. Zamecnik, *Proc. Natl. Acad. Sci. U. S. A.*, 1988, **85**, 7079–7083.
100. M. D. Benson, N. R. Dasgupta, S. M. Rissing, J. Smith and H. Feigenbaum, *Amyloid*, 2017, **24**, 217–223.
101. D. Gaudet, V. J. Alexander, B. F. Baker, D. Brisson, K. Tremblay, W. Singleton, R. S. Geary, S. G. Hughes, N. J. Viney, M. J. Graham, R. M. Crooke, J. L. Witztum, J. D. Brunzell and J. J. P. Kastelein, *N. Engl. J. Med.*, 2015, **373**, 438–447.
102. G. Hung, X. Xiao, R. Peralta, G. Bhattacharjee, S. Murray, D. Norris, S. Guo and B. P. Monia, *Nucleic Acid Ther.*, 2013, **23**, 369–378.
103. F. Rigo, S. J. Chun, D. A. Norris, G. Hung, S. Lee, J. Matson, R. A. Fey, H. Gaus, Y. Hua, J. S. Grundy, A. R. Krainer, S. P. Henry and C. F. Bennett, *J. Pharmacol. Exp. Ther.*, 2014, **350**, 46–55.
104. B. Leavitt, S. Tabrizi, H. Kordasiewicz, B. Landwehrmeyer, S. Henry, T. Zanardi, E. Swayze, D. Norris, A. Smith, R. Lane and F. Bennett, *Neurology*, 2016, **86**, PL01.002.
105. S. P. Henry, R. S. Geary, R. Yu and A. A. Levin, *Curr. Opin. Invest. Drugs*, 2001, **2**, 1444–1449.
106. N. J. Viney, J. C. van Capelleveen, R. S. Geary, S. Xia, J. A. Tami, R. Z. Yu, S. M. Marcovina, S. G. Hughes, M. J. Graham, R. M. Crooke, S. T. Crooke, J. L. Witztum, E. S. Stroes and S. Tsimikas, *The Lancet*, 2016, **388**, 2239–2253.
107. T. P. Prakash, J. Yu, M. T. Migawa, G. A. Kinberger, W. B. Wan, M. E. Ostergaard, R. L. Carty, G. Vasquez, A. Low, A. Chappell, K. Schmidt, M. Aghajan, J. Crosby, H. M. Murray, S. L. Booten, J. Hsiao, A. Soriano, T. Machemer, P. Cauntay, S. A. Burel, S. F. Murray, H. Gaus, M. J. Graham, E. E. Swayze and P. P. Seth, *J. Med. Chem.*, 2016, **59**, 2718–2733.
108. S. Obika, K.-I. Morio, D. Nanbu and T. Imanishi, *Chem. Commun.*, 1997, 1643–1644.
109. E. M. Straarup, N. Fisker, M. Hedtjarn, M. W. Lindholm, C. Rosenbohm, V. Aarup, H. F. Hansen, H. Orum, J. B. Hansen and T. Koch, *Nucleic Acids Res.*, 2010, **38**, 7100–7111.
110. L. Pedersen, P. H. Hagedorn, M. W. Lindholm and M. Lindow, *Mol. Ther. Nucleic Acids*, 2014, **3**, e149.
111. T. Yamamoto, H. Yasuhara, F. Wada, M. Harada-Shiba, T. Imanishi and S. Obika, *J. Nucleic Acids*, 2012, **2012**, 7.
112. E. E. Swayze, A. M. Siwkowski, E. V. Wancewicz, M. T. Migawa, T. K. Wyrzykiewicz, G. Hung, B. P. Monia and C. F. Bennett, *Nucleic Acids Res.*, 2007, **35**, 687–700.

113. P. H. Hagedorn, V. Yakimov, S. Ottosen, S. Kammler, N. F. Nielsen, A. M. Høg, M. Hedtjärn, M. Meldgaard, M. R. Møller, H. Ørum, T. Koch and M. Lindow, *Nucleic Acid Ther.*, 2013, **23**, 302–310.
114. P. P. Seth, A. Siwkowski, C. R. Allerson, G. Vasquez, S. Lee, T. P. Prakash, E. V. Wancewicz, D. Witchell and E. E. Swayze, *J. Med. Chem.*, 2009, **52**, 10–13.
115. R. Stanton, S. Sciabola, C. Salatto, Y. Weng, D. Moshinsky, J. Little, E. Walters, J. Kreeger, D. Dimattia, T. Chen, T. Clark, M. Liu, J. Qian, M. Roy and R. Dullea, *Nucleic Acid Ther.*, 2012, **22**, 344–359.
116. G. Melanie, Z. Andreas, K. Erich, M. Christine, M. B. Martina, W. Christian, S. Franz, S. Thomas and T. Yann, *Environ. Mol. Mutagen.*, 2017, **58**, 112–121.
117. A. D. Burdick, S. Sciabola, S. R. Mantena, B. D. Hollingshead, R. Stanton, J. A. Warneke, M. Zeng, E. Martsen, A. Medvedev, S. S. Makarov, L. A. Reed, J. W. Davis, 2nd and L. O. Whiteley, *Nucleic Acids Res.*, 2014, **42**, 4882–4891.
118. P. J. Kamola, J. D. A. Kitson, G. Turner, K. Maratou, S. Eriksson, A. Panjwani, L. C. Warnock, G. A. Douillard Guilloux, K. Moores, E. L. Koppe, W. E. Wixted, P. A. Wilson, N. J. Gooderham, T. W. Gant, K. L. Clark, S. A. Hughes, M. R. Edbrooke and J. D. Parry, *Nucleic Acids Res.*, 2015, **43**, 8638–8650.
119. S. A. Burel, C. E. Hart, P. Cauntay, J. Hsiao, T. Machemer, M. Katz, A. Watt, H. H. Bui, H. Younis, M. Sabripour, S. M. Freier, G. Hung, A. Dan, T. P. Prakash, P. P. Seth, E. E. Swayze, C. F. Bennett, S. T. Crooke and S. P. Henry, *Nucleic Acids Res.*, 2016, **44**, 2093–2109.
120. S. Kakiuchi-Kiyota, P. H. Koza-Taylor, S. R. Mantena, L. F. Nelms, A. E. Enayetallah, B. D. Hollingshead, A. D. Burdick, L. A. Reed, J. A. Warneke, L. O. Whiteley, A. M. Ryan and N. Mathialagan, *Toxicol. Sci.*, 2014, **138**, 234–248.
121. S. Kakiuchi-Kiyota, L. O. Whiteley, A. M. Ryan and N. Mathialagan, *Nucleic Acid Ther.*, 2015, **26**, 93–101.
122. S. Sewing, F. Boess, A. Moisan, C. Bertinetti-Lapatki, T. Minz, M. Hedtjaern, Y. Tessier, F. Schuler, T. Singer and A. B. Roth, *PLoS One*, 2016, **11**, e0159431.
123. A. Dieckmann, P. H. Hagedorn, Y. Burki, C. Brügmann, M. Berrera, M. Ebeling, T. Singer and F. Schuler, *Mol. Ther. Nucleic Acids*, 2018, **10**, 45–54.
124. P. J. Kamola, K. Maratou, P. A. Wilson, K. Rush, T. Mullaney, T. McKevitt, P. Evans, J. Ridings, P. Chowdhury, A. Roulois, A. Fairchild, S. McCawley, K. Cartwright, N. J. Gooderham, T. W. Gant, K. Moores, S. A. Hughes, M. R. Edbrooke, K. Clark and J. D. Parry, *Mol. Ther. Nucleic Acids*, 2017, **8**, 383–394.
125. A. R. MacLeod and S. T. Crooke, *J. Clin. Pharmacol.*, 2017, **57**, S43–S59.
126. D. Hong, R. Kurzrock, Y. Kim, R. Woessner, A. Younes, J. Nemunaitis, N. Fowler, T. Zhou, J. Schmidt, M. Jo, S. J. Lee, M. Yamashita, S. G. Hughes,

L. Fayad, S. Piha-Paul, M. V. P. Nadella, M. Mohseni, D. Lawson, C. Reimer, D. C. Blakey, X. Xiao, J. Hsu, A. Revenko, B. P. Monia and A. R. MacLeod, *Sci. Transl. Med.*, 2015, 7, 314ra185.
127. S. J. Ross, A. S. Revenko, L. L. Hanson, R. Ellston, A. Staniszewska, N. Whalley, S. K. Pandey, M. Revill, C. Rooney, L. K. Buckett, S. K. Klein, K. Hudson, B. P. Monia, M. Zinda, D. C. Blakey, P. D. Lyne and A. R. Macleod, *Sci. Transl. Med.*, 2017, **9**, eaao4188.
128. S. K. Pandey, T. M. Wheeler, S. L. Justice, A. Kim, H. S. Younis, D. Gattis, D. Jauvin, J. Puymirat, E. E. Swayze, S. M. Freier, C. F. Bennett, C. A. Thornton and A. R. MacLeod, *J. Pharmacol. Exp. Ther.*, 2015, **355**, 329–340.
129. T. P. Prakash, M. J. Graham, J. Yu, R. Carty, A. Low, A. Chappell, K. Schmidt, C. Zhao, M. Aghajan, H. F. Murray, S. Riney, S. L. Booten, S. F. Murray, H. Gaus, J. Crosby, W. F. Lima, S. Guo, B. P. Monia, E. E. Swayze and P. P. Seth, *Nucleic Acids Res.*, 2014, **42**, 8796–8807.
130. N. H. Skotte, A. L. Southwell, M. E. Ostergaard, J. B. Carroll, S. C. Warby, C. N. Doty, E. Petoukhov, K. Vaid, H. Kordasiewicz, A. T. Watt, S. M. Freier, G. Hung, P. P. Seth, C. F. Bennett, E. E. Swayze and M. R. Hayden, *PLoS One*, 2014, **9**, e107434.
131. A. L. Southwell, N. H. Skotte, H. B. Kordasiewicz, M. E. Ostergaard, A. T. Watt, J. B. Carroll, C. N. Doty, E. B. Villanueva, E. Petoukhov, K. Vaid, Y. Xie, S. M. Freier, E. E. Swayze, P. P. Seth, C. F. Bennett and M. R. Hayden, *Mol. Ther.*, 2014, **22**, 2093–2106.
132. K. Nishina, W. Piao, K. Yoshida-Tanaka, Y. Sujino, T. Nishina, T. Yamamoto, K. Nitta, K. Yoshioka, H. Kuwahara, H. Yasuhara, T. Baba, F. Ono, K. Miyata, K. Miyake, P. P. Seth, A. Low, M. Yoshida, C. F. Bennett, K. Kataoka, H. Mizusawa, S. Obika and T. Yokota, *Nat. Commun.*, 2015, **6**, 7969.
133. R. S. Geary, *Expert Opin. Drug Metab. Toxicol.*, 2009, **5**, 381–391.
134. C. S. Shemesh, R. Z. Yu, H. J. Gaus, P. P. Seth, E. E. Swayze, F. C. Bennett, R. S. Geary, S. P. Henry and Y. Wang, *Nucleic Acid Ther.*, 2016, **26**, 223–235.
135. S. T. Crooke, S. Wang, T. A. Vickers, W. Shen and X.-H. Liang, *Nat. Biotechnol.*, 2017, **35**, 230–237.
136. C. Beltinger, H. U. Saragovi, R. M. Smith, L. LeSauteur, N. Shah, L. DeDionisio, L. Christensen, A. Raible, L. Jarett and A. M. Gewirtz, *J. Clin. Invest.*, 1995, **95**, 1814–1823.
137. M. A. Guvakova, L. A. Yakubov, I. Vlodavsky, J. L. Tonkinson and C. A. Stein, *J. Biol. Chem.*, 1995, **270**, 2620–2627.
138. M. Butler, R. M. Crooke, M. J. Graham, K. M. Lemonidis, M. Lougheed, S. F. Murray, D. Witchell, U. Steinbrecher and C. F. Bennett, *J. Pharmacol. Exp. Ther.*, 2000, **292**, 489–496.
139. C. M. Miller, A. J. Donner, E. E. Blank, A. W. Egger, B. M. Kellar, M. E. Ostergaard, P. P. Seth and E. N. Harris, *Nucleic Acids Res.*, 2016, **44**, 2782–2794.
140. A. K. Hare and E. N. Harris, *Biochem. Biophys. Res. Commun.*, 2015, **456**, 257–261.

141. E. N. Harris, B. A. Baggenstoss and P. H. Weigel, *Am. J. Physiol. Gastrointest. Liver Physiol.*, 2009, **296**, G1191–G1199.
142. A. J. Donner, E. V. Wancewicz, H. M. Murray, S. Greenlee, N. Post, M. Bell, W. F. Lima, E. E. Swayze and P. P. Seth, *Nucleic Acid Ther.*, 2017, **27**, 209–220.
143. M. Tanowitz, L. Hettrick, A. Revenko, G. A. Kinberger, T. P. Prakash and P. P. Seth, *Nucleic Acids Res.*, 2017, **45**, 12388–12400.
144. K. Schmidt, T. P. Prakash, A. J. Donner, G. A. Kinberger, H. J. Gaus, A. Low, M. E. Ostergaard, M. Bell, E. E. Swayze and P. P. Seth, *Nucleic Acids Res.*, 2017, **45**, 2294–2306.
145. C. M. Miller, M. Tanowitz, A. J. Donner, T. P. Prakash, E. E. Swayze, E. N. Harris and P. P. Seth, *Nucleic Acid Ther.*, 2018, **28**, 119–127.
146. M. J. Graham, S. T. Crooke, D. K. Monteith, S. R. Cooper, K. M. Lemonidis, K. K. Stecker, M. J. Martin and R. M. Crooke, *J. Pharmacol. Exp. Ther.*, 1998, **286**, 447–458.
147. A. Sehgal, A. Vaishnaw and K. Fitzgerald, *J. Hepatol.*, 2013, **59**, 1354–1359.
148. E. A. Biessen, H. Vietsch, E. T. Rump, K. Fluiter, J. Kuiper, M. K. Bijsterbosch and T. J. van Berkel, *Biochem. J.*, 1999, **340**, 783–792.
149. A. L. Schwartz, D. Rup and H. F. Lodish, *J. Biol. Chem.*, 1980, **255**, 9033–9036.
150. L. M. Steirer, E. I. Park, R. R. Townsend and J. U. Baenziger, *J. Biol. Chem.*, 2009, **284**, 3777–3783.
151. R. J. Stockert, *Physiol. Rev.*, 1995, **75**, 591–609.
152. Z. Katzir, N. Nardi, I. Geffen, C. Fuhrer and Y. I. Henis, *J. Biol. Chem.*, 1994, **269**, 21568–21575.
153. D. A. Wall, G. Wilson and A. L. Hubbard, *Cell*, 1980, **21**, 79–93.
154. H. J. Geuze, J. W. Slot, G. J. A. M. Strous, H. F. Lodish and A. L. Schwartz, *Cell*, 1983, **32**, 277–287.
155. Y. C. Lee and R. T. Lee, in *Carbohydrates in Chemistry and Biology*, ed. B. Ernst, G. W. Hart and P. Sinaý, Wiley-VCH Verlag GmbH, Weinheim, Germany, 2008, vol. 4, ch. 72, pp. 549–561.
156. G. A. Kinberger, T. P. Prakash, J. Yu, G. Vasquez, A. Low, A. Chappell, K. Schmidt, H. M. Murray, H. Gaus, E. E. Swayze and P. P. Seth, *Bioorg. Med. Chem. Lett.*, 2016, **26**, 3690–3693.
157. A. M. Krieg, *Biochim. Biophys. Acta*, 1999, **1489**, 107–116.
158. A. M. Krieg, A.-K. Yi, S. Matson, T. J. Waldschmidt, G. A. Bishop, R. Teasdale, G. A. Koretzky and D. M. Klinman, *Nature*, 1995, **374**, 546.
159. M. Rutz, J. Metzger, T. Gellert, P. Luppa, G. B. Lipford, H. Wagner and S. Bauer, *Eur. J. Immunol.*, 2004, **34**, 2541–2550.
160. S. Henry, K. Stecker, D. Brooks, D. Monteith, B. Conklin and C. F. Bennett, *J. Pharmacol. Exp. Ther.*, 2000, **292**, 468–479.
161. A. M. Krieg, *Annu. Rev. Immunol.*, 2002, **20**, 709–760.
162. X.-H. Liang, H. Sun, W. Shen and S. T. Crooke, *Nucleic Acids Res.*, 2015, **43**, 2927–2945.
163. X.-H. Liang, W. Shen, H. Sun, G. A. Kinberger, T. P. Prakash, J. G. Nichols and S. T. Crooke, *Nucleic Acids Res.*, 2016, **44**, 3892–3907.

164. W. Shen, X.-H. Liang and S. T. Crooke, *Nucleic Acids Res.*, 2014, **42**, 8648–8662.
165. J. K. Bailey, W. Shen, X.-H. Liang and S. T. Crooke, *Nucleic Acids Res.*, 2017, **45**, 10649–10671.
166. S. Wang, H. Sun, M. Tanowitz, X.-H. Liang and S. T. Crooke, *Nucleic Acids Res.*, 2016, **44**, 7314–7330.
167. C. M. Miller, W. B. Wan, P. P. Seth and E. N. Harris, *Nucleic Acid Ther.*, 2018, **28**, 86–96.
168. S. Wang, H. Sun, M. Tanowitz, X.-H. Liang and S. T. Crooke, *Nucleic Acids Res.*, 2017, **45**, 5309–5322.

CHAPTER 4

Antisense Technology: Liver Targeting and Beyond for Drug Discovery

RICHARD S. GEARY*, BRENDA F. BAKER AND BRETT P. MONIA

Ionis Pharmaceuticals, 2855 Gazelle Court, Carlsbad, 92010, USA
*E-mail: rgeary@ionisph.com

4.1 Introduction

Single-stranded phosphorothioate antisense oligonucleotides (ASOs) represent the most advanced RNA therapeutic technology today, with three RNA-targeted ASOs now approved by the US Food and Drug Administration (FDA) and two more under regulatory review for market authorization in 2018. Early in the process of exploring bio-distribution of systemically administered ASOs it was shown that, while distribution of the ASOs was rapid and broad,[1–6] a relatively large fraction of the administered dose was taken up by the liver across different species.[7,8] This raised the likelihood that the innate pharmacokinetics of phosphorothioate-containing ASOs would provide preferential pharmacodynamics in this organ. Early work further documented that the distribution was homogenous throughout the liver, but was not equal for all cell types, with 80–90% localized in non-parenchymal cells, Kupffer and endothelial cells and the remainder within

Drug Discovery Series No. 68
Advances in Nucleic Acid Therapeutics
Edited by Sudhir Agrawal and Michael J. Gait

Published by the Royal Society of Chemistry, www.rsc.org

hepatocytes.[9,10] This early work provided clues to how targeted delivery to specific cell types, such as hepatocytes, may optimize the pharmacokinetic (PK) and pharmacodynamic (PD) properties of ASOs in the liver. Several PK and PD publications followed, showing that the pharmacokinetics within the liver was associated directly with ASO-mediated activity on the target mRNA.[11–13] These early studies also showed that both parenchymal and non-parenchymal cells retained the ASOs for days to weeks depending on the chemical modifications. These studies solidified the potential to regulate virtually any expressed RNA through ASO-mediated mechanisms, particularly in the hepatocyte.

Hepatocytes are primary producers of key proteins that support or affect a broad range of physiological processes, including key metabolic and immunological functions, in both health and disease.[14–19] The broad array of proteins produced by the liver[15,18] include apolipoproteins (apo) involved in lipid metabolism and homeostasis,[19,20] G-protein-coupled receptors involved in glucose metabolism and glycogen storage,[21] coagulation components involved in hemostasis,[22] complement factors involved in immune functions and transmembrane proteases involved in regulation of iron metabolism.[23–25] In addition, many genes expressed by hepatocytes have been identified that are predisposed to mutations that cause a variety of human diseases, including systemic amyloidosis.[26–28]

One of the first ASO liver targets to advance to clinical testing was apoB-100,[11,29–31] the carrier of cholesterol in the atherogenic lipid particle known as low-density lipoprotein–cholesterol (LDL–C). All previous and subsequent therapeutics approved for lowering LDL–C worked by increasing clearance of LDL–C *via* upregulation of the LDL receptor expressed in the liver.[32,33] ApoB-100, the lipoprotein carrier of LDL-cholesterol, is required for assembly of circulating LDL–C by the liver and thus represented the ideal approach to lowering atherogenic particles associated with cardiovascular disease by inhibiting its production. Antisense inhibition of apoB-100 was validated in clinical settings with mipomersen,[34–39] a 2′-*O*-methoxyethyl (2′MOE)-modified gapmer ASO sometimes referred to as a second-generation (Gen 2.0) ASO. Mipomersen (Kynamro) became the first systemically administered (SC) ASO approved by the US Food and Drug Administration (FDA; January 2013). Mipomersen was approved specifically for treating homozygous familial hypercholesterolemia (HoFH) which is caused by loss of function mutations in the LDL receptor, thus disabling the mechanism of action of statins as well as pro-protein convertase subtilisin/kexin type 9 (PCSK9) inhibitors, and highlighting the need for novel approaches such as inhibition of apoB100 production.

Advances in nucleic acid chemical modifications were followed by advances in screening methods to identify optimally active and tolerable ASO sequences, which have vastly improved the drug properties of ASOs compared with earlier versions. The earliest phosphorothioate oligodeoxynucleotide ASOs (first-generation ASOs) were limited in their broad systemic use due to reduced RNA binding affinity, relative biological instability[40] and pro-inflammatory properties.[41,42] In addition, optimized screening procedures were not in place during the evaluation period of first-generation ASOs, which limited identification of optimal first-generation ASOs. When chimeric ("gapmer") ASO designs were introduced in which 2′-alkoxy chemical modifications were incorporated at the 3′ and 5′ ends of the molecule, thereby enhancing the potency (through improved binding affinity) and stability (through increased nuclease resistance) while maintaining RNaseH capacity, biological activity was substantially improved (see Chapter 2).[43–46] Mipomersen was a product of this chemical innovation, specifically with the optimized 2′MOE modification.

Based on the lessons learned during the development of mipomersen, more intensive screening and optimizing selection of active human candidate sequences (Gen 2+) in this chemical class of 2′MOE gapmers was implemented. These optimizations in candidate selection resulted in further improvements in tolerability, primarily by reduction in pro-inflammatory properties that manifested at the site of injection or as constitutional symptoms shortly after injection.[47,48]

More recent chemical modifications involving conjugation of a targeting ligand, triantennary *N*-acetyl galactosamine (GalNAc), to the 5′ end of the 2′MOE ASO molecule have greatly increased uptake by hepatocytes *via* asialoglycoprotein receptor 1 (ASGPR1)[49–53] which substantially increases the potency of hepatocyte-targeted ASOs by more than 30-fold over the unconjugated ASOs.[54,55] Ultimately, this results in much lower doses required to deliver benefit and the potential for less frequent low-volume SC injections.[55] Further 2′-bicyclic sugar modifications have now been introduced into development utilizing the (*S*-)constrained ethyl (cEt) furanose chemistry (Generation 2.5)[56,57] with enhanced potency of a shortened ASO construct (*e.g.* 14–16 nucleotides in length), now enabling yet additional potency gains in the liver, but importantly, enhancing antisense activity well beyond the liver, including tumor[58–60] and muscle.[61] Figure 4.1 demonstrates the progression in advances of medicinal chemistry associated with improved activity in the liver and beyond.

4.2 Liver Targeting

4.2.1 The Beginning

In early pharmacokinetic and biodistribution studies it was consistently observed that the highest fraction of a 2′MOE PS ASO administered dose was in the liver.[4–6] Furthermore, the bioavailability of a subcutaneous (SC) administered dose was essentially identical to that of a dose administered by intravenous (IV) infusion for these compounds. Based on the known role of proteins produced by the liver, initial drug discovery and development activities were primarily focused on cardiometabolic targets known to play important roles in cardiovascular disease, such as lipids, or in glucose control important in diabetes (Table 4.1).

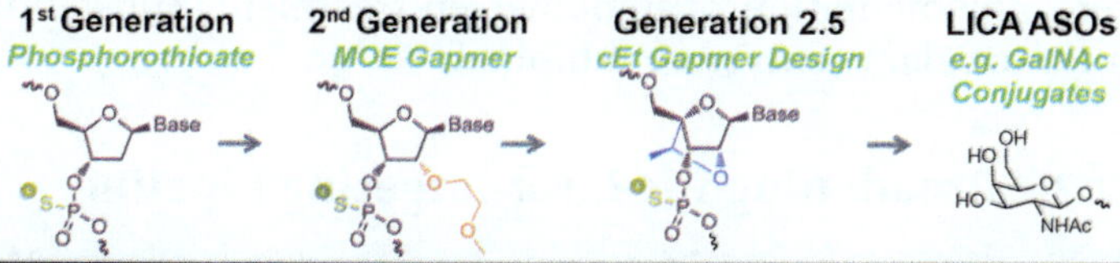

	1st Generation	2nd Generation	Generation 2.5	LICA ASOs
Chemistry Attributes	✓ Adds stability ✓ Improves distribution to tissues	✓ Increases potency ✓ Increases stability ✓ Reduces non-specific toxicities	✓ Improves potency and therapeutic index ✓ Expands range of targets and tissues	✓ Improve potency and therapeutic index in liver ✓ Specific for hepatocyte targets
Potency	1200 to 3500 mg/week	~100 to 400 mg/week in liver	20 to 100 mg/week in liver	5 to 40 mg/week
Dose Frequency	Daily to 3x/week	Weekly to biweekly	Weekly to biweekly	Weekly to monthly
Routes of Administration	I.V., enema, intravitreal	Sub Q, I.V., inhalation, topical, intrathecal, IVT	Same + Oral	Same as Generation 2.5

Figure 4.1 Medicinal chemistry innovation continues to improve drug-like properties of ASOs.

Table 4.1 Early liver target programs.

Drug	Chemistry	Target	Disease	Development status
Mipomersen (ISIS 301012)[37,39,63]	2′MOE	ApoB100	Homozygous familial hypercholesterolemia	Commercial FDA approval 2013
Miravirsen[64]	LNA	microRNA miR-122	Hepatitis C virus	Discontinued post-Phase 2
Atesidorsen (ATL1103; ISIS 227452)	2′MOE	Growth hormone receptor	Acromegaly	Phase 2 ongoing
ISIS 113715[65]	2′MOE	Protein tyrosine phosphatase 1B	Type 2 diabetes	Discontinued post-Phase 2 to bring forward higher potency backup sequence
ISIS 325568[66]	2′MOE	Glucagon receptor	Type 2 diabetes	Discontinued post-Phase 2 to bring forward higher potency backup sequence

The apoB-100-targeted 2′MOE ASO, mipomersen, was the first to move into late-stage testing and ultimately to be approved for treating high LDL–C levels in patients with HoFH by the FDA.[62] ApoB-100 targeting resulted in an ideal lipid profile with decreases in very low-density lipoprotein–cholesterol (VLDL–C), LDL–C, lipoprotein a [Lp(a)] and triglycerides, coupled with increases in high-density lipoprotein–cholesterol (HDL–C) in HoFH patients.[37] While it is not possible to design large outcome trials in rare disease populations, in retrospective exploration comparing events prior to and after implementation of mipomersen treatment, substantial reductions in cardiovascular events were exhibited.[63]

4.2.2 Broadening the Liver-targeting Pipeline

The knowledge obtained from mipomersen development allowed the clinical validation of the ASO approach for liver targets, which led to new strategies to improve ASO profiles, focused primarily on potency and tolerability. The incidence of unwanted side effects at the injection site or shortly following injection was not sustainable for larger populations. Thus the discovery process implemented greater stringency, broader and deeper sequence analysis and human cell-based assay screens to ensure that the very best sequences were selected for human candidacy. This optimized system has resulted in improved side-effect profiles and improved potency for optimally selected sequences.[48]

Table 4.2 lists some of the improved compounds identified using these improved strategies including follow-on sequences for protein tyrosine phosphatase 1B (PTP1b) and glucagon receptor (GCGR) that exhibited significant improvements in both potency and tolerability compared with the first sequences taken into clinical use prior to implementation of the enhanced screening practice. Additionally, the pipeline was further broadened to include other high-value liver-expressed cardiometabolic targets, *e.g.*, ApoC-III, Factor XI, Apo(a) and angiopoietin-like 3 (ANGPTL3) as well as additional targets outside of the initial cardiometabolic focus, *e.g.*, growth hormone receptor (GHR), diacylglycerol acyltransferase 2 (DGAT2), transthyretin (TTR), fibroblast growth factor receptor 4 (FGFR4) and prekallikrein (PKK).

Two of these optimized 2′MOE gapmer ASOs, volanesorsen and inotersen, have successfully completed Phase 3 clinical registration studies and are filed with both the FDA and European Medicines Agency (EMA) for market authorization in two distinct rare disease populations. In early Phase 2 proof-of-concept work with volanesorsen it was shown that targeting apoC-III dramatically lowered triglycerides in familial chylomicronemia syndrome (FCS) patients generally lacking functional lipoprotein lipase due to loss of function mutations.[67] Following this learning, two successful volanesorsen Phase 3 studies (COMPASS and APPROACH) reported that a weekly dose of 300 mg volanesorsen resulted in robust lowering of triglycerides (TG) of −77% from baseline in familial chylomicronemia syndrome (FCS) patients.[68] The absolute TG levels were thus reduced from a baseline average of >2000 mg dl^{-1} to less than 750 mg dl^{-1} in a majority of treated patients. This reduction in TGs resulted in a reduced risk of pancreatitis and daily symptoms of gastrointestinal pain and discomfort in patients exhibiting these symptoms early in study.[68] The SC-administered ASO was well tolerated, with few discontinuations due to injection-related side-effects in the 6–12 month controlled registration trial for FCS (APPROACH). For inotersen, a single Phase 3 study was conducted in patients with hereditary TTR (hATTR) amyloidosis with Stage

Table 4.2 Evolving liver targets and optimized back-up sequences.

Drug[a]	Target	Disease	Development status
Volanesorsen (ISIS 304801)[67,68,70]	APOC-III	Familial chylomicronemia syndrome (FCS) and familial partial lipodystrophy (FPL)	Phase 3 FCS complete and filed (FDA/EMA); Phase 3 (FPL ongoing)
Inotersen (ISIS 420915)[71]	Transthyretin (TTR)	hATTR amyloidosis	Phase 3 hATTR complete and filed (FDA/EMA)
IONIS-PTP1B$_{Rx}$ (ISIS 404173)[70]	Protein tyrosine phosphatase 1B (PTP-1B)	Diabetes	Phase 2 completed
IONIS-GCGR$_{Rx}$ (ISIS 449884)	Glucagon receptor	Type 2 diabetes	Phase 2 ongoing
IONIS-GCCR$_{Rx}$ (ISIS 449884)	Glucocorticoid receptor	Type 2 diabetes and Cushing's syndrome	Phase 2 completed
IONIS-FXI$_{Rx}$ (BAY 2306001; ISIS 416858)[72]	Factor XI	Thrombosis	Phase 2b ongoing
IONIS-APO(a) $_{Rx}$ (ISIS 494372)[73]	APO (a)	Coronary artery disease	Replaced with GalNac conjugate, Phase 2
IONIS-ANGPTL3$_{Rx}$ (ISIS 563580)[55]	ANGPTL3	Dyslipidemias	Replaced with GalNac conjugate, Phase 1
IONIS-PKK$_{Rx}$ (ISIS 546254)	Kallikrein B1	Hereditary angioedema and chronic migraine	Replaced with GalNAc after Phase 1
IONIS-DGAT2$_{Rx}$ (ISIS 484137)	DGAT2	Non-alcoholic steatohepatitis NASH	Phase 2 ongoing
IONIS-FGFR4$_{Rx}$ (ISIS 463588)	FGFR4	Obesity	Discontinued post Phase 2
IONIS-HBV$_{Rx}$ (ISIS 505358)	HBV surface AG	HBV infection	Phase 2 ongoing

[a]Optimized second generation ASO (Gen 2+), all 2′MOE gapmers.

1 or 2 peripheral neuropathy in the presence or absence of cardiomyopathy. High statistical significance was achieved for two primary endpoints, the neuropathy impairment score (mNIS+7) paired with the quality of life instrument, the Norfolk QoL-DN.[69] Inotersen also exhibited excellent tolerability, with no study drop-outs for injection-related adverse events in this 15 month placebo-controlled study. The improved tolerability for these two sequences verified an improved profile attributed to the increased screening employed in the human candidate selection process.[69]

4.3 Innovations in Liver Targeting

While great improvements in tolerability for 2′MOE second-generation ASOs were being made involving improved screening processes, knowledge obtained about the distribution in the liver and cell uptake was also being gained that led to new strategies to further improve ASO potency and widen safety margins by significantly reducing the active dose. Results from early longitudinal distribution kinetic studies indicated that the unconjugated ASO rapidly associated with proteins on the surface of cells in the liver[74] within minutes of administration. Within hours the ASOs were taken into the cells, later understood to be primarily, if not predominantly, through endocytotic mechanisms.[75] This mechanism of ASO uptake was relatively rapid, but much of the intracellular-trafficked ASO was in a non-functional or non-active compartment, namely the endosome or lysosome or the non-parenchymal cell compartments in the liver. Ultimately, antisense activity was a function largely of escape of a small proportion of ASO from the endosome by an, as yet, unknown mechanism. Further work to understand liver distribution showed that non-parenchymal cells were the first to take up ASOs and a large fraction of the delivered ASO resided in these cells with much less in the hepatocytes.[9] Further exploration of liver cell uptake demonstrated that the non-active pathway(s) were saturable and could be competitively inhibited by non-active or "dummy" oligonucleotides.[7,76] Much of the inactive component in the liver was due to a large fraction of the delivered dose in liver being taken up in Kupffer cells and endothelial cells (non-parenchymal) and a much smaller fraction taken up by the hepatocytes themselves. Competitively reducing the saturable uptake in the non-parenchymal fraction or endocytotic non-functional pathways with "dummy oligos" was suggested to be the likely explanation for a more preferred uptake and observed increase in antisense activity in the hepatocytes.

These exploratory mechanistic data indicated that more selective uptake in hepatocytes may indeed greatly improve activity for ASOs targeting mRNA primarily expressed in hepatocytes. The scientific literature is replete with information regarding targeting ASGPR1 on hepatocytes using *N*-acetyl galactose sugars.[77,78] Study of this literature led to the realization that receptor-mediated uptake may provide selective cell uptake. Ligand coupling of triantennary *N*-acetyl galactose sugars to antisense molecules was reported, and a new approach to liver targeting was brought forward for ASOs.[49] Initial preclinical work showed that this approach consistently delivered ASOs preferentially to hepatocytes and increased the potency of a 2′MOE ASO by approximately tenfold in mice and non-human primates when compared with the unconjugated ASO.[50–52]

Table 4.3 lists the current ligand-conjugated GalNAc 2′MOE ASOs that have now moved into clinical development. The first GalNAc conjugated ASO to achieve clinical validation was AKCEA-APO(a)-L_{Rx},[54] followed rapidly by additional targeted sequences that have now consistently shown substantial improvements in potency, tolerability and safety.[55] In regard to potency, the estimated dose causing 50% of the maximum effect (ED_{50}) in man was shown to be less than 5 mg per week (SC) compared with 130–150 mg per week for the unconjugated form, a remarkable 30-fold or greater increase in potency in man (Figure 4.2). Since the half-life of the ASO remains at approximately one month, once monthly or even less frequent dosing is supported. Multiple

Table 4.3 Most recent ASOs active in clinical development.

Drug[a]	Target	Disease	Development status
AKCEA-APOCIII-L_{Rx} (ISIS 678354)	APOC-III	Hypertriglyceridemia	Phase 2
AKCEA-APO(a)-L_{Rx} (ISIS 681257)[54]	Apo(a)	Cardiovascular disease (CVD)	Phase 2
AKCEA-ANGPTL3-L_{Rx} (ISIS 703802)[55]	Angiopoietin-like 3	CVD	Phase 2
GSK3389404 (IONIS-HBV-L_{Rx}) (ISIS 712408)	HBV surface antigen	Hepatitis B	Phase 2
IONIS-FB-L_{Rx} (ISIS 696844)	Complement factor B	Dry age-related macular degeneration (AMD)	Phase 2
IONIS-PKK-L_{Rx} (ISIS 721744)	Kallikrein B1	Hereditary angioedema	Phase 1
IONIS-GHR-L_{Rx} (ISIS 766720)	Growth hormone receptor	Acromegaly	Phase 1
IONIS-TMPRSS6-L_{Rx} (ISIS 702843)	Transmembrane protease, Serine 6	β-Thalassemia	Phase 1
IONIS-AGT-L_{Rx} (ISIS 757456)	Angiotensinogen	Resistant hypertension	Phase 1

[a]All 5′ GalNAc conjugated 2′MOE gapmer ASO.

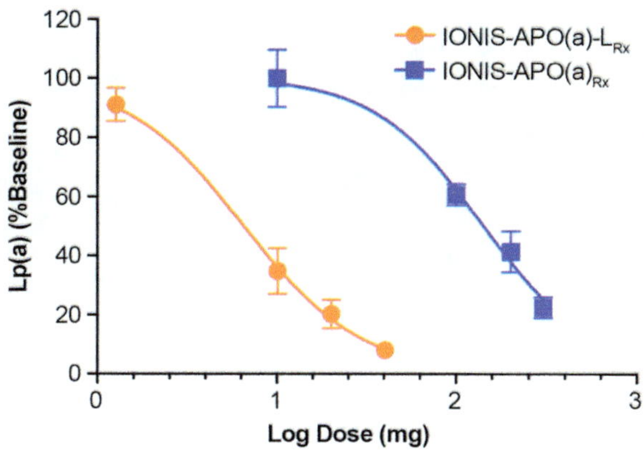

Figure 4.2 Greater than 30-fold increase in potency with a Ligand-Conjugated Antisense (LICA) ASO compared with the parent ASO.[54] Dose–response curves of IONIS-APO(a)-LRx and IONIS-APO(a)Rx in healthy human volunteers after four weeks of subcutaneous dosing shown by the percentage change from baseline in serum Lp(a) *vs.* Dose (mg). The weekly ED_{50} was 4.5 and 145 mg, respectively, using placebo data as the low-dose data points in each curve for modeling purposes.

Phase 2 studies are now underway testing multiple doses and less frequent dosing regimens. The advances in tolerability and improved safety margins provided by this strategy bode well for the future of liver-targeted ASOs in both rare and less rare diseases, like cardiovascular disease.

Additional advances in potency are likely given the advances in Gen 2.5 chemistry (bicyclic sugar chemistry, *e.g.* (*S*)-constrained ethyl (cEt)), combined with ligand-conjugated GalNAc. Gen 2.5 chemistry already delivers fivefold to tenfold greater potency due to increased binding affinity to the cognate mRNA. Adding GalNAc targeting to the hepatocyte has been shown preclinically to similarly improve the potency of this chemistry thus adding another tenfold increase in potency (Figure 4.3). This advance moves the technology to predicted ED_{50s} of less than 1 mg per week, further broadening the safety margin for this approach, and opens up the possibility of oral delivery based on such marked increases in potency.

4.4 Beyond the Liver

ASOs that contain Gen 2.5 chemistry in sequences that are lacking GalNAc ligands for hepatocyte targeting, broadly distribute to multiple tissues, including the liver. Coupled with the improved binding affinity to the cognate binding site on the targeted mRNA, these compounds have shown greatly improved antisense activity in tissues beyond the liver. This was first shown to be the case in cancer models, in which multiple tumor models were examined and compared with Gen 2.0 ASOs.[58–60] This improvement

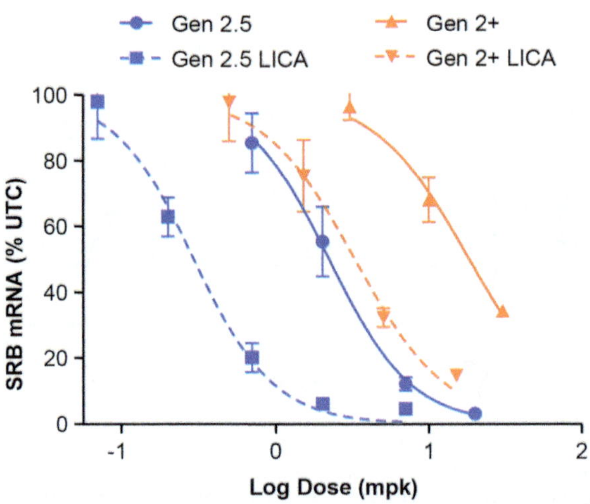

Figure 4.3 Generation 2.5 ASO chemistry plus LICA further enhances potency in mouse model studies.[79] Percentage change of scavenger receptor B1 mRNA levels from untreated control (UTC) *vs.* dose (mg kg^{-1}). Gen 2.5 LICA ASO ED_{50} = 0.3 mg kg^{-1} *vs.* 2.2 mg kg^{-1} for parent ASO; Gen 2+ LICA ASO ED_{50} = 3.3 mg kg^{-1} *vs.* 18.3 mg kg^{-1} for parent ASO.

Table 4.4 SC/IV administration of generation 2.5 (cEt) ASOs.

Drug	Target	Disease	Development status
Danvatirsen (AZD9150; IONIS-STAT3-2.5$_{Rx}$; ISIS 481464)[58,81]	Signal transducer and activator of transcription 3 (STAT3)	Advanced cancers and lymphoma	Phase 2
IONIS-AR-2.5$_{Rx}$ (AZD5312; ISIS 560131)[82]	Androgen receptor (AR)	Cancer	Phase 1
IONIS-KRAS-2.5$_{Rx}$ (AZD4785; ION 651987)[59]	Kirsten rat sarcoma virus oncogene homolog (KRAS)	Lung cancer, pancreatic cancer, colorectal cancer	Phase 1

in tumor targeting has been responsible for the advancement of multiple Gen 2.5 ASOs into development for oncology targets. Table 4.4 lists the Gen 2.5 cEt ASOs that have subsequently moved into development and their current status. Muscle became another opportunity for Gen 2.5 ASOs when initial work in mouse models for myotonic dystrophy targeting dystrophia myotonica protein kinase (DMPK) exhibited promising efficacy, which was followed by similar proof-of-concept in models of centronuclear myopathies.[61,80]

While promising systemic efforts to move beyond the liver, to tumors and muscle, are ongoing, alternative routes of administration have opportunistically opened the door to RNA targets in additional tissues. For example, while systemic administration of ASOs excludes central targets in the CNS due to the inability of ASOs to sufficiently cross the blood–brain barrier, intrathecal administration directly to the cerebrospinal fluid has yielded substantial distribution and antisense activity centrally.[83] Another advantage of direct administration to the CNS is the prolonged half-life of ASOs in brain and spinal cord.[8,84] Nusinersen (Spinraza) the first FDA approved intrathecally administered ASO therapy (see Chapter 6) and also the first and only treatment to date for spinal muscular atrophy (SMA)[85] is administered intrathecally every 4 months, consistent with the prolonged half-life. Nusinersen was also the first fully modified 2′MOE ASO to enter development designed to alter pre-mRNA splicing.[86–88] In this case, the ASO, upon binding to the survival motor neuron 2-(SMN2) pre-mRNA, promotes inclusion of an exon in the mature mRNA, which ultimately results in an increase in the SMN protein required for a healthy neuromuscular junction. Nusinersen represents just one more example of how transformative ASO technology can be when targeting the right target in the right tissue and/or cell type. While there are many opportunities being explored to do just that in the liver hepatocyte, there is much to do beyond the liver, including in the CNS. Table 4.5 lists the CNS-targeting ASOs that have disclosed targets currently in development. Many more with, as yet, undisclosed targets have entered non-clinical development, but have not yet reached clinical settings.

Table 4.5 Intrathecal administration in neurodegenerative disease.

Drug	Chemistry	Disease	Development status
Nusinersen (ISIS 396443)[86,88,89]	2′MOE, fully modified	Spinal muscular atrophy	Commercialized
IONIS-HTT$_{Rx}$ (ISIS 443139)	2′MOE, gapmer, mixed backbone	Huntington's disease	Phase 2
IONIS-SOD1$_{Rx}$ (ISIS 666853)	2′MOE, gapmer, mixed backbone	Amyotrophic lateral sclerosis	Phase 1
IONIS-MAPT$_{Rx}$ (ISIS 814907)	2′MOE, gapmer, mixed backbone	Alzheimer's and frontal temporal dementia	Phase 2

Table 4.6 Beyond the liver *via* additional alternative routes of administration.

Drug	Chemistry	Disease	Route	Development status
Vitravene (ISIS 2922)[96]	2′H	Cytomegalovirus retinitis	Intravitreal	Commercial
ISIS-IL4ra (ISIS-369645)[97]	2′MOE	Asthma	Inhalation	Phase 2
TPI ASM8[94]	2′OMe	Asthma	Inhalation	Phase 2
IONIS-ENAC-2.5$_{Rx}$ (ION-443139)[95]	cEt/2′MOE	Cystic fibrosis	Inhalation	Preclinical

Additional locally administered ASOs focused on inhibition of scarring of the skin (intradermal injection),[90] autoimmune disease in the gastrointestinal tract (oral formulation),[91] diseases of the eye (intravitreal injection) (Rho mutation, CMV retinitis) and chronic diseases of the lung (such as cystic fibrosis) have moved into development (Table 4.6). Once again, limitations in systemic distribution can be overcome by use of local routes of administration of ASOs. For example, the safety and delivery to the lung *via* inhalation of ASOs has been substantiated clinically[92,93] with an interleukin 4 (IL4) receptor alpha inhibitor indicated for asthma and for a dual ASO approach.[94] Preclinical work in cystic fibrosis models has shown promising results using Gen 2.5 ASOs delivered to the lung by inhalation.[95] Additional lung targets are nearing development, given the success in delivering active ASOs to the lung by inhalation.

4.5 Conclusions

Antisense technology has now been clinically and commercially validated as an important new drug discovery platform for human therapeutics. Based on the novel concept of targeting RNA instead of protein, the RNA therapeutic strategy, now validated with antisense technology, has emerged as the third platform for drug discovery alongside small molecules and protein therapeutics (protein replacement and monoclonal antibodies). The key to success for antisense technology was basic research across multiple areas including subcellular, cellular and whole-animal pharmacokinetic studies, coupled with pharmacodynamic correlations, medicinal chemistry, the exploitation of multiple mechanisms of action for ASO drugs and through the testing and validation of novel routes of delivery. Through this research, the hepatocyte has emerged as the most exploited cell type for antisense therapeutics to date for a broad range of human diseases, resulting in multiple commercial approvals for severe diseases with high unmet medical needs and with many more expected in the future. Advancements in the understanding of ASO subcellular trafficking have led to ligand-receptor targeting strategies, which have substantially enhanced further the hepatocyte as a target cell type opportunity for ASO therapeutics. Furthermore, advancements in medicinal chemistry and through the validation of novel routes of delivery have opened up many new organ systems and cell types for human therapeutics, including local delivery strategies in the CNS, eye and the lung, as well as for systemic applications, such as for cancer and muscular disorders.

The progress made in the development of antisense technology for human therapeutics has been remarkable, yet new breakthroughs are occurring at a rapid pace and these are expected to further enhance the potential therapeutic applications for this technology. Such advances include RNA biology research, which will continue to result in a better understanding of the connection between RNA and human disease, providing new targets and strategies that can be exploited with antisense technology. Furthermore, advances in ligand-receptor targeting that go beyond the liver, to open new opportunities in new cell types and organ systems are expected based on the mechanistic advancements made using such strategies in the liver. Finally, medicinal chemistry will continue to evolve and improve the properties of antisense drugs, leading to molecules that distribute to cells more efficiently, are better tolerated and are more potent. Antisense technology has already had a very meaningful and important impact on the lives of many patients with incurable diseases. Yet, we have only begun to take full advantage of the great potential antisense technology offers for human therapeutics. This potential will be realized through ongoing and future investments in basic research as it has in the past.

References

1. S. T. Crooke, M. J. Graham, J. E. Zuckerman, D. Brooks, B. S. Conklin, L. L. Cummins, M. J. Greig, C. J. Guinosso, D. Kornbrust, M. Manoharan, H. M. Sasmor, T. Schleich, K. L. Tivel and R. H. Griffey, *J. Pharmacol. Exp. Ther.*, 1996, **277**, 923–937.
2. J. A. Phillips, S. J. Craig, D. Bayley, R. A. Christian, R. Geary and P. L. Nicklin, *Biochem. Pharmacol.*, 1997, **54**, 657–668.
3. R. Z. Yu, R. S. Geary, J. M. Leeds, T. Watanabe, M. Moore, J. Fitchett, J. Matson, T. Burckin, M. V. Templin and A. A. Levin, *J. Pharm. Sci.*, 2001, **90**, 182–193.

4. R. S. Geary, R. Z. Yu, T. Watanabe, S. P. Henry, G. E. Hardee, A. Chappell, J. Matson, H. Sasmor, L. Cummins and A. A. Levin, *Drug Metab. Dispos.*, 2003, **31**, 1419–1428.
5. R. Z. Yu, R. S. Geary, D. K. Monteith, J. Matson, L. Truong, J. Fitchett and A. A. Levin, *J. Pharm. Sci.*, 2004, **93**, 48–59.
6. R. Z. Yu, T. W. Kim, A. Hong, T. A. Watanabe, H. J. Gaus and R. S. Geary, *Drug Metab. Dispos.*, 2007, **35**, 460–468.
7. R. S. Geary, E. Wancewicz, J. Matson, M. Pearce, A. Siwkowski, E. Swayze and F. Bennett, *Biochem. Pharmacol.*, 2009, **78**, 284–291.
8. R. S. Geary, D. Norris, R. Yu and C. F. Bennett, *Adv. Drug Delivery Rev.*, 2015, **87**, 46–51.
9. M. J. Graham, S. T. Crooke, D. K. Monteith, S. R. Cooper, K. M. Lemonidis, K. K. Stecker, M. J. Martin and R. M. Crooke, *J. Pharmacol. Exp. Ther.*, 1998, **286**, 447–458.
10. M. J. Graham, S. T. Crooke, K. M. Lemonidis, H. J. Gaus, M. V. Templin and R. M. Crooke, *Biochem. Pharmacol.*, 2001, **62**, 297–306.
11. R. Z. Yu, K. M. Lemonidis, M. J. Graham, J. E. Matson, R. M. Crooke, D. L. Tribble, M. K. Wedel, A. A. Levin and R. S. Geary, *Biochem. Pharmacol.*, 2009, **77**, 910–919.
12. R. Z. Yu, H. Zhang, R. S. Geary, M. Graham, L. Masarjian, K. Lemonidis, R. Crooke, N. M. Dean and A. A. Levin, *J. Pharmacol. Exp. Ther.*, 2001, **296**, 388–395.
13. H. Zhang, J. Cook, J. Nickel, R. Yu, K. Stecker, K. Myers and N. M. Dean, *Nat. Biotechnol.*, 2000, **18**, 862–867.
14. V. T. Samuel and G. I. Shulman, *Cell Metab.*, 2018, **27**, 22–41.
15. C. Ding, Y. Li, F. Guo, Y. Jiang, W. Ying, D. Li, D. Yang, X. Xia, W. Liu, Y. Zhao, Y. He, X. Li, W. Sun, Q. Liu, L. Song, B. Zhen, P. Zhang, X. Qian, J. Qin and F. He, *Mol. Cell. Proteomics*, 2016, **15**, 3190–3202.
16. B. Gao, *Cell. Mol. Immunol.*, 2016, **13**, 265–266.
17. M. W. Robinson, C. Harmon and C. O'Farrelly, *Cell. Mol. Immunol.*, 2016, **13**, 267–276.
18. C. Kampf, A. Mardinoglu, L. Fagerberg, B. M. Hallstrom, K. Edlund, E. Lundberg, F. Ponten, J. Nielsen and M. Uhlen, *FASEB J.*, 2014, **28**, 2901–2914.
19. R. A. Davis and T. Y. Hui, *Arterioscler. Thromb. Vasc. Biol.*, 2001, **21**, 887–898.
20. M. Mietus-Snyder, F. M. Sladek, G. S. Ginsburg, C. F. Kuo, J. A. Ladias, J. E. Darnell, Jr. and S. K. Karathanasis, *Mol. Cell. Biol.*, 1992, **12**, 1708–1718.
21. M. C. Petersen, D. F. Vatner and G. I. Shulman, *Nat. Rev. Endocrinol.*, 2017, **13**, 572–587.
22. M. DeSancho, in *Textbook of Hepatology: From Basic Science to Clinical Practice*, ed. J. Rodés, J.-P. Benhamou, A. Blei, J. Reichen and M. Rizzetto, Blackwell Publishing Ltd, Massachusetts, 3rd edn, 2007, p. 255.
23. D. F. Wallace, *Clin. Biochem. Rev.*, 2016, **37**, 51–62.
24. S. Guo, C. Casu, S. Gardenghi, S. Booten, M. Aghajan, R. Peralta, A. Watt, S. Freier, B. P. Monia and S. Rivella, *J. Clin. Invest.*, 2013, **123**, 1531–1541.

25. K. E. Finberg, M. M. Heeney, D. R. Campagna, Y. Aydinok, H. A. Pearson, K. R. Hartman, M. M. Mayo, S. M. Samuel, J. J. Strouse, K. Markianos, N. C. Andrews and M. D. Fleming, *Nat. Genet.*, 2008, **40**, 569–571.
26. J. D. Sipe, M. D. Benson, J. N. Buxbaum, S. I. Ikeda, G. Merlini, M. J. Saraiva and P. Westermark, *Amyloid*, 2016, **23**, 209–213.
27. S. Valleix, G. Verona, N. Jourde-Chiche, B. Nedelec, P. P. Mangione, F. Bridoux, A. Mange, A. Dogan, J. M. Goujon, M. Lhomme, C. Dauteuille, M. Chabert, R. Porcari, C. A. Waudby, A. Relini, P. J. Talmud, O. Kovrov, G. Olivecrona, M. Stoppini, J. Christodoulou, P. N. Hawkins, G. Grateau, M. Delpech, A. Kontush, J. D. Gillmore, A. D. Kalopissis and V. Bellotti, *Nat. Commun.*, 2016, **7**, 10353.
28. V. Plante-Bordeneuve and G. Said, *Lancet Neurol.*, 2011, **10**, 1086–1097.
29. R. M. Crooke, M. J. Graham, K. M. Lemonidis, C. P. Whipple, S. Koo and R. J. Perera, *J. Lipid Res.*, 2005, **46**, 872–884.
30. E. Merki, M. J. Graham, A. E. Mullick, E. R. Miller, R. M. Crooke, R. E. Pitas, J. L. Witztum and S. Tsimikas, *Circulation*, 2008, **118**, 743–753.
31. A. E. Mullick, W. Fu, M. J. Graham, R. G. Lee, D. Witchell, T. A. Bell, C. P. Whipple and R. M. Crooke, *J. Lipid Res.*, 2011, **52**, 885–896.
32. C. Stancu and A. Sima, *J. Cell. Mol. Med.*, 2001, **5**, 378–387.
33. J. D. Horton, J. C. Cohen and H. H. Hobbs, *Trends Biochem. Sci.*, 2007, **32**, 71–77.
34. J. J. Kastelein, M. K. Wedel, B. F. Baker, J. Su, J. D. Bradley, R. Z. Yu, E. Chuang, M. J. Graham and R. M. Crooke, *Circulation*, 2006, **114**, 1729–1735.
35. F. Akdim, E. S. Stroes, E. J. Sijbrands, D. L. Tribble, M. D. Trip, J. W. Jukema, J. D. Flaim, J. Su, R. Yu, B. F. Baker, M. K. Wedel and J. J. Kastelein, *J. Am. Coll. Cardiol.*, 2010, **55**, 1611–1618.
36. F. Akdim, D. L. Tribble, J. D. Flaim, R. Yu, J. Su, R. S. Geary, B. F. Baker, R. Fuhr, M. K. Wedel and J. J. Kastelein, *Eur. Heart J.*, 2011, **32**, 2650–2659.
37. F. J. Raal, R. D. Santos, D. J. Blom, A. D. Marais, M. J. Charng, W. C. Cromwell, R. H. Lachmann, D. Gaudet, J. L. Tan, S. Chasan-Taber, D. L. Tribble, J. D. Flaim and S. T. Crooke, *Lancet*, 2010, **375**, 998–1006.
38. R. D. Santos, P. B. Duell, C. East, J. R. Guyton, P. M. Moriarty, W. Chin and R. S. Mittleman, *Eur. Heart J.*, 2015, **36**, 566–575.
39. R. D. Santos, F. J. Raal, J. M. Donovan and W. C. Cromwell, *J. Clin. Lipidol.*, 2015, **9**, 201–209.
40. J. M. Leeds, M. J. Graham, L. Truong and L. L. Cummins, *Anal. Biochem.*, 1996, **235**, 36–43.
41. S. P. Henry, T.-W. Kim, K. Kramer-Stickland, T. A. Zanardi, R. A. Fey and A. A. Levin, in *Antisense Drug Discovery: Principles, Strategies, and Applications*, ed. S. T. Crooke, CRC Press, Taylor & Francis Group, Florida, 2nd edn, 2008, p. 327.
42. T. J. Kwoh, in *Antisense Drug Discovery: Principles, Strategies, and Applications*, ed. S. T. Crooke, CRC Press, Taylor & Francis Group, Florida, 2nd edn, 2008, p. 365.

43. B. P. Monia, E. A. Lesnik, C. Gonzalez, W. F. Lima, D. McGee, C. J. Guinosso, A. M. Kawasaki, P. D. Cook and S. M. Freier, *J. Biol. Chem.*, 1993, **268**, 14514–14522.
44. K. H. Altmann, D. Fabbro, N. M. Dean, T. Geiger, B. P. Monia, M. Muller and P. Nicklin, *Biochem. Soc. Trans.*, 1996, **24**, 630–637.
45. S. M. Freier and K. H. Altmann, *Nucleic Acids Res.*, 1997, **25**, 4429–4443.
46. C. F. Bennett, in *Antisense Drug Discovery: Principles, Strategies, and Applications*, ed. S. T. Crooke, CRC Press, Taylor & Francis Group, Florida, 2nd edn, 2008, p. 273.
47. K. Hovingh, J. Besseling and J. Kastelein, *Expert Opin. Drug Saf.*, 2013, **12**, 569–579.
48. S. T. Crooke, B. F. Baker, T. J. Kwoh, W. Cheng, D. J. Schulz, S. Xia, N. Salgado, H. H. Bui, C. E. Hart, S. A. Burel, H. S. Younis, R. S. Geary, S. P. Henry and S. Bhanot, *Mol. Ther.*, 2016, **24**, 1771–1782.
49. T. P. Prakash, J. Yu, M. T. Migawa, G. A. Kinberger, W. B. Wan, M. E. Ostergaard, R. L. Carty, G. Vasquez, A. Low, A. Chappell, K. Schmidt, M. Aghajan, J. Crosby, H. M. Murray, S. L. Booten, J. Hsiao, A. Soriano, T. Machemer, P. Cauntay, S. A. Burel, S. F. Murray, H. Gaus, M. J. Graham, E. E. Swayze and P. P. Seth, *J. Med. Chem.*, 2016, **59**, 2718–2733.
50. R. Z. Yu, R. Gunawan, N. Post, T. Zanardi, S. Hall, J. Burkey, T. W. Kim, M. J. Graham, T. P. Prakash, P. P. Seth, E. E. Swayze, R. S. Geary, S. P. Henry and Y. Wang, *Nucleic Acid Ther.*, 2016, **26**, 372–380.
51. R. Z. Yu, M. J. Graham, N. Post, S. Riney, T. Zanardi, S. Hall, J. Burkey, C. S. Shemesh, T. P. Prakash, P. P. Seth, E. E. Swayze, R. S. Geary, Y. Wang and S. Henry, *Mol. Ther. Nucleic Acids*, 2016, **5**, e317.
52. K. Schmidt, T. P. Prakash, A. J. Donner, G. A. Kinberger, H. J. Gaus, A. Low, M. E. Ostergaard, M. Bell, E. E. Swayze and P. P. Seth, *Nucleic Acids Res.*, 2017, **45**, 2294–2306.
53. M. Tanowitz, L. Hettrick, A. Revenko, G. A. Kinberger, T. P. Prakash and P. P. Seth, *Nucleic Acids Res.*, 2017, **45**, 12388–12400.
54. N. J. Viney, J. C. van Capelleveen, R. S. Geary, S. Xia, J. A. Tami, R. Z. Yu, S. M. Marcovina, S. G. Hughes, M. J. Graham, R. M. Crooke, S. T. Crooke, J. L. Witztum, E. S. Stroes and S. Tsimikas, *Lancet*, 2016, **388**, 2239–2253.
55. M. J. Graham, R. G. Lee, T. A. Brandt, L. J. Tai, W. Fu, R. Peralta, R. Yu, E. Hurh, E. Paz, B. W. McEvoy, B. F. Baker, N. C. Pham, A. Digenio, S. G. Hughes, R. S. Geary, J. L. Witztum, R. M. Crooke and S. Tsimikas, *N. Engl. J. Med.*, 2017, **377**, 222–232.
56. P. P. Seth, A. Siwkowski, C. R. Allerson, G. Vasquez, S. Lee, T. P. Prakash, E. V. Wancewicz, D. Witchell and E. E. Swayze, *J. Med. Chem.*, 2009, **52**, 10–13.
57. P. P. Seth, G. Vasquez, C. A. Allerson, A. Berdeja, H. Gaus, G. A. Kinberger, T. P. Prakash, M. T. Migawa, B. Bhat and E. E. Swayze, *J. Org. Chem.*, 2010, **75**, 1569–1581.
58. D. Hong, R. Kurzrock, Y. Kim, R. Woessner, A. Younes, J. Nemunaitis, N. Fowler, T. Zhou, J. Schmidt, M. Jo, S. J. Lee, M. Yamashita, S. G. Hughes,

L. Fayad, S. Piha-Paul, M. V. Nadella, M. Mohseni, D. Lawson, C. Reimer, D. C. Blakey, X. Xiao, J. Hsu, A. Revenko, B. P. Monia and A. R. MacLeod, *Sci. Transl. Med.*, 2015, **7**, 314ra185.

59. S. J. Ross, A. S. Revenko, L. L. Hanson, R. Ellston, A. Staniszewska, N. Whalley, S. K. Pandey, M. Revill, C. Rooney, L. K. Buckett, S. K. Klein, K. Hudson, B. P. Monia, M. Zinda, D. C. Blakey, P. D. Lyne and A. R. Macleod, *Sci. Transl. Med.*, 2017, **9**, eaao4188.
60. A. R. MacLeod and S. T. Crooke, *J. Clin. Pharmacol.*, 2017, **57**(Suppl 10), S43–S59.
61. D. Jauvin, J. Chretien, S. K. Pandey, L. Martineau, L. Revillod, G. Bassez, A. Lachon, A. R. MacLeod, G. Gourdon, T. M. Wheeler, C. A. Thornton, C. F. Bennett and J. Puymirat, *Mol. Ther. Nucleic Acids*, 2017, **7**, 465–474.
62. http://www.accessdata.fda.gov/drugsatfda_docs/label/2016/203568s008lbl.pdf (last accessed January 2018).
63. P. B. Duell, R. D. Santos, B. A. Kirwan, J. L. Witztum, S. Tsimikas and J. J. P. Kastelein, *J. Clin. Lipidol.*, 2016, **10**, 1011–1021.
64. H. L. Janssen, S. Kauppinen and M. R. Hodges, *N. Engl. J. Med.*, 2013, **369**, 878.
65. B. A. Zinker, C. M. Rondinone, J. M. Trevillyan, R. J. Gum, J. E. Clampit, J. F. Waring, N. Xie, D. Wilcox, P. Jacobson, L. Frost, P. E. Kroeger, R. M. Reilly, S. Koterski, T. J. Opgenorth, R. G. Ulrich, S. Crosby, M. Butler, S. F. Murray, R. A. McKay, S. Bhanot, B. P. Monia and M. R. Jirousek, *Proc. Natl. Acad. Sci. U. S. A.*, 2002, **99**, 11357–11362.
66. K. W. Sloop, J. X. Cao, A. M. Siesky, H. Y. Zhang, D. M. Bodenmiller, A. L. Cox, S. J. Jacobs, J. S. Moyers, R. A. Owens, A. D. Showalter, M. B. Brenner, A. Raap, J. Gromada, B. R. Berridge, D. K. Monteith, N. Porksen, R. A. McKay, B. P. Monia, S. Bhanot, L. M. Watts and M. D. Michael, *J. Clin. Invest.*, 2004, **113**, 1571–1581.
67. D. Gaudet, D. Brisson, K. Tremblay, V. J. Alexander, W. Singleton, S. G. Hughes, R. S. Geary, B. F. Baker, M. J. Graham, R. M. Crooke and J. L. Witztum, *N. Engl. J. Med.*, 2014, **371**, 2200–2206.
68. D. Gaudet, A. Digenio, V. Alexander, M. Arca, A. Jones, E. Stroes, J. Bergeron, F. Civeira, L. Hemphill, D. Blom, J. Flaim, S. Hughes, R. S. Geary, S. Tsimikas, J. Witztum and E. Bruckert, *Atherosclerosis*, 2017, **263**, e10.
69. T. C. A. K. Wang, M. Waddington Cruz, M. J. Polydefkis, P. J. Dyck, M. Scheinberg, V. Plante-Bordeneuve, J. Berk, F. Barroso, D. Adams, T. Brannagan, C. Whelan, G. Merlini, B. M. Drachman, S. B. Heitner, I. Conceicao, H. Schmidt, G. Vita, J. M. Campistol, J. Gamez, E. Gane, P. Gorevic, A. S. Bulle Oliveira, B. P. Monia, S. G. Hughes, J. Kwoh, B. W. McEvoy, B. F. Baker, E. J. Ackermann and M. D. Benson, *Ann. Neurol.*, 2017, **82**, S108–S109.
70. A. Digenio, R. L. Dunbar, V. J. Alexander, M. Hompesch, L. Morrow, R. G. Lee, M. J. Graham, S. G. Hughes, R. Yu, W. Singleton, B. F. Baker, S. Bhanot and R. M. Crooke, *Diabetes Care*, 2016, **39**, 1408–1415.

71. E. J. Ackermann, S. Guo, M. D. Benson, S. Booten, S. Freier, S. G. Hughes, T. W. Kim, T. Jesse Kwoh, J. Matson, D. Norris, R. Yu, A. Watt and B. P. Monia, *Amyloid*, 2016, **23**, 148–157.
72. H. R. Buller, D. Gailani and J. I. Weitz, *N. Engl. J. Med.*, 2015, **372**, 1672.
73. S. Tsimikas, N. J. Viney, S. G. Hughes, W. Singleton, M. J. Graham, B. F. Baker, J. L. Burkey, Q. Yang, S. M. Marcovina, R. S. Geary, R. M. Crooke and J. L. Witztum, *Lancet*, 2015, **386**, 1472–1483.
74. M. Butler, K. Stecker and C. F. Bennett, *Lab. Invest.*, 1997, **77**, 379–388.
75. E. Koller, T. M. Vincent, A. Chappell, S. De, M. Manoharan and C. F. Bennett, *Nucleic Acids Res.*, 2011, **39**, 4795–4807.
76. A. J. Donner, E. V. Wancewicz, H. M. Murray, S. Greenlee, N. Post, M. Bell, W. F. Lima, E. E. Swayze and P. P. Seth, *Nucleic Acid Ther.*, 2017, **27**, 209–220.
77. R. J. Stockert, *Physiol. Rev.*, 1995, **75**, 591–609.
78. A. A. D'Souza and P. V. Devarajan, *J. Controlled Release*, 2015, **203**, 126–139.
79. T. P. Prakash, M. J. Graham, J. Yu, R. Carty, A. Low, A. Chappell, K. Schmidt, C. Zhao, M. Aghajan, H. F. Murray, S. Riney, S. L. Booten, S. F. Murray, H. Gaus, J. Crosby, W. F. Lima, S. Guo, B. P. Monia, E. E. Swayze and P. P. Seth, *Nucleic Acids Res.*, 2014, **42**, 8796–8807.
80. H. Tasfaout, S. Buono, S. Guo, C. Kretz, N. Messaddeq, S. Booten, S. Greenlee, B. P. Monia, B. S. Cowling and J. Laporte, *Nat. Commun.*, 2017, **8**, 15661.
81. S. Odate, V. Veschi, S. Yan, N. Lam, R. Woessner and C. J. Thiele, *Clin. Cancer Res.*, 2017, **23**, 1771–1784.
82. Y. Yamamoto, Y. Loriot, E. Beraldi, F. Zhang, A. W. Wyatt, N. Al Nakouzi, F. Mo, T. Zhou, Y. Kim, B. P. Monia, A. R. MacLeod, L. Fazli, Y. Wang, C. C. Collins, A. Zoubeidi and M. Gleave, *Clin. Cancer Res.*, 2015, **21**, 1675–1687.
83. R. A. Smith, T. M. Miller, K. Yamanaka, B. P. Monia, T. P. Condon, G. Hung, C. S. Lobsiger, C. M. Ward, M. McAlonis-Downes, H. Wei, E. V. Wancewicz, C. F. Bennett and D. W. Cleveland, *J. Clin. Invest.*, 2006, **116**, 2290–2296.
84. K. T. Luu, D. A. Norris, R. Gunawan, S. Henry, R. Geary and Y. Wang, *J. Clin. Pharmacol.*, 2017, **57**, 1031–1041.
85. M. Wadman, *Science*, 2016, **354**, 1359–1360.
86. C. A. Chiriboga, K. J. Swoboda, B. T. Darras, S. T. Iannaccone, J. Montes, D. C. De Vivo, D. A. Norris, C. F. Bennett and K. M. Bishop, *Neurology*, 2016, **86**, 890–897.
87. K. J. Swoboda, *N. Engl. J. Med.*, 2014, **371**, 1752–1754.
88. R. S. Finkel, E. Mercuri, B. T. Darras, A. M. Connolly, N. L. Kuntz, J. Kirschner, C. A. Chiriboga, K. Saito, L. Servais, E. Tizzano, H. Topaloglu, M. Tulinius, J. Montes, A. M. Glanzman, K. Bishop, Z. J. Zhong, S. Gheuens, C. F. Bennett, E. Schneider, W. Farwell, D. C. De Vivo and E. S. Group, *N. Engl. J. Med.*, 2017, **377**, 1723–1732.
89. R. S. Finkel, C. A. Chiriboga, J. Vajsar, J. W. Day, J. Montes, D. C. De Vivo, M. Yamashita, F. Rigo, G. Hung, E. Schneider, D. A. Norris, S. Xia, C. F. Bennett and K. M. Bishop, *Lancet*, 2016, **388**, 3017–3026.

90. M. Sisco, Z. B. Kryger, K. D. O'Shaughnessy, P. S. Kim, G. S. Schultz, X. Z. Ding, N. K. Roy, N. M. Dean and T. A. Mustoe, *Wound Repair Regener.*, 2008, **16**, 661–673.
91. G. Monteleone and F. Pallone, *N. Engl. J. Med.*, 2015, **372**, 2461.
92. S. A. Gregory and J. Karras, in *Antisense Drug Discovery: Principles, Strategies, and Applications*, ed. S. T. Crooke, CRC Press, Taylor & Francis Group, Florida, 2nd edn, 2007, p. 665.
93. J. G. Karras, J. R. Crosby, M. Guha, D. Tung, D. A. Miller, W. A. Gaarde, R. S. Geary, B. P. Monia and S. A. Gregory, *Am. J. Respir. Cell Mol. Biol.*, 2007, **36**, 276–285.
94. H. Imaoka, H. Campbell, I. Babirad, R. M. Watson, M. Mistry, R. Sehmi and G. M. Gauvreau, *Clin. Exp. Allergy*, 2011, **41**, 1740–1746.
95. J. R. Crosby, C. Zhao, C. Jiang, D. Bai, M. Katz, S. Greenlee, H. Kawabe, M. McCaleb, D. Rotin, S. Guo and B. P. Monia, *J. Cystic Fibrosis*, 2017, **16**, 671–680.
96. B. Roehr, *J. Int. Assoc. Physicians AIDS Care*, 1998, **4**, 14–16.
97. R. A. Fey, M. V. Templin, J. D. McDonald, R. Z. Yu, J. A. Hutt, A. P. Gigliotti, S. P. Henry and M. D. Reed, *Inhalation Toxicol.*, 2014, **26**, 452–463.

CHAPTER 5

Oligonucleotide-based Toll-like Receptor Antagonists and Therapeutic Applications

EKAMBAR R. KANDIMALLA[†] AND SUDHIR AGRAWAL*

ARNAY Sciences LLC, Shrewsbury, MA 01545, USA
*E-mail: sagrawal@arnaysciences.com

5.1 Introduction

Endosomal Toll-like receptors (TLRs) are part of the innate immune system that sense molecular signatures derived from pathogens and protect against invading infections. However, under the conditions of breakdown of self-tolerance, the immune system recognizes self-antigens and/or endogenous nucleic acids and produces inflammatory responses and antibodies that attack the body's own tissues, leading to autoimmune and inflammatory diseases.[1] Immune complexes containing self-nucleic acids act as endogenous ligands to TLRs, namely TLR7, 8, and 9, stimulating interferon-α (IFN-α) and other proinflammatory cytokines and thus exacerbating autoimmune diseases (Figure 5.1).[1–3] Uncontrolled induction of TLR-mediated IFN-α and

[†]Present address: Exicure, Inc., 8045 Lamon Ave., Skokie, IL 60077, USA

Drug Discovery Series No. 68
Advances in Nucleic Acid Therapeutics
Edited by Sudhir Agrawal and Michael J. Gait

Published by the Royal Society of Chemistry, www.rsc.org

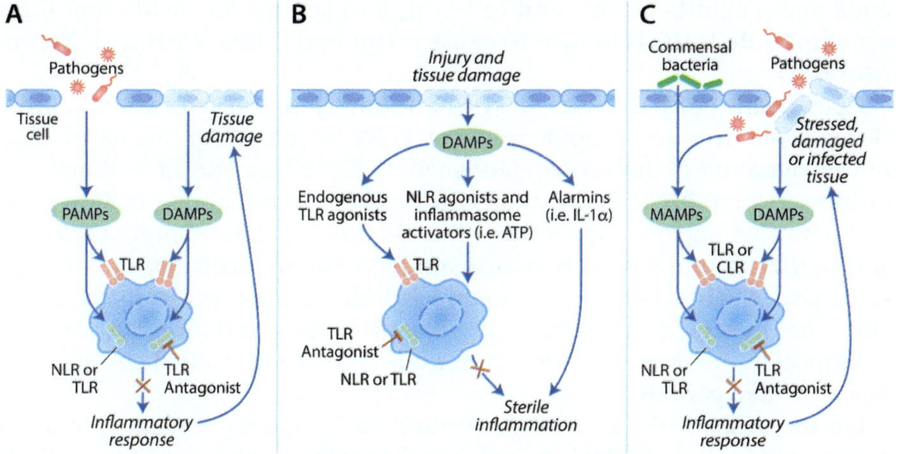

Figure 5.1 Autoimmune diseases are generally defined as the dysregulation of the immune system. (A) Chronic infection and inflammation leads to generation of endogenous nucleic acid-based pathogen-associated molecular patterns (PAMPs) and damage-associated molecular patterns (DAMPs). These PAMPs and DAMPs activate TLR7, 8, and 9 and other innate immune receptors resulting in induction of the inflammatory response. (B) Uncontrolled chronic activation of these pathways leads to elevated levels of inflammatory cytokines further exacerbating the inflammation. (C) Microbe-associated molecular patterns (MAMPs) and DAMPs could also be generated from chronic inflammation caused by commensal bacteria. Inhibition of TLR 7, 8, and 9 by TLR antagonist candidates provides a novel approach for the treatment of autoimmune and inflammatory diseases.

other proinflammatory cytokines by endogenous ligands has been implicated in several autoimmune and inflammatory diseases, including rheumatoid arthritis (RA), myositis, systemic lupus erythematosus (SLE), psoriasis, and atherosclerosis.[1–6] Monoclonal antibodies targeting a number of downstream elements, such as tumor necrosis factor-α (TNF-α), interleukin-12 (IL-12), IL-6, IL-17A, interferon-γ induced protein-10 (IP-10), IFN-α, and B lymphocyte stimulator (BLyS), are currently being developed or are in use for the treatment of autoimmune diseases, including RA, SLE, and psoriasis.[7] In the immune cascade, TLRs are upstream of these cytokine factors and targeting TLRs with oligonucleotide-based antagonists is a rational approach to mitigate induction of multiple inflammatory cytokines that are central to autoimmune pathophysiology.

The antimalarial agent hydroxychloroquine (HQ) is commonly used for the treatment of SLE, RA, and other autoimmune diseases.[8–10] HQ-treated SLE-patient immune cells do not produce cytokines, including IFN-α and TNF-α, when stimulated with TLR7 and TLR9 agonists, indicating that HQ inhibits endosomal TLR-mediated immune responses.[11] HQ suppression of TLR-mediated immune responses is a result of neutralization of endosomal acidification[12] and/or binding to self-nucleic acids that interact with TLRs.[13] HQ is known to cause retinopathy, neuromyotoxicity, and cardiotoxicity.[14]

It has been shown that certain synthetic oligodeoxynucleotides containing G-rich sequences act as antagonists of TLR9 by inhibiting factors involved in the signaling pathways of TLR9 such as signal transducer and activator of transcription (STAT)-1, -3, and -4 and/or other downstream factors proximal to nuclear factor kappa B (NF-κB) activation.[15–23] These compounds have shown therapeutic activity in mouse models of lupus, arthritis, and multiple sclerosis.[24–30] However, G-rich oligonucleotides are known to form quartet and other hyperstructures, lack drug-like properties, and have not advanced to clinical development,[31–33] except for one compound studied encapsulated in a virus-like particle.

On the basis of the above observations, it is suggestive that agents that act as antagonists of TLR7, 8, and 9 and thereby inhibit the TLR-mediated immune cascade triggered by endogenous nucleic acids or self-antigens could provide a novel approach for the treatment of autoimmune and inflammatory diseases. In this chapter we describe a novel class of synthetic oligonucleotides that act as antagonists of TLR7, 8, and 9, studies of these in preclinical models of autoimmune and inflammatory diseases, and proof of concept clinical trials.

5.2 Oligonucleotide-based TLR Antagonists

During the early clinical development of a phosphorothioate (PS) ASO, referred to as GEM91, targeted to human immunodeficiency virus-1 (HIV-1), subcutaneous administration led to the observation of flu-like symptoms in patients. This was the first observation of immune activation by an antisense molecule in humans. There were several reports on the immunostimulatory properties of palindromic sequences containing CG dinucleotides.[34–36] The GEM91 sequence (5′-CTCTCGCACCCATCTCTCTCCTTCT-3′) contained a CG dinucleotide motif and our studies delineated that this dinucleotide was responsible for immune activation by GEM91.[37–39] At that time, it was our intention to mitigate immune responses of PS-ASOs containing CG dinucleotides with chemical modifications.[37,38] During these studies a detailed understanding was developed of how the chemical modifications introduced at or around a CG dinucleotide within an oligonucleotide modulate immune responses.[37,38] These studies led us to identify a novel class of oligonucleotides that acted as antagonists of TLRs.

It is important to note that TLR7, 8, and 9 are expressed primarily in endosomal compartments of immune cells. The ligands for TLR7 and 8 are single-stranded viral or bacterial RNA molecules or their degradation products, and ligands for TLR9 include bacterial or synthetic DNA containing unmethylated CG dinucleotides.[31] Under pathological conditions TLR7, 8, and 9 could also be activated by endogenous nucleic acids.[1] While these TLRs have specifically defined ligands and signaling cascades, activation of a specific TLR may result in alteration of the function of other TLRs through cross-talk. For example, knockout of TLR7 or 9 in lupus-prone mice has provided evidence supporting the idea that TLR7 and TLR9 play pathogenic and protective roles.[40] On the basis of these observations it is important to identify antagonists that would inhibit TLR7-, 8-, and 9-mediated immune cascades for better therapeutic outcome.

5.2.1 Structure–Activity Relationship Studies

PS-oligodeoxynucleotides (PS-oligos) containing CG dinucleotides activate immune responses by binding to TLR9, activating the downstream signaling cascade and thus leading to NF-κB activation, and induction of cytokines (Chapter 14).[31] Structure–activity relationship studies showed that site-specific incorporation of 2′-*O*-methylribonucleotides at positions adjacent to CG dinucleotides on the 5′-side led to mitigation of immune-stimulatory activity.[41] This observation raised the possibility that these substitutions abolished the recognition of CG dinucleotide-containing PS-oligos by TLR9 and thereby prevented them acting as immune stimulators. When studies were conducted in the presence of a TLR9 agonist, it was noted that modified CG containing PS-oligos inhibited the immune-stimulatory activity of the agonist by binding to TLR9, providing evidence that modified CG-containing PS-oligos were being recognized by TLR9 and acted as antagonists.[31]

Detailed studies have highlighted key features of the oligonucleotide-based TLR antagonists. These oligonucleotides contain two key components, an immune-stimulatory motif and an immune-regulatory region positioned on the 5′-side adjacent to the immune-stimulatory motif. Extensive structure–activity relationship studies have been carried out with systematic incorporation of modifications in both of these regions. Modifications which have been studied in the immune regulatory region included 2′-*O*-methylribonucleotide,[41] 2′-*O*-methoxyethyl or 3′-*O*-methyl (which also results in a 2′–5′-linkage),[42] methylphosphonate linkages, which result in non-ionic internucleotide linkages,[43,44] 2′–5′-linkages,[45] non-nucleotide linkers,[46] or abasic nucleotides.[47,48] Similarly modifications of the immune stimulatory motif included natural 2′-*O*-methylribo-C, 5-methyl-dC, 5-methyl-2′-*O*-methylribo-C for C and 2′-*O*-methylribo-G, 7-deaza-dG, or araG for G.[31,49–51] These studies have provided a repertoire of chemical modifications which could be used to design TLR antagonist candidates.[49–51]

5.2.2 TLR Antagonists

On the basis of the insights gained from the detailed structure–activity relationship studies, several TLR antagonists have been designed. Figure 5.2 shows examples of four TLR antagonists. TLR antagonist example 1 is

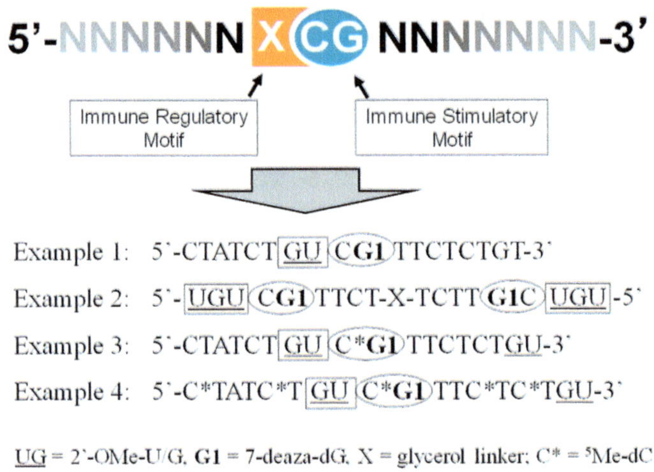

Figure 5.2 TLR antagonists consist of an immune-stimulatory motif and an immune-regulatory motif. Examples of four TLR antagonists are shown. In the examples, the immune-regulatory motif consists of 2′-*O*-methyl-ribonucleotide modifications (shown in the boxes on the 5′-end of the immune-stimulatory motifs), and immune-stimulatory motifs consists of 7-deazadG or, 5-methyl-dC-7deazadG (shown in ovals).

an 18-mer PS-oligo with a modified immune-stimulatory motif CG1 (G1 is 7-deaza-dG, which is recognized by TLR9)[52] and the immune-regulatory motif consists of two 2′-*O*-methylribonucleotides on the 5′-side adjacent to CG1. TLR antagonist example 2 is an 18-mer PS-oligo with two identical 9-mers attached *via* their 3′-ends. The presence of a 3′–3′-linkage provides increased stability against nucleases and two accessible 5′-ends for TLR9 recognition.[53–57] The immune-regulatory motif of TLR antagonist example 2 consists of three contiguous 2′-*O*-methylribonucleotides adjacent to the 5′-side of the immune-stimulatory motif. TLR antagonist examples 3 and 4 are 18-mer PS-oligos with modified immune-stimulatory motifs C*G1, wherein C* is 5-methyl-dC. In both the examples 3 and 4, the immune-regulatory motif consists of two 2′-*O*-methylribonucleotides on the 5′-side of each immune-stimulatory motif. In addition, all the dCs are substituted with 5-methyl-dCs in TLR antagonist example 4. Further details of these TLR antagonist examples have been published.[49–51]

The four examples described in Figure 5.2 represent general sequence and chemical modifications used in immunostimulatory and immunoregulatory motifs of a broad class of TLR7 and 9 or TLR7, 8, and 9 antagonists developed in our laboratory.

From the portfolio of compounds studied, three novel TLR antagonist candidates, IMO-3100, which inhibits TLR7 and 9 activation, and IMO-8400 and IMO-9200, which inhibit TLR7, 8, and 9 activation, have advanced to clinical evaluation. Sequence compositions of these clinical candidates have not been disclosed.

5.2.3 Inhibitory Activity of TLR Antagonists

In general, TLR antagonists as single agents do not induce immune responses in mouse and human cell cultures, or *in vivo* in mice, non-human primates (NHPs), and humans. TLR antagonists can be categorized into two classes. One class of TLR antagonists inhibits TLR7 and 9 and the other class of antagonists inhibits TLR7, 8, and 9. TLR antagonist examples 1 and 2 shown in Figure 5.2 have been shown to inhibit TLRs 7 and 9 and TLR antagonist examples 3 and 4 have been shown to inhibit TLRs 7, 8, and 9.

TLR antagonist example 1 inhibited TLR7 and TLR9 agonist-induced immune activity in cell lines, mouse spleen cells, and in human peripheral blood mononuclear cells (PBMCs), *in vivo* in mice, and NHPs. TLR antagonist example 1 showed inhibition of TLR7 and 9 agonist-induced NF-κB activation in HEK293, J774 macrophage cell lines and in human PBMCs.[49,58] The TLR inhibition was through competitive binding to the receptor and was reversible with increasing concentrations of TLR agonist.[49] In human PBMC cultures, TLR antagonist inhibited TLR9 and TLR7 and 8 agonists-induced cytokines and TLR9 agonist-induced B cell proliferation when co-incubated.[49] In mice, TLR antagonist inhibited TLR9 agonist-induced cytokine secretion for up to 5 days post dosing, and the inhibition was dose-dependent.[49] The TLR antagonist also inhibited immune responses induced by an RNA-based TLR7 and 8 agonist, but not TLR2, 3, 4, or 5 agonist, in mice.[49] Additionally, TLR antagonist inhibited TLR9 agonist-induced IL-6 in lupus-prone MRL-lpr mouse spleen cell cultures.[49]

TLR antagonist example 3 showed potent inhibition of TLR 7- and 9-mediated immune responses in murine cell-based assays and *in vivo* in mice.[50] Since TLR8 is not functional in mice, example 3 was evaluated in human cells. Figure 5.3 shows TLR7, 8, and 9 agonist-induced cytokine responses in human PBMCs. When the PBMCs were co-cultured with TLR agonist and TLR antagonist example 3, this resulted in the inhibition of TLR7-, 8-, and 9 agonists-mediated production of a broad range of cytokines, including TNF-α, IL-12, IL-6, IFN-α, IL-1β, and IP-10 in healthy human PBMCs (Figure 5.3). Additionally, TLR antagonist example 3 showed inhibition of TLR7 and TLR9 agonists-induced cytokines in human plasmacytoid dendritic cell (pDC), and TLR8 agonist-induced cytokines in myeloid dendritic cell (mDC) cultures.[50] In PBMCs obtained from SLE patients, TLR antagonist example 3 showed inhibition of TLR7 and 9 agonists-induced cytokines, including IFN-α, TNF-α, IL-6, IP-10, macrophage inflammatory protein (MIP)-1α, and MIP-1β.[50] Similarly, TLR antagonist example 3 showed TLR7 and TLR9 agonists-induced cytokine inhibition in mouse primary spleen cell cultures as well.[50]

5.2.4 Inhibitory Activity of Clinical Candidates

Gene expression profiling of TLR7 and 9 antagonist IMO-3100 in the absence and presence of TLR7 and 9 agonists in human PBMCs was studied using human inflammatory response and autoimmunity microarray.[59] IMO-3100

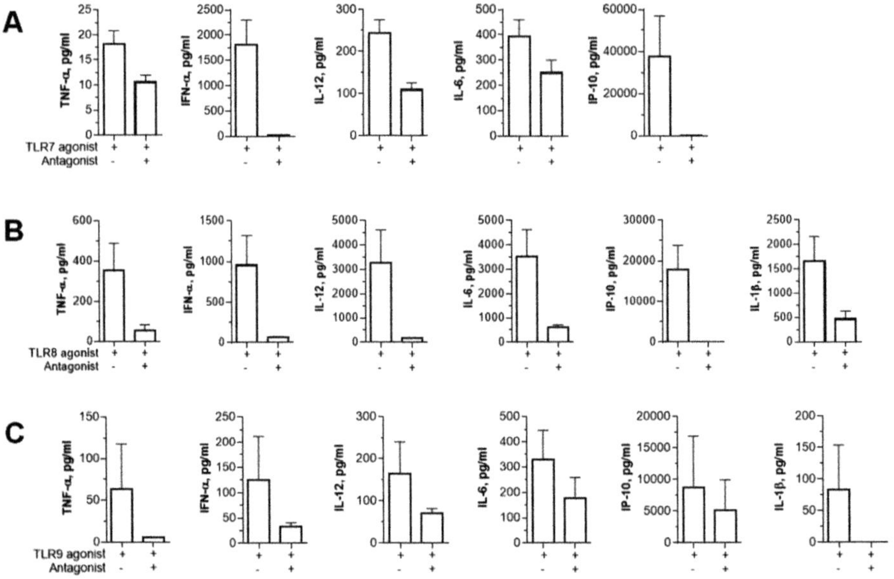

Figure 5.3 Antagonist activity of TLR antagonist example 3 (Figure 5.2) in freshly isolated PBMCs from healthy human donor blood. (A) Inhibition of TLR7 agonist-induced cytokines, (B) inhibition of TLR8 agonist-induced cytokines, and (C) inhibition of TLR9 agonist-induced cytokines. Data were adapted from Figure 4 of ref. 50. PBMCs were cultured with 50 μg ml^{-1} TLR7 agonist, or 50 μg ml^{-1} TLR8 agonist, or 3 μg ml^{-1} TLR9 agonist in the absence or presence of 10 μg ml^{-1} TLR antagonist example 3 for 24 h. Supernatants were collected and analyzed for cytokine levels by Luminex multiplex assay. Each bar represents mean + SEM of four donors. Details of TLR7, 8, and 9 agonists used in the study are published in ref. 50.

alone showed negligible regulation of gene expression. As expected, TLR9 agonist significantly upregulated a number of genes, including IP-10, myeloid differentiation primary response 88 (MyD88), monocyte chemoattractant protein-1 (MCP-1), IL-6, interferon regulatory factor 7 (IRF7), IL-17A, NF-κB1, and receptor-interacting serine/threonine-protein kinase 2 (RIPK2), which regulate immune responses, NF-κB signaling, inflammation, and chemotaxis networks. TLR7 agonist significantly upregulated genes including IP-10, IRF7, IFN-α1, IFN-α4, IFN-γ, IL-15Ra, IL-17A, IL-6, and toll-interleukin 1 receptor (TIR) domain-containing adaptor protein (TIRAP), which regulate host defense response, chemotaxis, and cytokine and chemokine signaling pathways. Treatment of PBMCs with IMO-3100 in the presence of either TLR7 or 9 agonist resulted in the inhibition of genes that were upregulated with either TLR agonist alone. These results indicate that IMO-3100 is a potent inhibitor of the immune responses induced by TLR7 and 9 agonists.[59]

In vivo activity of TLR antagonists was also evaluated in NHPs.[50,60–63] TLR antagonists were administered subcutaneously to NHPs and blood samples were collected at various post dosing time points. Then PBMCs isolated from these samples were stimulated *ex vivo* with agonists of TLR7, 8, and 9 and also with an agonist of TLR4, and the resulting cytokines were analyzed by multiplex analysis. For example, PBMCs isolated from NHPs treated with TLR antagonist candidate IMO-3100 as a single dose[62] or once a week for four weeks[63] showed reduced cytokine induction *ex vivo* in response to TLR7 and TLR9 agonists stimulation and there was no effect on TLR4 agonist-induced cytokines. In these assays, PBMCs from NHPs treated with IMO-3100 showed inhibition of IL-1β, IP-10, IL-6, IL-12, and IFN-α, stimulated by both TLR7 and TLR9 agonists.[63] In addition, MIP-1α and MIP-1β and regulated upon activation normal T cell expressed and secreted (RANTES) induced by TLR7 agonist and MCP-1 induced by TLR9 agonist were also inhibited.[63]

PBMCs collected from NHPs treated with IMO-8400 produced low levels of cytokines when stimulated *ex vivo* with TLR7, 8, or 9 agonists compared with pre-dose PBMCs from the same animals, indicating that IMO-8400 treatment inhibited TLR7, 8, and 9 agonist-induced cytokines.[50] Additionally, no inhibition of TLR4 agonist-induced cytokines was observed.[50] Similar results were observed with TLR7, 8, and 9 antagonist candidate IMO-9200 in NHPs.[61]

5.3 Studies of TLR Antagonists in Disease Models

The TLR antagonist examples shown in Figure 5.2 and the clinical candidates IMO-3100, IMO-8400, and IMO-9200 have been evaluated in multiple preclinical models of autoimmune and inflammatory diseases, including psoriasis, SLE, RA, Duchenne muscular dystrophy (DMD), inflammatory bowel disease (IBD), atherosclerosis, and HIV-1.

5.3.1 Psoriasis

Psoriasis is a complex chronic inflammatory skin disorder characterized by epidermal hyperplasia with leukocyte infiltration into epidermis and dermis resulting in T helper (Th)1–Th2–Th17 imbalance leading to aberrant expression of proinflammatory cytokines. Evidence indicates that the antimicrobial cathelicidin, LL37, produced by keratinocytes complexes with self-DNA, binds to TLR9 on pDCs, leading to excessive production of IFN-α that is implicated in the initiation of psoriasis lesions.[4] Additionally, LL37–self-RNA complexes bind to TLR7 and TLR8 on pDCs and mDCs, respectively, stimulating the production of IFN-α, TNF-α, IL-6, and IL-23 leading to activation of an inflammatory cascade initiating psoriatic leasions.[64,65] Moreover, topical application of the TLR7 agonist imiquimod has been shown to trigger psoriasis in the presence of dermal pDCs.[66] These studies provide strong evidence that abnormal activation of TLR7, 8, and 9 contributes to the initiation and maintenance of psoriasis.

IMO-3100 and IMO-8400 have been studied as therapeutic agents in an IL-23-induced mouse model of psoriasis.[67] For example, IMO-8400 inhibited IL-23-induced ear thickness in a dose-dependent fashion associated with reduction of epidermal hyperplasia in the histology of ear and dorsal skin tissues.[68] Treatment with IMO-8400 led to significant reduction of Th1 and Th17 cytokines in the skin and/or serum, including the expression of inflammasome components, nucleotide-binding oligomerization domain, leucine rich repeat and pyrin domain containing protein 3 (NLRP3) and absent in melanoma 2 (AIM2) and antimicrobial peptides in the skin.[68]

Gene expression profiling of IL-23-induced lesional skin of mice revealed that in the absence of treatment with antagonists, IL-23 affected about 5100 genes implicated in psoriasis.[69] Treatment with the TLR antagonist candidate IMO-8400, which inhibits TLR7, 8, and 9, led to modulation of more than 2300 genes compared with 1800 genes modulated by the TLR antagonist candidate IMO-3100, which inhibits TLR7 and 9 only. Treatment with either IMO-8400 or IMO-3100 strongly reduced (over 17-fold) IL-17A expression, normalized IL-17-induced genes, including beta-defensin and chemokine (C-X-C motif) ligand 1 (CXCL1), and keratin 16.[69] These results support the use of TLR antagonist candidates for the potential treatment of psoriasis.

5.3.2 Systemic Lupus Erythematosus (SLE)

SLE is an autoimmune disease characterized by production of autoantibodies against nucleic acid-associated antigens. In SLE patients, immune complexes of endogenous DNA- and RNA-associated antigens stimulate inflammatory responses through TLRs and exacerbate disease pathology.

The TLR antagonists were studied in lupus prone MRL-lpr and/or NZBW/F1 mouse models.[60,70–72] Treatment with TLR antagonists prevented development of 'butterfly-like skin rash' on the face, reduced serum anti-DNA antibody levels and improved lesional skin and renal histopathology in MRL-lpr mice compared with mice administered phosphate-buffered saline (PBS). As an example, Figure 5.4A shows prevention of development of 'butterfly rash' in MRL-lpr mice treated with TLR antagonist example 1.[73] Figure 5.4B shows improvement in lesional skin histology following treatment with the TLR antagonist example 1.[73]

In lupus prone NZBW/F1 mice, treatment with IMO-8400 reduced serum levels of autoantibodies targeting DNA, RNP (ribonuclear protein), Smith antigen, SSA (Sjogren's Syndrome antigen A), and SSB (Sjogren's Syndrome antigen B) compared with untreated mice.[74] In addition, treatment with IMO-8400 also led to reduction in blood urea nitrogen and proteinuria and improved kidney histopathology in mice compared with PBS-treated mice. Serum cytokine analysis of mice treated with IMO-8400 showed reduced serum IL-12 and IL-1β and increased IL-10 levels compared with PBS-treated mice. Levels of IL-6, inducible nitric oxide synthase (iNOS), and IL-1β mRNA were lower in the kidneys and spleens of mice treated with IMO-8400 compared with PBS-treated mice. Furthermore, IP-10, tumor necrosis factor

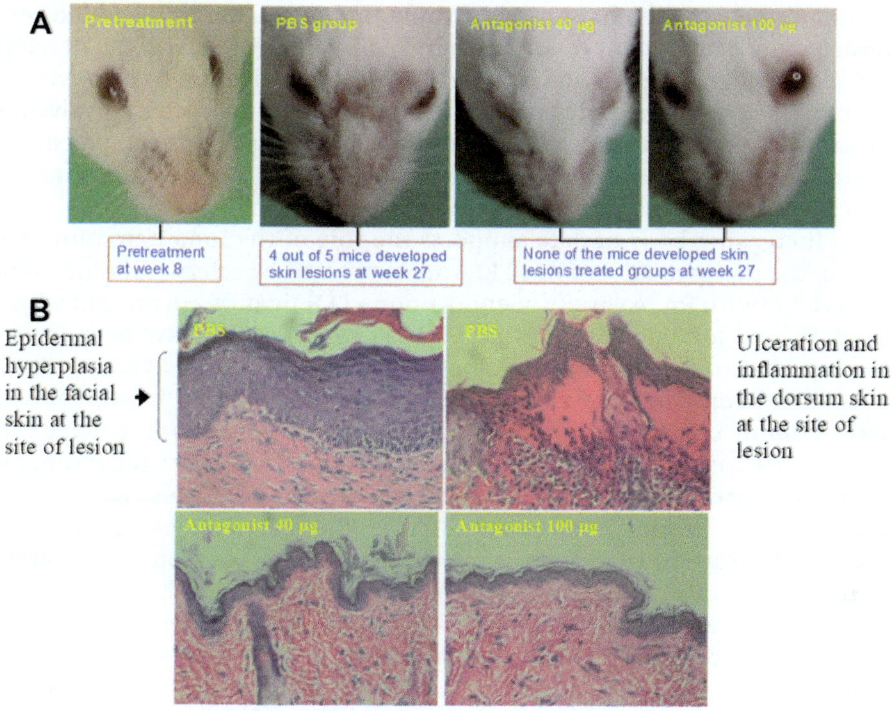

Figure 5.4 (A) Treatment of MRL-lpr mice with TLR antagonist example 1 blocks 'butterfly-like rash' on the faces of treated mice. PBS treated MRL-lpr mice developed 'butterfly-like' skin lesions. (B) Treatment of MRL-lpr mice with TLR antagonist example 1 reduced epidermal hyperplasia, ulceration, and inflammation in the lesional skin compared with PBS-treated mice.[73]

receptor superfamily member 9 (TNFRSF9), and apoptosis-mediating surface antigen FAS ligand (FASL) mRNA levels were lower and IL-4 mRNA levels were higher in the spleens of mice treated with IMO-8400 compared with the spleens of PBS-treated mice. While the inflammasome component NLRP3 mRNA was reduced, the antioxidant enzymes, catalase, and glutathione peroxidase 1 mRNAs were higher in the kidneys of mice treated with IMO-8400 compared with PBS-treated mice.[74] These results demonstrate that IMO-8400 effectively inhibited inflammatory pathways in lupus-prone NZBW/F1 mice and provide evidence of a potential novel therapeutic approach for the treatment of lupus.

5.3.3 Rheumatoid Arthritis (RA)

RA is a severely disabling disease with no known cure. It is characterized by destructive inflammation of the joints, leading to progressive disability and reduced life expectancy. The synovial membrane is infiltrated by activated immune cells that produce proinflammatory cytokines and matrix metalloproteases, which contribute to inflammation of the cartilage and bone destruction.[75] Treatment of RA has significantly improved with the development of biological agents, such as anti-TNF-α antibodies and B cell-depleting therapies, which neutralize components of the immune system involved in disease development.[76] A large proportion of patients are not responsive to current treatments, and new approaches with novel mechanisms of action and better therapeutic efficacy are still needed.

An increasing body of data supports the role of the innate immune system in the pathogenesis of RA. TLRs are expressed by cells within the joints of patients with RA. A variety of endogenous TLR ligands are present within the inflamed joints of patients with RA. TLR7 and TLR9 have been shown to play a pathogenic role in rheumatoid arthritis.[77,78] IMO-3100 and IMO-8400 have been evaluated in mouse models of arthritis.[79,80] Treatment with IMO-3100 in collagen antibody-induced arthritis (CAIA) and collagen-induced arthritis (CIA) mice models showed dose-dependent reductions in arthritic scores and decreases in inflammation, bone destruction, and cartilage loss in foot joint tissues. These changes were associated with increased serum IL-10. A decrease in complement deposition was observed in the CAIA model while lower serum anti-collagen antibody titer and IFN-γ levels in spleen cell cultures were detected in the CIA model compared with untreated mice.[79] Treatment with a combination of a fixed dose of etanercept, an anti-TNF drug, and escalating doses of IMO-3100 resulted in a greater reduction in inflammation and bone destruction induced in the CIA model.[81] These results demonstrate the therapeutic utility of TLR antagonists in models of arthritis and provide a rationale for the use of TLR antagonist candidates alone or in combination with etanercept for the potential treatment of arthritis.

5.3.4 Duchenne Muscular Dystrophy (DMD)

DMD is a genetic disorder characterized by progressive muscle degeneration and weakness due to loss of dystrophin protein (see Chapter 6). Recent studies have shown that dystrophin-deficient mouse muscle cells express higher levels of several TLRs, and aberrant activation of TLRs by endogenous single-stranded RNAs (ssRNAs) may play a role in degeneration and regeneration of dystrophic skeletal muscle in the mdx mouse model.[82] Treatment of *mdx* mice with IMO-8400 significantly reduced skeletal muscle inflammation and increased muscle strength, along with suppression of many inflammatory cytokines.[82] These observations provide a rationale for a potential novel therapeutic approach for DMD.

5.3.5 MyD88 L265P-positive B Cell Lymphoma (BCL)

Constitutive activation of NF-κB and Janus kinase/signal transducer and activator of transcription (JAK/STAT) signaling has been shown to promote cell survival in the activated B-cell-like (ABC) subtype of diffuse large

B-cell lymphoma (DLBCL). A highly recurrent oncogenic mutation, L265P, affecting MyD88 in ABC-DLBCL tumors has been identified.[83] The L265P mutation in MyD88 was reported in 29% of patients with ABC-DLBCL, in over 90% of patients with Waldenström's macroglobulinemia (WM), and in other B-cell lymphomas.[84] The MyD88 L265P oncogenic mutation has been shown to promote tumor cell survival by initiating signaling through IL-1 receptor-associated kinase 1 (IRAK-1), IRAK-4, NF-κB, and JAK/STAT, and secretion of IL-6 and IL-10. The knockdown of TLR7 and/or TLR9 expression, but not other TLRs, has been shown to block NF-κB signaling and decrease survival of cells expressing the MyD88 L265P oncogenic mutation.[85]

Therefore, we proposed that a TLR antagonist that inhibits TLR7, 8, and 9 activation would be a good candidate for treating patients who harbor the L265P mutation. Treatment of OCI-Ly10 and TMD8 cells, which express the L265P mutation, with IMO-8400 showed dose-dependent inhibition of cell growth, antitumor activity, and an increase in survival of tumor-bearing mice. Mice treated with IMO-8400 also showed increases of tumor-associated cytokines, IL-10, IL-2R, IP-10, and monokine induced by gamma interferon (MIG) compared with PBS-treated mice.[86] Furthermore, treatment of Waldenström macroglobulinemia cells expressing the L265P mutation with IMO-8400 led to similar inhibitory effects on cell signaling and survival. In contrast, no activity was observed with IMO-8400 treatment in a GCB (Germinal Center B cell)-DLBCL cell line, SU-DHL-6, lacking the L265P mutation, *in vitro* or *in vivo*.[86] Treatment with a combination of IMO-8400 and rituximab, an anti-CD20 agent, showed greater antitumor activity compared with either agent alone in MWCL-1 and OCI-Ly10 tumor models.[86]

Furthermore, IMO-8400 was studied in three cell lines with the MYD88 L265P mutation (OCI-Ly3, OCI-Ly10, and TMD8), primary bone marrow cells from a WM patient with the mutation (MWCL-1), and a control GCB-DLBCL cell line SU-DHL-6, with wild-type MYD88, with similar results to those described above.[87] IMO-8400 showed dose-dependent tumor growth inhibition and increased animal survival in a preclinical tumor model of disseminated OCI-LY10. In a subcutaneous tumor model, growth of large tumor nodules of approximately 500 mm^3 was significantly inhibited by IMO-8400 treatment.[87] The observed antitumor effect was correlated with decreased inhibitor of NK-κB alpha (IκBα) phosphorylation and IL-10 expression (gene and protein) in tumor cells and decreased IL-10 in the serum of the mice. In contrast, IMO-8400 treatment had no effects on control SU-DHL-6 cells *in vitro* or *in vivo*.[87]

5.3.6 Restenosis and Atherosclerosis

Post-interventional remodeling is a critical determinant of long-term efficacy of percutaneous coronary interventions. Restenosis is characterized by acute elastic recoil and intimal hyperplasia attributable to inflammation, smooth muscle cell (SMC) proliferation, and extracellular matrix turnover.[88] Under hypercholesterolemic conditions, this is accompanied by influx and accumulation of low-density lipoprotein (LDL) cholesterol in the vessel wall that becomes oxidized and taken up by macrophages, which then become foam cells and initiate a process of accelerated atherosclerosis.[89]

We studied the effects of TLR antagonist example 1 on post-interventional remodeling and accelerated atherosclerosis in a mouse model.[90] Treatment with TLR antagonist example 1 led to reduced macrophage activation and foam cell formation, which was accompanied by an increase in IL-10 production. Additionally, treatment with TLR antagonist example 1 led to a reduction in neointima formation, reduced macrophage presence in the lesions, as well as decreased high mobility group box 1 (HMGB1) release. These observations indicate the therapeutic potential of TLR antagonists in post-interventional remodeling to reduce restenosis and accelerated atherosclerosis.[90]

5.3.7 Inflammatory Bowel Disease (IBD)

IBD, which includes Crohn's disease (CD) and ulcerative colitis (UC), is a severe and debilitating autoimmune disorder characterized by chronic inflammation in the gastrointestinal tract. Several studies have shown that TLRs play a role in disease pathogenesis of IBD.[91–93]

TLR antagonist example 1 and IMO-9200 have been evaluated in models of IBD.[94,95] In 2,4,6-trinitrobenzene sulphonic acid (TNBS)- and Dextran sulfate sodium (DSS)-induced colitis models, oral administration of IMO-9200 improved animal body weight and survival. IBD mice treated with IMO-9200 showed improvements in colon length and histology.[95] Treatment with IMO-9200 also reduced inflammatory mediators and cytokines in the colon and serum, and restored the transforming growth factor beta/small and mothers against decapentaplegic 3 (TGF-β/Smad3) signaling pathway and modulated gene expression, including downregulation of genes associated with pro-inflammatory cytokines, IL-1β, IL-6, IL-12, IFN-γ, MCP-1, and TNF-α, tissue repair and modeling, antimicrobial peptides, and acute phase proteins and upregulation of genes involved in the maintenance of intestinal barrier integrity.[95] Currently a phase 2 clinical trial of IMO-9200 is ongoing in IBD patients.

5.3.8 HIV-1

As described in Section 5.2, subcutaneous administration of GEM91 to HIV patients led to flu-like symptoms associated with swollen lymph nodes. In these studies, GEM91 produced dose-dependent increases in viral loads in the plasma of HIV-positive patients compared with placebo control.[96] Though it was not clear at that time, later studies have shown that bacterial and viral DNA containing unmethylated CpG motifs *trans*-activate HIV-long terminal repeat (LTR) and HIV replication *via* TLR9.[97] Additionally, it has been shown that IFN-α induction following TLR7 stimulation by HIV RNA results in stronger $CD8^+$ T cell activation, which is a strong predictor of HIV-1 disease progression and transmission.[98] Together, these studies indicate that TLR-mediated inflammation plays a role in HIV replication and disease progression and, therefore, antagonists could be a choice of treatment for HIV-1.

An antagonist of TLR7 and 9 significantly inhibited humanized BLT mice-derived human pDC responses to stimulation with TLR7 agonist *in vitro*.[99,100] HIV-1 infection of BLT mice led to increased expression of immune activation marker CD38 on human T cells *in vivo*, and the treatment of infected mice with a TLR7 and 9 antagonist led to a significant reduction in CD38 expression.[99,100] These results indicate that treatment of HIV-1-infected humanized BLT mice with TLR7 and 9 antagonist resulted in a significant reduction of HIV-1-associated immune activation, indicating an important role of the TLR7 and 9 pathway in mediating immune activation in HIV-1 and possible application of TLR antagonists for the treatment of HIV-1.

5.4 Clinical Development of Lead TLR Antagonist Candidates

Clinical development of TLR antagonists was challenging, as it was difficult to know the optimal dose for use in humans for conducting a proof of concept trial. To evaluate safety and establish a dose at which TLR antagonist would inhibit the targeted TLR immune cascade, a clinical trial was conducted in healthy volunteers.[101,102] In this trial, IMO-3100 was administered subcutaneously at escalating single dose levels starting from 0.04 to 0.64 mg kg^{-1}. Treatment was safe and well tolerated. To evaluate pharmacodynamic activity of IMO-3100, blood was collected from treated subjects and subjects that had received placebo at various post-dosing time points. PBMCs were isolated from blood and stimulated *ex vivo* with agonists of TLR7 and 9. Cell culture supernatants were assayed for cytokines induced, and compared among various groups. Suppression of cytokine induction was observed in subjects treated with IMO-3100 at doses of 0.08 mg kg^{-1} or higher compared with placebo-treated subjects. At higher doses, of 0.32 and 0.64 mg kg^{-1}, plateauing of suppressive activity was noted. Importantly, suppressive activity of IMO-3100 was observed for up to five days post-dosing.[101,102] Similar results were observed with IMO-8400 in a phase 1 clinical trial.[103]

5.4.1 Moderate-to-severe Plaque Psoriasis

Having established the pharmacologically active doses of IMO-3100 in healthy subjects, a clinical proof of concept phase 1 trial was conducted in patients with moderate to severe plaque psoriasis. The clinical trial was a randomized, placebo-controlled, double-blind, 4 week study in 44 patients enrolled at 11 sites in the USA in 2012. IMO-3100 was administered subcutaneously once a week at 0.16 and 0.32 mg kg^{-1} doses. IMO-3100 treatment was well tolerated. Improvement in the Psoriasis Area and Severity Index (PASI), a disease-associated parameter, compared with pre-treatment were observed at both dose levels starting from day 15 and this was sustained through day 57 post-treatment. Patients treated with a 0.16 mg kg^{-1} dose of IMO-3100 showed better treatment-related changes. Improvements in PASI in treated patients were also associated with erythema plaque induration, epidermal thickness, changes in skin histology and immune cells and improvement in the expression profile of meta-analysis derived transcriptome-3 (MAD-3) psoriasis and IL-17-related genes (Figure 5.5).[104]

It has been shown that TLR8 also plays a role in autoimmune and inflammatory diseases and it would be important to block TLR7, 8, and 9 for optimal therapeutic outcome. Therefore, a follow-up trial in patients with moderate to severe plaque psoriasis was conducted with IMO-8400.[105] A placebo-controlled, randomized phase 2a trial of IMO-8400 in 46 patients was conducted at four dose levels, 0.075, 0.15, 0.3, and 0.6 mg kg^{-1}, or placebo, weekly for 12 weeks.[105] IMO-8400 across all dose levels did not cause any serious or severe adverse events. The most common treatment-related adverse events were dose-dependent injection-site reactions. All IMO-8400 groups showed clinical improvement, but a clear dose–response relationship and statistically significant differences with placebo were not observed (P = 0.26). Eleven (38%) out of 29 subjects on IMO-8400 treatment achieved ≥50% PASI-reduction, compared with 1 (11%) out of 9 subjects on placebo. Five (17%) and 2 (7%) IMO-8400-treated subjects achieved PASI-75 and PASI-90, respectively, compared with none in the placebo group.[105]

It is important to note that most of the patients showing the changes in PASI were in the two lower-dose groups. Four of the five patients showing PASI 50, and all four patients showing PASI 75 improvements were in the dose groups treated with 0.075 and 0.15 mg kg^{-1} dose. This observation, together with observations made with IMO-3100, highlights that lower doses of TLR antagonists are more effective than higher doses and use of higher doses may lead to local injection-site reactions, thereby creating a counter-productive mechanism of action.

5.4.2 Waldenström's Macroglobulinemia

In a phase 1/2, open label, dose escalation trial in relapsed or refractory Waldenström's Macroglobulinemia (WM) patients, IMO-8400 was well tolerated at dose levels of 0.6, 1.2, and 2.4 mg kg^{-1} (1.2 mg kg^{-1} twice) per week. In the highest-dose cohort studied (2.4 mg kg^{-1}), there were three responses and two showing stable disease out of six IMO-8400-treated patients with evaluable disease. One of these responders was refractory to ibrutinib. The median time to first response was ~10.5 weeks. In addition, improvements

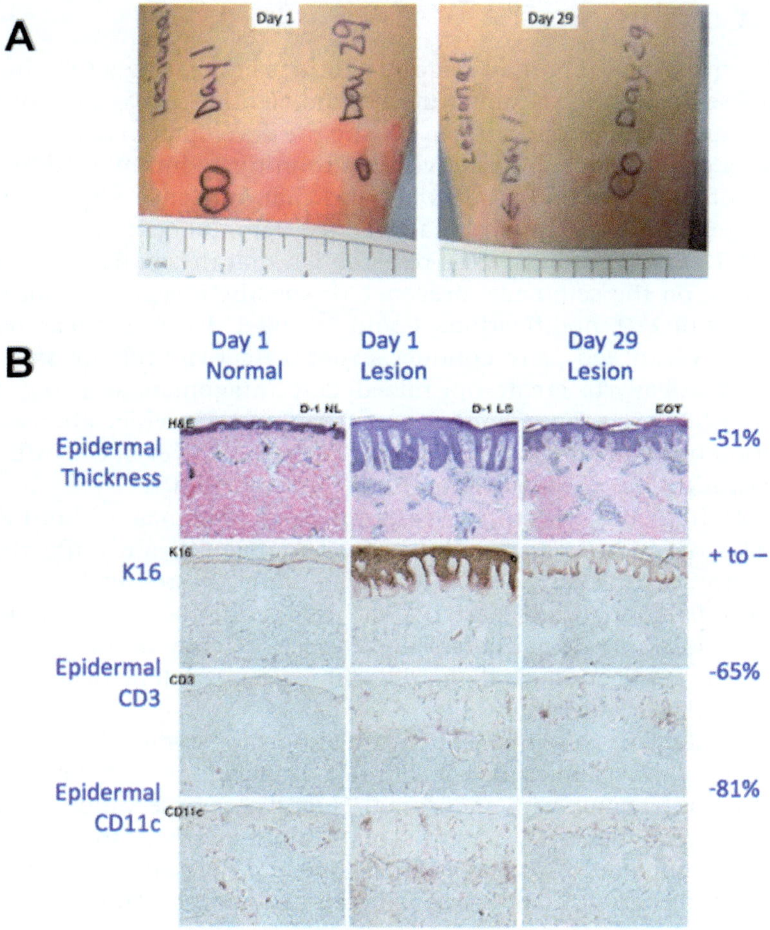

Figure 5.5 (A) Photograph of a representative patient showing the psoriatic plaques before (day1) and after (day 29) treatment with a dose of 0.16 mg kg^{-1} $week^{-1}$ of IMO-3100. (B) Treatment-related histological changes in the lesional skin on day 29 compared with day 1 in a representative patient treated with a dose of 0.16 mg kg^{-1} $week^{-1}$ of IMO-3100. The figure shows at day 29 K16 staining (marker of keratinocyte proliferation), reverting towards normal and decreasing infiltrates of $CD3^+$ lymphocytes and $CD11c^+$ mononuclear cells (Reproduced from ref. 104 with permission from Dr T. Sullivan).

in symptoms, hemoglobin and bone marrow were observed.[106] These data for WM patients provided the first clinical evidence supporting the conclusion that blocking the TLR pathway is a potential therapeutic approach for B-cell lymphoma harboring the MYD L265P mutation. Additionally, the data indicate that in non-autoimmune disease conditions a higher dose of TLR antagonist may be effective.

5.5 Conclusions

The discovery that modified oligonucleotides containing a CG dinucleotide and segments of 2′-*O*-substituted ribonucleotides act as antagonists of TLR7, 8, and 9 was serendipitous. During the development of second-generation ASOs, one of the objectives was to mitigate immune-stimulatory activity of ASOs. Several studies showed that antisense sequences with second-generation chemistry had reduced immune-stimulatory activity.[107–109] Detailed studies led us to establish that these oligonucleotides, depending on the sequence, presence of specific nucleotide motifs and placement of 2′-*O*-modifications, could inhibit TLR7, 8, and 9 activation. This observation led us to conduct structure–activity relationship studies systematically to create optimized TLR antagonists and to evaluate these candidates as therapeutic agents for diseases in which aberrant TLR activation by endogenous nucleic acids is the key driver of the inflammatory cascade. One of the advantages of this approach is that by blocking the TLRs, induction of multiple inflammatory cytokines, including TNF-α, IL-12, IL-6, and IFN-α, are inhibited without significantly affecting the constitutive levels. Early clinical proof of concept studies have yielded very promising results in patients with moderate-to-severe plaque psoriasis and also in WM patients. These classes of compounds have broad therapeutic applications and ongoing clinical trials will further shed light on their utility.

These extensive structure–activity relationship studies have provided insights into how chemical and structural modifications of oligonucleotides affect their interaction with endosomal TLRs. These observations also indicate a potential aberrant TLR activation by build-up of metabolically stable antisense compounds upon repeated dosing and their effects on safety signals. These insights should be considered in designing optimal antisense compounds.

Acknowledgements

The authors thank all of their colleagues for their contributions as cited in the references.

References

1. A. Marshak-Rothstein and I. R. Rifkin, *Annu. Rev. Immunol.*, 2007, **25**, 419.
2. M. B. Uccellini, A. M. Avalos, A. Marshak-Rothstein and G. A. Viglianti, *Methods Mol. Biol.*, 2009, **517**, 363.
3. A. N. Theofilopoulos, R. Gonzalez-Quintial, B. R. Lawson, Y. T. Koh, M. E. Stern, D. H. Kono, B. Beutler and R. Baccala, *Nat. Rev. Rheumatol.*, 2010, **6**, 146.

4. R. Lande, J. Gregorio, V. Facchinetti, B. Chatterjee, Y. H. Wang, B. Homey, W. Cao, Y. H. Wang, B. Su, F. O. Nestle, T. Zal, I. Mellman, J. M. Schröder, Y. J. Liu and M. Gilliet, *Nature*, 2007, **449**, 564.
5. A. Niessner, K. Sato, E. L. Chaikof, I. Colmegna, J. J. Goronzy and C. M. Weyand, *Circulation*, 2006, **114**, 2482.
6. L. Ronnblom and V. Pascual, *Lupus*, 2008, **17**, 394.
7. F. Conti, F. Ceccarelli, L. Massaro, E. Cipriano, M. Di Franco, C. Alessandria, R. R. Spinelli and R. Scrivo, *Clin. Ter.*, 2013, **164**, e413.
8. I. Ben-Zvi, S. Kivity, P. Langevitz and Y. Schoenfeld, *Clin. Rev. Allergy Immunol.*, 2012, **42**, 145.
9. S. J. Lee, E. Silverman and J. M. Bargman, *Nat. Rev. Nephrol.*, 2011, **7**, 718.
10. S. J. Katz and A. S. Russell, *Curr. Opin. Rheumatol.*, 2011, **23**, 278.
11. K. Sacre, L. A. Criswell and J. M. McCune, *Arthritis Res. Ther.*, 2012, **14**, R155.
12. R. Fox, *Lupus*, 1996, **5**(Suppl. 1), S4.
13. A. Kuznik, M. Bencina, U. Svajger, M. Jeras, B. Rozman and R. Jerala, *J. Immunol.*, 2011, **186**, 4794.
14. C. Tang, T. Godfrey, R. Stawell and M. Nikpour, *Intern. Med. J.*, 2012, **42**, 968.
15. A. M. Krieg, T. Wu, R. Weeratna, S. M. Efler, L. Love-Homan, L. Yang, A. K. Yi, D. Short and H. L. Davis, *Proc. Natl. Acad. Sci. U. S. A.*, 1998, **95**, 12631.
16. P. Lenert, L. Stunz, A. K. YI, A. M. Krieg and R. F. Ashman, *Antisense Nucleic Acid Drug Dev.*, 2001, **11**, 247.
17. I. Gursel, M. Gursel, H. Yamada, K. J. Ishii, F. Takeshita and D. M. Klinman, *J. Immunol.*, 2003, **171**, 1393.
18. F. J. Barrat, T. Meeker, J. Gregorio, J. H. Chan, S. Uematsu, S. Akira, B. Chang, O. Duramad and R. L. Coffman, *J. Exp. Med.*, 2005, **202**, 1131.
19. O. Duramad, K. L. Fearon, B. Chang, J. H. Chan, J. Gregorio, R. L. Coffman and F. J. Barrat, *J. Immunol.*, 2005, **174**, 5193.
20. R. F. Ashman, J. A. Goeken, J. Drahos and P. Lenert, *Int. Immunol.*, 2005, **17**, 411.
21. P. Lenert, *Clin. Exp. Immunol.*, 2005, **140**, 1.
22. R. F. Ashman, J. A. Goeken, E. Latz and P. Lenert, *Int. Immunol.*, 2011, **23**, 203.
23. D. Bayik, I. Gursel and D. M. Klinman, *Pharmacol. Res.*, 2016, **105**, 216.
24. P. Lenert, K. Yasuda, L. Busconi, P. Nelson, C. Fleenor, R. S. Ratnabalasuriar, P. L. Nagy, R. F. Ashman, I. R. Rifkin and A. Marshak-Rothstein, *Arthritis Res. Ther.*, 2009, **11**, R79.
25. L. Dong, S. Ito, K. J. Ishii and D. M. Klinman, *Arthritis Rheum.*, 2005, **52**, 651.
26. P. S. Patole, D. Zecher, R. D. Pawar, H. J. Grone, D. Schlondorff and H. J. Anders, *J. Am. Soc. Nephrol.*, 2005, **16**, 3273.
27. F. J. Barrat, T. Meeker, J. H. Chan, C. Guiducci and R. L. Coffman, *Eur. J. Immunol.*, 2007, **37**, 3582.

28. L. Dong, S. Ito, K. J. Ishii and D. M. Klinman, *Arthritis Rheum.*, 2004, **50**, 1686.
29. P. P. Ho, P. Fontoura, M. Platten, R. A. Sobel, J. J. DeVoss, L. Y. Lee, B. A. Kidd, B. H. Tomooka, J. Capers, A. Agrawal, R. Gupta, J. Zernik, M. K. Yee, B. J. Lee, H. Garren, W. H. Robinson and L. Steinman, *J. Immunol.*, 2005, **175**, 6226.
30. C. Guiducci, M. Gong, Z. Xu, M. Gill, D. Chaussabel, T. Meeker, H. J. Chan, T. Wright, M. Punaro, S. Bolland, V. Soumelis, J. Banchereau, R. L. Coffman, V. Pascual and F. J. Barrat, *Nature*, 2010, **465**, 937.
31. E. R. Kandimalla and S. Agrawal, *Adv. Polym. Sci.*, 2012, **249**, 61.
32. S. Agrawal, P. L. Iadorola, J. Temsamani, Q. Zhao and D. R. Shaw, *Bioorg. Med. Chem. Lett.*, 1996, **6**, 2219.
33. S. Agrawal, W. Tan, Q. Cai, X. Xie and R. Zhang, *Antisense Nucleic Acid Drug Dev.*, 1997, **7**, 245.
34. S. Yamamoto, E. Kuramoto, S. Shimada and T. Tokunaga, *Jpn. J. Cancer Res.*, 1988, **79**, 866.
35. J. P. Messina, G. S. Gilkeson and D. S. Pisetsky, *J. Immunol.*, 1991, **147**, 1759.
36. T. Tokunaga, O. Yano, E. Kuramoto, Y. Kimura, T. Yamamoto, T. Kataoka and S. Yamamoto, *Microbiol. Immunol.*, 1992, **36**, 55.
37. Q. Zhao, J. Temsamani, P. L. Iadarola, J. Ziang and S. Agrawal, *Biochem. Pharmacol.*, 1996, **51**, 173.
38. S. Agrawal and Q. Zhao, *Curr. Opin. Chem. Biol.*, 1998, **2**, 519.
39. S. Agrawal and E. R. Kandimalla, *Nat. Biotechnol.*, 2004, **22**, 1533.
40. Y. Liu, N. L. Seto, C. Carmona-Rivera and M. J. Kaplan, *Lupus Sci. Med.*, 2018, **5**, e000259.
41. Q. Zhao, D. Yu and S. Agrawal, *Bioorg. Med. Chem. Lett.*, 1999, **9**, 3453.
42. Q. Zhao, D. Yu and S. Agrawal, *Bioorg. Med. Chem. Lett.*, 2000, **10**, 1051.
43. S. Agrawal and E. R. Kandimalla, *Curr. Cancer Drug Targets*, 2001, **1**, 197.
44. D. Yu, E. R. Kandimalla, Q. Zhao, Y. Cong and S. Agrawal, *Bioorg. Med. Chem.*, 2001, **9**, 2803.
45. D. Yu, E. R. Kandimalla, Q. Zhao, Y. Cong and S. Agrawal, *Nucleic Acids Res.*, 2002, **30**, 1613.
46. D. Yu, E. R. Kandimalla, Y. Cong, J. Tang, J. Y. Tang, Q. Zhao and S. Agrawal, *J. Med. Chem.*, 2002, **45**, 4540.
47. D. Yu, E. R. Kandimalla, Q. Zhao, Y. Cong and S. Agrawal, *Bioorg. Med. Chem. Lett.*, 2001, **11**, 2263.
48. D. Yu, E. R. Kandimalla, Q. Zhao, L. Bhagat, Y. Cong and S. Agrawal, *Bioorg. Med. Chem.*, 2003, **11**, 459.
49. D. Wang, L. Bhagat, D. Yu, F. G. Zhu, J. X. Tang, E. R. Kandimalla and S. Agrawal, *J. Med. Chem.*, 2009, **52**, 551.
50. E. R. Kandimalla, L. Bhagat, D. Wang, D. Yu, T. Sullivan, N. La Monica and S. Agrawal, *Nucleic Acids Res.*, 2013, **41**, 3947.
51. D. Yu, D. Wang, F.-G. Zhu, L. Bhagat, M. Dai, E. R. Kandimalla and S. Agrawal, *J. Med. Chem.*, 2009, **52**, 5108.

52. E. R. Kandimalla, L. Bhagat, Y. Li, D. Yu, D. Wang, Y. P. Cong, S. S. Song, J. X. Tang, T. Sullivan and S. Agrawal, *Proc. Natl. Acad. Sci. U. S. A.*, 2005, **102**, 6925.
53. S. Agrawal and E. R. Kandimalla, *Biochem. Soc. Trans.*, 2007, **35**, 1461.
54. D. Yu, Q. Zhao, E. R. Kandimalla and S. Agrawal, *Bioorg. Med. Chem. Lett.*, 2000, **10**, 2585.
55. E. R. Kandimalla, L. Bhagat, D. Yu, Y. Cong, J. Tang and S. Agrawal, *Bioconjugate Chem.*, 2002, **13**, 966.
56. D. Yu, E. R. Kandimalla, L. Bhagat, J. Y. Tang, Y. Cong, J. Tang and S. Agrawal, *Nucleic Acids Res.*, 2002, **30**, 4460.
57. E. R. Kandimalla, L. Bhagat, D. Wang, D. Yu, F. G. Zhu, J. Tang, H. Wang, P. Huang, R. Zhang and S. Agrawal, *Nucleic Acids Res.*, 2003, **31**, 2393.
58. L. Bhagat, D. Yu, M. Dai, E. R. Kandimalla and S. Agrawal, DNA-based antagonists of Toll-like receptors: Insights into the mechanism of action presented at the 4th Annual Meeting of the Oligonucleotide Therapeutics Society, October 15–18, 2008, Boston, MA.
59. W. Jiang, L. Bhagat, D. Yu, E. R. Kandimalla and S. Agrawal, *J. Immunol.*, 2009, **182**(1 Supplement), 48.25.
60. L. Bhagat, D. Wang and W. Jiang, IMO-9200, a novel clinical-stage TLR antagonist for the treatment of autoimmune diseases, inhibits TLR-mediated immune responses and shows activity in animal models, presented at the Annual Meeting of the Oligonucleotide Therapeutics Society, October 12–15, 2014, San Diego, CA.
61. T. Sullivan and L. Bhagat, IMO-9200, a novel clinical-stage TLR antagonist for the treatment of autoimmune diseases, suppresses TLR-mediated immune responses in non-human primates following systemic administration, presented at the Annual Meeting of the Oligonucleotide Therapeutics Society, October 12-15, 2014, San Diego, CA.
62. L. Bhagat, D. Yu, J. X. Tang, T. Sullivan, E. R. Kandimalla and S. Agrawal, IMO-3100, a novel antagonist, suppresses TLR7- and TLR9-mediated immune responses in non-human primates presented at Keystone Symposium on Tolerance and Autoimmunity (B7), February 21–26, 2010, Taos, NM (Poster # 216).
63. L. Bhagat, J. X. Tang, T. Sullivan, E. R. Kandimalla and S. Agrawal, *Arthritis Rheum.*, 2010, **62**(supplement), 856.
64. D. Ganguly, G. Chamilos, R. Lande, J. Gregorio, S. Meller, V. Facchinetti, B. Homey, F. J. Barrat, T. Zal and M. Gilliet, *J. Exp. Med.*, 2009, **206**, 1983.
65. E. Lee, W. L. Trepicchio, J. L. Oestreicher, D. Pittman, F. Wang, F. Chamian, M. Dhodapkar and J. G. Krueger, *J. Exp. Med.*, 2004, **199**, 125.
66. M. Gilliet, C. Conrad, M. Geiges, A. Cozzio, W. Thürlimann, G. Burg, F. O. Nestle and R. Dummer, *Arch. Dermatol.*, 2004, **140**, 14090.
67. F.-G. Zhu, D. Yu, E. R. Kandimalla and S. Agrawal, A Toll-like receptor antagonist prevents development of IL-23-induced psoriasis-like dermal changes in mice, presented at Federation of Clinical Immunology Societies, FOCiS Meeting, June 5–9, 2008, Boston, MA.

68. W. Jiang, F.-G. Zhu, L. Bhagat, D. Yu, J. X. Tang, E. R. Kandimalla, N. La Monica and S. Agrawal, *J. Invest. Dermatol.*, 2013, **133**, 1777.
69. M. Suarez-Farinas, R. Arbeit, W. Jiang, F. S. Ortenzio, T. Sullivan and J. G. Krueger, *PLoS One*, 2013, **8**, e84634.
70. F.-G. Zhu, L. Bhagat, D. Yu, E. R. Kandimalla and S. Agrawal, A novel class of DNA-based TLR antagonists ameliorates collagen-induced arthritis in mice, presented at FASEB Summer Research Conference on Autoimmunity, July 14–19, 2007, Saxtons River, VT.
71. F.-G. Zhu, D. Yu, E. R. Kandimalla and S. Agrawal, Evaluation of a DNA-based Toll-like receptor antagonist for treatment of collagen-induced arthritis in DBA/1 mice, presented at the 4th Annual Meeting of the Oligonucleotide Therapeutics Society, October 15–18, 2008, Boston, MA.
72. F.-G. Zhu, D. Yu, E. Kandimalla, N. La Monica and S. Agrawal, Treatment with IMO-3100, a novel TLR7 and TLR9 dual antagonist, inhibits disease development in lupus prone NZBW/F1 mice, presented at Keystone Symposium on Dendritic cells and the initiation of adaptive immunity, February 12–17, 2011, Santa Fe, NM (poster #233).
73. F.-G. Zhu, E. R. Kandimalla, D. Yu, D. Wang and S. Agrawal, Novel class of DNA-based compounds act as antagonists for TLR7 and 9: In vitro and in vivo studies in MRL-lpr and NZBW/F1 mouse models, presented at Keystone Meeting on Biology of B cells in health and disease, February 10, 2007, Banff, Alberta, Canada.
74. F.-G. Zhu, W. Jiang, L. Bhagat, D. Yu, E. R. Kandimalla, N. La Monica and S. Agrawal, *Autoimmunity*, 2013, **46**, 419.
75. M. Feldmann, F. M. Brennan and R. N. Maini, *Cell*, 1996, **85**, 307.
76. S. Shankar and R. Handa, *J. Postgrad. Med.*, 2004, **50**, 293.
77. L. Svelander, H. Erlandsson, J. C. Lorentzen, C. Trollmo, L. Klareskog and A. Bucht, *Arthritis Rheum.*, 2004, **50**, 297.
78. M. F. Roelofs, L. A. Joosten, S. Abdollahi-Roodsaz, A. W. van Lieshout, T. Sprong, F. H. Van den Hoogen, W. B. van den Berg and T. R. Radstake, *Arthritis Rheum.*, 2005, **52**, 2313.
79. N. La Monica, F.-G. Zhu, M. Reardon, M. Precopio, D. Yu, E. R. Kandimalla and S. Agrawal, Treatment with IMO-3100, a novel TLR7 and TLR9 dual antagonist, inhibits disease development in a mouse model of collagen antibody-induced arthritis (CAIA), presented at FOCIS Annual Meeting, June 23–26, 2011, Washington, DC.
80. F.-G. Zhu, E. R. Kandimalla and S. Agrawal, Antagonists of Toll-like receptors 7 and 9 and potential application for treatment of rheumatoid arthritis, presented at the American Chemical Society National Meeting, August 16–20, 2009, Washington, DC.
81. F.-G. Zhu, D. Yu, E. R. Kandimalla and S. Agrawal, *Proceedings of the Annual Meeting American College of Rheumatology*, 2009, Ab # 569.
82. A. Henriques-Pons, Q. Yu, S. Rayavarapu, T. Cohen, B. Ampong, H. J. Cha, V. Jahnke, J. VanderMeulen, D. Wang, W. Jiang, E. R. Kandimalla, S. Agrawal, C. F. Spurney and N. Kanneboyina, *Hum. Mol. Genet.*, 2014, **23**, 2604.

83. V. N. Ngo, R. M. Young, R. Schmitz, S. Jhavar, W. Xiao, K. H. Lim, H. Kohlhammer, W. Xu, Y. Yang, H. Zhao, A. L. Shaffer, P. Romesser, G. Wright, J. Powell, A. Rosenwald, H. K. Muller-Hermelink, G. Ott, R. D. Gascoyne, J. M. Connors, L. M. Rimsza, E. Campo, E. S. Jaffe, J. Delabie, E. B. Smeland, R. I. Fisher, R. M. Braziel, R. R. Tubbs, J. R. Cook, D. D. Weisenburger, W. C. Chan and L. M. Staudt, *Nature*, 2011, **470**, 115.
84. L. Xu, Z. R. Hunter, G. Yang, Y. Zhou, Y. Cao, X. Liu, E. Morra, A. Trojani, A. Greco, L. Arcaini, M. Varettoni, J. R. Brown, Y. T. Tai, K. C. Anderson, N. C. Munshi, C. J. Patterson, R. J. Manning, C. K. Tripsas, N. I. Lindeman and S. P. Treon, *Blood*, 2013, **121**, 2051.
85. J. Q. Wang, Y. S. Jeelall, L. L. Ferguson and K. Horikawa, *Front. Immunol.*, 2014, **5**, 367.
86. W. Jiang, D. Wang and L. Bhagat, presented at 8th International Workshop on Waldenström's Macroglobulinemia and Symposium on Advances in Multiple Myeloma, August 13–17, 2014, London, UK.
87. L. Bhagat, D. Wang, W. Jiang and S. Agrawal, *Cancer Res.*, 2014, **74**(19 Suppl), 2570, Abstract nr 2570.
88. N. M. Pires, J. W. Jukema, M. J. Daemen and P. H. Quax, *Vasc. Pharmacol.*, 2006, **44**, 257.
89. J. H. Lardenoye, D. J. Delsing, M. R. de Vries, M. M. Deckers, H. M. Princen, L. M. Havekes, V. W. van Hinsbergh, J. H. van Bockel and P. H. Quax, *Circ. Res.*, 2000, **87**, 248.
90. J. C. Karper, M. M. Ewing, K. L. Habets, M. R. de Vries, E. A. Peters, A. M. van Oeveren-Rietdijk, H. C. de Boer, J. F. Hamming, J. Kuiper, E. R. Kandimalla, N. La Monica, J. W. Jukema and P. H. Quax, *Arterioscler. Thromb. Vasc. Biol.*, 2012, **32**, e72.
91. D. Berkowitz, R. Peri, R. Lavy and A. Kessel, *Hum. Immunol.*, 2013, **74**, 1519.
92. C. Steenholdt, L. Andresen, G. Pedersen, A. Hansen and J. Brynskov, *Scand. J. Gastroenterol.*, 2009, **44**, 195.
93. B. Zheng, M. E. Morgan, H. J. van de Kant, J. Garssen, G. Folkerts and A. D. Kraneveld, *Biochim. Biophys. Acta*, 2013, **1832**, 2162.
94. D. Wang, D. Yu, E. R. Kandimalla and S. Agrawal, A synthetic DNA-based antagonist of TLR7 and TLR9 protects mice against TNBS-induced colitis, presented at the 4th Annual Meeting of the Oligonucleotide Therapeutics Society, October 15–18, 2008, Boston, MA.
95. F.-G. Zhu, D. Wang, W. Jiang, L. Bhagat and S. Agrawal, *Gastroenterology*, 2015, **148**, S324, abstract Sa1757.
96. S. Agrawal and R. R. Martin, *J. Immunol.*, 2003, **171**, 1621.
97. O. Equils, M. L. Schito, H. Karahashi, Z. Madak, A. Yarali, K. S. Michelsen, A. Sher and M. Arditi, *J. Immunol.*, 2003, **170**, 5159.
98. A. Meier, J. J. Chang, E. S. Chan, R. B. Pollard, H. K. Sidhu, S. Kulkarni, T. F. Wen, R. J. Lindsay, L. Orellana, D. Mildvan, S. Bazner, H. Streeck, G. Alter, J. D. Lifson, M. Carrington, R. J. Bosch, G. K. Robbins and M. Altfeld, *Nat. Med.*, 2009, **15**, 955.

99. J. Chang, R. J. Lindsay, E. H. Doyle, V. Vrbanac, E. N. Seung, T. Dudek, R. J. Bosch, M. Precopio, E. Kandimalla, A. Tager and M. Altfeld, *Retrovirology*, 2012, **9**(Suppl. 2), 172.
100. J. Chang, R. J. Lindsay, E. H. Doyle, V. Vrbanac, E. N. Seung, T. Dudek, M. Precopio, R. J. Bosch, E. Kandimalla, A. Tager and M. Altfeld, *J. Immunol.*, 2012, **188**, 113.19.
101. T. Sullivan, G. Atiee, L. Bhagat, W. Jiang, E. Kandimalla, J. Murphy, M. Precopio, M. Putta and R. Arbeit, Safety and pharmacodynamics of IMO-3100, a novel Toll-like receptor antagonist for autoimmune and inflammatory diseases, in a rising single-dose phase 1 clinical trial, presented at 6th Annual meeting of the Oligonucleotide Therapeutic Society, Oct 20–23, 2010, Dana Point, CA.
102. T. Sullivan, G. Atiee, L. Bhagat, W. Jiang, J. Murphy, N. Ostavnenko, M. Precopio, M. R. Putta, E. R. Kandimalla and R. Arbeit, IMO-3100, a novel Toll-like receptor antagonist for autoimmune and inflammatory diseases: Safety and pharmacodynamics in multiple-dose phase 1 clinical trial, presented at Keystone Symposium on Immunoregulatory Networks, April 1–6, 2011, Breckenridge, CO.
103. L. Bhagat, W. Jiang, D. Yu, R. D. Arbeit and T. Sullivan, presented at FOCIS Annual Meeting, June 27-30, 2013, Boston, MA.
104. A. B. Kimball, J. Krueger, T. Sullivan and R. D. Arbeit, *J. Invest. Dermatol.*, 2013, S26.
105. D. M. W. Balak, M. B. A. van Doorn, R. D. Arbeit, R. Rijneveld, E. Klaassen, T. Sullivan, J. Brevard, H. B. Thio, E. P. Prens, J. Burggraaf and R. Rissmann, *Clin. Immunol.*, 2017, **174**, 63.
106. S. K. Thomas, W. A. Harb, J. T. Beck, G. Nashat, M. L. Palomba, S. M. Ansell, H. Eradat, E. N. Libby III, R. H. Advani, J. Hajdenberg, L. T. Heffner, J. Hoffman, D. H. Vesole, L. Simov, N. Wyant, J. Brevard, J. O'Leary and S. Agrawal, *Blood*, 2015, **126**, 1540.
107. S. Agrawal, Z. Jiang, Q. Zhao, D. Shaw, Q. Cai, A. Roskey, L. Channavajjala, C. Saxinger and R. Zhang, *Proc. Natl. Acad. Sci. U. S. A.*, 1997, **94**, 2620.
108. S. Agrawal, *Biochim. Biophys. Acta*, 1999, **1489**, 53.
109. S. Agrawal and E. R. Kandimalla, *Mol. Med. Today*, 2000, **6**, 72.

CHAPTER 6

Splicing Modulation for Therapeutics

ELENA DAOUTSALI AND ANNEMIEKE AARTSMA-RUS*

Department of Human Genetics, Leiden University Medical Center, Einthovenweg 20, 2333 ZC, Leiden, The Netherlands
*E-mail: A.M.Aartsma-Rus@lumc.nl

6.1 Introduction

6.1.1 RNA Splicing

The coding sequence of a gene is contained usually within a number of small exons that are interspersed by much longer introns. Genes exhibit large heterogeneity in the number and length of their introns. After transcription genes are expressed as precursor mRNAs (pre-mRNAs). Introns are removed from the pre-mRNA and exons are ligated together, a procedure called splicing, forming the mRNA that is translated into a protein. Pre-mRNA splicing should be precise and accurate in order to produce an open reading frame, which is crucial for efficient protein generation during translation. The precision and accuracy of this procedure relies on sequences and structures within the pre-mRNA transcript that guide splicing proteins and help in the proper processing of the pre-mRNA into mRNA by assuring the correct cleavage and ligation reactions of RNA splicing.

Drug Discovery Series No. 68
Advances in Nucleic Acid Therapeutics
Edited by Sudhir Agrawal and Michael J. Gait

Published by the Royal Society of Chemistry, www.rsc.org

6.1.1.1 Splicing Machinery and Sequence Motifs Required for RNA Splicing

RNA splicing is carried out by a ribonucleoprotein complex, the spliceosome, which consists of five small nuclear ribonucleoproteins (U1, U2, U4, U5 and U6 snRNPs) and numerous other splicing factors. Proteins and splicing factors of the spliceosome guide and position the RNA in order for the snRNPs to bind to the correct splicing sites and begin the catalytic steps of splicing. Splice sites are special sequences at the intron–exon junction and they are usually surrounded by conserved motifs. Analysis of intron–exon junctions in the GENBANK databank has allowed identification of these conserved motifs. If one looks further into the normal splicing mechanism one can see that most splice sites obey the GT–AG rule. The 5′ splice site (donor) almost invariably includes a GT dinucleotide at the first two positions of the intron, and is surrounded by a larger and less conserved sequence. The 3′ splice site (acceptor), that is located at the other end of the intron contains three conserved sequences: a branchpoint (an A nucleotide generally located between 20 and 50 nucleotides upstream of the 3′ (acceptor) splice site), an intronic polypyrimidine tract (PPT) between the branchpoint and the 3′ splice site and an AG dinucleotide in the last two positions of the intron. Recognition of the conserved sequences on the 3′ and 5′ (donor) splicing sites by the spliceosome are important for correct splicing (Figure 6.1).

6.1.1.2 Other Sequence Motifs that Influence RNA Splicing

Splicing is a complex process that requires the interplay of multiple factors. Generally, pre-mRNA transcripts have many motifs that look like splice sites. Splicing factors have to recognize the true exons and splice sites from all possible splice site-like motifs. Especially when exons contain weaker donor and/or acceptor splice sites, the consensus motifs may not provide sufficient information for the splicing machinery to determine whether a site is suitable for spliceosome assembly and therefore splicing. Splicing then relies on additional motifs (non-splice-site regulatory sequences) that affect spliceosome assembly in a positive or negative manner. RNA elements that stimulate spliceosome assembly are called splicing enhancers. Splicing enhancers can be exonic (ESEs) or intronic (ISEs). ESEs are binding sites for a subgroup of splicing factors, the serine/arginine-rich (SR) proteins. SR proteins contain RNA recognition motif sequences (RRM) and carboxy-terminal arginine/serine (RS) domains. When SR proteins bind to an ESE they recruit splicing factors, therefore promoting exon definition and/or the inclusion of an exon that is weakly defined.[1] As for the intronic variety of splicing enhancers, they appear to be mechanistically different from ESEs, but some of them are SR protein-dependent as well. They bind to proteins of the heterogeneous nuclear ribonucleoprotein (hnRNP) complex. Sequence analyses have identified several ISEs near alternative exons.

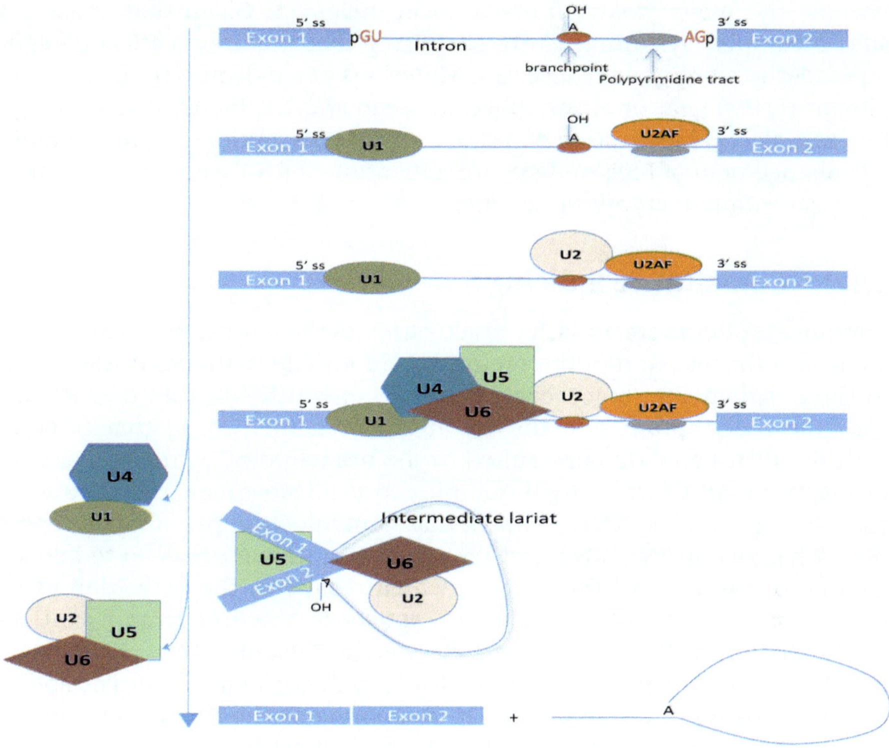

Figure 6.1 In the first phase of spliceosome assembly, U1 snRNP binds to the 5′ splice site (ss) and U2 snRNP is recruited to the branchpoint with the help of an additional splicing factor, U2AF that binds in the polypyrimidine tract. This early spliceosome assembly is called the A complex. Later binding of the tri-SNP U4–U5–U6 gives rise to the B complex. Structural rearrangements and conformational changes of the snRNPs lead to its activation. U6 dissociates from U4 and replaces U1. U1 and U4 are dissociated from the spliceosome and U6 binds with U2 creating the intermediate lariat (C complex). Splicing is completed by two transesterification reactions; the first step is cleavage in the 5′ss and formation of the lariat intron–exon 2 and the second step is cleavage at the 3′ss and ligation of the two exons.

For example, G-rich sequences containing at least one G triplet have been shown to enhance splicing by recruitment of hnRNP H and hnRNP F to introns.[2] On the other hand some RNA elements block spliceosome assembly and act as silencers or repressors. They can be of the exonic (ESSs) or intronic (ISSs) variety. Both of them serve as binding sites for splicing repressors, such as hnRNP proteins, or can have unfavourable secondary structures.

A strong match between a splice site and a consensus sequence is a major factor in the selection of a particular splice site. When the splice sites are weak, the splicing regulatory elements (SREs), ESEs, ISEs, ESSs and ISSs become the major contributors to exon inclusion. Given that splicing is such a complex system, it is not surprising that mutations affecting splicing underlie many genetic diseases. Mutations or single nucleotide polymorphisms (SNPs) near or at the splice sites can affect splicing in several ways, for example: (a) inactivation of a splice site, (b) disruption of ESEs and ISEs, (c) enhancement of ESSs and ISSs, (d) alteration of RNA secondary structures or (e) activation of cryptic splice sites.

6.1.2 Constitutive Splicing

Canonical splicing starts with the activation of the mature catalytic spliceosome and the excision of the pre-mRNA donor site. U1 snRNP is recruited to the 5′ splice site, together with some non-snRNP factors that interact with the branchpoint and the U2 auxiliary factor (U2AF) that interacts with the PPT. Later U2 is recruited to the branchpoint by the 65k subunit of the U2AF and forms the A complex (prespliceosome). U4, U5 and U6 snRNPs form a tri-snRNP complex that generates the precatalytic B complex. Changes in RNA–RNA and RNA–protein interactions lead to destabilization of the U1 and U4 snRNPs, which results in the activation of the spliceosome, or the B^{act} complex. U6 snRNP is now free to replace U2 at the branchpoint. The spliceosome cleaves the 5′ exon–intron junction and the cleavage is carried out by a branchpoint adenosine residue. The spliced intron forms an intermediary lariat that connects the 5′G of the intron to the 2′A at the branchpoint sequence. Next, the phosphodiester bond at the 3′ exon–intron junction is cleaved by the free hydroxyl group of the 5′ exon, the lariat is released and the downstream exon is joined to the upstream exon. The mature mRNA transcript is now released from the spliceosome (Figure 6.1).[3,4]

Overall, constitutive splicing shapes the process of intron removal and exon ligation. Constitutive exons are always included in the transcripts, whilst alternative exons are not always included in a transcript. Furthermore, mutations can activate splice sites, leading to the inclusion of cryptic exons, *i.e.* parts of introns being aberrantly included in transcripts.

6.1.3 Alternative Splicing

Compared with constitutive splicing where all exons in a specific gene are included in the transcript, alternative splicing is a variation where some exons are excluded, included or trimmed by the splicing machinery, resulting in various mRNA transcripts of the same gene, thus giving rise to different protein isoforms. Alternative splicing allows, for example, humans to produce over 100 000 proteins from approximately 20 000 genes. Alternatively spliced exons are defined by weaker splice sites and usually require additional splicing factors to be recognized by the spliceosome.[5] In order to determine which exons are removed and which are included in the

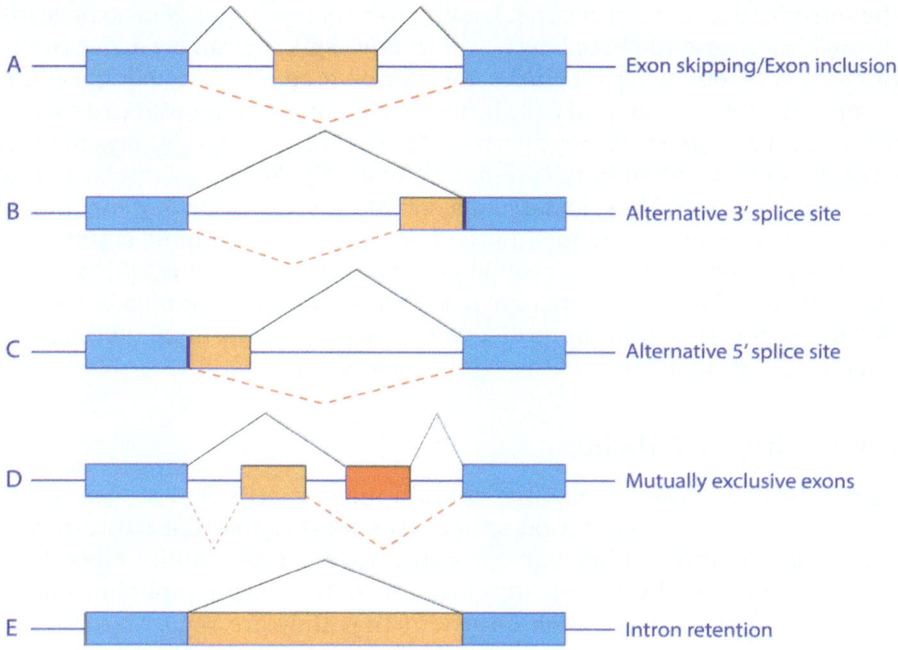

Figure 6.2 Schematic depiction of five main types of alternative splicing. (A) Exon skipping/Exon inclusion. Exon skipping accounts for approximately 40% of alternative splicing events in higher eukaryotes. (B), (C) alternative 3′ or 5′splice site. This happens when more than one splice site is recognised at the end of the exon. (D) The existence of mutually exclusive exons is a less frequent event that results in alternative transcript variants. (E) Intron retention is a more frequent alternative splicing event in plants, fungi and protozoa. Blue lines indicate the normal procedure and red dashed lines the alternative splicing event.

mRNA, different RNA sequences and splicing factors (RNA–protein complexes) play a role. Consequently binding of these inhibitory or enhancing factors actually determines whether a splice site is used or not. Alternative splicing events can be categorized as exon skipping, intron retention, alternative first and last exon, alternative 3′ and 5′ untranslated regions (UTRs), mutually exclusive exons, alternative promoter usage or alternative polyadenylation (Figure 6.2).[6,7]

Alternative splicing is an important mechanism for regulating tissue-specific and developmentally specific isoform expression. If the mechanism is disrupted, it can lead to disease, *e.g.* in myotonic dystrophy where certain splicing factors are unable to function, leading to a shift towards embryonic isoforms for multiple gene products.[8] This results in a multi-system disease where patients suffer from, amongst other effects, myotonia, heart problems, cataracts, and diabetes. Besides the generation of different protein isoforms from the same gene, alternative splicing can also result in the introduction of stop codons, leading to suppression of gene expression through nonsense-mediated decay. Alternative splicing can even give rise to proteins with opposing functions, *e.g.* skipping an exon encoding a transmembrane domain can lead to a shift from a membrane receptor to a soluble protein. An example is tumour necrosis factor receptor superfamily, member 6 (*FAS*), a gene involved in cell apoptosis. When the FAS transcript is alternatively spliced, exon 6 is skipped and a soluble protein instead of membrane receptor is produced. The membrane receptor has an immune-suppressive function, whilst the soluble protein enhances inflammations. Shifts in the ratio between the two isoforms can lead to disease, as exemplified by the fact that patients with autoimmune diseases, such as systemic lupus erythematosus (SLE), have high levels of the soluble protein.[6]

6.1.4 Cryptic Splicing

As mentioned above, transcripts contain many motifs that are similar to splice sites but are not actual splice sites. Mutations that cause inactivation of a genuine splice site can cause these cryptic splice sites to be used instead. Furthermore, mutations near the cryptic splicing motifs can strengthen them to such an extent that they are recognized by the splicing machinery as a genuine splice site.[9] Eukaryotic genomes contain many cryptic splicing sites that once activated result in various genetic diseases.[10] A well-known example of cryptic splicing is the thalassemia-associated mutations that occur in the 5′ splice site of intron 1 of the β-globin gene. The mutations lead to the activation of cryptic splice sites, which result in the inclusion of aberrant exons into the β-globin transcript, inducing defective beta-globin expression. Since β-globin is an essential part of haemoglobin, this lack results in beta-thalassemia, a severe disease where patients suffer from anaemia and toxic accumulation of iron.[9–11]

6.2 Therapeutic Exon Skipping Options

6.2.1 Antisense Oligonucleotides (ASOs) and Chemical Modifications

The mechanism of alternative splicing offers great diversity in the human proteome and has been proposed to be a contributor to the complexity of organisms. Expression of different splicing products in certain cell types and tissues indicates that cells can use and tolerate different mRNA transcripts and protein isoforms. Even though the process of splicing is not fully elucidated yet, approximately 50% of human genetic diseases have been found to be caused by mutations that lead to splicing abnormalities. Furthermore, the latest report of the human gene mutation database shows that 10% of genetic diseases are a result of a single base pair substitution in a splice site.[6,12] Taking into account that splicing is such a universal process, there is a broad opportunity for research on drugs that can manipulate splicing.

Splicing modulation can be achieved by use of antisense oligonucleotides (ASO), which can target and hybridize to a specific region in the pre-mRNA. In that way, they can interfere with RNA–RNA or RNA–protein interactions that are involved in splicing. These molecular tools can be exploited in different ways; to prevent the inclusion of an aberrant exon, to restore an open reading frame, to generate less protein or a non-toxic protein or to include an exon that is skipped.

ASOs are synthetic molecules consisting of nucleotides or chemically modified analogues. Splice-modulating ASOs are usually 15–30 nucleotides long. Target binding creates a steric block that prevents splicing factors from binding, thus inducing exon skipping or exon inclusion. Like other therapeutic ASOs, splice-modulating ASOs need to be chemically modified to improve stability and pharmacokinetic properties. For more in-depth information on chemical modifications see Chapter 1. Most splice-modulating ASOs have a phosphorothioate (PS) backbone. The PS backbone modification increases ASO stability, improves cellular uptake and also improves serum half-life *in vivo* due to binding of the backbone to serum protein, thus preventing renal clearance. Splice-modulating ASOs need an additional chemical modification to prevent RNase H-mediated cleavage (see Chapter 2). This can be achieved by modifications in the 2′ position of the ribose sugar, such as the 2′-*O*-methyl (2OMe) and 2′-*O*-methoxyethyl (MOE). Another type of modified oligonucleotide, the phosphorodiamidate morpholino (PMO) has been used almost exclusively to modify splicing. PMOs have a morpholine ring in place of the furanose ring found in natural nucleic acids and a neutral phosphorodiamidate backbone in place of the negatively charged phosphodiester backbone. PMOs do not bind to plasma proteins and, as such, are rapidly cleared by the kidney upon systemic treatment. However, they are very stable and inert and appear to be tolerated at very high doses.[13–15]

6.2.2 Restoring Cryptic Splicing

6.2.2.1 β-Thalassemia

β-Thalassemia is a severe, genetic blood disorder, caused by lack or reduction of the β-globin subunit of the protein hemoglobin. It affects a large group of people in the Mediterranean countries, Middle East, South East Asia and Africa. Close to 200 mutations have been reported to lead to defective β-globin expression and deficiency, but only a few of them account for approximately 90% of the thalassemia cases worldwide.[16] The most common mutations activate cryptic splice sites that induce aberrant splicing of β-globin pre-mRNA, which prevents normal translation of β-globin protein. β-thalassemia was the first disease for which ASOs were employed to redirect

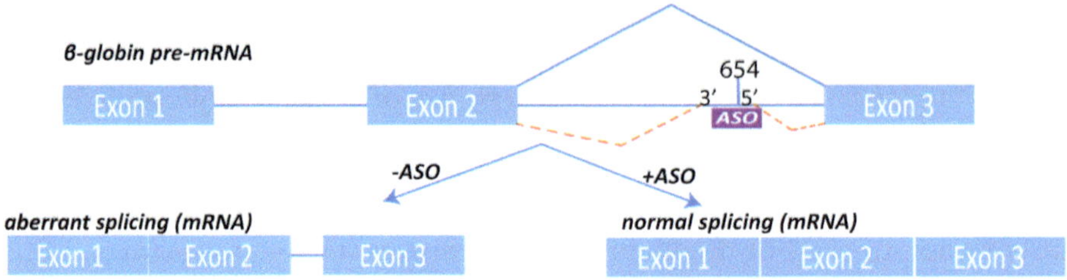

Figure 6.3 Splicing modulation for β-thalassaemia. IVS2-654 is a mutation in intron 2 of β-globin. The mutation creates an intronic 5′ splice site and activates a 3′ cryptic splice site that leads to the inclusion of a part of intron 2 in the mRNA transcript and to defective β-globin expression. ASOs that target the cryptic splice site hide it from the splicing machinery, therefore preventing inclusion of the intronic region into the transcript. Blue lines indicate the normal procedure and red dashed lines the alternative splicing event.

cryptic splicing in model systems. Mutation IVS2-654 (C>T) activates a 3′ cryptic splice site in intron 2, thus creating an aberrant 5′ splice site (Figure 6.3). While the genuine splice site is unaffected and functional, the cryptic splice site outcompetes it, resulting in partial intron retention in the spliced mRNA. ASOs were designed to hybridize to the aberrant splice site in order to prevent the binding of splicing factors, with the aim of restoring normal splicing. Tests in *in vitro* splicing systems and in patient-derived cell cultures indeed showed that this was feasible.[17,18] Furthermore, tests with ASOs in transgenic mice that carry the mutation showed successful restoration of cryptic splicing *in vivo*.[19] A limitation of this transgenic mouse model is that is does not recapitulate the fact that in patients with β-thalassemia the α-globin chain is not affected, and due to the lack of β-globin, α-globin accumulates in erythroid precursors. In a model that does recapitulate this phenomenon, it was shown that a combination of small interfering RNA (siRNA), targeting alpha-globin, and a splice-modulating ASO, targeting beta-globin, restored the balance between α and β-globin biosynthesis. This increased erythropoiesis but also increased the therapeutic effect of ASO treatment of β-globin.[20]

6.2.2.2 Ataxia Telangiectasia

Ataxia Telangiectasia (AT) is a rare, autosomal recessive disease, with poor prognosis. AT is characterized by progressive cerebellar ataxia, neurological impairment, an increased risk of cancer and chromosome instability. It is caused by mutations in the Ataxia Telangiectasia Mutated (*ATM*) gene that is important for cell cycle control, response to oxidative stress, apoptosis and chromatin changes and repair of DNA double-strand breaks. Close to 600 *ATM* mutations have been identified. Most of the mutations affect splicing, either by creating cryptic splice sites or by interfering

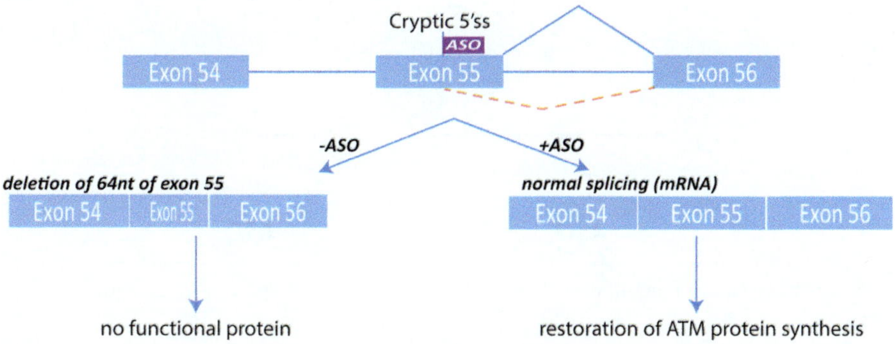

Figure 6.4 Splicing modulation for ataxia telangiectasia. Example of a mutation in the *ATM* gene. A mutation in exon 55 creates a cryptic splice donor and results in deletion of 64 nucleotides at the end of the exon. Targeting this site with an ASO leads to the restoration of exon 55 in the mRNA transcript and the ATM protein synthesis. Blue lines indicate the normal procedure and red dashed lines the alternative splicing event.

with splicing regulatory elements.[21] ATM protein is not detected in most of the patients, however, a subset of patients exhibiting some protein expression has a milder disease phenotype. Taking this fact into consideration, AT patients might benefit even from modest increases in ATM protein levels. ASOs have been used to correct splicing for three different cryptic splice sites mutations in AT (Figure 6.4). ASOs were designed to target and hide strong cryptic splice sites, aiming to restore normal splicing. For each mutation, ASOs restored the ATM splicing and resulted in production of the functional full-length ATM protein in patient-derived cultured cells.[22]

6.2.3 Reading Frame Restoration

6.2.3.1 Duchenne Muscular Dystrophy

Duchenne muscular dystrophy (DMD) is a neuromuscular, X-linked disease that has an early onset, where patients progressively loose muscle function from early life, become wheelchair-dependent around the age of 12 and need assisted ventilation at the age of 20. On average, with good care, life expectancy is around 30. The disease is caused by mutations, generally large deletions or duplications of one or more exons, which disrupt the open reading frame of dystrophin. This protein is involved in providing muscle fibers with stability during contraction through the connection of the cytoskeleton F-actin to proteins in the extracellular matrix surrounding muscle fibers. DMD muscle fibers that lack dystrophin are continuously damaged during normal exercise, resulting in chronic loss of muscle tissue and function. Similar to DMD, in Becker muscular dystrophy (BMD) patients have mutations in the *DMD* gene, mainly deletions of one or more exons. However, BMD has a

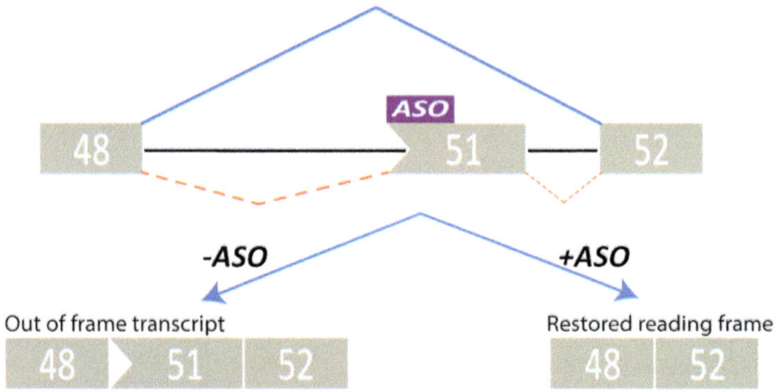

Figure 6.5 Exon skipping for DMD. A deletion of exons 49 and 50 results in an out of frame transcript. An ASO that targets exon 51 leads to exon skipping, restoration of the reading frame and a partially functional dystrophin protein. Red dashed lines illustrate the splicing for the out of frame transcript. Blue lines indicate the splicing of the in-frame transcript.

milder progression because the deletions do not disrupt the reading frame, allowing the production of a shorter but partially functional dystrophin, so the link between actin and extracellular matrix can still be made. The difference between the two diseases gave rise to the concept of using ASOs to restore the reading frame and thus convert the absence of functional dystrophin of DMD patients into the presence of a partially functional dystrophin, just as in BMD patients (Figure 6.5).[23]

In order to restore the reading frame, and counterintuitively, the DMD mutations have to be made larger. ASOs can be used to hide exons from the spliceosome in order to restore the reading frame. Which exon to skip depends on the size and location of the mutation.[24] Exon 45 is the most frequently observed single-exon deletion in DMD, whereas a deletion of both exon 45 and 46 is associated with BMD. Thus the skipping of exon 46 was anticipated to restore the reading frame and dystrophin protein production for DMD patients with an exon 45 deletion. Exon 46 skipping was tested in patient-derived muscle cell cultures carrying an exon 45 deletion, where ASO treatment indeed induced exon 46 skipping and restored dystrophin synthesis.[25] In a follow-up study 11 additional DMD exons, the skipping of which would apply to the most frequent deletions in DMD, were successfully skipped in wild-type cells.[26] Exon skipping and dystrophin restoration was further confirmed for numerous mutations and mutation types[23,27] in mouse[28] and dog models.[29] In an attempt to decrease mutation specificity multi-exon skipping was also studied by use of a combination of ASOs in patient cultured myotubes that contained different DMD deletions (Figure 6.6). Even though multiple exon skipping could cover a higher percentage of DMD patients and it appears feasible in a mouse model,[30] its applicability could be challenging since a combination of many ASOs is required.[31]

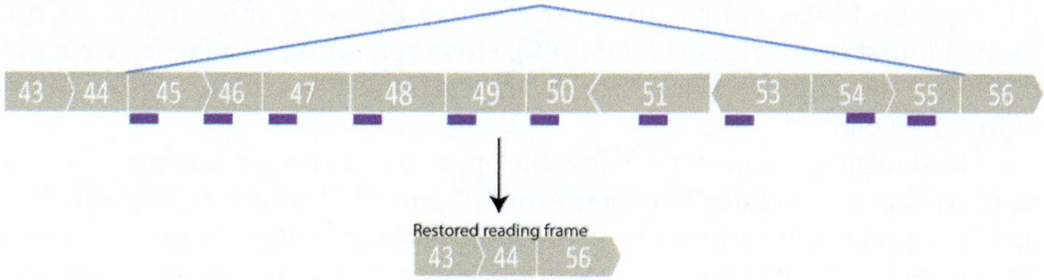

Figure 6.6 Multi-exon skipping for DMD. The approach was used in an exon 52 deletion (frame-shift) mouse model. A mix of 10 PMOs targeting exons 45–51 and 53–55 is able to induce skipping of these exons and restore dystrophin expression and improve muscle function. The blue line indicates the exon 45–55 multi-exon skipping event.

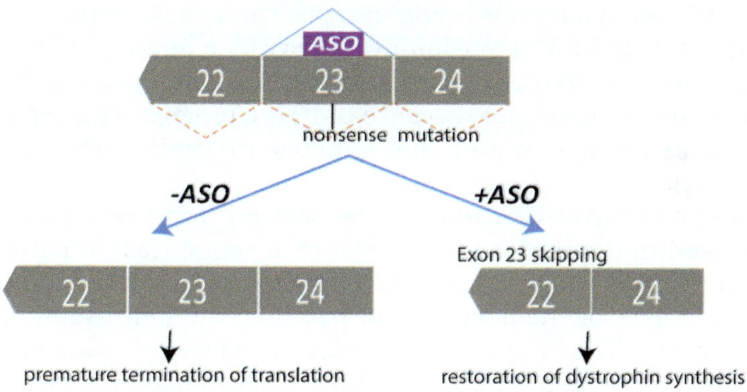

Figure 6.7 Exon skipping for the *mdx* mouse model for DMD. A nonsense mutation at nucleotide 3185 of the mRNA causes a premature termination of translation in exon 23. By skipping exon 23 the mutation is bypassed without disturbing the reading frame and dystrophin production was restored in the *mdx* mouse model. The red dashed lines indicate the splicing of exons 22, 23 and 24, while the blue lines indicate the exon 23 skip event.

Animal models of DMD played an important role in the clinical development of ASOs for DMD exon skipping. The *mdx* mouse model is the most widely used and studied animal model of DMD. It carries a nonsense mutation in exon 23 of the mouse *Dmd* gene. In this mouse model the consequences of lack of dystrophin are not as severe as in humans since muscle function and survival are only mildly impaired. Nevertheless, histological examination of the muscles shows a clear dystrophic phenotype.[32] Since exon 23 is an in-frame exon, skipping it would bypass the nonsense mutation without affecting the reading frame (Figure 6.7). Intramuscular injection of ASO targeting exon 23 skipping in the *mdx* mouse restored dystrophin production and improved function of the injected muscle.[28] Different chemistries have been tested systemically in this model and for exon

23 skipping PMOs appear to show a better skipping efficiency than 2′-*O*-methyl phosphorothioates (2OMePS). However, both chemistries were able to induce dystrophin production is skeletal muscles and also heart, albeit at lower levels.[13,33]

Exon skipping is a mutation-specific approach. However, because the deletion breakpoints cluster between exon 45 and 55,[34] targeting certain exons applies to relatively large subsets of DMD patients.[24] ASOs targeting exon 44 (applicable to 6% of patients), 45 (applicable to 8% of patients), 51 (applicable to 13% of patients) and 53 (applicable to 8% of patients) have been or are being tested in clinical trials. Exon 51 skipping (Figure 6.5) has been developed most thoroughly and for both the 2OMePS chemistry (drisapersen) and the PMO chemistry (eteplirsen). Drisapersen, developed by Prosensa–GSK–BioMarin, has been tested in three phase 2 and one phase 3 clinical trial, most often using subcutaneous delivery of 6 $mg\ kg^{-1}\ week^{-1}$ and the 6 minute walk test as a primary endpoint.[35–37] This test measures the distance someone is able to walk at maximum capacity for 6 minutes. It was not originally developed for DMD, but borrowed from the cardiovascular field, where it was used to test endurance. Because DMD is a progressive disease, the distance patients are able to walk in 6 minutes will decline when the disease progresses.

In placebo-controlled studies, there was an improvement in the distance walked in 6 minutes for younger (6–8 years) treated patients compared with those that received placebo, whereas in a larger phase 3 trial with a more heterogeneous group of patients, the improvement was not statistically significant or clinically relevant. Moreover, side effects were observed in the treated group, including injection-site reactions and subclinical proteinuria in almost all patients and severe thrombocytopenia in a few patients.[35–37] Drug registration was applied for with the Food and Drug Administration (FDA) and the European Medicines Agency (EMA). The FDA did not approve drisapersen, and the application to EMA was withdrawn before the official evaluation.[38] At this point, the clinical development of drisapersen, but also of 2OMePS ASOs targeting exon 44, 45 and 53 was stopped.

Eteplirsen has been tested less extensively in clinical trials.[39–42] Eteplirsen was administered intravenously weekly at doses of 30 or 50 $mg\ kg^{-1}$ in an open-label study that is currently ongoing for over 4 years, involving 12 patients. Since no placebo-group was used beyond the first 6 months, data were compared to matched untreated patients from natural history studies, revealing a slower disease progression when looking at the distance walked in 6 minutes[41] and at respiratory decline.[42] Dystrophin restoration was confirmed in the treated patients after 48 weeks of treatment,[40] although levels were very low (less than 1% on average). The FDA gave eteplirsen accelerated approval in 2016, based on the observed increases in dystrophin expression, explicitly stating that functional effects have yet to be confirmed and that data on this need to be provided by the company no later than 2021.[38] Evaluation by the EMA is currently pending.[43]

6.2.3.2 *Dystrophic Epidermolysis Bullosa*

Dystrophic epidermolysis bullosa is a severe skin disorder in which mild trauma induces severe blistering and scarring of skin and blistering of the mucosal surfaces of the respiratory and gastrointestinal tracts. It can be inherited in both dominant and recessive autosomal patterns. The disease is caused by mutations in *COL7A1* gene that encodes collagen VII.[44] Type VII collagen is the major component of fibrils, which ensure the attachment of the epidermis to the dermis, and is expressed by epidermal keratinocytes and dermal fibroblasts. Most patients have small mutations. Patients with in-frame mutations show a milder phenotype. This indicates that, just as in DMD, restoration of the reading frame can have a therapeutic effect. Interestingly, *COL7A1* exons are mostly in-frame and as such, skipping small mutations is anticipated to bypass the mutation without affecting protein function. Skipping of exon 70, which contains a recurrent mutation, was successfully induced *in vitro* in cultured patient keratinocytes and *in vivo* in a patient skin graft-transplanted nude rat model (Figure 6.8).[45] Exon 70 skipping in an animal model resulted in the formation of anchoring fibrils, indicating that the collagen is functional. Whereas in the previous *in vivo* studies the antisense oligonucleotide (ASO) administration was local to patient skin grafts, a recent study tried systemic delivery of oligonucleotides. Systemic delivery of ASOs at the tail base of mice induced exon skipping and restoration of collagen type VII in a skin graft placed close to the head. These data provide the first proof of concept of ASO systemic treatment in recessive dystrophic epidermolysis bullosa.[46]

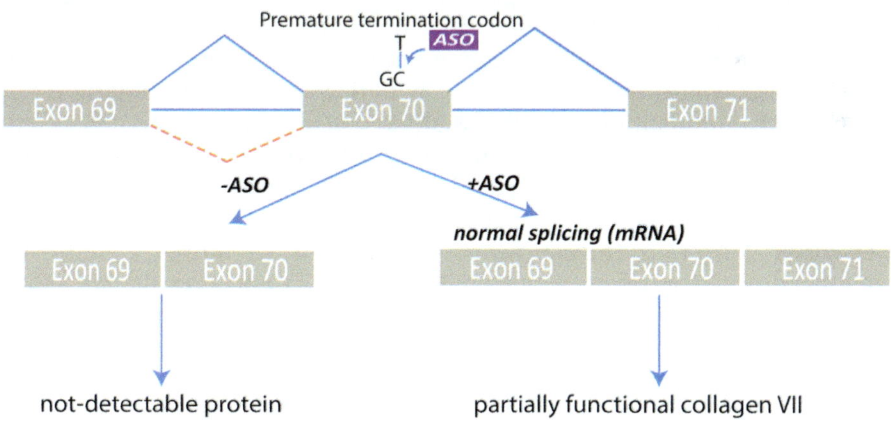

Figure 6.8 Exon skipping for dystrophic epidermolysis bullosa. A small deletion is localized in exon 70 of the *COL7A1* gene. An ASO targeting exon 70 can successfully induce exon 70 skipping, thus bypassing the mutation and allowing the production of a shorter partially functional collagen VII. The red dashed lines indicate splicing of the transcript including the mutated exon 70, the blue lines indicate the exon 70 skipping event.

6.2.4 Exon Inclusion

6.2.4.1 Spinal Muscular Atrophy

Spinal muscular atrophy (SMA) is an autosomal recessive neuromuscular disorder characterized by degeneration of motor neurons in the spinal cord that leads to progressive muscle weakness and paralysis. It is classified into four categories depending on the age of onset and severity (SMAI, SMAII, SMAIII and SMAIV). SMA type I, or Werdnig–Hoffmann disease, is the most common and severe type. Infants with SMA type I have a very early onset, before 6 months of age and never achieve the ability to sit unsupported. SMA type I patients require permanent ventilation *via* a tracheotomy and often die before they are 2 years of age. All types of SMA are caused by a defect in the *SMN* gene that prevents production of the ubiquitously expressed SMN protein. SMN protein is most abundantly expressed in the motor neurons of the spinal cord. Complete lack of SMN protein is embryonically lethal. However, deletion of *SMN1* is not lethal in humans due to the presence of *SMN2*, a homologous gene that has the potential to produce an identical protein.[47] *SMN1* and *SMN2* differ in their coding sequence only by one nucleotide. However, this silent variation results in skipping of exon 7 in the majority of SMN2 transcripts, which are translated into truncated and unstable proteins. Only about 10% of *SMN2* gene transcripts include exon 7, resulting in the production of low amounts of SMN proteins. Even though the pathogenesis of the disease relies on the absence of *SMN*, the severity of the disease is related to the number of copies of *SMN2*. SMA type I patients usually have two copies, whereas SMA type II ones usually have three copies and SMA types III and IV have three or four copies.[47] The presence of the homologous *SMN2* gene and its importance in disease severity led to the idea that the level of exon 7 inclusion in the SMN2 transcripts could be increased, and then a larger amount of SMN protein would be produced which would be anticipated to benefit SMA patients (Figure 6.9).[48] Methoxyethyl phosphorothioate (MOEPS) ASOs targeting an ISS within intron 7 successfully led to the inclusion of exon 7 and increased SMN protein levels in patient-derived fibroblasts.[49] Similar results were also obtained with a PMO ASO.[50] Since ASOs do not cross the blood–brain barrier, direct delivery to the central nervous system is required. Cerebroventricular administration of the ASO in SMA mouse models resulted in exon 7 inclusion and corrected the cellular and functional phenotypes (including survival in more severe models) by increasing SMN protein levels in the motor neurons of the spinal cord and survival.[51–53]

The successful animal studies encouraged the conducting of clinical trials with nusinersen (trade name Spinraza), a 2MOEPS ASO targeting the ISS in intron 7. The clinical development of nusinersen was sponsored by Biogen and Ionis Pharmaceuticals. The first study involved a dose-escalation study where type II/III patients received intrathecal injections of nusinersen. The trial showed that intrathecal delivery was well tolerated.[54]

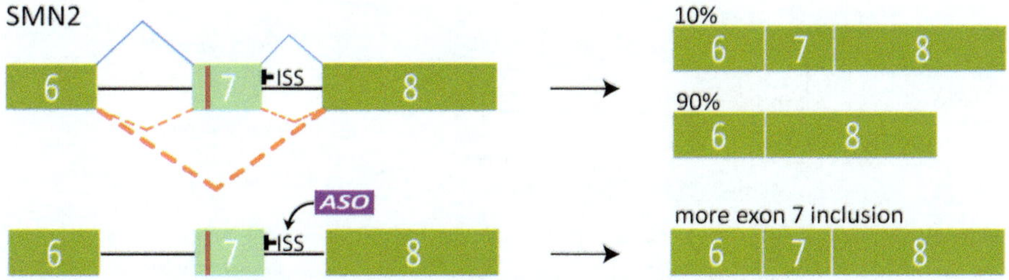

Figure 6.9 Exon inclusion in SMA. The majority of SMN2 transcripts lack exon 7, due to a silent mutation that makes this exon poorly recognized by the splicing machinery, resulting in a truncated version of the SMN protein that is less functional and unstable. By targeting an ISS located in intron 7 with an ASO the inclusion of exon 7 in the mRNA transcript can be increased, leading to higher levels of SMN protein. Red dashed lines indicate the splicing without ASO treatment and blue lines the increased exon 7 inclusion after ASO treatment.

A follow-up study in SMA type I patients, revealed the feasibility of intrathecal delivery also in this fragile group of patients.[55] Four infants died, but this was thought to be caused by the disease and not the drug. Tissue examination of autopsy material revealed up-to sixfold increase of full-length SMN protein, confirming that intrathecal delivery resulted in restoration of SMN protein in target tissues. This study also revealed that ASOs have a long half-life in the nervous system, of 4–6 months, meaning that, unlike drisapersen and eteplirsen, dosing does not have to happen weekly and can be much lower (12 mg per injection). In follow-up studies, patients received three or four loading doses in the first 2–3 months, followed by maintenance doses every 6 months.

Then several Phase 3 trials were initiated in symptomatic SMA type 1 patients, pre-symptomatic SMA type 1 patients and SMA type 2 patients. Results for the symptomatic SMA type 1 patients have been published.[56] This trial involved 121 patients who took part in a randomized multicentre sham-procedure-controlled double-blind study. Following interim analysis in 82 of the patients who were treated for 1 year, 40% of treated patients showed achievement of the motor milestones normally not achieved by type 1 SMA patients. Afterwards, unblinding took place and all patients received treatment. Most of the adverse events that were noticed were due to the disease and not to nusinersen.[57,58] A trial in type 2 SMA patients confirmed the therapeutic effect of nusinersen treatment in type II/III patients, where motor-function was increased compared with that in patients receiving placebo, when treatment was initiated early after the onset of symptoms, and disease progression was halted or slowed compared with placebo when treatment was initiated later.[59] Also in SMA type 1 patients, early initiation of treatment resulted in a larger effect. More patients responded when treated before 12 weeks after the onset of disease, and even more patients appear to

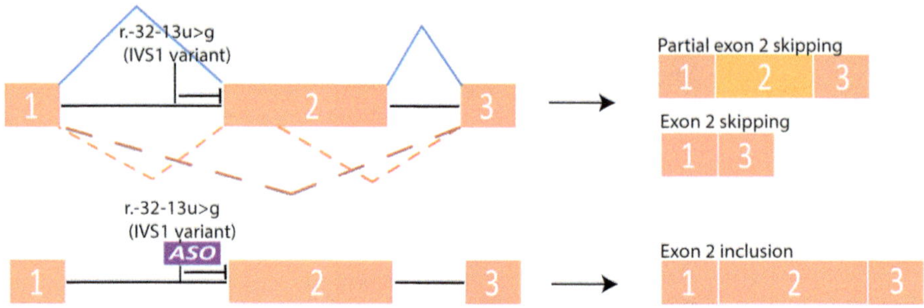

Figure 6.10 Splicing modulation for Pompe disease. The IVS1 variant (c.-32-13U>G) is located in the PPT of a pseudoexon upstream of exon 2. It makes the acceptor splice site weak, resulting in partial or complete exon 2 skipping. Skipping of exon 2 results in mRNA degradation and reduced GAA protein synthesis and production. ASOs targeting the acceptor and donor sites of the pseudoexon promote inclusion of exon 2 and restoration of the mRNA transcript and translation of the protein. Red dashed lines indicate the aberrant splicing events. The blue lines indicate the normalized splicing events.

respond in the pre-symptomatic trial, for which results are not yet published, but interim analyses have been presented at meetings and in press releases. Nusinersen was approved in December 2016 by FDA and EMA June 2017.[60]

6.2.4.2 Glycogen Storage Disease Type II or Pompe Disease

Pompe disease is a monogenic autosomal recessive lysosomal disease caused by a deficiency of acid alpha-glucosidase (GAA) enzyme, which leads to progressive wasting of skeletal muscle cells. It is a quite heterogeneous disease regarding the age of onset, severity of symptoms and extent of organ involvement and is categorized in three forms: infantile, childhood and adult.[61] The IVS1 (c.-32-13T>G) *GAA* variant is the most common mutation for the childhood and adult disease form and promotes both partial and complete skipping of exon 2 during pre-mRNA splicing. ASOs were designed to target a splicing regulatory element in GAA intron 1 in patient fibroblasts and resulted in exon 2 inclusion and increased GAA enzyme activity (Figure 6.10).[62] A follow-up study on iPSC-derived multinucleated myotubes showed exon 2 skipping at similar levels to those in patient fibroblasts. Blocking of 3′ and 5′ cryptic splice sites also led to restoration of canonical splicing and GAA enzyme activity.[63]

6.2.5 Generating Less Protein or Non-toxic Protein

6.2.5.1 Hutchinson–Gilford Progeria Syndrome

Hutchinson–Gilford Progeria is a rare autosomal dominant paediatric disorder that is characterized by premature aging, growth impairment and shortened lifespan. In most of the patients, progeria is caused by a

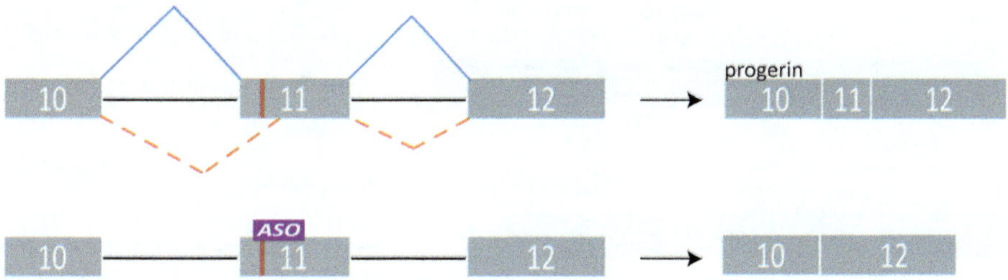

Figure 6.11 Restoration of cryptic splicing in progeria. A mutation in exon 11 of the *LMNA* gene introduces a cryptic splice site leading to loss of 150nt from the end of the exon and production of a truncated protein, progerin. Targeting the mutation in exon 11 results in decreased levels of progerin. Red dashed lines indicate the aberrant splicing events. The blue lines indicate the normalized splicing events.

mutation in exon 11 of the *LMNA* gene. The *LMNA* gene encodes A-type lamins that are intermediate filament proteins of the inner nuclear lamina, which is involved in maintenance of nuclear shape and structure, DNA replication and gene expression. The mutation in *LMNA* creates a cryptic splice site that leads to production of a shortened protein called progerin.[64] ASOs designed to target the donor splice site of exon 10 and/or the mutation in exon 11 of the *LMNA* gene, successfully reduced progerin production in mouse fibroblasts. Similar results have been achieved in patient-derived fibroblasts (Figure 6.11). *In vivo* testing of ASO in a progeria mouse model showed amelioration of the symptoms and improvement of survival.[65] In addition to being therapeutic tools, ASOs can also be used to disrupt normal gene expression and therefore create disease models for research on the mechanisms of the disease. In progeria, ASOs have been used to redirect processing of exon 11 and enhance production of progerin in human myogenic cells, thus generating an *in vitro* model of premature muscle ageing.[66] Another study also showed that multiple ASOs led to increased progerin production in patient fibroblasts, which is helpful for investigating cellular and molecular hallmarks.[67]

6.2.5.2 Huntington Disease

Huntington disease (HD) is an autosomal dominant neurodegenerative disorder that is caused by an expansion in the CAG repeat region of the huntingtin (*HTT*) gene and belongs to the broad group of polyglutamine (polyQ) disorders (see also Chapter 7). The pathological hallmark of polyQ disorders is the accumulation of polyQ proteins in the brain that leads to motor, cognitive and psychiatric symptoms. HD has been an attractive target for DNA- and RNA-based therapies because it results from mutation in one allele of *HTT*, indicating that elimination of the mutant HTT or reduction of its expression can prevent or slow down the disease pathology. When mutant Htt is cleaved by caspases, smaller toxic N-terminal fragments that contain the polyQ tail are produced and form insoluble aggregates. ASOs

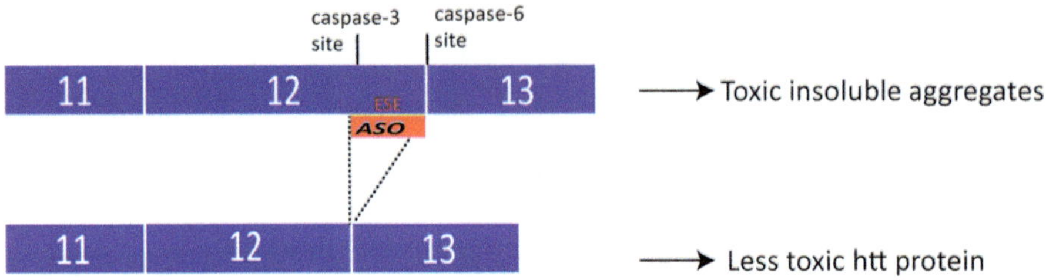

Figure 6.12 Generation of a less toxic protein in Huntington disease. The ASO is targeting an ESE in the 3′ part of exon 12 of the *HTT* gene. Partial skipping of exon 12 results in the formation of a new modified Htt protein that lacks the caspase-3 site and part of the caspase-6 site, which are involved in cleavage of Htt into smaller toxic proteins that lead to insoluble aggregates.

have been used to induce partial skipping of exon 12 where the caspase cleavage sites are located, in patient-cultured fibroblasts and *in vivo* in wild type mice (Figure 6.12).[68] In this way a modified, shorter mutant Htt is produced, which is an advantage when compared with Htt-lowering approaches, since the Htt transcript and protein levels remain unchanged. The same ASO was used in a recent *in vivo* study that used the YAC128 mouse model of HD and resulted in exon skipping. However behavioural experiments have not been conducted yet.[69]

6.3 Future Perspectives, Towards Additional Approved Splice-modulating ASOs

6.3.1 The Challenge of Personalized Medicine Development

It will be clear from the preceding sections that splice-modulating ASOs have a lot of potential as therapeutic entities, especially for rare genetic diseases, which often have an unmet medical need. However, herein also lie several challenges. Because the diseases are rare, only a limited number of patients are available for the conduct of clinical trials, which makes it harder to obtain sufficient evidence on drug efficacy needed to apply for regulatory approval. Whilst for SMA nusinersen applies to all subtypes, for most other genetic diseases, splicing modulation is a mutation-specific approach where a single splice-modulating ASO applies to only a subset of patients, or in extreme cases to individual patients.[70]

Another challenge is that for diseases with unmet medical need, there is often limited experience with conducting clinical trials. As such, outcome measures to assess clinical benefit for patients with that specific disease are often lacking.[71] This has been the case for DMD when drisapersen and eteplirsen were first tested in trials and the 6 minute walk test was adopted from the cardiovascular field as a functional endpoint. While this measure can in fact measure disease progression for DMD, it can do this only in ambulant patients (a majority of patients are non-ambulant) and there are several caveats, *e.g.* the distance walked in 6 minutes will increase in very young patients due to growth. Work is currently ongoing to develop new outcome measures – also in non-ambulant patients.[72] However, ideally this work should be carried out before clinical trials are initiated.

Both for DMD and SMA, ASOs have now been approved.[73] It is anticipated that other fields will benefit from the development path taken for these ASOs and the close collaboration and trilateral education that occurred between patient representatives, regulators and academics.[71]

6.3.2 ASO Delivery

As with most therapies, delivery to the target cells in the target tissue is a challenge also for oligonucleotides.[74] ASOs are small molecules and without a PS backbone, upon systemic treatment they will be rapidly cleared by the kidney, while large amounts of ASOs with a PS backbone will be taken up by both kidney and liver. Uptake by other tissues is limited and most ASOs will not cross the blood–brain-barrier. It should also be born in mind that disease pathology may improve delivery to that tissue (*e.g.* dystrophic muscle will take up more ASOs than healthy muscle) but can also result in lower uptake (*e.g.* due to severe fibrosis). When possible, local delivery is probably preferred, since this will result in a higher effective dose in the target cells and a lower load on liver and kidney, and thus less chance of safety concerns. In fact, local delivery to the nervous system has turned out to be very effective, as underlined by the success of nusinersen.[55] Upon local delivery (intrathecal or intraventricular), neurons take up ASOs very efficiently and distribution occurs throughout the central and peripheral nervous system. Furthermore, the nervous system is a very benign environment for ASOs, leading to half-lives of months.

6.3.3 Future Perspective

Two splice-modulating ASOs have been approved and it is anticipated that more will follow. Given that local delivery to the nervous system has proven well tolerated and very effective, it is expected that there will be a focus on splice-modulating ASOs (and ASOs in general) as therapeutics for neurological diseases. Furthermore, given that liver and kidney take up large amounts of ASOs, diseases of these organs are also likely candidates for ASO therapeutics. However, here one has to bear in mind that delivery to the target organ does not necessarily mean that the ASO is also taken up by the target cells within that tissue.[74]

It is now clear how to best design and test splice-modulating ASOs in proof of concept studies and these studies have been carried out for many diseases (only a subset is showcased in this chapter). The field has now moved to the next generation of challenges, involving how to test splice-modulating ASOs in clinical trials (as discussed in Section 6.3.1), how to develop additional chemical modifications and how to improve delivery for diseases where only systemic delivery is an option. These latter two topics are discussed in other chapters of this book and relate not only to splice-modulating ASOs but also to other therapeutic oligonucleotide modalities. Regarding future development of additional splice-regulating ASOs, on one hand more focus should be on the requirements for clinical development of ASOs and this should be an ongoing dialogue with patients and regulators.[71] On the other hand, there should also be more focus on basic science to facilitate more optimal chemical modifications, to increase knowledge of AON delivery (*e.g.* cellular uptake) and to improve our understanding of splicing.[74]

References

1. D. L. Black, *Annu. Rev. Biochem.*, 2003, **72**, 291.
2. Y. Wang, M. Ma, X. Xiao and Z. Wang, *Nat. Struct. Mol. Biol.*, 2012, **19**, 1044.
3. C. L. Will and R. Lührmann, *Cold Spring Harbor Perspect. Biol.*, 2011, **3**, 1.
4. H. Chen and S. Cheng, *Biosci. Rep.*, 2012, **32**, 345.
5. Y. Wang, J. Liu, B. Huang, Y. Xu, J. Li and L. Huang, *et al.*, *Biomed. Rep.*, 2014, 152.
6. L. M. Gallego-Paez, M. C. Bordone, A. C. Leote, N. Saraiva-Agostinho, M. Ascensão-Ferreira and N. L. Barbosa-Morais, *Hum. Genet.*, 2017, **136**, 1015.
7. H. Keren, G. Lev-Maor and G. Ast, *Nat. Rev. Genet.*, 2010, **11**, 345.
8. M. Nakamori, K. Sobczak, A. Puwanant, S. Welle and K. Eichinger, *Ann. Neurol.*, 2013, **74**, 862.
9. V. Arechavala-Gomeza, B. Khoo and A. Aartsma-Rus, *Appl. Clin. Genet.*, 2014, **7**, 245.
10. Y. Kapustin, E. Chan, R. Sarkar, F. Wong, I. Vorechovsky and R. M. Winston, *et al.*, *Nucleic Acids Res.*, 2011, **39**, 5837.
11. X. Roca, R. Sachidanandam and A. R. Krainer, *Nucleic Acids Res.*, 2003, **31**, 6321.
12. A. J. Matlin and M. J. Moore, *Adv. Exp. Med. Biol.*, 2007, **623**, 14.
13. B. Wu, P. Lu, E. Benrashid, S. Malik, J. Ashar and T. J. Doran, *et al.*, *Gene Ther.*, 2010, **17**, 132.
14. M. M. Evers, L. J. A. Toonen and W. M. C. van Roon-Mom, *Adv. Drug Delivery Rev.*, 2015, **87**, 90.
15. M. A. Havens and M. L. Hastings, *Nucleic Acids Res.*, 2016, **44**, 6549.
16. H. Sierakowska, M. J. Sambade, S. Agrawal and R. Kole, *Proc. Natl. Acad. Sci. U. S. A.*, 1996, **93**, 12840.
17. G. Lacerra, H. Sierakowska, C. Carestia, S. Fucharoen, J. Summerton and D. Weller, *et al.*, *Proc. Natl. Acad. Sci. U. S. A.*, 2000, **97**, 9591.
18. Z. Dominski and R. Kole, *Proc. Natl. Acad. Sci. U. S. A.*, 1993, **90**, 8673.

19. S. Svasti, T. Suwanmanee, S. Fucharoen, H. M. Moulton, M. H. Nelson and N. Maeda, *et al.*, *Proc. Natl. Acad. Sci. U. S. A.*, 2009, **106**, 1205.
20. S. Y. Xie, Z. R. Ren, J. Z. Zhang, X. B. Guo, Q. X. Wang, S. Wang, D. Lin, X. L. Gong, W. Li, S. Z. Huang, F. Zeng and Y. T. Zeng, *et al.*, *Hum. Mol. Genet.*, 2007, **16**, 2616.
21. L. Chessa, R. Micheli and A. Molinaro, *J. Rare Disord. Diagn. Ther.*, 2016, **2**, 1.
22. L. Du, J. M. Pollard and R. A. Gatti, *Proc. Natl. Acad. Sci. U. S. A.*, 2007, **104**, 6007.
23. A. Aartsma-Rus, *RNA Biol.*, 2010, **7**, 453.
24. A. Aartsma-Rus, I. Fokkema, J. Verschuuren, I. Ginjaar, J. Van Deutekom and G. J. Van Ommen, *et al.*, *Hum. Mutat.*, 2009, **30**, 293.
25. J. C. van Deutekom, M. Bremmer-Bout, A. A. Janson, I. B. Ginjaar, F. Baas, J. T. den Dunnen and G. J. van Ommen, *Hum. Mol. Genet.*, 2001, **10**, 1547.
26. A. Aartsma-Rus, M. Bremmer-Bout, A. A. M. Janson, J. T. Den Dunnen, G. J. B. Van Ommen and J. C. T. Van Deutekom, *Neuromuscular Disord.*, 2002, **12**, S71–S77.
27. Y. Takeshima, H. Wada, M. Yagi, Y. Ishikawa, Y. Ishikawa and R. Minami, *et al.*, *Brain Dev.*, 2001, **23**, 788.
28. C. J. Mann, K. Honeyman, A. J. Cheng, T. Ly, F. Lloyd and S. Fletcher, *et al.*, *Proc. Natl. Acad. Sci. U. S. A.*, 2001, **98**, 42.
29. T. Yokota, E. Hoffman and S. Takeda, *Methods Mol. Biol.*, 2011, **709**, 299.
30. Y. Echigoya, Y. Aoki, B. Miskew, D. Panesar, A. Touznik and T. Nagata, *et al.*, *Mol. Ther. - Nucleic Acids*, 2015, **4**, e225.
31. A. Aartsma-Rus, A. A. M. Janson, W. E. Kaman, M. Bremmer-Bout, G.-J. B. van Ommen and J. T. den Dunnen, *et al.*, *Am. J. Hum. Genet.*, 2004, **74**, 83.
32. P. Sicinski, Y. Geng, A. Ryder-Cook, E. Barnard, M. Darlison and P. Barnard, *Science*, 1989, **244**, 1578.
33. H. Heemskerk, C. De Winter, P. Van Kuik, N. Heuvelmans, P. Sabatelli and P. Rimessi, *et al.*, *Mol. Ther.*, 2010, **18**, 1210.
34. A. Aartsma-Rus, J. C. T. Van Deutekom, I. F. Fokkema, G. J. B. Van Ommen and J. T. Den Dunnen, *Muscle Nerve*, 2006, **34**, 135.
35. N. Goemans, E. Mercuri, E. Belousova, H. Komaki, A. Dubrovsky and C. M. McDonald, *et al.*, *Neuromuscular Disord.*, 2017, **28**, 4.
36. N. M. Goemans, M. Tulinius, M. van den Hauwe, A.-K. Kroksmark, G. Buyse and R. J. Wilson, *et al.*, *PLoS One*, 2016, **11**, e0161955.
37. T. Voit, H. Topaloglu, V. Straub, F. Muntoni, N. Deconinck and G. Campion, *et al.*, *Lancet Neurol.*, 2014, **13**, 987.
38. A. Aartsma-Rus and A. M. Krieg, *Nucleic Acid Ther.*, 2017, **27**, 1.
39. S. Cirak, V. Arechavala-Gomeza, M. Guglieri, L. Feng, S. Torelli and K. Anthony, *et al.*, *Lancet*, 2011, **378**, 595.
40. J. R. Mendell, L. R. Rodino-Klapac, Z. Sahenk, K. Roush, L. Bird and L. P. Lowes, *et al.*, *Ann. Neurol.*, 2013, **74**, 637.
41. J. R. Mendell, N. Goemans, L. P. Lowes, L. N. Alfano, K. Berry and J. Shao, *et al.*, *Ann. Neurol.*, 2016, **79**, 257.

42. T. B. Kinane, O. H. Mayer, P. W. Duda, L. P. Lowes, S. L. Moody and J. R. Mendell, *J. Neuromuscular Dis.*, 2017, **5**, 47.
43. A. Aartsma-Rus, V. Straub, R. Hemmings, M. Haas, G. Schlosser-Weber and V. Stoyanova-Beninska, *et al.*, *Nucleic Acid Ther.*, 2017, **27**, 251–259.
44. N. Hammami-Hauasli, D. U. Kalinke, H. Schumann, U. Kalinke, B. F. Pontz and I. Anton-Lamprecht, *et al.*, *J. Invest. Dermatol.*, 1997, **109**, 384.
45. M. Goto, D. Sawamura, W. Nishie, K. Sakai, J. R. McMillan and M. Akiyama, *et al.*, *J. Invest. Dermatol.*, 2006, **126**, 2614.
46. J. Bremer, O. Bornert, A. Nyström, A. Gostynski, M. F. Jonkman and A. Aartsma-Rus, *et al.*, *Mol. Ther. - Nucleic Acids*, 2016, **5**, e379.
47. A. D'Amico, E. Mercuri, F. D. Tiziano and E. Bertini, Spinal muscular atrophy, *Orphanet J. Rare Dis.*, 2011, **6**, 1.
48. Y. Hua, K. Sahashi, F. Rigo, G. Hung, G. Horev and C. F. Bennett, *et al.*, *Nature*, 2011, **478**, 123.
49. L. Cartegni and A. R. Krainer, *Nat. Struct. Biol.*, 2003, **10**, 120.
50. L. A. Skordis, M. G. Dunckley, B. Yue, I. C. Eperon and F. Muntoni, *Proc. Natl. Acad. Sci. U. S. A.*, 2003, **100**, 4114.
51. M. A. Passini, J. Bu, A. M. Richards, C. Kinnecom, S. P. Sardi and M. Stanek, *et al.*, *Sci. Transl. Med.*, 2011, **3**, 72ra18.
52. Y. Hua, K. Sahashi, G. Hung, F. Rigo, M. A. Passini and C. F. Bennett, *et al.*, *Genes Dev.*, 2010, **24**, 1634.
53. Y. Hua, T. A. Vickers, H. L. Okunola, C. F. Bennett and A. R. Krainer, *Am. J. Hum. Genet.*, 2008, **82**, 834.
54. C. A. Chiriboga, K. J. Swoboda, B. T. Darras, S. T. Iannaccone, J. Montes and D. C. De Vivo, *et al.*, *Neurology*, 2016, **86**, 890.
55. R. S. Finkel, C. A. Chiriboga, J. Vajsar, J. W. Day, J. Montes and D. C. De Vivo, *et al.*, *Lancet*, 2016, **388**, 3017.
56. R. S. Finkel, E. Mercuri, B. T. Darras, A. M. Connolly, N. L. Kuntz and J. Kirschner, *et al.*, *N. Engl. J. Med.*, 2017, **377**, 1723.
57. D. R. Corey, *Nat. Neurosci.*, 2017, **20**, 497.
58. C. A. Chiriboga, *Expert Rev. Neurother.*, 2017, **17**, 955.
59. for the C. S. G., E. Mercuri, B. T. Darras, C. A. Chiriboga, J. W. Day, C. Campbell, A. M. Connolly, S. T. Iannaccone, J. Kirschner, N. L. Kuntz, K. Saito, P. B. Shieh, M. Tulinius, E. S. Mazzone, J. Montes, K. M. Bishop, Q. Yang, R. Foster, S. Gheuens, C. F. Bennett and W. Farwell, *et al.*, *N. Engl. J. Med.*, 2018, **378**, 625.
60. A. Aartsma-Rus, P. Balabanov, L. Binetti, M. Haas, M. Haberkamp and J. Mitchell, *et al.*, *Lancet Neurol.*, 2017, **16**, 264.
61. M. Di Rocco, D. Buzzi and M. Tarò, *Acta Myol.*, 2007, **26**, 42.
62. E. van der Wal, A. J. Bergsma, J. M. Pijnenburg, A. T. van der Ploeg and W. W. M. P. Pijnappel, *Mol. Ther. - Nucleic Acids*, 2017, **7**, 90.
63. E. van der Wal, A. J. Bergsma, T. J. M. van Gestel, S. L. M. in 't Groen, H. Zaehres and M. J. Araúzo-Bravo, *et al.*, *Mol. Ther. - Nucleic Acids*, 2017, **7**, 101.

64. S. Rodriguez, F. Coppedè, H. Sagelius and M. Eriksson, *Eur. J. Hum. Genet.*, 2009, **17**, 928.
65. P. Spitali and A. Aartsma-Rus, *Cell*, 2012, **148**, 1085.
66. Y. B. Luo, C. Mitrpant, A. M. Adams, R. D. Johnsen, S. Fletcher and F. L. Mastaglia, *et al.*, *PLoS One*, 2014, **9**, e98306.
67. L. G. Fong, T. A. Vickers, E. A. Farber, C. Choi, U. J. Yun and Y. Hu, *et al.*, *Hum. Mol. Genet.*, 2009, **18**, 2462.
68. M. M. Evers, H.-D. Tran, I. Zalachoras, O. C. Meijer, J. T. den Dunnen and G.-J. B. van Ommen, *et al.*, *Nucleic Acid Ther.*, 2014, **24**, 4.
69. J. Casaca-Carreira, L. J. A. Toonen, M. M. Evers, A. Jahanshahi, W. M. C. van-Roon-Mom and Y. Temel, *Biomed. Pharmacother.*, 2016, **84**, 93.
70. A. Aartsma-Rus, A. Ferlini, N. Goemans, A. M. G. Pasmooij, D. J. Wells and K. Bushby, *et al.*, *Hum. Gene Ther.*, 2014, **25**, 885.
71. V. Straub, P. Balabanov, K. Bushby, M. Ensini, N. Goemans and A. De Luca, *et al.*, *Lancet Neurol.*, 2016, **15**, 882.
72. A. Mayhew, E. S. Mazzone, M. Eagle, T. Duong, M. Ash and V. Decostre, *et al.*, *Dev. Med. Child Neurol.*, 2013, **55**, 1038.
73. A. Aartsma-Rus, *Nucleic Acid Ther.*, 2017, **27**, 67.
74. C. Godfrey, L. R. Desviat, B. Smedsrød, F. Piétri-Rouxel, M. A. Denti and P. Disterer, *et al.*, *EMBO Mol. Med.*, 2017, **9**, 545.

CHAPTER 7

Targeting Toxic Repeats

M. LEONTIEN VAN DER BENT[†a,b], REMCO T. P. VAN CRUCHTEN[†a] AND DERICK G. WANSINK[a*]

[a]Radboud Institute for Molecular Life Sciences (RIMLS), Radboud University Medical Center, Department of Cell Biology, Geert Grooteplein 26-28, 6525 GA Nijmegen, The Netherlands; [b]Radboud Institute for Molecular Life Sciences (RIMLS), Radboud University Medical Center, Department of Biochemistry, Geert Grooteplein 26-28, 6525 GA Nijmegen, The Netherlands
*E-mail: rick.wansink@radboudumc.nl

7.1 Introduction

Repeat-expansion diseases form a special class of human disorders characterized by the presence of an unstable, expanded microsatellite in the genome of the patient.[1,2] The human genome contains several different microsatellite sequences, but only some of these repeats cause disease once they become unstable and expand above a certain threshold length. The associated molecular disease mechanism and the symptoms in patients depend on the expression and function of the mutated gene, the length of the repeat and its location in the gene.

Owing to their repetitive nature, expanded DNA sequences and their repeat RNA products show unusual functional and structural properties.[3] It is these atypical, repeat-associated characteristics that not only play a central role in

[†]These authors contributed equally to this work.

Drug Discovery Series No. 68
Advances in Nucleic Acid Therapeutics
Edited by Sudhir Agrawal and Michael J. Gait

Published by the Royal Society of Chemistry, www.rsc.org

disease manifestation, but also shape the development of therapeutic strategies and the identification of active compounds.

In this chapter, we will illustrate the uniqueness of repetitive DNA and RNA sequences, their roles in molecular mechanisms in repeat-expansion diseases and the therapeutic strategies and tools that are currently being developed against them. We will focus on the development of active compounds targeted directly or indirectly at the toxic microsatellites, not at protein products or molecular players further downstream in the pathological cascades of the various diseases. Along with antisense strategies [such as oligonucleotides or RNA interference (RNAi)], the more recent approaches using small molecules and gene-editing technology will also be discussed. There is no cure for any of the repeat-expansion diseases yet, but recent developments have moved extremely fast and the first clinical trials have been initiated.

7.2 Expanded Repetitive Sequences and Human Disease

7.2.1 Repeat Instability

Currently, more than 40 disorders are linked to unstable repeat sequences in the human genome.[1,2] The first microsatellite repeats were reported in 1991 and were made of trinucleotide units: a CGG repeat relating to fragile X syndrome [FXS; fragile X mental retardation 1 (*FMR1*) gene] and a CAG repeat causing spinal bulbar muscular atrophy (SBMA). Soon thereafter, the discovery of the mutations causing myotonic dystrophy type 1 (DM1; CTG repeat) and Huntington's disease (HD; CAG repeat) followed. Later on, an expanding tetranucleotide repeat was identified for DM2 (CCTG repeat) and a pentanucleotide repeat associated with spinocerebellar ataxia type 10 (SCA10; ATTCT repeat). Hexanucleotide repeats were more recently discovered to cause SCA36 (GGCCTG repeat)[4] and *C9orf72*-linked amyotrophic lateral sclerosis and frontotemporal dementia (*C9orf72*-ALS/FTD; GGGGCC repeat).[5,6] The latest disease added to the family was Fuchs' endothelial corneal dystrophy (FECD), linked to an expanded CTG repeat, in 2012.[7]

Repeat-instability diseases mostly belong to the group of neurodegenerative or neuromuscular diseases. Brain and skeletal muscle are generally the most affected organs, which corresponds to the expression of the mutated genes in these tissues. Genetic anticipation is frequently observed in families, which means that the disease symptoms occur earlier and become more severe in successive generations.[8] Anticipation is linked to expansion of the unstable repeat from one generation to the next, whereby a longer repeat correlates with more severe symptoms. Intergenerational contraction can also occur, but is rare. It should be noted, however, that not all repeat-expansion diseases show a clear correlation between repeat length and disease manifestation.

Repeats in the general population are polymorphic, around 5–35 repeat units in length depending on the disease, but stably inherited.[1,2] The threshold length above which a repeat becomes unstable and typically expands is usually in the region of 30–40 units. Repeat instability not only occurs between generations, but also during a patient's life. This so-called somatic instability can be dramatic and is thought to be responsible for the progressive nature of the symptoms. Hence, next to the inherited repeat length in the progenitor allele, the rate of somatic expansion is a determinant of the course of disease.

A typical repeat length in patients with a CAG/polyglutamine (polyQ) repeat disease [*e.g.* HD, SCA, DRPLA (dentato-rubro-pallido-luysian atrophy) and SBMA] is medium size and contains 35–80 repeat units (200 in extreme cases).[1,2] Hence, the general length difference between healthy and disease-causing alleles in these patients is only small. In contrast, many of the repeats outside the open reading frame (ORF) of the mutated gene, like those in DM1/2, *C9orf72*-ALS/FTD, Friedreich's ataxia (FRDA), FXS and SCA10, can expand to several thousand units. Long repeats sometimes already occur in progenitor alleles at birth, but predominantly develop due to cell-type-specific somatic expansion. The large difference between the healthy and pathogenic alleles in this second group of diseases provides a therapeutic window and makes the repeats and their unusual secondary structures logical targets for allele-selective therapy.

The third factor contributing to the complicated correlation between phenotype and unstable repeat is the occurrence of repeat interruptions, often at one end of the expansion. Complex repeats have been reported for DM2, FRDA, FXS/FXTAS, several SCAs and more recently also for DM1.[9,10] In DM1, repeat interruptions are thought to lower the rate of somatic expansion and thereby decrease disease severity and progression.

7.2.2 Molecular Mechanisms of Disease

For 17 repeat-expansion diseases, therapeutic studies based on the targeting of toxic repeats - directly or indirectly – have been reported (Figure 7.1). These diseases relate to the following repeat units: CAG, CGG, CTG, GAA, CCTG, ATTCT, GGCCTG and GGGGCC. The best-studied examples in terms of therapy development are DM1/2, HD, *C9orf72*-ALS/FTD and the group of CAG/polyQ-related SCAs. Only one or few therapy-oriented publications currently exist for DRPLA and SBMA (both CAG), FECD (CTG), FXS (CGG), SCA10 (AATTCT) and SCA36 (GGCCTG). To the best of our knowledge, no therapeutic studies have been published on repeat targeting in oculopharyngeal muscular dystrophy (OPMD; imperfect GCN repeat) or any of the other polyalanine diseases.

When it comes to disease mechanisms, the family of repeat-expansion diseases is highly diverse. For example, even within the group of CAG/polyQ expansion diseases (more than 10 different genes), each expanded

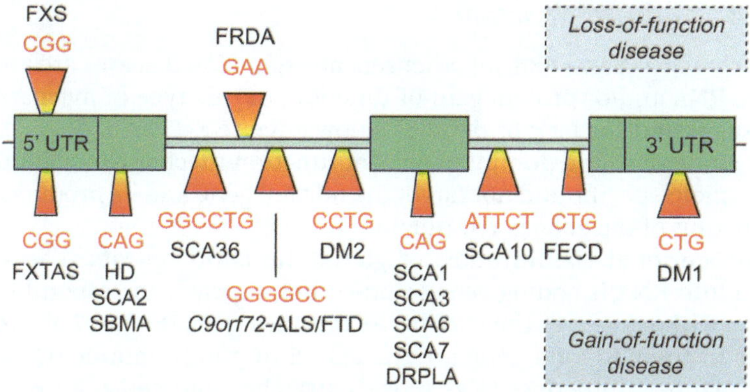

Figure 7.1 Repeat sequences related to human diseases and their locations in the gene. Only repeat-expansion diseases for which therapeutic strategies have been published are shown. Eight different expanded repeat sequences (in red) in 17 diseases (see list of abbreviations) have been targeted by therapeutic compounds at the DNA or RNA level. The unstable repeats can be located across the gene, in exons [dark green; 5′ untranslated region (UTR), open reading frame (ORF) or 3′ UTR] or in introns (light green). Depending on their location and length, repeats may cause disease *via* loss-of-function (top) or gain-of-function mechanisms (bottom).

gene elicits a distinct molecular mechanism in the cell types in which it is expressed. The expanded gene thus gives rise to a unique phenotype in patients, even though several symptoms are shared within the disease group (*e.g.* the SCAs). We do not intend to give a comprehensive overview of all disease mechanisms related to repeat instability or downstream cellular effects thereof, but refer to reviews for more information on these topics.[11–17] Here we will discuss common mechanistic concepts in disease mechanisms initiated by expanded repeats.

7.2.2.1 Loss-Of-Function

Of the repeat diseases discussed here, only two are related to a loss-of-function mechanism: FRDA and FXS (Figure 7.1). FRDA is an autosomal recessive disease caused by an unstable GAA repeat in the first intron of *FXN*. An expanded GAA repeat leads to transcriptional repression, resulting in reduced levels of *FXN* mRNA and frataxin protein. FXS is typically caused by >200 CGG triplets ("full mutation") in the 5′ untranslated region (UTR) of *FMR1* on the X-chromosome, which causes abnormal DNA methylation, histone deacetylation and silencing of the gene. Intriguingly, repeat lengths of between 55 and 200 triplets ("premutation") in *FMR1* result in fragile X tremor/ataxia syndrome (FXTAS). An *FMR1* allele carrying such a medium-sized CGG repeat is not silenced and causes an RNA gain-of-function.

7.2.2.2 Gain-Of-Function

Apart from FRDA and FXS, all other repeat-expansion diseases are associated with an RNA and/or protein gain-of-function, or this type of mechanism has been suggested for lack of detailed knowledge (*e.g.* SCA10, SCA36). Therapeutic strategies for a dominant gain-of-function mechanism should preferably be allele-specific and not target the normal gene and its products, unless the function of the gene is redundant.

A conventional classification of gain-of-function repeats that are transcribed into RNA is coding *versus* non-coding repeats, corresponding to the location with respect to the ORF in the mutated gene (Figure 7.1). Expanded repeats in the ORF of a gene are translated to polyglutamine (CAG repeat) or polyalanine (imperfect GCN repeat) stretches and cause a toxic protein gain-of-function, like in HD and most SCAs. Unstable repeats outside the ORF, a highly diverse group, are located in introns or UTRs and usually lead to a toxic RNA gain-of-function. In these latter cases the gained function generally consists of aberrant protein recruitment and nuclear retention of the expanded transcript, visualized by RNA fluorescence *in situ* hybridization (FISH) as so-called foci.

However, recent findings have shown that the division into coding *versus* non-coding repeats may actually not be so easy to make. In fact, it is most likely that several repeat-expansion diseases are caused by a combination of complex pathogenic mechanisms directed by abnormal RNA as well as toxic protein products. First, for the group of coding CAG repeats, it has been shown that CAG-expanded transcripts in SCA3 and HD not only encode abnormal polyQ proteins, but also initiate RNA gain-of-function effects.[18,19] Secondly, many, if not all, expanded repeats in the genome appear to be bi-directionally transcribed, which produces antisense transcripts along with the well-known sense transcripts. Antisense transcripts are generally low in copy number and bear the complementary repeat sequence.[20] The structure and function of the antisense genes is still largely unknown. Thirdly, both sense and antisense repeat transcripts may produce toxic polymeric proteins *via* a mechanism termed repeat-associated non-AUG (RAN) translation, regardless of the repeat location. RAN translation products have indeed been reported in *C9orf72*-ALS/FTD, DM, HD and FXTAS.[21] In conclusion, for all gain-of-function diseases, the therapeutic goal is to prevent transcription or to neutralize transcripts, since these may be toxic and also serve as templates for the production of toxic proteins.

7.3 Why are Expanded Repeats so Special as Therapeutic Targets?

Expanded repeats are not only special in terms of their primary sequences, but they are also prone to forming higher-order structures in DNA and RNA. Aberrant structures can form in RNA and in single-stranded genomic DNA during replication, DNA repair, transcription and other processes that involve separation of the complementary strands. The driving force behind this is the high GC-content and the internal complementarity in many repeat sequences. The occurrence of abnormal structures in expanded genes and transcripts – most notably R-loops, triple helices, hairpins and G-quadruplexes (G4s) (Figure 7.2) – strongly relates to mechanisms of repeat instability as well as to pathological mechanisms of disease[1,2] and also has important implications for toxic repeats as therapeutic targets.

7.3.1 Structures Formed by Expanded Repeats

7.3.1.1 R-Loop

R-loops occur in many repeat-expansion disorders and may arise during transcription, when the nascent RNA hybridizes to its template DNA strand (Figure 7.2A). The RNA–DNA duplex leaves the non-template strand unbound and free to form secondary structures. R-loops have physiological functions, but when formed by repeats, deleterious effects, such as epigenetic remodelling and repression of expression, can arise.[22]

7.3.1.2 Triple Helix

Next to R-loops, GAA·TTC repeats in FRDA have been found to form also unusual triple helical DNA structures. Triple helix structures are composed of three DNA strands, in which one strand folds back into the major groove of a double helix in its flanking region, leaving its partner strand unbound (Figure 7.2A). A triple helix is highly stable and is thought to contribute to FRDA's disease aetiology by inhibiting transcription.[23]

7.3.1.3 Imperfect Hairpin

In single-stranded DNA and RNA, most repeat expansions can form a stem-loop structure referred to as an imperfect hairpin, or simply hairpin (Figure 7.2).[24] The characteristics of such hairpins are most extensively studied in RNA *in vitro*. The CNG sequence of many disease-causing repeats provides two G–C base pairs per triplet when forming a CNG·CNG helical stem, resulting in structures varying in stability depending on the mismatched nucleotide. Using $(CNG)_{20}$ model RNAs, the following order of hairpin stability has been determined: CCG < CUG < CAG < CGG.[25,26] This can be explained by C·C, U·U and A·A mismatches forming an interaction with one hydrogen bond, while a G·G pair is thought to form two.[24] These types of hairpin occur in small independent clusters on each strand in the genomic DNA of the CTG·CAG repeat in dystrophia myotonica protein kinase (*DMPK*) in DM1 (Figure 7.2A).[27] In addition, next to hairpins CGG and CCG repeats are also thought to form G4 structures (see below).

The DM2-linked CCUG repeat also forms hairpins, but these are likely to be less stable than equally long CUG hairpins because of the CU·CU double mismatch.[28] The hairpin formed by the SCA10 AUUCU repeat is relatively unstable as it contains two A·U base pairs interrupted by UCU·UCU mismatches.[29]

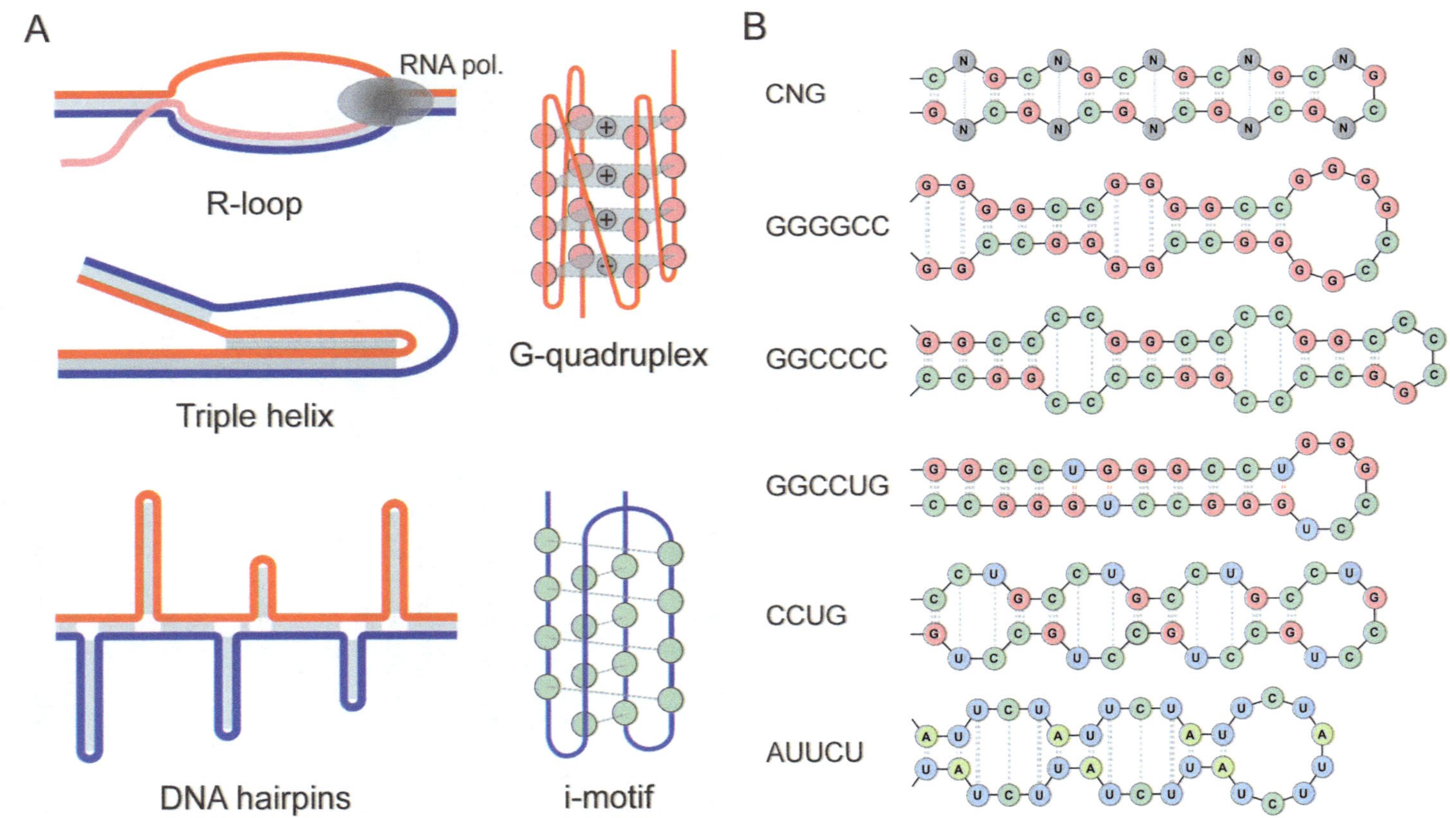

Figure 7.2 Implications of repeat expansion for DNA and RNA structure. (A) Structures that can be formed in the genome during strand separation and examples of quadruplexes. Complementary DNA strands are in dark red and blue; the nascent transcript in light red. For simplicity only a parallel, unimolecular G-quadruplex (G4) consisting of four planar G-quadruplets is shown, although several variations are possible. An i-motif consisting of two times four C–C interactions potentially formed by a GGCCCC repeat is shown as an example. (B) Hairpins that can be formed by expanded repeats in RNA. Dashed lines indicate hydrogen bonds and are red for the G–U wobble pairs. Dotted lines in the CNG hairpin indicate that this interaction depends on the N nucleotide. For illustration purposes only one potential loop configuration is shown; this can vary depending on the flanking region, repeat type and length. Structures in (B) were generated using Forna[140] and edited.

Repeat interruptions have implications for the folding of expanded RNA and possibly also DNA. Interruptions in FXTAS,[30] SCA1[31] and SCA2[32] transcripts appear to have a protective function in the healthy allele by creating branched hairpin structures, whilst in DM1 they stabilize the hairpin.[10] Thus one expects that targeting the repeat interruptions is not a feasible approach in these diseases.

7.3.1.4 *Quadruplexes*

C9orf72-ALS/FTD GGGGCC expansions can form highly stable intramolecular and intermolecular G4s (Figure 7.2A).[33–35] A G4 is a cation-stabilized structure characterized by the canonical motif $G_{3+}N_{1-7}G_{3+}N_{1-7}G_{3+}N_{1-7}G_{3+}N_{1-7}$, but variations are possible. In cells, G4s are found in RNA as well as in single-stranded DNA and form *via* Hoogsteen interactions between four guanine bases in a plane. Naturally occurring G4s are involved in cellular processes, such as telomere function and RNA processing, localization and translation.[36] Interestingly, GGGGCC as well as CGG repeat model molecules have been reported to fold as a G4 in equilibrium with a hairpin.[35,37–39] G4 folding by these repeats has been supported by immunofluorescence experiments in *C9orf72*-ALS/FTD cells[40] and by interaction of CGG repeats with G4-binding ligands and proteins.[41,42] Simulation studies have shown that GGCCUG RNA in SCA36 can also form G4 structures,[43] but this most likely competes with a highly stable hairpin that can be formed by purely G·C and U·G (wobble) base pairs. G4 motifs generally do not in fact assume a G4 structure in cells,[44] but the natural occurrence of a few adjacent G4 motifs is different from the dozens of successive motifs in a long repeat.

The GGCCCC repeat in the *C9orf72*-ALS/FTD antisense DNA strand is able to form intercalated motifs (i-motifs):[39] structures reminiscent of G4s but formed by two intercalating C-rich parallel duplexes in antiparallel fashion (Figure 7.2A).[45] In addition, the GGCCCC repeat in the antisense DNA strand has been shown to form intramolecular quadruplexes that consist of G and C bases, unlike its sense counterpart.[34,46] Similar types of quadruplexes and i-motifs have also been suggested for *FMR1* DNA antisense CCG repeats.[47] Whether such structures actually form in patients' cells remains to be determined.

7.3.2 Structures of Repeat-Expanded Transcripts

In the living cell, transcripts are actively unfolded, leading to differences between structures found *in vitro* and in a physiological environment.[48] It should be noted, however, that the influence of the cellular environment on repeat structure has only been minimally addressed. In addition to the cellular environment, flanking regions in the natural transcript, which are usually not present in the model molecules described above, can be important also.

The CAG repeats in androgen receptor (*AR*; SMBA),[49] ataxin 3 (*ATXN3*; SCA3)[50] and huntingtin (*HTT*; HD)[49,51] transcripts and the CGG repeat in *FMR1* (FXS, FXTAS)[30] transcripts are able to interact with these flanking regions, which base pair more efficiently than the repeat itself in a hairpin configuration. For *HTT* transcripts this means that a healthy repeat does not form a hairpin. Notably, this might explain the specificity for the pathogenic transcripts when the repeat is targeted in HD and SCA3 cell models by antisense oligonucleotides (ASOs).[52] For the CUG repeat in *DMPK* transcripts (DM1),[53] and the CAG repeats in *ATXN2* (SCA2),[32] calcium voltage-gated channel subunit alpha1 A (*CACNA1A*; SCA6)[50] and atrophin 1 (*ATN1*; DRPLA)[50] transcripts, the repeat appears to base pair only with itself. However, except for the *HTT* transcript, only short flanking regions of a few dozen to 74 nucleotides were studied for transcripts containing pathogenic repeats, limiting these conclusions to close-range interactions. In contrast, we found that the CUG repeat does interact with other regions in *DMPK* transcripts, which stresses the importance of examining long–range interactions as well (R. T. P. van Cruchten, B. Wieringa and D. G. Wansink, submitted).

The structures of antisense repeat transcripts have only been studied specifically for *C9orf72*-ALS/FTD. The GGCCCC repeat RNA does not fold as a G4 but as a hairpin, reminiscent of the sense transcript, with only CC·CC mismatches (Figure 7.2B).[34,54] For a hint on folding of other antisense repeats, knowledge about the HD CAG repeat structure, for example, can be extrapolated to the DM1 antisense repeat and *vice versa*, keeping in mind that flanking regions certainly may play a role.

7.3.3 Structural Implications for Therapeutic Targeting

For therapeutic compounds that rely on base pairing, the accessibility of the target sequence can influence the efficacy of the therapy. Strand displacement capacity of ASOs can compensate for this, but stable secondary structures are generally more difficult to engage than more available ones. So as to determine whether to target the body of the gene or transcript, the structure of the repeat, its accessibility and that of other unique parts of the transcript should be taken into account. This can be tested by screening[55] or by computational methods.[56] For structure disruption or splice modulation, there is often little choice in what region to target, but for RNA breakdown and blocking of transcription or translation there is usually more freedom. For example, imperfectly paired hairpins are, in principle, easy targets for perfectly matched ASOs. A G4 sequence, on the other hand, may be more difficult to disrupt and bind, since this structure competes with base pairing. Sequences capable of forming a G4 may therefore be more attractive for targeting with small molecules. For gene-targeting ASOs, disruption of DNA structure may be the goal and structural information may provide clues as to which strand is most likely to be vulnerable to ASO binding.

7.4 Therapeutics to Target Expanded Repeats

7.4.1 Therapeutic Strategies

The first publications on therapeutic intervention in repeat-expansion diseases originate from the late 1990s. However, the last decade has seen a major increase in the number of studies related to the targeting of toxic repeats (Figure 7.3). Depending on the loss- or gain-of-function effect caused by the repeat under study, the strategic principle may be activating or inhibitory and the molecular target can be the repeat in the gene (*i.e.* intervention at the DNA level), the expanded transcript (*i.e.* RNA level) or both. An overview of the main strategic principles along with their targets is given in Figure 7.4. Toxic translation products, such as the polyQ proteins or RAN translation polypeptides, might be targeted too, but these types of approach are beyond the scope of this chapter.

We now discuss three classes of therapeutics against nucleic acids: small molecules, ASOs and protein effectors. The small molecules mainly comprise compounds that stabilize or disrupt repeat structures and thus inhibit protein binding. The other two classes of therapeutics comprise (i) blocking-type compounds that modulate transcription, inhibit repeat expansion, or block protein interactions, and (ii) compounds with enzymatic or enzyme-recruiting activity in order to excise, degrade or repair a gene or its products. Additional types of therapeutics, not further covered here, are viruses and artificial mRNAs in the classical gene-replacement approaches designed to compensate for loss of gene activity (*e.g.* for FRDA and FXS).

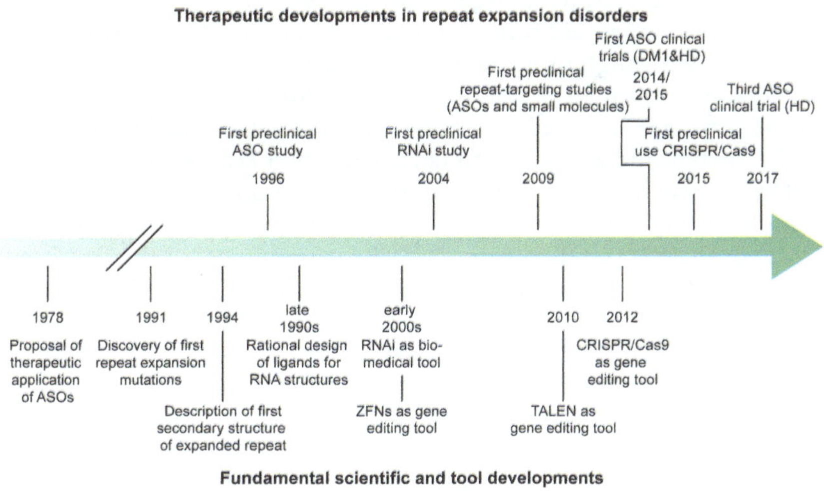

Figure 7.3 Overview of the key therapeutic developments in repeat expansion disorders (top) and their preceding fundamental scientific and tool developments (bottom).

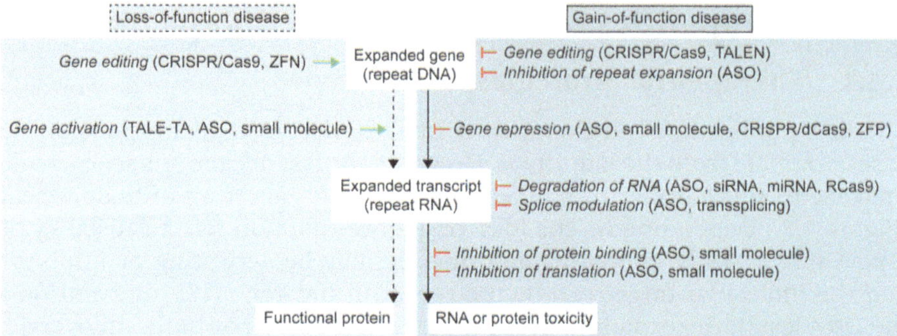

Figure 7.4 Therapeutic strategies for repeat-expansion diseases. The scheme shows the main reported therapeutic strategies and active compounds to target repeat DNA and RNA in loss-of-function (left) and gain-of-function repeat expansion diseases (right). TALE-TA = transcription activator-like effector-transcription activator; ZFP = zinc finger protein.

7.4.2 Small Molecules Targeting Repeat Structures

The principle of targeting repeats with small molecules is based on the recognition and modulation of higher-order structures, such as hairpins and G4s. Small molecules that selectively bind repetitive DNA or RNA have been identified through trial and error, but mostly through medium- or high-throughput screening or rational design combined with the use of DNA- and RNA-focused chemical libraries. Small molecules may bind a repeat only once or many times, depending on their binding characteristics and the repetitive nature of the target. Also larger modular compounds have been identified, basically consisting of multiple repeated small molecules, which bind with high affinity through multivalent interactions.

The largest number of studies designed for the targeting of higher-order repeat structures by small molecules has been published for DM, especially by the Disney and Zimmerman labs. Their first studies originate from 2009 and by now each lab has published more than a dozen papers on a diverse set of high-affinity small, multimeric and modular compounds that bind CUG (DM1) or CCUG (DM2) hairpins.[57,58] The general goal of the development of these potential therapeutics was the inhibition of muscleblind-like protein 1 (MBNL1) binding to the repeats, a hallmark of DM pathogenesis, followed by the dispersal of MBNL1 nuclear aggregates (so-called foci in immunofluorescence microscopy) and the restoration of aberrant DM-typical alternative splicing. Important determinants of biological activity of CUG/CCUG-selective small molecules *in vivo* in animal models appear to be solubility in water, cell permeability and cytotoxicity.[59,60] More recent small-molecule approaches include targeting of the CTG expansion in *DMPK* in order to interfere with transcription[61,62] or to induce CUG RNA cleavage.[61] Well-known DNA-binding molecules pentamidine and actinomycin D also preferentially inhibit transcription of the expanded *DMPK* gene,[63,64] but the high cytotoxicity of these compounds remains a concern.

For FXTAS, in addition to compounds that bind expanded CGG RNA and reduce its gain-of-function,[65,66] the Disney group designed a molecule that reacts covalently with the repeat.[67] As a result, RAN translation was inhibited, but not downstream canonical translation. In another study, the activity of a modularly assembled small molecule was compared with that of repeat-targeting ASOs.[68] Both types of molecules inhibit RAN translation as intended, however the ASO also inhibits canonical translation.

The GGGGCC repeat in *C9orf72*-ALS/FTD can be bound by small aromatic molecules.[37,69] These compounds reduce the number of repeat RNA foci and the expression of RAN translation products in repeat-expressing cells[37,69] and improve survival in a *Drosophila* model.[69] To what extent these compounds also interfere with the *C9orf72*-ALS/FTD repeat in the gene is not yet clear. Another study indicated that the porphyrin tetra-(*N*-methyl-4-pyridyl)porphyrin (TMPyP4) binds $(GGGGCC)_8$ RNA *in vitro*, disrupts its structure and blocks protein interactions, but this compound was not further tested in cells for functional activity.[70]

Only a few reports on small-molecule targeting exist for transcripts with CAG repeats, such as in HD and SCAs. The polyphenolic compound myricetin was identified in a screen for small molecules to bind an expanded CAG hairpin *via* base stacking at the A·A mismatch.[71] A positive effect of myricetin on cell viability of polyQ-expressing cells was demonstrated, but whether this effect relates to the repeat interaction or is due to myricetin's cytoprotective activity[72] is not clear. Similar to the studies discussed above for DM1, modular Hoechst 33 258-derived compounds can inhibit formation of abnormal MBNL1-CAG repeat RNA complexes.[73] Zhang *et al.* reported on peptides that directly interact with expanded CAG RNA in a repeat-length-dependent manner.[74,75] Although these peptides with a molecular weight of >1.5 kDa may not qualify as small molecules, they suppressed repeat toxicity *in vitro* and *in vivo* in a *Drosophila* SCA model. Finally, for SCA10, a bioactive, dimeric bis-benzamidine compound was identified that binds A·U base pairs in the hairpin structures formed by AUUCU-expanded transcripts, which constitute the toxic entity in this RNA-mediated disease.[76]

Finally, epigenetic therapy for the reactivation of expanded, silenced *FMR1* and *FXN* has been examined in cells of FXS and FRDA patients.[12,77–79] Positive effects were reported, but this type of therapy remains an indirect approach, not directly targeted at the expanded triplet repeats, and the risk of genome-wide effects needs to be carefully investigated. Also, a potentially toxic side effect in the case of FXS would be the expression of long CGG-expanded *FMR1* transcripts, which may be subject to RAN translation. A more direct approach for reversal of *FXN* inactivation, was demonstrated by the Gottesfeld lab, who used β-alanine-linked pyrrole–imidazole polyamides, which specifically bind the GAA·TTC repeat, thereby altering the DNA structure or opening the chromatin for transcription.[80] A similar type of polyamide coupled to a small molecule that recruits transcription elongation machinery acts as a gene-specific stimulator and promotes transcription across the expanded GAA repeat in *FXN*.[81]

7.4.3 Antisense Strategies

In contrast to the approaches with small molecules, antisense strategies are based on base-pairing interactions at the nucleotide level. Hence, these methods are very specific to the sequence that is targeted. ASOs are short stretches of DNA or RNA, or chemically modified versions thereof,[82] which have a sequence that is usually fully complementary to their nucleotide target. The development and function of nucleotide analogues is described in detail in Chapter 1, while their mechanisms of action are covered in Chapters 2 and 3.

In short, ASOs function either by activating degradation through the recruitment of RNase H, or through steric hindrance and blocking of protein interaction. RNase H-recruiting ASOs bind to their target and form a DNA–RNA hybrid, which is recognized by RNase H, leading to cleavage of the RNA strand in the hybrid. Steric blocking-type ASOs can be directed against DNA to block transcription or interfere with higher-order structures, or to RNA to displace RNA binding proteins and thus relieve protein sequestration, block canonical or RAN translation or direct alternative splicing. Degradation of target RNA has also been induced by a few blocking-type ASOs, but the cellular mechanisms through which this occurs have not been elucidated yet.

7.4.3.1 ASOs Directed Against DNA

By targeting DNA, ASOs can inhibit transcription and thus prevent formation of toxic RNA and protein products from expanded repeat-containing genes. Zaghloul *et al.* showed that gene-targeting ASOs directed against the repeat can block transcription in HD fibroblasts.[83] The authors used locked nucleic acid (LNA)–DNA mixmers (see Chapters 1 and 2), which are capable of strand invasion. Additionally, they showed that phosphorothioate (PS) linkages are necessary and that the minimum effective ASO length in their system is 12 nucleotides. The translatability of this approach to other repeat-expansion disorders might depend on the location of the repeat in the gene, which is exon 1 in the case of HD. Using the same approach against the CTG repeat in the 3′ UTR for DM1, we ourselves have not been able to affect *DMPK* expression (L. van der Bent, R. Brock and D. G. Wansink, unpublished data).

Conversely, gene-targeting ASOs can also be used to increase transcription, for example for FRDA, where the GAA repeat causes transcriptional silencing. Repeat-targeting ASOs prevented the formation of higher-order structures that inhibit gene expression, which led to increased *FXN* expression.[84,85] In contrast to the findings of Zaghloul *et al.*, a PS backbone did not appear to be essential for these ASOs.

Finally, gene-targeting ASOs may also suppress repeat instability.[86] LNA-containing ASOs targeting the CTG repeat in DM1 reduced repeat expansion in cell culture and in a DM1 mouse model. As a possible mechanistic explanation, it was proposed that the ASOs reduce the formation of R-loops.

7.4.3.2 ASOs Directed Against RNA

Most ASO strategies are directed against RNA targets. In some disorders, such as HD, certain haplotypes co-segregate with the expanded repeat, such that single-nucleotide polymorphisms SNPs can be used to selectively target the expanded allele and its RNA products in a large proportion of the patient population.[87] The prevalence of certain haplotypes depends on the patient population, and SNP-targeting ASOs are not a treatment option for every patient. The first study that reported the use of ASOs that target SNPs dates from 2011.[88] In this study, the authors showed the feasibility of this approach in HD patient fibroblasts, in mouse neurons and in two transgenic HD mouse models. Since then, many studies have reported SNP-targeting ASO approaches, and WAVE Life Sciences is currently recruiting for clinical trials with SNP-targeting ASOs against HD (http://www.clinicaltrials.gov; March 2018).

Another way to increase specificity for the expanded allele is by targeting the repeat itself. A long repeat contains many more binding sites for a repeat ASO than a healthy repeat, and in many cases expanded transcripts are retained in the nucleus, to which PS-containing ASOs preferentially localize.[89] Steric-blocking ASOs can cause dissolution of structured RNAs and release of RNA-binding proteins, which is thought to be one of the major drivers of disease in RNA gain-of-function disorders, such as DM1. Ever since the first repeat-targeting ASO strategies with steric blocking ASOs were reported in 2009 for DM1, HD and SCA3,[52,90,91] a variety of different chemistries have been applied. RNase H-recruiting ASOs directed against repeats have also been described, for example for DM1 and for *C9orf72*-ALS/FTD.[92,93] Some reports related to *C9orf72*-ALS/FTD, HD and SCA3 describe ASOs that partially overlap the repeat, which reduces the risk of off-target effects on other repeat-containing transcripts.[52,93–95]

Exon skipping is sometimes an option to prevent protein toxicity, while maintaining at least some functionality in the protein product. This approach was described for example for SCA3, where the mutated ATXN3 protein is proteolytically cleaved, which results in toxic polyQ-containing fragments. Skipping of exon 10 in *ATXN3* transcripts leads to the production of a truncated protein that still contains functional domains, but no polyQ stretch.[95,96] Alternatively, skipping of exons 8 and 9 leads to the production of a truncated protein with improved stability due to the absence of certain proteolytic cleavage sites. This leads to a reduction of toxic polyQ-containing fragments.[97] Similarly, for HD, a partial skip of exon 12 leads to the removal of caspase-3 and -6 motifs, which results in improved stability of the protein and reduced formation of the toxic N-terminal protein fragments.[98]

Although many groups try to achieve allele-specificity, there have been studies that indicate that, at least in HD and DM1, downregulation of the healthy allele does not lead to safety issues in mice and nonhuman primates.[99–102] Since it had previously been described that knock-out of *Dmpk* in a DM1 mouse model led to deleterious effects and that inactivation of wild-type *Htt* in the mouse forebrain led to neurodegeneration,[103,104] this is a point of controversy in the field. If downregulation of functional protein derived from the healthy allele is not expected to cause problems, ASOs may also be directed against other parts of the transcript in a non-allele-specific fashion. IONIS Pharmaceuticals initiated clinical trials using such ASOs for DM1 and for HD (http://www.clinicaltrials.gov; March 2018). The former trial has been concluded and this ASO has not been pushed forward to later phase clinical trials, because the intracellular ASO concentrations that were reached in the target tissues were too low. The HD clinical trial is still ongoing. In this trial, the ASO is administered intrathecally, which can be expected to lead to better delivery, similar to Nusinersen (Spinraza), the now approved IONIS ASO for spinal muscular atrophy.[105]

7.4.3.3 RNA Interference

RNA interference leads to degradation of the transcript and can be induced in various ways: direct use of short interfering RNAs (siRNAs), through viral transduction with short hairpin RNAs (shRNAs) or artificial microRNAs (miRNAs), or by using antisense transcripts that are complementary to the target transcript. Only a few studies have made use of antisense gene fragments to induce RNAi. shRNAs have been described as being more toxic compared with artificial miRNAs[106] and a shift towards the use of miRNA-like molecules seems apparent from the articles that have been published over the last few years.

As for ASO strategies, RNAi strategies targeting both repeat and non-repeat sequences have been described. The first reports of use of RNAi for repeat-expansion disorders date from 2004 and targeted non-repeat sequences in SCA1 and SCA3. In the SCA1 study, the authors showed the efficacy of AAV-delivered shRNA in cell culture and in SCA1 mice.[107] The study on SCA3 evaluated the use of siRNAs that target a SNP that immediately flanks the CAG repeat.[108] Since then, a large number of RNAi studies have been published for a range of repeat expansion disorders. Most of these studies target non-repeat sequences, some in an allele-specific manner.

Expanded repeats are notoriously difficult to target using RNAi. An extreme case is the hexanucleotide repeat in *C9orf72*-ALS/FTD, because the target is intronic, retained in the nucleus and has 100% GC-content. Nonetheless, Hu *et al.* showed that it can be done using both duplex siRNAs and single stranded siRNAs (ss-siRNAs).[109] Duplex siRNAs were only effective when more than two mismatches were introduced, whereas ss-siRNAs also functioned when fully complementary to the repeat. This is likely to be due to the fact that ss-siRNAs do not have to dissociate to be loaded into the RNA-induced silencing complex (RISC). The duplex siRNAs target both sense and antisense strands, both of which are implicated in the molecular pathology, but less efficiently than ss-siRNA.[109,110] Targeting (parts of) CAG and CUG repeats also appears possible. This strategy was first published in 2011 and was carried out in HD and SCA3 fibroblasts.[111] The authors showed that fully complementary siRNAs against the CAG repeat also silenced transcripts other than *HTT* and *ATXN3*. By introduction of two mismatches and an asymmetrical design, a higher degree of gene- and allele-selectivity was achieved.

Notwithstanding the large body of preclinical literature on RNAi for repeat-expansion disorders, to our knowledge, clinical trials with RNAi agents for repeat-expansion disorders have not been initiated yet.

7.4.3.4 *Other Antisense Strategies*

A few additional antisense strategies have been reported in the context of repeat-expansion disorders, including the use of ribozymes, DNA enzymes and *trans*-splicing, which can be induced by providing a template for a splice switch. A couple of studies around the year 2000 used DNA enzymes or ribozymes to excise the repeat in HD and DM1.[112–115] To the best of our knowledge, there have been no follow-up studies in this area. Artificial *trans*-splicing has been reported as a strategy for HD and DM1.[116–118] The paucity of reports using these alternative strategies indicates that they are either not developed far enough, or not efficient enough to be translated into more advanced preclinical settings.

7.4.4 CRISPR/Cas9, TALEN, ZFN and Other Protein Effectors

The development of genome editing technology using zinc finger nucleases (ZFNs), transcription activator-like effector nucleases (TALENs) and the clustered regularly interspaced short palindromic repeats–CRISPR-associated protein (CRISPR–Cas) system has revolutionized the possibilities of targeted gene editing, correction and interference.[119] Cleavage by a single CRISPR–Cas9 complex close to a CGG·CCG or CTG·CAG expansion induced spontaneous repeat contraction.[120,121] Complete excision of the 5′ UTR CGG repeat in *FMR1*,[122] the intronic GAA repeat in *FXN*[123,124] and the 3′ UTR CTG repeat in *DMPK*[120,125] was achieved through dual ZFN or CRISPR–Cas9 cleavage on both sides of the repeat in a more predictable fashion. In essentially all these reports, removal of the expanded repeat resulted in reversal of the cellular phenotype in corrected patient cells. As mentioned above, it is unclear if and when complete loss of the normal protein products would be tolerated (*e.g.* during development), indicating the importance of allele-specific silencing.[126] Allele-specific editing of CAG-expanded *HTT* was based on SNPs within protospacer adjacent motif (PAM) sequences, resulting in specific removal of a large 5′ part of the mutated gene, thereby silencing the formation of toxic gene products only.[127,128] A correction of the expanded CAG repeat through homologous recombination in SCA2 induced pluripotent stem cell (iPSC) lines was reported by Marthaler *et al.*[129]

A second set of strategies was designed to reduce the actual transcription of the repeat segment, or activate it in the case of loss-of-function diseases. Xia and colleagues used TALENs to insert a premature polyadenylation site in intron 9 of *DMPK*, in order to induce early termination of transcription, 5′ of the repeat.[130,131] As a result, DM1-typical biomarkers were normalized, but the effect of the reduced protein expression is still uncertain as mentioned above. Catalytically inactive, endonuclease Cas9 (dCas9) protein that was directly targeted to expanded repeats reduced disease markers in DM1, DM2 and *C9orf72*-ALS/FTD cells, presumably by perturbing transcription elongation.[132] In a similar type of study, rationally designed zinc finger proteins recognized the CAG repeat in HD cells and repressed *HTT* transcription by steric hindrance.[133] The fact that long expanded repeats are indeed difficult to transcribe was also illustrated by the inhibitory effect of elongation factor suppressor of Ty 4 homolog 1 (*SUPT4H*) silencing on the expression of repeat-containing transcripts.[134,135] In addition, activation of transcription was shown by a TALE-VP64 transcriptional activator domain targeted to the *FXN* promoter, which caused a two-fold increase of mRNA and frataxin protein level.[136] A hypermethylated *FMR1* gene was reactivated by a dCas9–ten-eleven translocation 1 (Tet1) demethylase fusion protein targeted to the expanded CGG repeat, which rescued the FXS phenotype in neurons.[137]

Finally, repeat RNA can also be directly targeted and degraded by enzymatic complexes. RCas9, a dCas9 fusion to the suppressor with morphological effect on genitalia homolog 6 (*SMG6*) protease inhibitor I2 N-terminal (PIN) RNA endonuclease domain, was able to degrade CUG, CCUG and GGGGCC repeat-containing RNAs.[138] An artificial site-specific RNA endonuclease, a fusion of a Pumilio RNA binding family member 1 (*PUM1*) Pumilio and feminization 3 protein binding factor (PUF) domain and the *SMG6* PIN domain was designed to recognize and degrade CUG repeat RNA in DM1.[139]

7.5 Challenges for Multisystemic Repeat Diseases

The field of therapeutics for repeat-expansion diseases has made some impressive advances, and many new exciting tools have expanded the range of possibilities. At present, we believe that there are three main issues that slow down further clinical development of these drugs: targeting, delivery and specificity.

As we have described above, some clinical trials have already been completed or are in progress. Especially with the success of Nusinersen in mind, it would appear that ASOs that are injected directly into the CNS by intrathecal administration have a high chance of succeeding in clinical use. This is a promising prospect for those diseases that are limited to the CNS, such as HD. However, many repeat-expansion disorders affect multiple tissues, some of which are difficult to target systemically. One example is DM1, which not only affects the CNS, but also skeletal muscles, the heart and multiple other organs. Systemic administration, for example by intravenous or subcutaneous injection, often leads to a less favourable pharmacokinetic profile due to clearance by liver and kidneys and can lead to off-target effects. Different targeting strategies are being applied to increase uptake by target tissues, such as use of liposomes and/or peptides. Viral delivery can also confer tissue specificity, due to the tropism of the virus, but here immune responses can limit the applicability.

At the cellular level, the efficiency of uptake of compounds by cells and subsequent endosomal release and trafficking to their site of action introduces another hurdle for delivery of therapeutic compounds. Large, hydrophilic molecules, such as RNA or proteins, are often not taken up efficiently by cells, unlike small hydrophobic molecules, which are generally more cell-permeable. Ideally, formulations would be such that they not only increase targeting to specific tissues, but also increase cellular uptake and release of the compounds.

Specificity of treatments is a third important consideration. The various approaches that we have described have different ways of achieving specificity. Small molecules, for example, might be specific for certain higher-order structures that are formed by expanded repeats. ASOs, RNAi agents and gene-editing strategies are specific for certain DNA or RNA sequences. Still, higher-order structures and target sequences might be shared with the healthy allele as well as with other transcripts, so allele- and gene-specific targeting remains a challenge.

In conclusion, these are exciting times for the development of therapeutics for toxic repeat-expansion diseases. Fundamental knowledge on the genetic and structural background of these diseases can now be used to apply novel, promising tools for alleviating the disease burden or to remove the cause of disease altogether. Although many preclinical questions remain to be answered, the first three clinical trials provide valuable insights for applying cutting edge technology from the last decade to its full potential.

Abbreviations

AR	Androgen receptor
ASO	Antisense oligonucleotide
ATN1	Atrophin 1
ATXN	Ataxin
CACNA1A	Calcium voltage-gated channel subunit alpha1 A
C9orf72-ALS/FTD	*C9orf72*-linked amyotrophic lateral sclerosis and frontotemporal dementia
(d)Cas9	(Catalytically inactive) CRISPR-associated protein
CNS	Central nervous system
CRISPR	Clustered regularly interspaced short palindromic repeats
DM1/2	Myotonic dystrophy type 1/2
DMPK	Dystrophia myotonica protein kinase
DRPLA	Dentato-rubro-pallido-luysian atrophy
FECD	Fuchs' endothelial corneal dystrophy
FMR1	Fragile X mental retardation 1
FRDA	Friedreich's ataxia
FXS	Fragile X syndrome
FXTAS	Fragile X tremor/ataxia syndrome
G4	G-quadruplex

HD	Huntington's disease
HTT	Huntingtin
i-motif	Intercalating motif
iPSC	Induced pluripotent stem cell
LNA	Locked nucleic acid
MBNL1	Muscleblind-like protein 1
miRNA	MicroRNA
OPMD	Oculopharyngeal muscular dystrophy
ORF	Open reading frame
PAM	Protospacer adjacent motif
PIN	Protease inhibitor I2 N-terminal
PUF	Pumilio and feminization 3 protein binding factor domain
PUM1	Pumilio RNA binding family member 1
RAN	Repeat associated non-AUG
SBMA	Spinal bulbar muscular atrophy
SCAn	Spinocerebellar ataxia type n
shRNA	Short hairpin RNA
siRNA	Small interfering RNA
SMG6	Suppressor with morphological effect on genitalia homolog 6
SNP	Single-nucleotide polymorphism
ss-siRNA	Single-stranded siRNA
SUPT4H	Suppressor of Ty 4 homolog 1
TALEN	Transcription activator-like effector nuclease
Tet1	Ten-eleven translocation 1
TMPyP4	Tetra-(*N*-methyl-4-pyridyl)porphyrin
UTR	Untranslated region
ZFN	Zinc finger nuclease

Acknowledgements

This work was supported by the Prinses Beatrix Spierfonds in combination with the Stichting Spieren voor Spieren under grants W.OR14-19 and W.OR16-09. We apologize to our colleagues whose work could not or could only partially be included due to size constraints. D. G. W. is inventor on a patent related to ASOs for treatment of DM1.

References

1. C. E. Pearson, K. N. Edamura and J. D. Cleary, *Nat. Rev. Genet.*, 2005, **6**, 729.
2. S. M. Mirkin, *Nature*, 2007, **447**, 932.
3. A. Ciesiolka, M. Jazurek, K. Drazkowska and W. J. Krzyzosiak, *Front. Cell. Neurosci.*, 2017, **11**, 97.
4. H. Kobayashi, K. Abe, T. Matsuura, Y. Ikeda, T. Hitomi, Y. Akechi and T. Habu, *et al.*, *Am. J. Hum. Genet.*, 2011, **89**, 121.

5. M. DeJesus-Hernandez, I. R. Mackenzie, B. F. Boeve, A. L. Boxer, M. Baker, N. J. Rutherford and A. M. Nicholson, *et al.*, *Neuron*, 2011, **72**, 245.
6. A. E. Renton, E. Majounie, A. Waite, J. Simón-Sánchez, S. Rollinson, J. R. Gibbs and J. C. Schymick, *et al.*, *Neuron*, 2011, **72**, 257.
7. E. D. Wieben, R. A. Aleff, N. Tosakulwong, M. L. Butz, W. E. Highsmith, A. O. Edwards and K. H. Baratz, *PLoS One*, 2012, **7**, e49083.
8. A. L. Castel, J. D. Cleary and C. E. Pearson, *Nat. Rev. Mol. Cell Biol.*, 2010, **11**, 165.
9. Z. Musova, R. Mazanec, A. Krepelova, E. Ehler, J. Vales, R. Jaklova and T. Prochazka, *et al.*, *Am. J. Med. Genet., Part A*, 2009, **149**, 1365.
10. C. Braida, R. K. A. Stefanatos, B. Adam, N. Mahajan, H. J. M. Smeets, F. Niel and C. Goizet, *et al.*, *Hum. Mol. Genet.*, 2010, **19**, 1399.
11. M. Boivin, R. Willemsen, R. K. Hukema and C. Sellier, *Eur. J. Med. Genet.*, 2018, **61**, 674.
12. E. Tabolacci and P. Chiurazzi, *Am. J. Med. Genet., Part A*, 2013, **161**, 2797.
13. C. Sandi, M. Sandi, S. A. Virmouni, S. Al-Mahdawi and M. A. Pook, *Front. Genet.*, 2014, **5**, 165.
14. M. Jimenez-Sanchez, F. Licitra, B. R. Underwood and D. C. Rubinsztein, *Cold Spring Harbor Perspect. Med.*, 2017, **7**, 1.
15. H. L. Paulson, V. G. Shakkottai, H. B. Clark and H. T. Orr, *Nat. Rev. Neurosci.*, 2017, **18**, 613.
16. G. Sicot, G. Gourdon and M. Gomes-Pereira, *Hum. Mol. Genet.*, 2011, **20**, 116.
17. T. G. Moens, L. Partridge and A. M. Isaacs, *Curr. Opin. Genet. Dev.*, 2017, **44**, 92.
18. L. B. Li, Z. Yu, X. Teng and N. M. Bonini, *Nature*, 2008, **453**, 1107.
19. M. Bañez-Coronel, S. Porta, B. Kagerbauer, E. Mateu-Huertas, L. Pantano, I. Ferrer and M. Guzmán, *et al.*, *PLoS Genet.*, 2012, **8**, e1002481.
20. R. Batra, K. Charizanis and M. S. Swanson, *Hum. Mol. Genet.*, 2010, **19**, R77.
21. J. D. Cleary and L. P. Ranum, *Curr. Opin. Genet. Dev.*, 2017, **44**, 125.
22. P. Richard and J. L. Manley, *J. Mol. Biol.*, 2017, **429**, 3168.
23. N. Sakamoto, P. D. Chastain, P. Parniewski, K. Ohshima, M. Pandolfo, J. D. Griffith and R. D. Wells, *Mol. Cell*, 1999, **3**, 465.
24. L. Błaszczyk, W. Rypniewski and A. Kiliszek, *Wiley Interdiscip. Rev.: RNA*, 2017, **8**, e1412.
25. M. Broda, E. Kierzek, Z. Gdaniec, T. Kulinski and R. Kierzek, *Biochemistry*, 2005, **44**, 10873.
26. K. Sobczak, G. Michlewski, M. De Mezer, E. Kierzek, J. Krol, M. Olejniczak and R. Kierzek, *et al.*, *J. Biol. Chem.*, 2010, **285**, 12755.
27. M. M. Axford, Y. H. Wang, M. Nakamori, M. Zannis-Hadjopoulos, C. A. Thornton and C. E. Pearson, *PLoS Genet.*, 2013, **9**, e1003866.
28. K. Sobczak, M. de Mezer, G. Michlewski, J. Krol and W. J. Krzyzosiak, *Nucleic Acids Res.*, 2003, **31**, 5469.

29. V. Handa, H. J. C. Yeh, P. McPhie and K. Usdin, *J. Biol. Chem.*, 2005, **280**, 29340.
30. M. Napierala, D. Michalowski, M. de Mezer and W. J. Krzyzosiak, *Nucleic Acids Res.*, 2005, **33**, 451.
31. K. Sobczak and W. J. Krzyzosiak, *J. Biol. Chem.*, 2004, **279**, 41563.
32. K. Sobczak and W. J. Krzyzosiak, *J. Biol. Chem.*, 2005, **280**, 3898.
33. P. Fratta, S. Mizielinska, A. J. Nicoll, M. Zloh, E. M. C. Fisher, G. Parkinson and A. M. Isaacs, *Sci. Rep.*, 2012, **2**, 1016.
34. K. Reddy, B. Zamiri, S. Y. R. Stanley, R. B. Macgregor and C. E. Pearson, *J. Biol. Chem.*, 2013, **288**, 9860.
35. A. R. Haeusler, C. J. Donnelly, G. Periz, E. A. J. Simko, P. G. Shaw, M. S. Kim and N. J. Maragakis, *et al.*, *Nature*, 2014, **507**, 195.
36. A. Cammas and S. Millevoi, *Nucleic Acids Res.*, 2017, **45**, 1584.
37. Z. Su, Y. Zhang, T. F. Gendron, P. O. Bauer, J. Chew, W. Y. Yang and E. Fostvedt, *et al.*, *Neuron*, 2014, **83**, 1043.
38. M. Malgowska, D. Gudanis, R. Kierzek, E. Wyszko, V. Gabelica and Z. Gdaniec, *Nucleic Acids Res.*, 2014, **42**, 10196.
39. A. Kovanda, M. Zalar, P. Šket, J. Plavec and B. Rogelj, *Sci. Rep.*, 2015, **5**, 17944.
40. E. G. Conlon, L. Lu, A. Sharma, T. Yamazaki, T. Tang, N. A. Shneider and J. L. Manley, *eLife*, 2016, **5**, e17820.
41. N. Ofer, P. Weisman-Shomer, J. Shklover and M. Fry, *Nucleic Acids Res.*, 2009, **37**, 2712.
42. P. Weisman-Shomer, E. Cohen, I. Hershco, S. Khateb, O. Wolfovitz-Barchad, L. H. Hurley and M. Fry, *Nucleic Acids Res.*, 2003, **31**, 3963.
43. Y. Zhang, C. Roland and C. Sagui, *ACS Chem. Neurosci.*, 2018, **9**, 1104.
44. J. U. Guo and D. P. Bartel, *Science*, 2016, **353**, aaf5371.
45. K. Gehring, J.-L. Leroy and M. Guéron, *Nature*, 1993, **363**, 561.
46. B. Zamiri, M. Mirceta, K. Bomsztyk, R. B. Macgregor and C. E. Pearson, *Nucleic Acids Res.*, 2015, **43**, 10055.
47. P. Fojtik, *Nucleic Acids Res.*, 2001, **29**, 4684.
48. S. Rouskin, M. Zubradt, S. Washietl, M. Kellis and J. S. Weissman, *Nature*, 2014, **505**, 701.
49. M. De Mezer, M. Wojciechowska, M. Napierala, K. Sobczak and W. J. Krzyzosiak, *Nucleic Acids Res.*, 2011, **39**, 3852.
50. G. Michlewski and W. J. Krzyzosiak, *J. Mol. Biol.*, 2004, **340**, 665.
51. S. Busan and K. M. Weeks, *Biochemistry*, 2013, **52**, 8219.
52. J. Hu, M. Matsui, K. T. Gagnon, J. C. Schwartz, S. Gabillet, K. Arar and J. Wu, *et al*, *Nat. Biotechnol.*, 2009, **27**, 478.
53. M. Napierała and W. J. Krzyzosiak, *J. Biol. Chem.*, 1997, **272**, 31079.
54. D. W. Dodd, D. R. Tomchick, D. R. Corey and K. T. Gagnon, *Biochemistry*, 2016, **55**, 1283.
55. S. P. Ho, Y. Bao, T. Lesher, R. Malhotra, L. Y. Ma, S. J. Fluharty and R. R. Sakai, *Nat. Biotechnol.*, 1998, **16**, 59.
56. Z. J. Lu and D. H. Mathews, *Nucleic Acids Res.*, 2008, **36**, W104.

57. M. D. Disney, *Drug Discovery Today*, 2013, **18**, 1228.
58. S. C. Zimmerman, *Beilstein J. Org. Chem.*, 2016, **12**, 125.
59. À. L. Gonzalez, P. Konieczny, B. Llamusi, E. Delgado-Pinar, J. I. Borrell, J. Teixidó and E. García-España, *et al.*, *PLoS One*, 2017, **12**, e0178931.
60. C. H. Wong, L. Nguyen, J. Peh, L. M. Luu, J. S. Sanchez, S. L. Richardson and T. Tuccinardi, *et al.*, *J. Am. Chem. Soc.*, 2014, **136**, 6355.
61. L. Nguyen, L. M. Luu, S. Peng, J. F. Serrano, H. Y. E. Chan and S. C. Zimmerman, *J. Am. Chem. Soc.*, 2015, **137**, 14180.
62. Y. Bai, L. Nguyen, Z. Song, S. Peng, J. Lee, N. Zheng and I. Kapoor, *et al.*, *J. Am. Chem. Soc.*, 2016, **138**, 9498.
63. L. A. Coonrod, M. Nakamori, W. Wang, S. Carrell, C. L. Hilton, M. J. Bodner and R. B. Siboni, *et al.*, *ACS Chem. Biol.*, 2013, **8**, 2528.
64. R. B. Siboni, M. Nakamori, S. D. Wagner, A. J. Struck, L. A. Coonrod, S. A. Harriott and D. M. Cass, *et al.*, *Cell Rep.*, 2015, **13**, 2386.
65. W. Y. Yang, F. He, R. L. Strack, S. Y. Oh, M. Frazer, S. R. Jaffrey and P. K. Todd, *et al.*, *ACS Chem. Biol.*, 2016, **11**, 2456.
66. M. D. Disney, B. Liu, W. Y. Yang, C. Sellier, T. Tran, N. Charlet-Berguerand and J. L. Childs-Disney, *ACS Chem. Biol.*, 2012, **7**, 1711.
67. W. Y. Yang, H. D. Wilson, S. P. Velagapudi and M. D. Disney, *J. Am. Chem. Soc.*, 2015, **137**, 5336.
68. T. Tran, J. L. Childs-Disney, B. Liu, L. Guan, S. Rzuczek and M. D. Disney, *ACS Chem. Biol.*, 2014, **9**, 904.
69. R. Simone, R. Balendra, T. G. Moens, E. Preza, K. M. Wilson, A. Heslegrave and N. S. Woodling, *et al.*, *EMBO Mol. Med.*, 2017, **10**, e201707850.
70. B. Zamiri, K. Reddy, R. B. Macgregor and C. E. Pearson, *J. Biol. Chem.*, 2014, **289**, 4653.
71. E. Khan, A. Tawani, S. K. Mishra, A. K. Verma, A. Upadhyay, M. Kumar and R. Sandhir, *et al.*, *ACS Chem. Biol.*, 2017, **13**, 180.
72. D. K. Semwal, R. B. Semwal, S. Combrinck and A. Viljoen, *Nutrients*, 2016, **8**, 90.
73. A. Pushechnikov, M. M. Lee, J. L. Childs-Disney, K. Sobczak, J. M. French, C. A. Thornton and M. D. Disney, *J. Am. Chem. Soc.*, 2009, **131**, 9767.
74. Q. Zhang, H. Tsoi, S. Peng, P. P. Li, K.-F. Lau, D. D. Rudnicki and J. C.-K. Ngo, *et al.*, *Dis. Models Mech.*, 2016, **9**, 321.
75. Q. Zhang, Z. S. Chen, Y. An, H. Liu, Y. Hou, W. Li and K.-F. Lau, *et al.*, *RNA*, 2018, **24**, 486.
76. W. Y. Yang, R. Gao, M. Southern, P. S. Sarkar and M. D. Disney, *Nat. Commun.*, 2016, **7**, 11647.
77. E. Soragni, W. Miao, M. Iudicello, D. Jacoby, S. De Mercanti, M. Clerico and F. Longo, *et al.*, *Ann. Neurol.*, 2014, **76**, 489.
78. Y. K. Chutake, C. C. Lam, W. N. Costello, M. P. Anderson and S. I. Bidichandani, *Nucleic Acids Res.*, 2016, **44**, 5095.
79. D. Kumari and K. Usdin, *Hum. Mol. Genet.*, 2015, **25**, 3689.
80. R. Burnett, C. Melander, J. W. Puckett, L. S. Son, R. D. Wells, P. B. Dervan and J. M. Gottesfeld, *Proc. Natl. Acad. Sci. U. S. A.*, 2006, **103**, 11497.

81. G. S. Erwin, M. P. Grieshop, A. Ali, J. Qi, M. Lawlor, D. Kumar and I. Ahmad, *et al.*, *Science*, 2017, **358**, 1617.
82. S. T. Crooke, *Nucleic Acid Ther.*, 2017, **27**, 70.
83. E. M. Zaghloul, O. Gissberg, P. M. D. Moreno, L. Siggens, M. Hallbrink, A. S. Jørgensen and K. Ekwall, *et al.*, *Nucleic Acids Res.*, 2017, **45**, 5153.
84. E. Grabczyk, *Nucleic Acids Res.*, 2000, **28**, 4930.
85. L. Li, M. Matsui and D. R. Corey, *Nat. Commun.*, 2016, **7**, 10606.
86. M. Nakamori, G. Gourdon and C. A. Thornton, *Mol. Ther.*, 2011, **19**, 2222.
87. C. Kay, J. A. Collins, N. H. Skotte, A. L. Southwell, S. C. Warby, N. S. Caron and C. N. Doty, *et al.*, *Mol. Ther.*, 2015, **23**, 1759.
88. J. B. Carroll, S. C. Warby, A. L. Southwell, C. N. Doty, S. Greenlee, N. Skotte and G. Hung, *et al.*, *Mol. Ther.*, 2011, **19**, 2178.
89. T. M. Wheeler, A. J. Leger, S. K. Pandey, A. R. Mac Leod, T. M. Wheeler, S. H. Cheng and B. M. Wentworth, *et al.*, *Nature*, 2012, **488**, 111.
90. T. M. Wheeler, K. Sobczak, J. D. Lueck, R. J. Osborne, X. Lin, R. T. Dirksen and C. A. Thornton, *Science*, 2009, **325**, 336.
91. S. A. M. Mulders, W. J. A. A. van den Broek, T. M. Wheeler, H. J. E. Croes, P. van Kuik-Romeijn, S. J. de Kimpe and D. Furling, *et al.*, *Proc. Natl. Acad. Sci. U. S. A.*, 2009, **106**, 13915.
92. J. E. Lee, C. F. Bennett and T. A. Cooper, *Proc. Natl. Acad. Sci. U. S. A.*, 2012, **109**, 4221.
93. C. J. Donnelly, P. W. Zhang, J. T. Pham, A. R. Heusler, N. A. Mistry, S. Vidensky and E. L. Daley, *et al.*, *Neuron*, 2013, **80**, 415.
94. J. Jiang, Q. Zhu, T. F. Gendron, S. Saberi, M. McAlonis-Downes, A. Seelman and J. E. Stauffer, *et al.*, *Neuron*, 2016, **90**, 535.
95. L. J. A. Toonen, F. Rigo, H. van Attikum and W. M. C. van Roon-Mom, *Mol. Ther.--Nucleic Acids*, 2017, **8**, 232.
96. M. M. Evers, H. D. Tran, I. Zalachoras, B. A. Pepers, O. C. Meijer, J. T. den Dunnen and G. J. B. van Ommen, *et al.*, *Neurobiol. Dis.*, 2013, **58**, 49.
97. L. J. A. Toonen, I. Schmidt, M. S. Luijsterburg, H. Van Attikum and W. M. C. Van Roon-Mom, *Sci. Rep.*, 2016, **6**, 1.
98. M. M. Evers, H.-D. Tran, I. Zalachoras, O. C. Meijer, J. T. den Dunnen, G.-J. B. van Ommen and A. Aartsma-Rus, *et al.*, *Nucleic Acid Ther.*, 2014, **24**, 4.
99. R. L. Boudreau, J. L. McBride, I. Martins, S. Shen, Y. Xing, B. J. Carter and B. L. Davidson, *Mol. Ther.*, 2009, **17**, 1053.
100. J. L. McBride, M. R. Pitzer, R. L. Boudreau, B. Dufour, T. Hobbs, S. R. Ojeda and B. L. Davidson, *Mol. Ther.*, 2011, **19**, 2152.
101. S. K. Pandey, T. M. Wheeler, S. L. Justice, A. Kim, H. S. Younis, D. Gattis and D. Jauvin, *et al.*, *J. Pharmacol. Exp. Ther.*, 2015, **355**, 310.
102. S. T. Carrell, E. M. Carrell, D. Auerbach, S. K. Pandey, C. F. Bennett, R. T. Dirksen and C. A. Thornton, *Hum. Mol. Genet.*, 2016, **25**, 4328.
103. S. Reddy, D. B. J. Smith, M. M. Rich, J. M. Leferovich, P. Reilly, B. M. Davis and K. Tran, *et al.*, *Nat. Genet.*, 1996, **13**, 325.

104. I. Dragatsis, M. S. Levine and S. Zeitlin, *Nat. Genet.*, 2000, **26**, 300.
105. A. Aartsma-Rus, *Nucleic Acid Ther.*, 2017, **27**, 67.
106. J. L. McBride, R. L. Boudreau, S. Q. Harper, P. D. Staber, A. M. Monteys, I. Martins and B. L. Gilmore, *et al.*, *Proc. Natl. Acad. Sci. U. S. A.*, 2008, **105**, 5868.
107. H. Xia, Q. Mao, S. L. Eliason, S. Q. Harper, I. H. Martins, H. T. Orr and H. L. Paulson, *et al.*, *Nat. Med.*, 2004, **10**, 816.
108. Y. Li, T. Yokota, R. Matsumura, K. Taira and H. Mizusawa, *Ann. Neurol.*, 2004, **56**, 124.
109. J. Hu, J. Liu, L. Li, K. T. Gagnon and D. R. Corey, *Chem. Biol.*, 2015, **22**, 1505.
110. J. Hu, F. Rigo, T. P. Prakash and D. R. Corey, *Nucleic Acid Ther.*, 2017, **27**, 87.
111. A. Fiszer, A. Mykowska and W. J. Krzyzosiak, *Nucleic Acids Res.*, 2011, **39**, 5578.
112. L. Yen, S. M. Strittmatter and R. G. Kalb, *Ann. Neurol.*, 1999, **46**, 366.
113. L. A. Phylactou, C. Darrah and M. J. A. Wood, *Nat. Genet.*, 1998, **18**, 378.
114. M. A. Bell, A. K. Johnson and S. M. Testa, *Biochemistry*, 2002, **41**, 15327.
115. M. A. Langlois, N. S. Lee, J. J. Rossi and J. Puymirat, *Mol. Ther.*, 2003, **7**, 670.
116. H. Rindt, P. F. Yen, C. N. Thebeau, T. S. Peterson, G. A. Weisman and C. L. Lorson, *Cell. Mol. Life Sci.*, 2012, **69**, 4191.
117. H. Rindt, C. M. Tom, C. L. Lorson and V. B. Mattis, *Front. Neurosci.*, 2017, **11**, 1.
118. H. Y. Chen, P. Kathirvel, W. C. Yee and P. S. Lai, *Gene Ther.*, 2009, **16**, 211.
119. C. E. Nelson, J. N. Robinson-Hamm and C. A. Gersbach, *Nat. Rev. Neurol.*, 2017, **13**, 647.
120. E. L. van Agtmaal, L. M. André, M. Willemse, S. A. Cumming, I. D. G. van Kessel, W. J. A. A. van den Broek and G. Gourdon, *et al.*, *Mol. Ther.*, 2017, **25**, 24.
121. C. Y. Park, T. Halevy, D. R. Lee, J. J. Sung, J. S. Lee, O. Yanuka and N. Benvenisty, *et al.*, *Cell Rep.*, 2015, **13**, 234.
122. N. Xie, H. Gong, J. A. Suhl, P. Chopra, T. Wang and S. T. Warren, *PLoS One*, 2016, **11**, e0165499.
123. Y. Li, U. Polak, A. D. Bhalla, N. Rozwadowska, J. S. Butler, D. R. Lynch and S. Y. R. Dent, *et al.*, *Mol. Ther.*, 2015, **23**, 1055.
124. D. L. Ouellet, K. Cherif, J. Rousseau and J. P. Tremblay, *Gene Ther.*, 2017, **24**, 265.
125. C. Provenzano, M. Cappella, R. Valaperta, R. Cardani, G. Meola, F. Martelli and B. Cardinali, *et al.*, *Mol. Ther.--Nucleic Acids*, 2017, **9**, 337.
126. S. Yang, R. Chang, H. Yang, T. Zhao, Y. Hong, H. E. Kong and X. Sun, *et al.*, *J. Clin. Invest.*, 2017, **127**, 2719.
127. J. W. Shin, K.-H. Kim, M. J. Chao, R. S. Atwal, T. Gillis, M. E. MacDonald and J. F. Gusella, *et al.*, *Hum. Mol. Genet.*, 2016, **25**, 4566.

128. A. M. Monteys, S. A. Ebanks, M. S. Keiser and B. L. Davidson, *Mol. Ther.*, 2017, **25**, 12.
129. A. G. Marthaler, A. Tubsuwan, B. Schmid, U. B. Poulsen, A. F. Engelbrecht, U. A. Mau-Holzmann and P. Hyttel, *et al.*, *Stem Cell Res.*, 2016, **16**, 202.
130. Y. Gao, X. Guo, K. Santostefano, Y. Wang, T. Reid, D. Zeng and N. Terada, *et al.*, *Mol. Ther.*, 2016, **24**, 1378.
131. G. Xia, Y. Gao, S. Jin, S. H. Subramony, N. Terada, L. P. W. Ranum and M. S. Swanson, *et al.*, *Stem Cells*, 2015, **33**, 1829.
132. B. S. Pinto, T. Saxena, R. Oliveira, H. R. Méndez-Gómez, J. D. Cleary, L. T. Denes and O. McConnell, *et al.*, *Mol. Cell*, 2017, **68**, 479.
133. M. Garriga-Canut, C. Agustín-Pavón, F. Herrmann, A. Sánchez, M. Dierssen, C. Fillat and M. Isalan, *Proc. Natl. Acad. Sci. U. S. A.*, 2012, **109**, E3136.
134. C. R. Liu, C. R. Chang, Y. Chern, T. H. Wang, W. C. Hsieh, W. C. Shen and C. Y. Chang, *et al.*, *Cell*, 2012, **148**, 690.
135. N. J. Kramer, Y. Carlomagno, Y. J. Zhang, S. Almeida, C. N. Cook, T. F. Gendron and M. Prudencio, *et al.*, *Science*, 2016, **353**, 708.
136. P. Chapdelaine, Z. Coulombe, A. Chikh, C. Gérard and J. P. Tremblay, *Mol. Ther.--Nucleic Acids*, 2013, **2**, e119.
137. X. S. Liu, H. Wu, M. Krzisch, X. Wu, J. Graef, J. Muffat and D. Hnisz, *et al.*, *Cell*, 2018, **172**, 979.
138. R. Batra, D. A. Nelles, E. Pirie, S. M. Blue, R. J. Marina, H. Wang and I. A. Chaim, *et al.*, *Cell*, 2017, **170**, 899.
139. W. Zhang, Y. Wang, S. Dong, R. Choudhury, Y. Jin and Z. Wang, *Mol. Ther.*, 2014, **22**, 312.
140. P. Kerpedjiev, S. Hammer and I. L. Hofacker, *Bioinformatics*, 2015, **31**, 3377.

CHAPTER 8

Research and Development of Oligonucleotides Targeting MicroRNAs (miRNAs)

A. L. MALINOWSKA AND J. HALL*

ETH Zürich, Department of Chemistry and Applied Biosciences, HCI H 437, Vladimir-Prelog-Weg 1-5/10, 8093 Zürich, Switzerland
*E-mail: jonathan.hall@pharma.ethz.ch

8.1 Introduction: MicroRNA Biogenesis and Functions

MicroRNAs (miRNAs) are small powerful endogenous regulators of many important physiological processes in a wide variety of organisms.[1] They were discovered independently in *Caenorhabditis elegans* by the Ambros and Ruvkun labs in 1993.[2,3] Human miRNAs are highly conserved, single-stranded RNAs of about 22 nucleotides (nt) in length. According to miRBase (http://www.mirbase.org; Release 22; March 2018), there are 2654 microRNAs in the human genome, most of which are encoded in introns or exons.[1]

miRNA biogenesis has been reviewed comprehensively (see for example Ha and Kim[4]) (Figure 8.1). Briefly, "canonical" miRNAs are transcribed by RNA polymerase II (Pol II) to yield long primary miRNAs (pri-miRNA) of more

Drug Discovery Series No. 68
Advances in Nucleic Acid Therapeutics
Edited by Sudhir Agrawal and Michael J. Gait

Published by the Royal Society of Chemistry, www.rsc.org

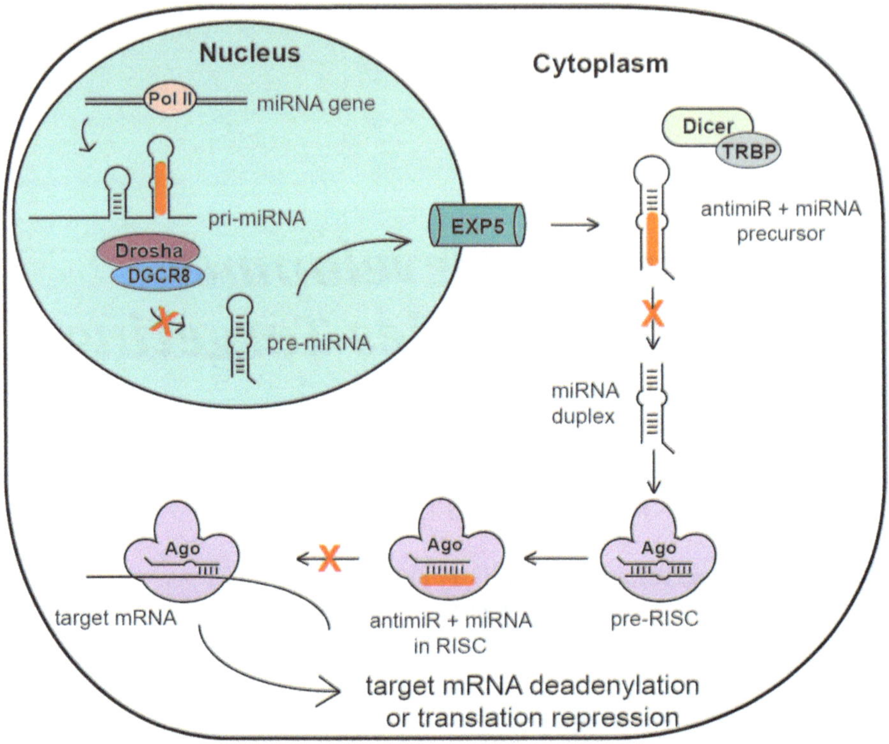

Figure 8.1 MicroRNA (miRNA) biogenesis showing the suppression of miRNA activity by an antimiR (orange). Silencing may follow a variety of mechanisms including: binding to miRNA precursors to disturb processing, sequestration of the mature strand in RISC, antimiR-mediated ejection of the miRNA from RISC, tailing and degradation of the miRNA mediated by the antimiR. Furthermore, a wide variety of chemistries are suitable for antimiR composition (see text for details). Image provided by Yuluan Wang.

than 1 kb in length: in some cases, the pri-miRNAs carries multiple miRNAs in a polycistronic cluster.[5] These are cleaved by a complex known as the Microprocessor, comprising the RNase III-type endonuclease Drosha and its cofactor DGCR8 (DiGeorge syndrome critical region 8) into an ~70 nt-long precursor miRNA (pre-miRNA) comprising a hairpin with a short overhang of a few nucleotides at the 3′-end and a monophosphate at the 5′-terminus.[6] The pre-miRNA is then shuttled to the cytoplasm by exportin 5 (EXP5)[7] where its terminal loop is excised by the RNase III Dicer, together with a cofactor, such as the transactivation-responsive RNA-binding protein (TRBP).[8,9] This yields an imperfect duplex of ~22 nt comprising the mature miRNA. Non-canonical mechanisms of maturation, *e.g.* those producing "mirtrons" have also been characterized which do not involve Drosha.[10,11] The mature miRNA-containing duplex is then associated with a protein of the Argonaute family (Ago) in the RISC (RNA-induced silencing complex). One of the mature miRNA strands – known as the "guide" – is maintained in the complex, whereas the "passenger" strand is ejected to give the complex which is responsible for miRNA-mediated post-transcriptional gene regulation:[12,13] both strands can act as a guide, possibly in a context-dependent fashion.[14]

Mature miRNAs engage their natural RNA targets in cells in several ways. The canonical mechanism involves binding to the 3′-untranslated region (3′UTR) of messenger RNAs (mRNAs) in order to trigger mRNA deadenylation and/or inhibition of translation.[15] Target recognition is mediated in most cases by complementary base-pairing between the so-called miRNA seed region (nucleotides 2 to 7 counting from the 5′-end) and binding sites in the mRNA, which are often conserved across species.[16] The seed region of the miRNA in RISC is pre-organized in an A-type helix for fast and stable binding to its target RNA.[17] In many cases, genes that are post-transcriptionally regulated by miRNAs are peppered with binding sites both for multiple miRNAs and/or the same miRNA, producing a high level of redundancy in the regulation. This can complicate pharmacological targeting, since the inhibition of a single molecular interaction may be insufficient to elicit a strong phenotypic effect. Base pairing between the central and terminal regions of the miRNA and mRNA may also contribute to binding, especially where the seed interaction is unusually short.[5] However, there are increasing accounts of non-canonical interactions that include: miRNAs binding in the 5′UTR of an mRNA[18,19] and the coding sequence of the target,[20,21] miRNAs that bind with complementarity *via* their central or 3′-regions,[22–24] miRNAs that bind to bulged mRNAs,[25–27] miRNA–mRNA interactions that result in enhanced translation of an mRNA[28] and miRNAs that interact with non-coding RNAs[29–31] including circular RNAs.[32,33]

miRNAs in the human genome that share the same seed sequence are classified as a miRNA "family".[34] Members of the same miRNA family regulate the same mRNA targets for the most part (because of their common seed sequence), resulting in a second level of regulatory redundancy that additionally complicates pharmacological approaches to targeting.[35]

A large proportion of nematode miRNAs do not have severe phenotypes, as assessed by genetic knockout studies,[36] leading to the conclusion that most conserved miRNAs are not essential for the organism's viability or development. For nematodes, it is likely that most of their miRNAs function simply to fine-tune the expression of genes as "rheostats" that maintain proper functioning of the cell.[37] However, in this respect nematodes are rather unusual since phenotypes have been observed for most of the mammalian miRNA families that are conserved across species.[1] This implies also, that loss- or gain-of-function mutants may critically affect a wide variety of processes, including those involved in disease. For example, point mutations in the seed region of the human miR-96 result in progressive hearing loss,[38] and germline deletions of the miR-17~92 cluster contribute to syndromic developmental defects in humans.[39]

8.2 miRNAs as Targets for Drugs

As miRNAs regulate so many important biological processes, it is hardly surprising that in some cases their dysregulation is linked causally to the development of diseases. These miRNAs represent potential targets for therapeutics. For instance, a large body of literature has described how miRNAs play important roles in the onset and progression of cancers,[40] including mechanisms of angiogenesis,[41] epithelial–mesenchymal transition,[42] the tumour microenvironment[43] and metastasis of cancers.[44] Many cancer-associated miRNAs or miRNA clusters were initially identified as such because they are located in chromosomal regions that are either amplified or deleted in cancer cells.[45,46] Some miRNAs, for example miR-34a and lethal 7 (let-7), are classified as tumour suppressors because they target the mRNAs that code for oncoproteins and their expression may be abrogated in certain tumours.[47] On the other hand, oncomiRs, including miR-21 and members of the miR-17~92 miRNA cluster, have tumour-promoting properties because they suppress mRNAs that code for tumour suppressor proteins.[47] Other disease areas besides cancer are equally touched by miRNAs. These include for example: miR-122 in hepatitis C infections,[48] miR-103/107 in diabetes[49] and miR-1, miR-133 and miR-208a in cardiac disorders,[50,51] to name a few.

8.3 AntimiR Oligonucleotides as Drugs

There are two opposite approaches for the correction of disease-related miRNAs' activities: miRNA replacement to correct the loss of activity of a given miRNA (family) and miRNA inhibition to attenuate an over-active miRNA. Hence, miRNAs can be considered as either therapeutic agents or therapeutic targets. In this chapter we focus on a selected set of examples of oligonucleotides and their derivatives as miRNA inhibitors.

8.3.1 The Development of AntimiR Medicinal Chemistry

Conceptually, the most straightforward way to inhibit the activity of an aberrant miRNA is by use of a complementary oligonucleotide, otherwise known as an antimiR, an anti-miRNA, or an AMO (anti-miRNA oligonucleotide). AntimiRs are chemically modified synthetic oligonucleotides of varying lengths with a sequence that is complementary to a part of or the complete miRNA mature guide strand. As with other RNA-targeting therapeutic technologies, antimiRs are highly potent and selective silencers of their complementary targets, but their use as drugs is complicated by tough technical challenges, particularly their problematic pharmacokinetic properties.[52] When delivered *in vivo*, antimiRs require protection against ubiquitous endonucleases and exonucleases present in serum, tissues and cells that readily degrade phosphodiester bonds. It is required that they avoid premature clearance *via* renal excretion and that they distribute to tissues

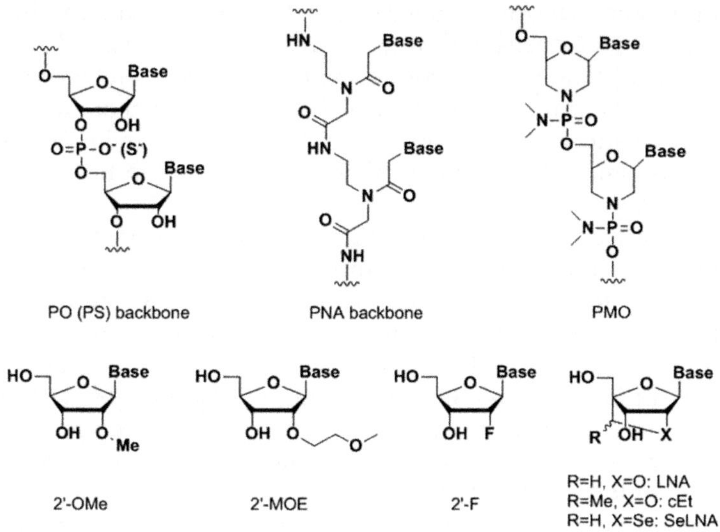

Figure 8.2 Commonly used chemical modifications in antimiRs.

via the bloodstream. They should not trigger an innate or adaptive immune response leading to toxic effects. Finally, these large, hydrophilic and negatively charged molecules should traverse cellular membranes to reach their intracellular targets, without being trapped in unproductive cellular compartments.

Activities in medicinal chemistry for antimiRs have followed similar paths to those for mRNA-targeting antisense oligonucleotides (Figure 8.2). Several comprehensive reviews as well as other chapters of this book (see Chapters 1 and 2) cover the field of antimiR medicinal chemistry in detail,[53–56] and therefore only a selection of prominent examples is presented here.

The main purpose of backbone and ribose modifications in antimiRs that are destined for use *in vivo* is to render them stable to exonucleases and endonucleases. As with other oligonucleotide drugs containing ribose rings, nucleotides are linked by phosphorothioate (PS) groups, in which one non-bridging oxygen in the natural phosphodiester (PO) backbone is replaced by a sulphur atom.[57] A large number of sulphur atoms in the antimiR greatly increases the hydrophobicity of the antimiR, which, in turn, improves its transport across cell membranes and its distribution *in vivo* thanks to a weak non-specific binding to circulating proteins.[52,58,59]

8.3.1.1 2′-O-Methyl-modified AntimiRs

The first reported efforts to inhibit miRNAs with oligonucleotides were carried out by direct microinjection of an unmodified antimiR DNA into *Drosophila* embryos.[60] This was quickly followed by experiments using antimiRs modified with 2′-*O*-methyl groups on the ribose (2′-OMe) (Figure 8.2).

The group of Zamore employed a 31 nt, 2′-OMe-modified antimiR to block let-7 activity in HeLa cells and in *C. elegans*,[61] whereas that of Tuschl used a shorter (24 nt) oligoribonucleotide against miR-21 in HeLa cells.[62] These early experiments provided the first conclusive evidence that miRNAs can be inhibited by an antisense-like mechanism and provided an important means to help functionalize the hundreds of newly discovered miRNAs in mammalian organisms.

It was a relatively small step thereafter to apply antimiRs *in vivo*, where 2′-OMe antimiRs require additional modifications or formulations for protection against nuclease degradation mechanisms. Recently, Wang and colleagues described a sophisticated delivery protocol in the form of a conjugate of polyethylene glycol (PEG)-functionalized nanographene oxide (NGO) with a biodegradable polyamidoamine (PAMAM) dendrimer (NGO–PEG-dendrimer) to deliver antimiR-21 into lung cancer-derived cells.[63] miR-21 regulates the tumour suppressor phosphatase and tensin homolog (*PTEN*). A 23 nt long antimiR-21 that was complexed with an NGO–PEG-dendrimer was found to be effective in A549 cells and in tumour xenografts in mice.[63] PTEN protein levels were increased in A549 cells, leading to reduced cell migration and invasion. Schnittert and associates reported delivery of 2′-OMe antimiRs into human-derived pancreatic stellate cells (hPSCs) with the help of a cell-penetrating peptide.[64] Conjugation with cell-penetrating peptides (CPPs) is often used to facilitate delivery of oligonucleotides *in vitro* and *in vivo*.[65] Short positively charged peptides associate with negatively charged oligonucleotides and protect them from degradation. miR-199a, a member of the miR-199 family, is thought to manifest pro-tumourigenic properties by transforming PSCs into cancer-associated fibroblasts (CAFs), a major cell type in the stroma of pancreatic ductal adenocarcinoma (PDA). In this example, a CPP consisting of two arginine-rich peptide chains was called upon to help deliver 2′-OMe antimiRs into cells. The improved cellular association and uptake[66] of the complexes was attributed to a macropinocytosis and a clathrin-mediated endocytosis. The delivery of antimiR-199a into hPSCs resulted in de-repression of mammalian target of rapamycin (mTOR) and reduction of characteristic activation transcription markers of CAFs [actin alpha 2 (Acta2), collagen 1 (Colla1) and platelet-derived growth factor receptor beta (PDGFR-β)], confirming the importance of miR-199a in hPSC-mediated tumour cell growth.

AntimiRs are increasingly used in combination with small-molecule drugs to identify new ways to address diseases with unmet medical need.[67] For example, the multi-kinase inhibitor sorafenib is a first-line drug treatment for advanced hepatocellular carcinoma (HCC), for which acquired sorafenib-resistance significantly complicates treatment. Increased levels of miR-494 in a subgroup of stem-like HCC cells correlates with poor survival prognosis and contributes to sorafenib resistance through activation of the mTOR pathway.[68] In an effort to determine the targetability of miR-494 a 2′-OMe antimiR-494 capped at both ends with PS linkages was delivered into HCC cells. It led to increased expression of tumour suppressors p27, p53 upregulated modulator of apoptosis (PUMA) and phosphatase and tensin homolog (PTEN) proteins, which are targets of miR-494. Use of the antimiR was extended to a diethylnitrosamine-induced HCC rat model, where enhanced anti-tumour effects of combined antimiR-494 and sorafenib treatment were confirmed.

8.3.1.2 Antagomirs

Antagomirs are a highly useful class of antimiR reagents for *in vivo* applications.[69] The original antagomir was a 2′-OMe-oligoribonucleotide, with two PS linkages at the 5′-end, four PS linkages at the 3′-end and a cholesterol group conjugated at the 3′-terminus to facilitate uptake into liver hepatocytes. Administration of three injections of 80 mg kg^{-1} antagomir-122 to mice produced a highly selective reduction in liver miR-122 activity, which correlated with an unexpected Ago2-independent reduction in miR-122 levels.[70] These experiments provided important insights into the role of miR-122 in the regulation of cholesterol biosynthesis and established antagomirs as a broadly-applicable class of antimiRs for loss-of-function studies *in vivo*.[71]

In a recent example, a miR-10b antagomir and paclitaxel (PTX) were used in a liposome formulation for inhibition of tumour growth and metastasis in a murine metastatic mammary tumour model.[72] miRNA-10b is engaged in the initiation of invasion and metastasis in breast cancer.[73] It was shown previously that treatments with high concentrations of antagomir-10b inhibited metastasis but did not prevent the growth of primary tumours,[74] hence the combination of a cytotoxic drug (paclitaxel) with antagomir-10b as a strategy to inhibit both processes. For improved endosomal escape by the antagomir in tumour cells, the liposomes were coated with a pH-responsive, anti-microbial-based peptide [D]-H_6L_9 (D-Lip).[75] In the slightly acidic environment of endosomes and lysosomes, the histidines of D-Lip are fully protonated, causing lysis of the endosomal membrane and release of their contents.[76] D-Lip facilitated the delivery of antagomir-10b, either with or without PTX, and de-repressed expression of a target of miR-10b – homeobox D10.[72] Furthermore, treatment of mice with D-Lip, antagomir-10b and PTX induced significant reductions of tumour volume and lung metastasis.

Park and colleagues characterized a 2′-OMe antimiR to miR-221 with two PS linkages at both ends for its ability to reduce cell proliferation of hepatocellular carcinoma (HCC) cells *in vitro*.[77] In *in vivo* experiments, this antimiR was conjugated to cholesterol to facilitate delivery and accumulation in the C57BL/6 mouse liver, similarly to the original report on antagomirs.[69] Administration of the antagomir-221 to mice with orthotopic tumours suppressed miR-221 levels in a concentration-dependent fashion and concomitantly increased expression of its targets $p27^{Kip1}$, $p57^{Kip2}$ and PTEN proteins. Tumour growth was impeded and survival of the antagomir-treated rodents increased.

8.3.1.3 2′-O-Methoxyethyl AntimiRs

The incorporation of 2′-*O*-methoxyethyl (MOE) modifications[78] (Figure 8.2) into antimiRs brought further improvements, with greater nuclease resistance and a higher binding affinity to their targets.[55] By use of PS-modified MOE antimiRs, Esau and associates studied selected miRNAs during adipocyte differentiation and discovered that targeting miR-143 in primary human pre-adipocytes hampered their differentiation and prevented adipocyte-specific gene expression.[79] A PS MOE antimiR was also employed to antagonize miR-122 and obtain information about its function in the liver.[80] Treatment of mice led to decreased levels of miR-122 and yielded a remarkable reduction in serum cholesterol levels with concomitant upregulation of miR-122 target genes. The data were in good agreement with those obtained during the original study on miR-122 using antagomirs[69] and confirmed that MOE oligonucleotides (without conjugated groups) are also effective antimiRs *in vivo*.

The presence of high levels of miRNA-10b is a hallmark of glioblastoma (GBM).[81] MOE-based antimiRs with PO or PS backbones have been employed to silence miR-10b, in order to help elucidate how miR-10b promotes cell survival in GBM tumours.[82] Studies conducted *in vitro* and in two mouse models showed that treatment with antimiR-10b results in attenuated growth and progression of GBM, without signs of toxicity. AntimiRs were administered to mice in several ways: jetPEI-formulated antimiR-10b with PO backbone for intratumoral injections, antimiR-10b with PS backbone for intravenous injections and antimiR-10b with PO backbone formulated with cationic lipid nanoparticles for continuous osmotic delivery. All delivery routes worked equally well in terms of silencing miR-10b activity in the orthotopic GBM8 xenograft mouse model. The data indicated that miR-10b affects cell cycle progression, as well as mRNA processing and alternative splicing, with muscleblind-like protein 2 (MBNL2), MBNL3, squamous cell carcinoma antigen recognized by T cells 3 (SART3) and arginine and serine rich coiled-coil 1 (RSRC1) identified as prominent targets of the miRNA. Surprisingly, miR-10b seems to bind mostly to the 5′UTRs of these targets using both its seed and its 3′-end.

8.3.1.4 2′-Fluoro-modified AntimiRs

In a search for alternative chemistries providing more potent antimiRs, oligonucleotides bearing 2′-fluoro (2′-F, FNA) modifications (Figure 8.2) have been investigated. The introduction of fluorine to the 2′-position of deoxyribose of oligonucleotides constrains the ribose to an RNA-like 3′-*endo*-conformation,[83] which enhances binding affinity to complementary RNAs.[84] Indeed, in a head-to-head comparison of PS antimiRs with various chemistries (2′-OMe, MOE, 2′-F and locked nucleic acid (LNA)–MOE) targeting miR-21 the affinity of 2′-F-modified oligonucleotide to the complementary RNA was higher than those of its 2′-OMe and MOE counterparts.[85] Moreover, when FNA antimiR-122 was capped using MOE modifications at both ends, increased de-repression of miR-122 target genes was observed *in vivo*.[86] The lack of observed reduction of miR-122 indicates that the miRNA is simply sequestered by the FNA antimiR.

miR-33a and miR-33b constitute the highly conserved miR-33 family and are miRNAs encoded within introns of the sterol-response-element-binding protein genes *SREBF2* and *SREBF1*, respectively.[87–89] Upregulation of miR-33 results in significant suppression of proteins involved in cholesterol transport [adenosine triphosphate-binding cassette transporter 1 (ABCA1)] and fatty acid metabolism [carnitine *O*-octanoyltransferase (CROT), carnitine palmitoyltransferase 1A (CPT1A)], which, in turn, leads to lowering of plasma high-density lipoprotein (HDL) levels and formation of atherosclerotic plaques. Rayner and colleagues showed that silencing of miR-33a or miR-33a/b in mice[90] and non-human primates,[91] respectively, with 2′-F–MOE-based antimiRs increases circulating HDL levels with concomitant enhancement of reverse cholesterol transport, thereby contributing to regression of atherosclerosis. These observations correlated with a reduction of very-low-density lipoprotein (VLDL) triglyceride levels in plasma (non-human primates). However, in a separate study, prolonged treatment of mice on a high-fat diet with antimiR-33 resulted also in increased levels of circulating triglycerides and lipid accumulation in the liver.[92]

8.3.1.5 Locked Nucleic Acids and Constrained Ethyl Derivatives

Locked nucleic acids (LNAs) are one of the most potent modifications used today in the antisense field[93,94] (Figure 8.2). LNAs are defined by 2′-*O*,4′-*C*-methylene-linked bicyclic ribonucleosides which, similarly to the 2′-fluoro modification, force a 3′-*endo* puckering of the sugar.[95] Compared with the aforementioned modifications, the LNAs have very high binding affinities to complementary RNA and provide a greatly increased stability against nucleases.

In the first study describing the use of LNA in an antimiR, an LNA–DNA chimera (an LNA central core flanked by seven DNA nucleotides) was used against miR-21 in glioblastoma cells.[96] Its activity was judged to be moderate, possibly due to nuclease degradation of the DNA wings. In contrast, a fully-modified LNA antimiR was an effective inhibitor of the *bantam* miRNA in HEK293 and *Drosophila* cells with high selectivity.[97] Today, it is known that the extreme affinity of LNAs for complementary sequences can be detrimental to their selectivity, but it can be tempered by incorporation of DNA or 2′-OMe monomers into the LNA antimiR.[56,98,99]

On the positive side, the high binding affinity of LNAs allows an general shortening of the antimiR length, to the extent that entire families of miRNAs can be silenced *in vivo* using "tiny LNAs", fully-modified LNA–PS octamers, which bind only to the seed region of a miRNA family.[100] Counter-intuitively, tiny LNAs are quite selective, despite the large number of 8 nt long complementary target sites in the transcriptome, though this is not the first time short, high-affinity oligonucleotides have provided sequence-selective targeting.[101] Thomas and associates have investigated short, seed-targeting LNA–PO oligonucleotides.[102] Using a luciferase reporter assay they determined that the minimum length of an antimiR needed to target let-7a with acceptable activity was 12–14 nt. Next, antiseed 14-mers against miR-17-5p and miR-20a, two members of the oncogenic miR-17~92 cluster, were electroporated into K562 cells,[103] thereby restoring expression of the anti-proliferative target p21.[104]

As emphasized previously, miRNA-21 is a well-characterized oncomiR that plays key roles in cell survival and is highly expressed in various cancers.[96,105] An LNA-based antimiR-21 encapsulated in stable nucleic acid lipid particles (SNALP) was administered to mice bearing brain tumours and showed delivery to the brain and accumulation in the tumour with concomitant inhibition of miR-21.[106] To improve glioblastoma tumour-specific targeting of the antimiR-21 *in vivo*, the SNALPs used for encapsulation of antimiR-21 were conjugated to chlorotoxin (CTX), a peptide known for specific binding to glioma cells.[107,108] Indeed, the SNALP-formulated antimiR-21 proved highly potent, causing de-repression of RhoB, a direct target of miR-21. When injections were followed by oral administration of the tyrosine kinase inhibitor sunitinib, a decline in tumour size was noted, together with reduced tumour cell proliferation, increased apoptosis and prolonged survival of the animals. Similar impressive results were obtained by targeting miR-21 in models of colorectal cancer.[109]

The use of LNA–DNA antimiRs *in vivo* is not restricted to models of cancer, and indeed they have also been used to great effect in the heart. In a study conducted by Montgomery *et al.*, subcutaneous injections of antimiR-208a targeted against cardiac-specific miR-208a resulted in improved cardiac function and survival of Dahl hypertensive rats with hypertension-induced heart failure.[110]

The role of the miR-15 family in cardiac ischemic injury and the subsequent remodelling process has also been investigated by Hullinger and colleagues.[111] miR-15a, miR-15b, miR-16, miR-195 and miR-497 are all members of the miR-15 family and their levels are elevated in the infarcted region of ischemic injury. To determine whether suppression of these miRNAs, and particularly miR-15b, contributes to the protection of cardiomyocytes from hypoxia-induced cell death antimiRs of different chemistries were investigated (Figure 8.3) including a full-length cholesterol-conjugated 2′-OMe antagomir, a 16 nt long LNA–DNA antimiR targeting miR-15b and a miR-15b tiny antimiR. Whereas the LNA–DNA antimiR showed selective inhibition of miR-15b, tiny-15b inhibited the entire miRNA family. The LNA-antimiRs were able to counteract stress-induced increases of miR-15b levels in murine cardiac tissue in a dose-dependent way when administered intravenously. Consistent results were obtained after testing in pigs (ear vein injection). Taken together, the study demonstrated the utility of LNA-based antimiRs for targeting single miRNAs (16-mer) or the entire miRNA families (octamer).

hsa-miR-15a-5p	5'-UAGCAGCACAUAAUGGUUUGUG
hsa-miR-497-5p	5'-CAGCAGCACACUGUGGUUUGU
hsa-miR-16-5p	5'-UAGCAGCACGUAAAUAUUGGCG
hsa-miR-195-5p	5'-UAGCAGCACAGAAAUAUUGGC
hsa-miR-15b-5p	5'-UAGCAGCACAUCAUGGUUUACA
antagomir-15b (2'-OMe)	chol-ATCGTCGTGTAGTACCAAATGT-5'
antimiR-15b (LNA–DNA)	TCGTCGTGTAGTACCA-5'
tiny-15b (LNA)	TCGTCGTG-5'

Figure 8.3 AntimiR strategies to target the miR-15/16 family: 8 nt LNA tiny-15b, 2′-OMe antagomir-15b with two PS groups at the ends and a 16 nt LNA–DNA chimera. The common seed sequence of the family is shown in yellow.

The constrained ethyl (cEt: 2′,4′-constrained 2′-*O*-ethyl)-modification is a derivative of the LNA structure (Figure. 8.2). This 2′,4′-bridged analogue (locked in the *N*-type conformation) carries a methyl group at position 6 of the 2,5-dioxabicyclo ring system as *R*- or *S*-stereoisomers. Oligonucleotides with this chemistry show similar binding affinity and mismatch discrimination when hybridized to RNA to that of LNAs and have also extreme resistance to nuclease degradation.[112] They have been incorporated into antimiRs, such as antimiR-21, and used effectively *in vivo*.[113] A number of cEt-containing oligonucleotides, including antimiRs are currently in clinical trials. The selenomethylene-locked nucleic acids (SeLNAs) are another variant of the LNAs, in which oxygen at the 2′-*O*-position has been replaced by a selenium atom (Figure 8.2). The antimiRs are characterized by high binding affinity to the target RNA and increased stability against nucleases.[114] A DNA oligonucleotide containing four thymine SeLNA monomers reduced miR-21 levels in cells, leading to an increase of the miR-21 target PDCD4 (programmed cell death 4) and reduced cell migration and invasiveness.

8.3.1.6 Peptide Nucleic Acids

Peptide nucleic acids (PNA) are electrostatically uncharged nucleic acid analogues with a peptide-like backbone composed of *N*-(2-aminoethyl)-glycine units with nucleobases conjugated *via* a carbonyl methylene linkage[115] (Figure 8.2). PNAs bind strongly to RNA and are not substrates for nucleases, however, unless a cationic amino acid is also added or the PNA is conjugated to a soluble peptide, they suffer from decreased aqueous solubility, with a tendency to aggregate. Nevertheless, PNAs have been used to great effect as antimiRs particularly when conjugated to groups that enhance cellular delivery *in vitro* and *in vivo*.[98,116–118] For example, PNA antimiRs conjugated to the pH (low) insertion peptide (pHLIP) were used to target miR-155 in a mouse model of lymphoma.[119] Being a derivative of bacteriorhodopsin C helix, pHLIP is a water-soluble polypeptide with a unique ability to insert across a lipid bilayer and form a stable transmembrane α-helix.[120,121] In acidic tumour environments, negatively charged residues of pHLIP are protonated and the peptide inserts into the cell membrane with the N-terminus in the extracellular space and the C-terminus placed in the cytosol. The pHLIP peptide was attached at its C-terminus to antimiR-155 molecule *via* a disulphide bond.[122] Intravenous injection of pHLIP–antimiR-155 significantly retarded tumour growth and metastasis of neoplastic lymphocytes in miR-155-addicted lymphoma subcutaneous tumour models. In transgenic *mir-155*^{LSLtTA} mice, systemically delivered pHLIP-antimiR accumulated in lymphoma nodes.

Other groups also have been conjugated to PNA antimiRs. A diethylene glycol moiety was introduced at the γ-position of PNA to create so-called miniPEG–γPNA ($^{MP}\gamma$PNA): this improved water solubility and increased target-binding affinity. The antimiR was used to antagonize miR-210 in tumour xenografts in immune-deficient nude mice.[123] Brognara and colleagues used a polyarginine-conjugated PNA molecule to target miR-221 expression in various cancer cells.[124] miR-221 belongs to the miR-221/222 cluster and is upregulated in breast cancer and glioma. The cyclin-dependent kinase inhibitor protein p27^{Kip1} is a validated target of miR-221. The initial study, carried out on MDA-MB-231 breast cancer cells, showed that fluorescein-labelled polyarginine-conjugated PNA antimiR was incorporated into target cells and efficiently antagonised miR-221,[125] thereby de-repressing p27^{Kip1} at the mRNA and protein levels.

8.3.1.7 Morpholino Oligonucleotides

A prominent modification in the oligonucleotide field is the phosphorodiamidate morpholino oligonucleotides (PMOs or morpholinos) (Figure 8.2). The PMO backbone consists of methylenemorpholine rings linked with phosphorodiamidate linkage groups. The backbone is completely resistant to nucleolytic enzymes.[126,127] Moreover, their lack of electrostatic charge diminishes interactions with proteins, which is reflected in low toxicity of morpholinos. Morpholinos show good aqueous solubility and high selectivity for complementary RNA, but a lower RNA-binding affinity compared with the aforementioned modifications. This modification has not been used extensively in antimiRs. However, in a study by Flynt and colleagues, a PMO antimiR was used to target miR-214 in zebrafish, uncovering one of its important functions in muscle cell development.[128]

8.3.2 New Insights into Mechanisms of Oligonucleotide-based miRNA Targeting

One of the important lessons from early antisense research was that a detailed understanding of the mechanisms by which oligonucleotides elicit their effects is important for the medicinal chemistry-based design of better reagents. This lesson was learned during the development of RNase-inducing antisense oligonucleotides and small interfering RNAs (siRNAs). Indeed, a small number of early works[85,129] indicated that the mechanism of miRNA inhibition by antimiRs is likely to be more complicated than mere stoichiometric sequestration of the miRNA. For example, in addition to binding mature miRNAs, antimiRs may also act on pri- or pre-miRNAs by invading these structures in the nucleus and cytoplasm, respectively. This leads to attenuation of their processing by Drosha and Dicer so that the synthesis of new mature miRNAs is interrupted. This principle was first shown by Kloosterman and colleagues through inhibition by morpholino antimiRs of miR-375 in zebrafish.[130] It was also shown that the activity of the antimiR-122 drug Miravirsen is partly due to inhibition of the processing of miR-122 precursors[131] (*vide infra*).

8.3.2.1 AntimiR-mediated Tailing and Trimming of miRNAs

In some of the early studies of miRNA inhibition, the target-binding affinity of an antimiR did not always correlate with its cellular activity (for example see Davis *et al.*[85]), implying that factors in addition to binding affinity were in play. Indeed, during their first experiments with antagomirs in cells, Krutzfeldt and colleagues observed a surprising reduction of miRNA levels by northern blotting,[69] indicating that the antagomir perhaps induced miRNA degradation.[70] This effect was also noted when using MOE-based antimiRs.[80,85] Although these effects were in some cases due to technical artefacts,[132] the hybridization of antimiRs to their miRNA targets does in some cases induce alterations in the miRNA, either a degradation of the miRNA (miR-122)[86] or the addition of uridines (miR-16 and miR-21: 2′-OMe antagomir).[133] The "tailing" of miRNAs in Ago-containing complexes results in their degradation *in vitro* by a recruited exonuclease, possibly as a natural protection against aberrant RNA interference.[134] Interestingly, antimiRs composed of 2′-F–MOE, LNA or cEt–DNA do not elicit this tailing effect and silence miRNAs by conventional heteroduplex formation, clearly demonstrating the modification-dependent effects.[86,99,113,132,135]

8.3.2.2 Targeting miRNAs in the RISC

Hogan and colleagues reported that antimiR activity is significantly affected by the nature and position of chemical modifications in the strand. They compared antimiRs against miR-21, which differed slightly in the placement of cEt modifications at two positions.[113] Despite similar binding affinity to a complementary RNA, only one of the antimiRs was able to associate with the miRNA isolated from liver lysates and de-repress miR-21 targets *in vitro* [ankyrin repeat domain 46 (ANKRD46)] and *in vivo* [ring finger protein 167 (Rnf167)]. Furthermore, the authors also observed that while strong hybridization of the antimiR to the miRNA seed region was essential for inhibitory activity, complementarity outside the seed region was surprisingly not required. This only partly overlaps with the findings of a comprehensive study showing that some complementarity in the 5′-end of 2′-OMe antimiRs was important for recognition, though these studies were carried out in cell culture only.[136] Taken together, Hogan and colleagues concluded that antimiRs associate with the target miRNA *inside* the RISC, *i.e.* in the presence of Ago proteins.[113]

A study by Brunschweiger and colleagues produced analogous findings. They investigated whether small drug-like fragments conjugated at internal positions of a short 2′-OMe antimiR could increase the cellular activity of antimiR-122.[137] Lennox and colleagues had previously described how small fragments conjugated to the terminal nucleotides of antimiRs, *e.g.* the *N*,*N*-diethyl-4-(4-nitronaphthalen-1-ylazo)-phenylamine (abbreviated ZEN) modifier (Figure 8.4), improves their binding affinity and nuclease resistance in serum and in cells.[138,139] Brunschweiger and associates prepared a small library of antimiR-122 conjugates and tested them in a luciferase reporter assay. Surprisingly, the potency of the reagents yielded a clear position-dependent activity pattern, *i.e.* modifications at the 2′-*O*-position opposite to nucleotide six of the miRNA seed completely abrogated antimiR activity. This trend was in accordance with previous unexplained observations.[85,113] After quantifying the antimiR in immune-precipitated Ago2 from treated cells and cross-comparing the results with a structural model of a miRNA–RNA duplex bound by Ago2,[140] the authors proposed that with hybridization of the antimiR to the miRNA, large 2′-*O*-modifications of the antimiR interfere with amino-acids Ile365 and Ser362 present in Ago2 in RISC. This study strongly indicated that in order to engage its target an antimiR needs to enter the RISC, where it may be subject to steric hindrance during hybridization with the miRNA.

Along the same lines, Schmidt and colleagues described chimeric oligonucleotide–small-molecule conjugates as more conventional drug-like structures for silencing miRNA activity.[141] These reagents are hybrids of a PNA tetranucleotide complementary to part of the miRNA seed region conjugated at the N-terminus to a small-molecule fragment that occupies the active site of Ago2. This novel design showed activity against miR-122 in *in vitro* assays, but these activities have not so far been reported in cells.

8.3.2.3 AntimiR-mediated Unloading of miRNAs from RISC

The MacRae group described how high complementarity between a miRNA and its mRNA target results in a fast release of the miRNA from Ago2 in RISC.[142] Moreover, while mismatches between the 5′-end of the miRNA and the complementary target in RISC were shown to support unloading, mismatches at the 3′-end hampered the release. As a miRNA needs to be loaded into RISC in order to carry out its function, it was proposed that modifications on an antimiR can be used to enhance release of the miRNA from the

RISC. In RISC the phosphate at the 5′-end of the miRNA contacts lysine residues in the PIWI-box.[143] By perturbing this interaction, dissociation of the miRNA from RISC is facilitated. For this purpose, Ariyoshi and colleagues designed and tested peptide-conjugated antimiRs called RINDA (RISC inhibitor disturbing active site of RISC),[144] which are antimiRs extended at their 3′-ends with three glutamic acid residues [RINDA(E3)-as] that bind the lysines and enhance release of the miRNA from RISC. Interestingly, some non-conjugated antimiRs also induced the release of the miRNA from RISC, indicating that this mechanism is not yet fully understood. The influence of conventional antimiR modifications on this mechanism was also investigated.[145] Six antimiRs with different chemistries (DNA, 2′-OMe–DNA, 2′-OMe, LNA–DNA or 2′-OMe–LNA) were tested in an unloading assay. The release of the target from RISC was observed in a time-dependent manner with the fastest dissociation [lowest half-life ($T_{1/2}$)] for the LNA antimiR and a good correlation between melting temperature (T_M) and $T_{1/2}$ values. This observation may partially explain the decrease of miRNA levels upon antimiR treatment reported by some groups.[142] Overall, it indicates that the regulation of RISC activity by blocking its active site combined with a targeting of the seed region of miRNA might significantly improve antimiR activity with an alternative pathway of miRNA silencing.

8.3.3 New Chemistries and Alternative Approaches for Targeting miRNAs

New classes of oligonucleotide modifications are constantly being designed in an overall effort to improve the properties of oligonucleotides either as tools or as therapeutic moieties. Their value is often assessed by incorporation into antimiRs where they can be compared directly with antimiRs composed of conventional chemistries.

Recently, serinol nucleic acids (SNAs) containing 2,6-diaminopurines were introduced[146] (Figure 8.4). SNAs are acyclic nucleic acids with a serinol backbone that is completely resistant to enzymatic digestion.[147] Furthermore,

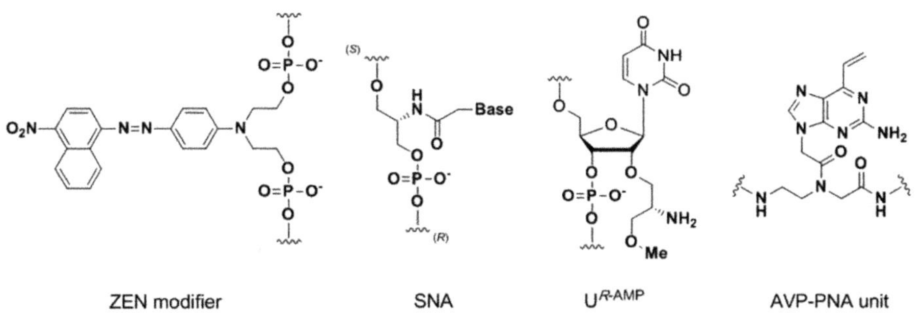

Figure 8.4 Recently introduced chemical modifications used to improve antimiR properties.

the substitution of adenine by 2,6-diaminopurine in SNAs improves their binding affinity to RNA thanks to a third hydrogen bond with uracil.[148] An SNA-based antimiR against miR-21 demonstrated similar activity to a 2′-OMe analogue in HeLa cells.[146] However, an antimiR in which all seven adenines were replaced by 2,6-diaminopurines displayed a much higher potency. Although the number of substitutions within the antimiR correlated with its affinity for a complementary RNA, there was no correlation between the number of substitutions and its cellular activity, which rather depended on the position of the modification within the strand, consistent with the aforementioned examples.[113,137]

A 2′-OMe antimiR interspersed with 2′-*O*-(*R*-2-amino-3-methoxypropyl) (*R*-AMP) uridine monomers and capped at the 5′- and 3′-ends with serinyl units has also been used to target miR-21 in cell culture systems[149] (Figure 8.4). The serinyl caps of the antimiR were introduced to provide protection against exonucleases, whereas the $U^{R\text{-AMP}}$ also served to maintain target-binding affinity. A serinyl-capped 2′-OMe antimiR containing three $U^{R\text{-AMP}}$ units in its sequence was a stronger inhibitor of miR-21 than unmodified 2′-OMe analogues in a luciferase assay in MCF-7 cells. A reduction of mature miR-21 levels (up to 95% after 24 h post-transfection) was observed, together with a de-repression of miR-21 targets PTEN and PDCD4 and reduced cell migration.

A variety of rather more esoteric oligonucleotide-based methods to inhibit miRNAs has been described. For instance, in the early years of antisense technology the covalent-targeting of RNAs with complementary oligonucleotides bearing cross-linkers was extensively investigated.[150] This strategy has now been tried in the antimiR field. A series of three PNA-modified antimiRs carrying a 2-amino-6-vinylpurine (AVP) monomer at their N-terminus (Figure 8.4) were designed to invade pre-miR-122 at different sites and inhibit maturation of the miRNA precursor.[151] Upon incubation with pre-miR-122 hairpin, three of the antimiRs crosslinked to the target with varying efficiencies. Two of them hampered pre-miR-122 processing in an *in vitro* Dicer assay to a higher degree than the parent antimiR control sequences bearing no cross-linker.

Vermeulen and colleagues described how the extension of a 2′-OMe-antimiR core with flanking double-stranded regions increases its potency. The authors proposed the hypothesis that this was most likely to be due to additional favourable interactions between the miRNA-containing RISC and flanking RNA structure of the antimiR. Moreover, this unusual approach was extended to the inhibition of multiple miRNAs simultaneously.[129]

Along similar lines, multiple-target AMOs (MTg-AMOs) are a type of synthetic sponge-like construct in which multiple DNA antisense elements are combined into a single construct separated by short linkers for combined inhibition of several miRNAs. MTg-AMOs targeting miR-21, miR-155 and miR-17-5p ($\text{MTg-AMO}_{21/155/17}$) and miR-1 and miR-133 ($\text{MTg-AMO}_{1/133}$) were used to demonstrate the principle. In comparison to antimiRs addressing single targets, MTg-AMO$_{21/155/17}$ showed stronger inhibitory effects on cancer cell growth.[152]

"Caged" PNAs are a novel class of antimiR that are activated by irradiation. Here, an antimiR PNA is embedded in a cyclic structure containing a 9,10-dialcoxyanthracene linker. Upon irradiation with red light, the linker is cleaved, thereby activating antimiR activity.[153] This principle was demonstrated using a 14 nt long PNA antimiR-92. Calling upon another strategy investigated in the antisense field,[154] Patutina and colleagues have designed and characterized miRNA-specific artificial ribonucleases – "miRNase" – to hybridize to and cleave miRNA sequences in a miRNA-selective manner.[155,156] The miRNase template comprises a peptide sequence acetyl-[(LeuArg)$_2$Gly]$_2$ conjugated *via* an aminohexyl linker to either the 5′- or 3′-end of the miRNA-targeting antimiR. miRNases of different design have been tested and shown to cleave site-selectively miR-21 at guanidine linkages.

8.4 AntimiRs in Clinical Trials

Ultimately, antimiRs are designed for use as drugs in humans. There are a number of topical reviews summarizing the progress in miRNA therapeutics.[47,157,158] In this section, examples of the most relevant advances from the last few years are presented (see Table 8.1).

8.4.1 AntimiRs Targeting miR-122 for Treatment of HCV

Miravirsen is a 15 nt LNA–DNA chimera antimiR with a PS backbone that entered clinical trials as a potential treatment for hepatitis C virus (HCV) infections. The development of miravirsen began with the discovery that miRNA-122 binds to and regulates three conserved binding sites at the 5′ (two binding sites) and 3′ (one binding site) ends of the noncoding region (NCR) in HCV RNA, thereby facilitating the replication of the virus in liver cells.[48,159] In contrast to conventional miRNA activity, binding of miR-122 to the 5′NCR of the viral RNA protects the latter from degradation by cellular exoribonucleases Xrn1 and Xrn2 [160,161] and prevents stimulation of innate immune responses.[162] miRNA-122 has a very high expression in liver hepatocytes because of its role in cholesterol and lipid metabolism.[69,80] Experiments carried out in mice demonstrated that a PS LNA–DNA antimiR complementary to miR-122 suppresses miR-122 activity dose-dependently when administered systemically,[99] leading to de-repression of miR-122 target mRNAs and a decline in cholesterol levels. The inhibitory effect is due to the formation of a stable heteroduplex between the antimiR and miR-122, though there is also evidence that miravirsen inhibits miR-122 biogenesis by targeting its precursors.[131] The potency of miravirsen was considerably higher than that of cholesterol-conjugated antagomirs.[69] Miravirsen was subsequently tested in African green monkeys and chronically-infected chimpanzees.[135,163] Treatment resulted in long-lasting suppression of HCV

Table 8.1 Current clinical trials with antimiR drugs (March 2018).

Name	Length	Targeted miRNA	Indicated disease	Chemistry	Company	Clinical trial phase
Miravirsen (SPC3649)	15 nt	miR-122	Hepatitis C virus (HCV)	LNA–DNA mixmer with full PS backbone	Santaris Pharma/ Roche	II (completed)
RG-101	—	miR-122	HCV	Mixed chemistry with *N*-acetyl galactosamine conjugate	Regulus Therapeutics	II (completed)
RG-012	—	miR-21	Alport syndrome	Undisclosed	Regulus Therapeutics/Sanofi Genzyme	I/II (recruiting participants)
RG-125 (AZD4076)	—	miR-103/107	Non-alcoholic steatohepatitis (NASH)	*N*-acetyl galactosamine conjugate with undisclosed chemistry	AstraZeneca/Regulus Therapeutics	I/II (active, not recruiting)
RGLS4326	—	miR-17	Autosomal dominant polycystic kidney disease (ADPKD)	Undisclosed, with a design allowing preferential targeting of the kidney	Regulus Therapeutics	I (ongoing)
Cobomarsen (MRG-106)	—	miR-155	Mycosis fungicides (MF) form of cutaneous T-cell lymphoma (CTCL)	LNA-based	miRagen Therapeutics	I (recruiting participants)

viremia with no rebound in viremia or adaptive mutations in miRNA-122 target sites. In 2010 miravirsen entered Phase 2a clinical trials, where as a monotherapy in saline it caused a dose-dependent and prolonged reduction in HCV levels in patients with chronic HCV genotype 1 infections.[164] There were no prominent adverse effects or resistance-associated mutations,[164] and plasma levels of miRNA-122 decreased.[165] However, a C3U nucleotide change in the 5′UTR of the viral RNA outside of the seed region has been observed in patients that relapse and may therefore be responsible for the development of drug resistance.[166]

Recently, a second antimiR (RG-101) targeting miR-122 has been tested during clinical trials.[167] RG-101 is a PS mixed chimera of undisclosed chemical composition, conjugated to *N*-acetyl galactosamine (GalNAc) for enhanced delivery to liver hepatocytes[168] through the asialoglycoprotein receptor.[169] Upon reaching hepatocytes, RG-101 is metabolized to its main active metabolite.[167] In mice, RG-101 demonstrated a 20-fold better activity in comparison with the unconjugated antimiR.[170] RG-101 lowered viral levels in mice and showed little toxicity in mice, rats and non-human primates. Phase 1b trials were carried out on patients with different chronic HCV genotypes and showed that a single subcutaneous injection (2 or 4 mg kg^{-1}) led to significant decreases in HCV RNA levels together with a decline in total plasma cholesterol levels, without serious adverse events.[167] However, in most of the patients, viral relapse was observed. In the patients with HCV genotype 1 this was mostly associated with the same C3U mutation as seen for miravirsen therapy.[166]

Both studies proved that miR-122 is an important host factor that contributes to HCV replication and that antimiR targeting of miR-122 is a highly promising therapeutic strategy, albeit with important unresolved issues: loss of miR-122 has been associated with the development of HCC;[171] the development of viral resistance and stiff competition for the antimiR from a new line of effective, recently approved small-molecule HCV protease inhibitors. Despite promising results, further development of RG-101 is uncertain (http://ir.regulusrx.com/news-releases/news-release-details/regulus-announces-pipeline-updates-and-advancements; Regulus Therapeutics press release from 12th of June 2017).

8.4.2 AntimiR-targeting of Other miRNAs in Clinical Studies

miRNAs have been highlighted for their roles in liver and kidney diseases and are promising targets for antimiR therapy, especially because of the tendency for single-stranded PS oligonucleotides to accumulate in kidney.[172] One of these antimiRs is RG-012, an antimiR of unspecified chemistry directed against miR-21 as a possible treatment for Alport Syndrome.[173] The second is a GalNAc-conjugated antimiR targeted to miR-103/107 (RG-125/AZD4076), which is being investigated in Phase 1/2 clinical trials as a potential treatment for non-alcoholic steatohepatitis (NASH) in patients with type 2 diabetes or pre-diabetes. Finally, the miR-17-5p family, and especially miR-17-5p, is implicated in the misregulation of mitochondrial metabolism and progression of polycystic kidney disease.[174] RGLS4326 (antimiR-17) is a potential treatment for autosomal dominant polycystic kidney disease (ADPKD).

miR-155 plays an important role in differentiation, function and proliferation of blood and lymph cells and has been proposed as a therapeutic target in lymphoma cells showing aberrant proliferation. Cobomarsen (MRG-106) is an LNA-based antimiR-155 in Phase 1 clinical testing for treatment of the mycosis fungoides (MF) form of cutaneous T-cell lymphoma (CTCL). Preliminary clinical data is reportedly promising.[175,176] However, since it has been suggested that miR-155 may also function as a tumour suppressor,[177] a therapeutic approach targeting miR-155 requires careful validation.[178]

8.5 Conclusions

AntimiRs are the newest class of oligonucleotide therapeutics, which simply reflects the young age of miRNA research. They share many of the major challenges that antisense oligonucleotides and siRNAs pose for medicinal chemists, including their demanding pharmacokinetic properties, the difficulties in identifying disease-critical (miRNA) targets and limitations in their medicinal chemistry caused by the mechanism by which antimiRs engage their targets. Undoubtedly, antimiR research has profited greatly from the previous experience with oligonucleotide therapeutics, especially because the best chemical modifications, which were designed and optimized over decades for other oligonucleotide drug classes, can be seamlessly mixed-and-matched in antimiR reagents. However, antimiRs present their own set of unique challenges, including a new RNA mechanism to unravel and the high level of redundancy that complicates targeting of many miRNA–mRNA interactions.

Successful drug discovery is a combination of the right target, the right drug and the right patient population. It is likely that antimiR projects will experience similarly high attrition rates to those of other therapeutic classes that address novel targets in clinical settings.[179] Although the speed of the development of miravirsen – from the first description of the role of miR-122 in HCV infections[48] to the antiviral activity of the antimiR in patients[164] – was thrilling, an antimiR-122 therapeutic has not yet been approved. Thus, the target is not yet formally validated. In our view, the major question for the field lies with the feasibility of miRNA targets: how many and which ones genuinely represent valid drug targets?

New chemistries for antimiRs and new miRNA-targeting strategies continue to be developed. A small number of these may even eventually be clinically applicable. Some of them seem rather esoteric at first glance, but then many working in the oligonucleotide field initially doubted that synthetically demanding chemistries, such as LNAs and the GalNAc conjugates, could be developed for use in humans. On a positive note, there are ample options for modified antimiRs available to researchers while new technologies are explored.

We predict that advances in three areas will influence and accelerate miRNA targeting using antimiRs. First is the continuing development of stereo-pure phosphorothioate oligonucleotides,[180] which were recently tested *in vitro*[181] and *in vivo*.[182] Second will be the development of new conjugated groups to aid the pharmacokinetic properties of antimiRs, following the example of the GalNAc conjugates. Third is the increasing use of the clustered regularly interspaced short palindromic repeats (CRISPR)–CRISPR-associated (Cas) system[183,184] as a means to edit miRNAs.[185] For sure, the coming years in anti-miR research will be exciting.

Acknowledgement

This work was supported by a grant from the Swiss National Science Foundation (205321_169612 to J.H.).

References

1. D. P. Bartel, *Cell*, 2018, **173**, 20–51.
2. R. C. Lee, R. L. Feinbaum and V. Ambros, *Cell*, 1993, **75**, 843–854.
3. B. Wightman, I. Ha and G. Ruvkun, *Cell*, 1993, **75**, 855–862.
4. M. Ha and V. N. Kim, *Nat. Rev. Mol. Cell Biol.*, 2014, **15**, 509–524.
5. D. P. Bartel, *Cell*, 2004, **116**, 281–297.
6. A. M. Denli, B. B. Tops, R. H. Plasterk, R. F. Ketting and G. J. Hannon, *Nature*, 2004, **432**, 231–235.
7. E. Lund, S. Guttinger, A. Calado, J. E. Dahlberg and U. Kutay, *Science*, 2004, **303**, 95–98.
8. G. Hutvagner, J. McLachlan, A. E. Pasquinelli, E. Balint, T. Tuschl and P. D. Zamore, *Science*, 2001, **293**, 834–838.
9. H. Y. Lee and J. A. Doudna, *RNA*, 2012, **18**, 2012–2019.
10. K. Okamura, J. W. Hagen, H. Duan, D. M. Tyler and E. C. Lai, *Cell*, 2007, **130**, 89–100.
11. J. G. Ruby, C. H. Jan and D. P. Bartel, *Nature*, 2007, **448**, 83–86.
12. A. Khvorova, A. Reynolds and S. D. Jayasena, *Cell*, 2003, **115**, 209–216.
13. D. S. Schwarz, G. Hutvagner, T. Du, Z. Xu, N. Aronin and P. D. Zamore, *Cell*, 2003, **115**, 199–208.
14. S. Ro, C. Park, D. Young, K. M. Sanders and W. Yan, *Nucleic Acids Res.*, 2007, **35**, 5944–5953.
15. S. Jonas and E. Izaurralde, *Nat. Rev. Genet.*, 2015, **16**, 421–433.
16. R. C. Friedman, K. K. Farh, C. B. Burge and D. P. Bartel, *Genome Res.*, 2009, **19**, 92–105.
17. N. T. Schirle and I. J. MacRae, *Science*, 2012, **336**, 1037–1040.
18. U. A. Orom, F. C. Nielsen and A. H. Lund, *Mol. Cell*, 2008, **30**, 460–471.

19. J. R. Lytle, T. A. Yario and J. A. Steitz, *Proc. Natl. Acad. Sci. U. S. A.*, 2007, **104**, 9667–9672.
20. Y. Tay, J. Zhang, A. M. Thomson, B. Lim and I. Rigoutsos, *Nature*, 2008, **455**, 1124–1128.
21. A. Brümmer and J. Hausser, *BioEssays: News Rev. Mol., Cell. Dev. Biol.*, 2014, **36**, 617–626.
22. J. Brennecke, A. Stark, R. B. Russell and S. M. Cohen, *PLoS Biol.*, 2005, **3**, e85.
23. C. Shin, J. W. Nam, K. K. Farh, H. R. Chiang, A. Shkumatava and D. P. Bartel, *Mol. Cell*, 2010, **38**, 789–802.
24. N. Cloonan, *BioEssays*, 2015, **37**, 379–388.
25. M. C. Vella, E.-Y. Choi, S.-Y. Lin, K. Reinert and F. J. Slack, *Genes Dev.*, 2004, **18**, 132–137.
26. D. Didiano and O. Hobert, *Nat. Struct. Mol. Biol.*, 2006, **13**, 849.
27. S. W. Chi, G. J. Hannon and R. B. Darnell, *Nat. Struct. Mol. Biol.*, 2012, **19**, 321–327.
28. S. Vasudevan, Y. Tong and J. A. Steitz, *Science*, 2007, **318**, 1931–1934.
29. L. Poliseno, L. Salmena, J. Zhang, B. Carver, W. J. Haveman and P. P. Pandolfi, *Nature*, 2010, **465**, 1033.
30. M. Cesana, D. Cacchiarelli, I. Legnini, T. Santini, O. Sthandier, M. Chinappi, A. Tramontano and I. Bozzoni, *Cell*, 2011, **147**, 358–369.
31. J. Imig, A. Brunschweiger, A. Brummer, B. Guennewig, N. Mittal, S. Kishore, P. Tsikrika, A. P. Gerber, M. Zavolan and J. Hall, *Nat. Chem. Biol.*, 2015, **11**, 107–114.
32. T. B. Hansen, T. I. Jensen, B. H. Clausen, J. B. Bramsen, B. Finsen, C. K. Damgaard and J. Kjems, *Nature*, 2013, **495**, 384.
33. S. Memczak, M. Jens, A. Elefsinioti, F. Torti, J. Krueger, A. Rybak, L. Maier, S. D. Mackowiak, L. H. Gregersen, M. Munschauer, A. Loewer, U. Ziebold, M. Landthaler, C. Kocks, F. le Noble and N. Rajewsky, *Nature*, 2013, **495**, 333.
34. D. P. Bartel, *Cell*, 2009, **136**, 215–233.
35. J. A. Vidigal and A. Ventura, *Trends Cell Biol.*, 2015, **25**, 137–147.
36. E. A. Miska, E. Alvarez-Saavedra, A. L. Abbott, N. C. Lau, A. B. Hellman, S. M. McGonagle, D. P. Bartel, V. R. Ambros and H. R. Horvitz, *PLoS Genet.*, 2007, **3**, e215.
37. D. P. Bartel and C. Z. Chen, *Nat. Rev. Genet.*, 2004, **5**, 396–400.
38. A. Mencia, S. Modamio-Hoybjor, N. Redshaw, M. Morin, F. Mayo-Merino, L. Olavarrieta, L. A. Aguirre, I. del Castillo, K. P. Steel, T. Dalmay, F. Moreno and M. A. Moreno-Pelayo, *Nat. Genet.*, 2009, **41**, 609–613.
39. L. de Pontual, E. Yao, P. Callier, L. Faivre, V. Drouin, S. Cariou, A. Van Haeringen, D. Genevieve, A. Goldenberg, M. Oufadem, S. Manouvrier, A. Munnich, J. A. Vidigal, M. Vekemans, S. Lyonnet, A. Henrion-Caude, A. Ventura and J. Amiel, *Nat. Genet.*, 2011, **43**, 1026–1030.
40. R. Gambari, E. Brognara, D. A. Spandidos and E. Fabbri, *Int. J. Oncol.*, 2016, **49**, 5–32.

41. W. Lou, J. Liu, Y. Gao, G. Zhong, D. Chen, J. Shen, C. Bao, L. Xu, J. Pan, J. Cheng, B. Ding and W. Fan, *Oncotarget*, 2017, **8**, 115787–115802.
42. M. Rokavec, M. G. Öner, H. Li, R. Jackstadt, L. Jiang, D. Lodygin, M. Kaller, D. Horst, P. K. Ziegler, S. Schwitalla, J. Slotta-Huspenina, F. G. Bader, F. R. Greten and H. Hermeking, *J. Clin. Invest.*, 2014, **124**, 1853–1867.
43. P. R. Kuninty, J. Schnittert, G. Storm and J. Prakash, *Front. Oncol.*, 2016, **6**, 3.
44. D. R. Hurst, M. D. Edmonds and D. R. Welch, *Cancer Res.*, 2009, **69**, 7495–7498.
45. K. A. O'Donnell, E. A. Wentzel, K. I. Zeller, C. V. Dang and J. T. Mendell, *Nature*, 2005, **435**, 839.
46. G. Di Leva, M. Garofalo and C. M. Croce, *Annu. Rev. Pathol.: Mech. Dis.*, 2014, **9**, 287–314.
47. R. Rupaimoole and F. J. Slack, *Nat. Rev. Drug Discovery*, 2017, **16**, 203.
48. C. L. Jopling, M. Yi, A. M. Lancaster, S. M. Lemon and P. Sarnow, *Science*, 2005, **309**, 1577–1581.
49. M. Trajkovski, J. Hausser, J. Soutschek, B. Bhat, A. Akin, M. Zavolan, M. H. Heim and M. Stoffel, *Nature*, 2011, **474**, 649–653.
50. E. M. Small, R. J. A. Frost and E. N. Olson, *Circulation*, 2010, **121**, 1022–1032.
51. S. Dangwal and T. Thum, *Annu. Rev. Pharmacol. Toxicol.*, 2014, **54**, 185–203.
52. R. S. Geary, *Expert Opin. Drug Metab. Toxicol.*, 2009, **5**, 381–391.
53. J. Weiler, J. Hunziker and J. Hall, *Gene Ther.*, 2005, **13**, 496–502.
54. K. A. Lennox and M. A. Behlke, *Pharm. Res.*, 2010, **27**, 1788–1799.
55. Z. Li and T. M. Rana, *Nat. Rev. Drug Discovery*, 2014, **13**, 622.
56. K. A. Lennox and M. A. Behlke, *Gene Ther.*, 2011, **18**, 1111–1120.
57. F. Eckstein, *Nucleic Acid Ther.*, 2014, **24**, 374–387.
58. D. A. Brown, S. H. Kang, S. M. Gryaznov, L. DeDionisio, O. Heidenreich, S. Sullivan, X. Xu and M. I. Nerenberg, *J. Biol. Chem.*, 1994, **269**, 26801–26805.
59. C. A. Stein, J. B. Hansen, J. Lai, S. Wu, A. Voskresenskiy, A. Høg, J. Worm, M. Hedtjärn, N. Souleimanian, P. Miller, H. S. Soifer, D. Castanotto, L. Benimetskaya, H. Ørum and T. Koch, *Nucleic Acids Res.*, 2010, **38**, e3.
60. A. Boutla, C. Delidakis and M. Tabler, *Nucleic Acids Res.*, 2003, **31**, 4973–4980.
61. G. Hutvágner, M. J. Simard, C. C. Mello and P. D. Zamore, *PLoS Biol.*, 2004, **2**, e98.
62. G. Meister, M. Landthaler, Y. Dorsett and T. Tuschl, *RNA*, 2004, **10**, 544–550.
63. F. Wang, B. Zhang, L. Zhou, Y. Shi, Z. Li, Y. Xia and J. Tian, *ACS Appl. Mater. Interfaces*, 2016, **8**, 9014–9021.
64. J. Schnittert, P. R. Kuninty, T. F. Bystry, R. Brock, G. Storm and J. Prakash, *Nanomedicine*, 2017, **12**, 1369–1384.

65. S. H. Lee, B. Castagner and J. C. Leroux, *Eur. J. Pharm. Biopharm.: Off. J. Arbeitsgem. Pharm. Verfahrenstech. e.V*, 2013, **85**, 5–11.
66. S. Aubry, F. Burlina, E. Dupont, D. Delaroche, A. Joliot, S. Lavielle, G. Chassaing and S. Sagan, *FASEB J.: Off. Publ. Fed. Am. Soc. Exp. Biol.*, 2009, **23**, 2956–2967.
67. J. R. van Beijnum, E. Giovannetti, D. Poel, P. Nowak-Sliwinska and A. W. Griffioen, *Angiogenesis*, 2017, **20**, 269–285.
68. D. Pollutri, C. Patrizi, S. Marinelli, C. Giovannini, E. Trombetta, F. A. Giannone, M. Baldassarre, S. Quarta, Y. P. Vandewynckel, A. Vandierendonck, H. Van Vlierberghe, L. Porretti, M. Negrini, L. Bolondi, L. Gramantieri and F. Fornari, *Cell Death Dis.*, 2018, **9**, 4.
69. J. Krutzfeldt, N. Rajewsky, R. Braich, K. G. Rajeev, T. Tuschl, M. Manoharan and M. Stoffel, *Nature*, 2005, **438**, 685–689.
70. J. Krutzfeldt, S. Kuwajima, R. Braich, K. G. Rajeev, J. Pena, T. Tuschl, M. Manoharan and M. Stoffel, *Nucleic Acids Res.*, 2007, **35**, 2885–2892.
71. C. S. Velu and H. L. Grimes, *Methods Mol. Biol.*, 2012, **928**, 185–195.
72. Q. Zhang, R. Ran, L. Zhang, Y. Liu, L. Mei, Z. Zhang, H. Gao and Q. He, *J. Controlled Release: Off. J. Controlled Release Soc.*, 2015, **197**, 208–218.
73. L. Ma, J. Teruya-Feldstein and R. A. Weinberg, *Nature*, 2007, **449**, 682.
74. L. Ma, F. Reinhardt, E. Pan, J. Soutschek, B. Bhat, E. G. Marcusson, J. Teruya-Feldstein, G. W. Bell and R. A. Weinberg, *Nat. Biotechnol.*, 2010, **28**, 341.
75. A. Makovitzki, A. Fink and Y. Shai, *Cancer Res.*, 2009, **69**, 3458–3463.
76. N. Papo, A. Braunstein, Z. Eshhar and Y. Shai, *Cancer Res.*, 2004, **64**, 5779–5786.
77. J. K. Park, T. Kogure, G. J. Nuovo, J. Jiang, L. He, J. H. Kim, M. A. Phelps, T. L. Papenfuss, C. M. Croce, T. Patel and T. D. Schmittgen, *Cancer Res.*, 2011, **71**, 7608–7616.
78. P. Martin, *Helv. Chim. Acta*, 1995, **78**, 486–504.
79. C. Esau, X. Kang, E. Peralta, E. Hanson, E. G. Marcusson, L. V. Ravichandran, Y. Sun, S. Koo, R. J. Perera, R. Jain, N. M. Dean, S. M. Freier, C. F. Bennett, B. Lollo and R. Griffey, *J. Biol. Chem.*, 2004, **279**, 52361–52365.
80. C. Esau, S. Davis, S. F. Murray, X. X. Yu, S. K. Pandey, M. Pear, L. Watts, S. L. Booten, M. Graham, R. McKay, A. Subramaniam, S. Propp, B. A. Lollo, S. Freier, C. F. Bennett, S. Bhanot and B. P. Monia, *Cell Metab.*, 2006, **3**, 87–98.
81. G. Gabriely, M. Yi, R. S. Narayan, J. M. Niers, T. Wurdinger, J. Imitola, K. L. Ligon, S. Kesari, C. Esau, R. M. Stephens, B. A. Tannous and A. M. Krichevsky, *Cancer Res.*, 2011, **71**, 3563–3572.
82. N. M. Teplyuk, E. J. Uhlmann, G. Gabriely, N. Volfovsky, Y. Wang, J. Teng, P. Karmali, E. Marcusson, M. Peter, A. Mohan, Y. Kraytsberg, R. Cialic, E. A. Chiocca, J. Godlewski, B. Tannous and A. M. Krichevsky, *EMBO Mol. Med.*, 2016, **8**, 268–287.
83. W. Guschlbauer and K. Jankowski, *Nucleic Acids Res.*, 1980, **8**, 1421–1433.

84. F. Guo, Q. Li and C. Zhou, *Org. Biomol. Chem.*, 2017, **15**, 9552–9565.
85. S. Davis, B. Lollo, S. Freier and C. Esau, *Nucleic Acids Res.*, 2006, **34**, 2294–2304.
86. S. Davis, S. Propp, S. M. Freier, L. E. Jones, M. J. Serra, G. Kinberger, B. Bhat, E. E. Swayze, C. F. Bennett and C. Esau, *Nucleic Acids Res.*, 2009, **37**, 70–77.
87. K. J. Rayner, Y. Suárez, A. Dávalos, S. Parathath, M. L. Fitzgerald, N. Tamehiro, E. A. Fisher, K. J. Moore and C. Fernández-Hernando, *Science*, 2010, **328**, 1570–1573.
88. S. H. Najafi-Shoushtari, F. Kristo, Y. Li, T. Shioda, D. E. Cohen, R. E. Gerszten and A. M. Näär, *Science*, 2010, **328**, 1566–1569.
89. T. J. Marquart, R. M. Allen, D. S. Ory and A. Baldan, *Proc. Natl. Acad. Sci. U. S. A.*, 2010, **107**, 12228–12232.
90. K. J. Rayner, F. J. Sheedy, C. C. Esau, F. N. Hussain, R. E. Temel, S. Parathath, J. M. van Gils, A. J. Rayner, A. N. Chang, Y. Suarez, C. Fernandez-Hernando, E. A. Fisher and K. J. Moore, *J. Clin. Invest.*, 2011, **121**, 2921–2931.
91. K. J. Rayner, C. C. Esau, F. N. Hussain, A. L. McDaniel, S. M. Marshall, J. M. van Gils, T. D. Ray, F. J. Sheedy, L. Goedeke, X. Liu, O. G. Khatsenko, V. Kaimal, C. J. Lees, C. Fernandez-Hernando, E. A. Fisher, R. E. Temel and K. J. Moore, *Nature*, 2011, **478**, 404.
92. L. Goedeke, A. Salerno, C. M. Ramírez, L. Guo, R. M. Allen, X. Yin, S. R. Langley, C. Esau, A. Wanschel, E. A. Fisher, Y. Suárez, A. Baldán, M. Mayr and C. Fernández-Hernando, *EMBO Mol. Med.*, 2014, **6**, 1133–1141.
93. S. Obika, D. Nanbu, Y. Hari, K.-i. Morio, Y. In, T. Ishida and T. Imanishi, *Tetrahedron Lett.*, 1997, **38**, 8735–8738.
94. J. Wengel, *Acc. Chem. Res.*, 1999, **32**, 301–310.
95. M. Petersen, K. Bondensgaard, J. Wengel and J. P. Jacobsen, *J. Am. Chem. Soc.*, 2002, **124**, 5974–5982.
96. J. A. Chan, A. M. Krichevsky and K. S. Kosik, *Cancer Res.*, 2005, **65**, 6029–6033.
97. U. A. Orom, S. Kauppinen and A. H. Lund, *Gene*, 2006, **372**, 137–141.
98. M. M. Fabani and M. J. Gait, *RNA*, 2008, **14**, 336–346.
99. J. Elmen, M. Lindow, A. Silahtaroglu, M. Bak, M. Christensen, A. Lind-Thomsen, M. Hedtjarn, J. B. Hansen, H. F. Hansen, E. M. Straarup, K. McCullagh, P. Kearney and S. Kauppinen, *Nucleic Acids Res.*, 2008, **36**, 1153–1162.
100. S. Obad, C. O. dos Santos, A. Petri, M. Heidenblad, O. Broom, C. Ruse, C. Fu, M. Lindow, J. Stenvang, E. M. Straarup, H. F. Hansen, T. Koch, D. Pappin, G. J. Hannon and S. Kauppinen, *Nat. Genet.*, 2011, **43**, 371–378.
101. W. M. Flanagan, R. W. Wagner, D. Grant, K.-Y. Lin and M. D. Matteucci, *Nat. Biotechnol.*, 1999, **17**, 48.
102. M. Thomas, K. Lange-Grunweller, E. Dayyoub, U. Bakowsky, U. Weirauch, A. Aigner, R. K. Hartmann and A. Grunweller, *RNA Biol.*, 2012, **9**, 1088–1098.

103. L. Venturini, K. Battmer, M. Castoldi, B. Schultheis, A. Hochhaus, M. U. Muckenthaler, A. Ganser, M. Eder and M. Scherr, *Blood*, 2007, **109**, 4399–4405.
104. I. Ivanovska, A. S. Ball, R. L. Diaz, J. F. Magnus, M. Kibukawa, J. M. Schelter, S. V. Kobayashi, L. Lim, J. Burchard, A. L. Jackson, P. S. Linsley and M. A. Cleary, *Mol. Cell. Biol.*, 2008, **28**, 2167–2174.
105. S. D. Selcuklu, M. T. Donoghue and C. Spillane, *Biochem. Soc. Trans.*, 2009, **37**, 918–925.
106. P. M. Costa, A. L. Cardoso, C. Custodia, P. Cunha, L. Pereira de Almeida and M. C. Pedroso de Lima, *J. Controlled Release: Off. J. Controlled Release Soc.*, 2015, **207**, 31–39.
107. J. Deshane, C. C. Garner and H. Sontheimer, *J. Biol. Chem.*, 2003, **278**, 4135–4144.
108. L. Soroceanu, Y. Gillespie, M. B. Khazaeli and H. Sontheimer, *Cancer Res.*, 1998, **58**, 4871–4879.
109. R. Nedaeinia, M. Sharifi, A. Avan, M. Kazemi, A. Nabinejad, G. A. Ferns, M. Ghayour-Mobarhan and R. Salehi, *Tumour Biol.: J. Int. Soc. Oncodev. Biol. Med.*, 2017, **39**, 1010428317692261.
110. R. L. Montgomery, T. G. Hullinger, H. M. Semus, B. A. Dickinson, A. G. Seto, J. M. Lynch, C. Stack, P. A. Latimer, E. N. Olson and E. van Rooij, *Circulation*, 2011, **124**, 1537–1547.
111. T. G. Hullinger, R. L. Montgomery, A. G. Seto, B. A. Dickinson, H. M. Semus, J. M. Lynch, C. M. Dalby, K. Robinson, C. Stack, P. A. Latimer, J. M. Hare, E. N. Olson and E. van Rooij, *Circ. Res.*, 2012, **110**, 71–81.
112. P. P. Seth, G. Vasquez, C. A. Allerson, A. Berdeja, H. Gaus, G. A. Kinberger, T. P. Prakash, M. T. Migawa, B. Bhat and E. E. Swayze, *J. Org. Chem.*, 2010, **75**, 1569–1581.
113. D. J. Hogan, T. M. Vincent, S. Fish, E. G. Marcusson, B. Bhat, B. N. Chau and D. G. Zisoulis, *PLoS One*, 2014, **9**, e100951.
114. S. Nahar, A. Singh, K. Morihiro, Y. Moai, T. Kodama, S. Obika and S. Maiti, *Biochemistry*, 2016, **55**, 7023–7032.
115. P. Nielsen, M. Egholm, R. Berg and O. Buchardt, *Science*, 1991, **254**, 1497–1500.
116. S. Y. Oh, Y. Ju and H. Park, *Mol. Cells*, 2009, **28**, 341–345.
117. M. M. Fabani, C. Abreu-Goodger, D. Williams, P. A. Lyons, A. G. Torres, K. G. Smith, A. J. Enright, M. J. Gait and E. Vigorito, *Nucleic Acids Res.*, 2010, **38**, 4466–4475.
118. A. G. Torres, M. M. Fabani, E. Vigorito, D. Williams, N. Al-Obaidi, F. Wojciechowski, R. H. Hudson, O. Seitz and M. J. Gait, *Nucleic Acids Res.*, 2012, **40**, 2152–2167.
119. C. J. Cheng, R. Bahal, I. A. Babar, Z. Pincus, F. Barrera, C. Liu, A. Svoronos, D. T. Braddock, P. M. Glazer, D. M. Engelman, W. M. Saltzman and F. J. Slack, *Nature*, 2014, **518**, 107.
120. O. A. Andreev, D. M. Engelman and Y. K. Reshetnyak, *Mol. Membr. Biol.*, 2010, **27**, 341–352.

121. Y. K. Reshetnyak, M. Segala, O. A. Andreev and D. M. Engelman, *Biophys. J.*, 2007, **93**, 2363–2372.
122. H. Østergaard, C. Tachibana and J. R. Winther, *J. Cell Biol.*, 2004, **166**, 337–345.
123. A. Gupta, E. Quijano, Y. Liu, R. Bahal, S. E. Scanlon, E. Song, W.-C. Hsieh, D. E. Braddock, D. H. Ly, W. M. Saltzman and P. M. Glazer, *Mol. Ther.–Nucleic Acids*, 2017, **9**, 111–119.
124. E. Brognara, E. Fabbri, E. Bazzoli, G. Montagner, C. Ghimenton, A. Eccher, C. Cantù, A. Manicardi, N. Bianchi, A. Finotti, G. Breveglieri, M. Borgatti, R. Corradini, V. Bezzerri, G. Cabrini and R. Gambari, *J. Neuro-Oncol.*, 2014, **118**, 19–28.
125. E. Brognara, E. Fabbri, F. Aimi, A. Manicardi, N. Bianchi, A. Finotti, G. Breveglieri, M. Borgatti, R. Corradini, R. Marchelli and R. Gambari, *Int. J. Oncol.*, 2012, **41**, 2119–2127.
126. J. Summerton and D. Weller, *Antisense Nucleic Acid Drug Dev.*, 1997, **7**, 187–195.
127. J. D. Moulton and Y. L. Yan, *Curr. Protoc. Mol. Biol.*, 2008, **83**, 26.8.1–26.8.29.
128. A. S. Flynt, N. Li, E. J. Thatcher, L. Solnica-Krezel and J. G. Patton, *Nat. Genet.*, 2007, **39**, 259–263.
129. A. Vermeulen, B. Robertson, A. B. Dalby, W. S. Marshall, J. Karpilow, D. Leake, A. Khvorova and S. Baskerville, *RNA*, 2007, **13**, 723–730.
130. W. P. Kloosterman, A. K. Lagendijk, R. F. Ketting, J. D. Moulton and R. H. Plasterk, *PLoS Biol.*, 2007, **5**, e203.
131. L. F. Gebert, M. A. Rebhan, S. E. Crivelli, R. Denzler, M. Stoffel and J. Hall, *Nucleic Acids Res.*, 2014, **42**, 609–621.
132. A. G. Torres, M. M. Fabani, E. Vigorito and M. J. Gait, *RNA*, 2011, **17**, 933–943.
133. S. L. Ameres, M. D. Horwich, J.-H. Hung, J. Xu, M. Ghildiyal, Z. Weng and P. D. Zamore, *Science*, 2010, **328**, 1534–1539.
134. P. Pisacane and M. Halic, *Nat. Commun.*, 2017, **8**, 15332.
135. J. Elmen, M. Lindow, S. Schutz, M. Lawrence, A. Petri, S. Obad, M. Lindholm, M. Hedtjarn, H. F. Hansen, U. Berger, S. Gullans, P. Kearney, P. Sarnow, E. M. Straarup and S. Kauppinen, *Nature*, 2008, **452**, 896–899.
136. B. Robertson, A. B. Dalby, J. Karpilow, A. Khvorova, D. Leake and A. Vermeulen, *Silence*, 2010, **1**, 10.
137. A. Brunschweiger, L. F. Gebert, M. Lucic, U. Pradere, H. Jahns, C. Berk, J. Hunziker and J. Hall, *Chem. Commun.*, 2016, **52**, 156–159.
138. K. A. Lennox, R. Owczarzy, D. M. Thomas, J. A. Walder and M. A. Behlke, *Mol. Ther.–Nucleic Acids*, 2013, **2**, e117.
139. K. A. Lennox, C. A. Vakulskas and M. A. Behlke, in *Drug Target miRNA: Methods and Protocols*, ed. M. F. Schmidt, Springer New York, New York, NY, 2017, pp. 51–69.
140. N. T. Schirle, J. Sheu-Gruttadauria and I. J. MacRae, *Science*, 2014, **346**, 608–613.

141. M. F. Schmidt, O. Korb and C. Abell, *ACS Chem. Biol.*, 2013, **8**, 2122–2126.
142. N. De, L. Young, P.-W. Lau, N.-C. Meisner, D. V. Morrissey and I. J. MacRae, *Mol. Cell*, 2013, **50**, 344–355.
143. J. S. Parker, S. M. Roe and D. Barford, *Nature*, 2005, **434**, 663–666.
144. J. Ariyoshi, D. Momokawa, N. Eimori, A. Kobori, A. Murakami and A. Yamayoshi, *Bioconjugate Chem.*, 2015, **26**, 2454–2460.
145. J. Ariyoshi, Y. Matsuyama, A. Kobori, A. Murakami, H. Sugiyama and A. Yamayoshi, *Nucleic Acid Ther.*, 2017, **27**, 303–308.
146. Y. Kamiya, Y. Donoshita, H. Kamimoto, K. Murayama, J. Ariyoshi and H. Asanuma, *ChemBioChem*, 2017, **18**, 1917–1922.
147. Y. Kamiya, J. Takai, H. Ito, K. Murayama, H. Kashida and H. Asanuma, *ChemBioChem*, 2014, **15**, 2549–2555.
148. C. Cheong, I. Tinoco and A. Chollet, *Nucleic Acids Res.*, 1988, **16**, 5115–5122.
149. S. Nahar, V. Kotikam, V. A. Kumar and S. Maiti, *Nucleic Acid Ther.*, 2016, **26**, 327–334.
150. C. Boiziau, A. S. Boutorine, N. Loreau, P. Verspieren, N. T. Thuong and J. J. Toulmé, *Nucleosides Nucleotides*, 1991, **10**, 239–244.
151. T. Akisawa, K. Yamada and F. Nagatsugi, *Bioorg. Med. Chem. Lett.*, 2016, **26**, 5902–5906.
152. Y. Lu, J. Xiao, H. Lin, Y. Bai, X. Luo, Z. Wang and B. Yang, *Nucleic Acids Res.*, 2009, **37**, e24.
153. S. G. Konig and A. Mokhir, *Bioorg. Med. Chem. Lett.*, 2013, **23**, 6544–6548.
154. J. Hall, D. Husken, U. Pieles, H. E. Moser and R. Haner, *Chem. Biol.*, 1994, **1**, 185–190.
155. O. A. Patutina, E. V. Bichenkova, S. K. Miroshnichenko, N. L. Mironova, L. T. Trivoluzzi, K. K. Burusco, R. A. Bryce, V. V. Vlassov and M. A. Zenkova, *Biomaterials*, 2017, **122**, 163–178.
156. O. A. Patutina, S. K. Miroshnichenko, A. A. Lomzov, N. L. Mironova and M. A. Zenkova, *Russ. J. Bioorg. Chem.*, 2017, **43**, 29–37.
157. M. Matsui and D. R. Corey, *Nat. Rev. Drug Discovery*, 2016, **16**, 167.
158. W. Cha, R. Fan, Y. Miao, Y. Zhou, C. Qin, X. Shan, X. Wan and T. Cui, *MedChemComm*, 2018, **9**, 396–408.
159. C. L. Jopling, S. Schutz and P. Sarnow, *Cell Host Microbe*, 2008, **4**, 77–85.
160. Y. Li, D. Yamane and S. M. Lemon, *J. Virol.*, 2015, **89**, 4857–4865.
161. C. D. Sedano and P. Sarnow, *Cell Host Microbe*, 2014, **16**, 257–264.
162. E. S. Machlin, P. Sarnow and S. M. Sagan, *Proc. Natl. Acad. Sci. U. S. A.*, 2011, **108**, 3193–3198.
163. R. E. Lanford, E. S. Hildebrandt-Eriksen, A. Petri, R. Persson, M. Lindow, M. E. Munk, S. Kauppinen and H. Orum, *Science*, 2010, **327**, 198–201.
164. H. L. Janssen, H. W. Reesink, E. J. Lawitz, S. Zeuzem, M. Rodriguez-Torres, K. Patel, A. J. van der Meer, A. K. Patick, A. Chen, Y. Zhou, R. Persson, B. D. King, S. Kauppinen, A. A. Levin and M. R. Hodges, *N. Engl. J. Med.*, 2013, **368**, 1685–1694.

165. M. H. van der Ree, A. J. van der Meer, A. C. van Nuenen, J. de Bruijne, S. Ottosen, H. L. Janssen, N. A. Kootstra and H. W. Reesink, *Aliment. Pharmacol. Ther.*, 2016, **43**, 102–113.
166. S. Ottosen, T. B. Parsley, L. Yang, K. Zeh, L. J. van Doorn, E. van der Veer, A. K. Raney, M. R. Hodges and A. K. Patick, *Antimicrob. Agents Chemother.*, 2015, **59**, 599–608.
167. M. H. van der Ree, J. M. de Vree, F. Stelma, S. Willemse, M. van der Valk, S. Rietdijk, R. Molenkamp, J. Schinkel, A. C. van Nuenen, U. Beuers, S. Hadi, M. Harbers, E. van der Veer, K. Liu, J. Grundy, A. K. Patick, A. Pavlicek, J. Blem, M. Huang, P. Grint, S. Neben, N. W. Gibson, N. A. Kootstra and H. W. Reesink, *Lancet*, 2017, **389**, 709–717.
168. Y. Huang, *Mol. Ther.–Nucleic Acids*, 2017, **6**, 116–132.
169. M. Spiess, *Biochemistry*, 1990, **29**, 10009–10018.
170. B. Bhat, S. Neben, J. Tay, K. Liu, N. Chau, D. Hogan, D. MacKenna and N. Gibson, *Hepatology*, 2013, **58**, 1377A–1395A.
171. W. C. Tsai, S. D. Hsu, C. S. Hsu, T. C. Lai, S. J. Chen, R. Shen, Y. Huang, H. C. Chen, C. H. Lee, T. F. Tsai, M. T. Hsu, J. C. Wu, H. D. Huang, M. S. Shiao, M. Hsiao and A. P. Tsou, *J. Clin. Invest.*, 2012, **122**, 2884–2897.
172. R. S. Geary, D. Norris, R. Yu and C. F. Bennett, *Adv. Drug Delivery Rev.*, 2015, **87**, 46–51.
173. I. G. Gomez, D. A. MacKenna, B. G. Johnson, V. Kaimal, A. M. Roach, S. Ren, N. Nakagawa, C. Xin, R. Newitt, S. Pandya, T. H. Xia, X. Liu, D. B. Borza, M. Grafals, S. J. Shankland, J. Himmelfarb, D. Portilla, S. Liu, B. N. Chau and J. S. Duffield, *J. Clin. Invest.*, 2015, **125**, 141–156.
174. S. Hajarnis, R. Lakhia, M. Yheskel, D. Williams, M. Sorourian, X. Liu, K. Aboudehen, S. Zhang, K. Kersjes, R. Galasso, J. Li, V. Kaimal, S. Lockton, S. Davis, A. Flaten, J. A. Johnson, W. L. Holland, C. M. Kusminski, P. E. Scherer, P. C. Harris, M. Trudel, D. P. Wallace, P. Igarashi, E. C. Lee, J. R. Androsavich and V. Patel, *Nat. Commun.*, 2017, **8**, 14395.
175. C. Querfeld, F. M. Foss, L. C. Pinter-Brown, P. Porcu, B. M. William, T. Pacheco, B. M. Haverkos, Y. H. Kim, J. Guitart, A. S. Halwani, J. DeSimone, A. G. Seto, L. A. Pestano, A. L. Jackson, P. J. Williams, B. A. Dickinson, J. Ruckman, G. Gordon, P. Rubin and B. S. Marshall, *Blood*, 2017, **130**, 820.
176. C. Querfeld, T. Pacheco, F. M. Foss, A. S. Halwani, P. Porcu, A. G. Seto, J. Ruckman, M. L. Landry, A. L. Jackson, L. A. Pestano, B. A. Dickinson, M. Sanseverino, D. M. Rodman, G. Gordon and W. Marshall, *Blood*, 2016, **128**, 1829.
177. C. Lei, Y. Wang, Y. Huang, H. Yu, Y. Huang, L. Wu and L. Huang, *PLoS One*, 2012, **7**, e52310.
178. T. Catela Ivkovic, G. Voss, H. Cornella and Y. Ceder, *Cancer Lett.*, 2017, **407**, 113–122.
179. H. L. Lightfoot and J. Hall, *Nucleic Acids Res.*, 2012, **40**, 10585–10595.
180. N. Oka and T. Wada, *Chem. Soc. Rev.*, 2011, **40**, 5829–5843.

181. M. Li, H. L. Lightfoot, F. Halloy, A. L. Malinowska, C. Berk, A. Behera, D. Schumperli and J. Hall, *Chem. Commun.*, 2017, **53**, 541–544.
182. N. Iwamoto, D. C. D. Butler, N. Svrzikapa, S. Mohapatra, I. Zlatev, D. W. Y. Sah, Meena, S. M. Standley, G. Lu, L. H. Apponi, M. Frank-Kamenetsky, J. J. Zhang, C. Vargeese and G. L. Verdine, *Nat. Biotechnol.*, 2017, **35**, 845–851.
183. Y. Zhao, Z. Dai, Y. Liang, M. Yin, K. Ma, M. He, H. Ouyang and C. B. Teng, *Sci. Rep.*, 2014, **4**, 3943.
184. L. S. Qi, M. H. Larson, L. A. Gilbert, J. A. Doudna, J. S. Weissman, A. P. Arkin and W. A. Lim, *Cell*, 2013, **152**, 1173–1183.
185. J. Yang, X. Meng, J. Pan, N. Jiang, C. Zhou, Z. Wu and Z. Gong, *RNA Biol.*, 2018, **15**, 35–43.

CHAPTER 9

Oligonucleotide Targeting of Long Non-coding RNAs

I. BLOKHIN[a], O. KHORKOVA[b], J. HSIAO[b] AND C. WAHLESTEDT*[a]

[a]Center for Therapeutic Innovation and Department of Psychiatry and Behavioral Sciences, University of Miami, Miller School of Medicine, 1501 NW 10th Avenue, Miami, FL 33136, USA; [b]OPKO Health Inc., 10320 USA Today Way, Miramar, FL 33025, USA
*E-mail: CWahlestedt@med.miami.edu

9.1 Introduction

Long non-coding RNAs (lncRNAs) are non-coding transcripts that are more than 200 nucleotides long. They are sometimes called "dark matter of the transcriptome" because of their pervasive expression and enormous but still unclear functional potential. Systematic research into lncRNAs began within the last 10–15 years, and both its path and timing somewhat resemble the progress in the study of microRNA (miR). miRs constitute relatively short (19–24 nucleotides long) non-coding molecules that act as post-transcriptional inhibitors (see Chapter 8). The first miR, lineage 4 (lin-4), was identified in *Caenorhabditis elegans* in 1993.[1] Seven years later, the first mammalian miR, lethal 7 (let-7), was discovered.[2] Currently, 25 years later, thousands of miRs have been characterized in virtually all species, and ~2600 miRs have been described in humans (miRBase, Release 21, accessed in February 2018).

Drug Discovery Series No. 68
Advances in Nucleic Acid Therapeutics
Edited by Sudhir Agrawal and Michael J. Gait

Published by the Royal Society of Chemistry, www.rsc.org

Many miRs are implicated in etiology and pathogenesis of medical conditions, and several miR-based therapeutics have reached the clinical development stage (reviewed by Rupaimoole and Slack[3]).

Characterization of lncRNAs, however, appears to be more challenging compared with miRs. One major reason is the structural and functional complexity of lncRNAs. While miRs function post-transcriptionally in the cytoplasm, lncRNAs act in both nucleus and cytoplasm and regulate transcriptional and post-transcriptional processes by interacting with DNA, RNA (both coding and non-coding) and proteins. It is still unclear how many lncRNAs are encoded in the human genome, but their number greatly surpasses the number of protein-coding genes and miRs combined. While there are approximately 20 000 protein-coding genes and 2600 miRs, the number of known lncRNAs already exceeds 50 000 and is constantly growing (LNCipedia, v. 5.0, accessed in February 2018).

The large number of lncRNAs, along with their vast functional potential, creates an opportunity for a novel class of therapies. Oligonucleotide-based compounds are an obvious choice for targeting lncRNAs, and advances in oligonucleotide chemistry and delivery platforms indicate the feasibility of this approach. As a therapeutic modality, oligonucleotides had a slow start but are gradually gaining momentum (see Chapters 1 and 2), with over 400 clinical trials of oligonucleotide-based drugs registered at http://ClinicalTrials.gov (accessed on February 5, 2018) and over 15 000 patients treated with oligonucleotide-based therapies by 2017. Notably, more oligonucleotide compounds reached clinical trials in the last 5 years than in the preceding decades. Compared with traditional therapeutics, such as small molecules and antibodies, oligonucleotides have several advantages for targeting lncRNA. Because of precise complementarity between oligonucleotides and lncRNAs, screening and testing are relatively fast. Furthermore, oligonucleotides can enter the cell and engage intracellular enzymatic machineries to specifically target lncRNAs that are otherwise inaccessible to antibodies (which act on the cellular surface) and are less specifically regulated by small molecules. As oligonucleotide-based drugs are more specific, the number of potential side effects is relatively limited. Recent progress in the chemistry of oligonucleotides also improved target specificity and delivery. However, the lncRNA field is still in its infancy and hard work is required to understand the lncRNA-dependent mechanisms of disease.

In this chapter, we discuss the history, biology and classification of lncRNAs. We also review the current status of lncRNA-based medicines and highlight future therapeutic applications.

9.2 History of lncRNAs

The central dogma of molecular biology, stating that RNA is an intermediate step in the flow of information from DNA to protein, was proposed by Crick[4] just a few years after his and Watson's discovery of structure of DNA. In the framework of the "one gene–one protein concept", this dogma flourished for a long time, and the only accepted types of RNA that had no coding potential were ribosomal RNA (rRNA) and transfer RNA (tRNA), components of the translational machinery. Throughout this period, non-coding sequences (which constitute approximately 98% of the genome) were considered non-functional. Few scientists, however, had the exceptional insight to be able to predict early that the genome may encode multiple non-coding features with functions yet to be conceived.[5,6] In the 1990s, several publications described non-coding transcripts that were clearly functional. For example, an essential role of lncRNAs in gene silencing and imprinting was uncovered at that time (reviewed recently by Carrel[7]). The lncRNA X-inactive specific transcript (Xist) was found to confer transcriptional silencing and inactivation of the X chromosome *via* formation of heterochromatin.[8] Furthermore, it was shown that lncRNA H19 participated in imprinting of the insulin-like growth factor 2 (IGF2) gene and that disruption of this mechanism was present in a large proportion of cervical cancers.[9] Another example is lncRNA potassium voltage-gated channel subfamily Q member 1-overlapping transcript1 (KCNQ1OT1), which was studied in 1999 with relation to Beckwith–Wiedemann syndrome, a pediatric overgrowth disorder involving a predisposition to tumor development.[10] KCNQ1OT1 is transcribed in the antisense direction to KCNQ1, a gene that encodes a potassium voltage-gated channel. KCNQ1OT1 is expressed only from the paternal allele and controls the expression of several imprinted genes. The maternal allele of KCNQ1OT1 is normally suppressed, in part *via* methylation of CpG islands in its promoter. In patients with Beckwith–Wiedemann syndrome, KCNQ1OT1 is expressed from both alleles, which impairs the imprinting and expression of KCNQ1OT1 target genes.

Substantial progress in understanding of lncRNAs has come from oncology. Amongst the first lncRNAs identified in cancer were prostate cancer antigen 3 (PCA3) and prostate cancer glandular epithelial mRNA-like 1 (PCGEM1) (in 1999 and 2000, respectively), both dysregulated in prostate cancer.[11,12] In 2003, the lncRNA metastasis-associated lung adenocarcinoma transcript 1 (MALAT1) was found as a prognostic marker in non-small cell lung cancer (reviewed by Gutschner *et al.*[13]). In 2007, lncRNA H19 was mechanistically linked to hepatocellular carcinoma and bladder cancer: H19 inhibition resulted in decreased tumorigenicity while ectopic expression of H19 led to higher tumorigenic potential of these cancers (reviewed by Raveh *et al.*[14]). Within the next several years, lncRNAs were shown to be altered and to regulate the progression of various malignancies such as neuroblastoma,[15] stomach cancer,[16] colorectal cancer,[17] head and neck squamous cell carcinoma,[18] esophageal cancer[19] and others. Furthermore, involvement of lncRNAs in some malignancies was dependent on splicing patterns. For example, metastases in patients with non-small-cell lung cancer were associated with specific splice variants of lncRNA MALAT1 (reviewed by Gutschner *et al.*[13]).

As a result, by the middle of 2000s, it had become clear that lncRNAs represent a functional entity and are involved in the development and progression of multiple diseases. Moreover, it was also becoming evident that lncRNAs constitute a large (probably the largest) fraction of the transcriptome. The first evidence of their pervasive expression was provided by tiling arrays[20] but results were not widely accepted. Skepticism persisted for a long time,[21] and the central dogma of molecular biology continued to hamper the progress of lncRNA research.

The real explosion of the lncRNA field came with the advent of RNA sequencing along with groundbreaking FANTOM, ENCODE and GENCODE projects[22,23] which for the first time seriously challenged the central dogma by demonstrating that while protein-coding genes constitute only about 2% of the genome, more than 80% of the genome is transcribed, generating an astonishing number of long non-coding transcripts. Furthermore, this plethora of non-coding transcripts appeared to be further diversified by alternative splicing. In addition, it was found that the phenomenon of antisense lncRNA transcripts, which was previously thought to be rare, is quite widespread, and 50–70% of protein-coding loci express antisense lncRNA.[24] In 2010s, the growth of the lncRNA field became exponential, with the number of publications referenced in PubMed increasing from 338 in 2012 to 2759 in 2017.

The new wave of lncRNA research revealed further important functions of lncRNAs. For example, a series of studies dissected and characterized the regulatory role of natural antisense transcripts (NATs). One of these was 'type 2 brain-derived neurotrophic factor (BDNF)-locus transcripts' encoded by a gene that overlaps the BDNF gene and is transcribed in a reverse orientation [later termed antisense 1 (BDNF-AS1)]. In 2012, a reciprocal depolarization-dependent pattern of expression of BDNF and BDNF-AS1 in human neocortex cultures was shown.[25] Likewise, a discordant regulation of BDNF and BDNF-AS1 was demonstrated *in vivo*.[26]

Furthermore, lncRNAs were shown to play an active role in nuclear body architecture. In 2009, Clemson *et al.*[27] identified nuclear enriched abundant transcript 1 (NEAT1), a highly abundant nuclear lncRNA, which co-localized with mRNAs on the periphery of paraspeckles [nuclear domains responsible for mRNA nuclear retention and adenosine to inosine (A–I) editing] and coordinated their assembly.

Characterization of the transcriptional landscape of the four human homeobox (HOX) loci identified 231 HOX ncRNAs. One of them, termed HOX transcript antisense RNA (HOTAIR), residing in the HOXC locus, repressed transcription of the HOXD locus, located on another chromosome, by interacting with Polycomb repressive complex 2 (PRC2) and leading to histone H3 lysine-27 trimethylation (H3K27m). HOTAIR was later shown to be involved in multiple cancers (reviewed by Cai *et al.*[28]).

Moreover, it was found that lncRNAs in sense and antisense orientation control transcription of rRNA genes collectively referred to as rDNA. Silencing of rDNA by nucleolar remodeling complex (NoRC), a chromatin-remodeling complex that triggers heterochromatin formation, requires a lncRNA forming a DNA–RNA triplex that is subsequently recognized by DNA methyltransferase DNMT3b.[29]

Another important mechanism was lncRNA involvement in histone modification. In 2012, Li *et al.*[30] identified antisense 1 short-chain dehydrogenase/reductase family member 4 (AS1DHRS4), a natural antisense transcript that inhibits its partner proteins in the dehydrogenase/reductase family member 4–dehydrogenase/reductase family member 4-like 2–dehydrogenase/reductase family member 4-like 1 (DHRS4–DHRS4L2–DHRS4L1) gene cluster. By pairing with ongoing sense transcripts, not only did AS1DHRS4 mediate deacetylation of histone H3 and demethylation of histone 3 lysine 4 (H3K4) in *cis* for the DHRS4 gene, but also it interacted physically *in trans* with the epigenetic modifiers [H3K9- and H3K27-specific histone methyltransferases G9a and enhancer of Zeste 2 (EZH2)], targeting the promoters of the downstream DHRS4L2 and DHRS4L1 genes and inducing local repressive H3K9me2 and H3K27me3 histone methylations.

Faced with the constantly growing numbers of lncRNAs, many research groups aimed to develop online tools that addressed the cellular localization of lncRNAs, their expression and their functional status computationally, and a number of "lncRNA Ontology" packages have emerged. As of the beginning of 2018, among the most popular are "lncAtlas",[31] "lnc2Meth",[32] "lncRNA Ontology",[33] "NeuraNetL2GO",[34] "LNCiPedia"[35] and "LnCeDB".[36]

9.3 Biology and Functions of lncRNA

The biogenesis of lncRNAs is similar to the biogenesis of mRNA from protein-coding genes. lncRNAs are transcribed by RNA polymerase II, capped at their 5′-ends, polyadenylated, spliced (both constitutively and alternatively) and exported into the cytoplasm by the same transport machineries. The promoters of lncRNAs are bound by the same transcription factors and are marked by typical histone marks (methylation and acetylation). lncRNAs have several unique features though. While some lncRNAs are exported into the cytoplasm, some are retained in the nucleus or reside in both cellular compartments. The mechanism targeting lncRNA to nucleus or cytoplasm is still incompletely understood but probably involves specific targeting sequences. lncRNAs are relatively weakly expressed compared with protein-coding genes, but their expression is more tissue-specific.[23] It is generally accepted that the sequences of lncRNA are less conserved than that of protein-coding genes. However, the weaker conservation of sequence is mostly the result of computational averaging, since some lncRNA are ultraconserved[37] while others are primate-specific.[38]

In contrast to miRs that target mRNA in the cytoplasm, lncRNAs act at both transcriptional and post-transcriptional levels in various cellular compartments *via* interactions with DNA, RNA and proteins (Figure 9.1). The same lncRNA can sometimes function *via* different mechanisms. It is not surprising, therefore, that functional ontology of lncRNA is hard to assemble. In this

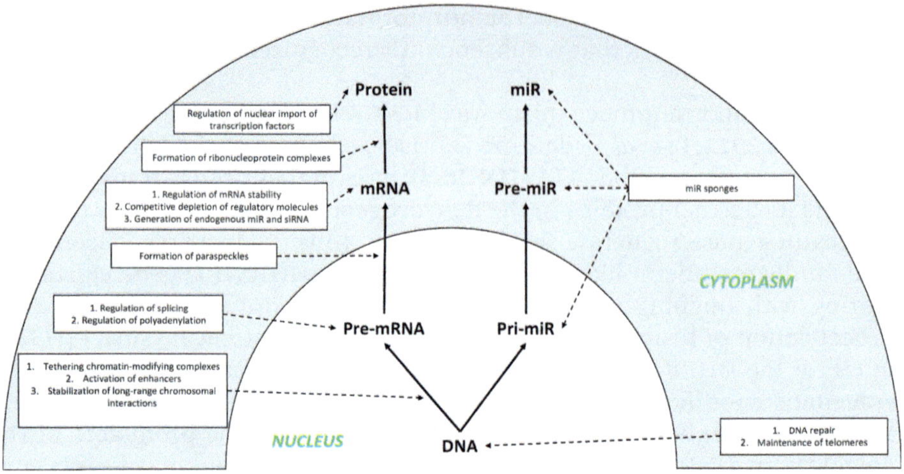

Figure 9.1 Functions of lncRNAs. lncRNAs act in both the nucleus and cytoplasm and regulate processing and functions of both protein-coding genes and microRNAs. In the nucleus, lncRNAs regulate DNA repair and maintenance of telomeres as well as participate in tethering chromatin-modifying complexes, activation of enhancers, stabilization of long-range chromosomal interactions, formation of paraspeckles, splicing and polyadenylation. In the cytoplasm, lncRNAs control mRNA stability, deplete molecules that competitively regulate mRNA levels, generate endogenous microRNAs and siRNAs, form ribonucleoprotein complexes and regulate nuclear import of transcription factors. In addition, lncRNAs serve as microRNA sponges in both nucleus and cytoplasm, thereby fine-tuning levels of pri-microRNAs, pre-microRNAs and mature microRNAs.

chapter, we will list and illustrate the main molecular functions of lncRNA starting with the role of lncRNAs in transcription, proceeding to post-transcriptional modifications and ending with translation.

9.3.1 lncRNAs as Regulators of Transcription

lncRNAs exert transcriptional control mainly *via* association with chromatin-modifying complexes, thus producing transcriptional modifications with different degrees of permanency. The key role of lncRNA in these cases is to tether 'general purpose' epigenetic machineries to specific genomic loci. An important aspect of engaging of epigenetic mechanisms by lncRNAs is the long duration of induced transcriptional changes. This mechanism underlies one of the first known functions of lncRNA, namely the regulation of allelic expression. lncRNA Xist tethers a repressive epigenetic complex PRC2 to the X chromosome in order to compensate for the difference in X-linked dosage between males and females. Regulation of transcription *via* recruitment of chromatin-modifying complexes later emerged as a common mechanism by which many lncRNAs function. For example, antisense non-coding RNA in the inhibitor of cyclin dependent kinase 4 locus (ANRIL) and HOTAIR also recruit PRC2.[39,40] Currently, approximately 20% of lncRNAs are expected to be bound by PRC2 to mediate PRC2-induced transcriptional inhibition.[41] In addition to PRC2, lncRNAs can interact with such epigenetic machineries as lysine-specific histone demethylase 1 (LSD1),[40] Growth arrest and DNA-damage-inducible protein GADD45 alpha (GADD45a),[42] WD repeat-containing protein 5–mixed lineage leukemia 1 (WDR5–MLL1)[43] and DNA (cytosine-5)-methyltransferase 1 (DNMT1).[44] Spatially, lncRNAs may serve as anchors of chromatin modifiers either in *cis* (in the vicinity of lncRNA synthesis) or/and in *trans* (at a remote location). Expression of lncRNA and its target protein-coding gene may be either concordant or discordant. Concordant regulation implies that expression of lncRNA and its target protein-coding gene coincides. For example, antisense lncRNA glypican 3-antisense 1 (GPC3-AS1) increases expression of its target protein-coding gene GPC3.[45] Discordant regulation causes an opposite effect: inhibition of lncRNA BDNF-AS1 enhances the expression of the neighboring gene BDNF.[26]

lncRNA may also influence transcription by mediating the function of enhancers.[46] A high correlation between enhancer RNA (eRNA) transcription and enhancer activation has been documented. The precise mechanism is largely unknown; it is possible that eRNAs interact with cohesin and mediator complex to stabilize the 3D structure that aligns enhancer and promoter (*i.e.* enhancer–promoter looping). Mediator is a large protein complex with variable subunit composition essential for many aspects of RNA polymerase II-mediated transcription, including the reorganization of chromatin architecture and regulation of such transcriptional processes as pre-initiation, initiation, re-initiation, pausing and elongation (reviewed by Allen and Taatjes[47]). eRNAs may also participate in the modulation of chromatin accessibility of the promoters of their target genes[48] and facilitate release of paused polymerase II by sequestering negative elongation factor (NELF) or through enabling enhancer–promoter looping.[49] Another set of lncRNAs associated with enhancers are super-lncRNAs that form DNA–RNA triplexes with super-enhancers. Super-enhancers are characterized by enrichment for mediator complex and clusters of binding sites for transcription factors (reviewed by Hu and Tee[50]). Soibam[51] has identified 442 unique super-lncRNAs that have common sequence features, including a single triplex-forming repeat domain that forms an RNA–DNA triplex with anchor DNA sites.

lncRNAs are also involved in the regulation of transcription *via* stabilization of long-range chromosomal interactions in which remote gene promoters are brought to the same transcription factory through long-range chromosomal loops and rosette-like structures (reviewed by Choi and Feeney[52]).

9.3.2 lncRNA as Regulators of Post-transcriptional Processing

lncRNAs have been shown to participate in multiple post-transcriptional processes such as splicing, polyadenylation, regulation of transcript stability and miR-induced repression. For example, Huang *et al.*[53] showed that lncRNA tropomyosin 1-antisense (TPM1-AS), transcribed from the fourth intron of the TPM1 gene, regulates alternative splicing of TPM1 in esophageal cancer through sponging the splicing factor RNA binding motif protein 4 (RBM4) and preventing its interaction with TPM1 pre-mRNA. TPM1-AS overexpression or RBM4 knockdown decreased inclusion of exon 2a of TPM1, resulting in downregulation of TPM1 variants V2 and V7 and inhibition of cell migration and filopodium formation.

9.3.2.1 *Splicing*

lncRNAs may bridge proteins to nascent pre-mRNAs *via* either base pairing or protein intermediates. The lncRNA Gomafu, also known as myocardial infarction associated transcript (MIAT), affects formation of spliceosomes and inhibits splicing factor SF1.[54] The lncRNA MALAT1 can regulate RNA processing *via* interaction with serine–arginine-rich splicing factors thereby influencing splicing in a genome-wide fashion.[55,56] Alternative splicing is also globally regulated by sno-lncRNAs,[57] lncRNAs flanked by small nucleolar RNA.

9.3.2.2 *Polyadenylation*

lncRNAs may be involved in regulation of polyadenylation. For example, overexpression of the antisense gene lysine acetyltransferase 5 (KAT5) promoted the usage of distal polyA site in the sense gene ribonuclease H2 Subunit C (RNASEH2C), which generated a longer 3′ untranslated region (3′UTR), reduced protein expression and inhibited cell growth.[58]

9.3.2.3 *mRNA Stability*

lncRNA interaction with sense transcripts has been shown to affect mRNA stability. This interaction most likely involves direct RNA–RNA pairing and/or tethering and sponging of stability-regulating miRs and proteins. For example, the long intergenic non-protein coding RNA regulator of reprogramming (linc-RoR)–heterogeneous nuclear ribonucleoprotein I (hnRNP I) complex displaces hnRNPD from c-Myc mRNA, leading to increased stability of c-Myc mRNA, and linc-RoR knockout reduces the cellular c-Myc mRNA and protein levels.[59] The antisense lncRNA Beta-secretase 1-antisense (BACE-AS1) increases the stability of its sense partner BACE.[60]

9.3.2.4 *Competing Endogenous RNAs*

Competing endogenous RNAs (ceRNAs) affect the fate of mRNAs in an indirect way by competitive depletion of regulatory molecules. ceRNAs have sequence similarity to specific mRNAs and dynamically control the binding of their regulatory factors.[61]

9.3.2.5 Interactions with miRs

lncRNAs are also known to regulate gene expression post-transcriptionally *via* interactions with miRs. On one hand, lncRNAs can inhibit miRs by harboring miR-binding sites and thus depleting (sponging) miRs from the cytoplasm. For example, H19 lncRNA is a sponge of the let-7 miR family.[62] The same function is also attributed to a specific subclass of lncRNA, circular RNA (circRNA), in which 5′ and 3′ ends are linked. circRNA CDE1as can sponge miR-7.[63,64] Another mechanism of inhibition of miRs is functional blockade. The lncRNA BACE1-AS1 prevents miR-induced repression of BACE1 by masking the binding site for miR-485-5p.[65]

9.3.2.6 Endogenous siRNA and miRNA Generation

lncRNAs are also involved in generation of endogenous small interfering RNAs (siRNAs) and miRNAs. Thus, lncRNA miR100HG has two embedded miRs (miR-100 and miR-125b), which confer a mechanistic feedback to the lncRNA progenitor.[66] The lncRNA H19 contains miR-675 necessary for the upregulation of Slug and downregulation of E-cadherin during the epithelial to mesenchymal transition (EMT) in breast cancer cells.[67] In another example, overexpression of sex determining region Y-box containing gene 4 (Sox4) natural antisense transcript (NAT) (known as Sox4ot1) led to the production of a novel small RNA, Sox4_sir3, which demonstrated Dicer1-dependent biogenesis and characteristics resembling P-element induced wimpy testis homolog (PIWI)-interacting RNA.[68]

9.3.3 lncRNA as Regulators of Translation

In addition to regulating the formation and processing of the transcriptome, lncRNAs are also involved in control of translation. For example, the lncRNA translation regulatory long non-coding RNA 1 (treRNA) [originally named noncoding RNA-activating 7 (ncRNA-a7)] promotes formation of a ribonucleoprotein complex consisting of hnRNPK, Fragile X mental retardation syndrome-related protein 1 (FXR1), FXR2, poly(U) binding splicing factor 60 (PUF60) and splicing factor 3b subunit 3 (SF3B3) which interacts with eukaryotic translation initiation factor 4G (eIF4G) and the 3′UTR of E-cadherin mRNA and promotes displacement of E-cadherin mRNA from high-molecular-weight polysomes, ultimately resulting in reduction of E-cadherin protein levels.[69] TreRNA is overexpressed in B-cells in chronic lymphocytic leukemia and correlates with poor prognosis. Overexpression of treRNA reduced DNA damage and desensitized the B-cell chronic lymphocytic leukemia (B-CLL) cell line OSU-CLL to fludarabine and mafosfamide.[70]

Short interspersed repetitive element 2B sequence to upregulate translation molecules (SINEUPs) are a family of antisense lncRNAs that are able to increase translation of partially overlapping mRNAs. SINEUPs have two functional domains: a sequence at the 5′ of the SINEUP complementary to the target coding mRNA and an inverted SINE2B-containing domain that triggers enhanced loading of the mRNA–SINEUP complex onto heavy polysomes.[71] Synthetic SINEUPs incorporating minimal versions of these two functional domains are capable of increasing the yield of the corresponding target protein *in vitro*.[72] Synthetic SINEUPs have also been shown to upregulate translation *in vivo* in a medaka fish model of microphthalmia with linear skin defects syndrome, caused by mutation in cyclooxygenase 7B (cox7B), leading to amelioration of the disease phenotype. The synthetic SINEUP approach is used by TransSINE Technologies, a venture of RIKEN Genome Science Center (Yokohama, Japan), to upregulate expression in industrial human-cell-based protein production systems.

9.4 Classification of lncRNA

As the transcriptional landscape continued to expand, a need arose to classify non-coding RNA. Non-coding transcripts were arbitrarily divided into short non-coding RNA (shorter than 200 nucleotides (nt), including miR, tRNA, small nuclear RNA, small nucleolar RNA, *etc.*) and long non-coding RNA (longer than 200 nt). The "200 nt cutoff" is convenient, since it excludes many other non-coding transcript classes, such as tRNAs, miRs, small nuclear RNAs, PIWI-interacting RNAs and others.

Finding an optimal classification of lncRNAs is an important objective in the lncRNA field. The question of classification has many practical applications. A wet-lab approach to studying the multitude of lncRNAs would be long and expensive, and thus computational analysis of lncRNA families, based on an optimal lncRNA classification, could significantly aid progress. Another application is to lay the foundation for the systematic nomenclature of the rapidly growing family of lncRNAs. Among currently known tens of thousands of lncRNAs, only a tiny fraction have a unique name, while most have number-based identifiers (such as RP11-250B2.6, CTD-3247F14.2 or LINC00844) which are hard to use. The same sequence may also be named differently by different authors. Subsequently, as genomes are constantly updated and sequencing data are generated, this leads to a lot of inconsistencies among current major databases (such as UCSC or Ensembl). A comprehensive classification would allow us to generate usable and informative names for tens of thousands of lncRNAs, which will facilitate learning and communication and help with consistency across scientific publications and online resources.

Currently, multiple overlapping subclasses of lncRNA have been defined, including natural antisense transcripts (NATs), long intergenic non-coding RNAs (lincRNAs), long intervening non-coding RNA (lincRNA), very long intergenic non-coding RNAs (vlincRNAs), competing endogenous RNAs (ceRNA), activating ncRNAs (ncRNA-a), promoter-associated lncRNA (pRNA), SINEUPs and others. The non-systematic nature of these classifications is not helpful and possibly even harmful to the development of the lncRNA field.

Arguably, the ideal classification should reflect the biology and function of each particular lncRNA. This is not an easy task, since, in contrast to protein-coding genes that have clear sequence-based motifs, the functional status of lncRNAs largely depends on their three-dimensional structure rather than their sequence. We still do not fully understand the laws by which long RNA molecules form internal base pairs and how spatial conformation responds to the cellular environment. The computational complexity of accounting for all these interactions makes the routine prediction of lncRNA 3D structures currently not feasible. Moreover, as structures and functions of multiple non-coding transcripts become better known, divisions between potential classes become less and less obvious, for example, in cases of miR-containing lncRNAs, or lncRNAs encoding polypeptides, or mRNAs that simultaneously function as regulatory lncRNA.

In this regard, we suggest two parallel classifications, functional and genomic, that may be currently useful.

9.4.1 Functional Classification

This is based on the molecular function in which a particular lncRNA is implicated. We refer the reader to Section 9.3, in which we discussed the functions of lncRNAs from the standpoint of their effects on transcription, post-transcriptional RNA processing and translation (also graphically shown in Figure 9.1).

9.4.2 Genomic Classification

This is based on the genomic coordinates of lncRNAs in relation to their protein-coding genes. Thus, lncRNA may be intronic (encoded by intronic sequences of protein-coding gene on the same strand), partially intronic (encoded by both intronic and exonic segments of a protein-coding gene on the same strand) and intergenic (not overlapping with any gene). These three classes of lncRNAs represent the most abundant non-coding transcripts, which constitute approximately 50% of all cellular RNA mass.[73] Another abundant class is natural antisense transcripts (NATs), which partly or fully overlap with protein-coding genes on an opposite strand. NATs also may be close but non-overlapping, and consensus is still needed regarding what distance will constitute the threshold for non-overlapping antisense lncRNAs. Bidirectional lncRNAs are lncRNAs transcribed from the same promoter as a protein-coding gene but in a different direction (on the same strand). lncRNAs may also relate to protein-coding genes as enhancer- or promoter-associated lncRNAs (reviewed by Rinn and Chang[74]).

9.5 Targeting of Long Non-coding RNA by Oligonucleotides

Targeting of lncRNA with oligonucleotides can be achieved through blocking or knockdown, either *via* use of antisense oligonucleotides (ASO) or of siRNA. There are also a few alternative approaches to target lncRNAs, such as the use of small molecules. Upregulation of lncRNAs can be implemented using the clustered regularly interspaced short palindromic repeats–catalytically dead-CRISPR-associated protein 9 (CRISPR–dCas9) system fused with a transcription factor.

9.5.1 Antisense Oligonucleotides

ASOs are single-stranded DNA-based compounds that are widely used to achieve specific mRNA knockdown, which either utilize RNaseH to recognize an RNA–DNA heteroduplex and cleave the RNA strand or alternatively sterically block lncRNA interactions with other molecules (see Chapter 2). ASOs usually include a phosphorothioate (PS) backbone and various sugar modifications [such as 2′-*O*-methyl (2OMe), 2′-*O*-methoxyethyl (2MOE), morpholinos, locked nucleic acids (LNA), *etc.*] to increase stability *in vivo* (see Chapters 1–3 for a complete description). Since RNaseH is not active against nucleotides with most types of sugar modifications, ASO-mediated knockdown is usually achieved by arranging the ASO in a 'gapmer' configuration, with a central 'gap' of 8–10 unmodified nucleotides and 3–5 nucleotide-long sugar-modified ends. One of the FDA-approved compounds, mipomersen, degrades apolipoprotein B100 (APOB100) mRNA through the RNaseH pathway.[75,76] Many ASO-based oligonucleotides are currently being developed by Ionis and other companies (Chapter 3).

9.5.2 siRNAs

lncRNA knockdown can also be mediated by use of siRNAs, which degrade mRNA through a pathway involving the RNA-induced silencing complex (RISC) (see Chapters 10–12). siRNAs are composed of two RNA strands complementary to each other. siRNA oligonucleotides may also have chemical modifications (such as PS, 2′-*O*-Methyl, or LNA) to increase their stability. siRNAs generally rely on liposomes to maximize delivery into the cell. Furthermore, conjugation with carrier molecules (PEG, peptides, lipids and glycosides) have also been used in specific cellular types to increase uptake (Chapter 10). siRNA-based compounds are being developed by several companies, including Alnylam. In December 2017, Alnylam completed a rolling submission of a New Drug Application (NDA) to the US Food and Drug Administration (FDA) for patisiran, a lipid nanoparticle formulation of an RNAi-based oligonucleotide targeting transthyretin for the treatment of hereditary transthyretin-mediated amyloidosis. Investigational new drug applications to the FDA have been filed by Alnylam for six other RNAi compounds using the Enhanced Stabilization Chemistry (ESC)–*N*-acetyl galactosamine (GalNAc) delivery platform.

Blockade of interaction with proteins or nucleic acids is yet another mechanism for modulating the function of lncRNAs that are known to associate with other molecules. This blockade could be achieved using ASOs, either through direct base pairing with the interaction sites in lncRNAs or through induction of a conformational change in lncRNA 3D structure. The blockade approach has been used in currently approved oligonucleotide drugs targeting mRNAs. Very recently, two drugs that induce skipping of damaged exons were approved. Ionis introduced nusinersen, which targets the interaction between pre-mRNA and splicing machinery proteins for the treatment of spinal muscular atrophy (see Chapter 6). Sarepta used a similar approach (eteplirsen) for the treatment of Duchenne muscular dystrophy (see Chapter 6).

Approximately 20% of lncRNAs expressed in various cell types are bound by PRC2 and are involved in inducing PRC2-mediated transcriptional inhibition.[41] Designing oligonucleotides or small-molecule compounds to block this interaction presents a valid therapeutic strategy, and this approach is being explored by Translate Bio (formerly RaNA Therapeutics) to selectively upregulate individual genes through steric blocking of interactions of PRC2 with regulatory lncRNA.

As many lncRNA are regulated by miRs and are involved in sponging miRs, blocking miR interaction sites or introducing miR mimics represent viable therapeutic approaches that can provide flexible context-specific upregulation or downregulation of lncRNAs' abundance and function.

Bifunctional oligonucleotides can also be designed to both block the interaction and degrade the lncRNA. For example, both RNaseH-mediated lncRNA destruction and blocking lncRNA access to the epigenetic proteins will lead to changes in gene regulation by lncRNA. Subramanian *et al.*[77] reported multi-targeting oligonucleotides with both RNase H-activating and steric-blocking domains linked by a cleavable bond.

9.5.3 CRISPR and Other Approaches

Upregulation of lncRNAs may be achieved by use of CRISPR–CRISPR-associated catalytically inactive dCas9 (CRISPR–dCas9) (see Chapter 17). The CRISPR–Cas9 system constitutes a platform for RNA-guided DNA targeting. Fusing of dCas9 to a transcription factor [such as nuclear factor kappa B (NFκB) or activator protein1 (AP-1)] may enable stable transcriptional activation of lncRNA of interest in humans,[78,79] with the site of delivery determined by short guide (sg) RNA and the rate of transcription determined by the particular transcription factor.

Finally, by use of high-throughput screening, small-molecule compounds could be selected to block lncRNA interaction with epigenetic proteins and other molecules.[80]

9.6 Therapeutic Applications

A major asset of lncRNA is their flexibility to affect either a single gene or multiple cellular processes simultaneously. The fact that there are currently no lncRNA-targeting drugs mainly reflects upon how laborious translational efforts in the lncRNA field are. Nevertheless, an increasing number of lncRNA-based approaches are in the validation, proof-of-concept, or pre-clinical stages. We will list examples of possible lncRNAs applications in neurology, psychiatry, oncology, cardiology and gastroenterology. Notably, a single lncRNA may be involved in multiple diseases. For example, the inhibitor of cyclin-dependent kinase 4b–alternate reading frame–inhibitor of cyclin-dependent kinase 4a (INK4b–ARF–INK4a) locus has been linked consistently to susceptibility to cardiovascular disease and cancer. This locus encodes three protein-coding genes that play a central role in cell-cycle arrest as well as an lncRNA ANRIL. ANRIL has 20 exons that are alternatively spliced and contain long intersperse element (LINEs), short interspersed elements (SINEs) and Alu repetitive elements. The isoforms may exert differential effects on products of the INK4b–ARF–INK4a locus and their downstream targets. Modulation of ANRIL expression and splicing results in diverse physiological outcomes in both cardiovascular disease and cancer.[81] It is likely that many lncRNAs may be implicated in seemingly unrelated medical conditions, further emphasizing the complex nature of lncRNAs and the care with which their functional potential should be employed for therapeutic purposes.

9.6.1 Neurology and Psychiatry

Although oligonucleotide-based compounds do not cross the blood–brain barrier, one of the first successful attempts to apply them *in vivo* was to inject them directly into cerebrospinal fluid (CSF), either intrathecally or intracerebroventricularly, to target specific mRNAs.[82,83] One of the currently approved oligonucleotide drugs was designed for the treatment of spinal muscular atrophy, a neurological disease. This could be due to the fact that CSF represents a protected environment with negligible nuclease activity, *i.e.* an immunoprivileged zone.[84] Moreover, long half-lives of advanced oligonucleotide compounds permit infrequent dosing. Thus, treatment of neurological and psychiatric conditions may be managed using epidural or intrathecal administration. In the following sections, we will review some of the examples of neurological and psychiatric diseases that could be addressed by targeting lncRNA.

9.6.1.1 *Dravet's Syndrome (DS)*

This disease is caused by heterozygous loss-of-function mutations in the SCN1A gene, which encodes the pore-forming alpha subunit of the voltage-gated sodium channel Nav1.1. Clinically, DS is characterized by seizures in the first year of life and progressive psychomotor retardation. As the majority of DS-causing mutations do not produce toxic protein, upregulation of SCN1A expression represents a desirable therapeutic target. Such upregulation has been achieved by blocking the inhibitory natural antisense transcript (NAT) from the SCN1A locus (SCN1ANAT) using single-stranded oligonucleotide-based compounds (*i.e.* AntagoNATs). In non-human primates and a knock-in mouse model of DS, intrathecal injection of AntagoNATs induced upregulation of Scn1a. A weekly administration of AntagoNAT for 4 weeks reduced the frequency and the duration of seizures and normalized the excitability of hippocampal interneurons in the mouse model of DS.[85] The AntagoNAT technology, developed at the Scripps Research Institute and its spin-out CuRNA (Jupiter, FL), is currently being advanced by the Miami-based company OPKO Health.

9.6.1.2 *Angelman's Syndrome (AS)*

This disease is characterized by microcephaly, developmental delay, seizures and ataxia and is caused by the maternal deficiency of the imprinted gene ubiquitin-protein ligase E3A (UBE3A). A lncRNA UBE3A-ATS represses paternal UBE3A and is expressed only from the paternal allele. Intracerebroventricular (ICV) treatment with anti-Ube3a-ATS oligonucleotides led to a reduction in UBE3A-ATS expression and upregulation of UBE3A in mice.[86] The resultant partial restoration of UBE3A protein in the mouse model of AS improved cognitive deficits. Targeting of UBE3A-ATS as a treatment for AS is currently being investigated by Ionis. Notably, knockdown of Ube3a-ATS is complicated since multiple small nucleolar RNAs (snoRNAs) are processed from the same precursor transcript, and mutations affecting these snoRNAs are known to cause Prader–Willi syndrome. However, expression of these RNAs was not altered by the oligonucleotide treatment (possibly due to a very fast rate of snoRNAs processing).[86]

9.6.1.3 *Spinal Muscular Atrophy (SMA)*

SMA is a rare neuromuscular disorder caused by mutations in the survival of motor neuron 1 (SMN1) gene (see also Chapter 6). Adequate expression of SMN1 is necessary for survival of motor neurons, and SMA is characterized by the progressive loss of motor neurons leading to early death. The human genome may also contain a number of damaged copies of SMN1, termed SMN2, that demonstrate aberrant splicing of exon 7, thereby producing an unstable mRNA transcript. A recently approved oligonucleotide-based drug, nusinersen, prevents splicing out of exon 7 of the SMN2 gene and increases SMN protein expression. This therapy could potentially be combined with another newly proposed approach involving lncRNA from the SMN2 locus (SMN-AS1), which suppresses SMN2 transcription. Anti-SMN-AS1 oligonucleotides delivered together with SMN2 splice-switching oligonucleotides additively increased SMN protein expression and improved survival in a mouse model of severe SMA.[87] As SMN-AS1 functions through recruiting PRC2 complex, oligonucleotides capable of blocking the SMN-AS1–PRC2 interaction represent a possible way to increase SMN2 expression and are being investigated by Translate Bio (RaNA) as a treatment for SMA.[88]

9.6.1.4 Alzheimer's Disease (AD)

Aβ plaques and neurofibrillary tangles (NFTs) have been viewed as the main histopathological features of AD. However, the full spectrum of molecular determinants that contribute to AD pathogenesis is yet to be known. As therapies targeted at amyloid precursor protein (APP) processing have not yet yielded a viable treatment for AD, the search for novel efficient drugs is warranted, and progress in the lncRNA field has already suggested a few promising targets.

One lncRNA of interest in AD is early B cell factor 3-antisense (EBF3-AS), transcribed from the opposite strand of EBF3, a gene involved in neuronal specification and migration.[89] EBF3-AS knockdown by siRNA inhibited the apoptosis induced by Aβ25–35 and okadaic acid in SH-SY5Y cells.[90] Furthermore, low expression of BDNF has been associated with AD and other disorders of the nervous system. Interfering with the function of lncRNA BDNF-AS1 by use of ICV administration of AntagoNATs resulted in increased expression of BDNF protein in mouse brain and proliferation and differentiation of neurons in mouse neurospheres.[26]

Another lncRNA, BACE1-AS, is elevated in subjects with Alzheimer's disease and in APP transgenic mice.[60] After exposure to various cell stressors including Abeta1–42, expression of lncRNA BACE1-AS becomes elevated, resulting in increased BACE1 mRNA stability. Notably, BACE1 also cleaves the β2-subunit of voltage-gated sodium channels (Navβ2). APP transgenic mice exhibit increased levels of BACE1 and accumulation of cleavage fragments of Navβ2 as well as associated alterations in SCN1A expression and localization, which is accompanied by aberrant electroencephalogram (EEG) activity and impairments in the Morris water maze.[91]

It is also possible that there is no single lncRNA capable of conferring a sizeable phenotypic effect in patients with AD. Instead, in accordance with principles of systems biology, there is a set of lncRNAs that align to affect clinical outcomes. For example, sequencing of RNAs from the hippocampus from patients with late-onset AD identified multiple differentially expressed lncRNAs, including upregulated transcripts AD-linc1 and AD-linc2.[89] Furthermore, lncRNA lipoprotein receptor-related protein 1-antisense (LRP1-AS) is upregulated in patients with AD, which is accompanied by downregulation of LRP1. LRP1-AS directly binds and inhibits the activity of high mobility group box 2 (HMGB2) protein, which is known to enhance Sterol regulatory element-binding protein 1a (SREBP1a)-dependent transcription of LRP1. LRP1-AS-targeted oligonucleotides inhibit its interaction with HMGB2 and increase LRP1 expression. As the SCN1A gene is known to be downregulated in AD, and overexpression of SCN1A ameliorates cognitive deficits in a mouse model of AD,[92] upregulation of SCN1A through oligonucleotide-mediated suppression of lncRNA SCN1ANAT could also be used for the treatment of the epileptiform activity in AD. It is possible that as the biology of AD and the chemistry of oligonucleotides develop and progress, multiplexing lncRNA-targeting may become a common approach.

9.6.1.5 Fragile X Mental Retardation 1 (FMR1)

Triplet-repeat expansions in the FMR1 gene result in FMR1 silencing, intellectual disability and neurodegeneration. The FMR1 locus expresses several lncRNAs, including FMR4, which is also silenced by repeat expansions. FMR4 regulates the chromatin state and the expression of several hundred genes in *trans*, including those that promote cellular proliferation of human neural precursor cells.[93] Allele-specific knockdown of the damaged FMR4 copy may represent a therapeutic strategy in FMR1.

9.6.1.6 Psychiatric Diseases

As psychiatric diseases represent biopsychosocial phenomena, such diseases are most complex at the molecular level. However, given the relative similarity of symptomatology for each psychiatric condition, this complexity is unlikely to be random. It is possible that lncRNAs, being able to regulate many molecular circuits simultaneously, can be core drivers of the complex pathogenesis of mental disorders. In fact, several lines of evidence indicate that lncRNAs may be especially important in psychiatric diseases. In contrast to protein-coding genes and small non-coding RNAs (for instance, miRs), a large proportion of lncRNAs are expressed only in the primate lineage[23] indicating that lncRNA evolution could be closely linked to the development of mental functions. Furthermore, bioinformatics analyses of genome-wide association studies indicated that while mutations in protein-coding genes are mainly related to metabolic diseases and anatomical abnormalities, mutations in lncRNAs are specifically associated with mental health disorders.[94] In addition, most lncRNAs are expressed in the brain,[95] probably reflecting the involvement of lncRNA's plasticity and ability to link environmental stimuli with gene expression[96] in information processing.

One potential implication of lncRNA research is schizophrenia (SZ). Although several lines of evidence indicate that SZ has a strong genetic component, there are only a few protein-coding genes that show significant association with this disease. Whilst lncRNA involvement in SZ has only recently gained attention, several lncRNAs of therapeutic interest have already been identified. The lncRNA Gomafu is downregulated in the cortical gray matter from patients with SZ and its depletion in mice results in increased dopamine levels in the brain. An increased level of dopamine is considered to be a major pathogenic force in SZ, and upregulation of Gomafu may therefore be employed to achieve a long-term decrease in dopamine concentrations.[97]

Furthermore, recent work showed that 125 lncRNAs were differentially expressed in peripheral blood mononuclear cells (PBMC) of SZ patients compared with healthy controls. Downregulation of one of them, NON-HSAT041499, correlated with a better response to antipsychotic treatment. Furthermore, PBMC levels of lncRNAs NONHSAT089447 and NONHSAT041499 correlated with a decrease in the Positive and Negative Syndrome Scale (PANSS) scores of patients after completion of the treatment.[98]

There has been an association of pro-inflammatory protein-coding genes and lncRNAs with the development of SZ. For example, expression of lncRNA interferon gamma-antisense 1 (IFNG-AS1) is decreased in SZ patients compared with controls and correlates with IFNG reduction in patients.[99]

9.6.2 Oncology

The role of lncRNA in cancer etiology and pathogenesis is currently under intense investigation. Involvement of lncRNAs in cancer seems to be broad. By use of CRISPR–Cas9 technology, genomic deletions were introduced into 700 lncRNAs, and 51 of these lncRNA were found to alter cancer growth.[100]

With its relatively low development effort and uniform toxicological profile, oligonucleotide therapeutic modality is highly amenable to treatment of cancers, since it can efficiently address mutations affecting a small number of patients. Not surprisingly, cancer treatment was one of the first applications of oligonucleotide drugs. For example, a clinical trial of custirsen (OGX-011, a second-generation antisense oligonucleotide targeting clusterin in patients with recurrent or metastatic prostate cancer) started in 2005 (http://ClinicalTrials.gov Identifier: NCT00258388). Given that lncRNAs also play a leading role in coordinating large sets of genes involved in cancer development and progression, lncRNA targets are likely to become dominant in oligonucleotide-mediated cancer treatment. As studies of lncRNAs in cancer are becoming more extensive (over 1500 publications listed in PubMed in 2017 alone), it is hard to give them justice in the limited space available in this review, and we will cite just a few recent examples to illustrate different possible lncRNA-based mechanisms and approaches to the treatment of cancers.

The initiating event in cancer development is frequently a mutation in one of the cell cycle control genes resulting from failure to repair DNA mismatches or double-stranded DNA breaks (DSB). The DNA damage can be caused either by environmental stimuli or by erroneous activation of cell division leading to initiation of replication forks that rapidly collapse. When DNA repair mechanisms are inadequate, DSBs accumulate and may eventually lead to cancer. At the same time, excessive activation of DSB repair genes is one of the reasons for radio- and chemo-resistance in cancer treatment.[101] DSB repair involves extensive reorganization of chromatin near break sites to allow access for repair factors and can be accomplished through two major pathways. The choice of DSB repair pathway is determined by the cell cycle phase: error-prone non-homologous end joining repair (NHEJ) during G1 or more precise homologous recombination during S/G2. Expression of TP53BP1 protein in G1 prevents accumulation of breast cancer type 1 susceptibility protein (BRCA1) at DSB sites, which would otherwise lead to homologous recombination. During DNA repair by homologous recombination, DSBs are detected by the meiotic recombination 11–RAD50–Nijmegen breakage syndrome 1 (MRN) complex, which activates ataxia telangiectasia mutated (ATM), which in turn phosphorylates the core histone variant H2AX (H2AFX), C-terminal-binding protein-interacting protein (CtIP) [Retinoblastoma-binding protein 8 (RBBP8)], BRCA1 and exonuclease EXO1. In NHEJ repair, DSBs are recognized by the Ku70–80 [X-ray repair cross complementing 6–5 (XRCC6–XRCC5) heterodimer which serves as a scaffold for DNA-PKcs (PRKDC), XRCC4, DNA ligase IV (LIG4), XRCC4-like factor (NHEJ1) and/or ATM and ataxia telangiectasia and Rad3-related protein (ATR)], followed by processing of DNA ends by polynucleotide kinase 3′-phosphatase (PNKP), Artemis [DNA cross-link repair 1C (DCLRE1C)] and Ku.[101] Multiple lncRNA have also been implicated in the DSB repair process. For example, damage-induced lncRNAs (dilncRNAs) synthesized from and towards damaged DNA ends at the sites of DSB associate with 53BP1 and are essential for DNA repair.[102]

Expression of a lncRNA damage-induced noncoding (DINO) is sufficient to activate DNA damage signaling, cell-cycle arrest and apoptosis in the absence of DNA damage. DINO is transcribed in the opposite direction from the cyclin dependent kinase inhibitor 1A (CDKN1A) (p21) promoter, is p53 (TP53)-inducible and can physically interact with the C-terminal RNA-binding region of p53 promoting p53 stabilization. The DINO–p53 pair co-localizes at multiple p53 target genes, including CDKN1A.[103]

Another lncRNA, [DNA damage-sensitive RNA1 (DDSR1)], was induced upon DNA damage in an ATM- and NFκB-dependent manner, did not require TP53 for induction after damage, but could regulate TP53 target genes. In cells lacking DDSR1, only the homologous recombination DSB repair pathway was compromised.[104]

TP53 is known to activate lncRNA TP53 Target 1 (TP53TG1) after DSB caused by ionizing irradiation or treatment with DNA-damaging agents. TP53TG1 binds to the DNA–RNA binding protein Y-Box binding protein 1 (YBX1) [nuclease-sensitive element-binding protein 1 (NSEP1)] and prevents its nuclear localization and YBX1-induced activation of oncogenes. TP53TG1 gene hypermethylation and inactivation in cancer cells releases the inhibition of YBX1 target genes, which leads to an increase in DSB and chemo-resistance.[105]

lncRNA in non-homologous end joining pathway 1 (LINP1) physically interacts with the Ku80–DNA–PKcs complex. LINP1 knockdown in triple-negative breast cancer cells led to reduced DSB repair, while overexpression of LINP1 increased NHEJ activity. Epidermal growth factor receptor (EGFR) activation upregulates LINP1 transcription, while TP53 activation stimulates miR-29 that downregulates LINP1. Oligonucleotide-mediated knockdown of LINP1 may reduce chemo-resistance in breast cancer.[48]

Events required for tumor development include activation of cell division and dedifferentiation. Appropriately, the INK4b–ARF–INK4a gene cluster, controlling cell cycle exit, is homozygously deleted or silenced in many human cancers. The lncRNA ANRIL is involved in the regulation of this locus and therefore represents one of the possible lncRNA targets in cancer. However, as the regulatory functions of ANRIL are mediated by its multiple isoforms and regulated by multiple proteins and ncRNAs, additional studies are needed to harness ANRIL as a therapeutic target.[81]

A critical step in cancer development is its dissemination from the original tumor site to multiple locations in the organism (*i.e.* metastasis). In the case of epithelial-derived cancers (carcinomas), this process is called epithelial to mesenchymal transition (EMT). During EMT, epithelial cells lose their polarity and cell–cell-extracellular matrix contacts and gain mesenchymal properties, including increased motility. Changes in the surrounding connective tissues enable cancer cells to intravasate into the bloodstream and disseminate. On the molecular level, EMT involves downregulation of epithelial markers, such as E-cadherin, claudins, desmosomes and occludins and upregulation of mesenchymal markers, such as N-cadherin (CDH1), vimentin (VIM) and fibronectin (FN1) as well as matrix metalloproteinases. These changes are mediated by transcription factors [TWIST1, TWIST2, SNAIL 1 (=SNAI1), SLUG (=SNAI2), zinc finger E-box binding homeobox 1 (ZEB1), ZEB2, Kruppel like factor 8 (KLF8), Forkhead box protein C2 (FOXC2), Goosecoid (GSC)]. Expression of these genes in turn can be induced by extracellular factors [Wingless/Integrated (Wnt), transforming growth factor (TGF)-β (=TGFB1), epidermal growth factor (EGF), fibroblast growth factor (FGF), hepatocyte growth factor (HGF)]. Hypoxic conditions inside tumors, arising due to insufficient vasculature, are known to induce EMT through hypoxia-inducible factor (HIF)-1α (=HIF1A) and HIF-2α [endothelial Period–aryl hydrocarbon receptor nuclear translocator protein–single-minded protein domain-containing protein 1 (=EPAS1)]. Negative regulation of EMT involves Sox2/octamer-binding transcription factor 4 (Oct4) [Pituitary-specific Pit-1–octamer transcription factor–uncoordinated-86 Class 5 homeobox 1 (POU5F1)], which suppress Snail, c-Myc, that downregulates TGF-β receptors, and Klf4 which induces a set of epithelial genes including E-cadherin. Since many of these processes involve chromatin remodeling and miR regulation, it is not surprising that lncRNAs play important roles in EMT.[106] The lncRNA MALAT-1 is overexpressed in tumors with higher risk of metastasis. MALAT-1 recruits PRC2 components suppressor of Zeste 12 homolog (Suz12) and EZH2 to the E-cadherin promoter and thus facilitates E-cadherin silencing. Another mechanism through which MALAT-1 induces EMT is by sponging miR-1, miR-204 and miR-205, which leads to derepression of their targets Slug, Zeb1 and Zeb2. Overexpression of lncRNA H19 is associated with the activation of EMT in numerous cancers. H19 silences E-cadherin by recruiting EZH2 to its promoter. H19 transcript contains a miR-675 necessary for downregulation of E-cadherin and upregulation of Slug. H19 is upregulated by TGF-β, hypoxia and HGF and Slug, which creates a positive feedback loop.[67] The lncRNA plasmacytoma variant translocation 1 (PVT1) is transcribed from the same chromosomal region as the c-Myc oncogene, frequently amplified in cancers. PVT1 promotes EMT through methylation of the promoter of Patched homolog 1 (PTCH1), a negative regulator of hedgehog signaling and by sponging miR-152.[106]

lncRNA treRNA is upregulated in clinical metastatic breast cancer samples and primary colon cancer samples. Overexpression of treRNA induces cell migration and xenograft metastases *in vivo*. Double-knockdown of hnRNP K and FXR2 or triple-knockdown of hnRNP K, FXR2 and FXR1 in treRNA-overexpressing xenografts abolished the metastasis-promoting effect of treRNA. TreRNA knockdown reduced the metastatic potential *in vivo*.[69] Additionally, in chronic lymphocytic leukemia, treRNA upregulation was associated with more aggressive disease and a poor response to chemotherapy through enhanced protection against DNA damage.[70]

One aspect that contributes to disease severity is the ability of cancer cells to undergo natural selection, which results in multiple aggressive phenotypes, including induction of tumor vascularization, that could be mediated by lncRNAs. Inhibition of vascularization is one of the possible targets in cancer treatment. The reverse of this strategy can be applied in wound healing and cardiovascular diseases, in which increased vascularization is desirable. For example, lncRNA small ubiquitin-related modifier 1 pseudogene 3 (SUMO1P3) is upregulated in colon cancer samples compared to normal epithelial cells. SUMO1P3 silencing reduced growth, metastases to liver and vascularization of colon cancer xenografts *in vivo*. SUMO1P3 knockdown decreased the levels of cyclin D1, vimentin and vascular endothelial growth factor A (VEGFA) and increased E-cadherin expression in tumors.[107]

9.6.3 Cardiology

lncRNAs are emerging as major players in the development of the cardiovascular system as well as in the pathogenesis of cardiovascular diseases. lncRNAs were the largest class of differentially regulated transcripts during differentiation of human pluripotent cells into early cardiovascular progenitor cells, committed vascular progenitor cells and vascular endothelial cells. Interestingly, expression of lncRNAs in this system was more tightly controlled than that of protein-coding genes.[108]

The INK4 locus on chromosome 9p21 containing the lncRNA ANRIL has the strongest linkage to genetic susceptibility for cardiovascular disease. The mechanism remains unclear but probably involves ANRIL-dependent regulation of inflammation underlying development of atherosclerosis. ANRIL has two isoforms, linear and circular, which are thought to play different roles in pathogenesis of cardiovascular diseases.[81]

Recent work showed that silencing of the lncRNA GATA binding protein 6-antisense (GATA6-AS) promotes blood vessel formation in a human endothelial cell xenograft model. Transfection of human umbilical vein endothelial cells with oligonucleotide-based compounds (gapmers) against GATA6-AS prior to transplantation to immunodeficient mice resulted in enhanced vascular growth[109] suggesting a therapeutic approach applicable to ischemic conditions.

9.6.4 Gastroenterology

Inflammatory bowel disease (IBD) is a multifactorial autoimmune disease with growing prevalence. Recent studies have identified multiple lncRNAs that are dysregulated in Crohn's disease and ulcerative colitis, including upregulated IFNG-AS1 and BC012900 and downregulated ANRIL. An oral oligonucleotide drug targeted against mRNA (alicaforsen) delivered through enema is currently undergoing clinical trials for IBD. The possibility of oral administration of lncRNA-targeting drugs may be particularly beneficial for the treatment of IBD and other diseases of the gastrointestinal tract.

9.7 Perspectives

Although functional lncRNAs were described as early as the 1990s, their potential and therapeutic applications were not proposed until the mid-2000s (reviewed by St Laurent and Wahlestedt and by Wahlestedt[96,110]). In recent years lncRNAs have been convincingly demonstrated to be indispensable in key cellular processes, including transcription, splicing, translation, DNA repair and regulation of miR levels. It was also shown that deregulation of lncRNA expression and/or activity results in perturbation of molecular circuits causing the development of multiple diseases. Furthermore, during recent years, significant progress has been made in optimization of the chemistry and delivery platforms for oligonucleotide-based therapeutics, including ASOs and siRNAs, which can be efficient tools in regulating the expression of lncRNAs. Because of consistently accelerating progress in both lncRNA biology and the development of oligonucleotide-based drugs, it is likely that lncRNA-dependent mechanisms of pathogenesis will soon be targeted in various medical conditions, including cardiovascular, infectious, psychiatric and neurological diseases as well as in cancer.

At the same time, we suggest that lncRNAs should be manipulated with caution. Despite significant progress having been achieved in understanding their biogenesis and functional profiles, their potential and scope of operations are still largely uncharted. We speculate that both a reductionist approach and a systems biology approach with the aid of bioinformatics and computational biology have to be implemented in order to dissect lncRNA-regulated networks. In parallel, hard work has to be continued to optimize the specificity and delivery of oligonucleotide-based therapies in order to ensure safety and efficacy.

References

1. R. C. Lee, *et al.*, *Cell*, 1993, **75**, 843–854.
2. B. J. Reinhart, *et al.*, *Nature*, 2000, **403**, 901–906.
3. R. Rupaimoole and F. J. Slack, *Nat. Rev. Drug Discovery*, 2017, **16**, 203–222.
4. F. H. Crick, *Cold Spring Harbor Symp. Quant. Biol.*, 1966, **31**, 1–9.
5. R. J. Britten and E. H. Davidson, *Q. Rev. Biol.*, 1971, **46**, 111–138.
6. L. E. Orgel and F. H. Crick, *Nature*, 1980, **284**, 604–607.
7. L. Carrel and C. J. Brown, *Philos. Trans. R. Soc. London B Biol. Sci.*, 2017, **372**, 20160355.
8. N. Brockdorff, *et al.*, *Nature*, 1991, **351**, 329–331.
9. S. Douc-Rasy, *et al.*, *Oncogene*, 1996, **12**, 423–430.
10. M. P. Lee, *et al.*, *Proc. Natl. Acad. Sci. U. S. A.*, 1999, **96**, 5203–5208.
11. M. J. Bussemakers, *et al.*, *Cancer Res.*, 1999, **59**, 5975–5979.
12. V. Srikantan, *et al.*, *Proc. Natl. Acad. Sci. U. S. A.*, 2000, **97**, 12216–12221.
13. T. Gutschner, *et al.*, *J. Mol. Med.*, 2013, **91**, 791–801.
14. E. Raveh, *et al.*, *Mol. Cancer*, 2015, **14**, 184.
15. G. K. Pandey, *et al.*, *Cancer Cell*, 2014, **26**, 722–737.
16. C. Yang, *et al.*, *Oncotarget*, 2015, **6**, 15311–15320.
17. J. F. Xiang, *et al.*, *Cell Res.*, 2014, **24**, 513–531.
18. N. Nohata, *et al.*, *Oral Oncol.*, 2016, **59**, 58–66.
19. C. Q. Li, *et al.*, *Oncogenesis*, 2017, **6**, e297.
20. P. Kapranov, *Science*, 2002, **296**, 916–919.
21. M. B. Clark and J. S. Mattick, *Semin. Cell Dev. Biol.*, 2011, **22**, 366–376.
22. The ENCODE Project Consortium, *Nature*, 2012, **489**, 57–74.
23. T. Derrien, *et al.*, *Genome Res.*, 2012, **22**, 1775–1789.
24. O. Khorkova, *et al.*, *Hum. Mol. Genet.*, 2014, **23**, R54–R63.
25. L. Lipovich, *et al.*, *Genetics*, 2012, **192**, 1133–1148.
26. F. Modarresi, *et al.*, *Nat. Biotechnol.*, 2012, **30**, 453–459.
27. C. M. Clemson, *et al.*, *Mol. Cell*, 2009, **33**, 717–726.
28. B. Cai, *et al.*, *Neoplasma*, 2014, **61**, 379–391.
29. H. Bierhoff, *et al.*, *Cold Spring Harbor Symp. Quant. Biol.*, 2010, **75**, 357–364.
30. Q. Li, *et al.*, *Proc. Natl. Acad. Sci. U. S. A.*, 2012, **109**, 14110–14115.
31. D. Mas-Ponte, *et al.*, *RNA*, 2017, **23**, 1080–1087.
32. H. Zhi, *et al.*, *Nucleic Acids Res.*, 2018, **46**, D133–d138.
33. Y. Li, *et al.*, *Oncotarget*, 2015, **6**, 39793–39805.
34. J. Zhang, *et al.*, *Bioinformatics*, 2018, **34**, 1750–1757.
35. P. J. Volders, *et al.*, *Nucleic Acids Res.*, 2015, **43**, D174–D180.
36. S. Das, *et al.*, *PLoS One*, 2014, **9**, e98965.
37. M. K. Iyer, *et al.*, *Nat. Genet.*, 2015, **47**, 199–208.
38. A. Necsulea, *et al.*, *Nature*, 2014, **505**, 635–640.
39. Y. Kotake, *et al.*, *Oncogene*, 2011, **30**, 1956–1962.
40. M. C. Tsai, *et al.*, *Science*, 2010, **329**, 689–693.
41. A. M. Khalil, *et al.*, *Proc. Natl. Acad. Sci. U. S. A.*, 2009, **106**, 11667–11672.

42. K. Arab, *et al.*, *Mol. Cell*, 2014, **55**, 604–614.
43. K. C. Wang, *et al.*, *Nature*, 2011, **472**, 120–124.
44. A. Di Ruscio, *et al.*, *Nature*, 2013, **503**, 371–376.
45. X. T. Zhu, *et al.*, *FEBS J.*, 2016, **283**, 3739–3754.
46. U. A. Orom, *et al.*, *Cell*, 2010, **143**, 46–58.
47. B. L. Allen and D. J. Taatjes, *Nat. Rev. Mol. Cell Biol.*, 2015, **16**, 155–166.
48. Y. Zhang, *et al.*, *Nat. Struct. Mol. Biol.*, 2016, **23**, 522–530.
49. H. Meng and B. Bartholomew, *J. Biol. Chem.*, 2017, **293**, 13786–13794.
50. Z. Hu and W.-W. Tee, *Biosci. Rep.*, 2017, **37**, BSR20160183.
51. B. Soibam, *RNA*, 2017, **23**, 1729–1742.
52. N. M. Choi and A. J. Feeney, *Front. Immunol.*, 2014, **5**, 49.
53. G. W. Huang, *et al.*, *Int. J. Biochem. Cell Biol.*, 2017, **90**, 59–67.
54. H. Tsuiji, *et al.*, *Genes Cells: Mol. Cell. Mech.*, 2011, **16**, 479–490.
55. D. Bernard, *et al.*, *EMBO J.*, 2010, **29**, 3082–3093.
56. V. Tripathi, *et al.*, *Mol. Cell*, 2010, **39**, 925–938.
57. K. L. McCann and S. J. Baserga, *Mol. Cell*, 2012, **48**, 155–157.
58. T. Shen, *et al.*, *Protein Cell*, 2017, **9**, 540–552.
59. J. Huang, *et al.*, *Nucleic Acids Res.*, 2016, **44**, 3059–3069.
60. M. A. Faghihi, *et al.*, *Nat. Med.*, 2008, **14**, 723–730.
61. R. V. Kartha and S. Subramanian, *Front. Genet.*, 2014, **5**, 8.
62. A. N. Kallen, *et al.*, *Mol. Cell*, 2013, **52**, 101–112.
63. T. B. Hansen, *et al.*, *Nature*, 2013, **495**, 384–388.
64. S. Memczak, *et al.*, *Nature*, 2013, **495**, 333–338.
65. M. A. Faghihi, *et al.*, *Genome Biol.*, 2010, **11**, R56.
66. Y. Lu, *et al.*, *Nat. Med.*, 2017, **23**, 1331–1341.
67. I. J. Matouk, *et al.*, *Oncotarget*, 2016, **7**, 3748–3765.
68. K. H. Ling, *et al.*, *Data Brief*, 2016, **7**, 282–290.
69. K. Gumireddy, *et al.*, *EMBO J.*, 2013, **32**, 2672–2684.
70. C. R. Miller, *et al.*, *Oncotarget*, 2017, **8**, 25942–25954.
71. C. Carrieri, *et al.*, *Nature*, 2012, **491**, 454–457.
72. S. Zucchelli, *et al.*, *Front. Cell. Neurosci.*, 2015, **9**, 174.
73. G. St Laurent, *et al.*, *BMC Genomics*, 2012, **13**, 504.
74. J. L. Rinn and H. Y. Chang, *Annu. Rev. Biochem.*, 2012, **81**, 145–166.
75. S. T. Crooke and R. S. Geary, *Br. J. Clin. Pharmacol.*, 2013, **76**, 269–276.
76. R. S. Geary, *et al.*, *Clin. Pharmacokinet.*, 2015, **54**, 133–146.
77. R. R. Subramanian, *et al.*, *Nucleic Acids Res.*, 2015, **43**, 9123–9132.
78. P. I. Thakore, *et al.*, *Nat. Methods*, 2015, **12**, 1143–1149.
79. H. Zhou, *et al.*, *Nat. Neurosci.*, 2018, **21**, 440–446.
80. R. Pedram Fatemi, *et al.*, *J. Biomol. Screening*, 2015, **20**, 1132–1141.
81. F. Aguilo, *et al.*, *Curr. Topics Microbiol. Immunol.*, 2016, **394**, 29–39.
82. C. Wahlestedt, *et al.*, *Nature*, 1993, **363**, 260–263.
83. C. Wahlestedt, *et al.*, *Science*, 1993, **259**, 528–531.
84. C. Wahlestedt, *Trends Pharmacol. Sci.*, 1994, **15**, 42–46.
85. J. Hsiao, *et al.*, *EBioMedicine*, 2016, **9**, 257–277.
86. L. Meng, *et al.*, *Nature*, 2015, **518**, 409–412.

87. C. d'Ydewalle, *et al.*, *Neuron*, 2017, **93**, 66–79.
88. C. J. Woo, *et al.*, *Proc. Natl. Acad. Sci. U. S. A.*, 2017, **114**, E1509–e1518.
89. M. Magistri, *et al.*, *J. Alzheimer's Dis.: JAD*, 2015, **48**, 647–665.
90. C. Gu, *et al.*, *DNA Cell Biol.*, 2018, **37**, 220–226.
91. B. F. Corbett, *et al.*, *J. Neurosci.: Off. J. Soc. Neurosci.*, 2013, **33**, 7020–7026.
92. L. Verret, *et al.*, *Cell*, 2012, **149**, 708–721.
93. V. J. Peschansky, *et al.*, *Mol. Cell. Neurosci.*, 2016, **74**, 49–57.
94. G. St Laurent, *et al.*, *BMC Med.*, 2014, **12**, 97.
95. T. R. Mercer, *et al.*, *Neuroscientist: Rev. J.*, 2008, **14**, 434–445.
96. G. St Laurent, 3rd and C. Wahlestedt, *Trends Neurosci.*, 2007, **30**, 612–621.
97. W. T. Carrick, *et al.*, *Front. Mol. Biosci.*, 2016, **3**, 69.
98. S. Chen, *et al.*, *Med. Sci. Monit.: Int. Med. J. Exp. Clin. Res.*, 2016, **22**, 3340–3351.
99. H. Ghafelehbashi, *et al.*, *Scand. J. Immunol.*, 2017, **86**, 479–485.
100. S. Zhu, *et al.*, *Nat. Biotechnol.*, 2016, **34**, 1279–1286.
101. Z. Wu and Y. Wang, *Oncotarget*, 2017, **8**, 102690–102704.
102. F. Michelini, *et al.*, *Nat. Cell Biol.*, 2017, **19**, 1400–1411.
103. A. M. Schmitt, *et al.*, *Nat. Genet.*, 2016, **48**, 1370–1376.
104. V. Sharma, *et al.*, *EMBO Rep.*, 2015, **16**, 1520–1534.
105. A. Diaz-Lagares, *et al.*, *Proc. Natl. Acad. Sci. U. S. A.*, 2016, **113**, E7535–e7544.
106. R. Heery, *et al.*, *Cancers*, 2017, **9**, 38.
107. L. M. Zhang, *et al.*, *Am. J. Transl. Res.*, 2017, **9**, 5461–5472.
108. L. Kurian, *et al.*, *Circulation*, 2015, **131**, 1278–1290.
109. P. Neumann, *et al.*, *Nat. Commun.*, 2018, **9**, 237.
110. C. Wahlestedt, *Nat. Rev. Drug Discovery*, 2013, **12**, 433–446.

CHAPTER 10

Conjugate-mediated Delivery of RNAi-based Therapeutics: Enhancing Pharmacokinetics–Pharmacodynamics Relationships of Medicinal Oligonucleotides

BRUNO M. D. C. GODINHO*[a], ANDREW H. COLES[a] AND ANASTASIA KHVOROVA[a,b]

[a]RNA Therapeutics Institute, University of Massachusetts Medical School, Worcester, MA, USA; [b]Program in Molecular Medicine, University of Massachusetts Medical School, Worcester, MA, USA
*E-mail: Bruno.godinho@umassmed.edu

10.1 Introduction

The discovery of the RNA interference (RNAi) pathway in the late 1990s–early 2000s[1] brought revitalizing optimism to the field of therapeutic gene silencing, previously dominated by antisense oligonucleotides (ASOs). The higher stability and potency of RNAi-based drugs and the rapid turnover of the

Drug Discovery Series No. 68
Advances in Nucleic Acid Therapeutics
Edited by Sudhir Agrawal and Michael J. Gait

Published by the Royal Society of Chemistry, www.rsc.org

RNA-induced silencing complex (RISC), generated significant interest in this new technology. With drug pipeline support from big pharma, RNAi therapeutics rapidly progressed and reached the clinical stage only 5–6 years after discovery. However, the surge of dose-limiting toxicities, the lack of effective delivery strategies and other stability issues hindered progression of those early clinical trials.[2,3] Two decades of research followed to gather crucial understanding of the pharmacokinetic (PK) and pharmacodynamic (PD) properties of siRNAs and allow the first RNAi product, Patisiran (Onpattro™), to receive approval from the Food and Drug Administration (FDA) for the treatment of liver disease.[2]

RNAi therapeutics has evolved and matured based on the understanding of key aspects that govern PK and PD of synthetic therapeutic oligonucleotides. The most advanced configuration to date consists of fully-modified small interfering RNAs (siRNAs) conjugated to *N*-acetylgalactosamine (GalNAc). GalNAc-conjugation enables targeted delivery to hepatocytes and allows potent and sustained gene silencing in the liver for several months after subcutaneous administration (see Chapter 11).[4] Given the revolutionizing success of this approach several clinical programs have now been initiated with this chemical configuration to treat other liver disorders. Furthermore, these accomplishments also spurred interest in the scientific community to explore the utility of other conjugated modalities for extrahepatic applications, which are likely to emerge in the near future.

This chapter provides an overview of the evolution of the field, focusing on the major cornerstones that determine productive PK and PD relationships for conjugated RNAi-based drugs: (i) the chemical stabilization of the oligonucleotide scaffold, as an essential pre-requisite that defines efficient and durable *in vivo* gene silencing; (ii) the nature of the conjugate as a strong determinant of tissue distribution; (iii) chemical and conjugation strategies to enhance endosomal release in order to achieve productive gene silencing; (iv) the influence of the route of administration, as a strategic approach to modulate clearance kinetics and sustained delivery; and (v) conjugation approaches to increase circulating times in the blood and improve tissue uptake.

10.2 Chemical Stabilization as a Prerequisite for Conjugate-mediated Delivery of siRNAs: Effects on Clearance, Distribution and Sustained Gene Silencing

Early clinical trials in the field were conducted using unmodified or terminally deoxy-modified 21 nucleotide (nt) siRNAs, that contained 3′-overhangs that mimicked natural Dicer substrates (Figure 10.1(a) and (b)).[5,6] Despite their double-stranded nature, these minimally modified molecules were easily degraded by exonucleases and endonucleases present in tissues and rapidly cleared from the body, resulting in poor

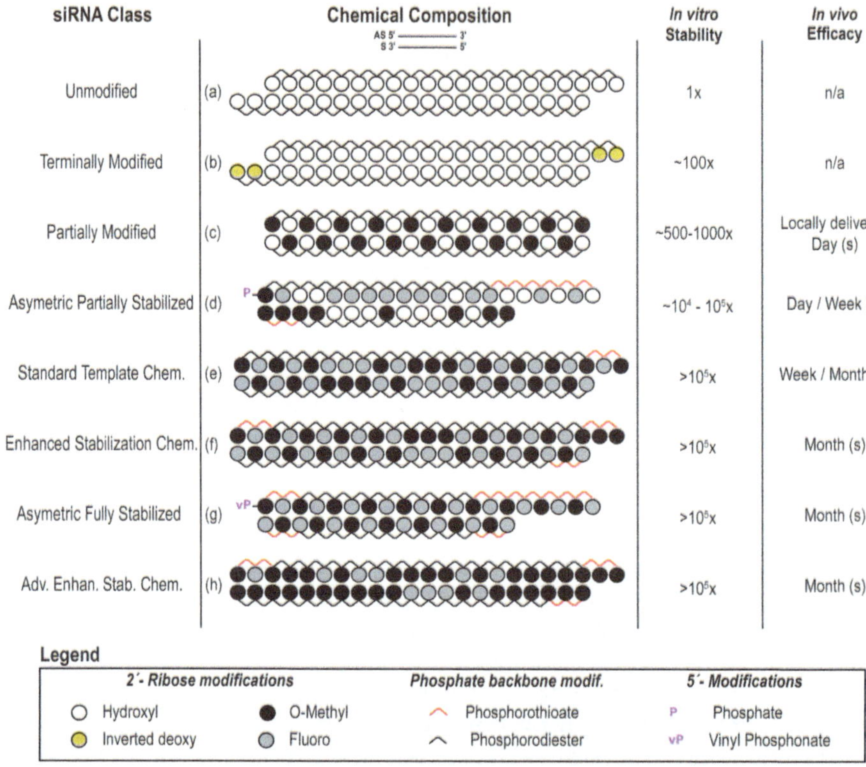

Figure 10.1 Chemical evolution of therapeutic siRNAs. Extensive chemical stabilization is a prerequisite for conjugate-mediated delivery of therapeutic siRNAs. Selected oligonucleotide stabilization patterns are displayed on the schematics above along with their *in vitro* stability and *in vivo* gene silencing efficacy relative to unmodified siRNAs.

gene silencing.[2] In addition, these oligonucleotides elicited unexpected interferon responses through activation of Toll-like receptors (TLRs), double-stranded RNA-dependent protein kinase (PKR) and retinoic acid-inducible gene I (RIG-I),[7] which contributed to the early termination of these trials. Formulation strategies, employing lipids or biocompatible polymers, were attempted as an alternative to mitigate some of the stability and immunostimulatory issues (reviewed by Yin *et al.* and Guzman-Villanueva *et al.*[8,9]). Although these systems provided substantial improvements in stability and enhanced cellular uptake, they were also associated with local and systemic toxicities[9] and never reached widespread success for oligonucleotide delivery in clinical settings.

These early failures led to significant efforts in medicinal chemistry to improve nuclease resistance and reduce toxicity of synthetic RNAi drugs.[10] Capitalizing on lessons learned from ASO technologies, chemical modifications have been introduced into different points of the oligonucleotide

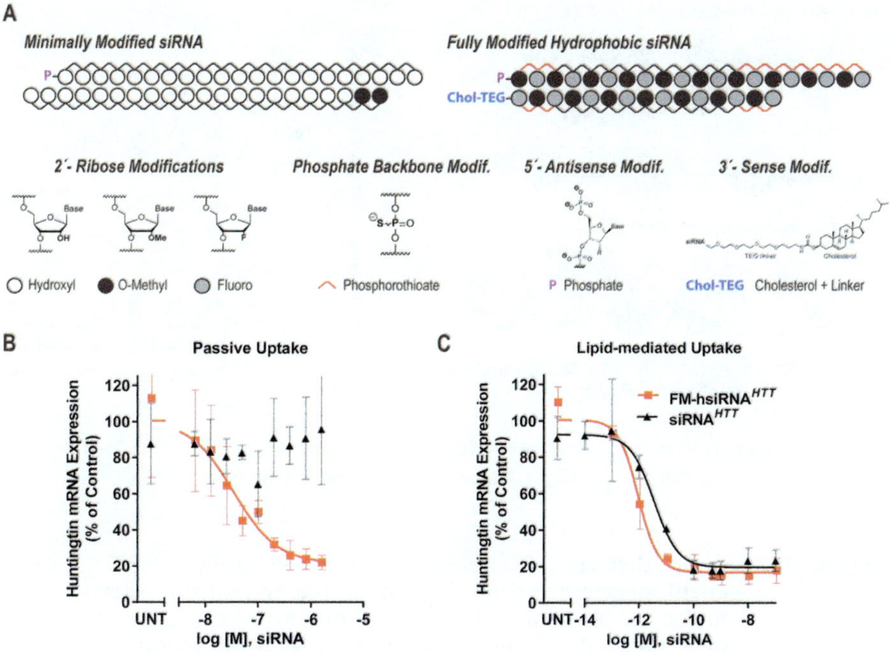

Figure 10.2 Fully modified hydrophobic siRNAs (hsiRNAs) are self-delivering and well-tolerated within the RISC complex. (A) Depicts the chemical composition of minimally modified canonical 21-mer siRNAs (siRNA^{HTT}) and asymmetric fully modified hydrophobic siRNAs (FM-hsiRNA^{HTT}) conjugated to cholesterol (Chol). (B, C) Comparison of gene silencing efficacy *in vitro* in the (B) absence or (C) presence of transfection reagent. Hela cells were treated with huntingtin (*Htt*) targeting siRNAs and *Htt* mRNA expression was assessed after 72 hours using Quantigene® branched DNA assay. Data was normalized to the peptidylprolyl isomerase B (PPIB) housekeeping gene and is expressed as a percentage of the value for the untreated control. (n = 3. Mean ± SD). (B, C) Adapted from ref. 15 with permission from Oxford University Press, Copyright 2018.

structure, including the phosphate backbone, sugar moieties and bases (reviewed by Khvorova and Watts, by Watts, Deleavey and Damha and by Wan and Seth,[2,10,11]). Groups at Atugen AG and Quark Pharmaceuticals experimented with 2′-ribose modifications, such as 2′-*O*-methyl (2′-OMe), to engineer partially modified siRNAs (Figure 10.1(c)).[12,13] While these studies showed that maintaining 2′-hydroxyl groups in key positions of the duplex was necessary to preserve A-helix formation and compatibility with RISC entry,[13] they also revealed that 3′-overhangs previously used in all canonical siRNAs were not necessary for efficacy.[13,14] Thus, both blunt and asymmetric constructs with more than 15 nt have been shown to load effectively into RISC and produce potent mRNA silencing (Figure 10.2). Partially modified siRNAs showed improved stability compared with

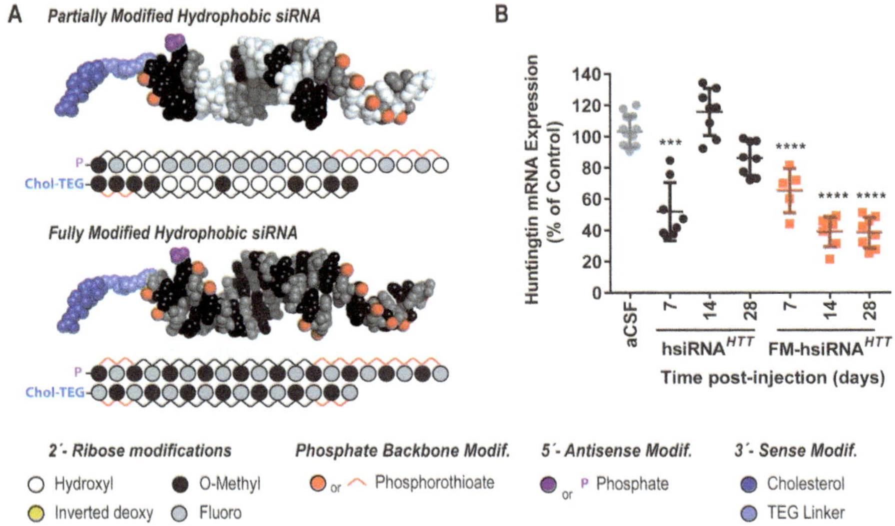

Figure 10.3 Local delivery of cholesterol-conjugated fully modified siRNAs enables sustained gene silencing in the brain. (A) Illustration of the molecular model and the chemical modification pattern of cholesterol-conjugated partially (top) and fully (bottom) modified hydrophobic siRNAs (hsiRNAs). (B) Comparison of gene silencing efficacy and duration of effect for cholesterol-conjugated partially (hsiRNA^{HTT}) and fully modified hydrophobic siRNAs (FM-hsiRNA^{HTT}) after local delivery in the brain. Stereotactic unilateral injections into the striatum (12 μg) were carried out in approximately 8 week-old mice. Animals were euthanized at different time points and huntingtin mRNA quantified by Quantigene® branched DNA assay. Data were normalized to the PPIB housekeeping gene and are expressed as percentages of the value for the untreated control. (n = 5–8. Mean ± SD). Kruskal–Wallis one-way ANOVA with Tukey's multi-comparison test. PyMol molecular models adapted from ref. 15 with permission from Oxford University Press, Copyright 2018.

unmodified siRNAs *in vitro*. However, their efficacy after *in vivo* administration was still limited.[12,15] This is most likely due to a more dynamic environment *in vivo*, with a small amount of oligo in a fluid rich in endonucleases, unlike the more static environment encountered *in vitro*. Local stereotactic delivery of a partially stabilized asymmetric siRNA to the mouse brain resulted in potent huntingtin (*Htt*) silencing in the striatum,[16] but the effect only lasted several days (Figure 10.3). Systemic delivery of partially modified siRNAs only resulted in transient knockdown of the p53 mRNA target in proximal tubules of the kidney,[12] or required very high doses (a cumulative dose of 150 mg kg^{-1}) for meaningful silencing of apolipoprotein B in the liver.[17] Recent studies tested partially modified siRNAs in the context of different conjugated modalities, including lipids and sugars, and demonstrated limited tissue accumulation and gene silencing after intravenous or subcutaneous administrations.[15] Despite the lack of robust *in vivo* efficacy data, these and other studies brought a greater understanding of which chemical modifications were well-tolerated by the RNAi machinery and which modifications significantly affected RNAi PD (*i.e.* gene silencing).[18]

In order to improve the *in vivo* utility of RNAi-based drugs, more extensive modifications have been evaluated (Figure 10.1(e)–(h)). Fully-modified siRNAs with alternating ribose modifications consisting of 2′-fluoro (2′-F) and 2′-OMe displayed increased resistance to nuclease degradation and were well-tolerated by RISC *in vitro*.[14] Similar chemical configurations have been used by Alnylam Pharmaceuticals and others in early versions of their stabilization chemistry (Figure 10.1(e)) to support productive gene silencing *in vivo*.[4] The introduction of terminal phosphorothioate (PS) backbone modifications, at the 5′-end and 3′-end of both guide and passenger strands, had profound effects on stability, *in vivo* efficacy and duration of effect. Indeed, the positioning of terminal PS linkages has been shown to have marked effects on leading GalNAc–siRNA delivery platforms[19] allowing reduction of the clinically effective dosing regimen from weekly administrations, for standard template chemistry (STC) (*e.g.* Revusiran yearly exposure ~28 g), to dosing every 3–6 months for enhanced stabilization chemistries (ESC) (*e.g.* Inclisiran yearly exposure ~0.6–1.2 g[20]).[21] In addition to these enhancements observed in stability and efficacy, the presence of PS modifications in the context of asymmetric constructs (Figure 10.1(d)–(g)), is thought to improve tissue distribution in a similar fashion to ASOs.[22–25] PS-containing ASOs have been shown to actively bind to serum proteins, which helps delay their rapid clearance through the kidney and increase their tissue distribution and retention.[23] The progress in this area has demonstrated that stabilization of sugar moieties and the phosphate backbone are of utmost importance to enhance PK and PD properties of these compounds.

Chemical addition of a 5′-phosphate (5′-P) to the antisense strand has been shown to significantly improve the silencing efficacy of modestly active sequences.[14,26,27] This feature mimics phosphorylation of Dicer products in heavily modified siRNA sequences, which may not be easily processed *in vivo* by relevant cleavage and polyadenylation factor I (CLP1) kinases.[28,29] While enhancing efficacy of some sequences *in vitro*, non-stabilized 5′-Ps are rapidly removed by phosphatases *in vivo*, limiting guide strand RISC entry (Figure 10.1(d)).[26] Thus, several strategies have been developed to chemically stabilize 5′-P.[26,30–33] The most studied synthetic 5′-P analog is 5′-vinyl phosphonate (5′-VP) (Figure 10.1(g)), where the bridging oxygen is substituted by carbon. Although direct comparison of 5′-VP, 5′-P and 5′-hydroxyl (5′-OH) confirmed that 5′-VP provides higher potency and duration of effect, interestingly, 5′-OH outperformed 5′-P indicating that the presence of a degradable phosphate might limit productive biological availability.[26] The improvement in efficacy granted by 5′-VP is most likely due to a combination of enhancement of stabilization and improved RISC entry. Indeed, Parmar *et al.* compared two different stereo configurations of this phosphate mimic, 5′-(*E*)-VP and 5′-(*Z*)-VP, where the former displayed higher RISC loading and *in vivo* efficacy.[33] These results were consistent with crystal structure analyses where 5′-(*E*)-VP enabled a better fit into Argonaute 2 (Ago2).[33] Thus, chemical stabilization of the 5′-P is required to enhance oligonucleotide stability and RISC loading, which, in turn, facilitates higher levels of tissue accumulation and extended duration of gene silencing upon systemic delivery.

The large majority of siRNA compounds currently being evaluated in clinical settings consist of fully modified constructs that include a combination of 2′-F and 2′-OMe ribose modifications [World Health Organization (WHO) database]. While early versions had a ratio of approximately 50 : 50 of both modifications, more advanced configurations (Figure 10.1(h)) have higher fractions of 2′-OMe.[4] Indeed, recent systematic studies have identified minimal positional requirements for 2′-F in order to maintain intrinsic activity allowing selection of 2′-OMe-rich constructs that enable highly potent gene silencing in rodents and non-human primates (NHPs).[4] The higher potency provided by these 2′-OMe-rich compounds is believed to result from the higher stability conferred by this chemical modification. The balance between 2′-F and 2′-OMe modifications is highly dependent on the combination of the modifications in both antisense and sense strands[4] and also dependent on target sequence. Going forward, the selection of optimized clinical candidates will require extensive primary and secondary screening processes to identify the ideal pattern of modification for a particular sequence.

In summary, chemical stabilization of siRNA and miRNA compounds is a strong pre-requisite for robust tissue delivery and sustained gene silencing efficacy *in vivo*. The impact of chemical stabilization is clearly much more pronounced in the context of conjugate-mediated than in formulated siRNA delivery. Chemical stabilization enhances the overall PK properties of siRNA compounds, as well as improving their biological activity. The high stability of these constructs results in the formation of intracellular depots (endosomal compartments), which through slow release yields a long-lasting source of siRNAs. In addition, it is likely that the long intracellular half-lives of these extensively modified siRNAs contribute to more sustained RISC activity, resulting in 9–12 months duration of gene silencing. Figure 10.1 summarizes the evolution of the chemical modification patterns used, with partial modification mostly increasing *in vitro* stability and only extensive chemical modification supporting sustained gene silencing *in vivo*. Thus, full chemical stabilization of the siRNA scaffold is a crucial requirement that establishes an appropriate foundation for testing and development of effective RNAi therapeutics.

10.3 Modulating Biodistribution of Therapeutic Oligonucleotides Using Conjugated Modalities: Targeted *versus* Broad Functional Delivery

Direct conjugation of ligands, such as lipids, carbohydrates and peptides, to the oligonucleotide entity has arisen as an effective alternative to formulation approaches to improve the PK properties of RNAi drugs. These modalities are generally coupled to the sense strand with attachment to the 3′-end being a preferred configuration. Modification of the guide strand at the 5′-end or 3′-end has been shown to interfere with RISC loading.[13] Bioconjugates that are currently under investigation can promote; (i) broad functional delivery of oligonucleotides, with overall distribution profile affected by the nature of the conjugate; (ii) or targeted delivery, where compounds are internalized through receptor-mediated endocytosis in specific tissues.

10.3.1 Broad Functional Delivery of Conjugated siRNAs

Early research with conjugated siRNAs focused on the attachment of a cholesterol moiety to improve cellular uptake and tissue retention.[17,34,35] Cholesterol-conjugated siRNAs (Chol–siRNAs) rapidly associate with cellular membranes and enable potent gene silencing after local[16,36] or systemic administration.[17,34] Due to their high hydrophobicity, when locally injected in the brain, Chol–siRNAs display a steep gradient of diffusion from the site of injection, limiting gene knockdown to defined structures around the injection site.[16,37,38] Fine tuning the hydrophobicity of conjugated oligonucleotides using other lipid modalities, such as docosahexaenoic acid (DHA), a polyunsaturated fatty acid, has enabled broader distribution from the site of injection while still allowing higher retention than unconjugated siRNA.[37,38] Although lipid conjugation offers a great means for modulating the spread of distribution within the central nervous system (CNS), it does not provide cell-type-specific delivery after local injections – oligos are internalized by both neurons and glia.[37,38] Similarly, lipid conjugates do not support targeted delivery after systemic administration, but instead allow for distinctive broad functional delivery (Figure 10.4).[39,40] The fact that most traditional small-molecule drugs also distribute to a variety of tissues and/or cell types in a non-specific fashion to achieve their desired therapeutic efficacy, suggests that functional delivery may also represent a valid strategy both for local and systemic applications with therapeutic oligonucleotides.

Lipid-conjugated oligonucleotides (reviewed by Osborn and Khvorova[41]) support distribution and potent gene silencing in a wide range of tissues including liver, kidneys, lungs, heart, spleen, muscle, fat and placenta, where the tissue of interest defines the choice of the hydrophobic

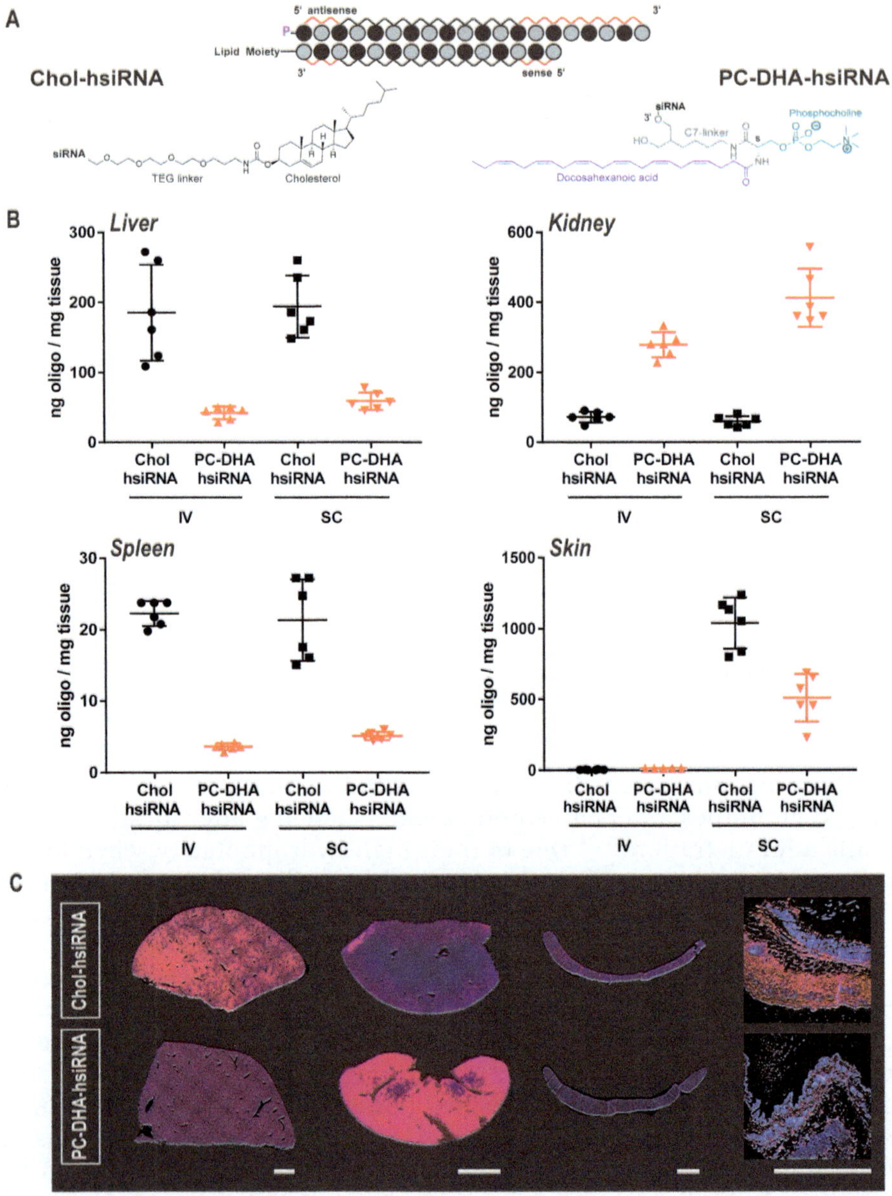

Figure 10.4 The nature of the conjugated modality is a strong determinant in the biodistribution of therapeutic oligonucleotides. Conjugation of cholesterol (Chol) or phosphocholine docosahexaenoic acid (PC-DHA) to fully chemically modified siRNAs results in a significantly different pattern of biodistribution after systemic delivery. (A) Chemical structures of Chol and PC-DHA moieties that were covalently attached to the 3′-end of the sense strand of a fully modified hsiRNA. (B, C) Comparison of oligonucleotide biodistribution after intravenous (IV) or subcutaneous (SC) administrations (20 mg kg^{-1}). Tissue distribution was evaluated by (B) peptide nucleic acid (PNA) hybridization assay and (C) fluorescent microscopy. For PNA assay, a single injection of unlabeled conjugated hsiRNA was performed. For microscopy assessments, two doses of fluorescently labeled lipid-conjugated hsiRNAs were given on consecutive days and all

modality (Figure 10.4)[39,40,42] (Turanov *et al.*[43]). Chol–siRNAs are known to mainly accumulate in the liver,[40,44] but functionally active delivery beyond the liver has been reported for other tissues, such as skeletal muscle.[45] Similarly, docosanoic acid (DCA)-conjugated siRNAs primarily accumulate in the liver, but also show enhanced delivery and silencing in muscle and heart.[39,40] Levels of muscle accumulation observed with DCA variants are significantly higher than with Chol–siRNAs, a good indication that fine tuning the conjugated modality can be used as a strategy to improve tissue delivery. In contrast, DHA–siRNAs and eicosapentaenoic acid-conjugated siRNAs (EPA–siRNAs) mainly distribute to the kidney but enable functional delivery to other tissues, including the lungs and adrenal glands.[39,40] While the mechanism of tissue uptake and cellular internalization of these compounds still warrants investigation, tissue distribution is thought to be dependent on their serum protein binding partners.[34,41]

Partitioning to different serum proteins is believed to be driven largely by the nature and hydrophobicity of the attached lipid moiety. For instance, relatively less hydrophobic DHA–siRNAs and lithocholic acid (LCA)–siRNAs mainly associate with high-density lipoproteins (HDL) and, to some extent, with albumin, whereas relatively more hydrophobic DCA–siRNAs and Chol–siRNAs primarily bind to low-density lipoproteins (LDL).[34,39] Thereafter, LDL-bound oligonucleotides preferentially distribute to LDL receptor (LDLR)-rich tissues, such as the liver, whereas HDL-associated oligonucleotides accumulate in tissues rich in scavenger receptor B1 (SR-BI), including in the ovary and adrenal glands.[39] Other studies using multivalent lipid conjugates have shown extensive binding to albumin, enabling long blood-circulating times and improving delivery to orthotopic breast cancer tumors.[46] In addition, binding to serum proteins allows lipid-conjugated siRNAs and ASOs to display significantly longer plasma half-lives than their unconjugated counterparts.[34,42,44,46,47] Given this feature, lipid-conjugation has also been used to facilitate pre-loading of siRNAs into recombinant human albumin particles, which exhibit improved plasma PK and increased delivery to the liver.[47] On the other hand, unconjugated siRNAs do not actively and/or tightly bind to plasma proteins and consequently exhibit very short blood residency times [half-life (t1/2) = ~2–5 min].[48–50] Together, this evidence indicates that binding to serum proteins not only influences tissue distribution, but also plays a major role in the clearance kinetics of therapeutic oligonucleotides.

The conjugated lipid modality is a strong determinant in the process of absorption from the site of administration. For instance, highly hydrophobic Chol–siRNAs and DCA–siRNAs are heavily retained at the site of subcutaneous injection, presenting delayed peak concentrations and shallower plasma concentration profiles when compared with less hydrophobic

---

animals were euthanized at 48 h. Order of tissues from left to right: liver, kidney, spleen and skin at the site of injection. Scale bar = 2 mm (liver, kidney spleen) and 1 mm (skin). n = 3–5. Mean ± SD. (B) PNA hybridization data reproduced from ref. 44 with permission from Mary Ann Liebert, Inc.

phosphocholine (PC)–DHA–siRNAs (Figure 10.4).[44] Together, this suggests that different conjugated modalities may be rationally selected to control the release and distribution from the site of administration, and to effectively modulate plasma concentration profiles of therapeutic oligonucleotides.

10.3.2 Targeted Delivery of Conjugated siRNAs

Efficient cell-type-specific delivery is considered one of the great challenges in the field and the holy grail for most therapeutics. The most clinically advanced chemical configuration that allows targeted delivery consists of a trivalent GalNAc–siRNA conjugate (see Chapter 11 and reviewed by Huang and by Springer and Dowdy[21,51]). When conjugated to fully chemically stabilized siRNA, it preferentially delivers to hepatocytes in the liver, with secondary but minor accumulation observed in kidneys and macrophages.[52] GalNAc–siRNAs actively bind to the asialoglycoprotein receptor (ASGPR) which is highly expressed in hepatocytes (500 000 per cell) and display rapid internalization and recycling (~15 minutes).[53] The high turnover and expression of the ASGPR, together with the increased stability of the oligonucleotide scaffold, results in very low doses of GalNAc–siRNAs being required for efficacious long-lasting gene silencing (>9 months).[4] This delivery strategy has been studied extensively using oligonucleotides with various chemical architectures, including symmetric and asymmetric siRNA constructs and ASOs.[4,15,32] Close to 40 clinical trials are now being conducted using the GalNAc delivery platform for different gene targets in the liver, including proprotein convertase subtilisin/kexin type 9 (PCSK9) for hypercholesterolemia (Inclisiran, NCT03399370),[20] delta-aminolevulinate synthase 1 (ALAS1) for acute hepatic porphyria (Givosiran, NCT03338816) and alpha-1 antitrypsin (AAT) for antitrypsin deficiency liver disease (ALN-AAT, NCT02503683)[21]. The great success of this technology has revitalized interest in other targeting moieties for extrahepatic delivery, such as peptides, antibodies and aptamers, many of which have been previously tested in the context of partially modified siRNAs without meeting efficacy expectations.

Cell-penetrating peptides (CPP) are the most well studied class of peptides used to enhance delivery of therapeutic oligonucleotides.[54,55] CPPs, such as TAT-transactivator[56] and Penetratin,[57] are cationic in nature, promoting cellular internalization of the oligonucleotide by electrostatic interaction and disruption of the cellular membrane. Although useful for *in vitro* applications, their high positive charge often leads to rapid aggregation after systemic administration. Furthermore, CPPs are more suited for functional delivery than for targeted delivery, since they are actively taken up by several cell types. In contrast, other peptides, such as arginylglycylaspartic acid (RGD), actively bind to integrins, which are membrane receptors that mediate interactions between proteins of the extracellular matrix and the cellular actin cytoskeleton.[58] Capitalizing on the differences in expression of certain subtypes of integrin receptors, RGD-conjugated siRNAs and ASOs were shown to be preferentially delivered to cancer cells through $\alpha v\beta 3$ integrin.[59–61] Similarly, therapeutic oligonucleotides conjugated to gastrin-releasing peptide

(GRP) or gonadotropin releasing hormone (GnRH) enabled delivery to malignant cells that overexpress the bombesin receptor 2 (BB2) or the GnRH receptor, respectively, with minimal delivery to control cells.[62,63] Several other moieties, including insulin-like growth factor (IGF1) peptide[64] and folate[65–67] have also been investigated for their potential application in oligonucleotide cancer therapeutics. Although these ligands presented encouraging results, the low efficiency and challenging reproducibility have spurred the search for more efficient alternatives.

Antibodies and aptamers bind to their cognate target or receptor with very high affinity [dissociation constant (Kd) $<10^{-9}$] and specificity and thus have also been considered to enhance delivery of therapeutic oligonucleotides.[68,69] Aptamers consist of DNA or RNA molecules that bind to their receptor through secondary- and tertiary-structure-based molecular recognition (see Chapters 15 and 16).[69,70] These have been applied successfully to deliver siRNA to prostate cancer tumors by targeting prostate-specific membrane antigen (PSMA) receptor,[71,72] but also to malignant T-cells through the cytotoxic T lymphocyte–associated antigen 4 (CTLA4) receptor.[73] Since aptamers are nucleic acids, their stabilization is paramount for the success of this strategy and, therefore, this has been the focus of new efforts in the field.[74] On the other hand, antibodies [or fragments of antibodies (Fab)] against the human insulin receptor (HIR),[75] epidermal growth factor receptor (EGFR)[76] and CD71 Fab[77] have also shown utility for delivery of siRNAs to the brain, tumors, and skeletal muscle respectively. The high affinity of aptamers and antibodies indicate that these strategies will be of great utility when used in the context of fully chemically modified siRNA constructs.

Other novel targeting moieties for systemic delivery include glucagon-like peptide-1 (GLP1) which has been investigated for delivery to pancreatic beta-cell islets. GLP1-conjugated ASOs enable efficient modulation of metastasis associated lung adenocarcinoma transcript 1 (MALAT1) and forkhead box protein O1 (FOXO1) targets in beta-cell islets, without major silencing effects reported in the liver.[78] Another new venture in the field is the use of centyrins for targeted siRNA delivery.[79] Centyrins consist of ~10 kDa protein scaffolds derived from fibronectin type III (FN3) domains which can be selected to target specific cell-surface receptors, such as EGFR and PSMA.[79] EGFR–Centyrin-conjugated siRNAs are effectively internalized by EGFR-positive tumors, where they evoke gene silencing effects without causing significant silencing in major clearance organs, such as liver and kidney. These scaffolds can be engineered to bind albumin, improving blood-circulating times and reducing renal clearance of conjugated oligonucleotides. The technology was originally optimized for siRNA delivery at Janssen (Pharmaceutical company of Johnson & Johnson) and is now being developed by a small spin-off company Aro Biopharmaceuticals. Interestingly, only a small fraction of selected Centyrins were active *in vivo*, indicating that geometry and affinity of receptor interactions is a major contributing factor. Also, most of these early studies were performed in the context of partially modified oligonucleotides, which most likely confounded the utility of this ligand. Thus, centyrins are now being evaluated using fully stabilized siRNA scaffolds to ascertain their utility for delivery of therapeutic oligonucleotides.

CNS-targeting ligands, such as transferrin and rabies virus glycoprotein (RVG), have also been extensively tested, but mostly in the context of formulated oligonucleotides. Transferrin-decorated delivery systems were shown to bind to the cognate receptor in the blood–brain barrier (BBB) allowing transvascular delivery of the oligonucleotide cargo to the brain.[80,81] Similarly, siRNAs formulated in RVG-containing delivery systems were able to bind to acetylcholine receptors and elicit gene silencing in the brain.[82] Alternatively, other research groups have conjugated siRNAs to small molecules, such as Sertraline[83,84] and Indatraline,[85] which targeted therapeutic oligonucleotides to sub-populations of neurons after intranasal administration. Again, these early promising studies have not yet gained major traction in the field due to a high degree of variability observed in results obtained by different research groups.

While all these approaches hold great promise for cell-type-specific delivery after local or systemic administration, without appropriate chemical stabilization of the oligonucleotide construct they only achieve sub-optimal results. In addition, other challenges, including endosomal entrapment and rapid clearance, may also significantly affect the assessment of the utility of tissue-targeting ligands and therefore need to be addressed to allow thorough analyses.

10.4 Productive Delivery of Therapeutic Oligonucleotides: Overcoming the Endosomal Barrier

Productive delivery of conjugated siRNAs is often hampered by entrapment of RNAi compounds within endosomal compartments. Indeed, it is believed that after distribution and internalization >99% of siRNAs localize to "intracellular oligonucleotide sinks", not actively contributing to the observed gene-silencing effect.[86] Thus, strategies that improve uptake and endosomal release hold great promise to improve potency and duration of gene-silencing effects, removing the need for high doses and reducing off-target effects.

Small-drug enhancers have been used experimentally to interfere with the processes of endocytic uptake and with the progression of the payload to late-stage lysosomal pathways. Guanabenz, an alpha-2 adrenergic receptor agonist, and Tetrandine, a calcium channel blocker, have been used successfully to increase endocytic uptake of Chol–siRNAs, but with no significant effects on endosomal release.[87,88] On the other hand, Lomatin is thought to alter trafficking of transferrin-containing vesicles affecting endosomal release of Chol–siRNAs, but not of lipid-based siRNA nanoparticles. Lysosomotrophic agents, such as chloroquine and hydroxychloroquine, were shown to enable endosomal release of siRNA in the context of nanoparticles,[89] but only showed limited effects in the case of conjugated Chol–siRNAs.[87] Thus, small-drug enhancers are likely to enable endosomal release through different mechanisms, which are dependent on the delivery strategy and pathway of cellular uptake, but also on the cell type. Extensive testing of this approach *in vivo* is now warranted to evaluate the feasibility of improving endosomal release for a diverse group of fully modified conjugated siRNAs.

Conjugation or co-administration of endosomolytic peptides, such as melittin, have also been considered as potential strategies to increase endosomal escape.[90] Melittin disrupts membrane bilayers by forming pore-like structures after reaching a critical concentration within the endosome.[90] Melittin or melittin-like peptide (MLP) polymers have been covalently attached to siRNAs containing GalNAc moieties for liver targeting.[91] These dynamic polyconjugates were able to efficiently deliver to hepatocytes *in vivo* and evoke potent gene silencing of two gene targets apolipoprotein B (*apoB*) and peroxisome proliferator-activated receptor alpha (*ppar-alpha*).[91] GalNAc-conjugated MLPs have also been used as an excipient to improve gene silencing of Chol–siRNAs.[92–94] Co-injections of Chol-siRNAs and GalNAc-MLPs resulted in up to a 500-fold increase in gene silencing for targets such as ApoB, coagulation factor VII and other hepatitis B virus (HBV)-related transcripts in the liver.[92–94] This strategy progressed to clinical settings, where it showed a reasonable tolerability profile in healthy volunteers.[95] Unfortunately, all clinical efforts with this platform have been stopped due to observation of long-term toxicity induced by the formulation in extended toxicology studies in NHPs.

While endosomal release agents can dramatically improve the potency and efficacy of therapeutic siRNAs, specially of partially stabilized oligonucleotides, the toxicity liabilities have reduced the enthusiasm of the field towards this approach. Alternatively, in the context of fully stabilized configurations, "oligonucleotide sinks" may support sustained release of therapeutic siRNAs over time, enabling long-lasting gene silencing and removing the requirement for endosomal release adjuvants.

10.5 The Effects of the Route of Administration: Local *versus* Systemic Delivery

The route of administration has a major effect on the key PK processes that govern tissue accumulation and retention, ultimately affecting the extent and potency of gene silencing. Once administered, compounds should be able to overcome relevant biological barriers on their path to reach the target site of action. Thus, selecting an adequate route of administration is dependent on the type of effect intended and on the intrinsic characteristics of specific conjugated oligonucleotide molecules. Here we give a summarized overview of the most commonly used local and systemic routes of administration for therapeutic oligonucleotides and their effects on PK and PD.

10.5.1 Local Delivery of Conjugated siRNAs

Local administrations have been used widely for intraocular and skin delivery, but also for delivery to the lungs and the CNS, in clinical-stage therapeutic oligonucleotide candidates. Administration at the target site of action, or on a membrane in close proximity, generally allows for the use of lower doses to achieve therapeutically meaningful gene silencing. Relevant examples from clinical settings include: intravitreal injections for reduction of vascular endothelial growth factor (VEGF) mRNA expression in age-related macular degeneration (NCT00557791);[96,97] intranasal delivery to the lung against the nucleocapsid of the respiratory syncytial virus (NCT00496821)[5] and intradermal injections of cholesterol-conjugated siRNAs against connective tissue growth factor (CTGF) for reduction of fibrosis and dermal scarring (NCT02246465).[98] These strategies avoid significant dilution, metabolism and elimination of the compound *en route*, as well as limiting distribution to untargeted tissues, hence reducing the likelihood of unwanted side effects. Rapid association with cellular membranes and/or interactions with components of the extracellular matrix enable local tissue retention and reduce immediate clearance.

Although local delivery of oligonucleotides is more immediately associated with easy-to-reach organs such as skin, eye and lungs, it has also been applied to less readily accessible tissues, such as the CNS. The invasiveness of the intervention is associated with the type of injection required, ranging from simple intrathecal infusions to intraparenchymal or intracerebroventricular (ICV) injections, which require highly skilled neurosurgical procedures. These injections result in markedly different distribution profiles, depending on the target site of administration. As an example, intraparenchymal injections of lipid-conjugated oligonucleotides in the brain result in focal distribution from the site of injection, whereas intracerebroventricular (ICV) injections lead to a more widespread distribution throughout the brain. The broad distribution attained through ventricular injections is thought to be driven by the bulk flow of the cerebrospinal fluid (CSF). Alternative less invasive approaches, which can be performed in out-patient settings, have also been investigated for direct administration into the CSF space. Intrathecal (IT) injections are used clinically to achieve meaningful concentrations of nusinersen, an approved antisense oligonucleotide (ASO) drug for spinal muscular atrophy (SMA) (see Chapter 6), in the CSF.[99] This approach is also being evaluated for the delivery of HTT- (NCT03342053) and superoxide dismutase (SOD1)-targeting (NCT01041222 and NCT02626699) ASOs now in clinical development.[100] Clearance of unconjugated ASOs from the CSF is known to occur rapidly with a relatively small fraction of the injected dose being retained in the CNS. Importantly, the secondary rate of CNS clearance after tissue uptake is very slow (months) and defined by the chemical architecture and stability of the oligonucleotide. For instance, nusinersen, a fully 2′-*O*-(2-methoxyethyl) and fully phosphorothioated ASO, is expected to have a significantly longer half-life in the CNS than mixed PS–PO backbone gapmers used in ongoing trials for Huntington's disease (HD) or amyotrophic lateral sclerosis (ALS) (see Chapters 2–4). Gapmers contain a central core, consisting of 8–10 nucleotides of unmodified DNA, which is necessary for RNaseH activity but which reduces the overall stability of the construct.

Despite the many challenges faced by local delivery approaches, direct administrations enable delivery of compounds to target tissues with precise control of doses and rates of infusion. Thus, local delivery strategies may still represent a logical choice for a subset of diseases with a well-defined and reachable tissue target, for which systemic approaches are deemed not suitable given the outstanding hurdles to delivery across relevant biological barriers, such as the BBB.

10.5.2 Systemic Delivery of Conjugated siRNAs

Systemic delivery of therapeutic oligonucleotides is required when the target tissue is not easily reachable or more generalized effects are desired. Given the developments seen in conjugate-mediated delivery of therapeutic oligonucleotides, the majority of ongoing RNAi clinical programs use systemic delivery approaches, aimed at targeting organs such as the liver. Nevertheless, these approaches impose several additional challenges since compounds need to overcome further biological barriers and avoid clearance by the reticulo-endothelial system (RES) before reaching the target organ. As a result, and in comparison with local administrations, higher doses or multiple administrations are generally required to attain therapeutic concentrations in target tissues.

The conventional and most broadly accepted form of administration for small molecules, oral administration, has found limited success in the clinical arena for therapeutic oligonucleotides.[101,102] The harsh biological environment of the gut poses great challenges for these drugs, which results in poor bioavailability (<1–10%).[101,102] Despite some recent successes with α-tocopherol-conjugated siRNAs,[103] oral administration generally requires adequate formulation in a drug-delivery system[104] or the use of permeation enhancers to improve delivery.[105] It is possible that recent advances in oligonucleotide stabilization chemistries can bring back enthusiasm to re-evaluate the potential of this route of administration.

Intravenous (IV) and subcutaneous (SC) injections are by far the most commonly used routes for systemic administrations in the therapeutic oligonucleotide field. They rely on the vascular system for distribution of the compounds. IV administrations afford the highest peak blood concentrations after injection, but also result in rapid renal clearance and elimination.[44] This effect is significant for unconjugated siRNAs ($t1/2$ = ~2–5 minutes),[49,50] but has also been reported for a wide variety of conjugated oligonucleotides for which the molecular weight is below the cut-off for passive glomerular filtration (<50 kDa).[44] Alternatively, SC and intramuscular (IM) injections have been used to delay this rapid initial clearance phase by establishing an oligonucleotide depot that provides sustained delivery over time. The intrinsic characteristics of the conjugated modality govern retention of the oligonucleotide at the site of injection and determine the rate of absorption into the blood compartment.[40,44] SC and IM injections result in lower peak concentrations compared with IVs, with highly lipophilic compounds presenting shallower concentration–time profiles than their hydrophilic counterparts. This results in reduced systemic exposure, with smaller areas under the curve (AUCs) in the blood, which contribute to a lower risk of dose-dependent systemic toxic effects. Despite the fact that these routes of administration display significantly different plasma concentration–time profiles, the pattern of oligonucleotide tissue distribution has been shown to remain similar (Figure 10.4).[15,44] In addition, given that SC injections may be carried out by the patients themselves, removing the costly burden of a requirement for specialized personnel for administration, this route of administration is one of the most widely popularized strategies for systemic delivery of oligonucleotides in clinical practice.

Systemic delivery of therapeutic oligonucleotides to well-perfused organs with permeable endothelia, such as the liver and kidney, has been achieved successfully in humans. Although the brain is also a hyper-vascularized organ, the presence of a highly specialized BBB precludes free diffusion of compounds from the blood to the brain parenchyma after regular IV or SC injections. Experimental intra-arterial administrations through the carotid artery have facilitated broad delivery of lipid-conjugated siRNAs (PC-DHA–hsiRNAs) to the brain when preceded by osmotic disruption of the BBB using mannitol.[123] Delivery through the carotid artery enables high concentrations of both mannitol and oligonucleotide at the brain vascular tree before hepatic and renal extractions take place. Although this approach significantly improves delivery to the brain, further evaluations are now warranted to establish long-term safety.

Other routes of administration include intraperitoneal[106] and rectal.[106] In both cases, compounds must be absorbed and enter circulation, where they partition and bind different plasma proteins, before reaching the target organ. The class and the extent to which the oligonucleotide binds to specific serum proteins will largely influence tissue distribution and uptake, as well as blood clearance kinetics. Although it is believed that protein binding during the distribution phase is mostly driven by the nature of the conjugate, recent studies indicate that the route of administration may also have an effect.[103] As an example, alpha-tocopherol-conjugated siRNA rapidly associates with HDL particles when administered IV,[107] but mostly associates with chylomicrons when administered enterically.[103] Given its strong influence on oligonucleotide PK and tissue accumulation, the selection of an appropriate route of administration is fundamental for the success of RNAi-based therapeutics.

The route of administration determines the fluids, tissues and barriers that the oligonucleotide will encounter while being absorbed, distributed and metabolized. Therefore, local and/or systemic approaches will present distinct advantages and drawbacks that must be weighted together with the intrinsic characteristics of the conjugated oligonucleotide when deciding which route of administration should be employed.

10.6 Enhancing PK Properties of Conjugated siRNAs: Reducing Clearance Kinetics and Accelerating Target Tissue Uptake

Fully chemically modified unconjugated siRNAs with an average molecular size of ~15 kDa are passively cleared by glomerular filtration within minutes of administration. Thus, different approaches have been adopted to rectify the poor PK properties of such highly hydrophilic molecules.

PS-modification and lipid-conjugation of siRNAs have helped to delay kidney clearance through active binding to serum proteins.[23,41] Indeed, serum protein binding is a common strategy utilized to improve the pharmacokinetic behavior of small molecules,[108] peptides and proteins.[109] Commercially available products, such as Abrazane (an albumin-based nanoparticle that encapsulates Taxol) and Levemir (a therapeutic peptide modified to associate non-covalently with endogenous albumin), rely on serum protein binding to avoid immediate elimination.[110] While simple, this is not without a cost. Serum protein binding probably dominates the tissue distribution of conjugated siRNAs, interfering with interactions with putative receptors, especially in tissues with low blood flow and non-fenestrated continuous endothelium.[39] Thus, to achieve meaningful target tissue uptake, the rate of siRNA clearance from the blood must be reduced. This is paramount in order to increase exposure and support productive uptake by low-abundance, slow-recycling receptors.

Enhancement of circulating time of therapeutic oligonucleotides and other biologics in the blood has been achieved through conjugation to polyethylene glycol (PEG) moieties, in a process called PEGylation.[111,112] PEGylation is currently used in many commercial biotherapeutics including proteins (*e.g.* Pegasys®,[113] PEG Intron®,[114] and Mircera®[115]), antibodies (*e.g.* Cimzia®[116]), liposomes (*e.g.* Doxil®[117]) and aptamer oligonucleotides (*e.g.* Macugen®[118]) to improve their delivery and therapeutic efficacy. In general, PEGylation acts by increasing the hydrodynamic size of the biopharmaceutical drug, reducing clearance by glomerular filtration and increasing plasma bioavailability. In the context of therapeutic oligonucleotides, this is effectively achieved with high-molecular-weight PEGs (>20–40 kDa) which enable considerable improvement in half-life (from 5 min to 1 hour, which represents 12-fold enhancement) of siRNAs.[119] In fact, it is believed that the long residency times provided by direct PEG conjugation surpass the reduction in protein and receptor binding affinity as a result of steric hindrance.[111] However, branched and linear PEG moieties have been reported to cause detrimental effects on gene silencing,[65,120] with longer PEGs affecting activity to a greater extent compared with shorter versions.[119,121] Although historically PEGs have been considered non-toxic and non-immunogenic, new evidence has confirmed the high occurrence of anti-PEG antibodies in clinical settings,[122] which may, in turn, have direct effects on the therapeutic utility of this strategy for therapeutic oligonucleotides. Thus, novel approaches to dynamically modulate the hydrodynamic size of conjugated siRNAs with low binding-affinities for their target tissue will most likely be required to improve bioavailability of these compounds.

Targeting high-abundance and rapid-recycling receptors favors productive delivery of conjugated siRNAs, despite the rapid clearance kinetics of these drugs *in vivo*.[52] Indeed, GalNAc–siRNAs are effectively internalized by hepatocytes even after IV administrations, where most of the dose (>90%) is rapidly eliminated.[19] Administration of GalNAc–siRNAs subcutaneously significantly delays the rapid initial clearance, improving delivery and silencing.[19] Since identifying ASGPR-type transport systems that allow meaningful delivery of siRNAs to extrahepatic tissues is a challenging task, selection of an appropriate route of administration may provide sufficient extension of circulating times in the blood to enable productive uptake in target cells.

10.7 Conjugation Chemistry for RNAi-based Therapeutics: Future Perspectives

While the first FDA-approved synthetic RNAi-based drug, Patisiran,[124] consists of a LNP-formulated siRNA, conjugate-mediated oligonucleotide delivery has evolved to be the clinically-dominant gene silencing paradigm.[2] Multivalent GalNAc conjugates, in the context of fully chemically modified siRNA scaffolds, enable efficient modulation of gene expression in hepatocytes and support up to 12 months of clinical efficacy in the liver.[20] The chemically defined manufacturing, ease of formulation in physiologically-relevant buffers, compatibility with SC injection and unprecedented duration of effect, together make this technology platform highly attractive for manipulation of gene expression. As a result, several companies, including Alnylam Pharmaceuticals, Arrowhead Pharmaceuticals, Dicerna Pharmaceuticals, Silence Therapeutics and many others, have initiated more than 40 clinical programs targeting liver diseases.

GalNAc-conjugated oligonucleotides revolutionized the field as the first true example of efficient targeted delivery. Optimization of the chemical architecture of the siRNA (dianophore) to enable excellent tissue distribution and PK–PD parameters rapidly expanded into multiple clinical programs for different liver diseases by simply modifying the targeting sequence (pharmacophore).[2] The success of the GalNAc approach has spurred interest in developing similar programs with ligands that enable cell-type-specific delivery of siRNAs beyond the liver. Designing such dianophores requires clear understanding of the key factors that affect absorption, distribution, metabolism and elimination (ADME) of conjugated siRNAs (Figure 10.5). All aspects, from selection of the adequate route of administration to every feature of the chemical design, should be carefully accounted for in order to enable and improve targeted delivery with long-lasting gene-silencing effects. In fact, in addition to the enhanced PK properties, these constructs should maintain features that enable the PD effect against the gene target of

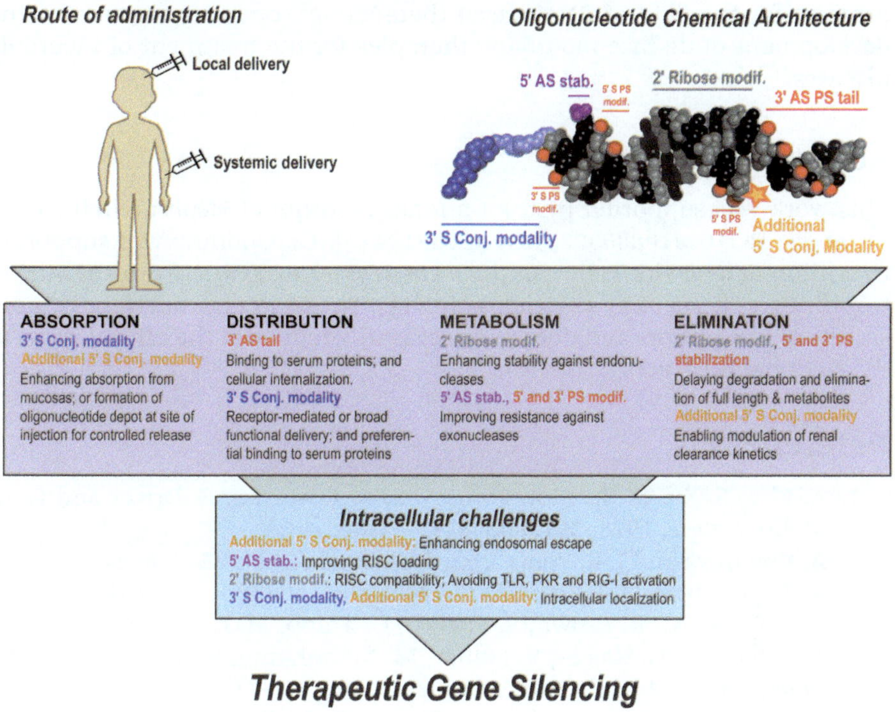

Figure 10.5 Key aspects governing the PK properties of conjugated RNAi-based compounds. A clear understanding of the key factors governing absorption, distribution, metabolism and elimination (ADME) of conjugated siRNAs, as well as of crucial intracellular challenges, is paramount for the clinical development of these compounds. The route of administration determines the need for absorption from the site of injection, as well as greatly influencing oligonucleotide kinetics in plasma. Local and systemic delivery approaches will largely depend on the gene target and on the nature of the conjugated modality. The chemical architecture of the oligonucleotides also has a major effect on the pharmacokinetic processes. On the molecular model above (top right) several prominent chemical features have been highlighted, and in the boxes below their effects on ADME and on the intracellular challenges have been summarized.

interest, including the rationally designed targeting sequence and features that enable RISC entry.

While targeted hepatic delivery of oligonucleotides has been solved with GalNAc conjugates, the utility of the conjugated-delivery approach remains to be demonstrated for extrahepatic tissues. Once a first clinical proof-of-concept study is established, it is expected that multiple programs will follow for different gene targets within the same tissue and cell-type. Alternatively, programs that enable broad functional delivery may offer the opportunity to treat other tissues beyond the liver, provided that the gene target has minimal significance in untargeted tissues. Together, both approaches hold great promise for the field of RNAi-based therapeutics, opening avenues for the development of disease-modifying therapies for the treatment of incurable diseases.

Acknowledgements

This work was supported by the National Institute of Health (NIH) [Grant numbers: RO1GM10880302, RO1NS03819415]. Dr. Godinho was supported by Milton-Safenowitz Fellowship (Grant ID# 17-PDF-363) from the Amyotrophic Lateral Sclerosis Association (ALSA). The content is solely the responsibility of the authors and does not necessarily represent the official views of the National Institutes of Health.

References

1. A. Fire, S. Xu, M. K. Montgomery, S. A. Kostas, S. E. Driver and C. C. Mello, *Nature*, 1998, **391**, 806–811.
2. A. Khvorova and J. K. Watts, *Nat. Biotechnol.*, 2017, **35**, 238–248.
3. K. Tiemann and J. J. Rossi, *EMBO Mol. Med.*, 2009, **1**, 142–151.
4. D. J. Foster, C. R. Brown, S. Shaikh, C. Trapp, M. K. Schlegel, K. Qian, A. Sehgal, K. G. Rajeev, V. Jadhav, M. Manoharan, S. Kuchimanchi, M. Maier and S. Milstein, *Mol. Ther.*, 2018, **26**, 708–717.
5. J. DeVincenzo, R. Lambkin-Williams, T. Wilkinson, J. Cehelsky, S. Nochur, E. Walsh, R. Meyers, J. Gollob and A. Vaishnaw, *Proc. Natl. Acad. Sci. U. S. A.*, 2010, **107**, 8800–8805.
6. P. K. Kaiser, R. A. Symons, S. M. Shah, E. J. Quinlan, H. Tabandeh, D. V. Do, G. Reisen, J. A. Lockridge, B. Short, R. Guerciolini, Q. Nguyen and Sirna-027-Study-Investigators, *Am. J. Ophthalmol.*, 2010, **150**, 33–39.e32.
7. M. Sioud, *Adv. Drug Delivery Rev.*, 2007, **59**, 153–163.
8. H. Yin, R. L. Kanasty, A. A. Eltoukhy, A. J. Vegas, J. R. Dorkin and D. G. Anderson, *Nat. Rev. Genet.*, 2014, **15**, 541–555.
9. D. Guzman-Villanueva, I. M. El-Sherbiny, D. Herrera-Ruiz, A. V. Vlassov and H. D. Smyth, *J. Pharm. Sci.*, 2012, **101**, 4046–4066.
10. J. K. Watts, G. F. Deleavey and M. J. Damha, *Drug Discovery Today*, 2008, **13**, 842–855.
11. W. B. Wan and P. P. Seth, *J. Med. Chem.*, 2016, **59**, 9645–9667.
12. B. A. Molitoris, P. C. Dagher, R. M. Sandoval, S. B. Campos, H. Ashush, E. Fridman, A. Brafman, A. Faerman, S. J. Atkinson, J. D. Thompson, H. Kalinski, R. Skaliter, S. Erlich and E. Feinstein, *J. Am. Soc. Nephrol.*, 2009, **20**, 1754–1764.
13. F. Czauderna, M. Fechtner, S. Dames, H. Aygun, A. Klippel, G. J. Pronk, K. Giese and J. Kaufmann, *Nucleic Acids Res.*, 2003, **31**, 2705–2716.
14. C. R. Allerson, N. Sioufi, R. Jarres, T. P. Prakash, N. Naik, A. Berdeja, L. Wanders, R. H. Griffey, E. E. Swayze and B. Bhat, *J. Med. Chem.*, 2005, **48**, 901–904.

15. M. R. Hassler, A. A. Turanov, J. F. Alterman, R. A. Haraszti, A. H. Coles, M. F. Osborn, D. Echeverria, M. Nikan, W. E. Salomon, L. Roux, B. M. D. C. Godinho, S. M. Davis, D. V. Morrissey, P. D. Zamore, S. A. Karumanchi, M. J. Moore, N. Aronin and A. Khvorova, *Nucleic Acids Res.*, 2018, **46**, 2185–2196.
16. J. F. Alterman, L. M. Hall, A. H. Coles, M. R. Hassler, M.-C. Didiot, K. Chase, J. Abraham, E. Sottosanti, E. Johnson, E. Sapp, M. Osborn, M. DiFiglia, N. Aronin and A. Khvorova, *Mol. Ther.—Nucleic Acids*, 2015, **4**, e266.
17. J. Soutschek, A. Akinc, B. Bramlage, K. Charisse, R. Constien, M. Donoghue, S. Elbashir, A. Geick, P. Hadwiger, J. Harborth, M. John, V. Kesavan, G. Lavine, R. Pandey, T. Racie, K. Rajeev, I. Röhl, I. Toudjarska, G. Wang, S. Wuschko, D. Bumcrot, V. Koteliansky, S. Limmer, M. Manoharan and H. Vornlocher, *Nature*, 2004, **432**, 173–178.
18. S. M. Elbashir, J. Martinez, A. Patkaniowska, W. Lendeckel and T. Tuschl, *EMBO J.*, 2001, **20**, 6877–6888.
19. J. K. Nair, H. Attarwala, A. Sehgal, Q. Wang, K. Aluri, X. Zhang, M. Gao, J. Liu, R. Indrakanti, S. Schofield, P. Kretschmer, C. Brown, S. Gupta, J. Willoughby, J. Boshar, V. Jadhav, K. Charisse, T. Zimmermann, K. Fitzgerald, M. Manoharan, K. Rajeev, A. Akinc, R. Hutabarat and M. Maier, *Nucleic Acids Res.*, 2017, **45**, 10969–10977.
20. K. K. Ray, U. Landmesser, L. A. Leiter, D. Kallend, R. Dufour, M. Karakas, T. Hall, R. P. Troquay, T. Turner, F. L. Visseren, P. Wijngaard, R. Wright and J. Kastelein, *N. Engl. J. Med.*, 2017, **376**, 1430–1440.
21. Y. Huang, *Mol. Ther.–Nucleic Acids*, 2017, **6**, 116–132.
22. M. Overhoff and G. Sczakiel, *EMBO Rep.*, 2005, **6**, 1176–1181.
23. R. S. Geary, D. Norris, R. Yu and C. F. Bennett, *Adv. Drug Delivery Rev.*, 2015, **87**, 46–51.
24. M. Tanowitz, L. Hettrick, A. Revenko, G. A. Kinberger, T. P. Prakash and P. P. Seth, *Nucleic Acids Res.*, 2017, **45**, 12388–12400.
25. K. Schmidt, T. P. Prakash, A. J. Donner, G. A. Kinberger, H. J. Gaus, A. Low, M. E. Østergaard, M. Bell, E. E. Swayze and P. P. Seth, *Nucleic Acids Res.*, 2017, **45**, 2294–2306.
26. R. A. Haraszti, L. Roux, A. H. Coles, A. A. Turanov, J. F. Alterman, D. Echeverria, B. M. Godinho, N. Aronin and A. Khvorova, *Nucleic Acids Res.*, 2017, **45**, 7581–7592.
27. W. F. Lima, H. Wu, J. G. Nichols, H. Sun, H. M. Murray and S. T. Crooke, *J. Biol. Chem.*, 2009, **284**, 26017–26028.
28. D. S. Schwarz, G. Hutvágner, B. Haley and P. D. Zamore, *Mol. Cell*, 2002, **10**, 537–548.
29. A. Nykänen, B. Haley and P. D. Zamore, *Cell*, 2001, **107**, 309–321.
30. E. Elkayam, R. Parmar, C. R. Brown, J. L. Willoughby, C. S. Theile, M. Manoharan and L. Joshua-Tor, *Nucleic Acids Res.*, 2016, **45**, 3528–3536.
31. R. Parmar, J. L. Willoughby, J. Liu, D. J. Foster, B. Brigham, C. S. Theile, K. Charisse, A. Akinc, E. Guidry, Y. Pei, W. Strapps, M. Cancilla, M.

Stanton, K. Rajeev, L. Sepp-Lorenzino, M. Manoharan, R. Meyers, M. Maier and V. Jadhav, *ChemBioChem*, 2016, **17**, 985–989.

32. T. P. Prakash, G. A. Kinberger, H. M. Murray, A. Chappell, S. Riney, M. J. Graham, W. F. Lima, E. E. Swayze and P. P. Seth, *Bioorg. Med. Chem. Lett.*, 2016, **26**, 2817–2820.
33. R. G. Parmar, C. R. Brown, S. Matsuda, J. L. S. Willoughby, C. S. Theile, K. Charissé, D. J. Foster, I. Zlatev, V. Jadhav, M. A. Maier, M. Egli, M. Manoharan and K. G. Rajeev, *J. Med. Chem.*, 2018, **61**, 734–744.
34. C. Wolfrum, S. Shi, K. N. Jayaprakash, M. Jayaraman, G. Wang, R. K. Pandey, K. G. Rajeev, T. Nakayama, K. Charrise, E. M. Ndungo, T. Zimmermann, V. Koteliansky, M. Manoharan and M. Stoffel, *Nat. Biotechnol.*, 2007, **25**, 1149–1157.
35. M. DiFiglia, M. Sena-Esteves, K. Chase, E. Sapp, E. Pfister, M. Sass, J. Yoder, P. Reeves, R. K. Pandey, K. G. Rajeev, M. Manoharan, D. Sah, P. D. Zamore and N. Aronin, *Proc. Natl. Acad. Sci. U. S. A.*, 2007, **104**, 17204–17209.
36. M. Byrne, R. Tzekov, Y. Wang, A. Rodgers, J. Cardia, G. Ford, K. Holton, L. Pandarinathan, J. Lapierre, W. Stanney, K. Bulock, S. Shaw, L. Libertine, K. Fettes, A. Khvorova, S. Kaushal and P. Pavco, *J. Ocul. Pharmacol. Ther.*, 2013, **29**, 855–864.
37. M. Nikan, M. F. Osborn, A. H. Coles, B. M. Godinho, L. M. Hall, R. A. Haraszti, M. R. Hassler, D. Echeverria, N. Aronin and A. Khvorova, *Mol. Ther.–Nucleic Acids*, 2016, **5**, e344.
38. M. Nikan, M. F. Osborn, A. H. Coles, A. Biscans, B. M. Godinho, R. A. Haraszti, E. Sapp, D. Echeverria, M. DiFiglia, N. Aronin and A. Khvorova, *Bioconjugate Chem.*, 2017, **28**, 1758–1766.
39. M. F. Osborn, A. H. Coles, A. Biscans, R. A. Haraszti, L. Roux, S. Davis, S. Ly, D. Echeverria, M. R. Hassler, B. M. Godinho, M. Nikan and A. Khvorova, *BioRxiv*, 2018, 288092.
40. A. Biscans, A. Coles, R. Haraszti, D. Echeverria, M. Hassler, M. Osborn and A. Khvorova, *BioRxiv*, 2018, 289439.
41. M. F. Osborn and A. Khvorova, *Nucleic Acid Ther.*, 2018, **28**, 128–136.
42. M. L. Hvam, Y. Cai, F. Dagnæs-Hansen, J. S. Nielsen, J. Wengel, J. Kjems and K. A. Howard, *Mol. Ther.*, 2017, **25**, 1710–1717.
43. A. A. Turanov, A. Lo, M. R. Hassler, A. Makris, A. Ashar-Patel, J. F. Alterman, A. H. Coles, R. Haraszti, L. Roux, B. M. D. C. Godinho, D. Echeverria, S. Pears, J. Iliopoulos, R. Shanmugalingam, R. Ogle, Z. K. Zsengeller, A. Hennessy, S. A. Karumanchi, M. J. Moore and A. Khvorova, *Nat. Biotechnol.*, 2018, **36**, 1164–1173.
44. B. M. Godinho, J. W. Gilbert, R. A. Haraszti, A. H. Coles, A. Biscans, L. Roux, M. Nikan, D. Echeverria, M. Hassler and A. Khvorova, *Nucleic Acid Ther.*, 2017, **27**, 323–334.
45. T. Khan, H. Weber, J. DiMuzio, A. Matter, B. Dogdas, T. Shah, A. Thankappan, J. Disa, V. Jadhav, L. Lubbers, L. Sepp-Lorenzino, W. Strapps and M. Tadin-Strapps, *Mol. Ther.–Nucleic Acids*, 2016, **5**, e342.

46. S. M. Sarett, T. A. Werfel, L. Lee, M. A. Jackson, K. V. Kilchrist, D. Brantley-Sieders and C. L. Duvall, *Proc. Natl. Acad. Sci. U. S. A.*, 2017, **114**, E6490–E6497.
47. K. Bienk, M. L. Hvam, M. M. Pakula, F. Dagnæs-Hansen, J. Wengel, B. M. Malle, U. Kragh-Hansen, J. Cameron, J. T. Bukrinski and K. A. Howard, *J. Controlled Release*, 2016, **232**, 143–151.
48. J. Christensen, K. Litherland, T. Faller, E. Van de Kerkhof, F. Natt, J. Hunziker, J. Krauser and P. Swart, *Drug Metab. Dispos.*, 2013, **41**, 1211–1219.
49. D. V. Morrissey, J. A. Lockridge, L. Shaw, K. Blanchard, K. Jensen, W. Breen, K. Hartsough, L. Machemer, S. Radka, V. Jadhav, N. Vaish, S. Zinnen, C. Vargeese, K. Bowman, C. S. Shaffer, L. B. Jeffs, A. Judge, I. MacLachlan and B. Polisky, *Nat. Biotechnol.*, 2005, **23**, 1002–1007.
50. B. M. Godinho, J. R. Ogier, A. Quinlan, R. Darcy, B. T. Griffin, J. F. Cryan and C. M. O'driscoll, *Int. J. Pharm.*, 2014, **473**, 105–112.
51. A. D. Springer and S. F. Dowdy, *Nucleic Acid Ther.*, 2018, **28**, 109–118.
52. T. S. Zimmermann, V. Karsten, A. Chan, J. Chiesa, M. Boyce, B. R. Bettencourt, R. Hutabarat, S. Nochur, A. Vaishnaw and J. Gollob, *Mol. Ther.*, 2017, **25**, 71–78.
53. A. Schwartz, S. Fridovich and H. Lodish, *J. Biol. Chem.*, 1982, **257**, 4230–4237.
54. P. Boisguérin, S. Deshayes, M. J. Gait, L. O'donovan, C. Godfrey, C. A. Betts, M. J. Wood and B. Lebleu, *Adv. Drug Delivery Rev.*, 2015, **87**, 52–67.
55. T. Lehto, K. Ezzat, M. J. Wood and S. E. Andaloussi, *Adv. Drug Delivery Rev.*, 2016, **106**, 172–182.
56. S. A. Moschos, S. W. Jones, M. M. Perry, A. E. Williams, J. S. Erjefalt, J. J. Turner, P. J. Barnes, B. S. Sproat, M. J. Gait and M. A. Lindsay, *Bioconjugate Chem.*, 2007, **18**, 1450–1459.
57. A. Muratovska and M. R. Eccles, *FEBS Lett.*, 2004, **558**, 63–68.
58. R. L. Juliano, X. Ming, O. Nakagawa, R. Xu and H. Yoo, *Theranostics*, 2011, **1**, 211–219.
59. M. R. Alam, V. Dixit, H. Kang, Z.-B. Li, X. Chen, J. Trejo, M. Fisher and R. L. Juliano, *Nucleic Acids Res.*, 2008, **36**, 2764–2776.
60. M. R. Alam, X. Ming, V. Dixit, M. Fisher, X. Chen and R. L. Juliano, *Oligonucleotides*, 2010, **20**, 103–109.
61. S. He, B. Cen, L. Liao, Z. Wang, Y. Qin, Z. Wu, W. Liao, Z. Zhang and A. Ji, *Drug Delivery*, 2017, **24**, 471–481.
62. X. Ming, M. R. Alam, M. Fisher, Y. Yan, X. Chen and R. L. Juliano, *Nucleic Acids Res.*, 2010, **38**, 6567–6576.
63. M. Sioud and A. Mobergslien, *Bioconjugate Chem.*, 2012, **23**, 1040–1049.
64. G. Cesarone, O. P. Edupuganti, C.-P. Chen and E. Wickstrom, *Bioconjugate Chem.*, 2007, **18**, 1831–1840.
65. C. Dohmen, T. Fröhlich, U. Lächelt, I. Röhl, H.-P. Vornlocher, P. Hadwiger and E. Wagner, *Mol. Ther.–Nucleic Acids*, 2012, **1**, 1–6.

66. E. A. Orellana, S. Tenneti, L. Rangasamy, L. T. Lyle, P. S. Low and A. L. Kasinski, *Sci. Transl. Med.*, 2017, **9**, eaam9327.
67. M. Thomas, S. A. Kularatne, L. Qi, P. Kleindl, C. P. Leamon, M. J. Hansen and P. S. Low, *Ann. N. Y. Acad. Sci.*, 2009, **1175**, 32–39.
68. J. Tushir-Singh, *Expert Opin. Biol. Ther.*, 2017, **17**, 325–338.
69. T. C. Chu, K. Y. Twu, A. D. Ellington and M. Levy, *Nucleic Acids Res.*, 2006, **34**, e73.
70. A. D. Keefe, S. Pai and A. Ellington, *Nat. Rev. Drug Discovery*, 2010, **9**, 537–550.
71. J. O. McNamara II, E. R. Andrechek, Y. Wang, K. D. Viles, R. E. Rempel, E. Gilboa, B. A. Sullenger and P. H. Giangrande, *Nat. Biotechnol.*, 2006, **24**, 1005–1015.
72. J. P. Dassie, X.-Y. Liu, G. S. Thomas, R. M. Whitaker, K. W. Thiel, K. R. Stockdale, D. K. Meyerholz, A. P. McCaffrey, J. O. McNamara II and P. H. Giangrande, *Nat. Biotechnol.*, 2009, **27**, 839–849.
73. A. Herrmann, S. J. Priceman, M. Kujawski, H. Xin, G. A. Cherryholmes, W. Zhang, C. Zhang, C. Lahtz, C. Kowolik, S. J. Forman, M. Kortylewski and H. Yu, *J. Clin. Invest.*, 2014, **124**, 2977–2987.
74. P. Röthlisberger and M. Hollenstein, *Adv. Drug Delivery Rev.*, 2018, **134**, 3–21.
75. C.-F. Xia, R. J. Boado and W. M. Pardridge, *Mol. Pharm.*, 2008, **6**, 747–751.
76. S. Bäumer, N. Bäumer, N. Appel, L. Terheyden, J. Fremerey, S. Schelhaas, E. Wardelmann, F. Buchholz, W. E. Berdel and C. Müller-Tidow, *Clin. Cancer Res.*, 2015, **21**, 1383–1394.
77. T. Sugo, M. Terada, T. Oikawa, K. Miyata, S. Nishimura, E. Kenjo, M. Ogasawara-Shimizu, Y. Makita, S. Imaichi, S. Murata, K. Otake, K. Kikuchi, M. Teratani, Y. Masuda, T. Kamei, S. Takagahara, S. Ikeda, T. Ohtaki and H. Matsumoto, *J. Controlled Release*, 2016, **237**, 1–13.
78. S. Andersson, TIDES: Oligonucleotide & Peptide Therapeutics, Boston, 2018.
79. S. D. Goldberg, R. M. Cardoso, T. Lin, T. Spinka-Doms, D. Klein, S. A. Jacobs, V. Dudkin, G. Gilliland and K. T. O'neil, *Protein Eng., Des. Sel.*, 2016, **29**, 563–572.
80. D. T. Wiley, P. Webster, A. Gale and M. E. Davis, *Proc. Natl. Acad. Sci. U. S. A.*, 2013, **110**, 8662–8667.
81. A. Cardoso, S. Simoes, L. De Almeida, N. Plesnila, M. P. de Lima, E. Wagner and C. Culmsee, *J. Controlled Release*, 2008, **132**, 113–123.
82. P. Kumar, H. Wu, J. L. McBride, K.-E. Jung, M. H. Kim, B. L. Davidson, S. K. Lee, P. Shankar and N. Manjunath, *Nature*, 2007, **448**, 39–43.
83. A. Bortolozzi, A. Castañé, J. Semakova, N. Santana, G. Alvarado, R. Cortés, A. Ferrés-Coy, G. Fernández, M. C. Carmona, M. Toth, J. Perales, A. Montefeltro and F. Artigas, *Mol. Psychiatry*, 2012, **17**, 612–623.
84. A. Ferrés-Coy, M. Galofré, F. Pilar-Cuéllar, R. Vidal, V. Paz, E. Ruiz-Bronchal, L. Campa, Á. Pazos, J. Caso, J. Leza, G. Alvarado, A. Montefeltro, E. Valdizán, F. Artigas and A. Bortolozzi, *Mol. Psychiatry*, 2016, **21**, 328–338.

85. D. Alarcón-Arís, A. Recasens, M. Galofré, I. Carballo-Carbajal, N. Zacchi, E. Ruiz-Bronchal, R. Pavia-Collado, R. Chica, A. Ferrés-Coy and M. Santos, *Mol. Ther.*, 2017, **26**, 550–567.
86. S. Ly, D. M. Navaroli, M.-C. Didiot, J. Cardia, L. Pandarinathan, J. F. Alterman, K. Fogarty, C. Standley, L. M. Lifshitz, K. D. Bellve, M. Prot, D. Echeverria, S. Corvera and A. Khvorova, *Nucleic Acids Res.*, 2016, **45**, 15–25.
87. J. Gilleron, P. Paramasivam, A. Zeigerer, W. Querbes, G. Marsico, C. Andree, S. Seifert, P. Amaya, M. Stöter, V. Koteliansky, H. Waldmann, K. Fitzgerald, Y. Kalaidzidis, A. Akinc, M. Maier, M. Manoharan, M. Bickle and M. Zerial, *Nucleic Acids Res.*, 2015, **43**, 7984–8001.
88. M. F. Osborn, J. F. Alterman, M. Nikan, H. Cao, M. C. Didiot, M. R. Hassler, A. H. Coles and A. Khvorova, *Nucleic Acids Res.*, 2015, **43**, 8664–8672.
89. S. R. Bhattarai, E. Muthuswamy, A. Wani, M. Brichacek, A. L. Castañeda, S. L. Brock and D. Oupicky, *Pharm. Res.*, 2010, **27**, 2556–2568.
90. K. K. Hou, H. Pan, P. H. Schlesinger and S. A. Wickline, *Biotechnol. Adv.*, 2015, **33**, 931–940.
91. D. B. Rozema, D. L. Lewis, D. H. Wakefield, S. C. Wong, J. J. Klein, P. L. Roesch, S. L. Bertin, T. W. Reppen, Q. Chu, A. V. Blokhin, J. Hagstrom and J. Wolff, *Proc. Natl. Acad. Sci. U. S. A.*, 2007, **104**, 12982–12987.
92. S. C. Wong, J. J. Klein, H. L. Hamilton, Q. Chu, C. L. Frey, V. S. Trubetskoy, J. Hegge, D. Wakefield, D. B. Rozema and D. L. Lewis, *Nucleic Acid Ther.*, 2012, **22**, 380–390.
93. C. I. Wooddell, D. B. Rozema, M. Hossbach, M. John, H. L. Hamilton, Q. Chu, J. O. Hegge, J. J. Klein, D. H. Wakefield, C. E. Oropeza, J. Deckert, I. Roehl, K. Jahn-Hofmann, P. Hadwiger, H. Vornlocher, A. McLachlan and D. Lewis, *Mol. Ther.*, 2013, **21**, 973–985.
94. V. S. Trubetskoy, J. B. Griffin, A. L. Nicholas, E. M. Nord, Z. Xu, R. M. Peterson, C. I. Wooddell, D. B. Rozema, D. H. Wakefield, D. L. Lewis and S. Kanner, *Nucleic Acids Res.*, 2016, **45**, 1469–1478.
95. T. Schluep, J. Lickliter, J. Hamilton, D. L. Lewis, C. L. Lai, J. Y. Lau, S. A. Locarnini, R. G. Gish and B. D. Given, *Clin. Pharmacol. Drug Dev.*, 2017, **6**, 350–362.
96. A. O. Garba and S. A. Mousa, *Ophthalmol. Eye Dis.*, 2010, **2**, 75–83.
97. L. Singerman, *Retina*, 2009, **29**, S49–S50.
98. R. Galiano, Plastic Surgery Foundation Industry Summit, Boston, 2015.
99. R. S. Finkel, E. Mercuri, B. T. Darras, A. M. Connolly, N. L. Kuntz, J. Kirschner, C. A. Chiriboga, K. Saito, L. Servais, E. Tizzano, M. Tulinius, J. Montes, A. Glanzman, K. Bishop, Z. Zhong, S. Gheuens, C. Bennett, E. Schneider, W. Farwell, D. De Vivo and ENDEAR-Study-Group, *N. Engl. J. Med.*, 2017, **377**, 1723–1732.
100. T. M. Miller, A. Pestronk, W. David, J. Rothstein, E. Simpson, S. H. Appel, P. L. Andres, K. Mahoney, P. Allred, K. Alexander, L. Ostrow, D. Schoenfeld, E. Macklin, D. Norris, G. Manousakis, M. Crisp, R. Smith, C. Bennett, K. Bishop and M. Cudkowicz, *Lancet Neurol.*, 2013, **12**, 435–442.

101. S. Akhtar, *J. Drug Targeting*, 2009, **17**, 491–495.
102. G. E. Hardee, *Ther. Delivery*, 2012, **3**, 143–145.
103. M. Murakami, K. Nishina, C. Watanabe, K. Yoshida-Tanaka, W. Piao, H. Kuwahara, Y. Horikiri, K. Miyata, N. Nishiyama, K. Kataoka, M. Yoshida, H. Mizusawa and T. Yokota, *Sci. Rep.*, 2015, **5**, 1–13.
104. M. Aouadi, G. J. Tesz, S. M. Nicoloro, M. Wang, E. Soto, G. R. Ostroff and M. P. Czech, *Nature*, 2009, **458**, 1180–1184.
105. L. G. Tillman, R. S. Geary and G. E. Hardee, *J. Pharm. Sci.*, 2008, **97**, 225–236.
106. S. D. Larson, L. N. Jackson, L. A. Chen, P. G. Rychahou and B. M. Evers, *Surgery*, 2007, **142**, 262–269.
107. K. Nishina, T. Unno, Y. Uno, T. Kubodera, T. Kanouchi, H. Mizusawa and T. Yokota, *Mol. Ther.*, 2008, **16**, 734–740.
108. M. Fasano, S. Curry, E. Terreno, M. Galliano, G. Fanali, P. Narciso, S. Notari and P. Ascenzi, *IUBMB Life*, 2005, **57**, 787–796.
109. M. S. Dennis, M. Zhang, Y. G. Meng, M. Kadkhodayan, D. Kirchhofer, D. Combs and L. A. Damico, *J. Biol. Chem.*, 2002, **277**, 35035–35043.
110. F. Kratz, *J. Controlled Release*, 2008, **132**, 171–183.
111. J. Winkler, *Future Med. Chem.*, 2015, **7**, 1721–1731.
112. P. Bailon and C.-Y. Won, *Expert Opin. Drug Delivery*, 2009, **6**, 1–16.
113. S. J. Hadziyannis and G. V. Papatheodoridis, *Expert Opin. Pharmacother.*, 2003, **4**, 541–551.
114. R. M. Bukowski, C. Tendler, D. Cutler, E. Rose, M. M. Laughlin and P. Statkevich, *Cancer*, 2002, **95**, 389–396.
115. J. M. Topf, *Expert Opin. Pharmacother.*, 2008, **9**, 839–849.
116. G. Y. Melmed, S. R. Targan, U. Yasothan, D. Hanicq and P. Kirkpatrick, *Nat. Rev. Drug Discovery*, 2008, **7**, 641–642.
117. Y. C. Barenholz, *J. Controlled Release*, 2012, **160**, 117–134.
118. D. Vavvas and D. J. D'Amico, *Ophthalmol. Clin.*, 2006, **19**, 353–360.
119. F. Iversen, C. Yang, F. Dagnæs-Hansen, D. H. Schaffert, J. Kjems and S. Gao, *Theranostics*, 2013, **3**, 201–209.
120. S. Jung, S. H. Lee, H. Mok, H. J. Chung and T. G. Park, *J. Controlled Release*, 2010, **144**, 306–313.
121. Z. Gaziova, V. Baumann, A.-M. Winkler and J. Winkler, *Bioorg. Med. Chem.*, 2014, **22**, 2320–2326.
122. J. K. Armstrong, G. Hempel, S. Koling, L. S. Chan, T. Fisher, H. J. Meiselman and G. Garratty, *Cancer*, 2007, **110**, 103–111.
123. B. M. D. C. Godinho, N. Henninger, J. Bouley, J. F. Alterman, R. A. Haraszti, J. W. Gilbert, E. Sapp, A. H. Coles, A. Biscans, M. Nikan, D. Echeverria, M. DiFiglia, N. Aronin and A. Khvorova, *Mol. Ther.*, **26**, 2580–2591.
124. O. B. Suhr, T. Coelho, J. Buades, J. Pouget, I. Conceicao, J. Berk, H. Schmidt, M. Waddington-Cruz, J. M. Campistol, B. R. Bettencourt, A. Vaishnaw, J. Gollob and D. Adams, *Orphanet J. Rare Dis.*, 2015, **10**, 109.

CHAPTER 11

Liver-targeted RNAi Therapeutics: Principles and Applications

KALLANTHOTTATHIL G. RAJEEV AND MUTHIAH MANOHARAN*

Alnylam Pharmaceuticals, 300 Third Street, Cambridge, Massachusetts, USA 02142
*E-mail: mmanoharan@alnylam.com

11.1 Introduction

The pioneering work that led to the development of transformative RNA interference (RNAi)-based therapeutic agents began in the late 1980s with a study by Napoli, Lemieux and Jorgensen on the anthocyanin biosynthesis pathways in petunias.[1] In 1992, Romano and Macino described quelling *Neurospora crassa*[2] and the observation of a similar phenomenon in *Caenorhabditis elegans* by Guo and Kemphus[3] led to a Nobel Prize-winning seminal publication by Fire, Mello and colleagues[4] in 1998, demonstrating that introduction of RNA into cells interfered with the function of an endogenous gene. The observed gene interference was caused by injected double-stranded, not single-stranded, RNA and occurred in a catalytic fashion.[4] A similar RNAi phenomenon exerted by double-stranded RNA in *C. elegans* was demonstrated in mammalian cells by use of synthetic small interfering RNAs (siRNAs) by

Drug Discovery Series No. 68
Advances in Nucleic Acid Therapeutics
Edited by Sudhir Agrawal and Michael J. Gait

Published by the Royal Society of Chemistry, www.rsc.org

Tuschl's laboratory.[5] Soon after the demonstration of RNAi-mediated target-specific gene silencing in mammalian cells, the enormous potential of this natural self-defense mechanism was recognized[6] and companies like Alnylam Pharmaceuticals took on the challenges of developing RNAi-based drugs, a new class of therapeutic modality with the potential to treat diseases previously thought to be undruggable.[7–9]

In this review we provide a historic brief overview of the development of technologies that allow delivery of RNAi therapeutics to liver and how chemical modifications have been used to avoid activation of the innate immune response, to ensure systemic stability of siRNA, and to enable delivery technologies. Readers are also referred to recent reviews covering clinical-stage RNAi therapeutics for both hepatic and extra-hepatic indications.[10–14]

11.2 The Role of Chemistry

In order to make drugs out of siRNAs, it is necessary to maintain robust pharmacodynamics with optimal pharmacokinetics, metabolic stability, and specificity to ensure drug safety. To make siRNAs safe for systemic application features that activate the innate immune response must be eliminated, siRNA seed-related off-target effects must be eliminated or minimized and potential metabolite-related chemical toxicity must be avoided. The siRNA engages with proteins associated with the RNA interference machinery to load preferentially one of the two strands of the double-stranded siRNA onto an RNase H-like endonuclease known as Argonaute 2 (Ago2). Ago2 is a core component of the RNA-induced silencing complex (RISC).[15] The strand that is loaded into the Ago2 is termed the guide (or antisense strand). The thermodynamic stability difference between the two ends of the siRNA that engage with the RISC – the thermodynamic asymmetry – is a critical determinant of selection of the guide strand. The 5′-end of the strand with less thermodynamic stability is preferentially loaded onto Ago2 to elicit RNAi-mediated gene silencing.[16]

The interaction of siRNA with certain classes of proteins such as Toll-like receptors (TLRs), RNA-dependent protein kinase (PKR) and double-stranded RNA-specific adenosine deaminases (ADARs) can trigger off-target effects.[17] Incorporation of the 2′-*O*-methyl (2′-OMe) ribosugar modification (Figure 11.1) can eliminate immune stimulatory effects to a great extent.[18] Incorporation of 2′-deoxy-2′-fluoro (2′-F) ribosugar modification (Figure 11.1) is well liked by the RISC machinery to a great extent, better than 2′-*O*-methyl (2′-OMe), locked nucleic acid (LNA), and 2′-*O*-methoxyethyl (2′-MOE) ribosugar modifications.[19,20] In addition, incorporation of minor-groove base modifications[17,21] and sterically demanding major-groove modifications[22] at critical sites, suppress off-target-related effects originating from interaction of the siRNA with PKR and ADAR, without inhibition of RNAi-activity. In certain instances, based on the sequence, certain chemically modified siRNAs, on the other hand, can trigger toxicity through non-specific

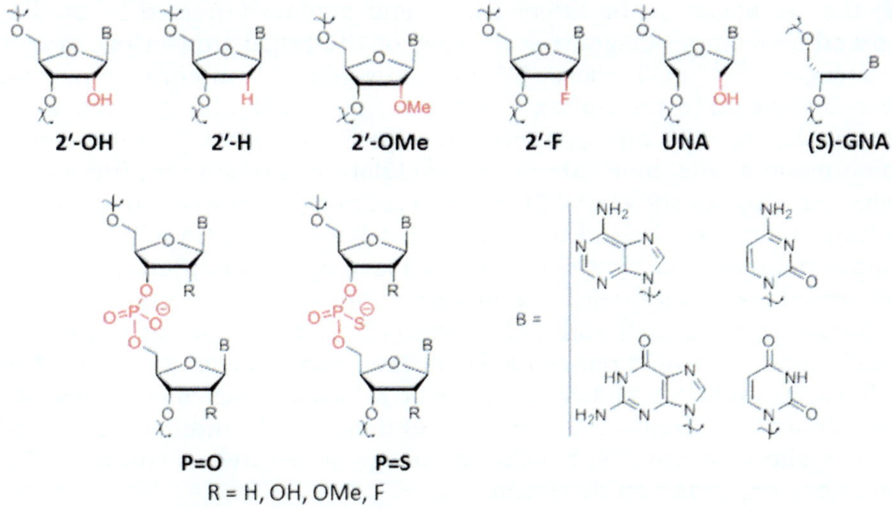

Figure 11.1 Ribosugar and backbone modifications used in designing siRNAs that are in clinical and pre-clinical developments. 2′-OH, 2′-H, 2′-OMe, and 2′-F are 2′-hydroxyl-, 2′-deoxy-, 2′-*O*-methyl-, and 2′-deoxy-2′-fluoro-ribosugar modifications, respectively. The acyclic sugar modifications UNA and (*S*)-GNA respectively are unlocked nucleic acid and (*S*)-glycol nucleic acids. P=O and P=S depict natural phosphate and phosphorothioate backbone, respectively.

binding to cellular proteins and the corresponding metabolites can trigger genotoxicity and mitochondrial toxicity by engaging with kinases and polymerases.[23]

The octamer short segment between nucleotide positions 1 and 9 from the 5′-end within both strands of the siRNA is analogous to the "seed" region of a microRNA (miRNA; see Chapter 8). The seed region is critical to the interaction with mRNA that results in inhibition of translation and/or mRNA degradation.[24] The Ago2 component of RISC that catalyzes siRNA-mediated mRNA cleavage is a member of the Argonaute protein superfamily of four proteins.[25] All Argonaute proteins participate in miRNA-mediated gene regulation. Non-specific interaction of the seed-region of the guide strand that is loaded onto any of the four Ago proteins can induce microRNA-like gene regulation by hybridizing to an mRNA that has perfect seed complementary to this region (although not to the rest of the siRNA guide strand). This results in so-called seed-related off-target gene regulation. Weakening of the thermodynamic stability of the duplex formed between the seed-region of siRNA and miRNA by incorporating acyclic nucleoside analogs, such as unlocked nucleic acid (UNA)[26,27] or glycol nucleic acid (GNA)[28,29] shown in Figure 11.1, at selected positions within the seed-region minimizes miRNA-type off-target effects of siRNAs.

In addition to mitigating immune response and off-target effects, the pharmacology of siRNA depends heavily on metabolic stability, delivery to the cytoplasm of the target tissue, and potency (affected by binding to and activation of Ago2 for cleavage of the target transcript). Several ribosugar,[28,30–33] nucleobase,[22,34] and phosphate[35–37] modifications have been evaluated in the context of improving or maintaining potency and metabolic stability and demonstrated their utility in a position-dependent manner, and hence are useful modulators of pharmacokinetics and pharmacology of siRNAs.[38–41] Both the type and position of modifications within the passenger and the guide strands of the siRNA affect potency and metabolic stability, and hence pharmacology of modified siRNAs.[42–44] Among the various nucleoside modifications evaluated, apart from the natural 2′-ribo (2′-OH) and 2′-deoxyribo (2′-H) modifications, only three nucleoside modifications namely 2′-OMe and 2′-deoxy-2′-fluoro (2′-F) ribosugars and UNA, shown in Figure 11.1, have been used in clinical and pre-clinical candidates thus far. Most of the clinical candidates also have a phosphorothioate (PS) backbone linkage at selected positions within both passenger and guide strands.

Chemical modifications can also inhibit degradation of siRNAs by exonucleases and endonucleases to increase metabolic stability and can facilitate delivery of siRNA to a desired tissue or cell type. The delivery mechanism determines the extent of chemical modification required on the siRNA. To date two modular delivery platforms – ionizable lipid nanoparticles (iLNPs)[45] and trivalent *N*-acetylgalactosamine (GalNAc) conjugates[46] – have been validated clinically for delivery of siRNAs to liver hepatocytes, as shown in Figure 11.2. In the LNP-based approach, the siRNA is formulated with lipid excipients and the formulation protects the siRNA from nucleases until it reaches the desired tissue, liver. In contrast, the GalNAc-conjugated siRNA is fully exposed to nucleases *en route* from the site of administration to the site of action. For LNP-formulated siRNAs, a minimal number of chemical

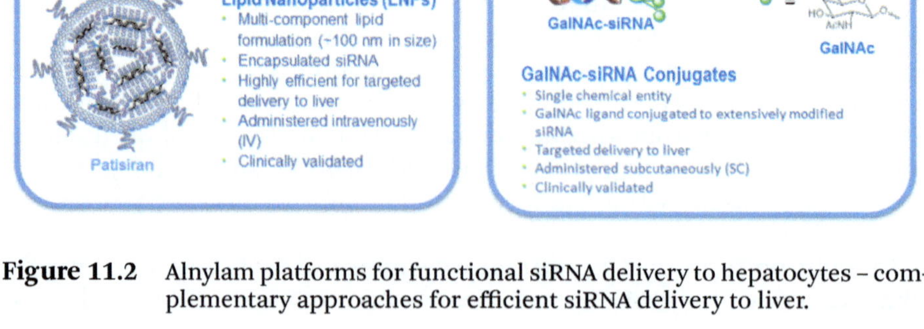

Figure 11.2 Alnylam platforms for functional siRNA delivery to hepatocytes – complementary approaches for efficient siRNA delivery to liver.

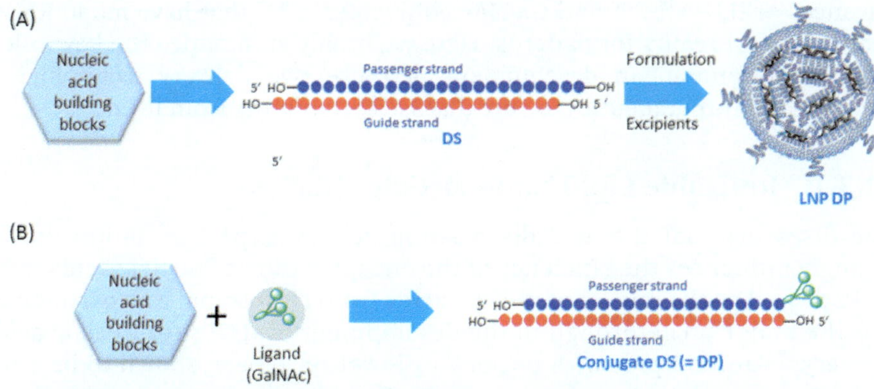

siRNA DS/DP Features	LNP	GalNAc Conjugate
Extent of chemical modifications	Light	Heavy
Shielding from nucleases	Formulation	Chemistry
Eliminate immune response	Chemistry	Chemistry
Delivery	LNP-mediated, ApoE-dependent	Receptor-ligand (ASGPR-GalNAc)-mediated
Administration	IV/IV infusion	SC

Figure 11.3 iLNP *vs.* GalNAc conjugate-based siRNA delivery. (A) iLNP approach – passenger and guide strands are separately synthesized from nucleic acid building blocks and annealing of equimolar mixture of the two strands yields the double-stranded siRNA, the drug substance (DS), which upon formulation with LNP excipients gives the LNP drug product (DP) in dosing buffer. (B) Ligand (GalNAc) approach – the trivalent GalNAc-conjugated passenger strand is synthesized from nucleic acid and ligand building blocks, and the corresponding guide strand from nucleic acid building blocks; annealing of an equimolar mixture of the two strands yields the GalNAc-conjugated siRNA DS, which in dosing buffer yields the DP. The table shows the distinguishing features of LNP-formulated and GalNAc-conjugated siRNA DPs.

modifications are used to suppress innate immune responses and potential off-target effects and to protect metabolic hot-spots,[14,47,48] but for GalNAc conjugates extensive chemical modification is required to achieve desirable clinical efficacy.[14,42–44,47] The design requirements and the differences between these two delivery approaches are summarized in Figure 11.3.

11.3 Liver-specific Delivery of siRNA

Despite improvements of drug-like properties of siRNA using chemical modifications, systemic delivery at therapeutically acceptable doses has proven to be a major hurdle for development of RNAi-based therapeutics. Now the field has two clinically validated modular delivery technologies –iLNPs and GalNAc conjugates – for delivery of siRNAs specifically to liver hepatocytes (Figure 11.2). Some recent reviews have summarized the unprecedented advances of iLNPs[14,45,48] and GalNAc conjugates[14,46,48] that have made RNAi therapeutics a reality for patients. Here we briefly summarize the key milestone achievements in development of these two delivery technologies enabling RNAi-therapeutics to cure diseases originating from liver.

11.3.1 Ionizable Lipid Nanoparticles (iLNPs)

The discovery that the acid dissociation constant (pK_a) of amino lipids strongly influences the efficiency of the encapsulation of nucleic acids into LNPs as well as the release of nucleic acids from endosomal compartments was the major breakthrough in the development of iLNPs for nucleic acid delivery.[49] Amino lipids with pK_a of 7 or lower have been shown to be critical for encapsulation, and subsequent efficient release, after uptake into cells. The low surface charge due to low pK_a translates to excellent tolerability of lipid nanoparticles. The ionizable lipid 1,2-dioleoyl-3-dimethylaminopropane (DODAP) was one of the early ionizable lipids shown to efficiently mediate nucleic acid delivery into cells.[50] After a lipid-chain structure–activity relationship (SAR) evaluation it was concluded that ionizable lipids with two or three double bonds with *cis* (*Z*)-configuration per acyl chain are optimal for encapsulation efficiency and that fully saturated acyl chains containing ionizable lipids are not suitable for formulating nucleic acids.[51] The discovery that a polyethylene glycol–lipid conjugate (PEG–lipid) can prevent aggregation and clearance of lipid vesicles by mononuclear phagocytes led to the development of stable lipid nanoparticles (SNALP) for nucleic acid encapsulation and delivery.[49,51,52] The major breakthrough in RNAi-mediated gene silencing in non-human primates (NHPs) was first demonstrated by Alnylam in 2006 by the use of *ApoB*-targeted siRNA encapsulated into SNALPs derived from another ionizable lipid called 1,2-dilinoleyloxy-*N*,*N*-dimethyl-3-aminopropane (DLinDMA), followed by intravenous administration.[53]

From mechanistic studies it was concluded that LNPs with a neutral surface charge bind to certain serum proteins, specifically apolipoprotein E (ApoE), which results in low-density lipoprotein receptor (LDLR)-mediated uptake into HepG2 and primary hepatocytes.[54,55] The ApoE-dependency of iLNPs to facilitate LDLR-mediated uptake into hepatocytes was demonstrated unambiguously by the use of ApoE-deficient mice, where a control cationic LNP (cLNP) with pK_a > 7 did not show any effect on activity.[56,57] The study findings also concluded that ApoE acts as an endogenous targeting ligand for iLNPs, but not cLNPs.[58]

Structure–activity guided lipid discovery with evaluation of nearly 60 lipids led to the synthesis of a novel, more potent ionizable lipid. DLin-MC3-DMA [MC3 or (6*Z*,9*Z*,28*Z*,31*Z*)-heptatriaconta-6,9,28,31-tetraen-19-yl-4-(dimethylamino)butanoate].[49,59] This ionizable lipid excipient has a pK_a of 6.44 and is used in patisiran formulation as shown in Figure 11.4. Patisiran is the most advanced RNAi therapeutic developed to

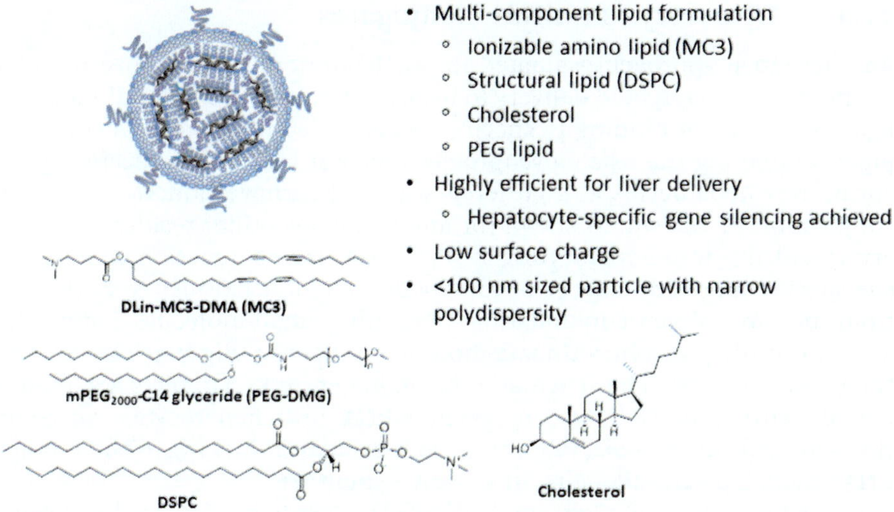

Figure 11.4 Ionizable lipid nanoparticles (iLNPs) for systemic RNAi therapeutics to liver describing the excipients of patisiran (ALN-TTR02) formulation and the characteristics of the formulation.

date. As described below the LNP–siRNA formulation patisiran[45,60–64] with trademark name ONPATTRO[65] is the only RNAi-based drug approved by the United States Food and Drug Administration (FDA). ONPATTRO™ was approved by the FDA to treat polyneuropathy of hereditary transthyretin amyloidosis (hATTR)[66,67] and European Medical Agency (EMA) approval is expected in 2018.[68]

11.3.2 Lipid-conjugated siRNA provided Proof of Concept for RNAi Therapeutics

Covalent conjugation of small molecules to siRNA has numerous advantages compared with non-viral vectors, particles, or excipient-based delivery systems.[44] In early efforts to deliver siRNA into cells, cholesterol conjugation was evaluated. A cholesterol-conjugated siRNA provided the first demonstration that conjugation of a pharmacokinetic modulating lipophilic small molecules to chemically modified siRNA could lead to silencing of endogenous genes. Treatment of mice with a cholesterol-conjugated siRNA targeting *ApoB* provided the first unambiguous proof of RNAi-mediated gene silencing driven by a synthetic agent *in vivo*.[69] The study was carried out at Alnylam in 2003, published in *Nature* in 2004 and was the first unambiguous proof of RNAi-mediated gene silencing *in vivo*. Administration of pharmacokinetic-modulating lipophilic small molecule-conjugated siRNA causes broad biodistribution and RNAi-mediated gene silencing in multiple tissues.[69,70]

11.3.3 Discovery of GalNAc Conjugates

An alternative approach evaluated in our laboratories was to use receptor ligand-mediated targeted delivery to liver. Exploration of the small-molecule ligands capable of binding to specific receptors led to the discovery of conjugates targeting the asialoglycoprotein receptor (ASGPR), a receptor present in liver hepatocytes at high levels with fast turnover kinetics. Covalent conjugation of GalNAc to siRNA (or any oligonucleotide) resulted in delivery specifically into hepatocytes.[44,71–73] The GalNAc conjugation enabled tissue-specific targeted delivery of therapeutic oligonucleotides, very distinct from pharmacokinetic-modulating lipophilic small-molecule conjugates and uniformly phosphorothioate-modified antisense oligonucleotides (see Chapters 2–4).[11,44,74] The schematic shown in Figure 11.5 depicts ASGPR-mediated delivery of GalNAc-conjugated siRNA into hepatocytes *via* clathrin-mediated endocytosis, followed by sub-cellular trafficking to elicit robust RNAi-mediated gene silencing in a catalytic manner.[36]

The achievement of a low single digit ED_{50} (median effective dose) using GalNAc–siRNA conjugates by subcutaneous (SC) administration in non-human primates and the need for very infrequent (monthly or quarterly) dosing by convenient SC administration has led to the advancement of several GalNAc-conjugated siRNAs targeting disease-causing genes in liver to early to late stage clinical development (Table 11.1, entries: 2, 3, 5, 6, 7, 9, 10, 11, 13, 15, 20 and 23).[11,46]

11.4 Clinical Candidates

A high-level overview of the liver-specific RNAi-based therapeutics that have advanced to various stages of clinical developments is provided in Table 11.1. The developmental candidates in Table 11.1 are designed to treat diseases that can be grouped into four categories: (I) orphan or rare (II) metabolic, (III) oncologic, and (IV) infectious. The table provides a high-level, up to date, overview of each drug entered into clinical development. The clinical trials of siRNA drugs intended for the treatment of liver-specific cancer[75] and infectious diseases[76,77] have been either completed or terminated at an early stage of clinical development due to safety concerns and/or poor clinical outcome. Readers are referred to the cited references for more details.[11,12,78] Several of the clinically advanced siRNA drugs from Alnylam are further discussed below.

11.4.1 ONPATTRO™ (Patisiran)

Hereditary transthyretin amyloidosis (hATTR) is a progressive multisystem, incapacitating and life-threatening disease with limited therapeutic options. hATTR amyloidosis can affect multiple parts of the body, including nerve, heart, and gastrointestinal systems. The mutant TTR-amyloid fibrils can deposit in the peripheral and central nervous system, gastrointestinal tract,

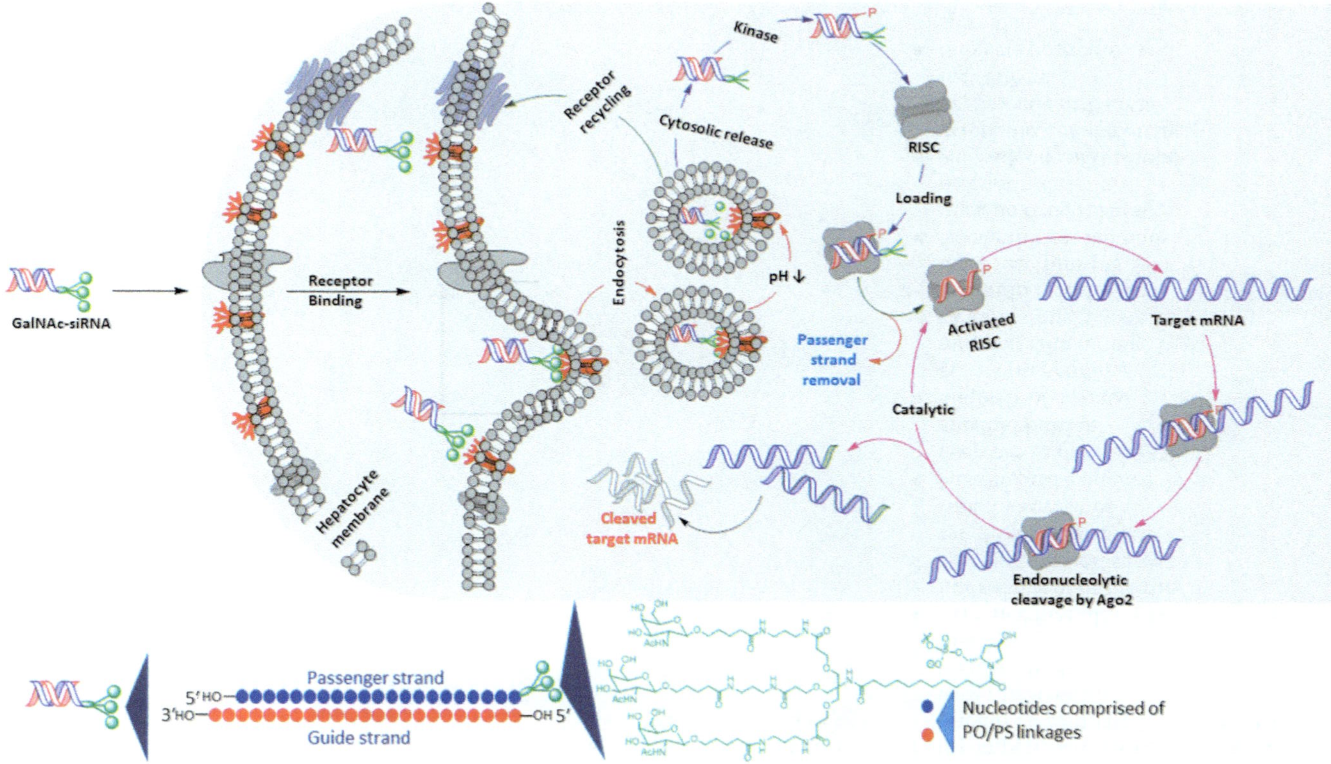

Figure 11.5 A schematic of ASGPR (red on the surface of hepatocyte membrane)-mediated endocytosis of GalNAc (green on the GalNAc–siRNA)-conjugated siRNA mediated by Ca^{2+} ions into early endosomes of hepatocytes followed by sub-cellular trafficking leading to RNAi-mediated silencing of target mRNA. After internalization, the low pH of the early endosomes results in detachment of the GalNAc–siRNA from the receptor with detachment of Ca^{2+} ions and concurrent cleavage of the anomeric linkage of the GalNAc moiety from the conjugate. A fraction of the siRNAs escapes into the cytosol. The 5′-end of the guide strand of siRNA that is released into the cytosol gets phosphorylated by kinases and then engages with RNA-induced silencing complex (RISC) to form the activated RISC. The activated RISC thus formed finds its target mRNA to produce endonucleolytic cleavage by RISC-associated Ago2 endonuclease. After product release, activated RISC engages with new mRNA molecules in a catalytic manner. With fully chemically modified siRNAs, the efficiency of 5′-binding of the guide strand can be improved by introducing a synthetic phosphate mimic such as 5′-(*E*)-vinylphosphonate.[36]

Table 11.1 Overview of liver-targeting siRNA-based therapeutics in clinical trials.

Orphan/Rare Diseases									
Entry	Drug	Drug substance/ Chemistry	Drug carrier	Target	Indication	Route, Dose	Key observations	Status	References
1	ONPATTRO™ (Patisiran, ALN-TTR02)	siRNA 2′-OH, 2′-OMe, 2′-H	LNP (DLin-MC3-DMA, PEG-DMG Lipid, Cholesterol, DSPC)	*TTR*	Hereditary ATTR (hATTR) Familial amyloid neuropathy	IV 0.3 mg kg^{-1} once every 3 weeks	• Generally well tolerated out to 25 months (*n*=27) • Mean 7.0 point decrease in modified neuropathy impairment score (mNIS+7), compared with the 30 point increase from natural history • Evidence for halting or improvement of neuropathy progression across range of baseline neuropathy severities • Histological evidence for decrease in dermal TTR amyloid burden • Majority of adverse events (AEs) mild to moderate, including mild flushing (22.2%) and mild infusion-related reactions (18.5%) • No significant lab findings; no drug-related discontinuations • No evidence of thrombocytopenia, renal toxicity, or systemic inflammatory effects • Subjects pretreated with steroids	Approved by FDA EMA approval awaited Phase 3	NCT01960348 NCT02510261 Adams *et al.*[61] Adams *et al*[62] Coelho *et al.*[79] Morrison[64] Suhr *et al.*[80] FDA[66] Alnylam[67,68]

2	ALN-TTRsc02	Enhanced stabilization chemistry (ESC) siRNA 2′-OMe, 2′-F, PS	GalNAc Conjugate	*TTR*	Hereditary ATTR (hATTR)	SC 5 to 300 mg total dose; 50 mg projected dose	• Generally well tolerated in healthy volunteers; no serious AEs (SAEs) and no discontinuations due to AEs • All AEs mild or moderate in severity • Max TTR knockdown of 98.4% with mean max of 97.1 ± 0.5% • Most potent Alnylam investigational RNAi therapeutic to date • Durability at least 4 months with a single dose	Phase 1, ongoing	NCT02797847 Alnylam[81]
3	Revusiran (ALN-TTRSC)	Standard Template Chemistry siRNA 2′-OMe, 2′-F, PS	GalNAc Conjugate	*TTR*	Transthyretin (TTR)-Mediated Familial Amyloidotic Cardiomyopathy (FAC)	SC	• TTR-mediated cardiac amyloidosis • Imbalance of mortality in the drug arm compared with placebo although durable mean maximum knockdown of TTR >88 % through 18 months was demonstrated earlier • Causes of the safety findings still not clear but may be related to high exposure levels of the drug (28 grams per year)	Phase 3 Terminated	NCT02319005 Zimmermann, *et al.*[82] K. Garber[83] Alnylam[84,85]
4	ALN-TTR01	siRNA, 2′-OH, 2′-OMe	LNP	*TTR*	Hereditary ATTR	IV 1 mg kg^{-1}	• Replaced with ALN-TT02	Phase 1 Completed	NCT01148953 Coelho *et al.*[79]

(*continued*)

Table 11.1 (*continued*)

Orphan/Rare Diseases									
Entry	Drug	Drug substance/ Chemistry	Drug carrier	Target	Indication	Route, Dose	Key observations	Status	References
5	Givosiran (ALN-AS1)	ESC siRNA 2′-OMe, 2′-F, PS	GalNAc Conjugate	*ALAS1*	Acute intermittent porphyria	SC	• Rapid, dose-dependent and durable lowering of aminolevulinic acid (ALA), ALA synthase 1 (ALAS1), and porphobilinogen (PBG) to near normal levels after two to four doses • Consistent clinical activity across three cohorts • 63% reduction in annualized attack rate • 73% reduction in annualized hemin doses • Attack rate reduction closely associated with the extent of ALA and PBG lowering • No drug-related SAEs or discontinuations due to AEs • Most common AEs were abdominal pain, headache, nasopharyngitis, nausea, vomiting	Phase 2, 3 ongoing	NCT03338816 NCT02949830 NCT03505853 NCT02452372 Chan *et al.*[86] Alnylam[87]

6	Fitusiran (ALN-AT3)	ESC siRNA 2′-OMe, 2′-F, PS	Conjugated trivalent GalNAc at 3′-end of sense strand	*AT3* encoding antithrombin	Hemophilia A or B, rare bleeding disorders	SC once a month, 80 mg fixed dose	• Mean maximal AT3 reduction 87% with increased thrombin generation at the lower end of the range observed in healthy participants • Median annualized bleed rate of 0 • No drug-related SAEs • No thromboembolic events • Mild ISRs • Transient increase in liver enzymes, possibly correlated with the hepatitis C virus (HCV) history of patients		NCT03549871 NCT03417102 NCT03417245 Pasi *et al.*[88] Sehgal *et al.*[89] Alnylam[90]
7	Lumasiran (ALN-GO1)	ESC siRNA 2′-OMe, 2′-F, PS	GalNAc Conjugate	*HAO1*	Primary hyperoxalurea 1 (PH1)	SC	• No drug related SAEs • Mild to moderate AEs • Up to eightfold increase in plasma glycolate in healthy volunteers	Phase 1, 2	NCT02706886 NCT03350451 Liebow *et al.*[91] Alnylam[92]
8	DCR-PH1	siRNA	Encore LNP	*PH1*	Primary hyperoxalurea 1	IV 0.05 mg kg^{-1} dose	• No data available	Phase 1 Terminated	NCT02795325

(*continued*)

Table 11.1 (*continued*)

Entry	Drug	Drug substance/ Chemistry	Drug carrier	Target	Indication	Route, Dose	Key observations	Status	References
					Orphan/Rare Diseases				
9	DCR-PHXC	Modified siRNA 2′-OMe, 2′-F, PS GalNAc conjugation through 2′-position of nucleoside	GalNAc Conjugated tetraloop (GalXC™ Technology)	*PH1*	Primary Hyperoxaluria	SC	• No data available	Phase 1 Ongoing	NCT03392896 Dicerna[93]
10	Cemdisiran (ALN-CC5)	ESC siRNA 2′-OMe, 2′-F, 2′-H, PS	GalNAc conjugate	*C5*	Compliment-mediated diseases [Atypical hemolytic uremic syndrome (aHUS), Paroxysmal nocturnal hemoglobinuria (PNH)]	SC	• No SAEs, no discontinuations due to AEs • 99% knockdown of *C5* • Sustained control of disease hemolysis with up to 67% reduction in eculizumab dose	Phase 1, 2	NCT03303313 NCT02352493 Alnylam[94]
11	ALN-AAT	ESC siRNA 2′-OMe, 2′-F, PS	GalNAc Conjugate	*SERPINA1*	Alpha-1 antitrypsin deficiency	SC	• Dose-dependent *AAT* target knockdown up to 89% • Liver enzyme elevation at high dose (15%)	Phase 1/2 terminated	NCT02503683 Alnylam[95]

12	ARC-AAT	UNA, Cholesterol	Cholesterol conjugated siRNA, melittin-like peptide–carboxy dimethylmaleic anhydride–GalNAc (MLP–CDM–GalNAc) delivery system	*SERPINA1*	Alpha-1 antitrypsin deficiency	IV 0.38–8.00 mg kg^{-1}	• Favorable tolerability in clinical settings with dose-dependent serum AAT reduction • MLP-related toxicity suspected in NHP	Phase 1 Terminated	NCT02363946 Arrowhead[96,97]
13	AROAAT1001 (ARO-AAT)	UNA, 2′-OMe, 2′-F, PS	GalNAc Conjugated TRiM™ Platform	*SERPINA1*	Alpha-1 antitrypsin deficiency	SC Single and Multi-dose	• No SAEs, No Severe AEs • AEs, one moderate (gastroenteritis) • Two cases of mild injection site erythema at 100 mg after first dose • No clinically meaningful adverse changes in blood urea nitrogen (BUN), creatinine, alanine aminotransferase (ALT), aspartate aminotransferase (AST) or total bilirubin • No pattern of adverse laboratory changes seen	Phase 1 Ongoing	NCT03362242 Arrowhead[98]

(continued)

Table 11.1 (*continued*)

Entry	Drug	Drug substance/ Chemistry	Drug carrier	Target	Indication	Route, Dose	Key observations	Status	References
					Orphan/Rare Diseases				
14	ND-L02-s0201 (BMS-986263)	Chemically modified siRNA	Vitamin A-conjugated LNP targeting	*HSP47*	Moderate to extensive hepatic fibrosis	IV	• No dose limiting toxicities • Infusion reactions	Phase 1b/2 ongoing; Phase 2 ongoing	NCT03538301 NCT03241264 NCT02227459 NCT01858935 BMS[99]
Metabolic Diseases									
15	Inclisiran ($ALN\text{-}PCS_{SC}$)	ESC siRNA 2′-OMe, 2′-F, 2′-H, PS	GalNAc conjugate	*PSCK9*	ASCVD heterozygous familial hypercholesterolemia; elevated cholesterol	SC 300 mg, day 1, day 90 then every 6 months.	• No liver function test (LFT) elevations related to study drug • No severe events related to drug and majority of AEs mild or moderate in severity • Maximum *PCSK9* inhibition of 95% with mean max reduction of LDL-cholesterol (LDL-C) of (64%) • All patients responded with significant LDL-C lowering: mean reduction of 52.6% (64 mg dl^{-1}) and up to 81% (122 mg dl^{-1}) at 6 months • Durability of silencing at least 6 months	Phase 3 ongoing	NCT03397121 NCT03400800 NCT03399370 NCT03159416 NCT03060577 NCT02963311 Ray *et al.*[100] Nishikido and Ray[101] Fitzgerald *et al.*[102] Ray *et al*[103] Bandyopadhyay *et al.*[104] Alnylam[105]

16	ALN-PCS02	siRNA 2′-OH, 2′-OMe, PS	LNP, MC3	*PCSK9*	Elevated LDL-C	IV 0.4 mg kg^{-1}	• 84 % and reductions in LDL cholesterol plasma levels of up to 50% following a single IV dose • No significant toxicities observed • Replaced by ALN-PCSsc	Phase 1 completed	NCT01437059 Fitzgerald *et al.*[106]
17	PRO-040201 (TKM-ApoB)	Unmodified siRNA	LNP (SNALP); DLinDMA	*Apo B*	Hypercholes-terolemia	IV	• Well tolerated with no evidence of liver toxicity. • Flulike symptoms at higher dose • ApoB protein and LDL cholesterol were transiently reduced by an average of 21.1% and 16.3%, respectively for patients received highest dose	Phase 1 Terminated	NCT00927459 Burnett *et al.*[107]

(*continued*)

Table 11.1 (*continued*)

Orphan/Rare Diseases									
Entry	Drug	Drug substance/ Chemistry	Drug carrier	Target	Indication	Route, Dose	Key observations	Status	References
Oncology									
18	TKM-080301 (TKM-PLK1)	siRNA chemistry undisclosed	LNP (DLinDMA)	*Polo-kinase 1*	Solid tumors with hepatocellular carcinoma, hepatoma liver cancer, adult liver cell carcinoma, adult cancer neuroendocrine tumors (NET), adrenocortical carcinoma (ACC), colorectal cancer with hepatic metastases, pancreas cancer with hepatic metastases, gastric cancer with hepatic metastases breast cancer with hepatic metastases ovarian cancer with hepatic metastases	IV 0.15–0.9 mg kg^{-1}	• Well tolerated at all dose levels • Most common drug-related AEs: mild to moderate levels of fever, chills, nausea, vomiting, fatigue • Maximum tolerated dose (MTD): 0.75 mg kg^{-1} body weight per week • Demonstrated tumor volume reduction along with overall stable disease or partial response with patients receiving >0.6 mg kg^{-1} dose • HCC – 43 subjects (12 subjects in the dose escalation arm, followed by 31 subjects in the expansion cohort, tumor response rate topline results: • 51% of subjects showed overall stable disease (SD) according to RECIST criteria; • 22% of subjects showed an overall partial response (PR) according to Choi response criteria; • Tumor density reduction of up to 59% was observed.	Phase 1/2 completed	NCT02191878 NCT01262235 NCT01437007 Lin *et al.*[108] Liu[109]

19	ALN-VSP02	Two siRNAs 2′-OH, 2′-OMe, 2′-H, PS	LNP (DLin DMA)	*KSP, VEGF*	Advanced solid tumors with liver involvement	IV	• Tumor reduction demonstrated, either complete remission or stable disease • 5′-rapid amplification of cDNA ends (RACE) confirmed the RNAi mechanism • Generally well-tolerated but reported infusion-related reaction that was reduced by slower infusion	Phase 1 completed	NCT01158079 NCT00882180 Tabernero *et al.*[110]
Infectious Diseases									
20	ALN-HBV	ESC siRNA 2′-OMe, 2′-F, PS	GalNAc conjugate	*HBV*	Non-cirrhotic patients with chronic HBV infection	SC 0.1–3.0 mg kg^{-1}	• Pre-clinical results up to 3.6 log_{10} HBsAg reduction	Being replaced by more potent compound	NCT02826018 Soriano[111] Alnylam[112]
21	ARC-520	Mixture of two choles-terol-conjugated siRNAs	Cholesterol-conjugated siRNA; MLP–CDM–GalNAc delivery system	*HBV*	Chronic hepatitis B (CHB)	IV	• siRNA targets covalently closed circular DNA (cccDNA) • MLP-related toxicity suspected	Phase 1 terminated	NCT02738008 NCT01872065 NCT02604199 NCT02604212 NCT02577029 NCT02065336 NCT02452528 NCT02535416 Schluep *et al.*[113] Wooddell *et al.*[114] Arrow-head[115,116]

(continued)

Table 11.1 (*continued*)

Orphan/Rare Diseases									
Entry	Drug	Drug substance/ Chemistry	Drug carrier	Target	Indication	Route, Dose	Key observations	Status	References
22	ARC-521	Mixture of two choles-terol-conjugated siRNAs	Cholesterol-conjugated siRNA MLP–CDM–GalNAc delivery system	*HBV*	Chronic hepatitis B	IV	• Optimized to silence integrated and cccDNA • Generally well tolerated in healthy volunteers and CHB patients. • One SAE of elevated transaminases occurred in a CHB patient 1 month after single dose, presumably due to viral flare secondary to fluc-tuating HBV DNA and antiviral therapy. • Mortality in non-human primates caused regula-tory hold and program termination. • Program discontinued due to EX-1 associated findings in non-clinical studies which were not seen in clinical trials. • MLP-related toxicity suspected	Phase 1 terminated	NCT02797522 Arrowhead[117]
23	ARO-HBV	Not known	GalNAc-Conjugated	*HBV*	Chronic hepatitis B	SC	No data	Phase 1/2 Ongoing	NCT03365947 Soriano[111] Arrowhead[118]

24	ARB-001467 (TKM-HBV)	Mixture of three siRNAs in combination with nucleot(s)ide analogues therapy. Chemistry not disclosed.	LNP	*HBV*	Hepatitis B/C infection	2 mg ml^{-1} IV Premedication to mitigate infusion-related reactions	• Greater reduction in serum hepatitis B surface antigen (HBsAg) levels with a favorable safety profile	Phase 2 Completed	NCT02631096 Mani *et al.*[119] Soriano[111] Arbutus[120,121]
25	TKM-130803	siRNA, chemistry not disclosed	LNP	*EBOV Kikwit*	Ebola virus disease (EVD)	IV 0.3 mg/kg per for 7 days	• No survival benefit for adult patients with severe EVD compared with historic controls	Single-arm Phase 2. Terminated	PACTR201501000997429 Dunning *et al.*[77] Rojek *et al.*[122]

eye, kidney, and/or heart. Deposition of TTR-amyloid fibrils in the peripheral and central nervous system leads to debilitating familial amyloid polyneuropathy (FAP), which eventually leads to the patient's death. The aggregation and deposition of mutant and wild-type transthyretin (TTR) protein in the heart leads to familial amyloid cardiomyopathy (FAC).[123,124]

Patisiran (Table 11.1, entry 1), an iLNP-formulated siRNA with MC3 lipid, which blocks production of the amyloid-forming TTR protein specifically in the liver, was tested in patients with FAP in a Phase 3 trial. The Phase 3 Apollo trial with 225 patients was a randomized, double-blind, placebo-controlled, global study designed to evaluate the efficacy and safety of patisiran in patients with hereditary ATTR amyloidosis with polyneuropathy.[61,79,80] The primary endpoint of the study was the change from baseline in modified Neurologic Impairment Score +7 (mNIS+7) relative to placebo at 18 months. Patisiran demonstrated benefit in patients with both early and advanced neuropathy at baseline. A post-hoc exploratory analysis has also been reported that demonstrated a significant decrease in the composite rate of all-causes hospitalization and mortality in patients receiving patisiran compared with those given placebo. Overall, patisiran improved multiple clinical manifestations of hereditary transthyretin amyloidosis.[62,125–129] Further attesting to the efficacy of patisiran, 99% of patients who completed the APOLLO study enrolled in the ongoing open-label extension study. The late-stage clinical study details can be found at ClinicalTrials.gov (NCT01960348, NCT02510261)

Very recent data analysis concluded that at 18 months, a larger proportion of patients on patisiran showed preservation (defined as no change in score) or improvement relative to baseline in each EuroQOL-5-dimension 5-level (EQ-5D-5L) analysis.[130] EQ-5D-5L is a standardized measure of health status based on five measures: mobility, self-care, usual activities, pain and/or discomfort, and anxiety and/or depression, and patients receiving patisiran showed significant improvement in all five measures compared with patients receiving placebo. In addition, overall health, measured by EuroQOL visual analogue scale (EQ-VAS), showed quality of life improvements from baseline in patients on patisiran whereas the patients that had received placebo showed even larger decline from the baseline.[130]

ONPATTRO™ (Patisiran, Table 11.1, entry 1) was approved by the United States FDA on August 10, 2018.[66,67] The regulatory approval of ONPATTRO™ (patisiran) for treating adult patients suffering from hTTR-mediated polyneuropathy (FAP) is a landmark achievement of RNAi-based therapeutic development.

11.4.2 Inclisiran

Inclisiran (Table 11.1, entry 15) is a long-acting GalNAc-conjugated siRNA modified with 2′-OMe, 2′-F and 2′-H, and PS at selected positions in a pattern that is referred to as enhanced stabilization chemistry or ESC.[42]

Inclisiran targets the mRNA encoding pro-protein convertase subtilisin/kexin type 9 (PCSK9). PCSK9 binds to the receptor for low density lipid (LDL) to direct the complex to the lysosome for degradation, allowing LDL to accumulate. Inhibition of expression of PCSK9 facilitates LDL clearance and reduces LDL levels in the bloodstream, thus ameliorating atherosclerosis progression. In the 500-patient ORION-1 Phase 2 trial, an average 41% reduction in LDL-cholesterol (LDL-C) was observed after a single 300 mg dose and a 50% reduction after two doses of 300 mg.[100,102,103] Inclisiran has an excellent safety profile. Multiple Phase 3 studies of inclisiran with different primary and secondary outcomes are ongoing. All the studies are placebo-controlled and the primary outcome measures include: percentage change in LDL-C from base line to day 510 and time-adjusted percentage change in LDL-C from baseline between day 90 and day 540. The secondary outcomes and other details can be found at ClinicalTrials.gov (NCT03399370, NCT03400800, NCT03397121)

11.4.3 Givosiran

Acute hepatic porphyrias (AHPs) are a group of ultra-rare diseases caused by genetic defects in components of the heme biosynthesis pathway.[131,132] Acute intermittent porphyria (AIP) is the most common subtype of AHP. AIP is characterized by disabling neurovisceral attacks and chronic symptoms caused by the upregulation of the first and rate-limiting step in heme synthesis catalyzed by ALA synthase 1 (ALAS1), and the resultant accumulation of the neurotoxic heme intermediates aminolevulinic acid (ALA) and porphobilinogen (PBG). The toxic heme intermediates cause acute incapacitating and potentially fatal attacks.

Givosiran (Table 11.1, entry 5) is another GalNAc-conjugated ESC siRNA. In a Phase 2 study, when givosiran was administered subcutaneously once a month or once quarterly, dose-dependent and prolonged urinary PBG and ALA lowering, after a single 2.5 mg kg^{-1} dose or multiple 1 mg kg^{-1} doses. Importantly, givosiran treatment of patients with AIP resulted in reductions in ALAS1, ALA, and PBG levels to levels similar to those in healthy volunteers. These decreases were associated with marked reductions in porphyria attacks and hemin usage, and significant increases in attack-free intervals.[87] Givosiran is generally well tolerated. There were no drug-related severe adverse effects and no discontinuations due to adverse effects in the Phase 2 trial. Reported mild to moderate adverse events possibly related to drug-included injection-site reactions, hypersensitivity, myalgia, headache, moderate renal impairment (in a patient with history of the same), and erythema. There were no clinically significant changes in vital signs, electrocardiogram (EKG), or clinical parameters. A Phase 3 clinical study of givosiran is currently underway and the primary outcome measure of the trial is the annualized rate of porphyria attacks in patients with AIP. The secondary outcomes and other details can be found at ClinicalTrials.gov (NCT03338816)

11.4.4 Fitusiran

Fitusiran (Table 11.1, entry 6), another GalNAc-conjugated ESC siRNA that targets the liver-expressed antithrombin 3 (*AT3*), which encodes an endogenous inhibitor of thrombin that is a genetically validated target for hemophilia.[88,89] Fitusiran is now in late-stage clinical testing. In the Phase 1 study and in the Phase 2 open-label extension OLE study of 32 patients, fitusiran was generally well tolerated during up to 14 months of dosing. Fitusiran given once a month subcutaneously robustly lowered AT3 levels by over 90% relative to patients treated with placebo, and there was clear evidence for restoration of hemostasis in severe hemophilia A and B patients. There were no thromboembolic events or laboratory evidence of pathological clot formation, even when bypassing agents were used to treat breakthrough bleeds, and no anti-drug antibodies were detected. There were two drug-related severe adverse events, and mild injection site reactions in 18% of patients. There were ALT increases in 33% patients, all of whom had a medical history of HCV, but no concurrent elevations of bilirubin were observed. Alnylam is transitioning fitusiran into a comprehensive Phase 3 program called ATLAS that began enrolling patients in January of 2018. The primary outcome measure of these studies is annualized bleeding rate (ABR). The secondary outcomes and other details can be found at ClinicalTrials.gov (NCT03549871, NCT03417102, NCT03417245).

11.4.5 TTRsc02

Due to the successes with other GalNAc-conjugated siRNAs, another *TTR*-targeting GalNAc-siRNA called TTRsc02 with ESC design[42] was evaluated in a Phase 1 randomized, placebo-controlled, single ascending-dose study in 48 healthy volunteers. Subcutaneous total doses ranged from 5 mg to 300 mg. The single doses resulted in robust TTR reductions of up to 98.4% (mean maximum of 97.1 ± 0.5%) with durability well over four months.[81] ALN-TTRsc02 at a dose of 50 mg in healthy volunteers produced robust TTR protein suppression (mean reduction of 86.2% on day 90 post dose). These results indicate that a once-quarterly dose could achieve potentially meaningful reductions in TTR in patients with ATTR amyloidosis. ALN-TTRsc02 was well tolerated in these healthy volunteers with no serious adverse events and no discontinuations. ALN-TTRsc02 is the most potent GalNAc conjugate developed so far. The clinical study details can be found at ClinicalTrials.gov (NCT02797847)

11.4.6 Revusiran

Revusiran was a first-generation conjugate with less metabolic stability and was the first GalNAc–siRNA conjugate evaluated in clinical trials, another landmark towards developing RNAi-based therapeutics.[47,82] Phase 1 and 2 clinical studies showed robust knockdown of TTR and the drug was generally well tolerated.[84,85] However, a Phase 3 clinical study of revusiran, called ENDEAVOUR, was discontinued in October 2016 due to an imbalance in mortality between patients that received placebo and drug-treated patients.[133] Extensive evaluation showed no clear reason for the observed mortality imbalance. While it is possible that the imbalance was a chance finding, a role for revusiran cannot be unambiguously excluded as drug exposure was very high (28 g cumulative annual dose or 500 mg chronic weekly dose).[134] The cumulative dose was 12–140 times greater than those used in other pipeline programs described above.

11.5 Conclusions and Outlook

In summary, the potential of siRNAs in treatment of otherwise undruggable diseases is real. The almost 30 drugs in various stages of clinical testing and the handful with proven efficacy indicate that this mode of gene silencing will find broad clinical application. Hurdles remain: a better understanding of the structural biology of Ago2 and the RISC and the interactions of these complexes with chemically modified siRNAs might enable design of more effective agents. The mechanisms of sequence-independent and sequence-dependent toxicities must be better understood. Identifying methods to improve the endosomal release of conjugates will improve the potency of RNAi therapeutics dramatically. Finally, GalNAc conjugation transformed the field, making possible subcutaneous, infrequent dosing, but methods for delivery into other tissues must be developed in order for RNAi-based therapeutics to broadly benefit patients.

References

1. C. Napoli, C. Lemieux and R. Jorgensen, *Plant Cell*, 1990, **2**, 279.
2. N. Romano and G. Macino, *Mol. Microbiol.*, 1992, **6**, 3343.
3. S. Guo and K. J. Kemphues, *Cell*, 1995, **81**, 611.
4. A. Fire, S. Xu, M. K. Montgomery, S. A. Kostas, S. E. Driver and C. C. Mello, *Nature*, 1998, **391**, 806.
5. S. M. Elbashir, J. Harborth, W. Lendeckel, A. Yalcin, K. Weber and T. Tuschl, *Nature*, 2001, **411**, 494.
6. G. L. Sen and H. M. Blau, *FASEB J.*, 2006, **20**, 1293.
7. A. R. de Fougerolles and J. M. Maraganore, *Antisense Drug Technology* (2nd edn), 2008, 465.
8. M. Manoharan and K. G. Rajeev, *Antisense Drug Technology* (2nd edn), 2008, 437.
9. D. Bumcrot, M. Manoharan, V. Koteliansky and D. W. Y. Sah, *Nat. Chem. Biol.*, 2006, **2**, 711.
10. C. Chakraborty, A. R. Sharma, G. Sharma, C. G. P. Doss and S.-S. Lee, *Mol. Ther.–Nucleic Acids*, 2017, **8**, 132.
11. S. T. Crooke, J. L. Witztum, C. F. Bennett and B. F. Baker, *Cell Metab.*, 2018, **27**, 714.

12. R. Titze-de-Almeida, C. David and S. S. Titze-de-Almeida, *Pharm. Res.*, 2017, **34**, 1339.
13. R. Titze-de-Almeida, C. David and S. S. Titze-de-Almeida, *Pharm. Res.*, 2018, **35**, 1.
14. J. C. Kaczmarek, P. S. Kowalski and D. G. Anderson, *Genome Med.*, 2017, **9**, 1.
15. I. J. Macrae, E. Ma, M. Zhou, C. V. Robinson and J. A. Doudna, *Proc. Natl. Acad. Sci. U. S. A.*, 2008, **105**, 512.
16. K. Ui-Tei, K. Nishi, T. Takahashi and T. Nagasawa, *Front. Non-Coding RNA*, 2012, **3**, 1.
17. H. Peacock, A. Kannan, P. A. Beal and C. J. Burrows, *J. Org. Chem.*, 2011, **76**, 7295.
18. R. Broering, C. I. Real, M. J. John, K. Jahn-Hofmann, L. M. Ickenstein, K. Kleinehr, A. Paul, K. Gibbert, U. Dittmer, G. Gerken and J. F. Schlaak, *Int. Immunol.*, 2014, **26**, 35.
19. M. Manoharan, A. Akinc, R. K. Pandey, J. Qin, P. Hadwiger, M. John, K. Mills, K. Charisse, M. A. Maier, L. Nechev, E. M. Greene, P. S. Pallan, E. Rozners, K. G. Rajeev and M. Egli, *Angew. Chem., Int. Ed.*, 2011, **50**, 2284.
20. P. S. Pallan, E. M. Greene, P. A. Jicman, R. K. Pandey, M. Manoharan, E. Rozners and M. Egli, *Nucleic Acids Res.*, 2011, **39**, 3482.
21. U. Ghanty, E. Fostvedt, R. Valenzuela, P. A. Beal and C. J. Burrows, *J. Am. Chem. Soc.*, 2012, **134**, 17643.
22. S. R. Suter, A. Ball-Jones, M. M. Mumbleau, R. Valenzuela, J. Ibarra-Soza, H. Owens, A. J. Fisher and P. A. Beal, *Org. Biomol. Chem.*, 2017, **15**, 10029.
23. M. M. Janas, M. K. Schlegel, C. E. Harbison, V. O. Yilmaz, Y. Jiang, R. Parmar, I. Zlatev, A. Castoreno, H. Xu, S. Shulga-Morskaya, K. G. Rajeev, M. Manoharan, N. D. Keirstead, M. A. Maier and V. Jadhav, *Nat. Commun.*, 2018, **9**, 1.
24. A. L. Jackson, J. Burchard, J. Schelter, B. N. Chau, M. Cleary, L. Lim and P. S. Linsley, *RNA*, 2006, **12**, 1179.
25. D. C. Swarts, K. Makarova, Y. Wang, K. Nakanishi, R. F. Ketting, E. V. Koonin, D. J. Patel and J. van der Oost, *Nat. Struct. Mol. Biol.*, 2014, **21**, 743.
26. J. B. Bramsen, M. M. Pakula, T. B. Hansen, C. Bus, N. Langkjaer, D. Odadzic, R. Smicius, S. L. Wengel, J. Chattopadhyaya, J. W. Engels, P. Herdewijn, J. Wengel and J. Kjems, *Nucleic Acids Res.*, 2010, **38**, 5761.
27. A. Pasternak and J. Wengel, *Org. Biomol. Chem.*, 2011, **9**, 3591.
28. M. K. Schlegel, D. J. Foster, A. V. Kel'in, I. Zlatev, A. Bisbe, M. Jayaraman, J. G. Lackey, K. G. Rajeev, K. Charisse, J. Harp, P. S. Pallan, M. A. Maier, M. Egli and M. Manoharan, *J. Am. Chem. Soc.*, 2017, **139**, 8537.
29. C. Theile, S. Milstein, M. K. Schlegel, M. Janas, V. R. Jadhav, D. Foster, M. Manoharan, K. G. Rajeev, M. Jayaraman, A. V. Kel'in, S. Matsuda, K. Charisse, J. K. Nair, M. A. Maier, A. Sehgal and C. Brown, *PCT Int. Appl.*, 2018, WO2018098328.

30. D. Zewge, F. Gosselin, D. M. Kenski, J. Li, V. Jadhav, Y. Yuan, S. S. Nerurkar, D. M. Tellers, W. M. Flanagan and I. W. Davies, *Bioconjugate Chem.*, 2014, **25**, 2222.
31. A. V. Kel'in, I. Zlatev, J. Harp, M. Jayaraman, A. Bisbe, J. O'Shea, N. Taneja, R. M. Manoharan, S. Khan, K. Charisse, M. A. Maier, M. Egli, K. G. Rajeev and M. Manoharan, *J. Org. Chem.*, 2016, **81**, 2261.
32. E. Malek-Adamian, D. C. Guenther, S. Matsuda, S. Martinez-Montero, I. Zlatev, J. Harp, M. Burai Patrascu, D. J. Foster, J. Fakhoury, L. Perkins, N. Moitessier, R. M. Manoharan, N. Taneja, A. Bisbe, K. Charisse, M. Maier, K. G. Rajeev, M. Egli, M. Manoharan and M. J. Damha, *J. Am. Chem. Soc.*, 2017, **139**, 14542.
33. T. Yamada, C. G. Peng, S. Matsuda, H. Addepalli, K. N. Jayaprakash, M. R. Alam, K. Mills, M. A. Maier, K. Charisse, M. Sekine, M. Manoharan and K. G. Rajeev, *J. Org. Chem.*, 2011, **76**, 1198.
34. H. Addepalli, Meena, C. G. Peng, G. Wang, Y. Fan, K. Charisse, K. N. Jayaprakash, K. G. Rajeev, R. K. Pandey, G. Lavine, L. Zhang, K. Jahn-Hofmann, P. Hadwiger, M. Manoharan and M. A. Maier, *Nucleic Acids Res.*, 2010, **38**, 7320.
35. G. Butora, N. Qi, W. Fu, T. Nguyen, H.-C. Huang and I. W. Davies, *Angew. Chem., Int. Ed.*, 2014, **53**, 14046.
36. R. G. Parmar, C. R. Brown, S. Matsuda, J. L. S. Willoughby, C. S. Theile, K. Charisse, D. J. Foster, I. Zlatev, V. Jadhav, M. A. Maier, M. Egli, M. Manoharan and K. G. Rajeev, *J. Med. Chem.*, 2018, **61**, 734.
37. T. P. Prakash, W. F. Lima, H. M. Murray, W. Li, G. A. Kinberger, A. E. Chappell, H. Gaus, P. P. Seth, B. Bhat, S. T. Crooke and E. E. Swayze, *Nucleic Acids Res.*, 2015, **43**, 2993.
38. S. A. Dar, A. Thakur, A. Qureshi and M. Kumar, *Sci. Rep.*, 2016, **6**, 20031.
39. G. F. Deleavey and M. J. Damha, *Chem. Biol.*, 2012, **19**, 937–954.
40. J. W. Engels, *New Biotechnol.*, 2013, **30**, 302.
41. M. Manoharan, *Curr. Opin. Chem. Biol.*, 2004, **8**, 570.
42. D. J. Foster, C. R. Brown, S. Shaikh, C. Trapp, M. K. Schlegel, K. Qian, A. Sehgal, K. G. Rajeev, V. Jadhav, M. Manoharan, S. Kuchimanchi, M. A. Maier and S. Milstein, *Mol. Ther.*, 2018, **26**, 708.
43. J. K. Nair, H. Attarwala, A. Sehgal, Q. Wang, K. Aluri, X. Zhang, M. Gao, J. Liu, R. Indrakanti, S. Schofield, P. Kretschmer, C. R. Brown, S. Gupta, J. L. S. Willoughby, J. A. Boshar, V. Jadhav, K. Charisse, T. Zimmermann, K. Fitzgerald, M. Manoharan, K. G. Rajeev, A. Akinc, R. Hutabarat and M. A. Maier, *Nucleic Acids Res.*, 2017, **45**, 10969.
44. J. K. Nair, J. L. S. Willoughby, A. Chan, K. Charisse, M. R. Alam, Q. Wang, M. Hoekstra, P. Kandasamy, A. V. Kel'in, S. Milstein, N. Taneja, J. O'Shea, S. Shaikh, L. Zhang, R. J. van der Sluis, M. E. Jung, A. Akinc, R. Hutabarat, S. Kuchimanchi, K. Fitzgerald, T. Zimmermann, T. J. C. van Berkel, M. A. Maier, K. G. Rajeev and M. Manoharan, *J. Am. Chem. Soc.*, 2014, **136**, 16958.

45. J. A. Kulkarni, P. R. Cullis and R. van der Meel, *Nucleic Acid Ther.*, 2018, **28**, 146.
46. A. D. Springer and S. F. Dowdy, *Nucleic Acid Ther.*, 2018, **28**, 109.
47. J. S. Butler, A. Chan, S. Fishman, J. L. S. Willoughby, T. D. Borland, S. Milstein, D. J. Foster, Q. Chen, J. Qin, B. R. Bettencourt, D. W. Sah, R. Alvarez, K. G. Rajeev, M. Manoharan, K. Fitzgerald, R. E. Meyers, S. V. Nochur, T. S. Zimmermann, S. Costelha, P. Goncalves and M. J. Saraiva, *Amyloid*, 2016, **23**, 109.
48. N. Shajari, B. Mansoori, S. Davudian, A. Mohammadi and B. Baradaran, *Curr. Drug Delivery*, 2017, **14**, 36.
49. Y. Y. C. Tam, S. Chen and P. R. Cullis, *Pharmaceutics*, 2013, **5**, 498.
50. S. C. Semple, S. K. Klimuk, T. O. Harasym, N. Dos Santos, S. M. Ansell, K. F. Wong, N. Maurer, H. Stark, P. R. Cullis, M. J. Hope and P. Scherrer, *Biochim. Biophys. Acta, Biomembr.*, 2001, **1510**, 152.
51. J. Heyes, L. Palmer, K. Bremner and I. MacLachlan, *J. Controlled Release*, 2005, **107**, 276.
52. I. Maclachlan, L. B. Jeffs, A. Judge, A. C. H. Lee, L. R. Palmer and V. Sood, *PCT Int. Appl.*, 2006, WO2006053430.
53. T. S. Zimmermann, A. C. H. Lee, A. Akinc, B. Bramlage, D. Bumcrot, M. N. Fedoruk, J. Harborth, J. A. Heyes, L. B. Jeffs, M. John, A. D. Judge, K. Lam, K. McClintock, L. V. Nechev, L. R. Palmer, T. Racie, I. Roehl, S. Seiffert, S. Shanmugam, V. Sood, J. Soutschek, I. Toudjarska, A. J. Wheat, E. Yaworski, W. Zedalis, V. Koteliansky, M. Manoharan, H.-P. Vornlocher and I. MacLachlan, *Nature*, 2006, **441**, 111.
54. C. L. Bisgaier, M. V. Siebenkas and K. J. Williams, *J. Biol. Chem.*, 1989, **264**, 862.
55. H. Jin, J. F. Lovell, J. Chen, Q. Lin, L. Ding, K. K. Ng, R. K. Pandey, M. Manoharan, Z. Zhang and G. Zheng, *Bioconjugate Chem.*, 2012, **23**, 33.
56. A. Akinc, W. Querbes, S. De, J. Qin, M. Frank-Kamenetsky, K. N. Jayaprakash, M. Jayaraman, K. G. Rajeev, W. L. Cantley, J. R. Dorkin, J. S. Butler, L. L. Qin, T. Racie, A. Sprague, E. Fava, A. Zeigerer, M. J. Hope, M. Zerial, D. W. Y. Sah, K. Fitzgerald, M. A. Tracy, M. Manoharan, V. Koteliansky, A. de Fougerolles and M. A. Maier, *Mol. Ther.*, 2010, **18**, 1357.
57. A. Akinc, A. Zumbuehl, M. Goldberg, E. S. Leshchiner, V. Busini, N. Hossain, S. A. Bacallado, D. N. Nguyen, J. Fuller, R. Alvarez, A. Borodovsky, T. Borland, R. Constien, A. de Fougerolles, J. R. Dorkin, K. Narayanannair Jayaprakash, M. Jayaraman, M. John, V. Koteliansky, M. Manoharan, L. Nechev, J. Qin, T. Racie, D. Raitcheva, K. G. Rajeev, D. W. Y. Sah, J. Soutschek, I. Toudjarska, H.-P. Vornlocher, T. S. Zimmermann, R. Langer and D. G. Anderson, *Nat. Biotechnol.*, 2008, **26**, 561.
58. P. R. Cullis and M. J. Hope, *Mol. Ther.*, 2017, **25**, 1467.
59. M. Jayaraman, S. M. Ansell, B. L. Mui, Y. K. Tam, J. Chen, X. Du, D. Butler, L. Eltepu, S. Matsuda, J. K. Narayanannair, K. G. Rajeev, I. M. Hafez, A. Akinc, M. A. Maier, M. A. Tracy, P. R. Cullis, T. D. Madden, M. Manoharan and M. J. Hope, *Angew. Chem., Int. Ed.*, 2012, **51**, 8529.
60. M. Rizk and S. Tuzmen, *Pharmgenomics Pers. Med.*, 2017, **10**, 267.

61. D. Adams, A. Gonzalez-Duarte, W. D. O'Riordan, C.-C. Yang, M. Ueda, A. V. Kristen, I. Tournev, H. H. Schmidt, T. Coelho, J. L. Berk, K.-P. Lin, G. Vita, S. Attarian, V. Plante-Bordeneuve, M. M. Mezei, J. M. Campistol, J. Buades, T. H. Brannagan, 3rd, B. J. Kim, J. Oh, Y. Parman, Y. Sekijima, P. N. Hawkins, S. D. Solomon, M. Polydefkis, P. J. Dyck, P. J. Gandhi, S. Goyal, J. Chen, A. L. Strahs, S. V. Nochur, M. T. Sweetser, P. P. Garg, A. K. Vaishnaw, J. A. Gollob and O. B. Suhr, *N. Engl. J. Med.*, 2018, **379**, 11.
62. D. Adams, O. B. Suhr, P. J. Dyck, W. J. Litchy, R. G. Leahy, J. Chen, J. Gollob and T. Coelho, *BMC Neurol.*, 2017, **17**, 181.
63. P. Kerschen and V. Plante-Bordeneuve, *Curr. Treat Options Neurol.*, 2016, **18**, 1.
64. C. Morrison, *Nat. Rev. Drug Discovery*, 2018, **17**, 156.
65. https://trademarks.justia.com/869/44/onpattro-86944919.html.
66. https://www.fda.gov/NewsEvents/Newsroom/PressAnnouncements/ucm616518.htm.
67. http://investors.alnylam.com/news-releases/news-release-details/alnylam-announces-first-ever-fda-approval-rnai-therapeutic.
68. http://investors.alnylam.com/news-releases/news-release-details/alnylam-receives-positive-chmp-opinion-onpattrotm-patisiran.
69. J. Soutschek, A. Akinc, B. Bramlage, K. Charisse, R. Constien, M. Donoghue, S. Elbashir, A. Geick, P. Hadwiger, J. Harborth, M. John, V. Kesavan, G. Lavine, R. K. Pandey, T. Racie, K. G. Rajeev, I. Roehl, I. Toudjarska, G. Wang, S. Wuschko, D. Bumcrot, V. Koteliansky, S. Limmer, M. Manoharan and H.-P. Vornlocher, *Nature*, 2004, **432**, 173.
70. C. Wolfrum, S. Shi, K. N. Jayaprakash, M. Jayaraman, G. Wang, R. K. Pandey, K. G. Rajeev, T. Nakayama, K. Charrise, E. M. Ndungo, T. Zimmermann, V. Koteliansky, M. Manoharan and M. Stoffel, *Nat. Biotechnol.*, 2007, **25**, 1149.
71. K. G. Rajeev, J. K. Nair, M. Jayaraman, K. Charisse, N. Taneja, J. O'Shea, J. L. S. Willoughby, K. Yucius, T. Nguyen, S. Shulga-Morskaya, S. Milstein, A. Liebow, W. Querbes, A. Borodovsky, K. Fitzgerald, M. A. Maier and M. Manoharan, *ChemBioChem*, 2015, **16**, 903.
72. S. Matsuda, K. Keiser, J. K. Nair, K. Charisse, R. M. Manoharan, P. Kretschmer, C. G. Peng, A. V. Kel'in, P. Kandasamy, J. L. S. Willoughby, A. Liebow, W. Querbes, K. Yucius, T. Nguyen, S. Milstein, M. A. Maier, K. G. Rajeev and M. Manoharan, *ACS Chem. Biol.*, 2015, **10**, 1181.
73. M. Manoharan, K. G. Rajeev, J. K. Narayanannair and M. Maier, *PCT Int. Appl.*, 2009, WO2009073809.
74. T. P. Prakash, M. J. Graham, J. Yu, R. Carty, A. Low, A. Chappell, K. Schmidt, C. Zhao, M. Aghajan, H. F. Murray, S. Riney, S. L. Booten, S. F. Murray, H. Gaus, J. Crosby, W. F. Lima, S. Guo, B. P. Monia, E. E. Swayze and P. P. Seth, *Nucleic Acids Res.*, 2014, **42**, 8796.
75. A. R. MacLeod and S. T. Crooke, *J. Clin. Pharmacol.*, 2017, **57**, S43.
76. R. Flisiak, J. Jaroszewicz and M. Lucejko, *Expert Opin. Biol. Ther.*, 2018, **18**, 609.

77. J. Dunning, F. Sahr, A. Rojek, F. Gannon, G. Carson, B. Idriss, T. Massaquoi, R. Gandi, S. Joseph, H. K. Osman, T. J. G. Brooks, A. J. H. Simpson, I. Goodfellow, L. Thorne, A. Arias, L. Merson, L. Castle, R. Howell-Jones, R. Pardinaz-Solis, B. Hope-Gill, M. Ferri, J. Grove, M. Kowalski, K. Stepniewska, T. Lang, J. Whitehead, P. Olliaro, M. Samai and P. W. Horby, *PLoS Med.*, 2016, **13**, e1001997/1001991.
78. F. Zahir-Jouzdani, F. Mottaghitalab, M. Dinarvand and F. Atyabi, *J. Drug Delivery Sci. Technol.*, 2018, **45**, 428.
79. T. Coelho, D. Adams, A. Silva, P. Lozeron, P. N. Hawkins, T. Mant, J. Perez, J. Chiesa, S. Warrington, E. Tranter, M. Munisamy, R. Falzone, J. Harrop, J. Cehelsky, B. R. Bettencourt, M. Geissler, J. S. Butler, A. Sehgal, R. E. Meyers, Q. Chen, T. Borland, R. M. Hutabarat, V. A. Clausen, R. Alvarez, K. Fitzgerald, C. Gamba-Vitalo, S. V. Nochur, A. K. Vaishnaw, D. W. Y. Sah, J. A. Gollob and O. B. Suhr, *N. Engl. J. Med.*, 2013, **369**, 819.
80. O. B. Suhr, T. Coelho, J. Buades, J. Pouget, I. Conceicao, J. Berk, H. Schmidt, M. Waddington-Cruz, J. M. Campistol, B. R. Bettencourt, A. Vaishnaw, J. Gollob and D. Adams, *Orphanet J. Rare Dis.*, 2015, **10**, 109.
81. http://www.alnylam.com/web/assets/ALN-TTRsc02_OTS_101415.pdf.
82. T. S. Zimmermann, V. Karsten, A. Chan, J. Chiesa, M. Boyce, B. R. Bettencourt, R. Hutabarat, S. Nochur, A. Vaishnaw and J. Gollob, *Mol. Ther.*, 2017, **25**, 71.
83. K. Garber, *Nat. Biotechnol.*, 2016, **34**, 1213.
84. http://www.alnylam.com/web/assets/Revusiran_Ph2-AHA_ABC_11152014.pdf.
85. http://www.alnylam.com/web/assets/ISA-2016_Revusiran-Ph2-OLE_070116.pdf.
86. A. Chan, A. Liebow, M. Yasuda, L. Gan, T. Racie, M. Maier, S. Kuchimanchi, D. Foster, S. Milstein, K. Charisse, A. Sehgal, M. Manoharan, R. Meyers, K. Fitzgerald, A. Simon, R. J. Desnick and W. Querbes, *Mol. Ther.–Nucleic Acids*, 2015, **4**, e263.
87. http://www.alnylam.com/wp-content/uploads/2018/04/EASL_Ph1-OLE_Capella_FINA'L04142018.pdf.
88. K. J. Pasi, S. Rangarajan, P. Georgiev, T. Mant, M. D. Creagh, T. Lissitchkov, D. Bevan, S. Austin, C. R. Hay, I. Hegemann, R. Kazmi, P. Chowdary, L. Gercheva-Kyuchukova, V. Mamonov, M. Timofeeva, C. H. Soh, P. Garg, A. Vaishnaw, A. Akinc, B. Sorensen and M. V. Ragni, *N. Engl. J. Med.*, 2017, **377**, 819.
89. A. Sehgal, S. Barros, L. Ivanciu, B. Cooley, J. Qin, T. Racie, J. Hettinger, M. Carioto, Y. Jiang, J. Brodsky, H. Prabhala, X. Zhang, H. Attarwala, R. Hutabarat, D. Foster, S. Milstein, K. Charisse, S. Kuchimanchi, M. A. Maier, L. Nechev, P. Kandasamy, A. V. Kel'in, J. K. Nair, K. G. Rajeev, M. Manoharan, R. Meyers, B. Sorensen, A. R. Simon, Y. Dargaud, C. Negrier, R. M. Camire and A. Akinc, *Nat. Med.*, 2015, **21**, 492.
90. http://www.alnylam.com/wp-content/uploads/2017/07/Geisen_ISTH-2017_Measurement-of-Thrombin-Generation_10July2017.pdf.

91. A. Liebow, X. Li, T. Racie, J. Hettinger, B. R. Bettencourt, N. Najafian, P. Haslett, K. Fitzgerald, R. P. Holmes, D. Erbe, W. Querbes and J. Knight, *J. Am. Soc. Nephrol.*, 2017, **28**, 494.
92. http://www.alnylam.com/wp-content/uploads/2018/06/Luma_OxalEurope_Slides_FINAL060820182.pdf.
93. http://investors.dicerna.com/static-files/6646d4f9-9dcf-48c6-8171-1096da2844c1.
94. http://www.alnylam.com/web/assets/ASH-2016_CC5_Ph-1-Part-C_05Dec2016_CAPELLA.pdf.
95. http://www.alnylam.com/web/assets/OTS_ALN-AAT_Phase-1_092816_Final2.pdf.
96. http://ir.arrowheadpharma.com/static-files/53d1fa15-0674-449d-a395-e6a5c2d4de54.
97. http://ir.arrowheadpharma.com/static-files/b01a866d-0655-460f-9489-1c51fbbdae48.
98. http://ir.arrowheadpharma.com/static-files/9a8cae92-6eb0-48b3-982c-de1680f34be3.
99. http://erj.ersjournals.com/content/50/suppl_61/PA881.
100. K. K. Ray, R. M. Stoekenbroek, D. Kallend, L. A. Leiter, U. Landmesser, R. S. Wright, P. Wijngard and J. J. P. Kastelstein, *Circulation*, 2018, **138**, 1304.
101. T. Nishikido and K. K. Ray, *Expert Opin. Invest. Drugs*, 2018, **27**, 287.
102. K. Fitzgerald, S. White, A. Borodovsky, B. R. Bettencourt, A. Strahs, V. Clausen, P. Wijngaard, J. D. Horton, J. Taubel, A. Brooks, C. Fernando, R. S. Kauffman, D. Kallend, A. Vaishnaw and A. Simon, *N. Engl. J. Med.*, 2017, **376**, 41.
103. K. K. Ray, U. Landmesser, L. A. Leiter, D. Kallend, R. Dufour, M. Karakas, T. Hall, R. P. T. Troquay, T. Turner, F. L. J. Visseren, P. Wijngaard, R. S. Wright and J. J. P. Kastelein, *N. Engl. J. Med.*, 2017, **376**, 1430.
104. D. Bandyopadhyay, A. Hajra, K. Ashish, A. Qureshi and S. Ball, *J. Cardiol.*, 2018, **71**, 523.
105. http://www.alnylam.com/wp-content/uploads/2017/08/ORION-1_ESC_presentation.pdf.
106. K. Fitzgerald, M. Frank-Kamenetsky, S. Shulga-Morskaya, A. Liebow, B. R. Bettencourt, J. E. Sutherland, R. M. Hutabarat, V. A. Clausen, V. Karsten, J. Cehelsky, S. V. Nochur, V. Kotelianski, J. Horton, T. Mant, J. Chiesa, J. Ritter, M. Munisamy, A. K. Vaishnaw, J. A. Gollob and A. Simon, *Lancet*, 2014, **383**, 60.
107. J. C. Burnett, J. J. Rossi and K. Tiemann, *Biotechnol. J.*, 2011, **6**, 1130.
108. P. J. C. Lin, Y. K. Tam and P. R. Cullis, *Clin. Lipidol.*, 2014, **9**, 317.
109. X. Liu, *Clin. Transl. Oncol.*, 2015, **8**, 185.
110. J. Tabernero, G. I. Shapiro, P. M. LoRusso, A. Cervantes, G. K. Schwartz, G. J. Weiss, L. Paz-Ares, D. C. Cho, J. R. Infante, M. Alsina, M. M. Gounder, R. Falzone, J. Harrop, A. C. S. White, I. Toudjarska, D. Bumcrot, R. E. Meyers, G. Hinkle, N. Svrzikapa, R. M. Hutabarat, V. A. Clausen, J.

Cehelsky, S. V. Nochur, C. Gamba-Vitalo, A. K. Vaishnaw, D. W. Y. Sah, J. A. Gollob and H. A. Burris, III, *Cancer Discovery*, 2013, **3**, 406.
111. V. Soriano, *AIDS Res. Rev.*, 2018, **20**, 125.
112. http://www.alnylam.com/web/assets/ALN-HBV_AASLD_111515.pdf.
113. T. Schluep, J. Lickliter, J. Hamilton, D. L. Lewis, C.-L. Lai, J. Y. N. Lau, S. A. Locarnini, R. G. Gish and B. D. Given, *Clin. Pharmacol. Drug Dev.*, 2017, **6**, 350.
114. C. I. Wooddell, M.-F. Yuen, H. L. Chan, R. G. Gish, S. A. Locarnini, D. Chavez, C. Ferrari, B. D. Given, J. Hamilton, S. B. Kanner, C.-L. Lai, J. Y. N. Lau, T. Shluep, Z. Xu, R. E. Lanford and D. L. Lewis, *Sci. Transl. Med.*, 2017, **9**, 1.
115. http://ir.arrowheadpharma.com/static-files/a1848345-85c7-4394-9e35-2599006fe8e3.
116. http://ir.arrowheadpharma.com/static-files/6bf371c3-2a4d-4f93-9171-8f03b07b4fc0.
117. http://ir.arrowheadpharma.com/static-files/80a0119d-5439-4125-9a83-b74e05941cda.
118. http://ir.arrowheadpharma.com/static-files/a574ef83-0235-4201-bad4-897b0eda3ca1.
119. N. Mani, A. G. Cole, J. R. Phelps, A. Ardzinski, K. D. Cobarrubias, A. Cuconati, B. D. Dorsey, E. Evangelista, K. Fan, F. Guo, H. Guo, T. O. Harasym, S. Kadhim, S. G. Kultgen, A. C. H. Lee, A. H. L. Li, Q. Long, S. A. Majeski, R. Mao, K. D. McClintock, S. P. Reid, R. Rijnbrand, N. M. Snead, H. M. M. Steuer, S. Tang, X. Wang, Q. Zhao and M. J. Sofia, *Antimicrob. Agents Chemother.*, 2018, **62**, 1.
120. https://investor.arbutusbio.com/static-files/ccae873d–79b9-48e0-b97f-2466920c52f0.
121. http://www.eatg.org/news/arbutus-announces-topline-results-for-arb-1467-phase-ii-cohort-4-for-the-treatment-of-hepatitis-b/.
122. A. M. Rojek, J. Dunning, A. Leliogdowicz, L. Castle, P. W. Horby, G. Carson, L. M. Van, F. Sahr and P. Olliaro, *Clin. Infect. Dis.*, 2018, **66**, 1454.
123. D. R. Jacobson, M. Ittmann, J. N. Buxbaum, R. Wieczorek and P. D. Gorevic, *Texas Heart Institute Journal*, 1997, **24**, 45.
124. D. R. Jacobson, R. D. Pastore, R. Yaghoubian, I. Kane, G. Gallo, F. S. Buck and J. N. Buxbaum, *N. Engl. J. Med.*, 1997, **336**, 466.
125. http://www.alnylam.com/wp-content/uploads/2018/07/1_ICNMD_Obici_FINAL.pdf.
126. http://www.alnylam.com/wp-content/uploads/2018/07/2_ICNMD_Coelho_FINAL.pdf.
127. http://www.alnylam.com/wp-content/uploads/2018/07/3_ICNMD_Mauermann_FINAL.pdf.
128. http://www.alnylam.com/wp-content/uploads/2018/07/4_ICNMD_Goel_FINAL.pdf.
129. http://www.alnylam.com/wp-content/uploads/2018/07/5_ICNMD_Zhang_FINAL.pdf.

130. http://www.alnylam.com/wp-content/uploads/2018/07/1.-Ajroud-Driss-APOLLO-Health-Status-PNS-2018.pdf.
131. P. Akhila, P. Sreethu and B. Vargheese, *World J. Pharm. Pharm. Sci.*, 2016, **5**, 1740–1750.
132. Z. Karim, S. Lyoumi, G. Nicolas, J.-C. Deybach, L. Gouya and H. Puy, *Clin. Res. Hepatol. Gastroenterol.*, 2015, **39**, 412.
133. http://investors.alnylam.com/static-files/509c77a7-c7cf-41ca-934f-f7928f925f47.
134. http://investors.alnylam.com/static-files/c8f6284e-841b-4f0e-846d-046aa73824b6.

CHAPTER 12

Advances and Challenges of RNAi-Based Anti-HIV Therapeutics

J. ZHOU[a] AND J. J. ROSSI*[a,b]

[a]Department of Molecular and Cellular Biology, Beckman Research Institute of City of Hope, 1500 E. Duarte Rd, Duarte, CA 91010, USA; [b]Irell and Manella Graduate School of Biological Sciences, Beckman Research Institute of City of Hope, 1500 E. Duarte Rd, Duarte, CA 91010, USA
*E-mail: jrossi@coh.org

12.1 Introduction

It has been almost two decades since the first description that double-stranded RNA (dsRNA) could functionally mediate a specific genetic interference in *Caenorhabditis elegans* to achieve post-transcriptional gene silencing (PTGS).[1] Mechanistically, PTGS, a natural-occurring RNA pathway, results in the degradation of the targeted mRNA in the cytoplasm *via* target sequence complementarity.[2] This phenomenon was termed RNA interference (RNAi). After this initial observation, the same phenomenon was documented in many other organisms, including mammalian cell lines and even animals. For instance, Elbashir *et al.* demonstrated that a synthetic 21-mer RNA duplex (small interfering RNA, siRNA) induced sequence-specific gene silencing in mammalian cells.[3] Song *et al.* showed the first *in vivo* evidence of

Drug Discovery Series No. 68
Advances in Nucleic Acid Therapeutics
Edited by Sudhir Agrawal and Michael J. Gait

Published by the Royal Society of Chemistry, www.rsc.org

siRNA-mediated gene silencing in a mouse model of hepatitis.[4] Soon thereafter, RNAi technology rapidly gained widespread attention, and it has since been established as a powerful tool for gene-function research as well as an emerging therapeutic strategy for various diseases.[5–8] A number of promising RNAi-based drugs against human diseases, including ophthalmic disorders, genetic diseases, renal diseases, liver cancer, and infectious diseases, have already been evaluated in clinical trials with encouraging safety profiles and efficacy.[9–11]

Natural RNAi mechanism involves two main classes of small regulatory RNAs: siRNA and microRNA (miRNA, ~21 nt single-stranded RNA).[12,13] The first class, siRNAs, are derived from long double stranded (ds) RNA molecules obtained upon cleavage mediated by the endonuclease Dicer. The resultant siRNAs are subsequently incorporated into the RNA-induced silencing complex (RISC) which degrades the target mRNA through site-specific cleavage. The second class, miRNAs encoded in the genome, are transcribed from endogenous primary miRNA transcripts (pri-miRNAs) that contain ~65–70 nucleotide stem-loop structures. The binding of miRNA to the mRNA target results in gene silencing, either through degradation of mRNA or through blockade of translation of mRNA. For therapeutic RNAi applications, various artificial RNAi effectors such as synthetic siRNA, miRNA, small hairpin RNA (shRNA), Dicer substrate RNA (DsiRNA) or different constructs have been developed to achieve effective gene silencing. For example, synthetic siRNAs can be transfected into cells through physical methods (*e.g.* electroporation) or polymer-based delivery vehicles (*e.g.* liposomes). The exogenously delivered siRNAs are loaded directly into RISC to trigger RNAi activity. Alternatively, endogenous expression of shRNA in cells can be achieved by transduction of a shRNA expression cassette using a viral delivery vector. In the latter case, the expression construct is transcribed by RNA polymerase I or III promoters into an shRNA that is subsequently exported from the nucleus to the cytoplasm by Exportin 5.[14] In the cytoplasm shRNA is processed by Dicer into siRNA for RISC loading.

The advent of RNAi technology for specific silencing of target genes provided new opportunities for treating HIV-1 infection. RNAi effectors, such as siRNA and shRNA, show unique therapeutic attributes for the control of HIV-1 infection, because the entire genome of the RNA virus is a target.[15–17] HIV infects all hematopoietic cells that express the CD4 receptor. These cells migrate to the lymphatic system, the thymus and even to the brain. Similarly combination antiretroviral therapy (cART), the simultaneous administration of multiple siRNAs or shRNAs against cellular HIV-1 dependency factors (HDFs) and the highly conserved HIV-1 RNA genome have been developed to prevent effectively the emergence of resistant variants, and to achieve enhanced and durable HIV-1 suppression.[18–20] In addition to their utility as stand-alone inhibitors, RNAi-based therapies can be used as an effective adjuvant to improve the therapeutic index by combining with other anti-HIV-1 therapy strategies, such as antiretroviral drugs, therapeutic RNAs (aptamer, decoy, ribozyme), HIV-1 vaccine, and cell-based therapy [chimeric antigen receptors (CAR)-T cell therapy].[21–23] The purpose of this chapter is to provide a description of the therapeutic landscape for HIV-1 *via* RNA interference. We first introduce the potential targets of anti-HIV-1 RNAi therapeutics and subsequently discuss potential challenges that limit effective RNAi activity. Next, we present current advances in the development of RNAi-based anti-HIV-1 therapeutics, with particular emphasis on *in vivo* delivery systems, by describing several representative applications. Finally, we discuss the prospects for future success.

12.2 Potential Targets for Anti-HIV-1 RNAi Therapeutics

12.2.1 Targeting the HIV-1 Viral Genome

HIV-1 is a chronic disease that attacks the human immune system.[24] Without therapeutic intervention, more than 90% of HIV-1-infected individuals will clinically progress to acquired immunodeficiency syndrome (AIDS).[25] Because of a greater understanding of the HIV-1 life cycle and pattern of gene expression, HIV-1 was the first infectious agent targeted by scientists using RNAi.[21] A schematic representation of HIV-1 genomic RNA is shown in Figure 12.1A. HIV-1 comprises a ~9 kbp genome with nine genes encoding 15 proteins. In addition to three indispensable structural genes [envelope (*env*), group specific antigen (*gag*), and polymerase (*pol*)], the genome also contains six accessory genes [transactivator (*tat*), regulator of expression of virion proteins (*rev*), viral protein R (*vpr*), viral infectivity factor (*vif*), viral protein U (*vpu*), and negative regulatory factor (*nef*)].

Highly specific siRNAs or shRNAs that target various regions of the HIV-1 genome, including *gag*-, *tat*-, *rev*-, *nef*-, and *vif*-coding sequences, have been demonstrated to degrade viral RNA and suppress HIV-1 replication.[18,26] These genes encode the HIV-1 core proteins, all of which are necessary for virion assembly and viral budding. Targeting of genes that encode two proteins (such as Tat and Rev) by one single siRNA or shRNA, has also been developed to obtain synergistic suppression.[27] Moreover, the viral RNA genome also contains two identical long terminal repeats (LTRs) located in the 5′ and 3′ terminal regions, which play important roles in viral gene expression. The 5′ LTR normally functions as an RNA polymerase II promoter and the 3′ LTR acts in transcription termination and polyadenylation. Therefore, the HIV-1 promoter regions have also been exploited as potential targets for siRNA-induced transcriptional gene silencing (TGS).[28–30] Several siRNAs targeting HIV-1 LTRs have been identified and shown to be capable of suppressing HIV-1 transcription through induction of epigenetic modifications in the HIV-1 promoter in both primary cells and *in vivo*.[31–34]

The life cycle of HIV-1 is divided into early and late stages (Figure 12.1B).[24,35] In the early stage, the HIV-1 envelope glycoproteins gp120 and

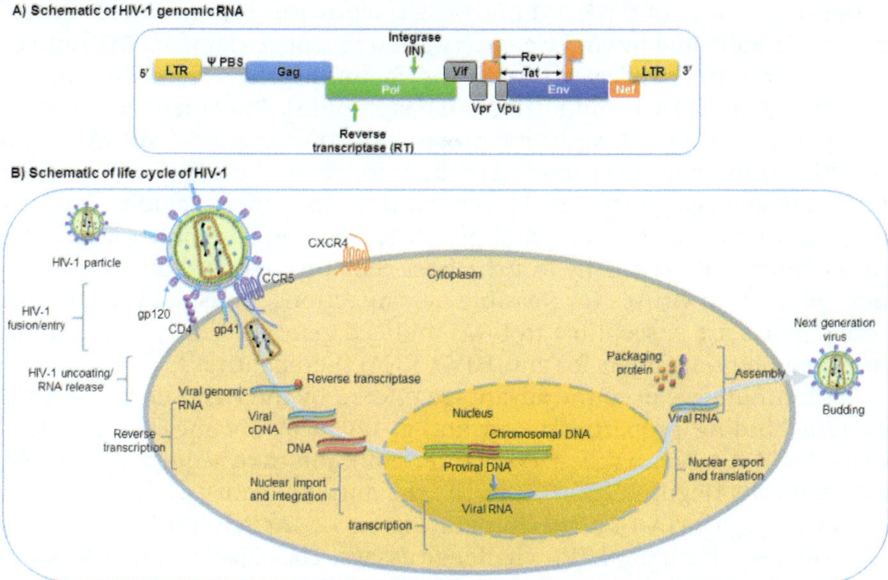

Figure 12.1 Schematic representation of the HIV-1 genome and HIV-1 life cycle. (A) HIV-1 genomic RNA contains nine genes (*gag*, *pol*, *env*, *tat*, *rev*, *vpr*, *vif*, *vpu*, and *nef*) and two long terminal repeats. PBS: primer-binding site that can be targeted by RNAi. (B) Life cycle of HIV-1. The first step is receptor binding and membrane fusion by HIV-1 envelope glycoproteins gp120 and gp41 to host receptors CD4 and either co-receptor C–C motif receptor (CCR5) or C-X-C-motif receptor 4 (CXCR4). After uncoating, viral content is released into the cytoplasm. The viral genome is reverse-transcribed by viral reverse transcriptase (RT) and integrated into the host genome by viral integrase (IN). Subsequently, the integrated viral DNA is transcribed and exported to the cytoplasm where viral RNA is translated. Finally, the viral protein and viral genome are packaged into new viral particles for budding.

gp41 recognize specific receptors on the surface of target cells to mediate entry and release of HIV-1 virions into the cytoplasm; gp120 is responsible for binding, whereas gp41 is essential for fusion of the virus to the cellular membrane.[36–38] Specifically, HIV-1 gp120 initially binds to the host cellular receptor CD4, and subsequently to host co-receptors, the C–C motif receptor (CCR5) in macrophages or the C-X-C-motif receptor 4 (CXCR4) in lymphocytes. Upon co-receptor binding, the HIV-1 gp41 transmembrane subunit interacts with the host cell membrane, resulting in the fusion of the viral and cellular membranes. After entry, the viral core is released into the cellular environment and uncoated. The HIV-1 viral RNA is reverse transcribed into cDNA in the cytoplasm by HIV-1 reverse transcriptase (RT), then imported into the host cell nucleus, and subsequently integrated into the host genome to become a provirus.[39] The viral RNA is available for RNAi targeting during uncoating and reverse transcription.

In the late stage of the HIV-1 life cycle, the provirus is transcribed, and this process is mediated by viral trans-activation response element (TAR) and the viral Tat protein as well as by several host factors [*e.g.* lens epithelium-derived growth factor (LEDGF), importin, and chaperonin]. Tat-dependence ensures the synthesis of a full-length viral progenomic transcript and mRNA.[40,41] The resulting transcripts are exported to the cytoplasm where they are translated into viral packaging proteins. The key viral components, including viral proteins and two copies of the viral genome, are then assembled into immature virions and eventually exit the host cells through budding, to become next-generation viruses for possible new infections. Thus, all the HIV-1-encoded genes are susceptible to RNAi-induced gene silencing, and therefore provide potential targets for anti-HIV-1 RNAi therapeutics.[21]

For example, we used the advantage offered by the HIV-1 envelope gp120 protein binding properties of an anti-gp120 RNA aptamer to specifically deliver siRNA targeting the HIV-1 *tat/rev* common exon sequence into HIV-1 infected cells (Figure 12.2).[42] To verify the anti-HIV activity of the aptamer–siRNA conjugates (A-1-tat/rev chimera, Figure 12.2A) in inhibiting HIV-1 replication, we carried out HIV-1 challenge assays. The experimental RNAs were incubated with HIV-1 IIIB infected-CEM cells (Figure 12.2B) or HIV-1 NL4-3 infected human peripheral blood mononuclear cells (PBMCs) (Figure 12.2C). At 7 days post-treatment, the cell-free supernatant was collected for HIV-1 p24 antigen by enzyme linked immunosorbent assay (ELISA) analysis, in which the strongest suppression was observed with the A-1-tat/rev chimera treatment. To confirm that the siRNA component was functioning along with the aptamer, we further carried out quantitative revers transcription-polymerase chain reaction (RT-PCR) assays to detect the target HIV-1 *tat/rev* gene expression (Figure 12.2D). We demonstrated that the treatment of infected cells with the A-1-tat/rev chimera is able to downregulate the HIV-1 *tat/rev* gene. However, the aptamer alone showed no effect on the expression of the HIV-1 *tat/rev* gene.

12.2.2 Targeting Host Factors

In addition to targeting the HIV-1 viral genome, host factors that are essential for viral entry and replication also represent attractive alternative targets to mitigate viral escape from RNAi.[43] Several aforementioned host factors responsible for viral entry have been successfully targeted by siRNAs or shRNAs for effective control of HIV-1 replication. For instance, siRNA or shRNA designed to knock down the CD4 receptor and the co-receptors CCR5 and CXCR4 in primary $CD4^+$ T cells, macrophages, and dendritic cells have been demonstrated to block viral binding/fusion, consequently reducing HIV-1 replication *in vitro* as well as *in vivo*.[43,44] However, effects on CD4 and CXCR4 receptors must be carefully assessed due to potential long-term safety concerns.[18] Because CD4 is an essential part of the human immune system and plays a central role in normal T cell receptor-induced activation of $CD4^+$ T cells, the arbitrary disruption of the CD4 gene may adversely affect signaling

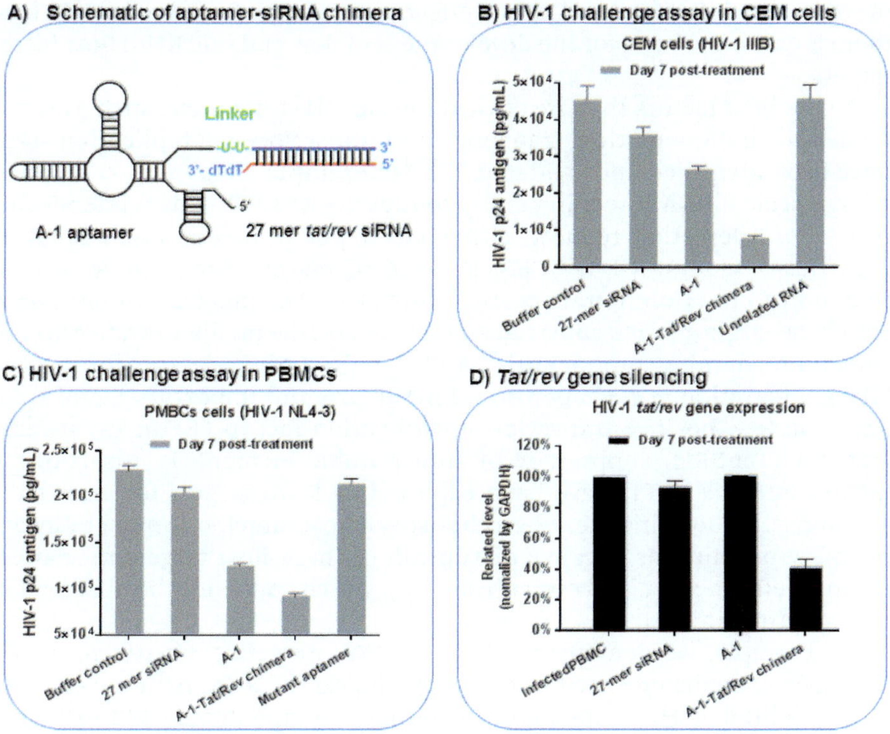

Figure 12.2 Inhibition of HIV-1 infection mediated by cell-specific aptamer-*tat*/*rev* siRNA conjugates. (A) Schematic representation of gp120 aptamer A-1-*tat*/*rev* siRNA chimera. The treatment of A-1-*tat*/*rev* siRNA chimera neutralized HIV-1 infection in (B) HIV-1 IIIB-infected CEM cells and (C) HIV-1 NL4-3-infected PBMCs culture, respectively. Data represent the average of triplicate measurements of HIV-1 p24 antigen by ELISA. (D) The siRNA delivered by the gp120 aptamer A-1 downregulated *tat*/*rev* gene expression in the HIV-1 infected PBMCs. Data represent the average of three replicates. Adapted from ref. 42 with permission from Oxford University Press, Copyright 2009.

to antigen-presenting cells and promote immunodeficiency.[45] Similarly, targeting CXCR4 may be problematic because CXCR4 is important in hematopoietic stem cell homing to the bone marrow and neuronal maturation.[46] In this regard, CCR5 is a promising target for RNAi because it is not required for the normal development and function of T-lymphocytes.[47] By artificially disruption of CCR5 (mutating or silencing the CCR5 gene), it is possible to generate HIV-1-resistant cells, and such an approach has been supported by a report on the "Berlin Patient" who received a CCR5-Δ32 homozygous allogeneic stem cell transplantation.[48–50] However, only approximately 50% of infection occurs with CCR5-tropic virus.[51,52] Furthermore, when mutated CCR5 is not available for HIV-1 entry, the virus can mutate to use an alternative co-receptor (CXCR4) to infect cells.[53] Additionally, CCR5 disruption only prevents new infections and has no effect on existing infected cells.[47,54] Thus, there is an unmet need for the development of new and safe RNAi host factor targets.

Others host factors that are critical for the HIV-1 life cycle, such as those related to viral integration, transport, and transcriptional replication steps have been identified and validated.[55–57] For example, Brass *et al.* conducted a large-scale siRNA screening study to identify 273 HIV-1 dependency factors whose depletion reduced either HIV-1 p24 production or viral gene activities, including CD4, CCR5, Rab6 (a regulator of retrograde protein transport to the Golgi), transportin 3-SR2 (TNPO3, a nuclear import factor for serine–arginine-rich substrates), and Med28 (the mediator transcription activation complex component).[55] Additionally, host co-factors involved in HIV-1 integration (*e.g.* chaperonin, LEDGF/p75 and importin-7), and transcription [*e.g.* positive transcription elongation fact (p-TEFb), Tat-specific factor 1 (Tat-SF1), suppressor of transposable element Ty insertions 5 (SPT5), and cyclin T1] have been explored as RNAi targets for anti-HIV-1 treatment.[18,58] In principle, when choosing a host target, one must be aware of safety precautions because disruption of these host targets may affect normal cell function, and even physiology, which could only be observed in a long-term clinical trial.

For example, we demonstrated that siRNAs targeting host factors CD4 and TNPO3 mediated specific gene silencing *via* the RNAi pathway and efficiently inhibited HIV-1 infection in human T-lymphoblast CEM cells and PBMCs.[58] We formulated cationic poly(amidoamine) (PAMAM) dendrimer (G5) with 27-mer DsiRNA as condensed spherical nanoparticles with diameters of approximately 100 nm (Figure 12.3A) and then applied the nanoparticles with cultured cells. The siRNA-mediated knockdown of CD4 mRNA was evaluated with quantitative reverse transcription-polymerase chain reaction (qRT-PCR) assay. The G5-CD4 siRNA nanoparticles effectively mediated gene silencing in CEM cells (Figure 12.3B) and primary PBMCs (Figure 12.3C). Next, we assessed the anti-HIV-1 activities of the G5 dendrimer-delivered siRNAs. The G5-siRNA nanoparticles were incubated with human PBMCs that had been previously infected with HIV-1 IIIB followed by ELISA for the HIV-1 p24 antigen release. Results presented in Figure 12.3D demonstrated that the single CD4 siRNA as well as a combination of CD4 siRNA and *tat/rev* siRNA all effectively suppressed HIV-1 replication. In contrast, treatment of infected cells with the empty G5 dendrimer or G5-scrambled siRNA nanoparticles had no effect on the HIV-1 p24 level.

12.3 Challenges in Obtaining Effective Anti-HIV-1 RNAi Activity

12.3.1 Key Barriers to *In Vivo* RNAi Efficacy

Although RNAi technology holds great therapeutic promise for human diseases, several early frustrations in clinical trials have tempered the initial excitement.[59–61] Before RNAi-based therapy can be realized as a general

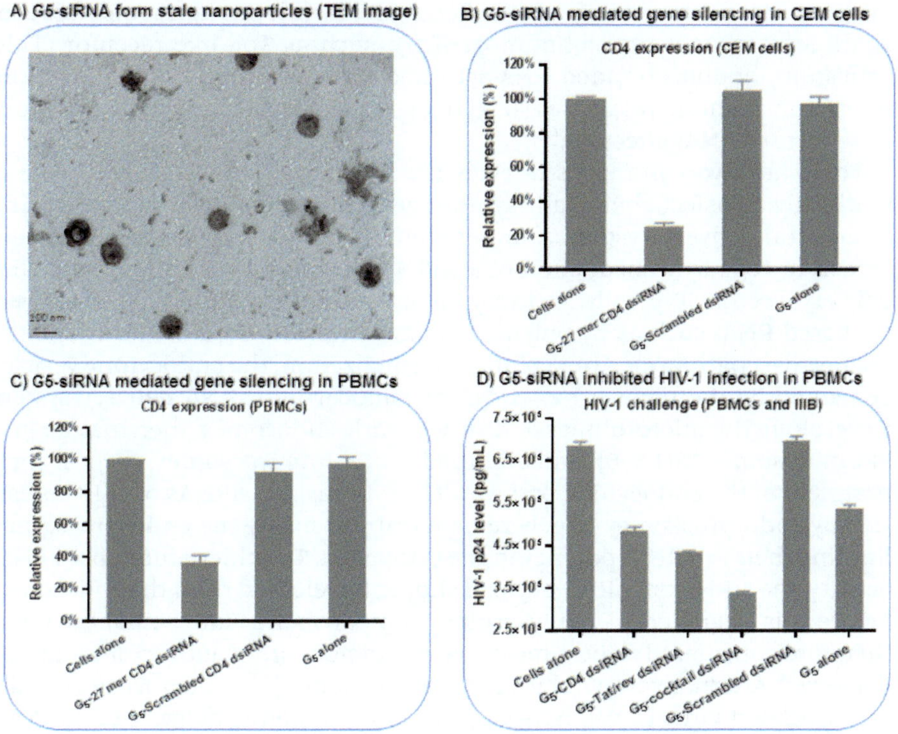

Figure 12.3 Inhibition of HIV-1 infection mediated by dendrimer-CD4 siRNA nanoparticles. (A) TEM image of dendrimer G5-27-mer dsiRNA complexes prepared in Opti-MEM transfection medium using dsiRNA and G5 at an amine to phosphate (N : P) ratio of 5. Dendrimer G5-CD4 siRNA nanoparticles downregulated CD4 expression levels in (B) CEM cells and (C) human PBMCs detected by qRT-PCR. Data represent the average of three replicates. (D) Dendrimer G5–siRNA nanoparticles inhibited HIV-1 infection in human PBMCs previously infected with HIV-1. Data represent the average of triplicate measurements of p24 and the average of three replicates. Adapted from ref. 58 with permission from American Society of Gene & Cell Therapy, Copyright 2011 The American Society of Gene & Cell Therapy. Published by Elsevier Inc.

strategy for the treatment of HIV-1, these RNA modalities must overcome several key challenges that derive from their intrinsic physiochemical properties as nucleic acid molecules.[10,62] Naked siRNA or shRNA is too negatively charged to diffuse passively across the lipid bilayer. When a synthetic siRNA or shRNA is administered systemically without a proper delivery formulation or chemical modification, it is vulnerable to nuclease degradation and is rapidly cleared from the blood by the kidneys and liver scavenger receptors due to its small size. The average diameter of siRNA and shRNA is less than 10 nm.[63] Moreover, unmodified siRNA or shRNA may trigger adverse off-target effects.[64,65] In theory, off-target effects mainly derive from sequence similarity of the guide or passenger strand of siRNAs or shRNAs to unintended mRNA targets. A passenger strand that is considered non-functional at the intended target may contribute to off-target effects. Meanwhile, certain sequences that could lead to unwanted immunogenicity, such as Toll-like receptor (TLR) activation, double-stranded RNA-activated protein kinase (PKR) induction, interferon immune responses (IFN) or cytotoxicity also contribute to off-target effects of RNAi effectors.[66]

There are two common strategies used to harness RNAi activity. One is to directly transfect chemically synthesized siRNA or shRNA into cells using a non-viral delivery system. Another alternative is to endogenously express shRNA in cells by transduction of a shRNA expression cassette using a viral delivery vector. Typically, delivery of a non-viral, synthetic delivery-vector-based RNA, such as by polymeric nanocarriers, is mediated through the endocytosis pathway.[67] After cellular internalization, the endocytotic vesicles generated by the nanocarrier-siRNA or nanocarrier-shRNA delivery system travel along the microtubules to fuse with early endosomes, then mature into late endosomes (pH 5–6), and eventually enter into lysosomes, the last compartment of the endocytotic pathway. The synthesized siRNAs or shRNAs carried by endocytosis may largely remain trapped inside the endosomes, thus limiting their full RNAi potency in the cytoplasm. To achieve functional RNAi activity, the siRNA or shRNA payload should be released from the endosomes before it is degraded in the lysosomal compartment. Endosomal escape is therefore considered to be a major barrier to non-viral nanovector-mediated delivery.[10] Alternatively, viral vectors, usually lentivirus, adenovirus, or adeno-associated virus (AAV), have been used for systemic delivery of shRNAs, and involve expression of shRNA driven by an RNA polymerase III promoter, such as U6.[68,69] In contrast to exogenous siRNAs or shRNAs delivered by non-viral vectors that must escape the endosome, endogenous shRNAs delivered by viral vectors are expressed in the nucleus and then exported into the cytoplasm where they are processed by the endoribonuclease Dicer to form mature siRNA for the RNAi pathway. Although endosomal escape is not problematic for shRNA, such shRNAs abundantly expressed by polymerase III promoters may induce toxicity by saturation of the nuclear transporter exportin 5, thereby blocking miRNA export.[70,71] In both cases, high or repeated dosage of siRNA or shRNA expression may also result in the saturation of the RNAi machinery, thus negatively affecting endogenous microRNA function.

12.3.2 The Need for Rational Design and for Chemical Modifications

To achieve efficient RNAi potency, extensive efforts are being made to surmount the barriers described above by optimizing siRNA and shRNA design and synthetic chemistry. Whether it is a synthetic or endogenous RNAi effector siRNA in the cytoplasm, one of its two strands is selected as a "guide" for the sequence-specific degradation of the complementary mRNA (defined as the "target"), while the other strand – the "passenger" – is rejected and/or degraded.[72] The intrinsic thermodynamic properties of each siRNA determine the selectivity and targeting ability of the guide strand. The 5′ end of the strand with the lowest stability is preferentially loaded into the RISC.[73–75] The two-base overhang at the 3′ end is recognized specifically by the Piwi Argonaut and Zwille (PAZ) domain (a single-stranded RNA-binding domain) and the Argonaute protein (a key RISC catalytic component).[72,76,77] The 3′ overhang also affects the polarity of Dicer entry into the complex, which consequently determines strand selectivity and influences overall RNAi efficiency.[78–80] Additionally, the accessibility of the primary sequence and secondary structure of the target mRNA should be taken into account, so that targeting of highly structured RNA motifs may be avoided.[81] A random siRNA design may result in suboptimal RNAi activity. Various sequence alignment algorithms (*e.g.* BLAST and Smith–Waterman) and online siRNA design software (*e.g.* siDESIGN from Dharmacon, siRNA Wizard from InvivoGen, BLOCK-iT RNAi designer from ThermoFisher) can be used to assist with siRNA and shRNA design and to minimize off-target effects.[82,83]

Precise chemical modifications in siRNA sequences not only improve stability but also minimize recipient immune responses.[63] Various positions in the siRNA sequence can be modified to improve physiochemical properties, confer nuclease resistance, and inhibit unwanted immune response, without affecting the original RNAi activity.[84] For example, 2′-fluoro (2′-F) and 2′-*O*-methyl (2′-OMe) modifications that protect siRNAs from RNase degradation are highly tolerated by the RNAi machinery, and prevent induction of innate immune responses. Thus far, and for the most part, synthetic siRNA therapeutics which have entered clinical trials have contained 2′-F and 2′-OMe modifications.[63] Recently, short-interfering ribonucleic neutral (siRNN) molecules have been developed, which demonstrate a reduced overall charge relative to siRNAs, while still maintaining robust RNAi activity.[85,86] As a result of modification of the backbone with charge-neutral phosphotriesters, the resulting siRNNs have shown superior physiochemical properties and improved deliverability and stability. However, selective modifications must be considered with caution because RNAi activity is highly position-sensitive. Hypermodification or improper positioning of a particular modification can compromise RNAi activity and even trigger cellular toxicity.[84,87] For example, although a boranophosphonate modification at the central position of the siRNA guide strand improved stability, it also reduced gene-silencing activity.[88] Universal rules may be applicable to specific sequences, but laborious evaluation and optimization are often needed for each individual siRNA. In-depth siRNA design and optimal design features, such as composition and length of the 3′ end, GC content, inverted repeats, DsiRNA, and chemical modifications, have been discussed previously in several reviews.[81,83,89]

12.3.3 The Need for Combinatorial RNAi

HIV-1 infects individuals chronically and thus requires long-term treatment. Development of viral resistance to drugs, including RNAi-based therapeutics, can result in the generation of viral escape mutants.[19,90] This is a common setback with HIV-1 therapies. Targeting the HIV-1 viral genome using RNAi can effectively suppress HIV-1 replication for some time. However, selective pressure may result in a point mutation in a critical position within the target sequence, consequently diminishing and even eliminating the RNAi inhibition.[90,91] HIV-1 is a rapidly evolving virus, due to the poor fidelity of HIV-1 reverse transcriptase, which markedly increases the genetic variation of HIV-1.[92,93] This poor fidelity may be responsible for the rapid emergence of RNAi-refractory viral mutants, even when the RNAi targets a highly conserved viral genome sequence. Therefore, a combinatorial RNAi approach that simultaneously targets different stages of the HIV-1 life cycle and/or host factors has been proposed to mitigate and even prevent viral escape, thus providing enhanced and durable viral suppression. For example, we have utilized a combination of three synthetic siRNAs targeting one highly conserved region of HIV-1 (*tat*/*rev*) and two host factors (CD4 and TNPO3).[58,94] By use of either a polymeric dendrimer-based nanoparticle or a gp120-specific aptamer-based targeted delivery system, we demonstrated that the combinatorial RNAi approach increased anti-HIV-1 activity effectively and prolonged viral suppression in an HIV-1-infected, humanized animal model. Similarly, Berkhout and colleagues used a single lentiviral vector to combine three shRNAs targeting three distinct and highly conserved regions of the HIV-1 genome (Pol1, Pol47, and RT5), which prevented virus escape even in the presence of unprotected cells.[95–97] There is a concern that multiple siRNAs may lead to competition for RISC, resulting in reduced gene silencing of some of the siRNAs within the combination. However, the use of DsiRNA may overcome this, as it has been shown to reduce competitive potential, presumably due to more efficient RISC loading.[98,99] Additionally, dual-targeting siRNAs that use a single duplex, wherein each strand is loaded separately into RISC, allowing the simultaneous downregulation of two different mRNA targets, may offer an efficient way to multiplex siRNA for multiple mRNA targets.[100]

12.4 Recent Progress and Clinical Development of Anti-HIV-1 RNAi

Because long-term or repeated administration of anti-HIV-1 RNAi therapeutics is required for the treatment of chronic HIV-1 infection, it is imperative to develop safe and efficient delivery systems. RNAi-based interventions for HIV-1 are primarily achieved using *in vivo* or *ex vivo* methods, in which various non-viral and viral vectors have been developed for effective RNAi activity *in vivo*. In the following sections, we review recent advances in the delivery and clinical development of RNAi-based anti-HIV-1 therapeutics.

12.4.1 *In Vivo* Delivery of Anti-HIV-1 RNAi Effectors

Although local administration of RNAi therapeutics may result in fewer cellular barriers being encountered and fewer pharmacokinetic concerns than systemic delivery, such a local delivery approach is not feasible for many diseases, in particular metastatic and malignant cancer, blood diseases, and viral infections, such as HIV-1. In these cases, the critical disease target sites can only be reached by systemic administration of siRNAs or shRNAs into the circulatory system. Because systemic delivery may ultimately result in a poor pharmacokinetic–pharmacodynamic (PK–PD) profile and inefficient gene silencing, improved systemic delivery methods are needed to overcome these challenges.

Non-viral siRNA or shRNA delivery may offer a flexible and feasible treatment for HIV-1 patients (Table 12.1). For example, synthetic polymers have been explored for delivering siRNAs *in vivo*. Liposomes (LP) functionalized with a lymphocyte function-associated antigen-1 (LFA-1)-specific monoclonal antibody have been formulated with anti-CCR5 siRNA for targeted siRNA delivery (Figure 12.4A).[101] The functionalized LP-siRNA nanoparticles were administered intravenously (IV injection) into humanized mice, and resulted in leukocyte-specific gene silencing that was sustained for 10 days. Following HIV-1 challenge, the treatment resulted in reduction of plasma viral load. In another example, our group used a generation 5 (G5) cationic polyamidoamine dendrimer (PAMAM) for functional delivery of combinatorial DsiRNAs against the HIV-1 *tat/rev* gene and host HIV-1 dependency factors CD4 and TNPO-3 (Figure 12.4B).[58] Systemic administration of G5 PAMAM dendrimer-multiplexed siRNA complexes by IV injection suppressed HIV-1 loads effectively and protected human $CD4^+$ T cell depletion, the hallmark of acute HIV-1 infection, in an HIV-1-infected humanized NSG mouse model. We also confirmed reductions in expression of the three targeted genes in PBMCs of treated mice, and observed no apparent IFN response or cytotoxicity. Importantly, in animals in which the viral load rebounded after cessation of treatment, retreatment with the G5 PAMAM dendrimer cocktail siRNA nanoparticles restored suppression. This resulted in a complete suppression of the HIV-1 viral load that persisted for an additional 3 weeks beyond the retreatment period.

Targeted anti-HIV-1 siRNA delivery *in vivo* has also been developed by use of either monoclonal antibody (mAb)-siRNA chimeras (Figure 12.4C) or nucleic acid aptamer–siRNA conjugates (Figure 12.4D). For example, Song *et al.* fused a mAb (F105) against HIV-1 envelope gp120 protein with protamine to deliver anti-siRNAs specifically to HIV-1-infected primary T cells and to viremic humanized mice.[102] In this targeted delivery system, F105 is responsible for targeting and internalization of siRNA cargoes into HIV-1-infected gp120-expressing cells *via* cell-surface receptor gp120, while the cationic protamine is responsible for binding and protecting the siRNAs from degradation. F105–protamine–anti-HIV-1 *gag* siRNA complexes induced specific gene silencing and efficiently inhibited HIV-1 replication *in vivo*. In a similar study, Kumar *et al.* conjugated a differentiation-7 (CD7)-specific antibody with a nine-arginine peptide to form a complex for targeted delivery of combinatorial siRNAs (ScFvCD7-9R).[103] CD7 is a human transmembrane protein expressed on thymocytes and the majority of human cells. ScFvCD7, a recombinant single-chain CD7 mAb, was used to promote receptor-specific recognition of and internalization into HIV-1-susceptible human cells.[104]

Table 12.1 *In vivo* delivery of anti-HIV-1 RNAi effectors.

Host or viral RNAi agents	Delivery vehicles	*In vivo* models	Route of administration	Results	Ref.
CCR5 siRNA	LFA-1 antibody coated lipid nanoparticles	HIV-1 infected humanized BLT (bone marrow, liver, thymus) mice model	Intravenous injection	Reduction in plasma viral loads; prevention of CD4 cells depletion	101
Cocktails of HIV-1 *tat*/*rev* siRNA, TNPO3 siRNA, and CD4 siRNA	Cationic polyamidoamine dendrimer G5 (PAMAM generation 5)	HIV-1 infected humanized recombination activating gene (Rag)–poly (rC) binding protein-deficient ($Rag2^{-/-}rc^{-/-}$) mice Model	Intravenous injection	Reduction in plasma viral loads; prevention of CD4 cells depletion	58
HIV-1 gag–c-myc, MDM2 and VEGF siRNA	HIV-1 envelope antibody F105 fused to protamine	WT mice model with HIV-1 env-expressing B6 melanoma tumors	Intratumoral or intravenous injection	Suppression of tumor progress and reduction in tumor size	102
HIV-1 Tat, vif siRNA and CCR5 siRNAs	Human transmembrane protein CD7 antibody fused to polyarginine peptide (CD7scFv-9R)	HIV-1 infected humanized BLT mice model	Intravenous injection	Reduction in plasma viral loads; prevention of CD4 cells depletion	103
HIV-1 *tat*/*rev* siRNA	Gp120 aptamer-siRNA chimeras	HIV-1 infected humanized rag mice model	Intravenous injection	Reduction in plasma viral loads; prevention of CD4 cells depletion	107
HIV-1 *tat*/*rev* siRNA, TNPO3 siRNA, and CD4 siRNA	Gp120 aptamer-stick bridge–siRNA conjugates	HIV-1 infected humanized rag mice model	Intravenous injection	Reduction in plasma viral loads; prevention of CD4 cells depletion	94
HIV-1 gag siRNA, vif siRNA and CCR5 siRNA	CD4 aptamer–siRNA chimeras	HIV-1 infected humanized BLT mice model	Intravaginal injection	Reduction in viremia; prevention of viral transmission to cervicovaginal explants and to the mice	109,110

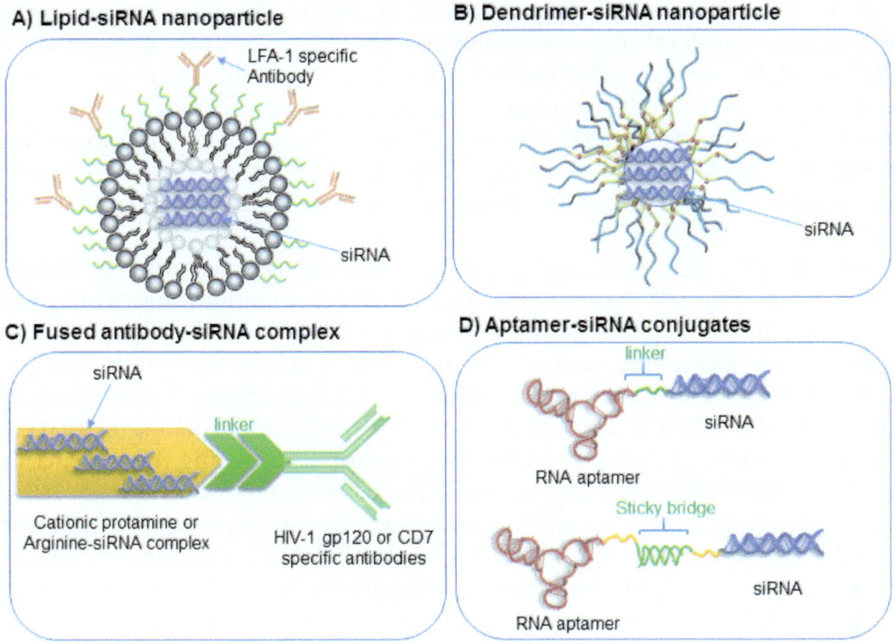

Figure 12.4 Schematic representation of *in vivo* anti-HIV-1 RNAi delivery. (A) LFA-1-specific antibody-functionalized lipid–siRNA nanoparticle; (B) G5 PAMAM dendrimer–siRNA nanoparticle; (C) fused antibody–siRNA complexes; and (D) cell-specific aptamer–siRNA conjugates.

Similar to protamine, the nine-arginine peptide is highly cationic and binds siRNA cargoes in the fusion protein. To improve RNAi potency and to reduce RNAi-resistant viral mutants, three siRNAs simultaneously targeting two different conserved HIV-1 genes (*vif* and *tat*) as well as the host CCR5 co-receptor were co-formulated with the ScFvCD7-9R delivery system. Following systemic administration, the complex suppressed HIV-1 viral loads and restored $CD4^+$ T-cell counts in viremic humanized mice.

Although cell-specific siRNA delivery can be achieved by using a mAb-fusion protein, recombinant protein fusion and formulation is laborious and time-consuming. In addition, costly production, combined with batch-to-batch variation and potential immunogenicity, may limit feasibility and utility in clinical translation. In this regard, cell-specific nucleic acid aptamers provide an alternative strategy for selective targeting and delivery of RNAi-based therapeutics *in vivo* (see Chapters 15 and 16).[105,106] Nucleic acid aptamers are single-stranded DNA or RNA molecules that specifically recognize and bind their targets with high affinity and specificity *via* folding into a unique three-dimensional structure that interacts with the target in a similar way to how an antibody binds to its antigen. Therefore, a nucleic acid aptamer is also known as a "chemical antibody." However, aptamers show many advantages over protein antibodies, including facile chemical synthesis and precise modification, lack of batch-to-batch variation during solid-phase production, *in vitro* selection procedure-unlimited target availability, high thermal stability, and minimal to no immunogenicity. An *in vitro* selection procedure called systematic evolution of ligands by exponential enrichment (SELEX) allows isolation of nucleic acid aptamers that target a desired cellular receptor, cell population, or even whole animals. Such aptamers can be used as targeted therapeutics as well as cell-specific delivery vehicles.[105]

Several RNA aptamers targeting HIV-1-infected or -susceptible cells have been isolated for HIV-1 treatment. These include aptamers specific to envelope glycoprotein gp120, host cell receptor CD4, or co-receptor CCR5. For example, we developed several 2′-fluoro-modified RNA aptamers that specifically bind HIV-1 gp120 with nanomolar binding affinity.[42] Upon binding to gp120, the HIV-1 gp120 aptamers were rapidly internalized into HIV-1-infected cells and neutralized HIV-1 replication by blocking the interaction of gp120 with the host CD4 receptor. We constructed two different HIV-1 gp120 aptamer-anti-HIV-1 siRNA conjugates through either covalent fusion or physical assembly (Figure 12.4D). In the covalent aptamer-siRNA chimera design, the gp120 aptamer was covalently fused to the sense strand of the siRNA *via* a short nucleotide linker (*e.g.* UU) through an *in vitro* T7 RNA polymerase-mediated transcription reaction. Subsequently, the transcribed aptamer-sense strand was annealed with the antisense strand of siRNA to form a chimeric RNA conjugate. In the non-covalent aptamer-stick-siRNA design, the synthetic aptamer and siRNA portions were hybridized using a GC-rich "sticky bridge" sequence, which provides a facile approach to conjugate various aptamers or siRNAs for combinatorial treatment. Systemic administration of both of the gp120 aptamer–anti-siRNA conjugates strongly targeting the conserved HIV-1 genomic *tat*/*rev* and the HIV-1 dependency factors CD4 and TNPO3 suppressed HIV-1 replication by several orders of magnitude and prevented the depletion of human $CD4^+$ T cells.[94] Importantly, the siRNA delivered by the gp120 aptamer was processed by Dicer and induced specific RNAi activity *in vivo*.[107] Compared with a single siRNA formulation, the combination of inhibitory gp120 aptamer and multiple anti-HIV siRNAs demonstrated a more effective and durable suppression of HIV-1. In a similar study, we recently isolated a cell-specific RNA aptamer against human CCR5 that specifically targets HIV-1-susceptible cells and inhibits CCR5-tropic HIV-1 infectivity.[108] Building on the aptamer-based RNAi technology, we constructed a dual inhibitory function CCR5 aptamer–anti-HIV-1 siRNA chimera. We demonstrated that CCR5 aptamer–siRNA chimeras specifically knocked down TNPO3 expression *via* the RNAi pathway and did not trigger a type I IFN response. In an additional example, an RNA aptamer specific to the human CD4 receptor was covalently fused with three different anti-HIV-1 siRNAs. The CD4 aptamer–siRNA chimeras specifically delivered the siRNAs to $CD4^+$ T cells and inhibited target gene expression *in vitro*.[109] When CD4 aptamer–siRNA chimeras bearing siRNAs targeting HIV-1 genes *gag* and *vif* and the host

CCR5 co-receptor were applied intravaginally to HIV-1-infected humanized mice, the RNA chimeras effectively prevented viral transmission to cervicovaginal explants and to the mice, indicating than the RNA conjugates can be used as an effective mucosal microbicide.[110] Despite these successes, neither the mechanism of endosomal escape nor the internalization kinetics of aptamer–siRNA conjugates have been elucidated. Continued efforts are required to enhance endosomal escape and to improve the bioavailability and PK–PD profile of these aptamer–siRNA conjugates *in vivo* by use of effective chemical modifications and formulation.

12.4.2 *Ex vivo* Delivery of Anti-HIV-1 shRNAs

Inspired by the "Berlin patient" who received an allogeneic hematopoietic stem cell transplant from a donor who had a homozygous deletion in the CCR5 gene (CCR5Δ32), the concept of repopulating the hematopoietic system with HIV-1-resistant cells has been exploited extensively for the potential cure of HIV-1.[54] Currently, one of the most common strategies is to manipulate autologous cells *ex vivo* by disrupting cellular factors essential for HIV-1 infection (*e.g.* CCR5), and then re-transplanting the HIV-1 resistant cells into the host.[111] *Ex vivo* delivery approaches typically require packaging of shRNAs into various viral vectors, followed by transduction into the host cells. Compared with non-viral vectors, viral vectors contain effective promoters that facilitate a high level of transgene expression. Targeted delivery and reduction of off-target effects can be achieved in some viral vectors (*e.g.* AAV) by introducing a tissue-specific promoter that ensures that RNA is expressed only in the target tissue.[112,113] Despite these advantages, it is important to consider safety issues, including potential immunogenicity and toxicity, low transgene capacity, induction of inflammation, risk of integration, and carcinogenicity.[68,114] Ongoing efforts are being made to develop a "new generation" of viral vectors with improved biosafety and efficiency.[115] The use of multiple plasmids allows *cis*- and *trans*-elements to be split, which improves the safety profile during viral preparation. To date, third-generation lentiviral vectors have been generated by use of four different plasmids and introduction of a self-inactivating (SIN) vector, thereby reducing the risks of integrated vector mobilization, oncogenesis, and recombinational generation of replication-competent viruses.[116–118] Generally, RNA polymerase III-driven expression cassettes in viral vectors are used for stable expression of shRNAs. However, an abundant amount of shRNA expressed by the U6 promoter may saturate exportin 5 and block miRNA export, thus inducing cellular toxicity.[70,119] In several recent studies, an endogenous miRNA backbone was adopted for shRNA expression, which avoided the toxicity caused by the saturation of the RNAi machinery.[120–122]

Due to their ability to transduce non-dividing cells, HIV-1-derived lentiviral vectors are used extensively in both basic research and clinical trials to engineer various human cells that are refractory to transfection, in particular primary PBMCs and $CD34^+$ human stem cells (HSCs). For example, a single SIN lentiviral vector was developed to stably express multiple shRNAs against *vif* and *tat/rev* viral transcripts and host CCR5, which were driven by three different RNA polymerase III promoters (U6, 7SK, and U1).[97] Its anti-HIV-1 efficacy was evaluated in human primary $CD4^+$ T cells, and indicated that the SIN lentiviral vector was safe and effective for the expression of multiple anti-HIV-1 siRNAs.[123] Moreover, strong and durable HIV-1 suppression was observed in cells transduced with a single lentiviral shRNA expression cassette (R3A) that targets HIV-1 integrase, protease, and *tat/rev* genes, indicating that an *ex vivo* combinatorial RNAi approach has the potential to prevent viral escape.[124] Recently, a preclinical *in vivo* evaluation demonstrated the safety of a lentiviral vector expressing three anti-HIV-1-shRNAs.[96] Additionally, as an adjuvant therapeutic, anti-HIV-1 shRNAs delivered by lentiviral vector were used to protect $CD8^+$ T cells modified with HIV-1-specific chimeric antigen receptor cells (CD4ζ CAR) from HIV-1 infection.[125] Despite highly potent anti-HIV-1 activities of HIV-1-specific CD4ζ CAR T cells, the poor survival and impaired functionality of the transduced cells in patients severely limited the therapeutic effects, because the ectopic expression of CD4ζ on $CD8^+$ T cells renders them susceptible to HIV-1 infection. In that study, Chen *et al.* co-expressed two anti-HIV-1 shRNAs targeting CCR5 and HIV-1 LTR which had been validated previously in HIV-1 CD4ζ CAR T cells. The results demonstrated that the numbers of CD4ζ CAR-modified $CD8^+$ T cells were maintained at similar levels relative to control cells, even under HIV-1 infection, thereby prolonging the immune function of HIV-1 CAR T cells.

12.4.3 Clinical Development of Anti-HIV-1 RNAi

So far, two RNAi-based clinical trials for HIV-1-infected patients have been evaluated in the early phase. Both trials relied on the lentiviral-vector-mediated *ex vivo* delivery of anti-HIV-1 shRNAs. HSCs survive for a long time and can develop into other cell types. Thus, through *ex vivo* modification, HIV-1-resistant cells, including monocytes, macrophages, and dendritic cells, can be generated *in vivo* from HIV-1-resistant HSCs.[126] We combined three different inhibitory genes into a single lentiviral vector, simultaneously expressing an shRNA that targets the conserved HIV-1 *tat/rev* gene, a nucleolus-localizing TAAR decoy that binds and sequesters the HIV-1 Tat protein, and a ribozyme that cleaves and downregulates the CCR5 co-receptor.[27,127] The triple combination has been applied *via ex vivo* gene delivery to HSCs in a first-in-human clinical trial conducted by the City of Hope and Benitec Inc. Compared with a single shRNA against HIV-1 *tat/rev* or double combinations of the shRNA and ribozyme or shRNA and decoy, this triple combination (lentiviral vector rHIV7-shI–TAR–CCR5RZ) was more effective and durable in suppression of HIV-1 replication. In this phase 0 safety and feasibility trial (ClinicalTrials.gov identifier NCT00569985), autologous hematopoietic progenitor cells collected from the patients were programmed with the lentiviral vector to produce HIV-1-resistant cells. These transduced cells were expanded *ex vivo*, purified, and then re-introduced into the patients. It was expected that the host would be repopulated with HIV-1-resistant HSCs. The triple combination delivered using lentiviral transduction of CD34$^+$ HSCs was shown to be safe and well tolerated, with no adverse events associated.[128] Although the delivered siRNA and ribozyme were detectable for up to 24 months after infusion in multiple peripheral blood cell lineages of all four patients who received the modified cells, no clinical benefit was observed. This may be attributed to the low frequency of modified cells in this study (less than 0.2% of the total circulating PBMCs). Further improvements that enhance vector design and transduction efficiency, and optimize transplantation, will be needed to reach a clinical endpoint.

In a different study, Calimmune, Inc. developed a lentiviral vector-expressing double-combination system of a CCR5 shRNA and C46 peptide (Cal-1) that is currently being evaluated in phase I/II clinical trials (ClinicalTrials.gov identifier NCT01734850).[129] The C46 peptide is derived from the HIV-1 gp41 envelope protein and inhibits viral fusion of both tropisms of HIV-1, acting as an essential inhibitory therapeutic in this double-combination design. Effective *ex vivo* delivery of Cal-1 was demonstrated in human PBMCs, primary CD4$^+$ T cells, and CD34$^+$ HSCs, in which both the CCR5 shRNA and C46 peptide were stably expressed, and the CCR5 gene was silenced *ex vivo*. In a humanized mouse model, the animals that received Cal-1-modified CD4$^+$ T cells significantly protected CD4$^+$ T cells after infection with CCR5-tropic HIV-1, and reduced viral loads in peripheral blood and lymphoid tissue, indicating that re-transplantation of the modified T cells conferred HIV-1 resistance.

12.5 Conclusions and Perspective

RNAi-based therapeutics offer an exquisite power to target undruggable human and viral genomes specifically with the ability to counter mutations and avoid the emergence of viral resistance. To date, more than 25 RNAi-based drugs have been evaluated in clinical trials for various indications, including age-related macular degeneration (AMD), respiratory syncytial virus, amyloidosis, renal diseases, hepatitis B virus, HIV-1 infection, pancreatic cancer, hepatocellular carcinoma, haemophilia, and amyloidosis. However, commercial success with RNAi-based therapeutics is still limited. The lessons learned from past failures provide valuable information. We cannot overlook the limitation and complexity of effective and well-controlled RNAi delivery. The need for optimizing chemistries and delivery vehicles remains imperative, in particular in the case of use of anti-HIV-1 RNAi in clinical settings. Clinical challenges include rapid viral escape and a latent HIV-1 reservoir. To achieve a realistic RNAi therapeutic for treating HIV-1 infection, further efforts should focus mainly on three areas: (1) the development of a safe and effective delivery formulation, by taking advantage of nanotechnology and conjugation chemistry, that improves the bioavailability and PD–PK profile *in vivo*; (2) the development of combinatorial strategies by use of multiple

RNAi, or by combining RNAi with other therapeutics, such as aptamers, ribozymes, peptides, antiviral drugs, mRNA-based vaccines, the clustered regularly interspaced short palindromic repeats-CRISPR associated protein 9 (CRISPR–Cas) gene editing system, or chimeric antigen receptor (CAR)-T cell therapy, resulting in synergistic therapeutic effects and increased breadth of strain coverage; and (3) the exploration of siRNA-directed TGS-mediated anti-HIV-1 strategies by targeting HIV-1 promoters or HIV-1-derived non-coding RNA (ncRNA).[30,130] siRNA-directed TGS induced a long-term stable epigenetic modification of gene expression that could be passed on to daughter cells, and thus may present possible opportunities for a functional cure of HIV-1 by targeting HIV-1 latency.[29]

Despite some major clinical setbacks in the past several years, Alnylam Pharmaceuticals, a leading company focused on RNAi therapeutics, is making great progress. Most recently, Alnylam announced that a breakthrough RNAi success had been achieved in a phase III study of Patisiran,[131,132] a lipid nanoparticle (LNP)-formatted siRNA drug targeting transthyretin (TTR) that is a protein produced in the liver and which facilitates the transport of vitamin A around the human body. In this treatment, Patisiran is administered intravenously into the patients through LNPs that deliver siRNA drug to target the TTR genes expressed in the liver. It has been demonstrated that the LNPs–siRNA system can mediate effective hepatocyte gene silencing *in vivo* with safe clinical profiles.[133] LNPs represent the leading delivery strategy for intravenously administering synthetic siRNA drugs into patients, particularly into patients who have liver-related disease.[134] Five LNP–siRNA formulations are currently being evaluated in clinical trials for the treatment of various liver-related diseases. The success of LNP systems in hepatic applications is due to several key contributing factors, including the liver's ideal physiology (highly perfused with fenestrated endothelium), the optimum acid-dissociation constant (pK_a) of the ionisable lipid, dissociable polyethylene glycol (PEG)–lipid and association of LNP with the endogenous ligand apolipoprotein E (ApoE).[135,136] Patisiran achieved all clinical endpoints in the treatment of rare cases of hereditary amyloid transthyretin (ATTR) amyloidosis with polyneuropathy. The company's first new drug application with the Federal Drug Administration (FDA) has been established. This major milestone is a well-deserved award for a 15 year journey of tireless work by numerous dedicated scientists. If approved, Patisiran would be the first commercial RNAi-based drug. We optimistically expect a new peak of RNAi therapeutics in the coming years.

Conflict of Interest Declaration

J. J. R. and J. Z. have an issued patent entitled "Cell-type specific aptamer-siRNA delivery system for HIV-1 therapy," USPTO, No. US 8, 222, 226 B2, issued date: July 17, 2012. J. J. R., J. Z., M. S. W, and K. V. M. have an issued patent entitled "Cell-specific internalizing RNA aptamers against human CCR5 and used therefore," USPTO, No. US 9 605 266, issued date:

March 28, 2017. The authors declare that they have no competing financial interests.

Acknowledgements

This work was supported by the US National Institutes of Health (grant numbers R01AI29329, R01AI42552, and R01HL07470 to J. J. R.). The content is solely the responsibility of the authors and does not necessarily represent the official views of the US National Institutes of Health. The authors thank Dr Sarah T. Wilkinson (Scientific Writer, City of Hope) for helpful advice on scientific writing.

References

1. A. Fire, S. Xu, M. K. Montgomery, S. A. Kostas, S. E. Driver and C. C. Mello, *Nature*, 1998, **391**, 806–811.
2. P. D. Zamore, *Cell*, 2006, **127**, 1083–1086.
3. S. M. Elbashir, J. Harborth, W. Lendeckel, A. Yalcin, K. Weber and T. Tuschl, *Nature*, 2001, **411**, 494–498.
4. E. Song, S. K. Lee, J. Wang, N. Ince, N. Ouyang, J. Min, J. Chen, P. Shankar and J. Lieberman, *Nat. Med.*, 2003, **9**, 347–351.
5. D. H. Kim and J. J. Rossi, *Nat. Rev. Genet.*, 2007, **8**, 173–184.
6. A. de Fougerolles, H. P. Vornlocher, J. Maraganore and J. Lieberman, *Nat. Rev. Drug Discovery*, 2007, **6**, 443–453.
7. D. Castanotto and J. J. Rossi, *Nature*, 2009, **457**, 426–433.
8. J. Kurreck, *Angew. Chem., Int. Ed. Engl.*, 2009, **48**, 1378–1398.
9. D. Scherman, A. Rousseau, P. Bigey and V. Escriou, *Gene Ther.*, 2017, **24**, 151–156.
10. S. F. Dowdy, *Nat. Biotechnol.*, 2017, **35**, 222–229.
11. B. Huggett and K. Paisner, *Nat. Biotechnol.*, 2017, **35**, 708–709.
12. H. Siomi and M. C. Siomi, *Mol. Cell*, 2010, **38**, 323–332.
13. J. L. Umbach and B. R. Cullen, *Genes Dev.*, 2009, **23**, 1151–1164.
14. Y. P. Liu, N. C. Schopman and B. Berkhout, *Nucleic Acids Res.*, 2013, **41**, 3723–3733.
15. L. Scherer, J. J. Rossi and M. S. Weinberg, *Gene Ther.*, 2007, **14**, 1057–1064.
16. S. K. Singh and R. K. Gaur, *BioDrugs*, 2009, **23**, 269–276.
17. J. Haasnoot, E. M. Westerhout and B. Berkhout, *Nat. Biotechnol.*, 2007, **25**, 1435–1443.
18. M. L. Bobbin, J. C. Burnett and J. J. Rossi, *Genome Med.*, 2015, **7**, 50.
19. D. Boden, O. Pusch, F. Lee, L. Tucker and B. Ramratnam, *J. Virol.*, 2003, **77**, 11531–11535.
20. J. J. Rossi, C. H. June and D. B. Kohn, *Nat. Biotechnol.*, 2007, **25**, 1444–1454.
21. A. Kwarteng, S. T. Ahuno and G. Kwakye-Nuako, *AIDS Res. Ther.*, 2017, **14**, 32.

22. V. Klemm, J. Mitchell, C. Cortez-Jugo, F. Cavalieri, G. Symonds, F. Caruso, A. D. Kelleher and C. Ahlenstiel, *Genes*, 2016, 7.
23. J. Zhou and J. J. Rossi, *BioDrugs*, 2012, **26**, 393–400.
24. Q. J. Sattentau and J. P. Moore, *Philos. Trans. R. Soc., B*, 1993, **342**, 59–66.
25. R. J. Pomerantz and D. L. Horn, *Nat. Med.*, 2003, **9**, 867–873.
26. J. Zhou and J. J. Rossi, *Gene Ther.*, 2011, **18**, 1134–1138.
27. M. J. Li, J. Kim, S. Li, J. Zaia, J. K. Yee, J. Anderson, R. Akkina and J. J. Rossi, *Mol. Ther.*, 2005, **12**, 900–909.
28. K. V. Morris and J. J. Rossi, *Curr. Opin. Mol. Ther.*, 2006, **8**, 115–121.
29. S. Saayman, T. C. Roberts, K. V. Morris and M. S. Weinberg, *Adv. Exp. Med. Biol.*, 2015, **848**, 169–189.
30. C. L. Ahlenstiel, K. Suzuki, K. Marks, G. P. Symonds and A. D. Kelleher, *Front. Immunol.*, 2015, **6**, 474.
31. K. Suzuki, S. Hattori, K. Marks, C. Ahlenstiel, Y. Maeda, T. Ishida, M. Millington, M. Boyd, G. Symonds, D. A. Cooper, S. Okada and A. D. Kelleher, *Mol. Ther.*, 2013, **2**, e137.
32. K. Suzuki, T. Shijuuku, T. Fukamachi, J. Zaunders, G. Guillemin, D. Cooper and A. Kelleher, *J. RNAi Gene Silencing*, 2005, **1**, 66–78.
33. S. Saayman, A. Ackley, A. M. Turner, M. Famiglietti, A. Bosque, M. Clemson, V. Planelles and K. V. Morris, *Mol. Ther.*, 2014, **22**, 1164–1175.
34. M. Yamagishi, T. Ishida, A. Miyake, D. A. Cooper, A. D. Kelleher, K. Suzuki and T. Watanabe, *Microbes Infect.*, 2009, **11**, 500–508.
35. C. Van Lint, S. Bouchat and A. Marcello, *Retrovirology*, 2013, **10**, 67.
36. S. Ugolini, I. Mondor and Q. J. Sattentau, *Trends Microbiol.*, 1999, **7**, 144–149.
37. J. M. Kilby and J. J. Eron, *N. Engl. J. Med.*, 2003, **348**, 2228–2238.
38. S. Jiang, Q. Zhao and A. K. Debnath, *Curr. Pharm. Des.*, 2002, **8**, 563–580.
39. D. F. Purcell and M. A. Martin, *J. Virol.*, 1993, **67**, 6365–6378.
40. S. K. Arya, C. Guo, S. F. Josephs and F. Wong-Staal, *Science*, 1985, **229**, 69–73.
41. B. Berkhout, R. H. Silverman and K. T. Jeang, *Cell*, 1989, **59**, 273–282.
42. J. Zhou, P. Swiderski, H. Li, J. Zhang, C. P. Neff, R. Akkina and J. J. Rossi, *Nucleic Acids Res.*, 2009, **37**, 3094–3109.
43. P. R. Gorry and P. Ancuta, *Curr HIV/AIDS Rep.*, 2011, **8**, 45–53.
44. H. J. Ryser and R. Fluckiger, *Drug Discovery Today*, 2005, **10**, 1085–1094.
45. G. Zauli, L. Catani, D. Gibellini, M. C. Re, D. Milani, P. Borgatti, A. Bassini, M. La Placa and S. Capitani, *Br. J. Haematol.*, 1995, **91**, 290–298.
46. Y. R. Zou, A. H. Kottmann, M. Kuroda, I. Taniuchi and D. R. Littman, *Nature*, 1998, **393**, 595–599.
47. G. Hutter, J. Bodor, S. Ledger, M. Boyd, M. Millington, M. Tsie and G. Symonds, *Viruses*, 2015, **7**, 4186–4203.
48. G. Hutter, D. Nowak, M. Mossner, S. Ganepola, A. Mussig, K. Allers, T. Schneider, J. Hofmann, C. Kucherer, O. Blau, I. W. Blau, W. K. Hofmann and E. Thiel, *N. Engl. J. Med.*, 2009, **360**, 692–698.

49. G. Hutter and S. Ganepola, *Sci. World J.*, 2011, **11**, 1068–1076.
50. K. Allers, G. Hutter, J. Hofmann, C. Loddenkemper, K. Rieger, E. Thiel and T. Schneider, *Blood*, 2011, **117**, 2791–2799.
51. Y. Lai, *Curr. Stem Cell Res. Ther.*, 2012, **7**, 310–317.
52. M. N. Swamy, H. Wu and P. Shankar, *Adv. Drug Delivery Rev.*, 2016, **103**, 174–186.
53. L. Kordelas, J. Verheyen, D. W. Beelen, P. A. Horn, A. Heinold, R. Kaiser, R. Trenschel, D. Schadendorf, U. Dittmer, S. Esser and H. I. V. A. G. Essen, *N. Engl. J. Med.*, 2014, **371**, 880–882.
54. K. Allers and T. Schneider, *Curr. Opin. Virol.*, 2015, **14**, 24–29.
55. A. L. Brass, D. M. Dykxhoorn, Y. Benita, N. Yan, A. Engelman, R. J. Xavier, J. Lieberman and S. J. Elledge, *Science*, 2008, **319**, 921–926.
56. H. Zhou, M. Xu, Q. Huang, A. T. Gates, X. D. Zhang, J. C. Castle, E. Stec, M. Ferrer, B. Strulovici, D. J. Hazuda and A. S. Espeseth, *Cell Host Microbe*, 2008, **4**, 495–504.
57. R. Konig, Y. Zhou, D. Elleder, T. L. Diamond, G. M. Bonamy, J. T. Irelan, C. Y. Chiang, B. P. Tu, P. D. De Jesus, C. E. Lilley, S. Seidel, A. M. Opaluch, J. S. Caldwell, M. D. Weitzman, K. L. Kuhen, S. Bandyopadhyay, T. Ideker, A. P. Orth, L. J. Miraglia, F. D. Bushman, J. A. Young and S. K. Chanda, *Cell*, 2008, **135**, 49–60.
58. J. Zhou, C. P. Neff, X. Liu, J. Zhang, H. Li, D. D. Smith, P. Swiderski, T. Aboellail, Y. Huang, Q. Du, Z. Liang, L. Peng, R. Akkina and J. J. Rossi, *Mol. Ther.*, 2011, **19**, 2228–2238.
59. P. K. Kaiser, R. C. Symons, S. M. Shah, E. J. Quinlan, H. Tabandeh, D. V. Do, G. Reisen, J. A. Lockridge, B. Short, R. Guerciolini and Q. D. Nguyen, *Am. J. Ophthalmol.*, 2010, **150**, 33–39.e32.
60. J. DeVincenzo, R. Lambkin-Williams, T. Wilkinson, J. Cehelsky, S. Nochur, E. Walsh, R. Meyers, J. Gollob and A. Vaishnaw, *Proc. Natl. Acad. Sci. U. S. A.*, 2010, **107**, 8800–8805.
61. J. DeVincenzo, J. E. Cehelsky, R. Alvarez, S. Elbashir, J. Harborth, I. Toudjarska, L. Nechev, V. Murugaiah, A. Van Vliet, A. K. Vaishnaw and R. Meyers, *Antiviral Res.*, 2008, **77**, 225–231.
62. K. A. Whitehead, R. Langer and D. G. Anderson, *Nat. Rev. Drug Discovery*, 2009, **8**, 129–138.
63. A. Khvorova and J. K. Watts, *Nat. Biotechnol.*, 2017, **35**, 238–248.
64. M. Robbins, A. Judge and I. MacLachlan, *Oligonucleotides*, 2009, **19**, 89–102.
65. O. Snove Jr and J. J. Rossi, *ACS Chem. Biol.*, 2006, **1**, 274–276.
66. M. Schlee, V. Hornung and G. Hartmann, *Mol. Ther.*, 2006, **14**, 463–470.
67. J. Zhou, K. T. Shum, J. C. Burnett and J. J. Rossi, *Pharmaceuticals*, 2013, **6**, 85–107.
68. K. V. Morris and J. J. Rossi, *Hum. Gene Ther.*, 2006, **17**, 479–486.
69. A. Schambach and M. Morgan, *Recent results in cancer research. Fortschritte der Krebsforschung. Progres dans les recherches sur le cancer*, 2016, vol. 209, pp. 17–35.

70. D. Grimm, K. L. Streetz, C. L. Jopling, T. A. Storm, K. Pandey, C. R. Davis, P. Marion, F. Salazar and M. A. Kay, *Nature*, 2006, **441**, 537–541.
71. O. Snove Jr and J. J. Rossi, *Genome Biol.*, 2006, **7**, 231.
72. C. Matranga, Y. Tomari, C. Shin, D. P. Bartel and P. D. Zamore, *Cell*, 2005, **123**, 607–620.
73. J. Liu, M. A. Carmell, F. V. Rivas, C. G. Marsden, J. M. Thomson, J. J. Song, S. M. Hammond, L. Joshua-Tor and G. J. Hannon, *Science*, 2004, **305**, 1437–1441.
74. G. Meister, M. Landthaler, A. Patkaniowska, Y. Dorsett, G. Teng and T. Tuschl, *Mol. Cell*, 2004, **15**, 185–197.
75. S. M. Hammond, S. Boettcher, A. A. Caudy, R. Kobayashi and G. J. Hannon, *Science*, 2001, **293**, 1146–1150.
76. G. L. Sen and H. M. Blau, *Nat. Cell Biol.*, 2005, **7**, 633–636.
77. J. B. Ma, Y. R. Yuan, G. Meister, Y. Pei, T. Tuschl and D. J. Patel, *Nature*, 2005, **434**, 666–670.
78. K. Sakurai, M. Amarzguioui, D. H. Kim, J. Alluin, B. Heale, M. S. Song, A. Gatignol, M. A. Behlke and J. J. Rossi, *Nucleic Acids Res.*, 2011, **39**, 1510–1525.
79. J. Zhou, M. S. Song, A. M. Jacobi, M. A. Behlke, X. Wu and J. J. Rossi, *Mol. Ther.*, 2012, **1**, e17.
80. S. D. Rose, D. H. Kim, M. Amarzguioui, J. D. Heidel, M. A. Collingwood, M. E. Davis, J. J. Rossi and M. A. Behlke, *Nucleic Acids Res.*, 2005, **33**, 4140–4156.
81. A. Vermeulen, L. Behlen, A. Reynolds, A. Wolfson, W. S. Marshall, J. Karpilow and A. Khvorova, *RNA*, 2005, **11**, 674–682.
82. M. Amarzguioui and H. Prydz, *Biochem. Biophys. Res. Commun.*, 2004, **316**, 1050–1058.
83. A. Reynolds, D. Leake, Q. Boese, S. Scaringe, W. S. Marshall and A. Khvorova, *Nat. Biotechnol.*, 2004, **22**, 326–330.
84. M. A. Collingwood, S. D. Rose, L. Huang, C. Hillier, M. Amarzguioui, M. T. Wiiger, H. S. Soifer, J. J. Rossi and M. A. Behlke, *Oligonucleotides*, 2008, **18**, 187–200.
85. B. R. Meade, K. Gogoi, A. S. Hamil, C. Palm-Apergi, A. van den Berg, J. C. Hagopian, A. D. Springer, A. Eguchi, A. D. Kacsinta, C. F. Dowdy, A. Presente, P. Lonn, M. Kaulich, N. Yoshioka, E. Gros, X. S. Cui and S. F. Dowdy, *Nat. Biotechnol.*, 2014, **32**, 1256–1261.
86. A. S. Hamil and S. F. Dowdy, *Methods Mol. Biol.*, 2016, **1364**, 1–9.
87. S. Shukla, C. S. Sumaria and P. I. Pradeepkumar, *ChemMedChem*, 2010, **5**, 328–349.
88. A. H. Hall, J. Wan, E. E. Shaughnessy, B. Ramsay Shaw and K. A. Alexander, *Nucleic Acids Res.*, 2004, **32**, 5991–6000.
89. J. P. DiNitto, L. Wang and J. C. Wu, *Biotechniques*, 2010, **48**, 303–311.
90. A. T. Das, T. R. Brummelkamp, E. M. Westerhout, M. Vink, M. Madiredjo, R. Bernards and B. Berkhout, *J. Virol.*, 2004, **78**, 2601–2605.
91. E. M. Westerhout, M. Ooms, M. Vink, A. T. Das and B. Berkhout, *Nucleic Acids Res.*, 2005, **33**, 796–804.

92. K. Bebenek, J. Abbotts, S. H. Wilson and T. A. Kunkel, *J. Biol. Chem.*, 1993, **268**, 10324–10334.
93. Y. Takeuchi, T. Nagumo and H. Hoshino, *J. Virol.*, 1988, **62**, 3900–3902.
94. J. Zhou, C. P. Neff, P. Swiderski, H. Li, D. D. Smith, T. Aboellail, L. Remling-Mulder, R. Akkina and J. J. Rossi, *Mol. Ther.*, 2013, **21**, 192–200.
95. E. Herrera-Carrillo, Y. P. Liu and B. Berkhout, *Mol. Ther.*, 2014, **22**, 596–606.
96. M. Centlivre, N. Legrand, S. Klamer, Y. P. Liu, K. Jasmijn von Eije, M. Bohne, E. S. Rijnstra, K. Weijer, B. Blom, C. Voermans, H. Spits and B. Berkhout, *Mol. Ther.*, 2013, **2**, e120.
97. O. ter Brake, K. t. Hooft, Y. P. Liu, M. Centlivre, K. J. von Eije and B. Berkhout, *Mol. Ther.*, 2008, **16**, 557–564.
98. D. H. Kim, M. A. Behlke, S. D. Rose, M. S. Chang, S. Choi and J. J. Rossi, *Nat. Biotechnol.*, 2005, **23**, 222–226.
99. M. Amarzguioui and J. J. Rossi, *Methods Mol. Biol.*, 2008, **442**, 3–10.
100. K. Tiemann, B. Hohn, A. Ehsani, S. J. Forman, J. J. Rossi and P. Saetrom, *RNA*, 2010, **16**, 1275–1284.
101. S. S. Kim, D. Peer, P. Kumar, S. Subramanya, H. Wu, D. Asthana, K. Habiro, Y. G. Yang, N. Manjunath, M. Shimaoka and P. Shankar, *Mol. Ther.*, 2010, **18**, 370–376.
102. E. Song, P. Zhu, S. K. Lee, D. Chowdhury, S. Kussman, D. M. Dykxhoorn, Y. Feng, D. Palliser, D. B. Weiner, P. Shankar, W. A. Marasco and J. Lieberman, *Nat. Biotechnol.*, 2005, **23**, 709–717.
103. P. Kumar, H. S. Ban, S. S. Kim, H. Wu, T. Pearson, D. L. Greiner, A. Laouar, J. Yao, V. Haridas, K. Habiro, Y. G. Yang, J. H. Jeong, K. Y. Lee, Y. H. Kim, S. W. Kim, M. Peipp, G. H. Fey, N. Manjunath, L. D. Shultz, S. K. Lee and P. Shankar, *Cell*, 2008, **134**, 577–586.
104. M. Peipp, H. Kupers, D. Saul, B. Schlierf, J. Greil, S. J. Zunino, M. Gramatzki and G. H. Fey, *Cancer Res.*, 2002, **62**, 2848–2855.
105. J. Zhou and J. Rossi, *Nat. Rev. Drug Discovery*, 2017, **16**, 181–202.
106. J. Zhou and J. J. Rossi, *Mol. Ther.*, 2014, **3**, e169.
107. C. P. Neff, J. Zhou, L. Remling, J. Kuruvilla, J. Zhang, H. Li, D. D. Smith, P. Swiderski, J. J. Rossi and R. Akkina, *Sci. Transl. Med.*, 2011, **3**, 66ra66.
108. J. Zhou, S. Satheesan, H. Li, M. S. Weinberg, K. V. Morris, J. C. Burnett and J. J. Rossi, *Chem. Biol.*, 2015, **22**, 379–390.
109. L. A. Wheeler, R. Trifonova, V. Vrbanac, E. Basar, S. McKernan, Z. Xu, E. Seung, M. Deruaz, T. Dudek, J. I. Einarsson, L. Yang, T. M. Allen, A. D. Luster, A. M. Tager, D. M. Dykxhoorn and J. Lieberman, *J. Clin. Invest.*, 2011, **121**, 2401–2412.
110. L. A. Wheeler, V. Vrbanac, R. Trifonova, M. A. Brehm, A. Gilboa-Geffen, S. Tanno, D. L. Greiner, A. D. Luster, A. M. Tager and J. Lieberman, *Mol. Ther.*, 2013, **21**, 1378–1389.
111. L. J. Scherer and J. J. Rossi, *Hum. Mol. Genet.*, 2011, **20**, R100–R107.
112. Y. Kashiwakura, K. Tamayose, K. Iwabuchi, Y. Hirai, T. Shimada, K. Matsumoto, T. Nakamura, M. Watanabe, K. Oshimi and H. Daida, *J. Virol.*, 2005, **79**, 609–614.

113. L. Yang, L. Bailey, D. Baltimore and P. Wang, *Proc. Natl. Acad. Sci. U. S. A.*, 2006, **103**, 11479–11484.
114. K. V. Morris and J. J. Rossi, *Gene Ther.*, 2006, **13**, 553–558.
115. M. Li and J. J. Rossi, *Methods Mol. Biol.*, 2008, **433**, 287–299.
116. D. Bonci, A. Cittadini, M. V. Latronico, U. Borello, J. K. Aycock, A. Drusco, A. Innocenzi, A. Follenzi, M. Lavitrano, M. G. Monti, J. Ross, Jr., L. Naldini, C. Peschle, G. Cossu and G. Condorelli, *Gene Ther.*, 2003, **10**, 630–636.
117. H. Miyoshi, U. Blomer, M. Takahashi, F. H. Gage and I. M. Verma, *J. Virol.*, 1998, **72**, 8150–8157.
118. T. Dull, R. Zufferey, M. Kelly, R. J. Mandel, M. Nguyen, D. Trono and L. Naldini, *J. Virol.*, 1998, **72**, 8463–8471.
119. D. Grimm, L. Wang, J. S. Lee, N. Schurmann, S. Gu, K. Borner, T. A. Storm and M. A. Kay, *J. Clin. Invest.*, 2010, **120**, 3106–3119.
120. P. Wu, M. A. Wilmarth, F. Zhang and G. Du, *Methods Mol. Biol.*, 2013, **936**, 195–207.
121. J. L. McBride, R. L. Boudreau, S. Q. Harper, P. D. Staber, A. M. Monteys, I. Martins, B. L. Gilmore, H. Burstein, R. W. Peluso, B. Polisky, B. J. Carter and B. L. Davidson, *Proc. Natl. Acad. Sci. U. S. A.*, 2008, **105**, 5868–5873.
122. R. L. Boudreau, I. Martins and B. L. Davidson, *Mol. Ther.*, 2009, **17**, 169–175.
123. F. Spanevello, A. Calistri, C. Del Vecchio, B. Mantelli, C. Frasson, G. Basso, G. Palu, M. Cavazzana and C. Parolin, *Mol. Ther.*, 2016, **5**, e312.
124. O. ter Brake, N. Legrand, K. J. von Eije, M. Centlivre, H. Spits, K. Weijer, B. Blom and B. Berkhout, *Gene Ther.*, 2009, **16**, 148–153.
125. C. Chen, P. Posocco, X. Liu, Q. Cheng, E. Laurini, J. Zhou, C. Liu, Y. Wang, J. Tang, V. D. Col, T. Yu, S. Giorgio, M. Fermeglia, F. Qu, Z. Liang, J. J. Rossi, M. Liu, P. Rocchi, S. Pricl and L. Peng, *Small*, 2016, **12**, 3604.
126. J. C. Burnett, J. A. Zaia and J. J. Rossi, *Curr. Opin. Immunol.*, 2012, **24**, 625–632.
127. M. Li and J. J. Rossi, *Methods Mol. Biol.*, 2005, **309**, 261–272.
128. D. L. DiGiusto, A. Krishnan, L. Li, H. Li, S. Li, A. Rao, S. Mi, P. Yam, S. Stinson, M. Kalos, J. Alvarnas, S. F. Lacey, J. K. Yee, M. Li, L. Couture, D. Hsu, S. J. Forman, J. J. Rossi and J. A. Zaia, *Sci. Transl. Med.*, 2010, **2**, 36ra43.
129. O. Wolstein, M. Boyd, M. Millington, H. Impey, J. Boyer, A. Howe, F. Delebecque, K. Cornetta, M. Rothe, C. Baum, T. Nicolson, R. Koldej, J. Zhang, N. Keech, J. Camba Colon, L. Breton, J. Bartlett, D. S. An, I. S. Chen, B. Burke and G. P. Symonds, *Mol. Ther.*, 2014, **1**, 11.
130. M. S. Weinberg and K. V. Morris, *Nucleic Acids Res.*, 2016, **44**, 6505–6517.
131. R. Titze-de-Almeida, C. David and S. S. Titze-de-Almeida, *Pharm. Res.*, 2017, **34**, 1339–1363.
132. M. Rizk and S. Tuzmen, *Pharmacogenomics Pers. Med.*, 2017, **10**, 267–278.

133. S. A. Barros and J. A. Gollob, *Adv. Drug Delivery Rev.*, 2012, **64**, 1730–1737.
134. J. Tabernero, G. I. Shapiro, P. M. LoRusso, A. Cervantes, G. K. Schwartz, G. J. Weiss, L. Paz-Ares, D. C. Cho, J. R. Infante, M. Alsina, M. M. Gounder, R. Falzone, J. Harrop, A. C. White, I. Toudjarska, D. Bumcrot, R. E. Meyers, G. Hinkle, N. Svrzikapa, R. M. Hutabarat, V. A. Clausen, J. Cehelsky, S. V. Nochur, C. Gamba-Vitalo, A. K. Vaishnaw, D. W. Sah, J. A. Gollob and H. A. Burris 3rd, *Cancer Discovery*, 2013, **3**, 406–417.
135. V. Kumar, J. Qin, Y. Jiang, R. G. Duncan, B. Brigham, S. Fishman, J. K. Nair, A. Akinc, S. A. Barros and P. V. Kasperkovitz, *Mol. Ther.*, 2014, **3**, e210.
136. Y. Y. Tam, S. Chen and P. R. Cullis, *Pharmaceutics*, 2013, **5**, 498–507.

CHAPTER 13

Nucleic Acid Innate Immune Receptors

SHIN-ICHIROH SAITOH* AND KENSUKE MIYAKE

Division of Innate Immunity, Department of Microbiology and Immunology, The Institute of Medical Science, The University of Tokyo, 4-6-1 Shirokanedai, Minatoku, Tokyo 108-8639, Japan
*E-mail: shinss@ims.u-tokyo.ac.jp

13.1 Introduction

In the innate immune system, pattern recognition receptors (PRRs) sense pathogen-associated molecular patterns (PAMPs) and host-derived damage-associated molecular patterns (DAMPs).[1–3] PAMPs are highly conserved motifs in pathogens, such as bacteria and viruses. Regarding PRRs, Toll-like receptors (TLRs) recognize highly conserved PAMPs in bacteria and viruses. TLRs are type I transmembrane proteins that contain leucine-rich repeats (LRRs) in the extracellular domain for ligand recognition and a Toll–interleukin-1 receptor (TIR) homology domain in the cytoplasmic region for activating signalling molecules, such as myeloid differentiation primary response 88 (MyD88) and TIR domain-containing adaptor-inducing interferon-β (TRIF).[4] Ten members of the TLR family are expressed in humans.[5] Among them, four TLRs respond to nucleic acids (NAs). TLR3, TLR7/8, and TLR9 recognize double-stranded RNA (dsRNA), single-stranded RNA (ssRNA), and single-stranded DNA (ssDNA), respectively. These NA-sensing

Drug Discovery Series No. 68
Advances in Nucleic Acid Therapeutics
Edited by Sudhir Agrawal and Michael J. Gait

Published by the Royal Society of Chemistry, www.rsc.org

TLRs are mainly expressed in immune cells, such as monocytes and macrophages, dendritic cells (DCs), and B cells. TLR3 exhibits broader expression in non-haematopoietic cells, such as fibroblasts and even neurons. NA-sensing TLRs are mainly localized in endosomes and lysosomes rather than on the cell surface.

In the cytoplasm, cytoplasmic dsRNA is recognized by retinoic acid-inducible gene-I (RIG-I) and by melanoma-associated gene-5 (MDA-5).[6,7] RIG-I and MDA-5 bind dsRNA in the cytosol to activate signalling for type I interferon induction *via* the mitochondrial antiviral signalling protein (MAVS).[8,9] Cytoplasmic dsDNA is recognized by the cyclic GMP–AMP synthase (cGAS)–stimulator of the interferon gene (STING) axis. Absent in melanoma 2 (AIM2)[10,11] gene products are components of the inflammasome for cytoplasmic DNA sensing.

13.2 Toll-like Receptors

13.2.1 TLR3 Recognizes dsRNA

TLR3 comprises a large N-terminal horseshoe-shaped ectodomain consisting of 23 conserved LRRs, a transmembrane region, and a C-terminal cytoplasmic TIR homology domain. TLR3 is broadly expressed in non-haematopoietic cells, macrophages, and DCs and recognizes dsRNA that is produced during viral replication.[12,13] TLR3 detects RNA viruses, such as influenza A,[14] West Nile virus,[15] and rhinovirus[16] and is activated by endogenous cellular RNAs, such as U1 RNA, or small interfering RNAs (siRNAs).[17,18] It can also detect DNA viruses, such as herpes simplex virus (HSV-1).[19] Human TLR3 deficiency predisposes individuals to herpes simplex encephalitis (HSE).[20–22] The structure of TLR3 has revealed that TLR3 dimer binds to one dsRNA.[23–26] To induce TLR3 dimerization, the length of the dsDNA must be longer than 40 bp.[26] TLR3 differs from other TLRs in the use of adaptor molecules. Dimerized TLR3 activates an adaptor molecule TRIF but not MyD88. TRIF activates downstream signalling molecules for nuclear factor kappa B (NF-κB) or interferon regulatory factor 3 (IRF3) activation, resulting in the production of inflammatory cytokines or type I interferon.[4]

13.2.2 TLR7 and TLR8 Recognize ssRNA and Guanosine or Uridine

TLR7 is expressed in B cells, plasmacytoid dendritic cells (pDCs), conventional dendritic cells (cDCs), and macrophages, whereas TLR8 is expressed in macrophages and granulocytes. TLR7 and TLR8 recognize guanosine- and uridine-rich ssRNAs, respectively.[27–30] Small synthetic ligands, such as imidazoquinolines and guanosine analogues, also activate TLR7 and TLR8. Mucolipin-1 is a lysosomal membrane protein that is essential for ssRNA trafficking to the lysosome in mouse DCs,[31] which indicates that a mucolipin-1-dependent vesicular trafficking pathway transports ssRNAs. TLR7 and

TLR8 detect exogenous viral RNAs derived from human immunodeficiency virus type-1,[27] hepatitis C virus,[32] Coxsackie virus[33] and influenza A virus.[28,34] TLR7 and TLR8 also detect endogenous ssRNAs derived from dead or dying cells.[35,36] Excessive activation of TLR7 and TLR8 by endogenous ssRNAs contributes to the progression of autoimmune diseases by promoting anti-nuclear antibody production. Crystal structure analyses have revealed that TLR7 has two ligand-binding sites that interact with small chemical ligands and oligoribonucleotides.[37] Both sites are essential for ssRNA-induced TLR7 activation. The binding site for small chemical ligands also binds guanosine, deoxyguanosine, 8-hydroxyguanosine, and other modified guanosines in the presence of ssRNA,[38] indicating that TLR7 functions as a sensor for guanosine or its derivatives.

Crystal structure analysis of TLR8 demonstrated that TLR8 also has two binding sites that are similar to those of TLR7.[39,40] The first binding site interacts with small chemical ligands or uridine. The second binding site binds to short oligonucleotides. Activated TLR8 reorganizes its structure to form a dimer. Human TLR8 is synergistically activated by uridine or uridine analogues in the presence of ssRNA. TLR8 senses the degradation products of ssRNA, such as uridine and short oligonucleotides,[39] which indicates that TLR8 acts as a sensor for uridine or its derivatives.[38] Both TLR7 and TLR8 are stimulated by RNA degradation products rather than by ssRNA fragments.

13.2.3 TLR9 Recognizes CpG-DNA

TLR9 recognizes ssDNAs containing an unmethylated CpG motif[41] and is broadly expressed in B cells, pDCs, cDCs, and macrophages. It senses infection by DNA viruses, such as HSV-1,[42,43] adenovirus,[44] and poxvirus.[45] TLR9 directly binds to CpG-DNA under the acidic-pH conditions of the endosome or lysosome. Isothermal titration calorimetry (ITC) analyses of the binding of CpG-DNA to TLR9 have demonstrated that the dissociation constant (K_d) values of binding are 20 nM at pH 6.0 and 2500 nM at pH 8.0.[46] TLR9 is proteolytically cleaved into N-terminal fragments and C-terminal fragments, which remain associated with each other after proteolytic cleavage. Crystal structure analyses of TLR9 have demonstrated that the N-terminal fragment of TLR9 recognizes the CpG motif in DNA, and the C-terminal fragment of TLR9 recognizes the DNA backbone.[46] CpG-DNA binds at the N-terminal fragment of one TLR9 molecule and the C-terminal fragment of another TLR9 molecule to form a TLR9 dimer.[47] Two CpG-DNAs act as molecular glue in the TLR9 dimer. In the case of methylated CpG-DNA, the binding affinity with TLR9 is decreased, and methylated CpG-DNA fails to serve as molecular glue for TLR9 dimerization.[46] The structural features of the TLR9 dimer indicate that ssDNAs, but not dsDNAs, can bind to TLR9. These results demonstrate that dsDNAs from bacteria, viruses, and dead cells must be processed into ssDNAs for TLR9 activation. Typically, DNases degrade self-DNAs to inhibit

TLR9 activation. However, DNase II, *i.e.*, the lysosomal DNase, is required for TLR9 activation because it processes DNA into appropriate short DNAs that can bind TLR9.

The processing of NAs and TLRs plays essential roles in innate immune responses to NAs.[48] Viral and bacterial DNA containing CpG motifs can activate TLR9. In addition, synthetic oligonucleotides including a CpG motif can activate TLR9 as mimics of viral and bacterial DNA. There are three different class of CpG-containing oligonucleotides, class A, B, and C. Class A, CpG-A, significantly induces type I interferon in pDCs, but weakly induces inflammatory cytokines. CpG-A contains a palindromic CpG motif in the middle of its sequence. Furthermore, phosphorothioated polyguanine (poly-G) sequences are attached to both ends. The 3′ poly-G tail is longer than the 5′ poly-G tail. Two CpG-A sequences form a duplex *via* a central palindromic sequence and their poly-G sequences form intermolecular G-tetrads. The tetrads induce self-assembly that results in a high-molecular-weight ordered structure. The large structure confers enhanced stability and increased endosomal uptake. These things contribute to the induction of a large amount of type I interferon. Class B, CpG-B, primarily induces large amounts of proinflammatory cytokines such as interleukin-6 (IL-6), IL-12 and tumour necrosis factor (TNF)-α. CpG-B consists of 18–28 base linear oligonucleotides containing fully phosphorothioated backbones with hexamer CpG motifs. Because CpG-B does not contain a central palindrome sequence and poly-G motif, it does not form high-molecular-weight ordered structures, but maintains a single-stranded structure. It is considered that the CpG-B internalization and trafficking pathways are different from those for CpG-A and that CpG-A and CpG-B activate TLR9 in different vesicles due to their structural differences. Their structural differences affect the resulting cytokine profile. Class C, CpG-C, strongly stimulates B cells and induces type I interferon. CpG-C combines characteristic features of both CpG-A and CpG-B. CpG-C contains fully phosphorothioated backbones with palindromic CpG motifs.

13.2.4 Chaperones Regulate the Maturation of NA-sensing TLRs

In humans, NAs that are internalized into the endosome or lysosome are recognized by NA-sensing TLRs, such as TLR3, TLR7, TLR8, and TLR9. These receptors recognize their ligands at low pH values. When acidification in the endolysosomal compartment is inhibited by chloroquine or bafilomycin A, NA-sensing by TLRs is impaired, which demonstrates that endosomal maturation is essential for NA recognition by TLRs. NA-sensing TLRs are synthesized and folded in the endoplasmic reticulum (ER). Several folding chaperones, such as gp96 and protein associated with TLR4 A (PRAT4A, also known as Canopy3) are required for maturation and correct folding of TLRs in the ER.[49,50] In contrast to other NA-sensing TLRs, TLR3 maturation in the ER is not dependent on gp96 or PRAT4A. TLR3 may require other chaperones for maturation in the ER. NA-sensing TLRs are glycosylated in the ER and are thereafter trafficked from the ER to the Golgi apparatus *via* coat protein II (COPII) vesicles.[51] NA-sensing TLRs are associated with uncoordinated protein 93 homolog B1 (Unc93B1), which is required for NA-sensing TLRs to exit from the ER and enter into the endosomes and lysosomes through the Golgi apparatus.[52] Mutation of the *Unc93b1* gene disrupts its physical association with NA-sensing TLRs and thus impairs the responses of NA-sensing TLRs by inhibiting their trafficking from the ER to the endosomes and lysosomes.[53,54]

13.2.5 Unc93B1 Regulates the Balance of TLR7 and TLR9 Responses

Unc93B1 differentially regulates its association with client TLRs, *i.e.*, TLR7 and TLR9, to balance their responses.[55] In the case of the Unc93B1 D34A mutation, TLR7 is preferentially associated with Unc93B1 D34A, and its trafficking to the endosomes or lysosomes increases at the expense of TLR9 association and trafficking. Consequently, *Unc93b1*$^{D34A/D34A}$ mice develop TLR7-dependent systemic inflammation.[56] Similarly, mutually contrasting roles of TLR7 and TLR9 in lupus-prone mice have been reported.[57,58] Deficiencies of TLR7 or TLR9 ameliorate or exacerbate, respectively, disease progression in lupus-prone MRL/lpr mice. TLR9 deficiency has been suggested to exacerbate diseases by enhancing TLR7 responses. These results indicate a competitive relationship between TLR7 and TLR9 in inflammatory disease models.

13.2.6 Proteolytic Cleavage of NA-sensing TLRs is Essential for Their Function

The extracellular domains of NA-sensing TLRs are cleaved in the endosomes and lysosomes into functional forms.[59] NA-sensing TLRs are cleaved by a variety of proteases in the endosomes or lysosomes, including cathepsins, asparagine endopeptidase, and furin-like proprotein convertases.[60–64] These proteases are active under acidic conditions, such as those in mature endosomes or lysosomes. Consequently, NA-sensing TLRs reach their functional form only in the endosome or lysosome. The compartmentalization and proteolytic cleavage of NA-sensing TLRs may prevent their inappropriate activation in the ER or the Golgi apparatus by self-DNA or self-RNA. Interestingly, after the proteolytic cleavage of the extracellular domain, the N-terminal fragments remain associated with the C-terminal fragments. Biochemical and structural analyses of TLR8 and TLR9 have revealed that the proteolytic cleavages of TLR8 and TLR9 are not required for ligand binding but are essential for ligand-dependent oligomerization.[46,65]

13.2.7 Trafficking of TLR7 and TLR9 is Essential for Type I Interferon Production in pDCs

Murine TLR7 specifically localizes in lysosome-associated membrane protein 2 (Lamp2)$^+$, ADP ribosylation factor-like 8b (Arl8b)$^+$ lysosomal vesicles in pDCs and is constitutively associated with Arl8b,[66] which is a lysosomal small G protein involved in lysosomal trafficking.[67] Arl8b binds to secreted effector protein *Salmonella*-induced filaments A (SifA) and the kinesin-interacting protein (SKIP), which contributes to the regulation of microtubule-mediated lysosomal trafficking towards the plus end. Arl8b–SKIP-dependent lysosomal trafficking is essential for TLR7-dependent type I interferon production in pDCs.[66] TLR7-containing lysosomal vesicles are trafficked to the signalling platform to activate signalling molecules for type I interferon induction (Figure 13.1). The signalling molecules include TNF receptor associated factor 3 (TRAF3), inhibitor of NF-κB kinase alpha (IKKα), and mammalian target of rapamycin complex 1 (mTORC1), which are constitutively associated with each other. Activated TLR7 first induces NF-κB activation for inflammatory cytokine production and secondly induces IRF7 activation by recruiting TRAF3 and IKKα to produce type I interferon.[66] Adaptor protein complex 3 (AP3) is a vesicular transport adaptor complex that can associate with clathrin and which is also required for type I interferon production upon activation of TLR7 and TLR9 in pDCs.[68] AP3 is involved in TLR9 trafficking to the lysosomes or lysosome-related organelles, which indicates that a type I interferon signalling platform exists in lysosome-related organelles. The

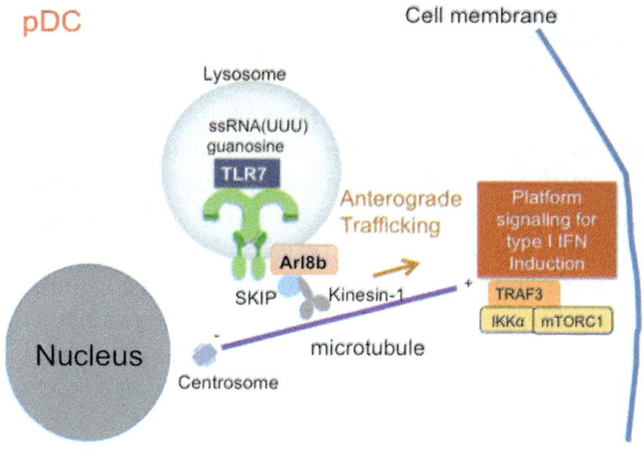

Figure 13.1 Arl8b-dependent TLR7 trafficking to the signalling platform for type I interferon production in pDCs. Upon TLR7 activation by recognition of guanosine and ssRNA (UUU), TLR7 traffics to the peripheral region by an Arl8b-dependent mechanism. TLR7 moves to the signalling platform to induce type I interferon production. The signalling platform includes the signalling molecules for type I interferon production, *i.e.*, TRAF3, IKKα, and mTORC1.

signalling platform for type I interferon production may be a target for medical treatment to control antiviral responses or TLR7-dependent autoimmune diseases that are caused by large amounts of type I interferon, such as systemic lupus erythematosus (SLE).

13.3 Nucleic Acids Sensing in the Cytoplasm

13.3.1 Cytosolic DNA Sensors Recognize dsDNA

NA-sensing TLRs sense NAs localized in the endosomal or lysosomal lumen and cannot directly sense cytosolic NAs. Surveillance of cytosolic NAs by cytosolic sensors is also important for innate immune responses against viruses. NA sensors of dsDNA or dsRNA are localized in the cytoplasm.

Cytosolic dsDNAs are sensed by cGAS, which can directly interact with dsDNA derived from DNA viruses and also interacts with RNA–DNA hybrids as an intermediate product of RNA virus replication.[69,70] After binding to cytosolic DNA, cGAS produces the cyclic dinucleotide second messenger 2′3′-cyclic guanosine monophosphate-adenosine monophosphate (cGAMP).[70–73] cGAMP binds to and activates STING [also called mitochondrial mediator of interferon regulatory factor 3 activation (MITA), methionine–proline–tyrosine–serine plasma membrane tetraspanner (MPYS) and endoplasmic reticulum interferon stimulator (ERIS)] to produce type I interferon. STING is activated on the surface of the ER and moves from the ER to vesicles through the Golgi apparatus.[10,74–77] STING interacts with the TNFR-associated NF-κB kinase (TANK)-binding kinase 1 (TBK1) on vesicles. Activated TBK-1 phosphorylates IRF3, which leads to type I interferon generation and activates NF-κB. In addition to self-derived cGAMP, STING also interacts with the cyclic dinucleotide (CDN) derived from bacteria and activates the downstream signalling pathways.[78–81]

The immune response to cytosolic DNA is not limited to type I interferon production. Cytosolic DNA stimulates inflammasome activity, which leads to the production of the pro-inflammatory cytokines IL-1β and IL-18.[82] These pro-inflammatory cytokines play an essential role in defence responses against bacterial, fungal, and viral pathogens.[83–85]

The human pyrin and hematopoietic interferon-inducible nuclear antigens (PYHIN) family member AIM2 recognizes cytosolic DNA, leading to caspase-1 activation and IL-1β secretion.[86–88] AIM2 binds cytosolic DNA, which includes bacterial, viral, and mammalian self-DNA, directly *via* the AIM2 HIN200 domain. Upon DNA binding, AIM2 is associated with the inflammasome adaptor apoptosis-associated speck-like protein (ASC), which contains a carboxy-terminal caspase activation and recruitment domain (CARD), *via* homotypic PYRIN domain interactions and which undergoes oligomerization to activate caspase-1.[89] *In vivo* analysis using AIM2-deficient mice has clearly demonstrated that the AIM2 inflammasome is essential for innate defensive responses to DNA viruses and intracellular bacterial infections.[90,91] Crystal structure analysis of the AIM2 HIN200 domains bound to dsDNA has demonstrated that AIM2 binds DNA in a sequence-independent manner because the interaction primarily involves electrostatic binding with the phosphate–sugar DNA backbone.[92]

13.3.2 RIG-I and MDA5 Recognize dsRNA and Activate MAVS to Induce Immune Responses

RIG-I and MDA5 are cytoplasmic RNA helicases that are expressed in almost all mammalian cell types[6,93,94] RIG-I and MDA5 are members of the RIG-I-like receptors (RLRs) and comprise two CARDs, *i.e.*, a DExH-box helicase domain and a C-terminal domain. In the steady state, CARDs bind to a helicase domain resulting in auto-inhibition. After binding of the C-terminal domain to an RNA molecule, the CARDs are released from a helicase domain by means of a conformational change. Subsequently, both the helicase domain and the C-terminal repressor domain interact with viral RNA.[95–97] CARDs are essential in the activation of signalling for type I interferon production. CARDs are activated by K63-polyubiquitination or polyubiquitin chain-binding. RIG-I responds to dsRNAs that are shorter than 1 kbp and have a 5′ triphosphate or a 3′ phosphate.[98,99] MDA5 responds to dsRNAs longer than 1 kbp.[100,101] RIG-I recognizes most negative-sense ssRNA viruses, including the Ebola virus of the filovirus family, measles virus, Sendai virus, Newcastle disease virus, respiratory syncytial virus, and influenza virus.[102] RIG-I also recognizes positive-sense ssRNA viruses, such as hepatitis C virus and Japanese encephalitis virus. MDA5 recognizes positive-sense ssRNA viruses, such as encephalomyocarditis virus, poliovirus and Coxsackie virus. Both RIG-I and MDA5 cross-detect the same viruses, including rotavirus, dengue virus, and West Nile virus. Some DNA viruses are also recognized by RIG-I because small dsRNAs are produced by RNA polymerase type III during viral replication.[103] RIG-I and MDA5 activate IRF3 and NF-κB through MAVS,[8,9] which has a CARD domain and a transmembrane domain that anchors MAVS to the mitochondrial membrane.[104] The CARD domain of MAVS interacts with RIG-1 and MDA5 through their CARD domains. Upon RNA binding, RIG-I and MDA5 form filaments and undergo oligomerization *via* the CARD domain. This oligomerization leads to MAVS polymerization on the mitochondrial membrane and activates TBK1 and IRF3 to induce production of type I interferon.[101,105–107]

13.4 Conclusions

Several NA sensors have been identified in the last 20 years. For pathogen sensing, NAs sensors are localized in endosomes, lysosomes, or the cytoplasm. Biochemical, molecular structure, and microscopic analysis have helped to unravel their recognition and activation mechanisms. We gradually have come to understand that NAs sensors are a double-edged sword. They play an essential role in antiviral responses. However, they recognize also their own NAs and induce the development of autoimmune disease, such as SLE. We must understand better the detailed NAs sensing mechanisms that control autoimmune diseases in order to utilize NAs for their effective treatments as drugs.

References

1. B. Beutler, *Mol. Immunol.*, 2004, **40**, 845–859.
2. O. Takeuchi and S. Akira, *Cell*, 2010, **140**, 805–820.
3. C. A. Janeway and R. Medzhitov, *Annu. Rev. Immunol.*, 2002, **20**, 197–216.
4. S. Akira and K. Takeda, *Nat. Rev. Immunol.*, 2004, **4**, 499–511.
5. S. Akira, S. Uematsu and O. Takeuchi, *Cell*, 2006, **124**, 783–801.
6. M. Yoneyama, M. Kikuchi, T. Natsukawa, N. Shinobu, T. Imaizumi, M. Miyagishi, K. Taira, S. Akira and T. Fujita, *Nat. Immunol.*, 2004, **5**, 730–737.
7. H. Kato, O. Takeuchi, S. Sato, M. Yoneyama, M. Yamamoto, K. Matsui, S. Uematsu, A. Jung, T. Kawai, K. J. Ishii, O. Yamaguchi, K. Otsu, T. Tsujimura, C. S. Koh, C. Reis e Sousa, Y. Matsuura, T. Fujita and S. Akira, *Nature*, 2006, **441**, 101–105.
8. T. Kawai, K. Takahashi, S. Sato, C. Coban, H. Kumar, H. Kato, K. J. Ishii, O. Takeuchi and S. Akira, *Nat. Immunol.*, 2005, **6**, 981–988.
9. L. G. Xu, Y. Y. Wang, K. J. Han, L. Y. Li, Z. Zhai and H. B. Shu, *Mol. Cell*, 2005, **19**, 727–740.
10. H. Ishikawa, Z. Ma and G. N. Barber, *Nature*, 2009, **461**, 788–792.
11. G. N. Barber, *Trends Immunol.*, 2014, **35**, 88–93.
12. L. Alexopoulou, A. C. Holt, R. Medzhitov and R. A. Flavell, *Nature*, 2001, **413**, 732–738.
13. N. Pirher, K. Ivicak, J. Pohar, M. Bencina and R. Jerala, *Nat. Struct. Mol. Biol.*, 2008, **15**, 761–763.
14. L. Guillot, R. Le Goffic, S. Bloch, N. Escriou, S. Akira, M. Chignard and M. Si-Tahar, *J. Biol. Chem.*, 2005, **280**, 5571–5580.
15. S. Daffis, M. A. Samuel, M. S. Suthar, M. Gale and M. S. Diamond, *J. Virol.*, 2008, **82**, 10349–10358.
16. C. A. Hewson, A. Jardine, M. R. Edwards, V. Laza-Stanca and S. L. Johnston, *J. Virol.*, 2005, **79**, 12273–12279.
17. K. Karikó, P. Bhuyan, J. Capodici and D. Weissman, *J. Immunol.*, 2004, **172**, 6545–6549.
18. K. Karikó, H. Ni, J. Capodici, M. Lamphier and D. Weissman, *J. Biol. Chem.*, 2004, **279**, 12542–12550.
19. S. Y. Zhang, E. Jouanguy, S. Ugolini, A. Smahi, G. Elain, P. Romero, D. Segal, V. Sancho-Shimizu, L. Lorenzo, A. Puel, C. Picard, A. Chapgier, S. Plancoulaine, M. Titeux, C. Cognet, H. von Bernuth, C. L. Ku, A. Casrouge, X. X. Zhang, L. Barreiro, J. Leonard, C. Hamilton, P. Lebon, B. Héron, L. Vallée, L. Quintana-Murci, A. Hovnanian, F. Rozenberg, E. Vivier, F. Geissmann, M. Tardieu, L. Abel and J. L. Casanova, *Science*, 2007, **317**, 1522–1527.

20. H. K. Lim, M. Seppänen, T. Hautala, M. J. Ciancanelli, Y. Itan, F. G. Lafaille, W. Dell, L. Lorenzo, M. Byun, E. Pauwels, Y. Rönnelid, X. Cai, S. Boucherit, E. Jouanguy, A. Paetau, P. Lebon, F. Rozenberg, M. Tardieu, L. Abel, A. Yildiran, A. Vergison, R. Roivainen, A. Etzioni, P. J. Tienari, J. L. Casanova and S. Y. Zhang, *Neurology*, 2014, **83**, 1888–1897.
21. Y. Guo, M. Audry, M. Ciancanelli, L. Alsina, J. Azevedo, M. Herman, E. Anguiano, V. Sancho-Shimizu, L. Lorenzo, E. Pauwels, P. B. Philippe, R. Pérez de Diego, A. Cardon, G. Vogt, C. Picard, Z. Z. Andrianirina, F. Rozenberg, P. Lebon, S. Plancoulaine, M. Tardieu, V. Doireau, E. Jouanguy, D. Chaussabel, F. Geissmann, L. Abel, J. L. Casanova and S. Y. Zhang, *J. Exp. Med.*, 2011, **208**, 2083–2098.
22. S. Y. Zhang and J. L. Casanova, *J. Exp. Med.*, 2015, **212**, 1342–1343.
23. L. Liu, I. Botos, Y. Wang, J. N. Leonard, J. Shiloach, D. M. Segal and D. R. Davies, *Science*, 2008, **320**, 379–381.
24. J. K. Bell, I. Botos, P. R. Hall, J. Askins, J. Shiloach, D. M. Segal and D. R. Davies, *Proc. Natl. Acad. Sci. U. S. A.*, 2005, **102**, 10976–10980.
25. J. Choe, M. S. Kelker and I. A. Wilson, *Science*, 2005, **309**, 581–585.
26. J. N. Leonard, R. Ghirlando, J. Askins, J. K. Bell, D. H. Margulies, D. R. Davies and D. M. Segal, *Proc. Natl. Acad. Sci. U. S. A.*, 2008, **105**, 258–263.
27. F. Heil, H. Hemmi, H. Hochrein, F. Ampenberger, C. Kirschning, S. Akira, G. Lipford, H. Wagner and S. Bauer, *Science*, 2004, **303**, 1526–1529.
28. S. S. Diebold, T. Kaisho, H. Hemmi, S. Akira and C. Reis e Sousa, *Science*, 2004, **303**, 1529–1531.
29. M. Jurk, F. Heil, J. Vollmer, C. Schetter, A. M. Krieg, H. Wagner, G. Lipford and S. Bauer, *Nat. Immunol.*, 2002, **3**, 499.
30. H. Hemmi, T. Kaisho, O. Takeuchi, S. Sato, H. Sanjo, K. Hoshino, T. Horiuchi, H. Tomizawa, K. Takeda and S. Akira, *Nat. Immunol.*, 2002, **3**, 196–200.
31. X. Li, S. Saitoh, T. Shibata, N. Tanimura, R. Fukui and K. Miyake, *Int. Immunol.*, 2015, **27**, 83–94.
32. Y. L. Zhang, Y. J. Guo, B. Li and S. H. Sun, *J. Hepatol.*, 2009, **51**, 29–38.
33. K. Triantafilou, G. Orthopoulos, E. Vakakis, M. A. Ahmed, D. T. Golenbock, P. M. Lepper and M. Triantafilou, *Cell. Microbiol.*, 2005, **7**, 1117–1126.
34. J. M. Lund, L. Alexopoulou, A. Sato, M. Karow, N. C. Adams, N. W. Gale, A. Iwasaki and R. A. Flavell, *Proc. Natl. Acad. Sci. U. S. A.*, 2004, **101**, 5598–5603.
35. V. Hornung, M. Guenthner-Biller, C. Bourquin, A. Ablasser, M. Schlee, S. Uematsu, A. Noronha, M. Manoharan, S. Akira, A. de Fougerolles, S. Endres and G. Hartmann, *Nat. Med.*, 2005, **11**, 263–270.
36. F. J. Barrat, T. Meeker, J. Gregorio, J. H. Chan, S. Uematsu, S. Akira, B. Chang, O. Duramad and R. L. Coffman, *J. Exp. Med.*, 2005, **202**, 1131–1139.
37. Z. Zhang, U. Ohto, T. Shibata, E. Krayukhina, M. Taoka, Y. Yamauchi, H. Tanji, T. Isobe, S. Uchiyama, K. Miyake and T. Shimizu, *Immunity*, 2016, **45**, 737–748.

38. T. Shibata, U. Ohto, S. Nomura, K. Kibata, Y. Motoi, Y. Zhang, Y. Murakami, R. Fukui, T. Ishimoto, S. Sano, T. Ito, T. Shimizu and K. Miyake, *Int. Immunol.*, 2016, **28**, 211–222.
39. H. Tanji, U. Ohto, T. Shibata, M. Taoka, Y. Yamauchi, T. Isobe, K. Miyake and T. Shimizu, *Nat. Struct. Mol. Biol.*, 2015, **22**, 109–115.
40. H. Tanji, U. Ohto, T. Shibata, K. Miyake and T. Shimizu, *Science*, 2013, **339**, 1426–1429.
41. H. Hemmi, O. Takeuchi, T. Kawai, T. Kaisho, S. Sato, H. Sanjo, M. Matsumoto, K. Hoshino, H. Wagner, K. Takeda and S. Akira, *Nature*, 2000, **408**, 740–745.
42. J. Lund, A. Sato, S. Akira, R. Medzhitov and A. Iwasaki, *J. Exp. Med.*, 2003, **198**, 513–520.
43. A. Krug, G. D. Luker, W. Barchet, D. A. Leib, S. Akira and M. Colonna, *Blood*, 2004, **103**, 1433–1437.
44. E. Basner-Tschakarjan, E. Gaffal, M. O'Keeffe, D. Tormo, A. Limmer, H. Wagner, H. Hochrein and T. Tüting, *J. Gene Med.*, 2006, **8**, 1300–1306.
45. C. Samuelsson, J. Hausmann, H. Lauterbach, M. Schmidt, S. Akira, H. Wagner, P. Chaplin, M. Suter, M. O'Keeffe and H. Hochrein, *J. Clin. Invest.*, 2008, **118**, 1776–1784.
46. U. Ohto, T. Shibata, H. Tanji, H. Ishida, E. Krayukhina, S. Uchiyama, K. Miyake and T. Shimizu, *Nature*, 2015, **520**, 702–705.
47. T. Shimizu, *Curr. Opin. Struct. Biol.*, 2017, **47**, 52–59.
48. K. Miyake, T. Shibata, U. Ohto and T. Shimizu, *J. Leukocyte Biol.*, 2017, **101**, 135–142.
49. Y. Yang, B. Liu, J. Dai, P. K. Srivastava, D. J. Zammit, L. Lefrançois and Z. Li, *Immunity*, 2007, **26**, 215–226.
50. K. Takahashi, T. Shibata, S. Akashi-Takamura, T. Kiyokawa, Y. Wakabayashi, N. Tanimura, T. Kobayashi, F. Matsumoto, R. Fukui, T. Kouro, Y. Nagai, K. Takatsu, S. Saitoh and K. Miyake, *J. Exp. Med.*, 2007, **204**, 2963–2976.
51. M. Tatematsu, K. Funami, N. Ishii, T. Seya, C. Obuse and M. Matsumoto, *J. Immunol.*, 2015, **195**, 4933–4942.
52. Y. M. Kim, M. M. Brinkmann, M. E. Paquet and H. L. Ploegh, *Nature*, 2008, **452**, 234–238.
53. K. Tabeta, K. Hoebe, E. M. Janssen, X. Du, P. Georgel, K. Crozat, S. Mudd, N. Mann, S. Sovath, J. Goode, L. Shamel, A. A. Herskovits, D. A. Portnoy, M. Cooke, L. M. Tarantino, T. Wiltshire, B. E. Steinberg, S. Grinstein and B. Beutler, *Nat. Immunol.*, 2006, **7**, 156–164.
54. M. M. Brinkmann, E. Spooner, K. Hoebe, B. Beutler, H. L. Ploegh and Y. M. Kim, *J. Cell Biol.*, 2007, **177**, 265–275.
55. R. Fukui, S. Saitoh, F. Matsumoto, H. Kozuka-Hata, M. Oyama, K. Tabeta, B. Beutler and K. Miyake, *J. Exp. Med.*, 2009, **206**, 1339–1350.
56. R. Fukui, S. Saitoh, A. Kanno, M. Onji, T. Shibata, A. Ito, M. Matsumoto, S. Akira, N. Yoshida and K. Miyake, *Immunity*, 2011, **35**, 69–81.
57. S. R. Christensen, J. Shupe, K. Nickerson, M. Kashgarian, R. A. Flavell and M. J. Shlomchik, *Immunity*, 2006, **25**, 417–428.

58. K. M. Nickerson, S. R. Christensen, J. Shupe, M. Kashgarian, D. Kim, K. Elkon and M. J. Shlomchik, *J. Immunol.*, 2010, **184**, 1840–1848.
59. S. E. Ewald, B. L. Lee, L. Lau, K. E. Wickliffe, G. P. Shi, H. A. Chapman and G. M. Barton, *Nature*, 2008, **456**, 658–662.
60. S. E. Ewald, A. Engel, J. Lee, M. Wang, M. Bogyo and G. M. Barton, *J. Exp. Med.*, 2011, **208**, 643–651.
61. B. Park, M. M. Brinkmann, E. Spooner, C. C. Lee, Y. M. Kim and H. L. Ploegh, *Nat. Immunol.*, 2008, **9**, 1407–1414.
62. F. E. Sepulveda, S. Maschalidi, R. Colisson, L. Heslop, C. Ghirelli, E. Sakka, A. M. Lennon-Duménil, S. Amigorena, L. Cabanie and B. Manoury, *Immunity*, 2009, **31**, 737–748.
63. A. Garcia-Cattaneo, F. X. Gobert, M. Müller, F. Toscano, M. Flores, A. Lescure, E. Del Nery and P. Benaroch, *Proc. Natl. Acad. Sci. U. S. A.*, 2012, **109**, 9053–9058.
64. M. M. Hipp, D. Shepherd, U. Gileadi, M. C. Aichinger, B. M. Kessler, M. J. Edelmann, R. Essalmani, N. G. Seidah, C. Reis e Sousa and V. Cerundolo, *Immunity*, 2013, **39**, 711–721.
65. H. Tanji, U. Ohto, Y. Motoi, T. Shibata, K. Miyake and T. Shimizu, *Proc. Natl. Acad. Sci. U. S. A.*, 2016, **113**, 3012–3017.
66. S. I. Saitoh, F. Abe, A. Kanno, N. Tanimura, Y. Mori Saitoh, R. Fukui, T. Shibata, K. Sato, T. Ichinohe, M. Hayashi, K. Kubota, H. Kozuka-Hata, M. Oyama, Y. Kikko, T. Katada, K. Kontani and K. Miyake, *Nat. Commun.*, 2017, **8**, 1592.
67. C. Rosa-Ferreira and S. Munro, *Dev. Cell*, 2011, **21**, 1171–1178.
68. M. Sasai, M. M. Linehan and A. Iwasaki, *Science*, 2010, **329**, 1530–1534.
69. D. Gao, J. Wu, Y. T. Wu, F. Du, C. Aroh, N. Yan, L. Sun and Z. J. Chen, *Science*, 2013, **341**, 903–906.
70. L. Sun, J. Wu, F. Du, X. Chen and Z. J. Chen, *Science*, 2013, **339**, 786–791.
71. A. Ablasser, M. Goldeck, T. Cavlar, T. Deimling, G. Witte, I. Röhl, K. P. Hopfner, J. Ludwig and V. Hornung, *Nature*, 2013, **498**, 380–384.
72. E. J. Diner, D. L. Burdette, S. C. Wilson, K. M. Monroe, C. A. Kellenberger, M. Hyodo, Y. Hayakawa, M. C. Hammond and R. E. Vance, *Cell Rep.*, 2013, **3**, 1355–1361.
73. P. Gao, M. Ascano, Y. Wu, W. Barchet, B. L. Gaffney, T. Zillinger, A. A. Serganov, Y. Liu, R. A. Jones, G. Hartmann, T. Tuschl and D. J. Patel, *Cell*, 2013, **153**, 1094–1107.
74. T. Saitoh, N. Fujita, T. Hayashi, K. Takahara, T. Satoh, H. Lee, K. Matsunaga, S. Kageyama, H. Omori, T. Noda, N. Yamamoto, T. Kawai, K. Ishii, O. Takeuchi, T. Yoshimori and S. Akira, *Proc. Natl. Acad. Sci. U. S. A.*, 2009, **106**, 20842–20846.
75. H. Konno, K. Konno and G. N. Barber, *Cell*, 2013, **155**, 688–698.
76. K. Mukai, H. Konno, T. Akiba, T. Uemura, S. Waguri, T. Kobayashi, G. N. Barber, H. Arai and T. Taguchi, *Nat. Commun.*, 2016, **7**, 11932.
77. H. Ishikawa and G. N. Barber, *Nature*, 2008, **455**, 674–678.
78. D. K. Karaolis, T. K. Means, D. Yang, M. Takahashi, T. Yoshimura, E. Muraille, D. Philpott, J. T. Schroeder, M. Hyodo, Y. Hayakawa,

B. G. Talbot, E. Brouillette and F. Malouin, *J. Immunol.*, 2007, **178**, 2171–2181.
79. S. M. McWhirter, R. Barbalat, K. M. Monroe, M. F. Fontana, M. Hyodo, N. T. Joncker, K. J. Ishii, S. Akira, M. Colonna, Z. J. Chen, K. A. Fitzgerald, Y. Hayakawa and R. E. Vance, *J. Exp. Med.*, 2009, **206**, 1899–1911.
80. J. J. Woodward, A. T. Iavarone and D. A. Portnoy, *Science*, 2010, **328**, 1703–1705.
81. D. L. Burdette, K. M. Monroe, K. Sotelo-Troha, J. S. Iwig, B. Eckert, M. Hyodo, Y. Hayakawa and R. E. Vance, *Nature*, 2011, **478**, 515–518.
82. V. Hornung and E. Latz, *Nat. Rev. Immunol.*, 2010, **10**, 123–130.
83. F. Geeraedts, N. Goutagny, V. Hornung, M. Severa, A. de Haan, J. Pool, J. Wilschut, K. A. Fitzgerald and A. Huckriede, *PLoS Pathog.*, 2008, **4**, e1000138.
84. T. D. Kanneganti, *Nat. Rev. Immunol.*, 2010, **10**, 688–698.
85. V. A. Rathinam and K. A. Fitzgerald, *J. Clin. Immunol.*, 2010, **30**, 632–637.
86. T. Bürckstümmer, C. Baumann, S. Blüml, E. Dixit, G. Dürnberger, H. Jahn, M. Planyavsky, M. Bilban, J. Colinge, K. L. Bennett and G. Superti-Furga, *Nat. Immunol.*, 2009, **10**, 266–272.
87. T. Fernandes-Alnemri, J. W. Yu, P. Datta, J. Wu and E. S. Alnemri, *Nature*, 2009, **458**, 509–513.
88. V. Hornung, A. Ablasser, M. Charrel-Dennis, F. Bauernfeind, G. Horvath, D. R. Caffrey, E. Latz and K. A. Fitzgerald, *Nature*, 2009, **458**, 514–518.
89. M. K. Atianand and K. A. Fitzgerald, *J. Immunol.*, 2013, **190**, 1911–1918.
90. V. A. Rathinam, Z. Jiang, S. N. Waggoner, S. Sharma, L. E. Cole, L. Waggoner, S. K. Vanaja, B. G. Monks, S. Ganesan, E. Latz, V. Hornung, S. N. Vogel, E. Szomolanyi-Tsuda and K. A. Fitzgerald, *Nat. Immunol.*, 2010, **11**, 395–402.
91. T. Fernandes-Alnemri, J. W. Yu, C. Juliana, L. Solorzano, S. Kang, J. Wu, P. Datta, M. McCormick, L. Huang, E. McDermott, L. Eisenlohr, C. P. Landel and E. S. Alnemri, *Nat. Immunol.*, 2010, **11**, 385–393.
92. T. Jin, A. Perry, J. Jiang, P. Smith, J. A. Curry, L. Unterholzner, Z. Jiang, G. Horvath, V. A. Rathinam, R. W. Johnstone, V. Hornung, E. Latz, A. G. Bowie, K. A. Fitzgerald and T. S. Xiao, *Immunity*, 2012, **36**, 561–571.
93. M. Yoneyama, M. Kikuchi, K. Matsumoto, T. Imaizumi, M. Miyagishi, K. Taira, E. Foy, Y. M. Loo, M. Gale, S. Akira, S. Yonehara, A. Kato and T. Fujita, *J. Immunol.*, 2005, **175**, 2851–2858.
94. J. Andrejeva, K. S. Childs, D. F. Young, T. S. Carlos, N. Stock, S. Goodbourn and R. E. Randall, *Proc. Natl. Acad. Sci. U. S. A.*, 2004, **101**, 17264–17269.
95. S. Cui, K. Eisenächer, A. Kirchhofer, K. Brzózka, A. Lammens, K. Lammens, T. Fujita, K. K. Conzelmann, A. Krug and K. P. Hopfner, *Mol. Cell*, 2008, **29**, 169–179.
96. K. Takahasi, H. Kumeta, N. Tsuduki, R. Narita, T. Shigemoto, R. Hirai, M. Yoneyama, M. Horiuchi, K. Ogura, T. Fujita and F. Inagaki, *J. Biol. Chem.*, 2009, **284**, 17465–17474.
97. K. Takahasi, M. Yoneyama, T. Nishihori, R. Hirai, H. Kumeta, R. Narita, M. Gale, F. Inagaki and T. Fujita, *Mol. Cell*, 2008, **29**, 428–440.

98. V. Hornung, J. Ellegast, S. Kim, K. Brzózka, A. Jung, H. Kato, H. Poeck, S. Akira, K. K. Conzelmann, M. Schlee, S. Endres and G. Hartmann, *Science*, 2006, **314**, 994–997.
99. K. Malathi, B. Dong, M. Gale and R. H. Silverman, *Nature*, 2007, **448**, 816–819.
100. H. Kato, O. Takeuchi, E. Mikamo-Satoh, R. Hirai, T. Kawai, K. Matsushita, A. Hiiragi, T. S. Dermody, T. Fujita and S. Akira, *J. Exp. Med.*, 2008, **205**, 1601–1610.
101. B. Wu, A. Peisley, C. Richards, H. Yao, X. Zeng, C. Lin, F. Chu, T. Walz and S. Hur, *Cell*, 2013, **152**, 276–289.
102. N. Vabret and J. M. Blander, *Front. Immunol.*, 2013, **4**, 468.
103. N. Chen, P. Xia, S. Li, T. Zhang, T. T. Wang and J. Zhu, *IUBMB Life*, 2017, **69**, 297–304.
104. R. B. Seth, L. Sun, C. K. Ea and Z. J. Chen, *Cell*, 2005, **122**, 669–682.
105. A. Peisley, C. Lin, B. Wu, M. Orme-Johnson, M. Liu, T. Walz and S. Hur, *Proc. Natl. Acad. Sci. U. S. A.*, 2011, **108**, 21010–21015.
106. A. Peisley, B. Wu, H. Yao, T. Walz and S. Hur, *Mol. Cell*, 2013, **51**, 573–583.
107. F. Hou, L. Sun, H. Zheng, B. Skaug, Q. X. Jiang and Z. J. Chen, *Cell*, 2011, **146**, 448–461.

CHAPTER 14

Synthetic Agonists of Toll-like Receptors and Therapeutic Applications

SUDHIR AGRAWAL* AND EKAMBAR R. KANDIMALLA†

ARNAY Sciences LLC, Shrewsbury, MA 01545, USA
*E-mail: sagrawal@arnaysciences.com

14.1 Introduction

The innate immune system is the evolutionarily conserved host's first line of defense mechanism against infections.[1] The innate immune system relies on pattern-recognition receptors (PRRs) that recognize a variety of pathogen-associated molecular patterns (PAMPs) as well as danger or damage-associated molecular patterns (DAMPs) to limit the early spread of infection and to prime delayed pathogen- or antigen-specific adaptive immunity, including memory T and B cell responses.[2,3] Among the PRRs, Toll-like receptors (TLRs), C-type lectin receptors (CLRs), retinoic acid-inducible gene-I (RIG-I)-like receptors (RLRs), nucleotide-binding domain, leucine-rich repeat-containing protein receptors (NLRs) and absent in melanoma (AIM)-like receptors have been well characterized. While PRRs recognize a wide variety of molecular structures derived from pathogen or host, the receptors that specifically recognize

†Present Address: Exicure, Inc., 8045 Lamon Avenue, Skokie, IL 60077, USA.

Drug Discovery Series No. 68
Advances in Nucleic Acid Therapeutics
Edited by Sudhir Agrawal and Michael J. Gait

Published by the Royal Society of Chemistry, www.rsc.org

molecular patterns of nucleic acids are of specific interest in the development of nucleic acid-based therapeutics, including antisense, small interfering RNA (siRNA), TLR modulation and aptamers. Specific nucleic acid or nucleotide molecular patterns are recognized by innate immune receptors that include RLRs, AIM2-like receptors (ALRs), stimulator of interferon genes (STING) and TLRs (Figure 14.1). Recent advances in oligonucleotide chemistries and understanding of their biological effects permitted therapeutic development of nucleic acid molecular patterns as a choice for the modulation of the innate immune system. In this chapter we describe structure–activity relationship studies of oligonucleotides, design of novel immunomodulatory compounds targeting TLR3, 7, 8, and 9 specifically, and their therapeutic applications.

14.1.1 RIG-I-like Receptors

The cytosolic retinoic acid-inducible gene 1-like (RIG-I-like) receptor family consists of RIG-I, melanoma differentiation-associated protein 5 (MDA5), and Laboratory of Genetics and Physiology 2 (LGP2), which are characterized by the presence of a conserved domain structure, consisting of a central DExD/H-box helicase domain and a C-terminal domain (CTD), which are responsible for recognizing viral RNAs that usually escape recognition by extracellular and endosomal TLRs.[4,5] In addition, RIG-I and MDA5 harbor two N-terminal caspase activation and recruitment domains (CARDs) which, upon virus sensing, initiate downstream signaling, leading to type I interferon (IFN) gene expression.[5] RLRs are constitutively expressed in a broad range of immune and nonimmune cells, including epithelial cells of the

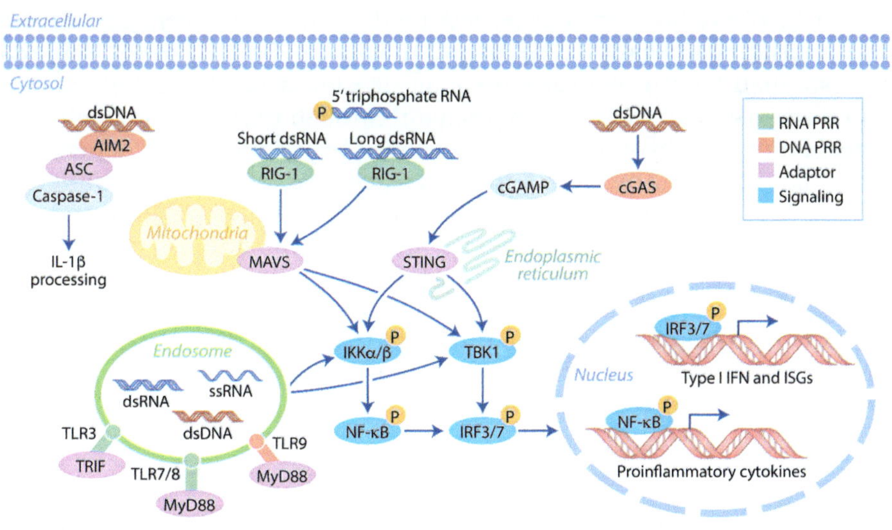

Figure 14.1 Nucleic acid sensing innate immune receptors and their signaling pathways.

central nervous system. Studies using synthetic or purified viral RNA revealed important molecular signatures that are required for RLR activation. A 5′-triphosphate (5′-ppp) moiety, present in the genomic RNA of many viruses, in concert with short blunt-end double-stranded RNA (dsRNA) stretches are critical for that ability of RIG-I to discriminate non-self from self RNA.[6,7] In contrast, MDA5 recognizes long dsRNA, such as poly(I–C).[8] Particular sequence compositions of the RNA ligands, such as poly(U/UC) motifs found in the genomic RNA, activate RIG-I more prominently. The three RLRs are broadly expressed in most tissues and signal innate immune activation in a variety of cell types. While they play a prominent role in triggering innate defenses within myeloid cells, epithelial cells, and cells of the central nervous system, their actions are not essential for IFN production in plasmacytoid dendritic cells despite their expression in this cell type. Activation of RIG-I-like receptors by RNA agonists induces IFN and proinflammatory cytokine responses that establish an antiviral state.[9]

14.1.2 AIM2-like Receptors (ALRs)

The family of cytosolic receptors expressed in hematopoietic cells, absent in melanoma 2 (AIM2)-like receptors (ALRs), comprises four members in humans and six in rodents. They are known to recognize double stranded (ds) DNA from bacteria and viruses and to protect against infections.[10–12] AIM2 consists of an N-terminal death and pyrin (DAPIN) or pyrin domain (PYD) and a C-terminal HIN-200 (hematopoietic interferon-inducible nuclear proteins with a 200-amino acid repeat) domain, which is known to have two oligonucleotide-binding folds.[12] Unlike the NLR inflammasomes, for which the exact ligand binding mechanisms are not well understood, AIM2 detects dsDNA by direct binding. The dsDNA binding occurs through its HIN-200 domain leading to oligomerization and subsequent binding to ASC [apoptosis-associated small protein effector of cell division cycle 42 (spec)-like protein containing a caspase recruitment domain] through PYD–PYD interactions.[13] Detection of DNA by AIM2 is not sequence-specific but rather depends on the length of DNA. Any dsDNA that is greater than 80 base pairs in length can provide a scaffold facilitating robust oligomerization of AIM2 and subsequent assembly of inflammasome complexes.[14] Inflammasome activation is initiated following DNA sensing by AIM2 leading to the secretion of interleukin (IL)-1β and IL-18 and pyroptosis.[12] Several members of the AIM2-like receptor gene families have been proposed to activate the interferon-stimulatory DNA pathway as well,[15] though, their role in the interferon-stimulatory DNA pathway is not well understood.

14.1.3 NOD-like Receptors, NLRP3 Inflammasome

Inflammasomes are multiprotein complexes that form in the cytoplasm in response to cellular damage and cytoplasmic pathogen-associated molecules during infection. Besides the AIM2-like receptor, the NOD-like receptor protein 3 (NLRP3) inflammasome that belongs to the NOD (nucleotide oligomerization domain)-like receptor family has been shown to be activated by DNA released after mitochondrial damage.[16,17] NLRP3 is localized in the cytoplasm and its activation leads to production of the proinflammatory cytokines, IL-1β and IL-18.

14.1.4 Cyclic GMP–AMP Synthase (cGAS) and the STING Pathway

While RLRs produce IFN responses to viral RNA, now it has been well established that beside AIM-2 receptors, a parallel pathway cGAS has recently emerged as a major sensor of cytosolic DNA of different origins including from viruses, self DNA and bacterial cyclic dinucleotides together with the adaptor protein stimulator of IFN genes (STING).[18,19] STING is localized in the endoplasmic reticulum and induces a type-1 IFN response upon activation by viral DNA.[20] cGAS is a cytoplasmic receptor that recognizes B-DNA and induces the synthesis of the cyclic dinucleotide cGAMP (cyclic guanosine monophosphate–adenosine monophosphate), which binds and activates STING to induce IFN gene expression.[21,22] Several synthetic dinucleotides have also been reported to activate the STING pathway.[23]

14.1.5 Toll-like Receptors (TLR)

Thirteen mammalian TLR family members, including ten present in humans and twelve present in rodents, have been identified. Of these thirteen, five TLRs referred to as endosomal TLRs are expressed in the membranes of the endosomes and recognize RNA and DNA molecular patterns of nucleic acids derived from viruses and bacteria and also synthetic oligonucleotides depending on nucleotide composition and structure.[24] TLR3 is the receptor for viral and synthetic double-stranded (ds) RNAs.[25] TLR7 and 8 are the receptors for viral single-stranded (ss) RNAs,[26,27] imidazoquinolines and nucleosides.[28,29] Bacterial and viral DNA molecules that contain unmethylated CpG motifs act as ligands for TLR9.[30] TLR13, which recognizes bacterial and viral rRNA, has been discovered in mice though a human analog has not been identified yet.[31]

TLRs are transmembrane receptors comprising extracellular leucine-rich repeats (LRRs) and a cytoplasmic Toll-IL-1R (TIR) domain connected through a transmembrane domain.[32,33] The TIR domain is structurally homologous to the IL-1 receptor. Binding of a PAMP to LRRs leads to TLR dimerization through the TIR domain and recruits an adapter protein to the TIR domain.[34] In general, TLR2–5, and 7–9 utilize myeloid differentiation factor 88 (MyD88) and TLR3 and TLR4 utilize Toll-interleukin 1 receptor homology (TIR)-domain-containing adaptor-inducing IFN-β (TRIF) as adapter molecules.[35] This phenomenon leads to the formation of a large helical oligomer called the myddosome,[36] which mediates engagement of IL-1-associated kinase (IRAK),

TLR-specific additional adapter proteins, and TNF receptor-associated factor 6 (TRAF6) in the signaling pathway, resulting in the activation of transcription factors activator protein1 (AP-1), E26 transformation specific like kinase 1 (ELK1), nuclear factor kappa B (NF-κB), and interferon regulatory factors (IRFs), depending on the PAMP encountered and TLR activated.[35]

14.2 Agonists of TLR3

TLR3 recognizes viral dsRNA and nucleic acid polymers such as poly I–C.[25] TLR3 is expressed in the endosomes of human myeloid dendritic cells (mDCs), monocytes and natural killer (NK) cells[37,38] and utilizes TRIF as an adaptor molecule for signaling.[39,40] The crystal structure of the TLR3 ectodomain bound to a 46-base pair dsRNA has been reported.[41]

Poly I–C, the most extensively studied dsRNA, induces type I interferons and other cytokines through activation of TLR3. Besides poly I–C, other enzymatically synthesized polymers, such as poly I–poly $C_{12}U$ (rintatolimod; Ampligen)[42] and poly A–U,[25] and poly I–C formulated with poly-L-lysine-carboxymethylcellulose (Hiltonol),[43] polyarginine,[44] or kanamycin–calcium ions,[45] have been studied as TLR3 agonists. Synthetic dsRNAs based on siRNA have also been shown to induce immune responses *via* activation of TLR3.[46,47] Though the major mode of action of poly I–C is through TLR3 stimulation, the other receptors, including protein kinase R (PKR), MDA5, RIG-I and inflammasome, have also been shown to be activated by poly I–C.[48,49] Since TLR3 agonists induce type I interferon production, they have been extensively studied preclinically and clinically as anti-infective agents, anticancer agents and as adjuvants with vaccines.[50,51] Though there has not been a TLR3-based agonist approved by the Food and Drug Administration (FDA), rintatolimod (Ampligen) that has been studied in severe myalgic encephalomyelitis–chronic fatigue syndrome (ME–CFS) patients[52] was recently approved for marketing in the Argentine Republic.

Enzymatic synthesis of poly I–C and other analogs leads to variations in length from batch to batch, which often contain single-stranded segments, which results in variability in solubility and biological activity, including toxicity, and inconsistent results[53–56] limiting their progress as therapeutic agents.

14.2.1 Synthetic Agonists of TLR3

We have designed and synthesized double-stranded synthetic oligoribonucleotides (dsORNs) of defined length to specifically activate TLR3 and with minimal off target activity, generated from the activation of other receptors as in the case of poly I–C.[57] Each strand of dsORN contains two distinct segments, namely an alignment segment composed of a heteronucleotide sequence and an oligo inosine (I) or an oligo cytidine (C) segment. dsORNs of 50 nt length with a 15-mer alignment segment and a 35-mer oligo I–oligo C segment form stable duplexes under physiological conditions and induce TLR3-mediated immune responses (Figure. 14.2).

dsORNs activate the IRF3 signaling pathway in J774 cells, induce production of cytokines, including IFN-β, IFN-α, interferon gamma-induced protein 10 (IP-10), IL-12 and IL-6, in murine and human cell-based assays and following systemic administration in mice and non-human primates.[57] These defined-length synthetic oligonucleotide-based agonists of TLR3 provide alternatives to poly I–C for use as therapeutic agents.

14.3 Agonists of TLR 7 and TLR 8

Virus-derived single-stranded RNAs (ssRNAs) activate TLR7-and/or TLR8-mediated immune responses.[58,59] Short synthetic ssRNAs that mimic viral RNA segments formulated in lipids have been shown to act as agonists of TLR7 and TLR8.[26,27] Additionally, certain small molecules, such as imidazoquinoline-based compounds and nucleoside analogs, such as loxoribine, 7-thia-8-oxo-guanosine, and 7-deazaguanosine, have also

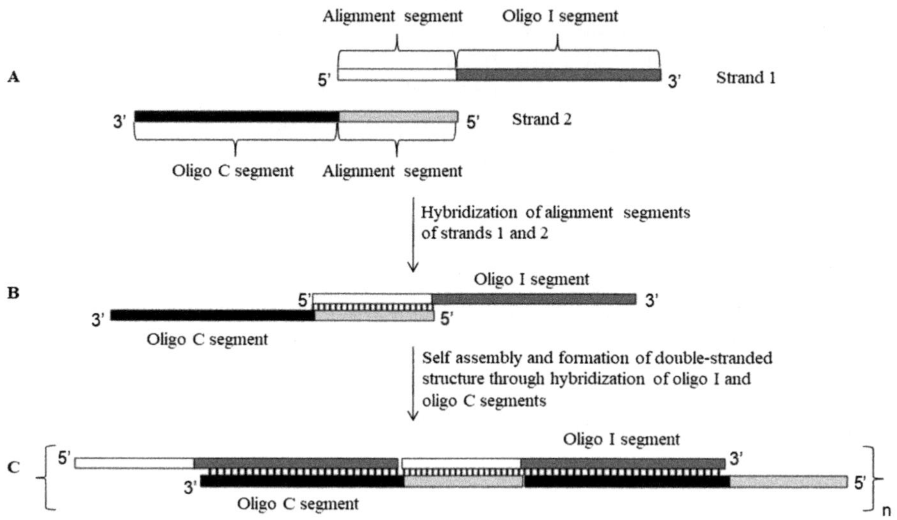

Figure 14.2 Design of oligoribonucleotides to form an extended double-stranded structure to act as potent TLR3 agonists. (A) Each ORN strand (1 and 2) is designed to consist of an alignment segment with heteronucleotide composition and an oligo I- or oligo C segment. Heteronucleotide alignment segments in strand 1 and 2 are complementary to each other. (B) The two strands are aligned through their complementary alignment segments forming a duplex structure with overhanging oligo I–oligo C segments. (C) The oligo I and oligo C segments self-assemble forming an extended duplex structure. Each strand is about 50 nt long with a 15 nt alignment segment and 35 nt oligo I- or oligo C strand. Reproduced from ref. 57 with permission from the Royal Society of Chemistry.

been shown to activate TLR7 and/or TLR8.[28,29] Imiquimod, an imidazoquinoline-based small molecule that acts as an agonist of TLR7 has been approved for the treatment of actinic keratosis, superficial basal cell carcinoma, and external genital warts by topical administration.[60] However, systemic administration of imiquimod and related compounds is known to cause toxicity.[60–62] A number of other nucleoside-based TLR7/8 agonists are currently under evaluation for infectious diseases and immuno-oncology applications.[63,64]

Recently, the crystal structure of unliganded TLR8 and its complex with a small molecule have been reported.[65] These studies elucidated that TLR8 exists in dimeric form and upon binding of the ligand a conformational change occurs in TLR8 ectodomains, leading to an activated configuration of the cytoplasmic TIR signaling domains. Interestingly, the later studies by the same group indicated that the natural ssRNA and nucleosides, which are generated from degradation of ssRNA, bind at two distinct sites and induce immune responses.[65] Imidazoquinoline-based compounds, however, have also been shown to induce immune responses *via* activation of other receptors, including A3 adenosine and the opioid growth factor receptors.[66,67]

TLR7 and TLR8 are expressed in different immune cell types. In primates, TLR7 is expressed in plasmacytoid dendritic cells (pDCs) and B cells, and TLR8 is expressed in mDCs and monocytes. In rodents, TLR8 is non-functional and TLR7 is expressed more widely in several types of immune cells than in primates and humans. In general, the activation of either TLR leads to a T-helper 1 (Th1)-type cytokine profile but the specific cytokine milieu depends on the type of cells within which these receptors are expressed and the signal transduction cascades they activate. For example, TLR7 agonists produce greater IFN-α secretion by pDCs and B cell activation compared with TLR8 agonists. In contrast, TLR8 agonists produce higher levels of IL-12 and other proinflammatory cytokines from mDCs compared with TLR7 agonists.[46]

14.3.1 Synthetic Agonists of TLR7 and TLR8

Synthetic ssRNAs have been shown to induce immune responses *via* activation of TLR7 and TLR8.[68–70] However, RNA is rapidly degraded by nucleases, and hence formulation of ssRNA with lipids is required to increase their nuclease stability and delivery for TLR7/8 activation.[70,71]

To develop synthetic RNAs as agonists of TLR7 and 8, we have stabilized the RNAs by linking two molecules of RNA *via* their 3′-ends. These compounds are referred to as stabilized immune modulatory RNA (SIMRA) compounds and are more stable in human serum than are ssRNA.[72,73] Our studies have shown that dinucleotides UA, UC and CA are highly susceptible to degradation by endonucleases compared with other dinucleotide combinations.[72,73] Avoiding these dinucleotide motifs in RNAs increases their stability against endonucleases.

We have carried out detailed studies of structure–activity relationships of a number of novel SIMRA compounds with varying sequence compositions, substituting 7-deaza-G for natural guanosine, and shown the optimal nucleotide compositions for selective TLR7 activation.[74] These TLR7-selective SIMRA compounds induce IFN-α production in isolated human pDC cultures (which express TLR7) with insignificant effects on mDCs (which express TLR8). The SIMRA compounds that activate TLR7 also show cytokine production in mice, both *in vitro* in splenocyte cultures and *in vivo* upon systemic administration.[72,74] The results of studies in TLR knockout mice indicate that both TLR7 and MyD88 are required for immune-stimulatory activity of TLR7-selective SIMRA compounds.[74] These are the first synthetic RNAs that have shown selective activation of TLR7 depending on the sequence composition and chemical modifications incorporated.

During the course of these studies, we have also identified SIMRA compounds that stimulate TLR8 selectively.[72] In general, TLR8 agonists induce high levels of IL-12 and TNF-α *via* activation of mDCs and insignificant amounts of IFN-α.[72,75] Furthermore, incorporation of non-nucleosidic linkers at the 5′-ends of SIMRA compounds increases their stability against nucleases and improves their ability to activate TLR8.[72,75]

Replacement of guanosine residues in certain SIMRA compounds that activate TLR8 by 7-deazaguanosine, shown to activate both TLR7 and TLR8, provides unique synthetic RNA-based compounds that act as dual TLR7 and TLR8 agonists.[72,75] TLR7-and 8-activating SIMRA compounds induce dose-dependent Th1-type cytokine production in human peripheral blood mononuclear cell (PBMC), and IFN-α production in human PBMC and pDC cultures.[72,75] pDCs, which express TLR7 and 9, but not TLR8, are the primary producers of IFN-α upon stimulation with appropriate TLR agonists.[76]

We have also studied substitution of a number of different ribonucleotide modifications in SIMRA compounds for their ability to activate TLR7 and TLR8 and found that arabino (ara)-G, ara-C, ara-U or ara-A substitutions in SIMRA compounds lead to activation of TLR8 in a transfected HEK293 cell reporter assay. SIMRA compounds containing arabinonucleotides induce a Th1-type cytokine profile and compounds containing ara-C also stimulate immune responses in human PBMC and pDC assays and *in vivo* in mice and non-human primates.[75,77] Phosphorothioate oligoribonucleotides (ORNs) with self-complementary sequences form secondary structures, and the thermal stability of the duplex depends on the length and GC composition of the duplex leading to greater nuclease stability.[78] ORNs containing a secondary structure show TLR8 activity in transfected HEK293 cells and induce cytokines and chemokines in human PBMCs.[78] In addition to TLR8 activity, certain ORNs containing a 'CUGAAUU' motif in the duplex-forming region activate TLR7 in transfected HEK293 cells and induce immune stimulation through TLR7 in human PBMCs and pDCs, and *in vivo* in mice.[78]

We have studied the immune response profiles induced by acute administration of naked SIMRA compounds that act as agonists of TLR7 and TLR8 in non-human primates without lipid formulation. Subcutaneous administration of a single-dose of SIMRA compounds to cynomolgus monkeys induces lymphocyte migration, CD69 expression on T lymphocytes, monocytes, and NK cells, and elevation of cytokine and chemokine levels in the plasma.[72,75] A SIMRA compound that activates both TLR7 and 8 induces higher levels of IFN-α than does a TLR8-only-activating SIMRA compound.[57,75] Based on the results of these structure–activity relationship studies we have elucidated a detailed understanding of the design of RNA-based TLR7, TLR8, and dual TLR7/8 agonists as summarized in Figure 14.3.

14.3.2 Preclinical Studies of Agonists of TLR7 and TLR8

Imiquimod, which acts as an agonist of TLR7, has been approved for the treatment of actinic keratosis, superficial basal cell carcinoma, and external genital warts by topical administration. However, systemic administration of imiquimod and related compounds is known to cause toxicity.[60–62] A number of other nucleoside-based TLR7/8 agonists are currently under evaluation for infectious diseases and immuno-oncology applications.[63,64]

An RNA-based dual TLR7/8 agonist SIMRA compound was evaluated for antitumor activity in preclinical tumor models of CT26.CL25 colon cancer and 3LL-C75 lung carcinoma.[79] The dual TLR7/8 agonist showed potent dose-dependent tumor growth inhibition following intraperitoneal administration in CT26.CL25 colon carcinoma-bearing mice. Treatment with a dual TLR7/8 agonist of CT26.CL25 colon carcinoma-bearing mice led to a marked decrease of cluster of differentiation (CD)4^{+} CD25^{+}, forkhead box

5'-NNNNNNNNNNN-3'-L-3'-NNNNNNNNNNN-5'

- Two 5'-ends provide increased *in vivo* activity
- Secondary structure at the 5'-end provides nuclease stability and activity
- Endonuclease susceptible CA, UA and UC dinucleotides are avoided in RNA
- 3'-3'-linkage provides stability against nuclease degradation
- Phosphorothioate backbone

- **Sequence characteristics of TLR7-specific SIMRA sequences**
  - Presence of about 50% G/U content
  - Presence of 7-deaza-G is critical
  - Presence of specific nucleotide motifs
- **Sequence characteristics of TLR7/8 dual SIMRA sequences**
  - Presence of >50% G/U content
  - Presence of 7-deaza-G or ara-C substitutions accepted
  - Certain secondary structures
- **Sequence characteristics of TLR8-specific SIMRA sequences**
  - Presence of high G/U content (>50%)
  - 5'-modifications increase potency
  - No U modification accepted

Figure 14.3 Some RNA sequence characteristics and chemical modifications identified for TLR7/8 recognition.

P3 (FoxP3)$^+$ T regulatory cells, and an increase of tumor-antigen-specific IFN-γ-secreting effector cells in splenocytes and peritoneal-infiltrating lymphocytes.[79] The same TLR7/8 SIMRA compound induced strong immune responses and tumor growth inhibition in 3LL-C75 tumor-bearing wild-type and TLR9 knockout mice, but not in TLR7 knockout or MyD88 knockout mice, indicating that the agonist induces an immune response *via* TLR7 and through the MyD88-dependent signaling pathway.[79] The SIMRA compound also prevented lung metastasis of CT26.CL25 tumors when administered subcutaneously.[79]

14.4 Agonists of TLR9

DNA fractions isolated from *Mycobacterium bovis* strain BCG elicit antitumor activity,[80,81] augment NK cell activity,[82,83] and induce interferon-α, β, and γ production.[83] Furthermore, DNA from different bacterial species, but not mammalian DNA, stimulates proliferation of murine lymphocytes.[84] Synthetic oligodeoxynucleotides containing certain palindromic sequences with CG dinucleotides, similar to those present in bacterial DNA, increase the cytolytic function of NK cells and induce secretion of IFN-γ.[85,86] Subsequent studies have shown that phosphorothioate oligodeoxynucleotides containing CpG motifs activate B cells and induce immune responses.[87]

During the mid-1990s a number of phosphorothioate antisense molecules were evaluated in human clinical trials for various disease indications.[88] Many of these antisense compounds inadvertently contained unmethylated CpG dinucleotides/motifs. We had designed an antisense sequence, referred to as GEM91 (5′-CTCTCGCACCCATCTCTCTCCTTCT-3′), targeting the *gag* region of HIV-1 for potential treatment of AIDS.[88] During the early clinical development, subcutaneous administration of GEM91 led to flu-like symptoms in patients, associated with swollen lymph nodes. This was the first observation of immune activation upon systemic administration of a synthetic oligonucleotide containing a TCG motif in humans. Later it was shown that bacterial and viral DNA containing unmethylated CpG motifs *trans*-activates human immunodeficiency virus long terminal repeat (HIV-LTR) and HIV replication *via* TLR9.[89] Consistent with these results, GEM91, produced dose-dependent increases in viral loads in the plasma of HIV-positive patients compared with placebo controls.[90] Though unintended, these results provided the first proof of concept for CpG oligonucleotide-mediated biological activity in humans. Later on it became clear that many of the antisense sequences containing the CG motif had largely off-target activity in preclinical disease models, and have been shown to elicit sequence-specific immune responses, including cytokine induction *in vitro* and in mice, including splenomegaly.[88,91–96] These effects are similar to those reported with bacterial DNA and palindromic sequences containing CpG dinucleotides.[82,84,87] The later studies have shown that TLR9 is the receptor that recognizes bacterial and synthetic DNA containing unmethylated CpG motifs and induces immune responses.[30]

The immunostimulatory effects of CpG oligonucleotides are dependent on the nucleotide sequence, the nature of the DNA backbone and the flanking sequences. On the basis of the type of immune-cell populations that are activated and the cytokine profiles produced, synthetic CpG oligonucleotides have been divided into three types, A, B, and C.

Type A CpG oligonucleotides comprise a CpG motif within a palindromic sequence, have a phosphodiester backbone[85] and are flanked by phosphorothioate poly(dG) sequences.[97,98] Type A CpG oligonucleotides activate pDCs and NK cells, leading to the secretion of type-I IFNs. However, the presence of poly(dG) sequences in an oligonucleotide is known to induce immune responses and bind to a number of proteins, thereby inhibiting their own action.[99] Moreover, problems associated with poly(dG) synthesis, purification, quality control, and the pharmacokinetics of poly(dG) sequence-containing oligonucleotides have limited their potential as therapeutic molecules.[100,101] A class A CpG oligonucleotide formulated in virus-like particles (CYT003) has been evaluated as an adjuvant in patients with allergies. In this trial, all groups, including placebo, showed a clinically important improvement in ACQ score. However, there was no significant difference between the CYT003 and placebo groups at week 12. No significant differences were seen in secondary outcomes as well. Due to a lack of efficacy, the study was prematurely terminated at the end of the treatment phase with no further follow-up.[102]

Single-stranded phosphorothioate oligonucleotides containing CpG dinucleotides in specific sequence contexts (CpG motifs) are referred to as type B. Type B CpG oligonucleotides activate B cells, macrophages, monocytes, and DCs, and produce proinflammatory cytokine secretion.[87] The two nucleotides flanking the CpG influence immunostimulatory activity and also species specificity.[87] The lack of secondary structure precludes long retention in early endosomes, unlike type A and C, and thereby they induce minimal levels of IFN-α by pDCs. A class B CpG oligonucleotide has been recently approved as an adjuvant in hepatitis B vaccine.[103]

Type C CpG oligonucleotides mimicking the effects of both type A and B CpG DNAs have been described.[104,105] Type C CpG oligonucleotides contain uniform phosphorothioate backbone modifications, as for type B CpG oligonucleotides, with both palindromic and non-palindromic segments for secondary structure formation. Type C CpG oligonucleotides induce IFN-α secretion and activate B cells *in vitro* in human cell-based assays. Class C CpG oligonucleotides have also been clinically evaluated and are discussed later in detail in this chapter.

14.4.1 Synthetic Agonists of TLR9

We have systematically studied the structure–activity relationships of oligonucleotides that act as agonists of TLR9.[106,107] The presence of a CpG motif in phosphodiester (PO) and phosphorothioate (PS) backbone oligonucleotides stimulate TLR9-mediated immune responses.[108,109] PO oligonucleotides, however, undergo rapid degradation *in vivo* and therefore, PS oligonucleotides, which are more resistant to degradation by ubiquitous nucleases, are commonly used.[110] Substitution of a sulfur atom by oxygen in the internucleotide PO-bond results in the formation of Rp and Sp diastereoisomers of the PS-bond. The Rp diastereoisomer of the PS-linkage stimulates a stronger TLR9-mediated immune response than does the Sp diastereoisomer but is rapidly degraded by nucleases in biological fluids.[111,112]

The negative charges of the internucleotide linkages between and adjacent to cytosine (C) and guanine (G) of the CpG motif are also required for TLR9 activation. Incorporation of a methylphosphonate (neutral charge) linkage at these positions leads to loss of TLR9 activation.[113–115] TLR9 activation is also dependent on the sequences flanking the CpG dinucleotide, the nature of the nucleotide backbone, and the secondary structures.[73,106,114,116,117]

14.4.1.1 Role of the 5′-end

In the above studies it was noted that the position of the CpG motif in an oligonucleotide also affected the immunostimulatory activity. To gain further insights we synthesized oligonucleotides with two CpG motifs, one placed towards the 5′-end and the other towards the 3′-end, incorporated 2′-*O*-methylribonucleotides selectively preceding the 5′- end or 3′-end of the CpG motifs or both (the presence of 2′-*O*-methylribonucleotides preceding a CpG dinucleotide suppresses TLR9 activation as described in Chapter 4) and studied for their TLR9-mediated immune responses in cell-based assays and *in vivo* in mice.[115] These studies indicated that a CpG motif placed closer to the 5′-end plays a critical role in TLR9 activation compared with the one present towards the 3′-end.[118] These results provide insights for the use of oligonucleotides with CpG dinucleotides when employed as TLR9 agonists and antagonists or antisense agents (Figure 14.4).

Two CpG oligonucleotides linked through their 5′-ends do not induce immune responses despite the presence of two CpG motifs.[119,120] The same oligonucleotides when linked through their 3′-ends produce higher levels of cytokines than the parent CpG oligonucleotide with a single 5′-end.[119–123] These studies provided the first evidence that an accessible or free 5′-end is required for TLR9-mediated immune stimulation. The presence of two accessible 5′-ends in a TLR9 agonist results in rapid activation of the transcription factor NF-κB, but similar activation of mitogen-activated protein (MAP) kinase, compared with a TLR9 agonist containing a single 5′-end in J774 cells.[121] Oligonucleotides with 11-nucleotides in each segment are more active than those containing shorter or longer segments.[106,121] The attachment of oligonucleotides through their 3′-ends not only provides two 5′-ends for optimal activation of TLR9, but also increases their stability against 3′-exonucleases. Oligonucleotides with a PO backbone and as short as five or six nucleotides linked through their 3′-ends act as potent TLR9 agonists to induce immune responses.[108,124] This design of TLR9 agonists allowed oral administration due to their increased stability in the gastrointestinal tract,

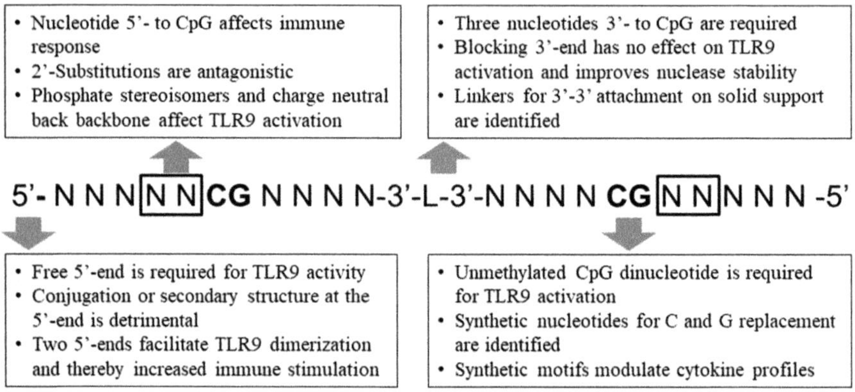

Figure 14.4 Role of flanking sequences and CpG for TLR9 activation.

with potent mucosal immune responses and adjuvant activity in the presence of antigens that prevents and reverses peanut-induced allergy in mouse models.[125–127]

Blocking the 5′-end of a TLR9 agonist with mononucleotides, dinucleotides or oligonucleotides or with small-molecule entities, such as fluorescein, abrogates TLR9-mediated immune-stimulatory activity, whilst conjugation of the same at the 3′-end retains the activity.[120] As an example, conjugation of a model peptide, a 28-amino acid β-amyloid peptide, to either the 5′- or 3′-end of a TLR9 agonist *via* C3 and C6 alkyl linkers results in loss or retention of immunostimulatory activity, respectively, compared with that of the parent TLR9 agonist without peptide conjugation *in vitro* and *in vivo* in mice.[128] Consistent with these findings, conjugation of ragweed allergen Amb a 1, to the 5′-end of a TLR9 agonist failed in phase 3 trial in patients for allergic asthma.[129]

The observation that the 5′-end is required for immune activation leads to a question about a compound (MGN1703, lefitolimod), which is being studied as a TLR 9 agonist. This compound is a dumbbell shaped DNA structure in which both ends have been linked, thereby containing no available ends. Evaluation of this compound in HEK293 TLR9 transfected cells failed to show any activity (Agrawal, unpublished data). It is possible that *in vivo* the compound is degraded, which allows the 5′-end to be accessible for immune activation.

These studies indicate that agonists containing two accessible 5′-ends facilitate rapid dimerization of the receptor, leading to greater activation of immune responses.[106] Moreover, TLR9 activation can be modulated through appropriate presentation of the free 5′-ends and synthetic immune-stimulatory motifs (described later in this chapter) leading to changes in the downstream cytokine induction profiles. In fact, recent studies have shown that TLR9 exists in monomeric form and that ligand binding facilitates dimerization, z-loop cleavage, and subsequent conformational changes in the endodomain.[130,131] However, only binding of oligonucleotides containing a CpG motif causes TLR9 dimerization and subsequent conformational changes in the endodomain, leading to the recruitment of an adaptor molecule and subsequent activation of immune-stimulatory signaling pathways.[130,131] Control oligonucleotides and antagonists bind to TLR9 but are incapable of inducing TLR9 dimerization and/or conformational changes.

Oligonucleotide sequences with two 5′-ends containing two identical segments that are linked through their 3′-ends *via* a non-nucleoside linker or a phosphorothioate linkage, referred to as immunomers, can be synthesized on an automated DNA synthesizer using two different synthetic strategies, linear and parallel synthesis.[121,132] We have studied the effects of length, nature, and stereochemistry of the linker incorporated in agonists of TLR9-mediated immune responses.[133,134] A number of linkers, including 1,2,3-propanetriol, (*S*)-(−)-1,2,4-butanetriol, 1,3,5-pentanetriol, *cis*,*cis*-1,3,5-cyclohexanetriol, *cis*,*trans*-1,3,5-cyclohexanetriol, 1,3,5-tris(2-hydroxyethyl)isocyanurate, tetraethyleneglycol, and hexaethyleneglycol were studied within TLR9 agonists.[134] These studies have provided a detailed understanding of the requirements for the distance between the two 5′-ends, rigid *versus* a flexible linker, and the spatial arrangement of the two branches for optimal immune responses. Results from these studies indicate that C3–C5 linkers, 1,2,3-propanetriol, (*S*)-(−)-1,2,4-butanetriol, or 1,3,5-pentanetriol, are optimal for stimulation of TLR9-mediated immune responses.[134] Rigid C3 linkers with different stereochemistries have little effect on immune stimulation, whilst linkers longer than C5 reduce TLR9-mediated immune stimulation.[134]

14.4.1.2 Synthetic Immune-stimulatory Motifs

Certain chemical modifications introduced within the CpG dinucleotide that alter its structure and conformation lead to the loss of immune-stimulatory activity of TLR9 agonists. One such modification is substitution with a methyl group at the 5-position of cytosine in the CpG motif of TLR9 agonists.[113] The presence of a methyl group at the 5-position of cytosine disrupts water-mediated hydrogen bonding in the TLR9 binding pocket, thereby destabilizing the 5-methyl-CpG oligonucleotide–TLR9 complex.[130] Vertebrates use this feature to distinguish self-DNA, which contains extensive methylated CpG motifs, from bacterial DNA. In later studies we observed that oligonucleotides containing 5-methyl-CpG oligonucleotides act as antagonists of TLR9.[135]

We have studied the effects of various pyrimidine analogs (Y), such as 5-OH-dC, dU, dP, 1-(2′-deoxy-β-D-ribofuranosyl)-2-oxo-7-deaza-8-methyl-purine, N^3-Me-dC, N^4–Et-dC, 5-propyne-dC, furano-dT, dF, 4-thio-dU, ψ-iso-dC, and arabinoC in place of cytosine and several purine nucleobases (R) such 7-deaza-dG (Figure 14.5), 7-deaza-8-aza-dG, 9-deaza-dG, N^1-Me-dG, N^2-Me-dG, 2-amino-D-purine, nebularine, 2-amino-dA, 7-deaza-D-xanthine, K-base 6-thio-dG, dI, 8-OMe-dG, 8-*O*-allyl-dG, and arabinoG in place of

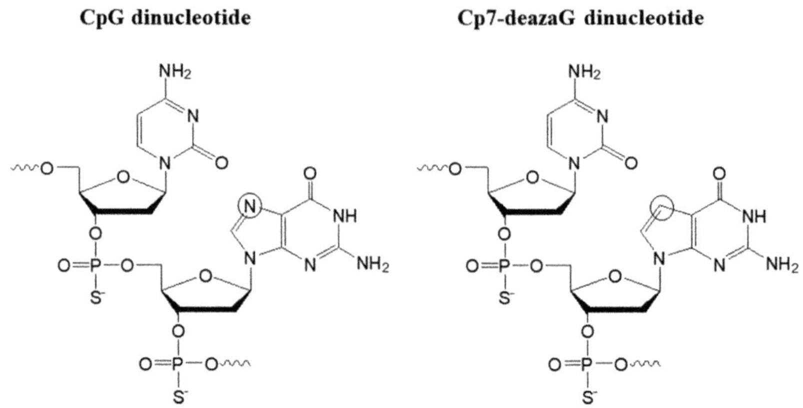

Figure 14.5 Structures of CpG and Cp7-deazaG dinucleotides. Cp7-deazaG dinucleotides represent a novel synthetic motif.

guanine in the CpG dinucleotide, respectively.[106,136,139–141] Agonists of TLR9 containing certain cytosine or guanine modifications show activity in TLR9 transfected HEK293 cells, mouse spleen, and human cell-based assays and *in vivo* in mice and provide insight as to which specific chemical modifications at C or G of the CpG motif are recognized by TLR9 and have the ability to modulate immune responses when substituted for natural C or G in TLR9 agonists. Moreover, our studies have shown that TLR9 recognizes arabino-C and arabino-G substitutions in C and G of a CpG dinucleotide, respectively. More importantly, TLR9 agonists containing specific substitutions induce a different immune response profile than a natural CpG dinucleotide.[106,107,117,122,123,134–141]

Modifications at the 2′-position of the sugar ring of a C or G nucleotide in the CpG motif of TLR9 agonists lead to loss of immune-stimulatory activity of TLR9 agonists.[113] In addition, other chemical modifications, such as methylphosphonate linkages,[114,115] 2′-alkyl or 3′-deoxy or -alkyl ribonucleosides,[133,142,143] non-nucleotide linkers,[144] or abasic nucleotides[145,146] in the flanking sequences incorporated at the fourth to sixth nucleotide positions 5′- to the CpG dinucleotide of TLR9 agonists increase immune-stimulatory activity. In contrast, substitutions incorporated adjacent to CpG on the 5′-side do not lead to immune stimulation.[114,142,143] In general, modifications incorporated in the 3′-flanking sequence distal to the CpG dinucleotide have effects dependent on the nature of modification.[73,107]

14.4.1.3 Secondary Structures

Secondary structures of oligodeoxynucleotides affect TLR9-mediated immune responses. TLR9 agonists containing a hairpin structure at the 3′-end show higher levels of IFN-α in human PBMC and pDC cultures compared with those without a 3′-end hairpin structure.[147,148] In contrast,

TLR9 agonists containing a hairpin structure at the 5′-end show loss of immune-stimulatory activity.[147,148] These results are consistent with the concept discussed earlier in this chapter that an accessible 5′-end is required for optimal activation of TLR9 and blocking the 5′-end with a secondary structure leads to loss of TLR9-mediated immune responses. However, *in vivo* studies of 3′-hairpin-containing TLR9 agonists in mice and non-human primates indicate that no immune responses are induced (Kandimalla and Agrawal unpublished results). This observation indicates that TLR9 agonists that can form intramolecular secondary structures, such as hairpins, are not effective inducers of immune responses *in vivo*, though they show TLR9-activation *in vitro*.[149–151]

Oligodeoxynucleotides containing a CpG motif in a palindromic sequence can form duplex structures, referred to as class C CpG oligos. These have been shown to act as agonists of TLR9 and induce higher levels of IFN-α compared with sequences without a secondary structure.[101,104,150] Depending on the length and base composition of the palindromic sequence, these oligodeoxynucleotides can preferentially form intramolecular hairpin structures rather than intermolecular duplex structures.[151] Formation of an intramolecular hairpin structure leads to loss of immune-stimulatory activity *in vivo* as discussed above. A class C CpG oligonucleotide (5′-TCGTCGTTTTC-GGCGCGCGCCG-3′) has been evaluated in patients with hepatitis C infection and failed to show significant hepatitis C reduction, largely due to potential formation of intramolecular structures.[152]

We have designed TLR9 agonists with self-complementary sequences that would preferentially form an intermolecular duplex structure, but not an intramolecular hairpin structure. Our studies have shown the ability of such self-complementary sequences containing TLR9 agonists to induce higher levels of IFN-α production along with other Th1-type cytokines *in vitro* in human cell-based assays and higher and sustained levels of IFN-α and IP-10 induction and NK cell activation *in vivo* in non-human primates.[73,107,123]

We have evaluated three different IFN-α-inducing agonists of TLR9 with different sequence compositions to determine the type of duplex structures formed and their ability to induce immune responses *in vitro* and *in vivo* in mice and non-human primates.[151] Thermal melting studies have indicated that, depending on the oligonucleotide sequence composition, they form intramolecular duplexes, intermolecular duplexes, or both type of structures at equilibrium. Although the oligonucleotides that form either type of structure induce production of Th1-type cytokines and chemokines, including high levels of IFN-α in human PBMC and pDC cultures, only the compounds that form intermolecular duplexes induce cytokine secretion in mice.[151] In non-human primates, the agonists that form intermolecular duplexes induce IFN-α and IP-10 secretion. In contrast, the agonist that forms an intramolecular duplex induces only lower levels of cytokines in non-human primates, indicating that this type of structure formation is less immune-stimulatory *in vivo* than the other structures. Consistent with these results, a TLR9 agonist containing a palindromic sequence with a potential to form an intramolecular hairpin structure induces low levels of IFN-α with no dose–response at a dose range of 1 to 20 mg in a clinical trial in healthy subjects.[152,153]

Based on the permutations and combinations of novel structures and synthetic immune-stimulatory motifs discussed above a number of novel synthetic agonists of TLR9, referred to as immune modulatory oligonucleotides (IMOs) have been synthesized and studied (Figure. 14.6).

Two synthetic agonists of TLR9 referred to as IMO-2055 (ref. 154) and IMO-2125 (ref. 151) have been evaluated extensively in preclinical and clinical studies. Both of these agents are comprised of two 5′-ends (3′–3′ linkage) and contain synthetic immunostimulatory motifs, Cp7-deazaG. All other TLR9 agonists studied to date by others contain CpG motifs.

14.5 Therapeutic Applications of Synthetic Agonists of TLR9

TLR9 agonists have been studied extensively as therapeutic agents for treatment of cancer, respiratory diseases, and infectious diseases, and for use as vaccine adjuvants.

14.5.1 Preclinical Studies in Cancer

A number of preclinical studies have been carried out with TLR9 agonists as anticancer agents either alone or in combination with chemotherapeutic agents, targeted therapies and monoclonal antibodies following systemic administration.[88,154–157] Our early studies indicated that these synthetic agonists of TLR9 induce strong immunopharmacological responses and antitumor effects *in vivo* in tumor models as monotherapy and in combination with chemotherapeutic agents.[155] Studies in cytokine knockout mice indicated that IL-12, but not IL-6, is required for antitumor effects. Treatment of tumor-bearing mice with synthetic agonists of TLR9 also led to the

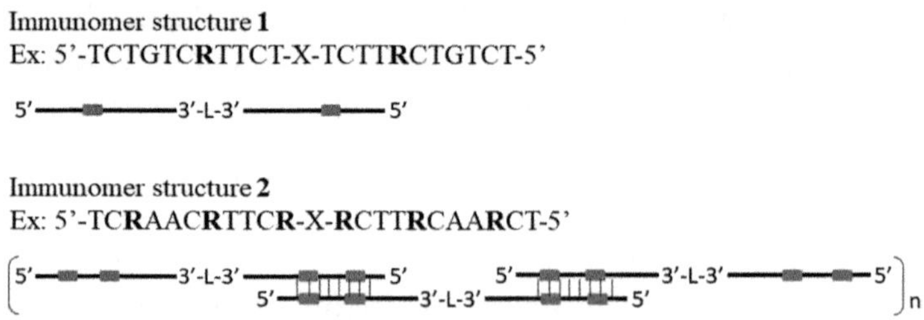

Figure 14.6 Two representative examples of synthetic agonists of TLR9 are shown. Both of these candidate examples contain synthetic immune stimulatory motifs, **CR**, wherein **R** represents 7-deaza-dG (structure shown in Figure 14.5). Structure **1** is a linear compound and structure **2** forms double-stranded structures.

development of strong tumor-specific adaptive immune responses that lead to rejection of the same tumor cell rechallenge.[155] The studies of TLR9 agonists in combination with epidermal growth factor receptor (EGFR) and vascular endothelial growth factor (VEGF) inhibitors have shown enhancement of antitumor effects *via* interference with the EGFR-related signaling pathways and reduction in angiogenesis in addition to TLR9-mediated immune response effects.[158,159] In addition, a cooperative effect of TLR9 agonist plus trastuzumab has been observed in a trastuzumab-resistant breast cancer model as a result of direct antitumor and anti-angiogenic activity of trastuzumab and antibody-dependent cell cytotoxicity (ADCC) enhancement by TLR9 agonist.[160] In mice, TLR9 agonist in combination with radiation therapy induced systemic antitumoral humoral responses, augmented tumor infiltration of NKDCs, reduced pulmonary metastases and improved survival of 3LL cancer-bearing mice,[161] and served as a prophylactic and therapeutic agent against acute radiation syndrome in adjuvant therapy in gastrointestinal malignancies.[162]

In the past few years immunotherapy of cancer by inhibiting key checkpoints in the immune system has led to durable responses in several tumor types, however, in only about 10–35% of patients.[163,164] It has been suggested that the tumor microenvironment (TME) is very important for the outcome, and patients with non-immunogenic "cold" TME have been shown to respond poorly. A number of approaches are being pursued to activate TME, including use of oncolytic viruses and synthetic innate immune modulators, with the goal of improving the outcome of immunotherapy. We and others have evaluated the use of intratumorally administered synthetic agonists of TLR9 to activate innate immune responses in the TME.[165–167]

The antitumor activity of intratumorally (i.t.) administered IMO-2125 has been evaluated in preclinical models of colon, lung, melanoma, and pancreas tumors. The i.t. delivery of IMO-2125 to a single tumor led to dose-dependent antitumor activity not only in the injected tumor but also at distal tumors, establishing systemic antitumor efficacy. Antitumor activity was associated with increased $CD3^+$ tumor-infiltrating lymphocytes (TILs), $CD8^+$ T cells, and cytotoxic T cell responses against tumor antigens. Depletion of $CD8^+$ T lymphocytes with CD8-specific antibodies abrogated the anti-tumor response, whereas depletion of $CD4^+$ lymphocytes with CD4-specific antibodies did not, indicating a key role of $CD8^+$ T-cells in exerting the antitumor responses. The mice that remained tumor-free after initial treatment with IMO-2125 rejected tumor growth after rechallenge with the same tumor cells, indicating the development of a tumor-specific adaptive immune response.[167,168]

Changes in TME were further supported by increased expression of several genes including indoleamine-pyrrole 2,3-dioxygenase (IDO-1), Programmed death-ligand 1 (PD-L1), T-cell immunoglobulin and mucin-domain containing-3 (TIM3), lymphocyte-activation gene 3 (LAG3) and cytotoxic T-lymphocyte antigen 4 (CTLA4), in both treated and distant tumors.[167,169] In these studies, elevation of IDO-1 levels following i.t. or subcutaneous (SC) administration of IMO-2125 possibly blunted the immune stimulation induced. A study was carried out in tumor-bearing mice treated with i.t. administration of IMO-2125, oral IDO-1 inhibitor, or a combination of the two agents. Mice treated with a combination of IMO-2125 and IDO-1 inhibitor showed the most potent antitumor activity, more than those treated with IMO-2125 alone, indicating that blocking the IDO-1 pathway potentiates antitumor activity of IMO-2125. In this study the IDO-1 inhibitor showed marginal antitumor activity.[170] In another study, i.t. administration of IMO-2125 in combination with an anti-PD1 antibody showed more potent antitumor activity than with either agent alone.[171] These observations led to the evaluation of a triple combination of i.t. administration of IMO-2125 plus anti-PD1 plus IDO-1 inhibitor, as well as these agents alone, and also combinations of the two agents. Treatment with the triple combination exerted the most potent antitumor activity compared with treatment with two agents. These outcomes guide us to appropriate combinations of agents which help to modulate the TME and expansion of tumor antigen-specific T cells ideal for immunotherapy.[172]

Similarly, i.t. administration of IMO-2125 in combination with a systemic anti-CTLA4 antibody showed very potent antitumor activity, associated with an increase in T-cell responses in the TME.[173] In another study both agents, IMO-2125 and anti-CTLA4 were administered intratumorally and exerted very potent and durable antitumor responses.[173]

Pancreatic cancer remains untreatable, with poor outcomes. The majority of pancreatic tumors are rich in stromal cells and are not the ideal tumors for immunotherapy. Studies with local treatment with IMO-2125 in murine syngeneic pancreatic ductal carcinoma (Panc02) and a metastatic pancreatic tumor model showed potent antitumor activity, associated with increased T-cell infiltration and upregulation of several checkpoint genes. This approach provides a novel strategy for treatment of pancreatic cancer.[174]

Clinical trials of intratumoral administration of IMO-2125 monotherapy and in combination with anti-CTLA4 are in progress, as discussed later in this chapter.

14.5.2 Treatment for Asthma and Allergies

TLR9 agonists have been shown to prevent and reverse allergen-induced Th2 immune responses that are commonly observed in asthma and allergic conditions in a number of preclinical models.[175,176] In ovalbumin (OVA)-sensitized mouse models of asthma, agonists of TLR9 prevent the development of allergic airway inflammation and airway hyper-responsiveness when co-administered during OVA-sensitization,[141] and reverse established allergic responses when administered to OVA-sensitized and OVA-challenged mice.[177] Furthermore, the agonists of TLR9 described herein have also been studied to prevent and treat OVA-induced Eustachian tube dysfunction in rat models, indicating the potential application in treating allergy-associated otitis media with effusion (OME), a chronic inflammatory condition of the middle ear.[178–180]

A synthetic agonist of TLR9 (IMO-2134) has been studied in a non-human primate model of *Ascaris suum* challenge. Administration of IMO-2134 by aerosol inhalation effectively attenuated inflammatory responses in the lungs after repeated allergen challenges. IMO-2134 inhibited IL-13 in BAL fluid after the allergen challenge, increased IP-10 levels in plasma, and induced expression of several IFN-dependent genes in BAL cells, indicating inhibition of Th2 responses and induction of a Th1-like immune response.[181] A ragweed allergen, Amb a 1, conjugated to the 5′-end of a TLR9 agonist has been the most extensively studied and evaluated in human clinical trials. However, the human trials have failed to meet the clinical end points as discussed in the previous section.[129]

14.5.3 Use as Vaccine Adjuvants

A number of preclinical studies have shown that TLR9 agonists provide protection against viruses, bacteria, parasites, and fungi through the induction of Th1-type cytokines (*e.g.*, IL-12, IFN-α, and IFN-γ).[182] TLR9 agonists also enhance antigen-specific Th1-type adaptive immune responses when used as adjuvants in combination with protein, peptide, and DNA vaccines in preclinical and clinical studies.[182,183] Recent studies have shown the potential application of synthetic agonists of TLR9 as adjuvants with hepatitis B surface antigen (HBsAg),[184,185] HIV gp-120-depleted whole killed HIV,[186] rabies vaccine,[187] dengue vaccine candidate,[188] human EGFR-2/neuregulin (HER-2/neu) DNA vaccine,[189] and telomerase reverse transcriptase (TERT) DNA vaccine[190,191] in animal models. Recently, a TLR9 agonist adjuvant containing HepB vaccine, Heplisav B, has been approved by the FDA.[103]

14.6 Clinical Development of Synthetic Agonists of TLR9

Several TLR9 drug candidates have advanced to clinical development. In this section we have focused on clinical development of two drug candidates IMO-2055 and IMO-2125, as these agents were extensively evaluated in our laboratory. A number of other candidates have been advanced to clinical development by others.

14.6.1 Clinical Trials in Hepatitis C Patients

Recombinant interferon used to be an important drug for treatment for the treatment of hepatitis C infection. It was proposed that induction of natural interferon may provide a better safety profile. Towards this goal IMO-2125 was studied as a candidate for the treatment of hepatitis C.

IMO-2125 was evaluated in phase-1 trials in hepatitis C patients alone and in combination with ribavirin.[192–194] Treatment with IMO-2125, administered subcutaneously once weekly for four weeks at four dose levels of 0.04 to 0.32 mg kg^{-1} was well tolerated. Dose-dependent increases in plasma IP-10, IFN-α, and 2′–5′-oligoadenylate synthetase at 24 h post IMO-2125 administration was observed. The induction of IFN-α was stronger after the fourth dose compared with the first dose.[192] Of the eight null-responder HCV patients treated with the highest dose level of IMO-2125, six showed maximum viral load reductions ranging from 1.0 to 3.5 $\log_{10}$ at least once during the treatment period.[192] The levels of plasma IFN-α induced by IMO-2125 correlated with viral load reduction.

14.6.2 Clinical Trials in Cancer

The safety and pharmacodynamic activity of subcutaneously administered IMO-2055 have been studied in phase 1 rising-dose trials in healthy subjects and in cancer patients. A dose-dependent transient migration of lymphocytes from peripheral blood, activation of surface markers on immune cells, and elevation of plasma cytokine levels have been observed, consistent with immune-stimulatory activity of IMO-2055. No significant acute phase or proinflammatory activity, including no effects on cortisol, C-reactive protein (CRP), or erythrocyte sedimentation rate (ESR) were observed.[88,195–198]

IMO-2055 was studied for the treatment of cancers in phase 1 and 2 clinical trials following subcutaneous administration either alone in renal cell carcinoma (RCC),[199] or in combination with approved anticancer agents in non-small-cell lung cancer (NSCLC),[200,201] relapsed and/or metastatic squamous cell carcinoma of head and neck (SCCHN),[202,203] and colorectal cancer patients.[204] Systemic administration of TLR9 agonists, including IMO-2055, CPG7909, or other agents, has not yielded promising results in cancer patients.[205]

While the above studies of IMO-2055 and other TLR9 agonists following systemic administration indicated that these were safe in combination with a number of anticancer agents, clinical benefit was not achieved.[206] Systemic administration of IMO-2055 or any other TLR9 agonist leads to induction of a host immune response without affecting the tumor microenvironment (TME). Intratumoral administration in one lesion led to induction of local immune responses which, in turn, would result in the attention of the host immune system to the injected lesion site allowing the migration of immune cells to distant tumor sites. This will allow conversion of the TME of injected tumors to become immunogenic and potentiation of the activity of systemic therapies with immune checkpoint blockade.

IMO-2125 is currently being evaluated in phase 1/2 clinical studies in anti-PD-(L)1 refractory melanoma patients in combination with anti-CTLA4 antibody (ipilimumab) or anti-PD-1 (pembrolizumab) in advanced metastatic melanoma patients.[207–210] IMO-2125 is administered intratumorally at four dose levels of 4, 8, 16, or 32 mg per injection once a week and ipilimumab or pembrolizumab is administered following a standard dosing regime. The combination of IMO-2125 with ipilimumab has been tolerated well at all dose levels studied. Importantly, dendritic cell activation associated with an elevated type 1 interferon signature in the injected lesion was detectable at 24 h post-dosing of IMO-2125, supporting the rationale of innate immune activation within the TME. Early signals of promising clinical efficacy have been confirmed in five out of ten patients based on response evaluation criteria in solid tumors (RECIST) response, including one durable complete response (CR) (>1 year), another two subjects had durable stable disease (SD) (>6 months). The overall clinical benefit rate was about 67%.[207–210] Importantly, the combination therapy induces CD8$^+$ T cell proliferation and activation that is preferential to the tumor inhibition. In addition, major T-cell clones, expanding on therapy in responding patients, are shared between local and distant lesions indicating that priming and reactivation is of a shared antigen.[207] IMO-2125 has been granted fast track designation in combination with ipilimumab to treat anti-PD-1 refractory melanoma patients and a phase 3 clinical trial is in progress.[172,211]

In addition to the ongoing combination trial, an IM0-2125 monotherapy trial is also in progress on multiple tumor types.[212]

14.7 Conclusions

Innate immune-receptor-meditated immunity is being recognized as a therapeutic modality in clinical settings. Natural ligands for each of these receptors have been identified and the signaling pathways have been well characterized. A number of synthetic oligonucleotides comprising various nucleotide sequences, lengths, and chemical modifications based on antisense, siRNA, aptamer, and immune modulatory mechanisms are currently being evaluated in human clinical trials for the treatment of various diseases.

Since innate immune receptors that recognize nucleic acids are the key components of our immune system, it is important to develop a thorough understanding of how these receptors would affect development of oligonucleotide therapeutics. Unintended activation of these receptors by an oligonucleotide drug candidate could affect the mechanism of action but also safety-related issues. Most of the phosphorothioate antisense oligonucleotides developed in the field failed largely due to unintended activation of TLR9 by the presence of CpG motifs, thereby affecting the intended mechanism of action and safety. It is important to note that many of the receptors described above recognize patterns of foreign 'exogenous' nucleic acids. However, these receptors also recognize endogenous nucleic acids, as is the case in many autoimmune diseases.[213] This is important in the context of oligonucleotide drug candidates with increased stability, which after repeated administrations, would result in increased retention and concentration in tissues, thereby creating conditions for unintended engagement of these receptors. This may lead to delayed inflammatory signal-mediated safety issues during chronic dosing.

Oligonucleotide drug candidates designed to recognize these receptors provide a very compelling approach to trigger receptor-mediated immune responses for the treatment of a broad range of diseases. Extensive studies on structure–activity relationships have provided insights into creating novel optimized agonists for nucleic acid-recognizing endosomal TLRs. Oligonucleotides targeting these receptors have shown encouraging results in preclinical models of infectious and respiratory diseases, cancer, and use as vaccine adjuvants. Recently Heplisav B, a hepatitis B vaccine containing a TLR9 agonist has been approved as an adjuvant.[103] Early data on intratumoral application of TLR9 agonists in immunotherapy of cancer are very encouraging, and a phase 3 trial is ongoing with IMO-2125.[211] This is the beginning of new era of application of oligonucleotides as innate immune receptor modulators and therapeutics. We have come a long way from observing an unintended immune activation in humans by a synthetic oligonucleotide containing CpG motif (GEM 91)[88,91] to now employing the mechanism for therapeutics.

References

1. C. A. Janeway Jr and R. Medzhitov, *Annu. Rev. Immunol.*, 2002, **20**, 197.
2. T. Kawai and S. Akira, *Semin. Immunol.*, 2007, **19**, 24.
3. A. Iwasaki and R. Medzhitov, *Science*, 2010, **327**, 291.
4. M. Yoneyama, M. Kikuchi, T. Natsukawa, N. Shinobu, T. Imaizumi, M. Miyagishi, K. Taira, S. Akira and T. Fujita, *Nat. Immunol.*, 2004, **5**, 730.
5. M. Yoneyama, M. Kikuchi, K. Matsumoto, T. Imaizumi, M. Miyagishi, K. Taira, E. Foy, Y. M. Loo, M. Gale Jr, S. Akira, S. Yonehara, A. Kato and T. Fujita, *J. Immunol.*, 2005, **175**, 2851.
6. V. Hornung, J. Ellegast, S. Kim, K. Brzózka, A. Jung, H. Kato, H. Poeck, S. Akira, K. K. Conzelmann, M. Schlee, S. Endres and G. Hartmann, *Science*, 2006, **314**, 994.
7. A. Baum, R. Sachidanandam and A. García-Sastre, *Proc. Natl. Acad. Sci. U. S. A.*, 2010, **107**, 16303.
8. H. Kato, O. Takeuchi, S. Sato, M. Yoneyama, M. Yamamoto, K. Matsui, S. Uematsu, A. Jung, T. Kawai, K. J. Ishii, O. Yamaguchi, K. Otsu, T. Tsujimura, C. S. Koh, C. Reis e Sousa, Y. Matsuura, T. Fujita and S. Akira, *Nature*, 2006, **441**, 101.
9. Y.-M. Loo and M. Gale Jr, *Immunity*, 2011, **34**, 680.
10. T. Fernades-Alnemri, J.-W. Yu, P. Datta, J. Wu and E. S. Alnemri, *Nature*, 2009, **458**, 509.
11. V. Hornung, A. Ablasser, M. Charrel-Dennis, F. Bauernfeind, G. Horvath, D. R. Caffrey, E. Latz and K. A. Fitzgerald, *Nature*, 2009, **458**, 514.
12. T. L. Roberts, A. Idris, J. A. Dunn, G. M. Kelly, C. M. Burnton, S. Hodgson, L. L. Hardy, V. Garceau, M. J. Sweet, I. L. Ross, D. A. Hume and K. J. Stacey, *Science*, 2009, **323**, 1057.

13. S. A. Schattgen and K. K. Fitzgerald, *Immunol. Rev.*, 2011, **243**, 109.
14. V. A. Rathinam, S. K. Vanaja and K. A. Fitzgerald, Nat Immunol, 2012, **13**, 333.
15. J. D. Hansen, L. N. Vojtech and K. J. Laing, *Dev. Comp. Immunol.*, 2011, **35**, 886.
16. K. Nakahira, J. A. Haspel, V. A. Rathinam, S. J. Lee, T. Dolinay, H. C. Lam, J. A. Englert, M. Rabinovitch, M. Cernadas, H. P. Kim, K. A. Fitzgerald, S. W. Ryter and A. M. Choi, *Nat. Immunol.*, 2011, **12**, 222.
17. K. Shimada, T. R. Crother, J. Karlin, J. Dagvadorj, N. Chiba, S. Chen, V. K. Ramanujan, A. J. Wolf, L. Vergnes, D. M. Ojcius, A. Rentsendorj, M. Vargas, C. Guerrero, Y. Wang, K. A. Fitzgerald, D. M. Underhill, T. Town and M. Arditi, *Immunity*, 2012, **36**, 401.
18. H. Ishikawa and G. N. Barber, *Nature*, 2008, **455**, 674.
19. S. Ouyang, X. Song, Y. Wang, H. Ru, N. Shaw, Y. Jiang, F. Niu, Y. Zhu, W. Qiu, K. Parvatiyar, Y. Li, R. Zhang, G. Cheng and Z. J. Liu, *Immunity*, 2012, **36**, 1073.
20. H. Ishikawa, Z. Ma and G. N. Barbe, *Nature*, 2009, **461**, 788.
21. L. Sun, J. Wu, F. Du, X. Chen and Z. J. Chen, *Science*, 2013, **339**, 786.
22. J. Wu, L. Sun, X. Chen, F. Du, H. Shi, C. Chen and Z. J. Chen, *Science*, 2013, **339**, 826.
23. F. Schwede, H. G. Genieser and A. Rentsch, *Handb. Exp. Pharmacol.*, 2017, **238**, 359.
24. L. A. O'Neill, D. Golenbock and A. G. Bowie, *Nat. Rev. Immunol.*, 2013, **13**, 453.
25. L. Alexopoulou, A. C. Holt, R. Medzhitov and R. A. Flavell, *Nature*, 2001, **413**, 732.
26. F. Heil, H. Hemmi, H. Hochrein, F. Ampenberger, C. Kirschning, S. Akira, G. Lipford, H. Wagner and S. Bauer, *Science*, 2004, **303**, 1526.
27. S. S. Diebold, T. Kaisho, H. Hemmi, S. Akira and C. Reis e Sousa, *Science*, 2004, **303**, 1529.
28. H. Hemmi, T. Kaisho, O. Takeuchi, S. Sato, H. Sanjo, K. Hoshino, T. Horiuchi, H. Tomizawa, K. Takeda and S. Akira, *Nat. Immunol.*, 2002, **3**, 196.
29. J. Lee, T. H. Chuang, V. Redecke, L. She, P. M. Pitha, D. A. Carson, E. Raz and H. B. Cottam, *Proc. Natl. Acad. Sci. U. S. A.*, 2003, **100**, 6646.
30. H. Hemmi, O. Takeuchi, T. Kawai, S. Sato, H. Sanjo, K. Hoshino, T. Horiuchi, H. Tomizawa, K. Takeda and S. Akira, *Nature*, 2000, **408**, 740.
31. M. Oldenburg, A. Krüger, R. Ferstl, A. Kaufmann, G. Nees, A. Sigmund, B. Bathke, H. Lauterbach, M. Suter, S. Dreher, U. Koedel, S. Akira, T. Kawai, J. Buer, H. Wagner, S. Bauer, H. Hochrein and C. J. Kirschning, *Science*, 2012, **337**, 1111.
32. M. V. Lasker and S. K. Nair, *J. Immunol.*, 2006, **177**, 11.
33. I. Botos, D. M. Segal and D. R. Davies, *Structure*, 2011, **19**, 447.
34. A. Bowie and L. A. O'Neill, *J. Leukocyte Biol.*, 2000, **67**, 508.
35. E. F. Kenny and L. A. O'Neill, *Cytokine*, 2008, **43**, 342.
36. J. C. Kagan, V. G. Magupalli and H. Wu, *Nat. Rev. Immunol.*, 2014, **14**, 821.

37. M. Matsumoto, K. Funami, M. Tanabe, H. Oshiumi, M. Shingai, Y. Seto, A. Yamamoto and T. Seya, *J. Immunol.*, 2003, **171**, 3154.
38. K. N. Schmidt, B. Leung, M. Kwong, K. A. Zarember, S. Satyal, T. A. Navas, F. Wang and P. J. Godowski, *J. Immunol.*, 2004, **172**, 138.
39. M. Yamamoto, S. Sato, H. Hemmi, K. Hoshino, T. Kaisho, H. Sanjo, O. Takeuchi, M. Sugiyama, M. Okabe, K. Takeda and S. Akira, *Science*, 2003, **301**, 640.
40. J. Brown, H. Wang, G. N. Hajishengallis and M. Martin, *J. Dent. Res.*, 2011, **90**, 417–427.
41. L. Liu, I. Botos, Y. Wang, J. N. Leonard, J. Shiloach, D. M. Segal and D. R. Davies, *Science*, 2008, **320**, 379.
42. C. Stahl-Hennig, M. Eisenblätter, E. Jasny, T. Rzehak, K. Tenner-Racz, C. Trumpfheller, A. M. Salazar, K. Uberla, K. Nieto, J. Kleinschmidt, R. Schulte, L. Gissmann, M. Müller, A. Sacher, P. Racz, R. M. Steinman, M. Uguccioni and R. Ignatius, *PLoS Pathog.*, 2009, **5**, e1000373.
43. M. S. Boukhvalova, T. B. Sotomayor, R. C. Point, L. M. Pletneva, G. A. Prince and J. C. Blanco, *J. Interferon Cytokine Res.*, 2010, **30**, 229.
44. D. D. Silva, M. Koopman, H. Brand, G. McKee and W. M. Kast, *J. Immunol.*, 2011, **186**, 53.
45. Y. Lau, L. Tang, A. McCall, E. Ooi and K. Subbarao, *J. Virol.*, 2010, **84**, 8639.
46. V. Hornung, W. Barchet, M. Schlee and G. Hartmann, *Handb. Exp. Pharmacol.*, 2008, **183**, 71.
47. M. E. Kleinman, K. Yamada, A. Takeda, V. Chandrasekaran, M. Nozaki, J. Z. Baffi, R. J. Albuquerque, S. Yamasaki, M. Itaya, Y. Pan, B. Appukuttan, D. Gibbs, Z. Yang, K. Karikó, B. K. Ambati, T. A. Wilgus, L. A. DiPietro, E. Sakurai, K. Zhang, J. R. Smith, E. W. Taylor and J. Ambati, *Nature*, 2008, **452**, 591.
48. S. Palchetti, D. Starace, P. De Cesaris, A. Filippini, E. Ziparo and A. Riccioli, *J. Biol. Chem.*, 2015, **290**, 5470.
49. M. Yu and S. J. Levine, *Cytokine Growth Factor Rev.*, 2011, **22**, 63.
50. T. M. Glavan and J. Pavelic, *Curr. Pharm. Des.*, 2014, **20**, 6555.
51. K. A. Martins, S. Bavari and A. M. Salazar, *Expert Rev. Vaccines*, 2015, **14**, 447.
52. W. M. Mitchell, *Expert Rev. Clin. Pharmacol.*, 2016, **9**, 755.
53. E. De Clercq, *Acta Microbiol. Immunol. Hung.*, 2005, **52**, 273.
54. M. Matsumoto and T. Seya, *Adv. Drug Delivery Rev.*, 2008, **60**, 805.
55. E. R. Homan, R. P. Zendzian, L. D. Schott, H. B. Levy and R. H. Adamson, *Toxicol. Appl. Pharmacol.*, 1972, **23**, 579.
56. R. A. Robinson, V. T. DeVita, H. B. Levy, S. Baron, S. Hubbard and A. S. Levine, *J. Natl. Cancer Inst.*, 1976, **57**, 599.
57. T. Lan, D. Wang, L. Bhagat, V. J. Philbin, D. Yu, J. X. Tang, M. R. Putta, T. Sullivan, N. La Monica, E. R. Kandimalla and S. Agrawal, *Org. Biomol. Chem.*, 2013, **11**, 1049.
58. S. S. Diebold, *Adv. Drug Delivery Rev.*, 2008, **60**, 813.
59. J. M. Lund, L. Alexopoulou, A. Sato, M. Karow, N. C. Adams, N. W. Gale, A. Iwasaki and R. A. Flavell, *Proc. Natl. Acad. Sci. U. S. A.*, 2004, **101**, 5598.

60. E. Hanna, R. Abadi and O. Abbas, *Int. J. Dermatol.*, 2016, **55**, 831.
61. P. Savage, V. Horton, J. Moore, M. Owens, P. Witt and M. E. Gore, *Br. J. Cancer*, 1996, **74**, 1482.
62. M. Gilliet, C. Conrad, M. Geiges, A. Cozzio, W. Thürlimann, G. Burg, F. O. Nestle and R. Dummer, *Arch. Dermatol.*, 2004, **140**, 1490.
63. J. P. Vasilakos and M. A. Tomai, *Expert Rev. Vaccines*, 2013, **12**, 809.
64. S. A. Patil, S. A. Patil, R. Patil and R. Hahizume, *Mini-Rev. Med. Chem.*, 2016, **16**, 309.
65. H. Tanji, U. Ohto, T. Shibata, K. Miyake and T. Shimizu, *Science*, 2013, **339**, 1426.
66. M. P. Schon, M. Schon and K. N. Klotz, *J. Invest. Dermatol.*, 2006, **126**, 1338.
67. I. S. Zagon, R. N. Donahue, M. Rogosnitzky and P. J. McLaughlin, *Exp. Biol. Med.*, 2008, **233**, 968.
68. A. D. Judge, V. Sood, J. R. Shaw, D. Fang, K. McClintock and I. MacLachlan, *Nat. Biotechnol.*, 2005, **23**, 457.
69. J. T. Marques and B. R. Williams, *Nat. Biotechnol.*, 2005, **23**, 1399.
70. M. Sioud, *J. Mol. Biol.*, 2005, **348**, 1079.
71. S. Dow, *Expert Opin. Drug Delivery*, 2008, **5**, 11.
72. T. Lan, E. R. Kandimalla, D. Yu, L. Bhagat, Y. Li, D. Wang, F. G. Zhu, J. X. Tang, M. R. Putta, Y. P. Cong, A. F. Trombino, T. Sullivan and S. Agrawal, *Proc. Natl. Acad. Sci. U. S. A.*, 2007, **104**, 13750.
73. S. Agrawal and E. R. Kandimalla, *Biochem. Soc. Trans.*, 2007, **35**, 1461.
74. T. Lan, M. Dai, D. Wang, F. G. Zhu, E. R. Kandimalla and S. Agrawal, *J. Med. Chem.*, 2009, **52**, 6871.
75. E. R. Kandimalla, M. Struthers, A. Bett, T. Wisniewski, S. Dubey, W. Jiang, M. Precopio, Z. Sun, H. Wang, T. Lan, S. Agrawal and D. R. Casimiro, *Cell. Immunol.*, 2011, **270**, 126.
76. M. Colonna, A. Krug and M. Cella, *Curr. Opin. Immunol.*, 2002, **14**, 373.
77. T. Lan, L. Bhagat, D. Wang, M. Dai, E. R. Kandimalla and S. Agrawal, *Bioorg. Med. Chem. Lett.*, 2009, **19**, 2044.
78. T. Lan, M. R. Putta, D. Wang, M. Dai, D. Yu, E. R. Kandimalla and S. Agrawal, *Biochem. Biophys. Res. Commun.*, 2009, **386**, 443.
79. D. Wang, M. Precopio, T. Lan, D. Yu, J. X. Tang, E. R. Kandimalla and S. Agrawal, *Mol. Cancer Ther.*, 2010, **9**, 1788.
80. S. Shimada, O. Yano, H. Inoue, E. Kuramoto, T. Fukuda, H. Yamamoto, T. Kataoka and T. Tokunaga, *J. Natl. Cancer Inst.*, 1985, **74**, 681.
81. T. Tokunaga, H. Yamamoto, S. Shimada, H. Abe, T. Fukuda, Y. Fujisawa, Y. Furutani, O. Yano, T. Kataoka and T. Sudo, *J. Natl. Cancer Inst.*, 1984, **72**, 955.
82. S. Shimada, O. Yano and T. Tokunaga, *Jpn. J. Cancer Res.*, 1986, **77**, 808.
83. S. Yamamoto, E. Kuramoto, S. Shimada and T. Tokunaga, *Jpn. J. Cancer Res.*, 1988, **79**, 866.
84. J. P. Messina, G. S. Gilkeson and D. S. Pisetsky, *J. Immunol.*, 1991, **147**, 1759.

85. T. Tokunaga, O. Yano, E. Kuramoto, Y. Kimura, T. Yamamoto, T. Kataoka and S. Yamamoto, *Microbiol. Immunol.*, 1992, **36**, 55.
86. S. Yamamoto, T. Yamamoto, T. Kataoka, E. Kuramoto, O. Yano and T. Tokunaga, *J. Immunol.*, 1992, **148**, 4072.
87. A. M. Krieg, A. K. Yi, S. Matson, T. J. Waldschmidt, G. A. Bishop, R. Teasdale, G. A. Koretzky and D. M. Klinman, *Nature*, 1995, **374**, 546.
88. S. Agrawal and E. R. Kandimalla, *Nat. Biotechnol.*, 2004, **22**, 1533.
89. O. Equils, M. L. Schito, H. Karahashi, Z. Madak, A. Yarali, K. S. Michelsen, A. Sher and M. Arditi, *J. Immunol.*, 2003, **170**, 5159.
90. S. Agrawal and R. R. Martin, *J. Immunol.*, 2003, **171**, 1621.
91. S. Agrawal and Q. Zhao, *Curr. Opin. Chem. Biol.*, 1998, **2**, 519.
92. T. Tanaka, C. C. Chu and W. E. Paul, *J. Exp. Med.*, 1992, **175**, 597.
93. R. F. Branda, A. L. Moore, L. Mathews, J. J. McCormack and G. Zon, *Biochem. Pharmacol.*, 1993, **45**, 2037.
94. K. W. McIntyre, K. Lombard-Gillooly, J. R. Perez, C. Kunsch, U. M. Sarmiento, J. D. Larigan, K. T. Landreth and R. A. Narayanan, *Antisense Res. Dev.*, 1993, **3**, 309.
95. C. F. Mojcik, M. F. Gourley, D. M. Klinman, A. M. Krieg, F. Gmelig-Meyling and A. D. Steinberg, *Clin. Immunol. Immunopathol.*, 1993, **67**, 130.
96. S. Agrawal and E. R. Kandimalla, *Mol. Med. Today*, 2000, **6**, 72.
97. D. Verthelyi, K. J. Ishii, M. Gursel, F. Takeshita and D. M. Klinman, *J. Immunol.*, 2001, **166**, 2372.
98. A. Krug, S. Rothenfusser, V. Hornung, B. Jahrsdörfer, S. Blackwell, Z. K. Ballas, S. Endres, A. M. Krieg and G. Hartmann, *Eur. J. Immunol.*, 2001, **31**, 2154.
99. R. F. Rando and M. E. Hogan, in *Applied Antisense Oligonucleotide Technology*, ed. C. A. Stein and A. M. Krieg, New York, NY, Wiley-Liss, 1998, p. 335.
100. S. Agrawal, P. L. Iadorola, J. Temsamani, Q. Zhao and D. R. Shaw, *Bioorg. Med. Chem. Lett.*, 1996, **6**, 2219.
101. S. Agrawal, W. Tan, Q. Cai, X. Xie and R. Zhang, *Antisense Nucleic Acid Drug Dev.*, 1997, **7**, 245.
102. T. B. Casale, J. Cole, E. Beck, C. F. Vogelmeier, J. Willers, C. Lassen, A. Hammann-Haenni, L. Trokan, P. Saudan and M. E. Wechsler, *Allergy*, 2015, **70**, 1160.
103. M. Abramowicz, G. Zuccotti and J.-M. Pflomm, *J. Am. Med. Assoc.*, 2018, **319**, 822.
104. G. Hartmann, J. Battiany, H. Poeck, M. Wagner, M. Kerkmann, N. Lubenow, S. Rothenfusser and S. Endres, *Eur. J. Immunol.*, 2003, **33**, 1633.
105. J. D. Marshall, K. Fearon, C. Abbate, S. Subramanian, P. Yee, J. Gregorio, R. L. Coffman and G. Van Nest, *J. Leukocyte Biol.*, 2003, **73**, 781.
106. E. R. Kandimalla and S. Agrawal, in *Toll and Toll Receptors an Immunologic Perspective*, ed. T. Rich, Kluwer Academic/Plenum Publishers, New York, NY, 2005, pp. 181–212.
107. E. R. Kandimalla and S. Agrawal, *Adv. Polym. Sci.*, 2012, **249**, 61.

108. D. Yu, F.-G. Zhu, L. Bhagat, H. Wang, E. R. Kandimalla, R. Zhang and S. Agrawal, *Biochem. Biophys. Res. Commun.*, 2002, **297**, 83.
109. A. H. Dalpke, S. Zimmermann, I. Albrecht and K. Heeg, *Immunology*, 2002, **106**, 102.
110. S. Agrawal, J. Temsamani and J. Y. Tang, *Proc. Natl. Acad. Sci. U. S. A.*, 1991, **88**, 7595.
111. D. Yu, E. R. Kandimalla, A. Roskey, Q. Zhao, L. Chen, J. Chen and S. Agrawal, *Bioorg. Med. Chem.*, 2000, **8**, 275.
112. N. Iwamoto, D. C. D. Butler, N. Svrzikapa, S. Mohapatra, I. Zlatev, D. W. Y. Sah, Meena, S. M. Standley, G. Lu, L. H. Apponi, M. Frank-Kamenetsky, J. J. Zhang, C. Vargeese and G. L. Verdine, *Nat. Biotechnol.*, 2017, **35**, 845.
113. Q. Zhao, J. Temsamani, P. L. Iadarola, J. Ziang and S. Agrawal, *Biochem. Pharmacol.*, 1996, **51**, 173.
114. S. Agrawal and E. R. Kandimalla, *Curr. Cancer Drug Targets*, 2001, **1**, 197.
115. D. Yu, E. R. Kandimalla, Q. Zhao, Y. Cong and S. Agrawal, *Bioorg. Med. Chem.*, 2001, **9**, 2803.
116. S. Agrawal and E. R. Kandimalla, *Ann. N. Y. Acad. Sci.*, 2003, **1002**, 30.
117. E. R. Kandimalla, F. G. Zhu, L. Bhagat, D. Yu and S. Agrawal, *Biochem. Soc. Trans.*, 2003, **31**, 654.
118. M. R. Putta, L. Bhagat, D. Wang, F. G. Zhu, E. R. Kandimalla and S. Agrawal, *ACS Med. Chem. Lett.*, 2013, **4**, 302.
119. D. Yu, Q. Zhao, E. R. Kandimalla and S. Agrawal, *Bioorg. Med. Chem. Lett.*, 2000, **10**, 2585.
120. E. R. Kandimalla, L. Bhagat, D. Yu, Y. Cong, J. Tang and S. Agrawal, *Bioconjugate Chem.*, 2002, **13**, 966.
121. D. Yu, E. R. Kandimalla, L. Bhagat, J. Y. Tang, Y. Cong, J. Tang and S. Agrawal, *Nucleic Acids Res.*, 2002, **30**, 4460.
122. E. R. Kandimalla, L. Bhagat, D. Wang, D. Yu, F. G. Zhu, J. Tang, H. Wang, P. Huang, R. Zhang and S. Agrawal, *Nucleic Acids Res.*, 2003, **31**, 2393.
123. E. R. Kandimalla, L. Bhagat, Y. Li, D. Yu, D. Wang, Y. P. Cong, S. S. Song, J. X. Tang, T. Sullivan and S. Agrawal, *Proc. Natl. Acad. Sci. U. S. A.*, 2005, **102**, 6925.
124. L. Bhagat, F.-G. Zhu, D. Yu, J. Tang, H. Wang, E. R. Kandimalla, R. Zhang and S. Agrawal, *Biochem. Biophys. Res. Commun.*, 2003, **300**, 853.
125. F. G. Zhu, E. R. Kandimalla, D. Yu, J. X. Tang and S. Agrawal, *Int. Immunopharmacol.*, 2004, **4**, 851.
126. D. Wang, E. R. Kandimalla, D. Yu, J. X. Tang and S. Agrawal, *Vaccine*, 2005, **23**, 2614.
127. F. G. Zhu, E. R. Kandimalla, D. Yu and S. Agrawal, *J. Allergy Clin. Immunol.*, 2007, **120**, 631.
128. M. R. Putta, F. G. Zhu, D. Wang, L. Bhagat, M. Dai, E. R. Kandimalla and S. Agrawal, *Bioconjugate Chem.*, 2010, **21**, 39.
129. P. S. Creticos, J. T. Schroeder, R. G. Hamilton, S. L. Balcer-Whaley, A. P. Khattignavong, R. Lindblad, H. Li, R. Coffman, V. Seyfert, J. J. Eiden,

D. Broide and Immune Tolerance Network Group, *N. Engl. J. Med.*, 2006, **355**, 1445.
130. U. Ohto, T. Shibata, H. Tanji, H. Ishida, E. Krayukhina, S. Uchiyama, K. Miyake and T. Shimizu, *Nature*, 2015, **520**, 702.
131. E. Latz, A. Verma, A. Visintin, M. Gong, C. M. Sirois, D. C. Klein, B. G. Monks, C. J. McKnight, M. S. Lamphier, W. P. Duprex, T. Espevik and D. T. Golenbock, *Nat. Immunol.*, 2007, **8**, 772.
132. M. R. Putta, D. Yu and E. R. Kandimalla, *Methods Mol. Biol.*, 2011, **764**, 263.
133. D. Yu, E. R. Kandimalla, Q. Zhao, Y. Cong and S. Agrawal, *Nucleic Acids Res.*, 2002, **30**, 1613.
134. M. R. Putta, D. Yu, L. Bhagat, D. Wang, F. G. Zhu and E. R. Kandimalla, *J. Med. Chem.*, 2010, **53**, 3730.
135. D. Yu, D. Wang, F.-G. Zhu, L. Bhagat, M. Dai, E. R. Kandimalla and S. Agrawal, *J. Med. Chem.*, 2009, **52**, 5108.
136. E. R. Kandimalla, D. Yu, Q. Zhao and S. Agrawal, *Bioorg. Med. Chem.*, 2001, **9**, 807.
137. M. R. Putta, F. Zhu, Y. Li, L. Bhagat, Y. P. Cong, E. R. Kandimalla and S. Agrawal, *Nucleic Acids Res.*, 2006, **34**, 3231.
138. D. Yu, M. R. Putta, L. Bhagat, Y. Li, F. G. Zhu, D. Wang, J. X. Tang, E. R. Kandimalla and S. Agrawal, *J. Med. Chem.*, 2007, **50**, 6411.
139. M. Struthers, A. Bett, T. Wisniewski, S. Dubey, M. Precopio, W. Jiang, Z. Sun, H. Wang, I. Nowak, M. R. Putta, D. Yu, J. X. Tang, E. R. Kandimalla, S. Agrawal and D. R. Casimiro, *Cell. Immunol.*, 2010, **263**, 105.
140. E. R. Kandimalla, L. Bhagat, F. G. Zhu, D. Yu, Y. P. Cong, D. Wang, J. X. Tang, J. Y. Tang, C. F. Knetter, E. Lien and S. Agrawal, *Proc. Natl. Acad. Sci. U. S. A.*, 2003, **100**, 14303.
141. D. K. Agrawal, J. Edwan, E. R. Kandimalla, D. Yu, L. Bhagat, D. Wang and S. Agrawal, *Int. Immunopharmacol.*, 2004, **4**, 127.
142. Q. Zhao, D. Yu and S. Agrawal, *Bioorg. Med. Chem. Lett.*, 1999, **9**, 3453.
143. Q. Zhao, D. Yu and S. Agrawal, *Bioorg. Med. Chem. Lett.*, 2000, **10**, 1051.
144. D. Yu, E. R. Kandimalla, Y. Cong, J. Tang, J. Y. Tang, Q. Zhao and S. Agrawal, *J. Med. Chem.*, 2002, **45**, 4540.
145. D. Yu, E. R. Kandimalla, Q. Zhao, Y. Cong and S. Agrawal, *Bioorg. Med. Chem. Lett.*, 2001, **11**, 2263.
146. D. Yu, E. R. Kandimalla, Q. Zhao, L. Bhagat, Y. Cong and S. Agrawal, *Bioorg. Med. Chem.*, 2003, **11**, 459.
147. Y. P. Cong, S. S. Song, L. Bhagat, R. K. Pandey, D. Yu, E. R. Kandimalla and S. Agrawal, *Biochem. Biophys. Res. Commun.*, 2003, **310**, 1133.
148. E. R. Kandimalla, L. Bhagat, Y. P. Cong, R. K. Pandey, D. Yu, Q. Zhao and S. Agrawal, *Biochem. Biophys. Res. Commun.*, 2003, **306**, 948.
149. K. Kapp, C. Kleuss, M. Schroff and B. Wittig, *Mol. Ther.--Nucleic Acids*, 2014, **3**, e270.
150. J. Vollmer, R. Weeratna, P. Payette, M. Jurk, C. Schetter, M. Laucht, T. Wader, S. Tluk, M. Liu, H. L. Davis and A. M. Krieg, *Eur. J. Immunol.*, 2004, **34**, 251.

151. D. Yu, M. R. Putta, L. Bhagat, M. Dai, D. Wang, A. F. Trombino, T. Sullivan, E. R. Kandimalla and S. Agrawal, *Antimicrob. Agents Chemother.*, 2008, **52**, 4320.
152. J. G. Mchutchison, B. R. Bacon, S. C. Gordon, E. Lawitz, M. Shiffman, N. H. Afdhal, I. M. Jacobson, A. Muir, M. Al Adhami, M. L. Morris, J. A. Lekstrom-Himes, S. M. Efler and H. L. Davis, *Hepatology*, 2007, **46**, 1341.
153. A. P. Vicari, T. Schmalbach, J. Lekstrom-Himes, M. L. Morris, M. J. Al-Adhami, C. Laframboise, P. Leese, A. M. Krieg, S. M. Efler and H. L. Davis, *Antiviral Ther.*, 2007, **12**, 741.
154. H. Wang, E. R. Rayburn, W. Wang, E. R. Kandimalla, S. Agrawal and R. Zhang, *Mol. Cancer Ther.*, 2006, **5**, 1585.
155. D. Wang, Y. Li, D. Yu, S. S. Song, E. R. Kandimalla and S. Agrawal, *Int. J. Oncol.*, 2004, **24**, 901.
156. D. Melisi, M. Frizziero, A. Tamburrino, M. Zanotto, C. Carbone, G. Piro and G. Tortora, *Biomedicines*, 2014, **2**, 211.
157. H. Wang, E. R. Rayburn, W. Wang, E. R. Kandimalla, S. Agrawal and R. Zhang, *Mol. Cancer Ther.*, 2006, **5**, 2106.
158. V. Damiano, R. Caputo, R. Bianco, F. P. D'Armiento, A. Leonardi, S. De Placido, A. R. Bianco, S. Agrawal, F. Ciardiello and G. Tortora, *Clin. Cancer Res.*, 2006, **12**, 577.
159. V. Damiano, R. Caputo, S. Garofalo, R. Bianco, R. Rosa, L. Racioppi, G. Merola, S. DePlacido, G. Fontanini, E. R. Kandimalla, S. Agrawal, F. Ciardiello and G. Tortora, *Proc. Natl. Acad. Sci. U. S. A.*, 2007, **104**, 12468.
160. V. Damiano, S. Garofalo, R. Rosa, R. Bianco, R. Caputo, T. Gelardi, G. Merola, L. Racioppi, C. Garbi, E. R. Kandimalla, S. Agrawal and G. Tortora, *Clin. Cancer Res.*, 2009, **15**, 6921.
161. H. Zhang, L. Liu, D. Yu, E. R. Kandimalla, H. B. Sun, S. Agrawal and C. Guha, *PLoS One*, 2012, **7**, e38111.
162. S. Saha, P. Bhanja, L. Liu, A. A. Alfieri, D. Yu, E. R. Kandimalla, S. Agrawal and C. Guha, *PLoS One*, 2012, **7**, e29357.
163. M. D. Hellmann, C. F. Friedman and J. D. Wolchok, *Adv. Immunol.*, 2016, **130**, 251.
164. E. J. Lipson, P. M. Forde, H. J. Hammers, L. A. Emens, J. M. Taube and S. L. Topalian, *Semin. Oncol.*, 2015, **42**, 587.
165. S. Wang, J. Campos, M. Gallotta, M. Gong, C. Crain, E. Naik, R. L. Coffman and C. Guiducci, *Proc. Natl. Acad. Sci. U. S. A.*, 2016, **113**, E7240.
166. F. Sato-Kaneko, S. Yao, A. Ahmadi, S. S. Zhang, T. Hosoya, M. M. Kaneda, J. A. Varner, M. Pu, K. S. Messer, C. Guiducci, R. L. Coffman, K. Kitaura, T. Matsutani, R. Suzuki, D. A. Carson, T. Hayashi and E. E. Cohen, *JCI Insight*, 2017, **2**, e93397.
167. D. Wang, W. Jiang, F. Zhu, X. Mao and S. Agrawal, *Int. J. Oncol.*, 2018, DOI: 10.3892/ijo.2018.4456.
168. D. Wang, F.-G. Zhu, X. Mao and S. Agrawal, *Cancer Immunol. Res.*, 2016, **4**(1 suppl.), abstract nr B094.
169. W. Jiang, D. Wang, F.-G. Zhu, L. Bhagat, J. DiMuzio and S. Agrawal, *Cancer Immunol. Res.*, 2016, **4**(1 suppl.), abstract nr B159.

170. D. Wang, W. Jiang, B. Lakshmi, J. DiMuzio, F.-G. Zhu and S. Agrawal, *Cancer Res.*, 2016, **76**(14 suppl.), abstract nr 3847.
171. D. Wang, F.-G. Zhu, J. DiMuzio and S. Agrawal, *Mol. Cancer Ther.*, 2015, **14**(12 suppl. 2), abstract nr B196.
172. E. K. Argon, F.-G. Zhu, S. Agrawal, J. Yingling and D. Wang, *Poster Presentation at AACR Annual Meeting 2018*, Chicago, IL, April 14-18, 2018, abstract no. 4704.
173. D. Wang, X. Mao, E. K. Argon, F.-G. Zhu and S. Agrawal, *Poster Presented at the Third CRI-CIMT-EATI-AACR International Cancer Immunotherapy Conference, Mainz, Germany*, September 6-9, 2017.
174. D. Wang, E. K. Argon, F.-G. Zhu and S. Agarwal, *Cancer Res.*, 2017, **77**(13 suppl.), abstract nr 5659.
175. C. Duez, P. Gosset and A. B. Tonnel, *Eur. J. Dermatol.*, 2006, **16**, 12.
176. S. Farrokhi, N. Abbasirad, A. Movahed, H. A. Khazaei, M. Pishjoo and N. Rezaei, *Immunotherapy*, 2017, **9**, 339.
177. F. G. Zhu, E. R. Kandimalla, D. Yu, J. X. Tang and S. Agrawal, *Int. Immunopharmacol.*, 2004, **4**, 851.
178. C. S. Ebert Jr, A. S. Rose, M. R. Patel, S. M. Hardy, E. R. Kandimalla, S. Agrawal, J. Prazma and H. C. Pillsbury 3rd, *Int. J. Pediatr. Otorhinolaryngol.*, 2006, **70**, 2019.
179. C. S. Ebert Jr, A. S. Rose, D. A. Blanks, R. P. Eapen, E. R. Kandimalla, S. Agrawal and J. Prazma, *Otolaryngol.--Head Neck Surg.*, 2007, **137**, 250.
180. D. A. Blanks, C. S. Ebert Jr, R. P. Eapen, E. R. Kandimalla, S. Agrawal and J. Prazma, *Otolaryngol.--Head Neck Surg.*, 2007, **137**, 321.
181. G. Dubois, J. Simmons, E. Martin, *et al*, *Presented at Toll 2008: Recent Advances in Pattern Recognition, Lisbon, Portugal*, September 24–27.
182. M. L. Mbow and R. T. Sarisky, *Drug News Perspect.*, 2005, **18**, 179.
183. J. Scheiermann and D. M. Klinman, *Vaccine*, 2014, **32**, 6377.
184. Y. Li, E. R. Kandimalla, D. Yu, J. X. Tang and S. Agrawal, *Int. Immunopharmacol.*, 2005, **5**, 981.
185. G. Swaminathan, E. A. Thoryk, K. S. Cox, S. Meschino, S. A. Dubey, K. A. Vora, R. Celano, M. Gindy, D. R. Casimiro and A. J. Bett, *Vaccine*, 2016, **34**, 110.
186. D. Trabattoni, A. Clivio, D. H. Bray, L. Bhagat, S. Beltrami, G. Maffeis, E. Cesana, P. Lowery, F. Lissoni, E. R. Kandimalla, T. Sullivan, S. Agrawal, R. Bartholomew and M. Clerici, *Vaccine*, 2006, **24**, 1470.
187. D. DiStefano, J. M. Antonello, A. J. Bett, M. B. Medi, D. R. Casimiro and J. ter Meulen, *Vaccine*, 2013, **31**, 4888.
188. G. Swaminathan, E. A. Thoryk, K. S. Cox, J. S. Smith, J. J. Wolf, M. E. Gindy, D. R. Casimiro and A. J. Bett, *Sci. Rep.*, 2016, **6**, 34215.
189. L. Aurisicchio, D. Peruzzi, A. Conforti, S. Dharmapuri, A. Biondo, S. Giampaoli, A. Fridman, A. Bagchi, C. T. Winkelmann, R. Gibson, E. R. Kandimalla, S. Agrawal, G. Ciliberto and N. LaMonica, *Clin. Cancer Res.*, 2009, **15**, 1575.
190. S. Dharmapuri, D. Peruzzi, C. Mennuni, F. Calvaruso, S. Giampaoli, G. Barbato, E. R. Kandimalla, S. Agrawal, E. Scarselli, G. Mesiti, G. Ciliberto, N. LaMonica and L. Aurisicchio, *Mol. Ther.*, 2009, **17**, 1804.

191. A. Conforti, B. Cipriani, D. Peruzzi, S. Dharmapuri, E. R. Kandimalla, S. Agrawal, F. Mori, G. Ciliberto, N. La Monica and L. Aurisicchio, *Vaccine*, 2010, **28**, 3520.
192. A. Muir, R. Ghalib, E. Lawitz, K. Patel, M. Rodriguez-Torres, A. Sheikh, S. Sapp, R. Taylor, A. Bexon, T. Sullivan, A. Dovholuk and J. McHutchison, *J. Hepatol.*, 2010, **52**, S14, abstract # 32.
193. *Study of Combination Treatment with IMO-2125 and Ribavirin in Naïve Hepatitis C-infected, Genotype 1 Patients*, NCT00990938.
194. *Placebo-controlled, Dose-escalation Study of the Safety of IMO-2125 in Hepatitis C-infected Patients*, NCT00728936.
195. J. Hwang, E. Fox-Sinclair, A. Bahrani, *et al.*, *Presented at the AACR-NCI-EORTC International Conference on Molecular Targets and Cancer Therapeutics, November 17-21*, Boston, MA, 2003, abstract # C111.
196. R. Martin, T. Sullivan, L. Bhagat, *et al*, *Presented at the AACR-NCI-EORTC International Conference on Molecular Targets and Cancer Therapeutics*, November 17-21, 2003, Boston, MA, abstract # C100.
197. T. Sullivan, L. Bhagat, E. Kandimalla *et al*, *Presented at the 2004 AACR Annual Meeting*, Orlando, Florida, 26-30 March, abstract # 4707 Proc Am Assoc Cancer Res, 2004, 45, 1087.
198. J. Hwang, S. Malik, S. Park, *et al*, *Presented at TOLL*, Taormina, Sicily, Italy, 2004, 8–11 May.
199. T. Kuzel, J. Dutcher, S. Ebbinghaus, *et al*, *Presented at the Eighth International Kidney Cancer Symposium*, Chicago, IL, September 25 and 26, 2009.
200. S. Malik, J. Hwang, I. Cotarla, T. Sullivan, R. Karr and J. Marshall, *Presented at the 12th World Conference on Lung Cancer, Seoul, Korea*, September 2-6, 2007.
201. D. A. Smith, P. Conkling, D. A. Richards, J. J. Nemunaitis, T. E. Boyd, A. C. Mita, G. de La Bourdonnaye, D. Wages and A. S. Bexon, *Cancer Immunol. Immunother.*, 2014, **63**, 787.
202. J. P. Michiels, M. C. Kaminsky, U. Keller, T. H. Brümmendorf, T. Goddemeier, U. Forssmann and J. P. Delord, *Invest. New Drugs*, 2013, **31**, 1207.
203. A. Ruzsa, M. Sen, M. Evans, L. W. Lee, K. Hideghety, S. Rottey, P. Klimak, P. Holeckova, J. Fayette, T. Csoszi, J. Erfan, U. Forssmann, T. Goddemeier, A. Bexon, C. Nutting and NA EMD 1201081 Study Group, *Invest. New Drugs*, 2014, **32**, 1278.
204. E. Chan, E. L. Kwak, J. Hwang, M. Heiskala, G. de La Bourdonnaye and M. Mita, *Cancer Chemother. Pharmacol.*, 2015, **75**, 701.
205. V. Hirsh, L. Paz-Ares, M. Boyer, R. Rosell, G. Middleton, W. E. E. Eberhardt, A. Szczesna, P. Reiterer, M. Saleh, O. Arrieta, E. Bajetta, R. T. Webb, J. Raats, R. J. Benner, C. Fowst, S. J. Meech, D. Readett and J. H. Schiller, *J. Clin. Oncol.*, 2011, **29**, 2667.
206. M. Guha, *Nat. Rev. Drug Discovery*, 2012, **11**, 503.
207. C. Haymaker, *Presentation at SITC 2017 Annual Meeting*, National Harbor, MD, November 8-12, 2017.

208. C. Haymaker, M. Uemura, R. Murthy, M. James, D. Wang, J. Brevard, S. Swann, J. Geib, M. Cornfeld, S. Chunduru, S. Agrawal, C. Yee, J. Wargo, R. Amaria, S. Patel, H. Tawbi, I. Glitza, S. Woodman, W.-J. Hwu, M. A. Davies, P. Hwu, W. Overwjik, C. Bernatchez and A. Diab, *Cancer Res.*, 2017, 77(13 suppl.), abstract nr 5652.
209. M. Uemura, C. L. Haymaker, R. Murthy, M. James, J. Geib, S. Swann, K. Lipford, C. Yee, J. Wargo, R. Amaria, S. Patel, H. Tawbi, I. Glitza, S. Woodman, W. Hwu, M. A. Davies, P. Hwu, W. Overwijk, C. Bernatchez and A. Diab, *Presented at 2017 ASCO-SITC Clinical Immuno-Oncology Symposium*, Orlando, FL, February 23-25, 2017.
210. A. Diab, C. Haymaker, W.-J. Hwu, M. Uemura, R. Murthy, M. James, J. Geib, M. Cornfeld, S. Swann, K. Lipford, S. Chunduru, R. Amaria, S. Patel, H. Tawbi, I. Glitza, M. A. Davies, W. Overwijk, C. Bernatchez and P. Hwu, *Presented at the European Society for Medical Oncology Congress*, September, 2017.
211. *A Study of IMO-2125 in Combination with Ipilimumab versus Ipilimumab Alone in Subjects with Anti-PD-1 Refractory Melanoma (ILLUMINATE 301); NCT03445533.*
212. *A Study of Intratumoral IMO-2125 in Patients with Refractory Solid Tumors (ILLUMINATE 101); NCT03052205 and A study of IMO-2125 in combination with ipilimumab versus ipilimumab alone in subjects with anti-PD-1 refractory melanoma (ILLUMINATE-301); NCT03445533.*
213. A. Marshak-Rothstein and I. R. Rifkin, *Annu. Rev. Immunol.*, 2007, **25**, 419.

CHAPTER 15

Prostate-specific Membrane Antigen (PSMA) Aptamers for Prostate Cancer Imaging and Therapy

SHAMBHAVI SHUBHAM[a], LI-HSIEN LIN[a], OFONIME UDOFOT[a,e], SVEN KRUPSE[a] AND PALOMA H. GIANGRANDE[a,b,c,d,e,f,g]*

[a]Department of Internal Medicine, University of Iowa, Iowa City, IA, 52242, USA; [b]Department of Radiation Oncology University of Iowa, Iowa City, IA, 52242, USA; [c]Molecular & Cellular Biology Program, University of Iowa, Iowa City, IA, 52242, USA; [d]Medical Scientist Training Program, University of Iowa, Iowa City, IA 52242, USA; [e]Abboud Cardiovascular Research Center, University of Iowa, Iowa City, IA 52242, USA; [f]Holden Comprehensive Cancer Center, University of Iowa, Iowa City, IA, USA; [g]Environmental Health Sciences Research Center (EHSRC), University of Iowa, Iowa City, IA, USA
*E-mail: paloma-giangrande@uiowa.edu

15.1 Introduction

15.1.1 Aptamers and SELEX

Aptamers are short, highly structured, synthetic DNA or RNA oligonucleotides that bind to their cognate targets with high affinity and specificity (Figure 15.1A). They are identified by an iterative *in vitro* selection process

Drug Discovery Series No. 68
Advances in Nucleic Acid Therapeutics
Edited by Sudhir Agrawal and Michael J. Gait

Published by the Royal Society of Chemistry, www.rsc.org

termed SELEX (systematic evolution of ligands by exponential enrichment) and were first described in 1990 by Ellington and Szostak and by Tuerk and Gold (Figure 15.1B).[1,2] The SELEX process begins with the generation of a randomized oligonucleotide (DNA or RNA) aptamer library ($\sim 10^{15}$ sequences), followed by incubation of the library with a target of interest. Those aptamers within the library that bind to the target are retained, while the non-binders are removed by stringent washes. Target-bound aptamers are subsequently amplified using polymerase chain reaction (PCR) (for DNA aptamer selection) or reverse transcription-polymerase chain reaction (RT-PCR) (for RNA aptamer selection). Typically, a minimum of 8 to 15 rounds of selection (including counter selections performed against control targets to eliminate non-specific binding sequences) are carried out to obtain individual aptamers that bind to their target with high affinity and specificity. During the early years of SELEX, individual candidate aptamer sequences were identified by cloning and sequencing the last round of selection. This method, however, is labor-intensive and translates in the analysis of a limited number of individual aptamer sequences (50 to 100 at best). Fortunately, in recent years, there have been major advancements in the SELEX process that have streamlined the aptamer selection process.[3] Notably, the combination of next-generation sequencing (NGS) and bioinformatics with the SELEX protocol have led to the effective development and thorough screening of aptamers against a vast repertoire of target molecules.[4–9] These technologies and their application to SELEX are thoroughly described in the following sections.[3,10]

Due to their intrinsic structural malleability, aptamers conform to the shape of the target molecule. Aptamer–target interactions involve hydrogen bonding, stacking of aromatic rings, electrostatic and hydrophobic interactions, and van der Waals forces.[11] These molecular interactions and forces impart to aptamers properties such as high binding affinity [picomolar to nanomolar range dissociation constants (Kds)] and selectivity to their cognate target molecules. Given their binding properties, aptamers are often compared to antibodies. However, aptamers offer notable advantages over antibodies and have been proposed as an alternative to antibodies for research, as well as for diagnostic and therapeutic applications.[3,12] Advantages include the following:

1. Owing to their small size, aptamers can penetrate tissues efficiently and are rapidly cleared from the circulation (an advantage for imaging applications).
2. Chemically-optimized aptamers are non-immunogenic when delivered systemically and studies have shown that they do not elicit an immune reaction even after repeated injections. This is a considerable advantage over protein-based drugs and antibodies, which can be immunogenic and require optimization prior to use in humans.
3. Unlike protein drugs, aptamers are thermally stable and can be subjected to several freeze–thaw cycles without affecting their activity.

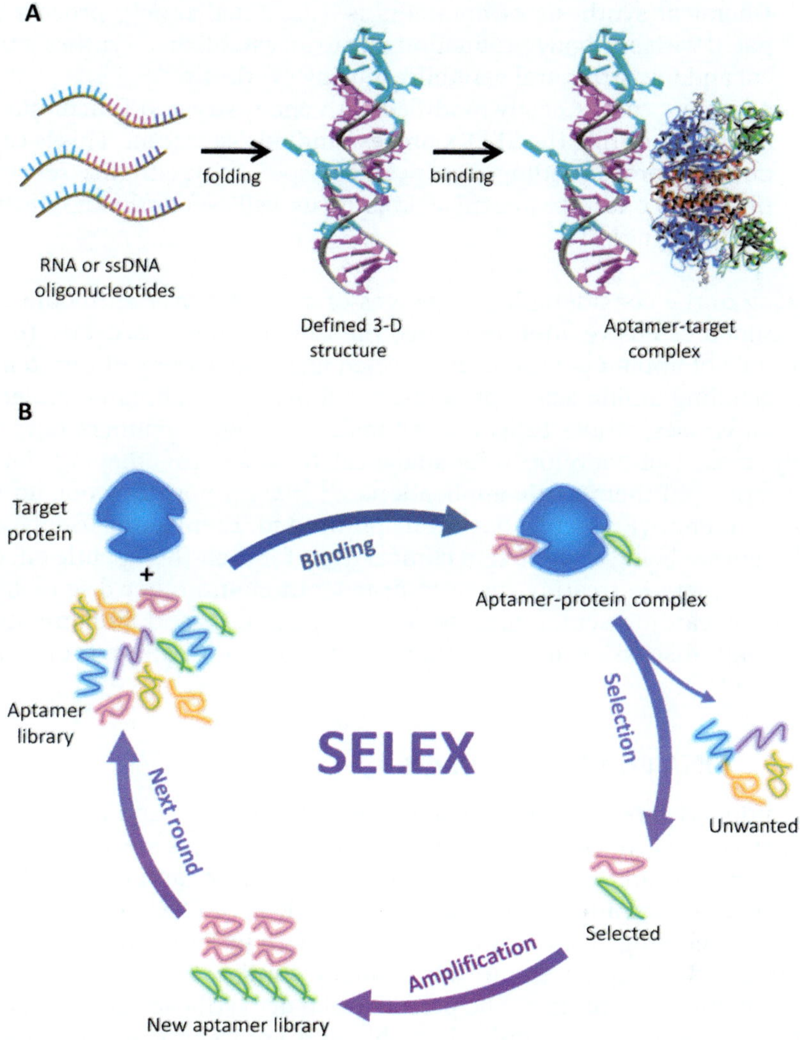

Figure 15.1 Aptamers and SELEX. (A) DNA and RNA aptamers fold into complex tertiary structures, resulting in high-affinity specific interactions with targets. (B) Schematic of the *in vitro* systematic evolution of ligands by exponential enrichment (SELEX) procedure. DNA or RNA aptamers are incubated with a target of interest (protein). Aptamers that do not bind the target are discarded (unwanted). Target-bound aptamers are retained, eluted from the target, and amplified using either PCR (for DNA aptamer selections) or RT-PCR (for RNA aptamer selections). The enriched aptamer pool is then used in a subsequent round of selection and the selection processed repeated until aptamers with the desired properties are enriched for.

4. Chemical synthesis of aptamers is a less challenging process compared with antibody production due to an established synthesis protocol and low structural variability during synthesis.
5. Aptamers can be easily modified with chemistry and functional moieties both during the SELEX process and post-selection. This is carried out to improve binding affinity and/or specificity, enhance serum stability, and optimize pharmacokinetics as well as for other applications (*e.g.* imaging).[13]

Owing to the considerable advantages of aptamers over antibodies, they have gained extensive attention since the time of their discovery. To date, thousands of aptamers have been selected against a variety of different targets, including amino acids, proteins, small metal ions, organic molecules, bacteria, viruses, whole cells, and animals.[14,15] These aptamers have been widely studied and developed for analytical, bioanalytical, imaging, delivery, diagnostic, and therapeutic applications.[7,15–18] Until now only one aptamer (Macugen, Pfizer/Eyetech)[19] has been approved for therapeutic use; however ten others are being evaluated in clinical trials for their therapeutic effectiveness and safety in treating macular degeneration and other diseases, such as cancer, cardiovascular disease (*e.g.* coagulation), and inflammation.[20] A thorough discussion of these therapeutic aptamers can be found in the literature.[10,15]

15.1.2 DNA *vs.* RNA Aptamers

As mentioned above, aptamers come in two flavors: DNA and RNA. In terms of the selection process, 'DNA SELEX' requires fewer steps since the selection is performed from the starting oligonucleotide library and only a PCR step is required for amplification. In contrast, for 'RNA SELEX', several steps are involved, including *in vitro* transcription followed by RT-PCR.

DNA and RNA aptamers are functionally similar, but possess different physiochemical properties. The presence of a deoxyribose sugar in the DNA makes DNA aptamers chemically stable and naturally resistant to serum nucleases (*e.g.* RNAse A). In contrast, RNA aptamers are prone to degradation owing to their scissile 2′-hydroxyl group (2′-OH) in the ribose.[15,21] To overcome the intrinsic chemical instability of RNA, several sugar modifications are added to RNA aptamers. Specifically, 2′ ribose sugar modifications such as 2′-fluoro (2′-F) or 2′-amino (2′-NH_2) on pyrimidine residues of RNA or 2′-*O*-methoxy motifs, and/or changes to the phosphodiester backbone with boranophosphate or phosphorothioate are regularly used modifications to furnish nuclease stability in RNA aptamers.[13,22] Although the hydroxyl group in the ribose sugar of RNA confers instability, it plays an important role in RNA folding. The 2′-OH groups on the ribose sugar engage in intra-strand hydrogen bond interactions and it provides the RNA with more structural diversity.[23] It is believed that the presence of the 2′-OH group and non-Watson–Crick base pairing in RNA enable RNA aptamers to fold into more complex three dimensional structures compared with single-stranded DNA (ssDNA) aptamers.[24] Despite physiochemical differences, DNA and RNA aptamers display similar affinities and specificities for cognate targets. Whether to proceed with a DNA *vs.* an RNA aptamer is dependent on the intended application. For example, DNA aptamers are favored for *ex vivo* diagnostic applications or as biosensors.[25] Meanwhile, RNA aptamers have been preferred for *in vivo* applications, including *in vivo* imaging, as therapeutic inhibitors, and for drug delivery.[26] Given their specific properties and preferred applications, due diligence is advised when choosing between DNA and RNA aptamers.

15.2 Prostate Specific Membrane Antigen (PSMA) Aptamers

15.2.1 PSMA

PSMA is a type II transmembrane glycoprotein that is expressed in normal prostatic epithelia and prostate cancer (PCa).[27] Expression of PSMA in PCa tissues positively correlates with the stage of disease and Gleason score.[28] PSMA expression is also higher in PCa cells from hormone-refractory patients[29,30] and increased PSMA expression is an independent marker of disease recurrence.[31,32] PSMA is different from prostate specific antigen (PSA), a soluble tryptic protease secreted by the prostate gland, which is used as a standard screening marker for PCa. However, results from large randomized trials indicate that PSA is not a reliable PCa biomarker.[33,34] Thus, PSMA has been recognized as a PCa biomarker. More recently, PSMA has also been suggested as a potential therapeutic target for cancers other than prostate, including cancers of the stomach, colon, bladder, kidney, breast, and brain.[35–37]

Given its enzymatic activities, PSMA is also referred to as folate hydrolase 1 (FOLH1), glutamate carboxypeptidase II (GCPII), and *N*-acetylated alpha-linked acidic dipeptides (Maladies).[28] PSMA was first cloned in 1993,[38] and found to be homologous in sequence to the human transferrin receptor (TfR 1) (Figure 15.2A and B). Because of the sequence similarity between TfR 1 (Figure 15.2A) and PSMA (Figure 15.2B), the cell uptake mechanisms of PSMA and TfR 1 are believed to be similar. The mechanism of TfR 1 cellular uptake proceeds by means of the cytosolic adaptor protein clathrin (Figure 15.2C).[39–41] Upon scission of the endocytotic vesicle containing the Tf ligand–receptor complex (Tf–TfR), the complex is directed to the early endosomal compartment. Sorting events initiated at this compartment determine the subsequent fate of internalized proteins and lipids, destining them either for recycling to the plasma membrane (as is the case for TfR), degradation in lysosomes, or delivery to the trans-Golgi network. Although the precise cellular uptake mechanism for PSMA is not known, it has been shown that internalization and recycling of PSMA is mediated through interactions with caveolin-1 and clathrin-coated pits.[42,43] Similarly to TfR, PSMA

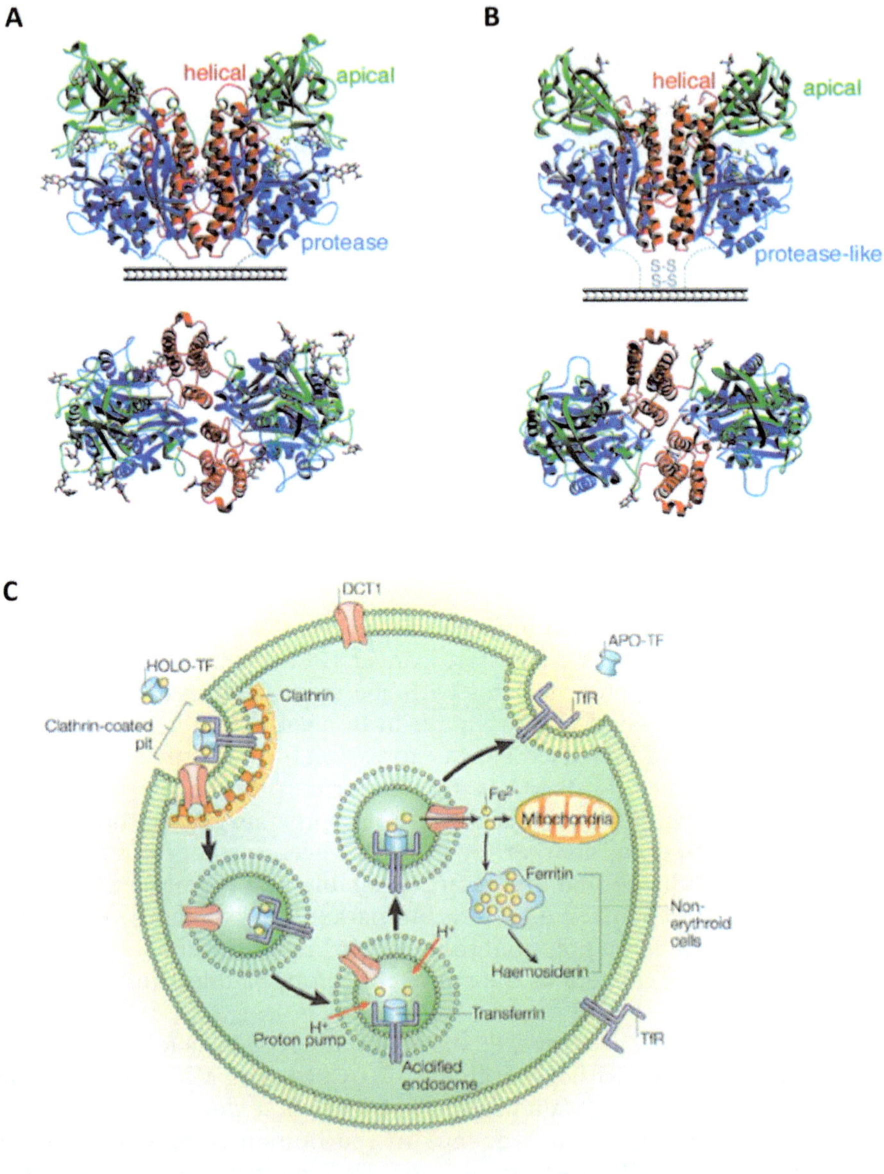

Figure 15.2 Similarities between PSMA and transferrin receptor 1 (TfR1). Crystal structures of PSMA and TfR 1 are shown in (A) and (B), respectively. Clathrin-dependent endocytosis of transferrin receptor is shown in (C). The human transferrin receptors display the best understood example of clathrin-dependent receptor-mediated endocytosis. The receptor uptake is mediated by various units of the triskelion shaped protein clathrin, coating a budding membrane vesicle from the cytoplasmic side. Clathrin recruitment is normally achieved through adaptor proteins binding the cytosolic terminus of the receptor protein. After scission from the cell membrane, the vesicle undergoes maturation to the sorting endosome (early endosome) where receptor-bound ligands are detached and the sole receptor can be recycled back to the cell surface. PSMA cellular uptake is believed to be similar to that of TfR 1. A and B adapted from ref. 127, Copyright (2005) National Academy of Sciences U.S.A. C adapted from ref. 128 with permission from Springer Nature, Copyright 2004.

is constitutively recycled from the plasma membrane to the cytoplasm *via* endocytosis, making it a promising target for pharmaceutical purposes such as imaging, diagnostics, and targeted drug delivery.

15.2.2 DNA and RNA PSMA Aptamers

To date, several DNA and RNA aptamers have been developed against PSMA either to specifically deliver imaging modalities and/or therapeutic agents for diagnosis and treatment of PCa.

15.2.2.1 PSMA RNA Aptamers

In 2004, Lupold and colleagues developed the first aptamers to PSMA.[44] These 2′ F-modified RNA aptamers were selected based on their ability to inhibit PSMA enzymatic (NAALADase) activity (Figure 15.3A).[44] Selection of the aptamers was performed using human recombinant purified PSMA (amino acids 1–706, comprising the extracellular domain of PSMA) protein. Two unique aptamer sequences, A9 and A10, were enriched for, following six rounds of positive and negative selection. The authors determined that aptamer A9 inhibited PSMA enzymatic activity (NAALADase) non-competitively with an average inhibitory constant (Ki) of 2.1 nM. In contrast, aptamer A10 behaved as a competitive inhibitor of PSMA enzymatic activity (NAALADase) with an average Ki of 11.9 nM. It was concluded that A9 and A10 probably bind to unique epitopes on PSMA since they exhibit different modes of inhibition. As expected, when tested for binding specificity on PSMA-positive (LNCaP) and -negative (PC-3) human PCa cell lines, the aptamers displayed strong preferential binding to LNCaP cells.

Typically, parent aptamer sequences are further trimmed down to shorter sequences to improve specificity and reduce cost of synthesis. 3′-Terminus base deletions were performed on both PSMA A9 and A10 aptamers and tested for function. Aptamer A10 was successfully truncated by 15 bases (from 71 down to 56 nucleotides), while retaining its function (truncated aptamer, A10-3, Ki = 20.5 nM) (Figure 15.3B). A10-3 was also tested for binding specificity and showed specific binding to PSMA-positive (LNCaP), but not to PSMA-negative (PC-3) cells. In contrast, using a similar 3′-terminus deletion strategy for aptamer A9 resulted in loss of function and it was concluded that the A9 aptamer sequence could not be reduced without enduring loss of function.

In a follow up study by our group,[45] a different truncation strategy was applied to the A9 aptamer that enabled reduction of the length of the parent aptamer, while retaining both its binding and inhibitory functions. Our truncation strategy was guided by RNA structural predictions and protein–RNA docking algorithms.[46–48] Base deletions from both the 5′- and 3′-termini of A9 were performed while retaining A9's original structural motifs. Seven truncated versions of the A9 aptamer were designed (A9a to A9g) ranging

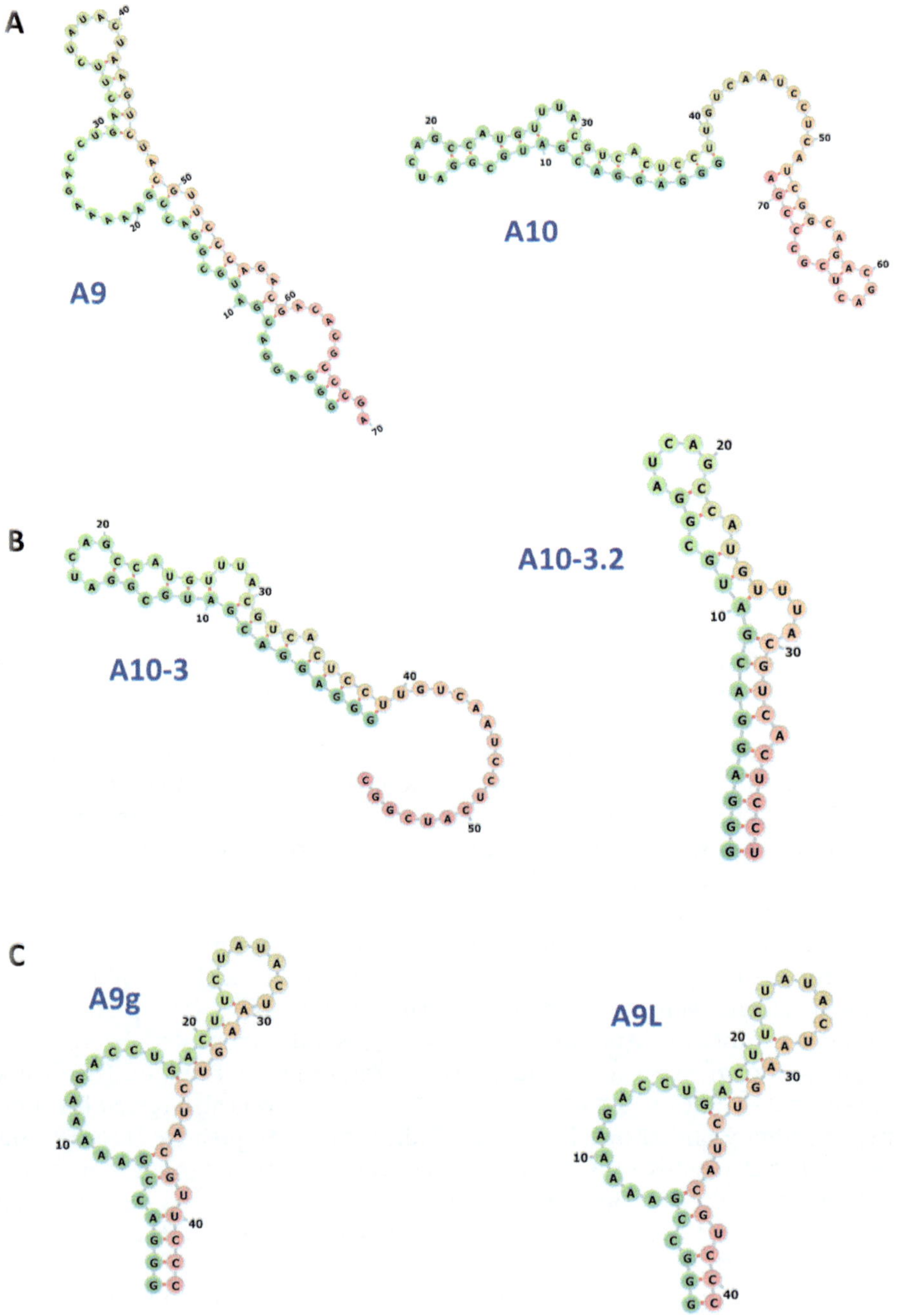

Figure 15.3 PSMA RNA aptamers. (A) Predicted secondary structures of parent (non-truncated) A9 RNA and A10 RNA aptamers. (B) Predicted secondary structures of the truncated versions of A10 aptamers, A10-3 aptamer was further truncated to A10-3.2. (C) Predicted secondary structures of the truncated forms of A9 RNA aptamers A9g and A9L. A9g was further truncated to A9L by deletion of two bases. Predicted RNA secondary structures were generated using ViennaRNA Package 2.0 (RNAfold).

from 66 bases (A9a) to 43 bases (A9g) in length. The NAALADase assay was used to assess inhibition of PSMA enzymatic activity of the various truncations. Remarkably, all seven truncations inhibited PSMA's enzymatic activity. The shortest of these truncated versions of A9 (A9g) exhibited inhibitory activity similar to that the parental A9 aptamer (Ki of 10 nM) (Figure 15.3C). These results led to further truncations of A9g to determine key binding and inhibitory elements within the aptamer. Truncations A9h (37 mer) and A9i (24 mer) retained sequence and structural loop elements of A9g, whereas A9j (30 mer) and A9k (21 mer) retained sequence and structural stem elements of A9g. Interestingly, unlike A9 and A9g, none of these additional truncations (A9h–A9k) retained functionality. When probed further, protein-docking studies with A9g and PSMA revealed key residues within A9g responsible for binding to PSMA. These studies helped guide the deletion of two more bases from A9g, resulting in an aptamer of 41 bases in length that was termed A9L (Figure 15.3C). A9L was functionally equivalent to A9 and A9g in terms of binding to human PSMA and inhibition of its enzymatic activity (NAALADase). Hence, it was concluded that A9g and A9L aptamers retain minimal essential sequence structure motifs required for function.

15.2.2.2 PSMA DNA Aptamers

Following the development and characterization of the PSMA RNA aptamers by Coffey's group, Almasi and colleagues[49] selected two DNA aptamers against human PSMA using a cell-based SELEX approach (cell-SELEX). To obtain sequences that bind to PSMA in the context of the cell membrane, initial rounds of selection were performed using PSMA-positive LNCaP cells. Counter selection using a PSMA-negative PCa cell line (PC-3) was performed at each round of selection to eliminate nonspecific sequences. Stringency during the cell-SELEX process was progressively increased by carrying out additional washes and reduction of the incubation times at subsequent rounds of selection. After 10 rounds of selection two aptamer sequences, A12 and B1, showed high affinity binding to PSMA-positive cells. Reported Kds for these sequences were 33.78 ± 3.77 and 57.4 ± 2.214 pmol, respectively.

Atabi and colleagues used a cell-SELEX approach similar to the one described above, to develop different DNA aptamer ligands targeting PSMA.[50] Similarly, cell-SELEX was carried out using the LNCaP cell line as the target and the PC-3 cell line as a nonspecific control for the counter selection. Following ten rounds of selection and counter-selection, one sequence was identified (referred to as Apt1) that was different in primary sequence from the PSMA DNA aptamers originally reported by Almasi and colleagues.[49] The apparent dissociation constant for Apt1 was determined to be 34 ± 3.8 pM. The DNA aptamer was linked to a myristilated chitosan nanogel (MCS) using glutaraldehyde. Thereafter, the aptamer–MCS complex was loaded with doxorubicin (DOX). In studies performed using PCa cells grown in culture, the aptamer–MCS–DOX complex demonstrated selective killing of PSMA-expressing PCa cells (LNCaP). In contrast, no toxicity was observed with PCa cells lacking PSMA expression (PC-3). Although encouraging, this study was limited to evaluation of cytotoxicity, selectivity, and efficacy of the aptamer–MCS–DOX complex *in vitro*. Further confirmation of its safety and efficacy in preclinical animal models is warranted.

In a subsequent study, Boyacioglu and colleagues[51] identified another DNA aptamer optimized for targeted delivery of DOX to PSMA-expressing PCa cells. The DNA aptamer was selected against the extracellular domain (amino acids 1–720) of human recombinant PSMA protein. To include DOX binding sites within the selected aptamer the DNA aptamer library was designed to include constant stem regions using sequences optimized for DOX binding flanking the variable regions. After 10 rounds of selection and counter-selection, the SZTI01 DNA aptamer was identified. To increase the avidity of the aptamer to PSMA and enhance its cell-internalization efficiency, aptamer SZTI01 was dimerized by using a poly A–T linker. The resultant dimeric aptamer–DOX conjugate was shown to selectively internalize and lead to the death of PSMA-positive PCa cells, but not to that of PCa cells lacking PSMA surface expression.

15.3 PSMA Aptamers Applications

15.3.1 Imaging and Diagnostic Applications

Aptamers are great candidates for targeted imaging applications due to their rapid blood clearance, target specificity and low immunogenicity.[52] The relatively small size of aptamers favors high permeability in solid tumors while their chemical versatility allows for easy conjugation to fluorophores and radio-imaging molecules.[52,53] Combining aptamers with several imaging modalities facilitates visualization of biological processes at the basic molecular and cellular levels.

15.3.1.1 Radio-imaging Applications

Nuclear imaging, including positron emission tomography (PET) and single-photon emission computed tomography (SPECT) can be used to monitor biological events deep in the body and to provide longitudinal assessment of the same patient with high detection sensitivity.[52,54] Labeling of single-stranded oligonucleotides with radioisotopes (*e.g.* ^{99m}Tc, ^{111}In, ^{125}I, for SPECT or ^{11}C, ^{18}F, ^{76}Br, ^{64}Cu for PET) can result in valuable radiopharmaceuticals with promising applications.[55,56] Previous studies demonstrated the use of aptamers in targeted imaging using gamma emitters, technetium-99m (^{99m}Tc) and indium-111 (^{111}In) for SPECT.[57–59] While these studies are encouraging, radiolabeling conditions used for gamma emitters can be harsh and often result in degradation of nucleic acid-based bioconjugates. Owing to the limitations of gamma emitters, researchers interested in developing nucleic acid-based imaging technologies have directed their focus to positron-emitting radionuclides.

In a study by our group,[60] the PSMA RNA aptamer (A10 3.2) was radiolabeled with a positron-emitting radionuclide (copper ^{64}Cu) using several different chelators. The A10-3.2 aptamer is a derivative of the A10 and A10-3 aptamers that was further truncated down to 39 bases (Figure 15.3B). Notably, A10-3.2 retains binding to PSMA but fails to inhibit PSMA's enzymatic activity.[61] In this study, the key parameters for labeling of the RNA aptamer were established. A10-3.2 RNA aptamer was conjugated to three bifunctional macrocyclic chelators followed by labeling with ^{64}Cu [half life ($t_{1/2}$) = 12.7 h]. The effects of different chelators on binding specificity of aptamer A10-3.2 against PSMA was evaluated along with establishing key parameters (pH, temperature, and reaction ratio) for optimal labeling. The stability and function (binding to PSMA) of the aptamer conjugates were investigated. However, while the initial characterization of the aptamer–chelator conjugates is promising, *in vivo* PET imaging studies were not performed, warranting evaluation of these conjugates in preclinical animal models of PCa.

15.3.1.2 Computed Tomography (CT) Applications

CT is one of the most frequently used and cost-effective biomedical imaging techniques. Currently available CT contrast agents are based on small iodinated molecules, which possess a number of limitations, including lack of targeted molecular imaging, short imaging times, and renal toxicity. Kim and colleagues[62] reported aptamer-conjugated gold nanoparticles (GNPs) for targeted CT imaging of PCa. Silver staining and CT imaging of PCa cell aggregates confirmed the selective accumulation of the PSMA RNA aptamer (A9) in PSMA-positive cancer cells (LNCaP) with minimal uptake into PSMA-negative PC-3 cells. The investigators demonstrated fourfold greater signal intensity in LNCaP *vs.* PC-3 cells using the aptamer-targeted GNP conjugate. While promising, these studies await further validation of the aptamer–GNP CT conjugate in an *in vivo* preclinical model of PCa.

15.3.1.3 Magnetic Resonance Imaging (MRI) Applications

MRI is a noninvasive medical test that physicians use to diagnose medical conditions. MRI uses a powerful magnetic field combined with radio-frequency pulses and a computer to produce detailed pictures of organs, soft tissues, bones, and other internal body structures.[54] Recent efforts have focused on super-paramagnetic iron oxide nanoparticles (SPIONs) as novel MRI contrast agents for imaging tumors. One advantage of SPIONS is that they can be used as theranostic agents[63] to image as well as to deliver therapeutic cargo to tumors. A recent study has combined aptamer technology with SPIONS as a way to develop a targeted theranostic for PCa.[64] In this study, the A9 PSMA RNA aptamer was conjugated to SPIONS. The aptamer-functionalized SPION conjugated was loaded with a cytotoxic drug cargo (DOX) and evaluated for efficacy in PCa cells grown in culture as well as in a preclinical animal model of PCa. Importantly, the authors demonstrated selective drug targeting to PSMA-expressing cancer cells both *in vitro* and *in vivo* with the aptamer-functionalized SPION. In addition, they were able to image the prostate tumors using MRI to determine treatment response. This approach is noteworthy as it allows for the simultaneous detection of prostate tumors by MRI and the targeted delivery of anti-cancer drugs to tumor sites with aptamers.

15.3.1.4 Optical Imaging Applications

Aptamers have been conjugated to fluorophores for small-animal imaging studies. Bagalkot and colleagues[65] described an aptamer-functionalized theranostic conjugate for PCa. The conjugate works based on the principle of fluorescence resonance energy transfer (FRET)[66,67] and takes advantage of the fluorescence properties of both DOX and Quantum dots (QD). DOX quenches the fluorescence of QDs since the emission spectra overlaps with the absorbance of DOX and, in parallel, DOX fluorescence is quenched due to intercalation within the bases of the aptamer. This leads to the formation of a Bi-FRET complex. In this study, the A10 PSMA RNA aptamer was conjugated to QD and DOX intercalated within the stem region of the aptamer. The resulting A10–QD–DOX conjugate is in an "OFF" state due to Bi-FRET (fluorescence of both QD and DOX is quenched). However, once the conjugate is internalized within the target cells, release of DOX from the aptamer results in fluorescence of DOX and the QD. The A10–QD–DOX conjugate can be used to simultaneously determine bio-distribution as well as to quantify the rate of internalization and release of the drug conjugate.

Imaging using near infrared (NIR) light has gained momentum for *in vivo* biodistribution studies due to improved penetrability and visibility of the signal.[68] In a recent study by our group,[69] NIR was used to determine the bio-distribution and pharmacokinetic properties of the PSMA A9g aptamer. In this study, the aptamer was coupled to the NIR fluorophore IRDye 800CW and the conjugate was administered systemically (by intravenous injection) to both tumor-bearing and control (non-tumor-bearing) mice. The distribution of the aptamer–NIR conjugate was monitored over time using bioluminescence imaging (BLI). In control mice, the A9g–NIR conjugate and a control non-targeting conjugate (A9g.6–NIR) were renally cleared within 8 h. In contrast, in tumor-bearing mice, the A9g–NIR conjugate was retained for 72 h, while non-targeting control conjugate was cleared within 8 h. Notably, in studies using xenograft mice bearing both PSMA-positive and PSMA-negative tumors implanted subcutaneously, we observed that the A9g–NIR conjugate was selectively retained in the PSMA-positive tumors. These studies highlight the power of NIR technology for monitoring the biodistribution of oligonucleotide-based drugs in real time *in vivo*.

Second-harmonic generation (SHG) imaging, which uses NIR to improve penetration depths of biological specimens, has been combined with aptamers for targeted imaging applications. In a recent study by Demeritte and colleagues, aptamer-functionalized gold nanocage assemblies were developed to image the ability of the PSMA A9 aptamer to bind to PCa cells.[70] The authors conjugated the A9 aptamer to a 'gold nanocage' and applied the conjugate to either PSMA-positive (LNCaP) or PSMA-negative (PC-3) cancer cell lines. Importantly, the aptamer SHG imaging probe was highly selective and was able to distinguish between target and non-target cells. While these results are encouraging, *in vivo* studies to help assess the efficiency and safety of the technology are awaited.

15.3.2 PSMA Aptamers as Therapeutic Inhibitors

Cell-targeted therapies that selectively target cancer cells have been developed to effectively treat some cancers, while reducing systemic toxicity.[18,71] In a study by Dassie and colleagues,[69] the authors used the A9g PSMA RNA aptamer to target PCa cells and inhibit their metastatic spread. As described above, the A9g RNA aptamer acts as an inhibitor of PSMA's enzymatic activity, which has been linked to carcinogenesis. The authors demonstrated that treatment of PCa cells with A9g aptamer results in reduced cell migration and invasion of PCa cells in culture and exhibited potent anti-metastatic activity *in vivo*. The efficacy of A9g *in vivo* was tested in a mouse model of metastatic PCa characterized by high osteotropism. Importantly, treatment of the animals by systemic injection of A9g significantly reduced the spread of PCa in the mouse model with only 10% of treated mice developing bone metastases compared with controls, thus, showing the efficacy of aptamer treatment. A9g was also shown to be non-toxic *in vivo* since no change in the appearance, weight or behavior of treated animals was observed during 4 weeks of treatment. Notably, the aptamer showed no signs of immunogenicity, even after repeated injections. Future work will be focused on combining the cytostatic properties (inhibition of cell migration and invasion) of the A9g aptamer with cytotoxic drug payloads for targeted combinatorial therapy of advanced PCa.

15.3.3 Targeted Delivery Applications

15.3.3.1 PSMA Aptamer RNAi Conjugates

RNA interference (RNAi) can be harnessed to silence virtually all genes in the human and viral genomes, making it a powerful tool for basic research and clinical applications.[72] Over the last three decades since its discovery, many efforts have been focused on optimizing RNAi modulators [*e.g.* small interfering RNAs (siRNAs), micro RNAs (miRNAs), short hairpin RNAs (shRNAs), and miRNA inhibitors (miR inhibitors)] for therapeutic applications.[73] These efforts have resulted in several RNAi-based drugs in advanced stages of clinical development.[74] One of the remaining challenges associated with the broad clinical translation of RNAi-based therapeutics is delivery to specific cell types. RNA aptamers have been proposed as excellent vehicles for targeted delivery of RNAi modulators to cells. RNA aptamer–siRNA conjugates are designed such that aptamers to cell-specific surface markers facilitate the delivery of RNAi modulators to the appropriate subcellular compartment (cytoplasm) of target cells. Aptamers have been conjugated to RNAi modulators (*e.g.* siRNAs) either by covalent linkage or complementation (annealing). The combination is typically referred to as aptamer–siRNA chimeras or AsiCs.

The concept of aptamer-mediated delivery of siRNAs was first described in 2006 in two independent studies.[75,76] These studies both took advantage of existing RNA aptamers to PSMA[44] to deliver the siRNA cargo. In a study by Ellington and colleagues,[76] two biotinylated A9 anti-PSMA aptamers and two biotinylated siRNAs directed against lamin A/C or glyceraldehyde 3-phosphate dehydrogenase (GAPDH) were joined with a modular streptavidin connector (Figure 15.4A). The resulting aptamer–streptavidin–siRNA conjugates were internalized by PSMA-positive cells and specific knockdown of target genes was observed.

In a study by McNamara *et al.*,[75] the authors harnessed a different PSMA RNA aptamer (A10) to deliver siRNAs directed against two genes overexpressed in human tumors [polo-like kinase 1 (Plk1) and B cell lymphoma 2 (bcl-2)] responsible for cancer cell survival. The authors designed AsiCs composed entirely of RNA by covalently linking the passenger strands of the siRNAs to the aptamer's 3′-terminus and annealing the respective guide strand to the aptamer–siRNA oligo (Figure 15.4B). Gene silencing by the siRNAs was confirmed by a reduction in the mRNA and protein levels of Plk1 and bcl-2, respectively. Importantly, silencing was observed only in the PSMA-positive PCa cells and required a functional PSMA targeting aptamer. The investigators demonstrated *in vivo* efficacy of the aptamer–siRNA conjugates following intratumoral injections in a human xenograft mouse model of PCa. This study was the first to describe *in vivo* efficacy of aptamer–siRNA conjugates.

In a subsequent study by the same group,[61] modifications were made to both the A10 aptamer and the siRNA cargo to enable efficacy upon systemic administration of the AsiCs and reduce potential off-target effects. In addition to modifications within the RNA, a 20 kDa polyethylene glycol (PEG) moiety was added to the AsiC to increase its half-life in plasma ($t_{1/2}$) and improve efficacy. The study demonstrated effective and pronounced tumor regression of PSMA-expressing tumors in athymic mice after systemic administration with no any indication of toxicity due to off-target or immunostimulatory effects.

In a seminal study by Gilboa and colleagues published in the journal *Nature*, AsiCs were developed for cancer immunotherapy applications.[77] Cancer cells often evade the immune system by downregulating the expression of cancer-specific antigens on the cell surface. One way

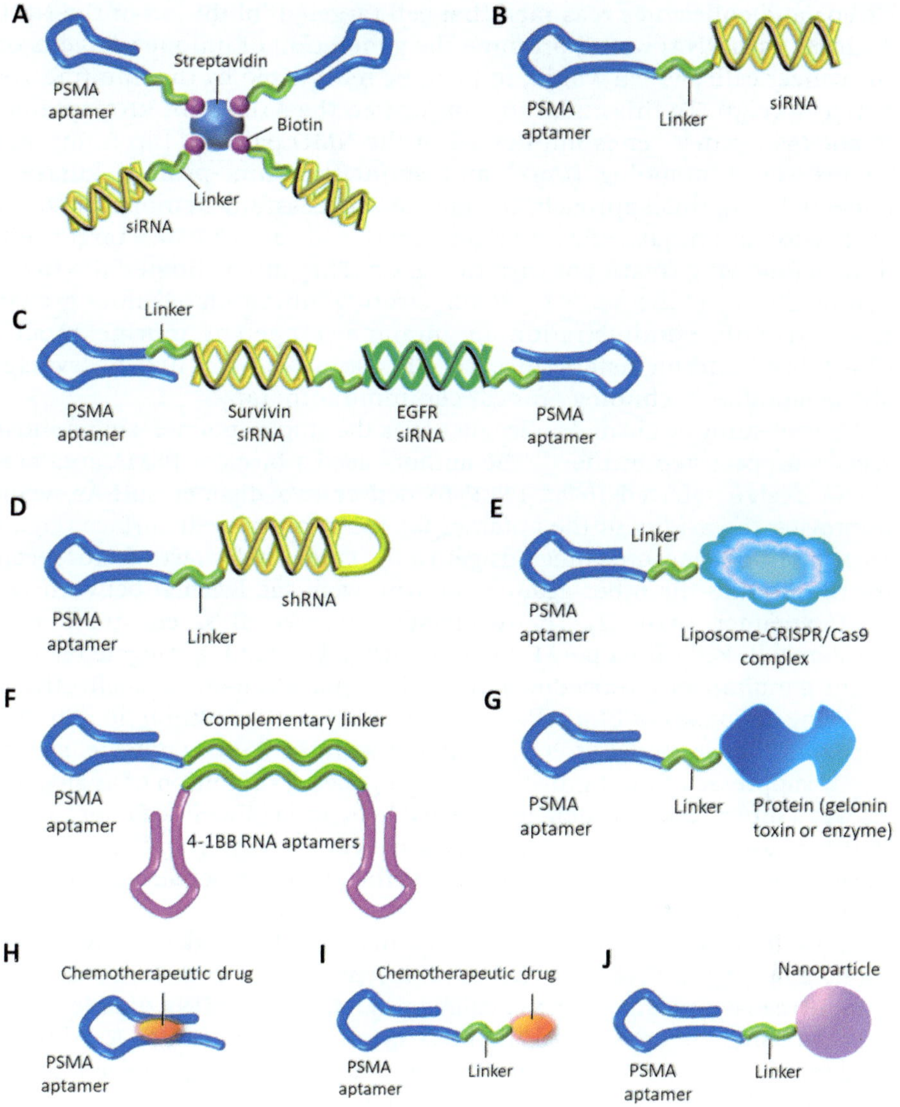

Figure 15.4 Schematic of PSMA aptamer conjugates and AsiCs. (A) PSMA aptamer–streptavidin–siRNA conjugate. (B) PSMA aptamer–siRNA chimera or AsiC. (C) Bivalent aptamer–dual siRNA *PSEP* conjugate. (D) PSMA aptamer–shRNA conjugate. (E) PSMA aptamer–CRISPR/Cas9 conjugate. (F) Bispecific PSMA aptamer conjugate. (G) PSMA aptamer protein toxin conjugate (H and I) PSMA aptamer small molecule conjugates. (J) PSMA aptamer functionalized nanoparticles.

the cancer cell can do this is by interfering with the nonsense-mediated mRNA decay (NMD) pathway, a surveillance pathway that exists in all eukaryotes.[78,79] Since no aberrant antigens are expressed on the cancer cell surface, cancer cells can shield themselves from an immune attack.

Gilboa and colleagues reasoned that cell-targeted inhibition of the NMD pathway with RNAi would promote the generation of unique antigens on the cancer cells, which would, in turn, be recognized by the immune system as foreign. To this end, they conjugated the A10 PSMA RNA aptamer to siRNAs against genes implicated in the NMD pathway [up-frameshift suppressor 2 homolog (*Upf2*) and serine/threonine-protein kinase 1 (*Smg1*)]. Using this approach, the authors successfully demonstrated the expression of unique cancer antigens on the surface of PSMA target cells *in vitro* following treatment with the AsiCs. They also validated the therapeutic efficacy of the AsiCs by demonstrating inhibition of tumor growth after systemic administration in immune-competent murine models of localized and metastatic disease. This study was the first to leverage siRNA–aptamer technology for cancer immunotherapy.

A recent study by Liu and colleagues took the single aptamer–single siRNA AsiC concept a step further.[80] The authors used a bivalent PSMA aptamer–siRNA design referred to as PSEP to deliver two distinct siRNAs, while improving the avidity of the aptamer for PSMA on the cell surface (Figure 15.4C). The siRNAs, one directed against the epidermal growth factor receptor (EGFR) and the other against survivin, were sandwiched between two PSMA aptamers (A10-3.2). The two PSMA aptamers–siRNA constructs were covalently linked with a poly U linker (UUUU). The PSEP conjugate resulted in the simultaneous knockdown of survivin and EGFR and was effective in inducing apoptosis of PCa cells *in vitro* and *in vivo*. In addition, in a human xenograft mouse model of PCa, treatment with the PSEP conjugate, significantly suppressed tumor growth and angiogenesis. Inhibition of angiogenesis was a direct result of silencing the EGFR–hypoxia inducible factor 1 alpha (HIF1α)–vascular endothelial growth factor (VEGF) pathway by the *PSEP* conjugate. The study by Liu and colleagues is the first to demonstrate efficacy of a bivalent AsiC conjugate *in vivo*.

In addition to cytotoxic siRNAs, aptamers can be used to deliver RNAi modulators (*e.g.* shRNAs) that act as radiosensitizing agents. Ionizing radiation (IR) causes multiple types of cellular injury, of which DNA double-strand breaks (DSBs) are considered the most cytotoxic. Chemical or RNAi-mediated inhibition of DNA repair proteins such as ataxia telangiectasia mutated (ATM) or Nijmegen breakage syndrome gene (NBS1) results in cellular hypersensitivity to IR.[81] To achieve enhanced radiation therapy for locally advanced PCa, a PSMA aptamer-shRNA was developed (Figure 15.4D).[82] The truncated A10 aptamer (A10-3)[44] was linked to a shRNA directed against a DNA repair pathway gene, DNA-activated protein kinase (DNAPK). Importantly, the DNAPK-A10 conjugate selectively reduced DNAPK in PCa cells, xenografts and human prostate tissues, resulting in sensitization of PSMA-positive PCa cells to ionizing radiation. Given that IR treatment is the standard-of-care for recurrent late-stage PCa, radio-sensitization *via* treatment with PSMA–DNAPK aptamer–shRNA chimeras has the potential to significantly improve clinical outcomes for patients with advanced disease.

When conjugating RNAi modulators to aptamers, investigators have favored two main strategies. The first strategy involves covalent linking of the siRNA, miRNA, or shRNA to the aptamer followed by annealing of the guide strand (in the case of siRNAs and miRNAs).[75,83] The second strategy involves the use of a linker or bridge sequence.[14] A less common approach has been to conjugate RNAi modulators, such as miRNAs, onto aptamer-functionalized dendrimers [*e.g.* polyamidoamine (PAMAM)].[84] In this study by Wu and colleagues, miR-15a and miR-16-1, which act as putative tumor suppressors by targeting the oncogene bcl-2, were conjugated to the PAMAM dendrimer. To enable selective delivery to PCa cells, the miRNA–dendrimer conjugate was functionalized with the A10-3 PSMA aptamer. The investigators confirmed cell-specific delivery of the miRNAs, downregulation of miRNA target genes (oncogenes), and subsequent death of the cancer cells. Importantly, the simultaneous delivery of the two tumor suppressor miRNAs resulted in synergistic induction of cell death compared with each miRNA alone. Although initial results are promising, the efficacy and safety of the dendrimer–aptamer targeted drug delivery system need to be validated in preclinical animal models of cancer.

15.3.3.2 PSMA Aptamer–CRISPR Conjugates

In addition to delivery of RNAi modulators to target cells, more recently, there has been an interest in exploring the possibility of using aptamers for cell-targeted delivery of genome-editing tools.[85] The clustered regularly-interspaced short palindromic repeats (CRISPRs)–CRISPR-associated protein 9 (Cas9) system is an emerging gene-editing technology. CRISPR was first discovered in bacteria and archaea as an immune surveillance mechanism that confers resistance to infection by viruses.[86] CRISPR is an array of DNA sequences in bacteria, which have been acquired over the course of time from invading viruses. These CRISPR arrays function as immune-memory for bacteria to recognize and remember viruses. In the event of a second infection by the virus, bacteria and archaea produce RNA segments from the CRISPR arrays to target the viral DNA.[87,88] Cas9 or a similar enzyme is then used by the bacteria to cleave the viral DNA. CRISPR–Cas9 has gained immense attention in the field of gene editing and is currently being researched for use in various applications, such as gene expression modulation and imaging.[89]

In a recent study by Zhen and colleagues a flexible aptamer-liposome–CRISPR–Cas9 chimera was designed using the A10 PSMA RNA aptamer (Figure 15.4E).[90] Aptamer-functionalized cationic liposomes were investigated for their ability to deliver CRISPR–Cas9 against Plk1 to tumor cells. The authors demonstrated that the A10 aptamer–liposome–CRISPR–Cas9 effectively silenced *plk1* expression *in vitro*. They also tested the complex in a xenograft mouse model of PCa and observed a conspicuous regression of tumor growth when treating with the aptamer–CRISPR–Cas9 Plk1 particle.

No significant immunogenicity was observed due to treatment with the complex, indicating that it can be safely used for therapeutics.

15.3.3.3 Bispecific PSMA Aptamer Conjugates

Deficiency of T cell co-stimulation at the tumor site has been proposed as a major reason why tumors are not responsive to vaccine-induced tumor immunity.[91,92] In an elegant study by Pastor and colleagues, a bispecific aptamer design was used to recruit activated (primed) T cells to the tumor site. Specifically, the authors linked the A10 PSMA aptamer to a bivalent 4-1BB aptamer directed against a co-stimulatory receptor expressed on T cells (murine 4-1BB). The two aptamers were conjugated using a complementary linker design. The bispecific conjugate was shown to promoted the survival and expansion of activated $CD8^+$ T cells while simultaneously recruiting the activated T cells to PSMA-positive tumors, leading to an enhanced immune response against these tumors (Figure 15.4F).[93] Significantly, systemic administration of the bispecific aptamer conjugate significantly increased vaccine-induce immunity and inhibited tumor growth in mice. Additionally, in contrast to mice treated with the 4-1BB antibody, mice treated with the bispecific aptamer conjugate did not display signs of toxicity (*e.g.* enlarged spleens and lymph nodes). From these studies, the authors concluded that the bispecific aptamer conjugate was a safe and effective cancer immunotherapy treatment for PSMA-expressing tumors. Future studies based on this approach await the development of a cross-species (humanized) conjugate that can be evaluated for efficacy and safety in preclinical animal models of cancer and in humans.

15.3.3.4 PSMA Aptamer Protein Toxin Conjugates

Antibodies directed against cancer cell surface antigens have been successfully used to deliver toxic payloads.[94] Given the many advantages of aptamers over antibodies (see above), several investigators have explored the use of aptamers for delivery of protein toxins to cancer cells. Towards this end, the A9 PSMA RNA aptamer and its derivatives were evaluated by two separate groups for their ability to deliver ribosome-inactivating protein drugs (gelonin and saporin) to PSMA-expressing cancer cells (Figure 15.4G).[6,95,96] Gelonin and saporin are potent toxins derived from plants, which act by a similar mechanism to the poison ricin (a potent toxin derived from castor beans).[97] Unlike ricin, gelonin and saporin lack a cell translocation domain required for delivery into a cell and thus are not active.[98] However, upon conjugation to a cell-penetrating ligand (*e.g.* PSMA aptamers), the drugs are delivered to the cytoplasm, where they engage with the ribosome and inactivate it, resulting in pronounced cell death. These studies demonstrate the ability of the A9 PSMA aptamer and its derivatives to deliver therapeutic cargo to the cytoplasm of target cells, rendering them ideal delivery vehicles for RNAi-based drugs.

15.3.3.5 PSMA Aptamer Small-molecule Conjugates

Chemotherapy is one of the main treatment lines for cancer. Although widely used, chemotherapy is limited by systemic toxicity due to the non-specific targeting of these cancer drugs. Conjugation of nonspecific cancer drugs with targeting ligands (*e.g.* aptamers) has been explored to improve drug safety profiles. In a recent study, Bagalkot and colleagues[99] explored the use of an aptamer to PSMA (RNA aptamer A10) for the targeted delivery of DOX to PSMA-expressing PCa cells (Figure 15.4H).[99] To assess delivery of DOX to target cancer cells, the investigators intercalated DOX within the RNA aptamer. When intercalated, DOX fluorescence is quenched. Once inside the cells, DOX is released from the aptamer. Upon release, DOX fluorescence is restored. The inherent fluorescence of DOX helps monitor the rate of uptake of the aptamer within target cells. The investigators were also able to determine potential off-target effects using this approach. Notably, the aptamer–DOX complex selectively internalized into target PSMA-expressing cells resulting in significant cytotoxicity. The concept of using PSMA aptamers to improve the cell-specificity of otherwise nonspecific chemotherapeutic drugs was corroborated in a later study, using a dimeric PSMA DNA aptamer–DOX conjugate.[51] In this latter study by Boyacioglu and colleagues, DOX was selectively delivered to PSMA-positive cells with minimal uptake and cytotoxicity to PSMA-null cells (Figure 15.4I).[100] While encouraging, studies evaluating the use of aptamers for delivery of chemotherapeutic drugs await validation in *in vivo* preclinical cancer models.

15.3.3.6 PSMA Aptamer Functionalized Nanoparticles

Nanoparticles are particles smaller than a few hundred nanometers.[101–103] They are several orders of magnitude smaller than human cells and exhibit properties distinct from both molecules and bulk solids. Due to their large surface area and loading capacity and versatile chemistry, nanoparticles can function as excellent carriers of anti-cancer drugs.[104,105] Several nanosystems have been developed for systemic cancer therapy such as liposomes, albumin-based particles, PEGylated proteins, biodegradable polymer–drug composites, and polymeric micelles.[106] Over the years, aptamer-functionalized nanoparticles have been evaluated for a broad range of applications, such as diagnostics and targeted delivery of drugs.[107,108]

In a 2004 proof of concept study by Farokhzad and colleagues,[109] the authors reported on the efficacy of an aptamer-functionalized nanoconjugate for the treatment of PCa. Nanoparticles formulated with biocompatible and biodegradable poly (D,L-lactic-*co*-glycolic acid)-*block*-poly (ethylene glycol) (PLGA-*b*-PEG) copolymer were loaded with rhodamine-labeled dextran and functionalized with the A10 PSMA RNA aptamer (Figure 15.4J).[109] The aptamer-functionalized nanoparticle bioconjugate was then evaluated for uptake into target (PSMA-expressing) PCa cells. As anticipated, the conjugate showed selective uptake in PSMA-expressing PCa cells (LNCaP) *vs.* PCa cells lacking PSMA expression (PC-3). In a later study by the same group published in 2006,[110] the A10 PSMA RNA aptamer-functionalized nanoconjugate was loaded with the cancer drug Docetaxel (DTX) and evaluated for efficacy in PCa cells in culture and *in vivo*, in a preclinical human xenograft mouse model of PCa. When added to cells in culture, the aptamer-functionalized DTX nanoconjugate (NP–Apt–DTX) resulted in the selective death of PSMA-expressing PCa cells *vs.* PCa cells lacking PSMA expression (PC-3). Similarly, when injected intratumorally into tumor bearing mice, NP–Apt–DXT resulted in selective killing of PSMA-expressing prostate tumors. These findings were later confirmed by an independent group who also reported on the pharmacokinetic and biodistribution profiles of the NP–Apt–DXT conjugate *vs.* free DTX.[111] While encouraging, these studies await evaluation of efficacy and safety of the aptamer-functionalized DTX nanoparticle following systemic administration. Additional considerations include efficiency of scale-up and cost due to the multi-component nature of the conjugate.

The broad application of the PSMA aptamer-functionalized nanoparticle technology for targeted delivery of chemotherapeutic drugs was later shown by Dhar and colleagues.[112] The investigators used similar PLGA nanoparticle conjugates functionalized with the A10 PSMA aptamer to deliver a different cancer drug, cisplatin (CDDP), to cancer cells. These findings are significant and provide a foundation for improving efficacy and reducing systemic toxicity of conventional cancer drugs.

More recently, aptamer-functionalized nanoparticles have been evaluated for theranostic applications. In a study by Wu and colleagues published in 2017, the authors investigated the use of A10-3.2 PSMA RNA aptamer-functionalized PLGA-based gas-filled nanobubbles (NBs) for their ability to (1) deliver a therapeutic drug cargo [Paclitaxel (PTX)] and (2) detect tumors using ultrasound technology.[113] Notably, the combination of drug-loaded, aptamer-functionalized NBs with low-frequency ultrasound resulted in high levels of drug release and significant cell apoptosis *in vitro*. Cell death was specific for PCa cells expressing PSMA on the cell surface. In a human xenograft mouse model of PCa, the aptamer-functionalized NBs in combination with low frequency ultrasound resulted in pronounced tumor inhibition, which was greater than that achieved with the NBs alone. Importantly, the combined treatment significantly prolonged the survival of tumor-bearing nude mice without obvious systemic toxicity due to PTX. This study is significant since the aptamer-functionalized multifunctional theranostic agent can be used for the purpose of obtaining ultrasound images and providing low-frequency ultrasound-triggered therapy of PCa. This is the first description of an aptamer-based strategy for both PCa imaging and chemotherapy.

In addition to directly killing tumor cells with cytotoxic drugs, angiogenesis inhibitors have been used to block blood supply to growing tumors. Within this framework, tumor endothelium-specific epitopes have been identified, and targeting molecules against these epitopes have been developed and evaluated for their ability to selectively target the tumor neovasculature.

One such epitope is PSMA.[114,115] In a study by Bandekar and colleagues, the investigators loaded ^{225}Ac (α-particle emitter) into liposomes functionalized with the A10 PSMA RNA aptamer.[116] Powerful therapeutics, such as α-particle emitters, have been suggested as ideal for cancer treatment because of their high energy and short range, resulting in high killing efficacies and low irradiation of the surrounding healthy tissues. The investigators demonstrated that the aptamer-targeted liposomes loaded with ^{225}Ac selectively bind, become internalized, and kill PSMA-expressing cells, including endothelial cells induced to express PSMA. These findings, combined with the unique ability of liposomes to be easily tuned (in terms of size and surface modification) for optimal biodistribution, indicate the potential of PSMA-targeting liposomes encapsulating α-particle emitters for selective anti-cancer and anti-vascular α radiotherapy.

15.4 Conclusions

Aptamers are being developed for basic research as well as diagnostic, imaging, and therapeutic applications. In this chapter, we have discussed the body of literature centered on PSMA DNA and RNA aptamers for the development of targeted approaches to drug delivery, imaging, and therapy of PCa. We highlighted the preferred application (niche) and clinical potential of each of these aptamers. We discussed advances in the SELEX process that have enabled the identification of PSMA aptamers with enhanced ability to bind to PSMA on the cell surface and undergo rapid uptake into PSMA-expressing PCa cells for targeted drug-delivery applications. We also discussed major challenges with respect to manufacturing and scale-up of long oligonucleotides and multi-component aptamer-functionalized nanoparticle drugs.

Over recent years, advances in chemical synthesis and the increased availability and lower costs of raw materials have significantly reduced costs of manufacturing. Similarly, use of modified nucleotides has dramatically improved both the stability and safety of therapeutic oligonucleotides. Like other therapeutic oligonucleotides entering clinical use, aptamers can be conjugated to high-molecular-weight PEG or nanoparticles to improve pharmacokinetics and biodistribution and deliver therapeutic cargos. Although chemically-modified oligonucleotide-based drugs seem to display minimal toxicity, multi-component aptamer drug formulations warrant further toxicity studies.[117,118] As more oligonucleotide-based drugs enter clinical use and more efforts are invested in optimizing the development of aptamers for clinical applications, we anticipate clinical testing to begin by the end of this decade.

15.5 Future Perspectives

Before aptamers can be broadly translated, continued efforts are warranted in several key areas. For therapeutic aptamers, one limiting factor is limited data on their safety, especially as it relates to aptamer drug formulations.[119]

In a recent Phase IIb trial (RADAR), an aptamer drug, Pegnivacogin, a 2′-fluoropyrimidine-modified RNA aptamer that inhibits coagulation factor IXa, coupled to an approximately 40 kDa branched molecule of methoxypolyethylene glycol (mPEG), caused allergic reactions in 3 out of 640 patients with acute coronary syndrome.[120] Two patients met criteria for anaphylaxis (one life-threatening) and one was an isolated dermal reaction; together they led to early termination of the RADAR trial.[117] A broad investigation into a cause for these three events identified an association between high levels of antibodies to PEG and the first-exposure allergic reactions. This investigation was the first to document the potential clinical significance of pre-existing antibodies to PEG, a component of numerous consumer and medicinal products.[119] These findings warrant further testing of aptamer drug formulations, including aptamer–nanoparticle conjugates, AsiCs, and aptamer–drug conjugates.

The road for developing aptamer-based diagnostics is not as torturous as the aptamer therapeutics one. This is especially true for DNA aptamers. Unlike RNA aptamers, synthesis of DNA aptamers is relatively inexpensive and can be easily integrated with a detection platform.[121] Aptamer–target interactions on a diagnostic platform can be captured through several different detection modes (*e.g.* fluorescent reporters, surface stress, electrical signal, chemiluminescence, and nanodevices).[122–125] The binding properties of DNA aptamers, which include high affinity and highly specific interactions with their cognate target, result in high signal-to-noise detection ratios. These properties lend DNA aptamers applications ranging from immunofluorescence or immunohistochemistry to immune-capture. For example, PSMA DNA aptamers are being evaluated for the capture and detection of circulating PCa cells in blood (our unpublished work in collaboration with Ana Zamay). Given the broad potential of PSMA as a biomarker for cancers of the prostate, breast, and brain, PSMA DNA aptamers could be developed into assays for diagnosing and staging these malignancies as well as for assessing response to therapy.

Alternatively, aptamers can be conjugated to a variety of molecules, including other therapeutic oligonucleotides, small-molecule drugs, nanocarriers, and chelators. This, coupled to the ability of many aptamers to bind to cell-surface receptors (*e.g.* PSMA) and internalize into specific cells, is likely to improve the safety profile of existing drugs and enable delivery to most organs and tissues. As many more oligonucleotide drugs enter clinical use,[126] aptamers to cell-surface receptors like PSMA can be developed to enhance the efficacy and safety profiles of this promising new class of drugs. In the meantime, we can learn from work performed on PSMA aptamers to better understand their specific niche market such as cancer diagnostics, cancer detection, cancer therapy and immunotherapy, and imaging. The PSMA aptamer field is still growing, since the description of the first PSMA RNA aptamers in 2002, and applications of these aptamers to other cancers besides prostate cancer, are rapidly expanding.

Acknowledgements

The authors would like to acknowledge Colette Galet, PhD (Department of Surgery, University of Iowa) for careful editing of the manuscript.

This work was supported by grants to P. H. G. from the Department of Defense Congressionally Directed Medical Research Programs (PR150627P1), University of Iowa Carver College of Medicine (CCOM) (Carver Collaborative Pilot Grant Award 2015) and University of Iowa Award from The Office of the Vice President for Research and Economic Development (OVPRED 2015). This work was also supported by a postdoctoral NIH training grant to O. U. (T32HL007121-41).

References

1. A. D. Ellington and J. W. Szostak, *Nature*, 1990, **346**, 818.
2. C. Tuerk and L. Gold, *Science*, 1990, **249**, 505–510.
3. Z. Zhuo, Y. Yu, M. Wang, J. Li, Z. Zhang, J. Liu, X. Wu, A. Lu, G. Zhang and B. Zhang, *Int. J. Mol. Sci.*, 2017, **18**, 2142.
4. A. Ozer, J. M. Pagano and J. T. Lis, *Mol. Ther.–Nucleic Acids*, 2014, **3**, e183.
5. W. H. Thiel, K. W. Thiel, K. S. Flenker, T. Bair, A. J. Dupuy, J. O. McNamara, 2nd, F. J. Miller and P. H. Giangrande, *Methods Mol. Biol.*, 2015, **1218**, 187–199.
6. D. D. Dickey, G. S. Thomas, J. P. Dassie and P. H. Giangrande, *Methods Mol. Biol.*, 2016, **1364**, 209–217.
7. S. Kruspe and P. H. Giangrande, *Biomedicines*, 2017, **5**, 45.
8. J. P. Dassie and P. H. Giangrande, *Ther. Delivery*, 2013, **4**, 1527–1546.
9. D. D. Dickey and P. H. Giangrande, *Methods*, 2016, **97**, 94–103.
10. J. Zhou and J. Rossi, *Nat. Rev. Drug Discovery*, 2017, **16**, 181–202.
11. J. C. Rohloff, A. D. Gelinas, T. C. Jarvis, U. A. Ochsner, D. J. Schneider, L. Gold and N. Janjic, *Mol. Ther.–Nucleic Acids*, 2014, **3**, e201.
12. H. Sun and Y. Zu, *Molecules*, 2015, **20**, 11959–11980.
13. C. Wilson and A. D. Keefe, *Curr. Opin. Chem. Biol.*, 2006, **10**, 607–614.
14. J. Zhou and J. Rossi, *Nat. Rev. Drug Discovery*, 2017, **16**, 181.
15. A. D. Keefe, S. Pai and A. Ellington, *Nat. Rev. Drug Discovery*, 2010, **9**, 537–550.
16. J. G. Bruno, *Molecules*, 2015, **20**, 6866–6887.
17. P. Röthlisberger, C. Gasse and M. Hollenstein, *Int. J. Mol. Sci.*, 2017, **18**, 2430.
18. H. Sun, X. Zhu, P. Y. Lu, R. R. Rosato, W. Tan and Y. Zu, *Mol. Ther.–Nucleic Acids*, 2014, **3**, e182.
19. E. W. Ng, D. T. Shima, P. Calias, E. T. Cunningham Jr, D. R. Guyer and A. P. Adamis, *Nat. Rev. Drug Discovery*, 2006, **5**, 123.
20. Y.-H. Lao, K. K. Phua and K. W. Leong, *ACS Nano*, 2015, **9**, 2235–2254.
21. J. Barciszewski and B. F. Clark, *RNA Biochemistry and Biotechnology*, Springer Science & Business Media, 2012.

22. S. Ni, H. Yao, L. Wang, J. Lu, F. Jiang, A. Lu and G. Zhang, *Int. J. Mol. Sci.*, 2017, **18**, 1683.
23. J. Fohrer, M. Hennig and T. Carlomagno, *J. Mol. Biol.*, 2006, **356**, 280–287.
24. B. Schneider, Z. Moravek and H. M. Berman, *Nucleic Acids Res.*, 2004, **32**, 1666–1677.
25. Q. Zhu, G. Liu and M. Kai, *Molecules*, 2015, **20**, 20979–20997.
26. K. Germer, M. Leonard and X. Zhang, *Int. J. Biochem. Mol. Biol.*, 2013, **4**, 27.
27. J. S. Horoszewicz, E. Kawinski and G. P. Murphy, *Anticancer Res.*, 1987, **7**, 927–935.
28. B. T. Ristau, D. S. O'Keefe and D. J. Bacich, *Urol. Oncol.*, 2014, **32**, 272–279.
29. A. Ghosh and W. D. Heston, *J. Cell. Biochem.*, 2004, **91**, 528–539.
30. A. Ghosh, X. Wang, E. Klein and W. D. Heston, *Cancer Res.*, 2005, **65**, 727–731.
31. G. P. Murphy, G. M. Kenny, H. Ragde, R. L. Wolfert, A. L. Boynton, E. H. Holmes, S. L. Misrock, G. Bartsch, H. Klocker and J. Pointner, *Urology*, 1998, **51**, 89–97.
32. G. P. Murphy, A. A. A. Elgamal, S. L. Su, D. G. Bostwick and E. H. Holmes, *Cancer*, 1998, **83**, 2259–2269.
33. B. J. Miles, M. Shalev, E. Aguilar-Cordova, T. L. Timme, H. M. Lee, G. Yang, H. L. Adler, K. Kernen, C. K. Pramudji, T. Satoh, Y. Gdor, C. Ren, G. Ayala, T. M. Wheeler, E. B. Butler, D. Kadmon and T. C. Thompson, *Hum. Gene Ther.*, 2001, **12**, 1955–1967.
34. F. H. Schroder, J. Hugosson, M. J. Roobol, T. L. Tammela, S. Ciatto, V. Nelen, M. Kwiatkowski, M. Lujan, H. Lilja, M. Zappa, L. J. Denis, F. Recker, A. Berenguer, L. Maattanen, C. H. Bangma, G. Aus, A. Villers, X. Rebillard, T. van der Kwast, B. G. Blijenberg, S. M. Moss, H. J. de Koning and A. Auvinen, *N. Engl. J. Med.*, 2009, **360**, 1320–1328.
35. M. C. Haffner, I. E. Kronberger, J. S. Ross, C. E. Sheehan, M. Zitt, G. Muhlmann, D. Ofner, B. Zelger, C. Ensinger, X. J. Yang, S. Geley, R. Margreiter and N. H. Bander, *Hum. Pathol.*, 2009, **40**, 1754–1761.
36. N. Nomura, S. Pastorino, P. Jiang, G. Lambert, J. R. Crawford, M. Gymnopoulos, D. Piccioni, T. Juarez, S. C. Pingle, M. Makale and S. Kesari, *Cancer Cell Int.*, 2014, **14**, 26.
37. A. G. Wernicke, S. Varma, E. A. Greenwood, P. J. Christos, K. S. Chao, H. Liu, N. H. Bander and S. J. Shin, *APMIS*, 2014, **122**, 482–489.
38. R. S. Israeli, C. T. Powell, W. R. Fair and W. D. Heston, *Cancer Res.*, 1993, **53**, 227–230.
39. L. Zecca, M. B. Youdim, P. Riederer, J. R. Connor and R. R. Crichton, *Nat. Rev. Neurosci.*, 2004, **5**, 863–873.
40. C. Harding, J. Heuser and P. Stahl, *J. Cell Biol.*, 1983, **97**, 329–339.
41. E. M. van Dam, T. Ten Broeke, K. Jansen, P. Spijkers and W. Stoorvogel, *J. Biol. Chem.*, 2002, **277**, 48876–48883.
42. C. Barinka, C. Rojas, B. Slusher and M. Pomper, *Curr. Med. Chem.*, 2012, **19**, 856–870.

43. Y. Su, L. Yu, N. Liu, Z. Guo, G. Wang, J. Zheng, M. Wei, H. Wang, A. G. Yang, W. Qin and W. Wen, *Cancer Lett.*, 2013, **338**, 282–291.
44. S. E. Lupold, B. J. Hicke, Y. Lin and D. S. Coffey, *Cancer Res.*, 2002, **62**, 4029–4033.
45. W. M. Rockey, F. J. Hernandez, S. Y. Huang, S. Cao, C. A. Howell, G. S. Thomas, X. Y. Liu, N. Lapteva, D. M. Spencer, J. O. McNamara, X. Zou, S. J. Chen and P. H. Giangrande, *Nucleic Acid Ther.*, 2011, **21**, 299–314.
46. S. Y. Huang and X. Zou, *Proteins: Struct., Funct., Bioinf.*, 2010, **78**, 3096–3103.
47. S. Cao and S.-J. Chen, *RNA*, 2005, **11**, 1884–1897.
48. S. Cao, B. Fürtig, H. Schwalbe and S.-J. Chen, *J. Phys. Chem. B*, 2010, **114**, 13609–13615.
49. F. Almasi, S. L. M. Gargari, F. Bitaraf and S. Rasoulinejad, *Avicenna J. Med. Biotechnol.*, 2016, **8**, 104.
50. F. Atabi, S. L. M. Gargari, M. Hashemi and P. Yaghmaei, *Iran. J. Pharm. Res.: IJPR*, 2017, **16**, 35.
51. O. Boyacioglu, C. H. Stuart, G. Kulik and W. H. Gmeiner, *Mol. Ther.–Nucleic Acids*, 2013, **2**, e107.
52. H. Hong, S. Goel, Y. Zhang and W. Cai, *Curr. Med. Chem.*, 2011, **18**, 4195–4205.
53. A. Z. Wang and O. C. Farokhzad, *J. Nucl. Med.*, 2014, **55**, 353–356.
54. S. S. Gambhir, *Nat. Rev. Cancer*, 2002, **2**, 683.
55. C. Younes, R. Boisgard and B. Tavitian, *Curr. Pharm. Des.*, 2002, **8**, 1451–1466.
56. A. Perkins and S. Missailidis, *Q. J. Nucl. Med. Mol. Imaging*, 2007, **51**, 292.
57. J. Charlton, J. Sennello and D. Smith, *Chem. Biol.*, 1997, **4**, 809–816.
58. B. J. Hicke, A. W. Stephens, T. Gould, Y.-F. Chang, C. K. Lynott, J. Heil, S. Borkowski, C.-S. Hilger, G. Cook and S. Warren, *J. Nucl. Med.*, 2006, **47**, 668–678.
59. D. Kryza, F. Debordeaux, L. Azéma, A. Hassan, O. Paurelle, J. Schulz, C. Savona-Baron, E. Charignon, P. Bonazza and J. Taleb, *PLoS One*, 2016, **11**, e0149387.
60. W. M. Rockey, L. Huang, K. C. Kloepping, N. J. Baumhover, P. H. Giangrande and M. K. Schultz, *Bioorg. Med. Chem.*, 2011, **19**, 4080–4090.
61. J. P. Dassie, X. Y. Liu, G. S. Thomas, R. M. Whitaker, K. W. Thiel, K. R. Stockdale, D. K. Meyerholz, A. P. McCaffrey, J. O. McNamara, 2nd and P. H. Giangrande, *Nat. Biotechnol.*, 2009, **27**, 839–849.
62. D. Kim, Y. Y. Jeong and S. Jon, *ACS Nano*, 2010, **4**, 3689–3696.
63. T. Lammers, F. Kiessling, W. E. Hennink and G. Storm, *Mol. Pharm.*, 2010, 7, 1899–1912.
64. M. K. Yu, Y. Y. Jeong, J. Park, S. Park, J. W. Kim, J. J. Min, K. Kim and S. Jon, *Angew. Chem.*, 2008, **120**, 5442–5445.
65. V. Bagalkot, L. Zhang, E. Levy-Nissenbaum, S. Jon, P. W. Kantoff, R. Langer and O. C. Farokhzad, *Nano Lett.*, 2007, **7**, 3065–3070.
66. R. B. Sekar and A. Periasamy, *J. Cell Biol.*, 2003, **160**, 629–633.

67. J. A. Brzostowski, T. Meckel, J. Hong, A. Chen and T. Jin, *Curr. Protoc. Protein Sci.*, 2009, 19.15.11–19.15.12.
68. X. Zhang, S. Bloch, W. Akers and S. Achilefu, *Curr. Protoc. Cytom.*, 2012, ch. 12, Unit 12 27, DOI: 10.1002/0471142956.cy1227s60.
69. J. P. Dassie, L. I. Hernandez, G. S. Thomas, M. E. Long, W. M. Rockey, C. A. Howell, Y. Chen, F. J. Hernandez, X. Y. Liu, M. E. Wilson, L. A. Allen, D. A. Vaena, D. K. Meyerholz and P. H. Giangrande, *Mol. Ther.*, 2014, **22**, 1910–1922.
70. T. Demeritte, Z. Fan, S. S. Sinha, J. Duan, R. Pachter and P. C. Ray, *Chemistry*, 2014, **20**, 1017–1022.
71. Y. H. Bae and K. Park, *J. Controlled Release*, 2011, **153**, 198–205.
72. D. H. Kim and J. J. Rossi, *Biotechniques*, 2008, **44**, 613.
73. L. Aagaard and J. J. Rossi, *Adv. Drug Delivery Rev.*, 2007, **59**, 75–86.
74. C. Chakraborty, A. R. Sharma, G. Sharma, C. G. P. Doss and S.-S. Lee, *Mol. Ther.–Nucleic Acids*, 2017, **8**, 132–143.
75. J. O. McNamara, 2nd, E. R. Andrechek, Y. Wang, K. D. Viles, R. E. Rempel, E. Gilboa, B. A. Sullenger and P. H. Giangrande, *Nat. Biotechnol.*, 2006, **24**, 1005–1015.
76. T. C. Chu, K. Y. Twu, A. D. Ellington and M. Levy, *Nucleic Acids Res.*, 2006, **34**, e73.
77. F. Pastor, D. Kolonias, P. H. Giangrande and E. Gilboa, *Nature*, 2010, **465**, 227–230.
78. S. Lykke-Andersen and T. H. Jensen, *Nat. Rev. Mol. Cell Biol.*, 2015, **16**, 665.
79. M. W. Popp and L. E. Maquat, *Curr. Opin. Genet. Dev.*, 2018, **48**, 44–50.
80. H. Y. Liu, X. Yu, H. Liu, D. Wu and J. X. She, *Sci. Rep.*, 2016, **6**, 30346.
81. M. D. Rainey, M. E. Charlton, R. V. Stanton and M. B. Kastan, *Cancer Res.*, 2008, **68**, 7466–7474.
82. X. Ni, Y. Zhang, J. Ribas, W. H. Chowdhury, M. Castanares, Z. Zhang, M. Laiho, T. L. DeWeese and S. E. Lupold, *J. Clin. Invest.*, 2011, **121**, 2383–2390.
83. X. Ni, Y. Zhang, J. Ribas, W. H. Chowdhury, M. Castanares, Z. Zhang, M. Laiho, T. L. DeWeese and S. E. Lupold, *J. Clin. Invest.*, 2011, **121**, 2383–2390.
84. X. Wu, B. Ding, J. Gao, H. Wang, W. Fan, X. Wang, W. Zhang, X. Wang, L. Ye and M. Zhang, *Int. J. Nanomed.*, 2011, **6**, 1747.
85. L. Cong, F. A. Ran, D. Cox, S. Lin, R. Barretto, N. Habib, P. D. Hsu, X. Wu, W. Jiang and L. Marraffini, *Science*, 2013, 1231143.
86. R. Barrangou, C. Fremaux, H. Deveau, M. Richards, P. Boyaval, S. Moineau, D. A. Romero and P. Horvath, *Science*, 2007, **315**, 1709–1712.
87. F. J. Mojica, C. Díez-Villaseñor, E. Soria and G. Juez, *Mol. Microbiol.*, 2000, **36**, 244–246.
88. K. S. Makarova, N. V. Grishin, S. A. Shabalina, Y. I. Wolf and E. V. Koonin, *Biol. Direct*, 2006, **1**, 7.
89. J. A. Doudna and E. Charpentier, *Science*, 2014, **346**, 1258096.

90. S. Zhen, Y. Takahashi, S. Narita, Y.-C. Yang and X. Li, *Oncotarget*, 2017, **8**, 9375.
91. M. Croft, *Nat. Rev. Immunol.*, 2009, **9**, 271–285.
92. T. H. Watts, *Annu. Rev. Immunol.*, 2005, **23**, 23–68.
93. F. Pastor, D. Kolonias, J. O. McNamara, 2nd and E. Gilboa, *Mol. Ther.*, 2011, **19**, 1878–1886.
94. A. M. Wu and P. D. Senter, *Nat. Biotechnol.*, 2005, **23**, 1137–1146.
95. T. C. Chu, J. W. Marks, 3rd, L. A. Lavery, S. Faulkner, M. G. Rosenblum, A. D. Ellington and M. Levy, *Cancer Res.*, 2006, **66**, 5989–5992.
96. L. Kelly, C. Kratschmer, K. E. Maier, A. C. Yan and M. Levy, *Nucleic Acid Ther.*, 2016, **26**, 156–165.
97. A. Shapira and I. Benhar, *Toxins*, 2010, **2**, 2519–2583.
98. A. Rust, L. J. Partridge, B. Davletov and G. M. Hautbergue, *Toxins*, 2017, **9**, 344.
99. V. Bagalkot, O. C. Farokhzad, R. Langer and S. Jon, *Angew. Chem. Int. Ed.*, 2006, **45**, 8149–8152.
100. O. Boyacioglu, C. H. Stuart, G. Kulik and W. H. Gmeiner, *Mol. Ther.-Nucleic Acids*, 2013, **2**, e107.
101. M. Ferrari, *Nat. Rev. Cancer*, 2005, **5**, 161.
102. R. F. Service, *Science*, 2003, **302**, 1310.
103. G. M. Whitesides, *Small*, 2005, **1**, 172–179.
104. W. Cai and X. Chen, *Small*, 2007, **3**, 1840–1854.
105. R. Sinha, G. J. Kim, S. Nie and D. M. Shin, *Mol. Cancer Ther.*, 2006, **5**, 1909–1917.
106. M. E. Davis and D. M. Shin, *Nat. Rev. Drug Discovery*, 2008, **7**, 771.
107. O. C. Farokhzad, S. Jon, A. Khademhosseini, T.-N. T. Tran, D. A. LaVan and R. Langer, *Cancer Res.*, 2004, **64**, 7668–7672.
108. E. Levy-Nissenbaum, A. F. Radovic-Moreno, A. Z. Wang, R. Langer and O. C. Farokhzad, *Trends Biotechnol.*, 2008, **26**, 442–449.
109. O. C. Farokhzad, S. Jon, A. Khademhosseini, T. N. Tran, D. A. Lavan and R. Langer, *Cancer Res.*, 2004, **64**, 7668–7672.
110. O. C. Farokhzad, J. Cheng, B. A. Teply, I. Sherifi, S. Jon, P. W. Kantoff, J. P. Richie and R. Langer, *Proc. Natl. Acad. Sci. U. S. A.*, 2006, **103**, 6315–6320.
111. Z. Chen, Z. Tai, F. Gu, C. Hu, Q. Zhu and S. Gao, *Eur. J. Pharm. Biopharm.*, 2016, **107**, 130–141.
112. S. Dhar, F. X. Gu, R. Langer, O. C. Farokhzad and S. J. Lippard, *Proc. Natl. Acad. Sci. U. S. A.*, 2008, **105**, 17356–17361.
113. M. Wu, Y. Wang, Y. Wang, M. Zhang, Y. Luo, J. Tang, Z. Wang, D. Wang, L. Hao and Z. Wang, *Int. J. Nanomed.*, 2017, **12**, 5313.
114. S. S. Chang, V. E. Reuter, W. Heston, N. H. Bander, L. S. Grauer and P. B. Gaudin, *Cancer Res.*, 1999, **59**, 3192–3198.
115. H. Liu, P. Moy, S. Kim, Y. Xia, A. Rajasekaran, V. Navarro, B. Knudsen and N. H. Bander, *Cancer Res.*, 1997, **57**, 3629–3634.
116. A. Bandekar, C. Zhu, R. Jindal, F. Bruchertseifer, A. Morgenstern and S. Sofou, *J. Nucl. Med.*, 2014, **55**, 107–114.

117. S. M. Nimjee, T. J. Povsic, B. A. Sullenger and R. C. Becker, *Nucleic Acid Ther.*, 2016, **26**, 147–155.
118. N. Que-Gewirth and B. Sullenger, *Gene Ther.*, 2007, **14**, 283.
119. N. J. Ganson, T. J. Povsic, B. A. Sullenger, J. H. Alexander, S. L. Zelenkofske, J. M. Sailstad, C. P. Rusconi and M. S. Hershfield, *J. Allergy Clin. Immunol.*, 2016, **137**, 1610–1613.e1617.
120. T. J. Povsic, J. P. Vavalle, L. H. Aberle, J. D. Kasprzak, M. G. Cohen, R. Mehran, C. Bode, C. E. Buller, G. Montalescot and J. H. Cornel, *Eur. Heart J.*, 2012, **34**, 2481–2489.
121. B. Strehlitz, N. Nikolaus and R. Stoltenburg, *Sensors*, 2008, **8**, 4296–4307.
122. C.-H. Leung, D. S.-H. Chan, H.-Z. He, Z. Cheng, H. Yang and D.-L. Ma, *Nucleic Acids Res.*, 2011, **40**, 941–955.
123. R. E Wang, Y. Zhang, J. Cai, W. Cai and T. Gao, *Curr. Med. Chem.*, 2011, **18**, 4175–4184.
124. S. Song, L. Wang, J. Li, C. Fan and J. Zhao, *TrAC Trends Anal. Chem.*, 2008, **27**, 108–117.
125. Y. S. Kim, N. H. A. Raston and M. B. Gu, *Biosens. Bioelectron.*, 2016, **76**, 2–19.
126. C. A. Stein and D. Castanotto, *Mol. Ther.*, 2017, **25**, 1069–1075.
127. M. I. Davis, M. J. Bennett, L. M. Thomas and P. J. Bjorkman, *Proc. Natl. Acad. Sci. U. S. A.*, 2005, **102**, 5981–5986.
128. L. Zecca, M. B. Youdim, P. Riederer, J. R. Connor and R. R. Crichton, *Nat. Rev. Neurosci.*, 2004, **5**, 863.

CHAPTER 16

Aptamers and Clinical Applications

G. ZON

TriLink BioTechnologies, 9955 Mesa Rim Road, San Diego, 92121, USA
*E-mail: gzon@trilinkbiotech.com

16.1 Introduction

By remarkable coincidence in 1990, two labs independently published the conceptually similar achievement of selecting RNA sequences that bind to specific target molecules, *i.e.* aptamers. Although somewhat different technical approaches were used, they nevertheless both started with a highly complex mixture of synthetic polyribonucleotides having "random sequence" (A/G/C/U) regions. The procedure described by Tuerk and Gold,[1] which was referred to as "systematic evolution of ligands by exponential enrichment" (SELEX), now has ~7500 citations in Google Scholar. The alternative strategy reported by Ellington and Szostak[2] has likewise had a major impact and is reflected in the ~6800 citations in Google Scholar to date. Indeed, the wide range of basic and applied topics related to aptamers covered here all follow from these two truly seminal publications and the thousands of subsequent investigations.

To put the present chapter into perspective, it is worth emphasizing that aptamers can function directly as therapeutic agents or serve as targeting agents (Figure 16.1), as reviewed by Poolsup and Kim.[3] Direct action

Drug Discovery Series No. 68
Advances in Nucleic Acid Therapeutics
Edited by Sudhir Agrawal and Michael J. Gait

Published by the Royal Society of Chemistry, www.rsc.org

of an aptamer drug can include direct binding to extracellular protein, cell-surface protein, or to viral surface protein, wherein each target is associated with a particular disease or clinical indication of interest for achieving therapeutic intervention. These three targets are also addressable by conventional small molecules or antibodies. However, it was recognized early on that aptamers might provide advantages in specificity or cost, respectively. The use of aptamers as targeting agents is depicted in Figure 16.1 as conjugates for delivery of either a nanoparticle loaded with drug, a therapeutic oligonucleotide, such as short-interfering RNA (siRNA), or a small-molecule drug, such as a cytotoxic agent. Not shown, but by conceptual extension, aptamer conjugates to fluorescent dyes or other types of detectable moieties can be used to target cells for detection or imaging *in vitro* or *in vivo* in either preclinical or clinical contexts, as well as in basic research.

Examples of each of these modalities are described in Section 16.2, which includes a variety of exemplary aptamer drugs reported recently in pre-clinical

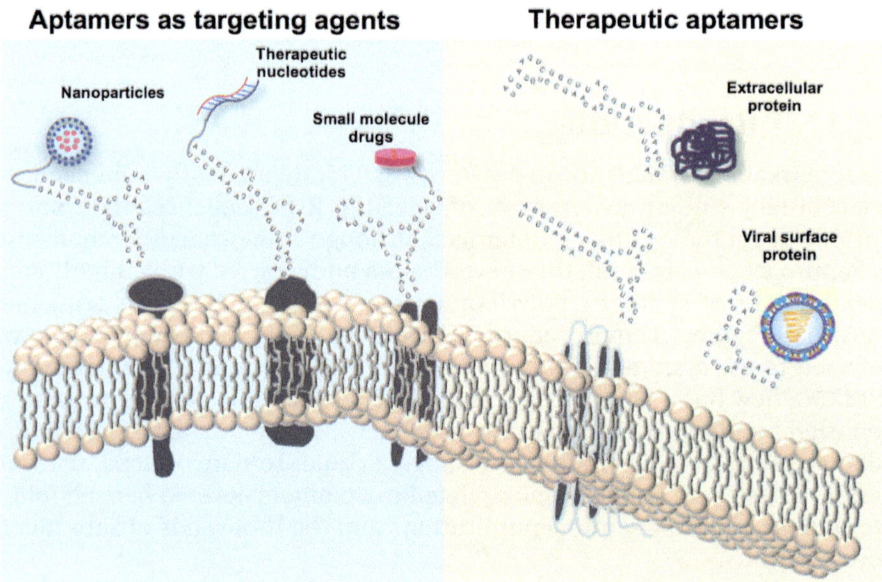

Figure 16.1 Cartoon depicting various modes for RNA aptamers (shown as stem–loop structures) functioning as therapeutic agents (left panel) or as cell-targeting agents (right panel). Modes for aptamers functioning as therapeutic agents involve aptamers binding to either extracellular protein, viral surface protein, or cell surface protein targets. Modes for aptamers functioning as cell-targeting agents involve aptamers linked to either a nanoparticle loaded with a drug, a therapeutic siRNA nucleotide, or a small-molecule drug for delivery by aptamer binding to cell surface proteins that trigger uptake. Reproduced from ref. 3 with permission from Elsevier, Copyright 2017.

development studies, and which indicate potential future directions of study. In Section 16.3, various recently published clinically relevant examples are given of aptamer-based targeting investigations, wherein aptamers are used to specifically target disease cells for efficient delivery of other types of drugs, such as cytotoxic agents or short interfering RNA (siRNA). In Section 16.4, there is an overview of the status of more than 30 clinical trials of aptamer drugs that were listed in 2017 in ClinicalTrials.gov,[4] which is a comprehensive registry and results database of publicly and privately supported clinical studies of human participants conducted around the world that is a service of the US National Institutes of Health. The final Section 16.5, describes Conclusions and Prospects.

In none of these cases is attention given to the method of aptamer selection, details for which are available in cited references, or in a number of lead ref. 5–9. However, it is important to note that requisite polydeoxyribonucleotide templates having "random sequence" regions with $(A/G/C/T)_n$ and $n = 20$, 30, *etc.* (for transcription into random RNA pools) have been shown by deep sequencing to vary substantially in the degree of randomness depending on the commercial supplier (TriLink BioTechnologies, unpublished data). Attendant consequences for deviation from "randomness" have been discussed by Ellington and coworkers.[9]

16.2 Recent Preclinical Studies of Aptamer Drugs

This section is devoted to published examples of preclinical investigations of potentially therapeutic aptamers. These representative examples selected from this literature are grouped into two broad categories, namely, "aptamer structures" and "non-ocular diseases" in subsections 16.2.1 and 16.2.2 that follow.

16.2.1 Aptamer Structures

16.2.1.1 SOMAmers and X-Aptamers

The recognition that there are examples of difficult protein targets for which standard RNA and DNA SELEX did not yield high-affinity aptamers led Gold and others at SomaLogic to create a new class of aptamer, the slow off-rate modified aptamer (SOMAmer). SOMAmers feature base modifications (Figure 16.2) that were found to be functionally compatible with enzymes used for SELEX enzymes, and which provide more structural diversity. The 2010 publication[10] of this innovation for efficient selection of aptamers for almost any protein has been viewed more than 33 000 times, thus indicating the importance of SOMAmers for proteomics assays, which are reported[11] to now include 3000 proteins. Not surprisingly, SOMAmers have also proved to be promising candidates for possible drug development, as exemplified by the following preclinical studies.

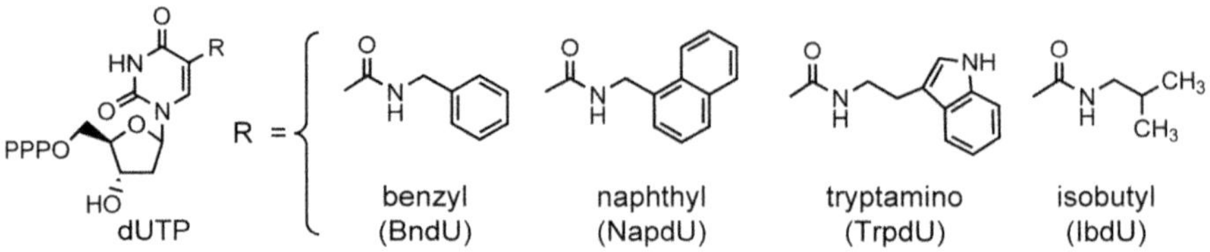

Figure 16.2 Structures in Gold *et al.*[10] for chemically modified 5′-triphosphate dU analogs bearing C5-position substituents R as shown. Reproduced from ref. 10 [https://doi.org/10.1371/journal.pone.0015004] under the terms of the Creative Commons Attribution License [https://creativecommons.org/licenses/by/4.0/].

A large team of investigators at SomaLogic and Otsuka Pharma reported[12] results aimed at developing a SOMAmer targeting interleukin-6 (IL-6), which is a pleiotropic cytokine that regulates immune and inflammatory responses, and for which overproduction is a hallmark of inflammatory diseases. Indirect inhibition of IL-6 signaling with the anti-IL-6-receptor antibody tocilizumab has provided some clinical benefit to patients. However, it was proposed that direct cytokine inhibition may be a more effective option. By use of SELEX, two classes of IL-6 SOMAmers were discovered starting from modified DNA libraries containing 40 random positions and either 5-(*N*-benzylcarboxamide)-2′-deoxyuridine (BndU) or 5-[*N*-(1-naphthylmethyl)carboxamide]-2′-deoxyuridine (NapdU) replacing dT. Post-SELEX optimization of one BndU and one NapdU SOMAmer led to improvements in IL-6 binding (tenfold) and inhibition activity (greater than 20-fold), resulting in lead SOMAmers with sub-nanomolar affinity [dissociation constant (K_d) = 0.2 nM] and potency [concentration giving 50% of maximum inhibition (IC_{50}) = 0.2 nM]. Although similar in inhibition properties and stability in human serum *in vitro* (>48 h), the two SOMAmers have unique sequences. Both SOMAmers prevented IL-6 signaling by blocking the interaction of IL-6 with its receptor and inhibited the proliferation of tumor cells *in vitro* as effectively as tocilizumab, which led to the conclusion that these SOMAmers may provide effective therapeutic alternatives for patients suffering from inflammatory diseases.

Two years later, follow-on work at SomaLogic and Otsuka Pharma with these anti-IL-6 SOMAmers led to a report[13] that one aptamer, SL1026, delays the onset and reduces the severity of rheumatoid symptoms in a collagen-induced arthritis model in cynomolgus monkeys. SL1026 (1 and 10 mg kg^{-1}), administered four-times daily delayed the progression of arthritis and the concomitant decrease in serum IL-6 levels compared with the untreated control group. Furthermore, SL1026 inhibited IL-6-induced STAT3 phosphorylation *ex vivo* in T lymphocytes from human blood and IL-6-induced C-reactive protein and serum amyloid A production in human primary hepatocytes. Importantly, SOMAmer treatment did not elicit an immune response, as evidenced by the absence of anti-SOMAmer antibodies in plasma of treated monkeys. It was concluded that SOMAmer antagonists of IL-6 may be attractive agents for the treatment of rheumatoid arthritis.

More recently, researchers at SomaLogic have published[11] a description of a major improvement of SOMAmer technology involving selection of DNA aptamers with two modified bases (Figure 16.3), which contrasts with only one modified base in first-generation SOMAmers[12,13] targeting IL-6. This recent study involved targeting human proprotein convertase subtilisin/kexin type 9 (PCSK9), which is of considerable therapeutic interest for autosomal dominant hypercholesterolemia.[14] Within the set of modifications tested, pairing of Tyr, 5-[*N*-(4-hydroxyphenyl-2-ethyl)carboxamide]-2′-deoxy (Tyr)-dU with either Nap-dC or 5-[*N*-(phenyl-3-propyl)carboxamide]-2′-deoxy (Pp)-dC produced the best-performing combinations. The synergistic benefit

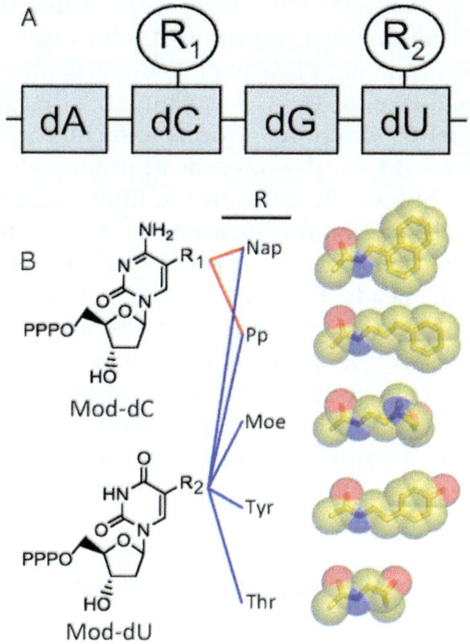

Figure 16.3 (A) Schematic of four bases in a DNA SOMAmer library showing double modifications with modified derivatives dC and dU. Modifications on dC are represented by the R_1 group and modifications on dU are represented by the R_2 group. (B) Structures of mod-dC and mod-dU triphosphates bearing a 5-(*N*-substituted-carboxamide) functional group R_1 and R_2, respectively, and space-filling models of R groups as follows: Nap, 5-[*N*-(1-naphthylmethyl)carboxamide]-2′-deoxy; Pp, 5-[*N*-(phenyl-3-propyl)carboxamide]-2′-deoxy; Moe, 5-[*N*-(1-morpholino-2-ethyl)carboxamide]-2′-deoxy; Tyr, 5-[*N*-(4-hydroxyphenyl-2-ethyl) carboxamide]-2′-deoxy; and Thr, 5-[*N*-(S-2-hydroxypropyl)carboxamide]-2′-deoxy. R_1 groups tested on dC were Nap and Pp (red lines), whereas R_2 groups tested on dU were Nap, Pp, Moe, Tyr, and Thr (blue lines). Reproduced from ref. 11 with permission from PNAS.

of combining two modified nucleotides in the same library was apparent from the fact that libraries with either Tyr-dU or Pp-dC alone were not sufficient to generate high-affinity ligands. Another advantage of using libraries with two modified nucleotides is enhanced epitope coverage, which is evidenced by the ability to identify ligands that are cross-reactive with PCSK9 from different species. Finally, dramatically enhanced nuclease resistance was observed for most SOMAmers with two modifications, without the need for sugar–phosphate backbone modifications, which adds an important feature for therapeutic uses.

In 2012, Gorenstein and collaborators published[15] the first report of a bead-based selection method for a new class of aptamers, X-Aptamers, wherein a small drug-like moiety was incorporated into an aptamer to achieve greatly enhanced binding attributed to increased structural diversity of the pendant moiety. This methodology, which has been recently described in detail,[16] has been expanded and commercialized by AM Biotech to X-Aptamers that now include 2′-deoxyuridine analogs bearing C5 substituents that are amino acid-like or a boronic acid (Figure 16.4). It is claimed[16] that X-Aptamers provide an unlimited number of functional groups and thus are superior targeting agents. In contrast to SELEX, the microbead process does not rely on PCR amplification of the oligonucleotides in the library and thus enables more diversity of chemical functionality that could achieve increased specificity and affinity. A recent publication[17] reports that an X-Aptamer library has approximately 2×10^9 beads, each bead carrying about 3×10^3 copies of a potential X-Aptamer consisting of a unique chemically modified strand of DNA. In this publication,[17] X-Aptamers were identified against two immune checkpoint proteins, namely recombinant programmed death 1 (PD-1) and programmed death ligand 1 (PD-L1). Selections were carried out by use of a bead-based X-Aptamer (XA) library containing several different amino acid functional groups attached to the C5 position of dU. The binding affinities

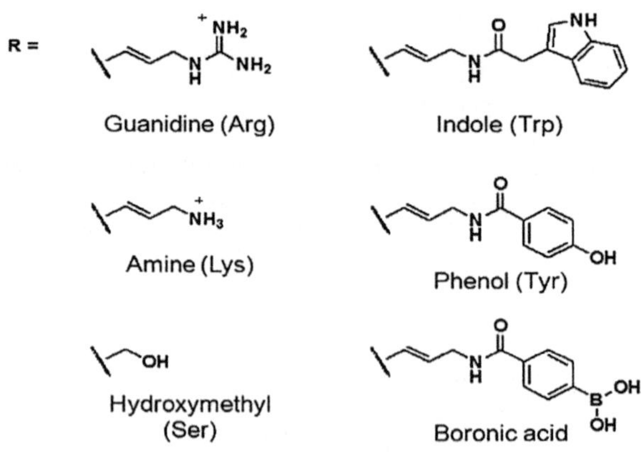

Figure 16.4 Structures of C5-position substituents in chemically modified dU nucleotides in X-Aptamer libraries. Reproduced with permission from AM Biotech.

and specificities of the selected XA-PD1 and XA-PDL1 were validated by human PD-1 and human PD-L1 expression cells, as well as by binding to human pancreatic ductal adenocarcinoma tissue, to demonstrate that XA-PD1 and XA-PDL1 can mimic antibody functions in *in vitro* assays.

16.2.1.2 Thioaptamers

Early development of P-chiral monothioaptamers and achiral dithioaptamers (Figure 16.5) has been chronicled by Thiviyanathan and Gorenstein.[18] Screening of P-chiral monothioaptamer libraries against E-selectin, elevated expression of which occurs in many types of inflammatory diseases, including diabetes, atherosclerosis, rheumatoid arthritis, and cancer, has been investigated by Mann *et al.*[19] This work led to discovery of a monothioaptamer that binds to E-selectin with nanomolar affinity on cultured endothelial cells and tumor-associated vasculature in human and mouse carcinomas. Since then there have been other reports and a recent review[20] that support potential clinical utility of thioaptamers, which are believed to exhibit tight binding due at least in part to favorable hydrophobic[21] interactions of sulfur moieties in these compounds.

Follow-on investigations of the above mentioned[19] E-selectin monothioaptamer (ESTA) have been reported by Morita *et al.*,[22] who sought to develop a PEGylated E-selectin targeting monothioaptamer with improved pharmaceutical properties. Serial deletion of stem–loops revealed a truncated version (ESTA7) as the minimally effective backbone structure necessary for inhibition of E-selectin at a level similar to ESTA. Chemical conjugation of 5 kDa and 10 kDa methoxy-polyethyleneglycol (PEG) did not interfere with ESTA7 binding. Importantly, *in vivo* experiments demonstrated that only 10 kDa PEG-conjugated ESTA7 (ESTA7-p10) inhibited metastases at a level similar to ESTA. Additionally, a single intravenous injection of ESTA7-p10 inhibited the development of lung, brain, and bone metastases through the blockade of E-selectin, and led to extension of elimination half-life, with a

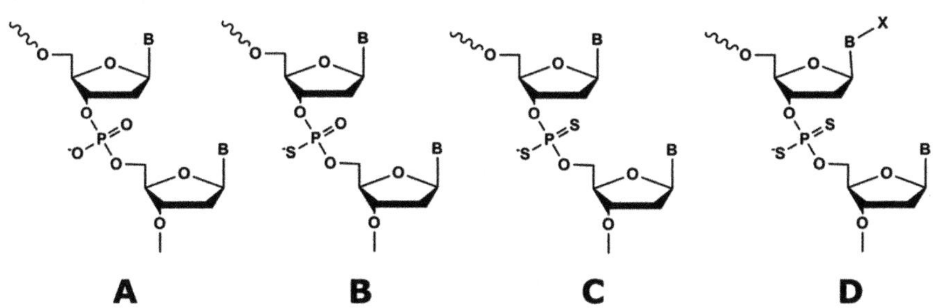

Figure 16.5 Partial structures for internucleotide linkages in aptamers that are unmodified (A), monophosphorothioate (B), or diphosphorothioate (C and D). Reproduced from ref. 18 with permission from John Wiley and Sons, Copyright 2012 WILEY-VCH Verlag GmbH & Co. KGaA, Weinheim.

longer interval between dosing in a spontaneous metastasis model, and was well tolerated.

HER2, which is a member of the human epidermal growth factor receptor family and which is overexpressed in certain aggressive types of breast cancer, has been investigated as an aptamer target by Hu *et al.*[23] They found a thioaptamer that could bind to the extracellular domain of HER2 with a K_d of 172 nM and which had minimal cross reactivity to trypsin or IgG. This thioaptamer was found to be more resistant to serum nucleases than the unmodified aptamer, and was capable of binding to the HER2-positive breast cancer cell lines SK-BR-3 and MDA-MB-453, but not to the HER2-negative cell line MDA-MB-231.

16.2.1.3 Photo-regulated Aptamers

Exogenous control of aptamer function can be achieved by optochemical approaches, such as introduction of photo-labile groups, so called cages, at strategic positions within an aptamer, as reviewed by Liu and Deiters.[24] These approaches expand the properties of aptamers far beyond those found in naturally occurring nucleic acids and allow precise spatiotemporal control over biological functions. Cages are light-sensitive groups, which can be attached to defined positions of nucleobases, thereby interfering, for example, with their Watson–Crick base pairing capabilities. The first example of a caged aptamer was reported in 2005 by Heckel and Mayer,[25] who introduced an *o*-nitrophenylpropyl group into thymidine residues of the thrombin binding DNA aptamer HD1.[26] Due to steric hindrance the caged aptamer showed a strongly reduced affinity towards its target. However, upon irradiation and deprotection the binding affinity to the target could be restored, thus enabling light-control of thrombin-dependent blood clotting. In a second study, Heckel *et al.*[27] used a similar approach to obtain an aptamer that was inhibited rather than activated upon irradiation, wherein light-dependent inactivation was achieved by extending the aptamer with a caged antisense region that blocked base-pairing properties. When the cage on the complementary antisense region was removed, base pairing was initiated resulting in the formation of a stable hairpin structure and thus leading to a loss of aptamer activity.

More recently, Li *et al.*[28] proposed that light triggering can be used to achieve spatiotemporal control of binding of an aptamer injected systemically to its target tissue *in vivo*, which would have implications for control of delivery of therapeutic aptamers and/or conjugated drugs or drug delivery systems. They designed a photo-triggerable system whereby the aptamer of interest is inactivated by hybridization to a photolabile complementary oligonucleotide (Figure 16.6). Upon irradiation, the complementary sequence breaks down, releasing the functional aptamer.

The aptamer of interest was the single-stranded DNA 26-mer aptamer AS1411, which is discussed in Section 16.4.2, and has the sequence 5′-GGT

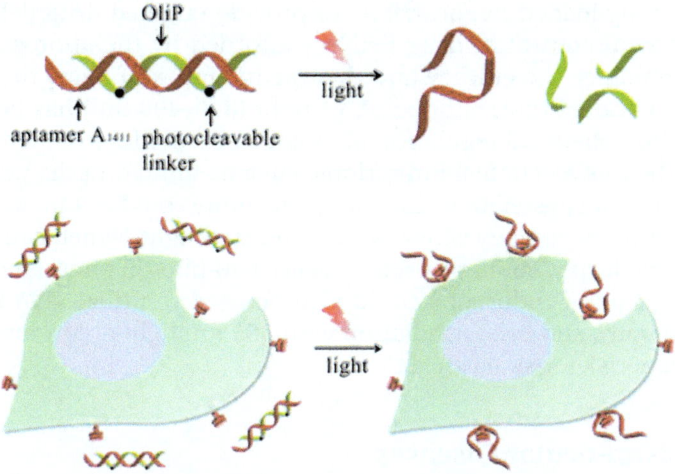

Figure 16.6 The AS1411 aptamer (A1411, red DNA strand) is hybridized to a complementary oligonucleotide (OliP, green DNA strand) containing photo-cleavable bonds (black dots). The hybridized complex cannot bind cells. A1411 can be released by light-triggered breakage of the OliP, allowing binding to cell surfaces. Reproduced from ref. 28 with permission from *PNAS*.

GGT GGT GGT TGT GGT GGT GGT GG-3′. This oligonucleotide folds and binds with high affinity and selectivity to nucleolin, which is overexpressed in certain cancer cells. A complementary photo-triggerable inhibitory oligonucleotide (OliP) was designed with the sequence 5′-CCA CCA L CCA CCA L CAA CCA C-3′, where L indicates an abasic, internucleotide linker[28] featuring a photo-labile *o*-nitrophenylethyl moiety. Intratumoral photoactivation of A1411/OliP was demonstrated by microscopy in nude mice with 4T1 tumors (diameters ~ 6–8 mm) on their mammary fat pads. Mice were administered A1411 or A1411/OliP (250 nmol kg^{-1}) intravenously, both labeled with the fluorescent dye Cy3. Injection of free Cy3-A1411 resulted in strong extravasated fluorescence in the tumor tissue 10 min after injection, whereas administration of Cy3-A1411–OliP resulted in minimal fluorescence in the same time frame. Irradiation of the tumor site (365 nm light at 200 mW cm^{-2} for 3 min) immediately after injection of Cy3-A1411–OliP resulted in more extravasated fluorescence throughout the tumor tissue compared with that obtained without light irradiation. Li *et al.*[28] concluded from these results that A1411–OliP could be dehybridized to liberate A1411 upon irradiation at the tumor site *in vivo*. Fluorescence imaging of tumor sections 2 h after administration showed distribution of Cy3-A1411 throughout the tumor tissue, whereas A1411/OliP was mainly localized around blood vessels, which Li *et al.*[28] presumed to be due to inhibition of the aptamer's binding to cells.

As discussed in Section 16.3, aptamers can be covalently bound to a wide variety of therapeutic agents, including chemotherapeutics or enzymes, siRNA, or drug-loaded nanocarriers, to provide targeted drug delivery. The spatiotemporal control of drug delivery afforded by the approach of Li *et al.*[28] may enhance the efficacy and therapeutic index of many drugs. These investigators noted that, although UVA light (320–400 nm) has been widely used for photochemical regulation of biological activities *in vitro* and *in vivo*, UVA light possesses clinical limitations, such as limited tissue penetration. However, they suggested that clinical applicability could be enhanced by use of two-photon technology whereby near-infrared light, which can penetrate tissues more deeply, could be used to cleave two-photon photolabile groups. Finally, fiber-optic endoscopy could also be used to utilize UVA light deep within the body, and brief irradiation with 365 nm light at low energy is not considered a risk for skin cancer.

16.2.2 Non-ocular Diseases

16.2.2.1 Cardiovascular and Cerebrovascular Diseases

Aptamers for treatment of cardiovascular diseases initially focused on myocardial infarction (MI) by selecting aptamers for binding to and inactivating proteins that are overproduced. For example, anticoagulant therapy is employed to prevent recurrent MI. Porcine and murine studies[29] utilized an aptamer against factor IXa to inhibit coagulation that was rapidly reversed by introduction of a complimentary oligonucleotide that inactivated the aptamer. Alternatively, an aptamer has been investigated for inhibition of von Willebrand factor (VWF), which drives the coagulation cascade, and the aptamer has been evaluated for the same purpose in a clinical trial.[30]

More recently, neointimal formation related to cardiovascular disease has been investigated in preclinical studies by Thiel *et al.*[31] using an RNA aptamer targeting vascular smooth muscle cells (VSMCs). These researchers point out that VSMC migration and proliferation is a key pathological process in vascular disorders, including arteriosclerosis, restenosis, vein-bypass graft disease, and allograft arteriopathy. The only available therapy to directly prevent VSMC activation is the local delivery of antiproliferative or antimigratory drugs (*e.g.*, paclitaxel) by percutaneous drug-eluting stents (DES). However, the safety of DES is limited by delayed re-endothelialization due to the drug's antiproliferative effect on endothelial cells (EC), contributing to late stent thrombosis and the need for prolonged dual platelet therapy. Thiel *et al.*[31] argue that, in contrast to changes in stent-delivery platforms to improve efficacy and safety, little progress has been made in optimizing the therapeutic drug cargo, which has been limited to derivatives of rapamycin. This improved efficacy compared with paclitaxel but still lacked cell-specificity. Consequently, Thiel *et al.*[31] considered aptamers as leading to a possible ideal therapy of hyperproliferative vascular disease that would specifically target VSMCs without impairing endothelial cell healing.

To develop a VSMC-targeted smart drug for the treatment of vascular disease, Thiel *et al.*[31] characterized an RNA aptamer (Apt 14) that selectively binds to VSMCs.[32] They showed that Apt 14 is a potent inhibitor of VSMC migration in response to multiple agonists by a mechanism that involves interruption of platelet-derived growth factor receptor-beta (PDGFR-β) activation and phosphatidylinositol-3-kinase (PI3K)/Ak strain transforming (Akt)-dependent signaling in VSMC. In addition, Apt 14 was shown to attenuate neointimal formation in a murine model of vascular injury. Thiel *et al.*[31] concluded that these findings, together with the long half-life in human serum, cross-reactivity with human VSMCs, and safety profile with human cells, justify additional preclinical testing of Apt 14 as a novel inhibitor of hyperproliferative vascular disease.

According to Varshney *et al.*,[33] pathological mechanisms operating in cerebrovascular diseases mediated *via* homocysteine (Hcy) have gained increased attention, and it has been known that Hcy causes endothelial dysfunction by inducing oxidative stress [correlated with the generation of reactive oxygen species (ROS)] in most cerebrovascular disorders. To slow down or prevent the onset of these diseases mediated by an increased level of Hcy, these researchers selected RNA aptamers targeting Hcy using SELEX. The best-binding Hcy RNA aptamers were further investigated for biological and functional characterization. A cell line (bEND3) treated with Hcy in the presence or absence of Hcy RNA aptamers for 24 h provided evidence that these aptamers ameliorate the effects of ROS, thereby reducing the oxidative stress in bEND cell lines. Apoptotic and autophagic cell death were measured, as well as ROS production and mitochondrial fusion and fission. Treatment of bEnd3 with Hcy RNA aptamer attenuated the production of free radicals and protected the cells from oxidative damage, which led Varshney *et al.*[33] to suggest that the Hcy-RNA aptamer has beneficial effects on cerebrovascular diseases and could be developed as a potential therapeutic agent against hyperhomocysteinemia.

16.2.2.2 Alzheimer's Disease

Currently, there are 44 million people suffering from Alzheimer's Disease (AD) or a related dementia globally, and it is estimated that by 2050 this will increase to over 135 million people worldwide, thereby becoming a global epidemic.[34] Liang *et al.*[35] noted that build-up of amyloid-β (Aβ) peptides in the brain has been linked to AD pathogenesis and may represent a key target for AD intervention. Furthermore, Aβ formation occurs *via* sequential proteolytic processing of amyloid precursor protein (APP) and is catalyzed by β- and γ-secretases. The β-site APP-cleaving enzyme 1 (BACE1) involved in cleavage of APP is upregulated in the brains and cerebrospinal fluid of AD patients. Liang *et al.*[35] suggested that BACE1 represents a promising target for mechanistic-based AD treatment, and added that to date no safe and effective BACE1 inhibitor has been used in clinical populations.

A purified human BACE1 extracellular domain was used by Liang *et al.*[35] as a target for SELEX, which yielded an unmodified DNA aptamer, A1, that exhibited a distinct inhibitory effect on BACE1 activity in an AD cell model. Furthermore, western blot analysis indicated that APPβ expression significantly decreased in the A1 treated group *vs.* the control group. Although these findings were considered to support preliminary feasibility of an aptamer evolved as a potential BACE1 inhibitor, several factors were discussed regarding *in vivo* applicability, namely, degradation of unmodified DNA, short shelf-life, and ability to penetrate the blood-brain barrier (BBB). Regarding the last of these factors, it is encouraging to note that development of systemically delivered oligonucleotide therapeutics that can cross the BBB has been recently demonstrated in a mouse model of spinal muscular atrophy (SMA) Shabanpoor *et al.*[36] investigated the ability of various BBB-crossing peptides for central nervous system (CNS) delivery of a splice-switching phosphorodiamidate morpholino oligonucleotide (PMO) targeting survival motor neuron 2 (*SMN2*) exon 7 inclusion (see Chapter 6). They identified a branched derivative of the well-known ApoE (141–150) peptide, which as a PMO conjugate was capable of exon inclusion in the CNS following systemic administration in newborn SMA mice and in adult mice, leading to an increase in the level of full-length *SMN2* transcript. A second cell-penetrating peptide, PMA internalization peptide 6a (Pip6a), as a PMO conjugate has also been shown to exhibit the same effects in the CNS of SMA mice.[37]

16.2.2.3 Antiviral Applications

Given the many years of the development of drugs against human immunodeficiency virus/acquired immunodeficiency syndrome (HIV/AIDS), it is not surprising that there have been numerous publications on aptamers that target various proteins known to be important in the life cycle or infectivity of HIV-1. Included among these targets are the CC chemokine receptor type 5 (CCR5),[38] HIV-1 protease (by aptamer delivery of siRNA),[39] and HIV-1 reverse transcriptase,[40] the latter of which was bound with picomolar affinity by selection of 2′-deoxy-2′-fluoroarabinonucleotide (FANA) aptamers.

In contrast, there are also examples of how aptamers have been developed for lesser-known viruses of medical importance. According to the US Centers for Disease Control and Prevention,[41] Rift Valley Fever (RVF) is an acute, fever-causing viral disease, most commonly observed in domesticated animals (such as cattle, buffalo, sheep, goats, and camels), with the ability to infect and cause illness in humans. The disease is caused by RVF virus (RVFV), a member of the genus *Phlebovirus* in the family Bunyaviridae and was first reported in livestock by veterinary officers in Kenya's Rift Valley in the early 1910s. RVF is generally found in regions of eastern and southern Africa where sheep and cattle are raised, but the virus exists in most of sub-Saharan Africa also, including West Africa and Madagascar. Outbreaks of

RVF can have major societal impacts, including significant economic losses, trade reductions, and epidemics of RVF in humans.

All viruses contain a protein (N) capsid that encapsidates the genomic polynucleotide. In a review of the mechanism for RNA encapsidation by negative-strand RNA viruses such as RVFV, Green *et al.*[42] state that the unique feature that distinguishes negative strand viruses from the rest of the virosphere is that the nucleocapsid, instead of the naked genome, is used as the template for viral nucleotide synthesis. For RVFV, the structure of the N protein has been elucidated, and within its NH_2-terminal helix the hydrophobic side chain of Trp24 fits into the pocket where the bases of the RNA are sequestered in the nucleocapsid. Specific binding of an aptamer to RVFV N-protein could be used for virus detection and/or blocking interaction of N with RVFV RNA as a possible mechanism of drug action. Ellenbecker *et al.*[43] have investigated this possibility by studying the RNA binding properties of N using both natural and artificially selected RNAs as ligands. They found that N binds to RNAs that resemble the terminal panhandle structures of RVFV genomic and antigenomic RNA. Furthermore, they used SELEX to isolate RNA aptamers that bound N with high affinity and determined that N specifically recognizes and binds to GAUU and pyrimidine or guanine motifs. The aptamer RNAs were used to construct a sensitive fluorescence-based sensor of N binding with potential applications for drug screening and imaging methodologies. Recent follow-on studies of this aptamer by Csik and Hornak[44] involved binding experiments in the presence and absence of suramin, a drug recently shown to inhibit RVFV replication. The results indicated that suramin acts as a competitive inhibitor that binds to the site on N where the aptamer RNA also binds. Therapeutic investigations of this potential anti-RVFV aptamer have yet to be reported.

The CDC states[45] that Japanese Encephalitis (JE) Virus (JEV) is the leading cause of vaccine-preventable encephalitis in Asia and the western Pacific. JEV is maintained in a cycle involving mosquitoes and vertebrate hosts, mainly pigs and wading birds. Humans can be infected when bitten by an infected mosquito. While only a small percentage of infected persons develop encephalitis, about one in four cases are fatal, and there is no specific treatment for JE. Han and Lee[46] have recently investigated aptamers to target JEV's single methyltransferase (MTase) domain located at the N-terminal region of the viral nonstructural protein NS5. JEV MTase is essential for viral replication and specifically catalyzes methylation of the viral RNA cap, which occurs exclusively in the cytoplasm. They found a JEV MTase-specific, tight-binding 2′-*O*-methyl pyrimidine-modified, truncated 24-mer RNA aptamer ($K_d \sim 16$ nM) that efficiently inhibited JEV production and replication in cells. They concluded that an RNA aptamer might be useful as an anti-JEV agent.

Recently, Jung *et al.*[47] have developed a 2′-fluoro 45-mer RNA aptamer that bound specifically and avidly ($K_d \sim 15$ nM) to dengue virus serotypes 2 and 3 methyltransferase (MTase), and which competitively blocks MTase methylation activity. These investigators concluded that this approach could be useful for development of an anti-dengue agent. According to the US CDC, with more than one-third of the world's population living in areas at risk for infection, dengue virus is a leading cause of illness and death in the tropics and subtropics.[48] As many as 400 million people are infected yearly. Dengue is caused by any one of four related viruses transmitted by mosquitoes. There are not yet any vaccines to prevent infection with dengue virus and the most effective protective measures are those that avoid mosquito bites. When infected, early recognition and prompt supportive treatment can substantially lower the risk of medical complications and death.

16.2.2.4 Antibacterial Applications

Relatively little effort to date has been applied to development of antisense or siRNA oligonucleotides as antibacterial agents. There are, however, publications describing targeting of various bacteria using aptamers. For example, Vivekananda *et al.*[49] investigated DNA aptamers as a novel approach to neutralize *Staphylococcus aureus* α-toxin, also known as alpha-hemolysin (Hla), which is the major cytotoxic agent released by the bacterium *S. aureus*. The pore-forming structure of Hla allows the toxin to carry out its major function, development of pores in the cellular membrane, eventually causing cell death. *S. aureus* causes a broad spectrum of diseases including life-threatening conditions such as septicemia, pneumonia, and toxic shock syndrome. Vivekananda *et al.*[49] reasoned that, because *S. aureus* has rapidly evolved resistance to many antibiotics, including methicillin, it is important to identify new therapeutic strategies, other than antibiotics, for inhibiting the harmful effects of this pathogen. After 10 rounds of SELEX, 49 potential anti-Hla aptamers were identified, and *in vitro* neutralization assays demonstrated that four of these aptamers significantly inhibited Hla-mediated cell death in Jurkat T cells. Furthermore, two of these four aptamers inhibited upregulation of inflammatory cytokines TNF-α and IL-17. Vivekananda *et al.*[49] concluded that, collectively, these data indicate the feasibility of generating functionally effective aptamers against Hla for treatment of *S. aureus* infections.

Although most *Escherichia coli* bacteria are harmless, some *E. coli* are pathogenic and can cause diarrhea, or illness outside of the intestinal tract, *via* contaminated water or food, or through contact with animals or people. Early investigations by Kolovskaya *et al.*[50] demonstrated that DNA aptamers could exhibit bacteriostatic effects against *Salmonella enteritidis* and *Salmonella typhimurium*, which are dangerous and common food-borne bacteria. More recent studies include neonatal *E. coli* meningitis, which is caused by a disease-causing strain known as *E. coli* K1 that has a capsular antigen shown to be immunochemically identical to the meningococcal group B polysaccharide.[51] Marton *et al.*[52] in Brazil, where meningitis/sepsis associated *E. coli* (MNEC) is especially problematic, used cell-based SELEX[53] (cell-SELEX) to isolate four DNA aptamers that bind strongly to the surface of *E. coli* cells with nanomolar K_d values. One of these aptamers is highly specific for MNEC from clinical isolates, and is the first aptamer found to have potential use in the diagnosis of MNEC-borne pathologies. Therapeutic utility of such an aptamer has yet to be investigated. Similar studies and results have been published by Duan *et al.*[54] for a DNA against *Shigella dysenteriae*, which is common in the developing world, and by Suh *et al.*[55] for *Listeria monocytogenes*, which is found globally.

16.2.2.5 Antiparasitic Applications

It has been estimated[56] that over 1 billion people worldwide are at risk of *Plasmodium falciparum* malaria, which led to over 400 000 deaths in 2015 according to WHO statistics cited by Nik Kamarudin *et al.*[57] These researchers note that there is grave concern that malaria parasites have developed a widespread resistance to anti-malarial drugs, especially in the endemic regions, and review strategies for how aptamers have been targeted and selected for possible malaria therapy. For example, Barfod *et al.*[58] selected RNA aptamers against a conserved region of the *P. falciparum* erythrocyte membrane protein 1 (PfEMP1) which is known to play a major role in the pathogenicity of the *P. falciparum* parasite. This protein enables the parasite to adhere to the endothelial linings of small blood vessels (cytoadherence) as well as to non-infected erythrocytes (rosetting), thus preventing clearance from the bloodstream. By use of SELEX against PfEMP1, they found RNA aptamers that showed 100% rosette-disrupting capacity at ~0.4 nM and concluded that RNA aptamers are promising candidates for adjunct therapy in severe malaria. In an analogous approach, Niles *et al.*[59] used SELEX to obtain heme-binding DNA aptamers that function in a manner akin to heme-binding anti-malarial compounds, such as chloroquine.

Visceral leishmaniasis, according to the World Health Organization in 2017, affects 300 000 people each year and causes 200 000 deaths annually. The utility of aptamers as antileishmanial drugs has been recently proposed by Keighobadi *et al.*,[60] who proposed that, since glycolysis is the only source of energy for leishmania trypanosome parasites, it could serve as a promising drug target *via* glycosome assembly that requires proteins called peroxins (PEX). If synthesis of PEX is specifically inhibited by an aptamer in the process of assembly or propagation of glycosome, parasite development and proliferation should be stopped.

16.3 Recent Studies of Aptamer-based Targeting of Drugs

As was noted in the introduction, aptamers have long been recognized as targeting agents for non-aptamer drugs (Figure 16.1). Recent reviews (*e.g.* Shigdar[61]) are available. The following examples of such delivery (excluding aptamer-based imaging[62]) are selected from publications in 2016–2017.

The nucleotide analogue 5-fluorouracil (5-FU) has been used as a cytotoxic cancer agent for many years, but it remains mostly ineffective against cancer stem cells, which are protected against apoptosis by survivin, according to AlShamaileh *et al.*[63] These researchers therefore proposed that an aptamer-siRNA chimera against survivin might lead to 5-FU eliminations of cancer stem cells. To do this, they leveraged their earlier development of the first RNA aptamers that target the cancer stem cell marker epithelial cell adhesion molecule (EpCAM) and, upon binding to target cells, these EpCAM aptamers undergo receptor-mediated endocytosis which allows for controlled release of drugs intracellularly. AlShamaileh *et al.*[63] studied an aptamer-siRNA construct (Figure 16.7) in a severe combined immunodeficiency (SCID) mouse model wherein HT-29 (human colorectal adenocarcinoma) tumors (60 mm^3) were present in left flanks. Mice received 2 nmol of aptamer-siRNA chimera on days 1, 3, and 5, and phosphate buffered saline (PBS) on days 7, 9, and 11, while mice with 5-FU treatment received intravenous injections of PBS on day 1 and 30 mg kg^{-1} of 5-FU *via* IV injections on days 3, 5, 7, 9, and 11. Combinatorial treatment with chimera and 5-FU led to a significantly slower rate of tumor growth compared with the mice that received 5-FU treatment with or without negative control chimera, and improved overall survival within the first month of treatment.

Mesoporous silica nanoparticles (MSNs) are reported[64] to be exceptionally promising drug carriers for controlled drug delivery systems because their morphology, pore structure, pore volume and pore size can be well tailored

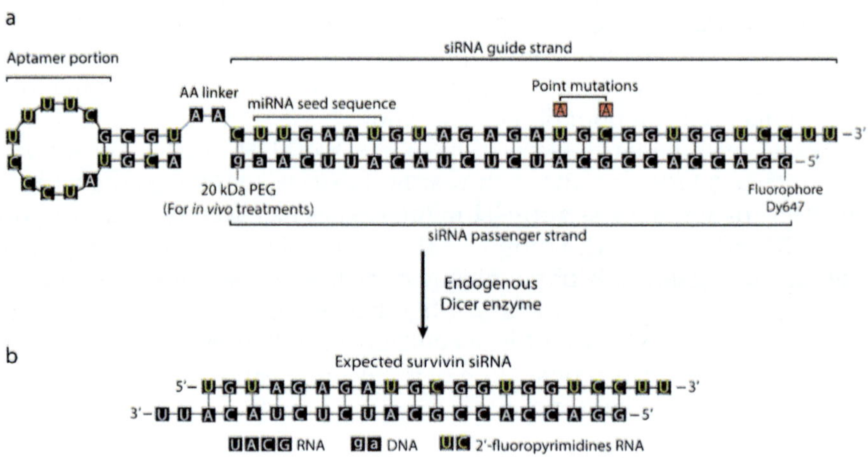

Figure 16.7 (a) Structure of the aptamer-siRNA chimera comprised of an aptamer (binding portion) and an RNA duplex (siRNA portion). (b) Upon cell binding and receptor-mediated endocytosis, the chimera is recognized by the endogenous Dicer enzyme and subsequently integrated into the RNAi machinery, leading to the release of the predicted 21-mer survivin siRNA. Reproduced from ref. 63 [https://doi.org/10.1038/s41598-017-05859-z] under the terms of a CC BY 4.0 license [https://creativecommons.org/licenses/by/4.0/].

to obtain certain drug release profiles. A review by She *et al.*[64] provides a comprehensive overview of the fabrication and modification of MSNs and their applications in tumor-targeted delivery. In addition, the characterization and analysis of MSNs with computer-aided strategies were described. The existing issues and future perspective concerning the applications of MSNs as drug carriers for controlled drug delivery systems were discussed. Pascual *et al.*[65] have used mucin 1 (MUC1), which is a cell-surface protein overexpressed in breast cancer, to direct MSNs loaded with safranin O, a biological stain, to tumor *vs.* non-tumor cells. Confocal microscopy showed that these nanoparticles (NPs) were internalized in breast cancer cells that overexpress MUC1 receptor, with subsequent pore opening and cargo release. A non-tumorigenic breast epithelial cell line that does not overexpress MUC1 showed reduced internalization, and negligible internalization was also found for these NPs with a scrambled-sequence control aptamer. Aptamer-guided NPs loaded with doxorubicin were internalized into the breast cancer cells and induced remarkable reductions in cell viability.

Taghavi *et al.*[66] have investigated Epirubicin (Epi) encapsulated NPs formulated with biocompatible and biodegradable poly(lactic-*co*-glycolic acid) (PLGA) modified with physically adsorbed chitosan (CS) to make efficient drug carriers for cancer cells. To improve anti-tumor efficacy, they used an aptamer against the MUC1 receptor. *In vivo* experiments using mice bearing a murine colon carcinoma cell line demonstrated that MUC1-targeted Epi-loaded NPs exhibited significant tumor growth inhibition compared with non-targeted NPs.

Taghdisi *et al.*[67] add that clinical use of Epi in the treatment of cancer has been limited due to its cardiotoxicity, and that aptamer-targeted delivery of chemotherapeutic agents could not only increase their efficacy but also reduce Epi's cardiotoxicity and other toxic side effects. They recognized that high drug loading and excellent stability of DNA dendrimers make these DNA nanostructures attractive candidates for clinical applications. Taghdisi *et al.*[67] therefore investigated three kinds of aptamers (MUC1, AS1411, and ATP aptamers) for targeted delivery of Epi to target cells, MCF-7 breast cancer cells and C26 murine colon carcinoma cells, as well as non-target CHO cells. Release profiles of Epi from these complexes were assessed at pH 5.4 and 7.4, and cytotoxicity was measured by the MTT assay. Internalization was evaluated by use of flow-cytometry analysis. These complexes were also evaluated for inhibition of tumor growth *in vivo*. More Epi was released from these complexes at pH 5.5, and flow cytometric analysis indicated that the designed complexes were efficiently internalized into target cells, but not into control cells. The complexes had more cytotoxicity in C26 and MCF-7 cells compared with Epi alone. Moreover, the complexes efficiently prevented tumor growth *in vivo*.

Harris *et al.*[68] proposed that the chemokine known as fractalkine can serve as a cancer molecular target, and therefore prepared aptamer-bearing micelles functionalized with an outer layer of PEG. They then investigated these micelles for targeting fractalkine-expressing colon adenocarcinoma cells. *In vitro* cell-binding results indicated that aptamer micelles bound to, and were internalized into, fractalkine-expressing cancer cells. Minimal surface binding was observed using normal cells. Even though partial PEGylation did not prevent serum adsorption, the micelles were highly resistant to endonuclease and exonuclease degradation. *In vivo* biodistribution studies and confocal studies demonstrated that even though both aptamer and control micelles showed tumor accumulation, only the aptamer-bearing micelles internalized into fractalkine-expressing cancer cells, thus demonstrating the potential of the approach and showing that fractalkine may serve as a specific target for nanoparticle delivery to cancer cells.

16.4 Clinical Studies of Aptamer Drugs Registered in ClinicalTrials.gov

Although there have been a number of relatively recent reviews[69–73] of various aspect of aptamers intended as therapeutics, this section deals exclusively with an overview of nearly 30 clinical studies of aptamer drugs listed in the ClinicalTrials.gov registry and database in 2017.[74] This broad scope and time frame contrasts with more limited and now outdated reviews[75,76] on clinical trials of aptamer drugs. The ClinicalTrials.gov website is relatively easy to navigate, and has built-in help "buttons" as well as dropdown menus as aids to find and filter the extensive amount of available information. Publications such as that by Tse *et al.*[4] can be consulted for additional guidance about this comprehensive registry and database provided by the U.S. National Institutes of Health.

A search of ClinicalTrials.gov in September 2017 using the term "aptamer" provided information for 32 different clinical studies, which when sorted by conditions related to these search results, gave 87 distinct medical conditions covering a wide range of diseases, with ocular being the most prevalent general category. A map view of clinical sites associated with these studies showed that, of the 60 total participating sites around the world, nearly half (29) were in the USA (19) and Europe (10). Reading procedural details for these 32 studies involving aptamers revealed that, in four studies, samples from subjects treated with non-aptamer drugs were to be used for *in vitro* investigations such as screening for aptamer biomarkers. The remaining 28 clinical studies of aptamer drugs *per se* were therefore available for summarizing herein, and can be compared numerically with the 151 clinical studies found by searching for "antisense," and the 82 clinical studies found by searching for "RNAi" or "siRNA," *i.e.* non-aptamer oligonucleotide drug studies.

The overarching format for summarizing these 28 clinical studies of aptamer drugs used herein is study "status" grouping as completed studies (Section 16.4.1), terminated or withdrawn studies (Section 16.4.2), and active studies (Section 16.4.3) based on the following verbatim definitions of study status provided by ClinicalTrials.gov.

- Completed: clinical study has ended normally, and participants are no longer being examined or treated (that is, the "last subject, last visit" has occurred).
- Terminated: clinical study has stopped recruiting or enrolling participants early and will not start again and participants are no longer being examined or treated.
- Withdrawn: clinical study stopped before enrolling its first participant.
- Active: clinical study is ongoing (that is, participants are receiving an intervention or being examined), but potential participants may not be being currently recruited.

These three groupings of clinical studies of aptamer drugs are presented in tabular form listing sponsor, study ID, aptamer drug, disease(s) and type of study. Study identification (ID) is a unique identification code given to each clinical study registered on ClinicalTrials.gov. The format is "NCT" followed by an eight-digit number (for example, NCT00000419). This identifier is also known as the NCT number, and can be searched directly *via* ClinicalTrials.gov or a web browser, such as Google Chrome.

16.4.1 Completed Clinical Studies

The majority of completed clinical studies involve ocular diseases (Table 16.1), of which age-related macular degeneration[77] (ARMD) is a frequently studied indication. ARMD is the leading cause of blindness for people over the age of 55 years, in the USA and Europe. There are two forms of ARMD, namely "dry" and "wet" ARMD. The "wet" form is characterized by the growth of new blood vessels into the central region of the retina. These new vessels cause severe visual loss due to retinal damage caused by subsequent leakage and scar formation. As of 2006, anti-vascular endothelial growth factor (VEGF) therapies and photodynamic therapies had been approved for "wet" ARMD, whereas for "dry" ARMD, which accounts for up to 90 percent of all cases of ARMD (8 million patients in the USA and an additional 8 million in Europe) there was no approved therapy. Visual loss in "dry" ARMD is typically not as severe as in "wet" ARMD, however, over time, dry ARMD can progress to the wet form of the disease.

Most notable among aptamers studied for treatment of ARMD is pegaptanib sodium (Macugen) (Figure 16.8), which originated at NeXstar Pharma and was later taken on by Eyetech Pharma. Pegaptanib sodium is a 5′-PEG (40 kDa)–3′-inverted dT blocked 2′-fluoro/2′-*O*-methyl modified 27-mer RNA aptamer targeted against VEGF, and is the only federally approved aptamer drug.[78] However, in an excellent review[70] by Zhou and Rossi, it is pointed out that Macugen has been superseded by VEGF-specific monoclonal antibodies such as bevacizumab (Avastin; Genentech) and ranibizumab (Lucentis; Genentech).

Another ARMD drug candidate, Ophthotech's ARC1905 (Anti-C5 Aptamer, Zimura®), is reported[79] to be a 38-mer aptamer sequence that was originally

Table 16.1 Completed studies listed in ClinicalTrials.Gov in 2017.

Sponsor (study ID)	Aptamer drug	Disease(s)	Type of study
Eyetech Pharma (NCT00215670)	Pegaptanib sodium (Macugen)	ARMD	Safety, tolerability and pharmacokinetics
Eyetech Pharma (NCT00321997)	Pegaptanib sodium (Macugen)	Wet ARMD	Safety and efficacy of doses *vs.* sham
Eyetech Pharma (NCT00040313)	Pegaptanib sodium (Macugen)	Diabetic macular edema	Pegaptanib sodium *vs.* sham
Retina Inst. of Hawaii (NCT01487070)	Pegaptanib sodium (Macugen)	Proliferative diabetic retinopathy (PDR)	Pegaptanib sodium <7 days before vitrectomy secondary to tractional retinal detachment in PDR
Ophthotech Corp (NCT00950638)	ARC1905 (Anti-C5 aptamer, Zimura®)	Dry ARMD	ARC1905 in subjects with dry ARMD
Ophthotech Corp (NCT00709527)	ARC1905 (Anti-C5 aptamer, Zimura®)	ARMD	ARC1905 given either with Lucentis® in subjects with ARMD
Ophthotech Corp (NCT01089517)	E10030 (Anti-PDGF pegylated aptamer, Fovista®)	ARMD	Safety and efficacy of E10030 plus Lucentis for ARMD
Ophthotech Corp (NCT00569140)	E10030 (Anti-PDGF pegylated aptamer, Fovista®)	ARMD	Safety, tolerability and pharmacokinetics
Eyetech Pharma (NCT00021736)	EYE001	ARMD/choroidal neovascularization	Phase 2/3
National Eye Institute (NCT00056199)	EYE001	Von Hippel–Lindau disease	Treating retinal tumors in patients with Von Hippel–Lindau disease
Ophthotech Corp (NCT02397954)	ARC1905 (Anti-C5 aptamer), Zimura®	Idiopathic polypoidal choroidal vasculopathy (IPCV)	Safety and tolerability
Archemix (NCT00632242)	ARC1779	Thrombotic thrombocytopenic/Von Willebrand disease Type-2b	Treating patients with von Willebrand factor-related platelet function
National Heart, Lung, and Blood Institute (NCT00113997)	REG1	Anticoagulation	Safety and dosing evaluation
NOXXON Pharma (NCT02079896)	NOX-H94	Anemia/end stage renal disease	Lexaptepid pegol (NOX-H94) in ESA-hyporesponsive anemia in dialysis patients
NOXXON Pharma (NCT00976729)	NOX-E36	Chronic inflammatory diseases/type 2 diabetes mellitus/systemic lupus erythematosus	First-in-human study
NOXXON Pharma (NCT00976378)	NOX-A12	Autologous stem cell transplantation	First-in-human study
NOXXON Pharma (NCT01194934)	NOX-A12	Hematopoietic stem cell transplantation	Multiple ascending dose study in healthy volunteers

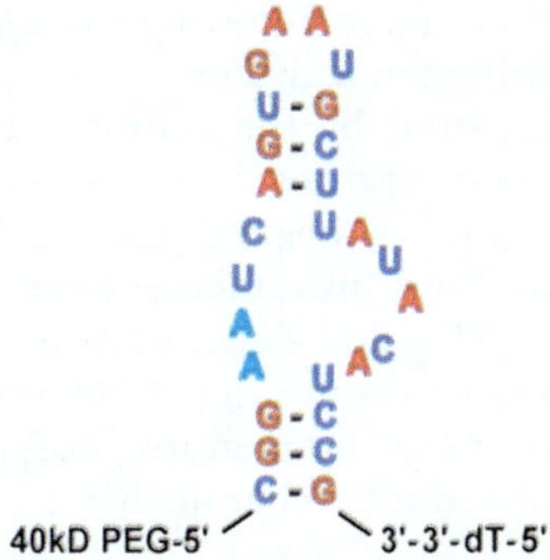

Figure 16.8 Structure in DrugBank[106] for chemically modified RNA aptamer pegaptanib sodium (Macugen) with 5′-PEG (40 kDa) and 3′-inverted dT blocking. Light blue nucleotides are 2′-hydroxyl, dark blue nucleotides are 2′-fluoro, and red nucleotides are 2′-*O*-methyl.

identified from a 2′-fluoro-pyrimidine-modified library, and all but three purines could be replaced by 2′-*O*-methyl purine nucleotides. For *in vivo* applications, an inverted dT and a 40 kDa PEG moiety were added to the 3′ and 5′ ends, respectively.[76]

Ophthotech's E10030, an anti-PDGF-BB aptamer (Fovista®) also known as ARC127 was likewise studied for treatment of ARMD. The sequence is reported[76] to be a 36-mer published in 1996 that was derived from a DNA library, and was only modified with a 3′-inverted dT moiety to inhibit exonuclease degradation. Further shortening and post-SELEX modifications (2′-fluoro pyrimidines, 2′-*O*-methyl purines, internal hexyl-linkers, all wherever possible) has been reported[76] to improve nuclease resistance and resulted in a 32-mer that was later named ARC126, and then ARC127 when pegylated at the 5′ terminus. An expert overview of the clinical failure of E10030 and the fate of anti-PDGF medications was published by Dunn *et al.*[80] in 2017.

Eyetech's EYE001 is described in a patent[81] application as a pegylated 28-mer RNA aptamer comprised of mostly 2′-fluoro and 2′-*O*-methyl ribonucleotides, and an inverted 3′-dT moiety. The 5′ end is conjugated to a Y-shaped 40 kDa PEG component comprised of two 20 kDa PEG chains attached to two amine groups on a lysine residue *via* carbamate linkages. This moiety is in turn linked to the oligonucleotide *via* the bifunctional amino linker, [HN–(CH2)$_5$O–], which is attached to the oligonucleotide through a standard phosphodiester bond.

Archemix's ARC1779 is a therapeutic aptamer antagonist of the A1 domain of Von Willebrand factor, the ligand for receptor glycoprotein 1b on platelets, and was studied as a novel antithrombotic agent for use in patients with acute coronary syndromes.[30] It has been reported[82] to be comprised of 2′-deoxy and 2′-*O*-methyl nucleotides with 3′-inverted dT and 5′-PEG (20 kDa) moieties, as well as one phosphorothioate linkage. Dornbos *et al.* reported[83] in 2017 that ARC 1779 inhibition of Von Willebrand factor is a potent thrombolytic and ameliorates stroke burden in mice following vascular injury compared with recombinant tissue plasminogen activator.

REG1, which was sponsored by the National Heart, Lung, and Blood Institute (NHLBI), has been reported[84] as a novel anticoagulation system consisting of pegnivacogin,[82] a pegaptanib-like RNA aptamer inhibitor of coagulation factor IXa, and anivamersen, the reverse complementary sequence oligonucleotide. This dual-function construct was used to test the hypothesis that near complete inhibition of factor IXa with pegnivacogin during percutaneous coronary intervention, followed by partial reversal with anivamersen, would reduce ischemic events compared with bivalirudin, without increasing bleeding. It was found[84] that REG1, as formulated, was associated with severe allergic reactions, leading to early termination, and there was no evidence that REG1 reduced ischemic events or bleeding compared with bivalirudin.

The last four clinical studies shown in Table 16.1 involve three aptamer drugs, namely NOX-H94, NOX-E36, and NOX-A12 that were sponsored by the German company NOXXON Pharma, and are each a Spiegelmer® (Figure 16.9), which is a registered trademark of NOXXON Pharma. Naturally occurring RNA or DNA oligomers are made of nucleotides with D-chirality, whereas Spiegelmers are made of mirror-image nucleotides with L-chirality that do not occur in nature, and can thus render Spiegelmers resistant to degradation by plasma and cellular nucleases, which are stereoselective for D- not L-chirality. The name Spiegelmer follows from Spiegel being the German word for mirror. Methods for generating these L-aptamers have been discussed in a review[76] of the evolution of Spiegelmer therapeutics by Vater and Klussmann. In brief, D-configured aptamers are selected against a synthetic target that is the mirror-image of the biological target of interest. The sequence of the identified D-aptamer is then chemically synthesized as an L-aptamer (Spiegelmer) that, based on the rules of symmetry, binds to the natural target of interest with the same affinity as the D-aptamer to the mirror-image selection target.

Schwoebel *et al.* in a preclinical study[85] of NOX-H94, which is conjugated to 40 kDa PEG,[82] point out that anemia of chronic inflammation is the most prevalent form of anemia in hospitalized patients, and that a hallmark of this disease is the intracellular sequestration of iron as a consequence of hepcidin-induced internalization and subsequent degradation of ferroportin, the hepcidin receptor and only known iron-export protein. They also report that NOX-H94, which binds human hepcidin with high affinity, blocked hepcidin-induced ferroportin degradation and ferritin expression in J774A.1 macrophages. Moreover, in an acute cynomolgus monkey model of interleukin 6 (IL-6)-induced hypoferremia, NOX-H94 inhibited serum iron reduction completely, and in a subchronic model of IL-6-induced anemia, NOX-H94 inhibited the decrease in hemoglobin concentration, which led them to conclude that NOX-H94 may represent a treatment option for patients suffering from anemia of chronic inflammation.

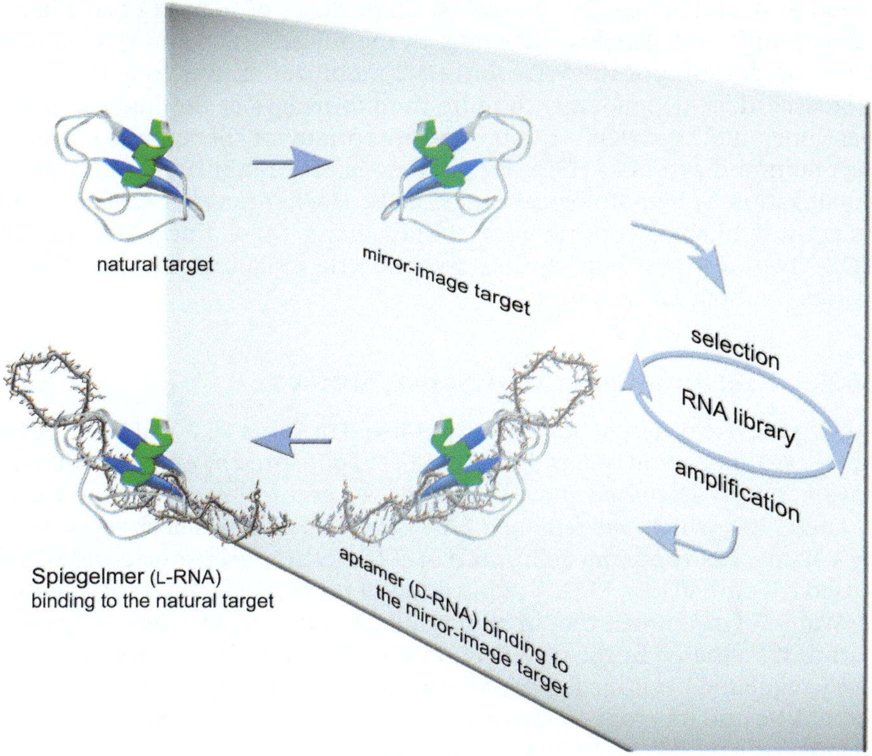

Figure 16.9 Cartoon depicting Spiegelmer (L-RNA) generation by starting with a natural target, synthesizing the mirror-image target for SELEX selection/amplification cycles to give D-aptamer that is then synthesized with L-phosphoramidites to give the Spiegelmer. Reproduced from ref. 76 [https://doi.org/10.1016/j.drudis.2014.09.004] under the terms of a CC BY 3.0 license [https://creativecommons.org/licenses/by/3.0/].

NOX-E36 (emapticap pegol), which is a 40-mer L-aptamer conjugated to 40 kDa PEG,[82] specifically binds and inhibits the pro-inflammatory chemokine C–C motif-ligand 2 (CCL2) (also called monocyte-chemotactic protein 1). NOX-E36 has been previously investigated in various preclinical studies that are discussed by Menne *et al.*,[86] who recently reported results from a randomized, double-blind, placebo-controlled phase IIa study in 75 albuminuric type 2 diabetics. They concluded that inhibition of the CCL2–CCL2 receptor axis with NOX-E36 was generally safe and well tolerated, and some beneficial effects on urinary albumin to creatinine ratio were observed in this exploratory study.

According to Vater *et al.*,[87] NOX-A12 is a 40 kDa PEGylated[82] L-aptamer that binds to CXCL12 (stromal cell-derived factor-1, SDF-1) with high affinity, thereby inhibiting CXCL12 signaling on both its receptors, CXCR4 and CXCR7. Furthermore, in animals, NOX-A12 mobilized white blood cells (WBCs) and hematopoietic stem and progenitor cells (HSCs) into peripheral blood (PB), and in healthy volunteers, single doses of NOX-A12 had a benign safety profile and also dose-dependently mobilized WBCs and HSCs into PB. Also, the duration of the WBC and HSC mobilization was long-lasting and increased dose-dependently to more than four days at the highest dose. It was concluded that NOX-A12 may be appropriate for therapeutic use in and beyond mobilization of HSCs, *e.g.*, in long-lasting mobilization and chemosensitization of hematological cancer cells. Hoellenriegel *et al.*[88] have more recently published confirmatory studies along these lines, showing that NOX-A12 interferes with chronic lymphocytic leukemia cell motility and causes chemosensitization.

16.4.2 Terminated or Withdrawn Studies

Among the five different aptamer drugs listed in Table 16.2, only two, namely ARC19499 sponsored by Baxalta and AS1411 sponsored by Antisoma Research have not been discussed above for completed studies listed in Table 16.1.

ARC19499, which was renamed BAX 499 by Baxalta, has been reported[70] to be a 32-mer RNA aptamer comprised of 2′-*O*-methyl purine nucleosides, 3′-inverted dT, and 40 kDa PEG. A publication of preclinical studies of ARC19499 by Waters *et al.*[89] noted that in hemophilia A and B the extrinsic coagulation pathway, mediated by factor VIIa and tissue factor (TF), is intact but is negatively regulated by tissue factor pathway inhibitor (TFPI), which inhibits both factor VIIa and its product, factor Xa. This inhibition limits clot initiation *via* the extrinsic pathway. ARC19499 blocks TFPI inhibition of both factor Xa and the TF-factor VIIa complex, corrects thrombin generation in hemophilia A and B plasma, and restores clotting in FVIII-neutralized whole blood. Waters *et al.*[89] then showed, using a monkey model of hemophilia, that ARC19499 restored thromboelastography clotting times to baseline levels and corrected bleeding time, thus supporting the hypothesis that ARC19499 inhibition of TFPI may be an effective alternative to then available treatments of bleeding associated with hemophilia.

AS1411 is reported[90] to be a 26-mer DNA aptamer that is G-rich and PEGylated. Stuart *et al.*[91] published results for an open-label randomized phase 2 trial comparing AS1411 in combination with high-dose cytarabine (HiDAC) with HiDAC alone in patients with primary refractory or relapsed acute myeloid leukemia (AML), who had received up to three previous lines of chemotherapy. Safety findings were similar across groups, except that grade 3 hypokalemia was more frequent with AS1411. It was concluded, based on the aggregate findings for this trial, that AS1411 in combination with HiDAC appeared to be well tolerated and shows promising signs of activity in patients with relapsed or refractory AML. Nevertheless, it was subsequently reported in non-scientific media[92] that these early data for candidate AS1411 in AML were judged by Antisoma to be inconclusive, and that the firm would terminate this study and look for other approaches to development of a drug.

Table 16.2 Terminated or withdrawn studies listed in ClinicalTrials.Gov in 2017.

Sponsor (study ID)	Aptamer drug	Disease(s)	Type of study
Eyetech Pharma (NCT00312351)	Pegaptanib sodium (Macugen)	Wet ARMD	Safety and efficacy of three different doses
Ophthotech Corp (NCT02387957)	E10030 (Anti-PDGF pegylated aptamer, Fovista®)	ARMD	18 Month phase 2a open label, randomized study of Avastin®, Lucentis®, or Eylea® (anti-VEGF therapy) administered with E10030
Ophthotech Corp (NCT01944839)	E10030 (Anti-PDGF pegylated aptamer, Fovista®)	ARMD	Phase 3 safety and efficacy study of E10030 in combination with Lucentis® *vs.* Lucentis®
Ophthotech Corp (NCT01940900)	E10030 (Anti-PDGF pegylated aptamer, Fovista®)	ARMD	Phase 3 safety and efficacy study of E10030 in combination with Lucentis® *vs.* Lucentis®
Ophthotech Corp (NCT01940887)	E10030 (Anti-PDGF pegylated aptamer, Fovista®)	ARMD	Phase 3 safety and efficacy study of E10030 in combination with either Avastin® or Eylea® *vs.* Avastin® or Eylea®
Archemix (NCT00694785)	ARC1779	Von Willebrand disease	Pharmacokinetics, pharmacodynamics, and safety of ARC1779 for von Willebrand disease type 2B; withdrawn
Baxalta US Inc (NCT01191372)	ARC19499	Hemophilia	First-in-human study and proof-of-mechanism study in hemophilia patients
Antisoma Research (NCT01034410)	AS1411	Acute myeloid leukemia (AML)	AS1411 with cytarabine for primary refractory or relapsed AML

16.4.3 Active Studies

There are only three active studies of aptamer drugs registered in ClinicalTrials.gov as of 2017. As listed in Table 16.3, these involve pegaptanib sodium sponsored by Eyetech, ARC1905 sponsored by Ophthotech, and E10030 also sponsored by Ophthotech. The first of these studies, which is a single-center trial at the Retina Institute of Hawaii, involves loading of intravitreal pegaptanib bi-weekly during the initial treatment period for diabetic macular edema (DME) when VEGF levels are the greatest, and then gradually extending the administration frequency to monthly as homeostasis ensues for the treatment of DME. This study was registered in 2011 and is currently recruiting patients.

Table 16.3 Active studies listed in ClinicalTrials.Gov in 2017.

Sponsor (study ID)	Aptamer drug	Disease(s)	Type of study
Eyetech Pharma (NCT01487044)	Pegaptanib sodium (Macugen)	Diabetic Macular Edema (DME)	Single-center trial of high frequency pegaptanib for rapid restoration of VEGF levels in DME
Ophthotech Corp (NCT02686658)	ARC1905 (Anti-C5 aptamer), Zimura®	Dry ARMD	Safety and efficacy of three different doses
Ophthotech Corp (NCT02591914)	E10030 (Anti-PDGF pegylated aptamer, Fovista®)	ARMD	Open-label safety, tolerability and development of subfoveal fibrosis for altering regimens of Fovista® and anti-VEGF therapy for ARMD

Opthotech's study of ARC1905 is a phase 2–3 trial to assess the safety and efficacy of intravitreous administration in subjects with geographic atrophy secondary to dry ARMD. This study was registered in 2016 and is currently recruiting participants in numerous locations in the USA as well as at some sites in Hungary. Ophthotech's trial of E10030 is an open-label trial located at Retinal Consultants of Arizona for safety, tolerability and development of subfoveal fibrosis by intravitreal administration of altering regimens of this aptamer drug and anti-VEGF therapy in subjects with neovascular ARMD. This study is ongoing but not recruiting patients.

In concluding this section, it should be emphasized that additional active studies may be in progress but not registered in this database.

16.5 Conclusions and Prospects

Perusal of the clinical studies information presented in Tables 16.1–16.3 reinforces several opinions expressed in recent reviews[82,90] of the clinical development of aptamer drugs. The first and most obvious is that only one drug, namely pegaptanib (Macugen), has been approved for use in humans, and that for ARMD. However, as was noted above, even this achievement has been diminished by subsequent approval of superior antibody-based drugs. Secondly, most of the other candidate aptamer drugs have also been, or still are, directed to ARMD or other ocular diseases.[93] It seems that many other non-ocular diseases, such as those discussed in Section 16.2, can be addressed by systemic administration, and repeated dosing, of longer-lived, chemically stabilized aptamers, which is technically and commercially feasible given what has been learned clinically and manufacturing-wise, over many years, in the antisense oligonucleotide field.[94,95] In this regard, extensive use of 2′-fluoro or 2′-*O*-methyl modified oligonucleotides represents, in effect, exploration of very limited structural diversity, which can now be expanded to hydrophobic SOMAmers and thioaptamers, or C5-modified dU

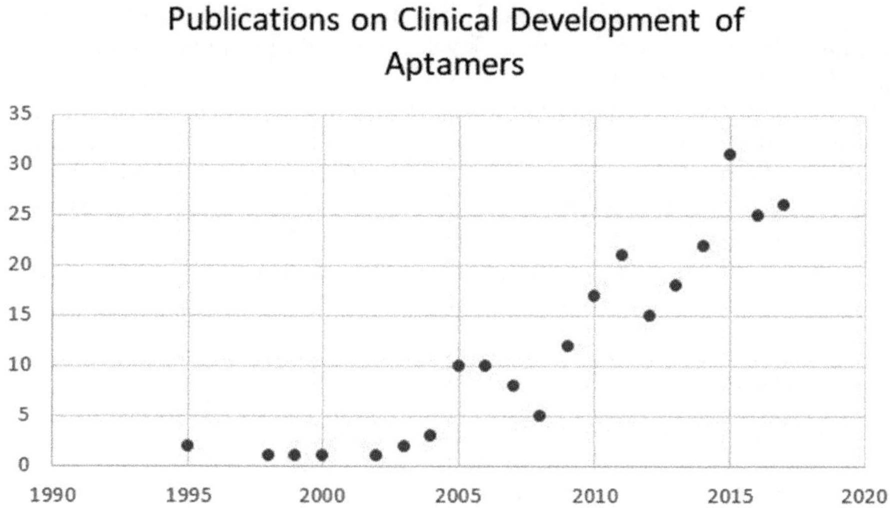

Figure 16.10 Chart of annual publications in PubMed (as of September 2017) found by searching for "clinical development of aptamers" in title or abstract.

X-Aptamers. Even greater structural diversity of aptamer libraries has been recently achieved by Kong *et al.*[96] using T4 DNA ligase-catalyzed DNA-templated oligonucleotide assembly. Also, expanded functionality for an L-RNA aptamer targeting miRNA biogenesis has been recently demonstrated by Kabza and Sczepanski.[97] Regarding chiral monothioaptamers, it is now possible to control phosphorothioate stereochemistry during synthesis[98] and thus examine the effect of chirality on binding to target. Additionally, an aptamer-optimization strategy recently published by Xu *et al.*[99] employs a structural computational modeling methodology that, when coupled to a standard functional assay, reveals key sequence and structural motifs of an RNA aptamer.

If prospects for the future of aptamer clinical development can be gleaned from retrospective analysis, then the number of annual publications for clinical development of aptamers (Figure 16.10) available in PubMed[100] indicates that there are likely to be continued increases over the next 5 or 10 years. Some clinical advances beyond aptamers summarized in Tables 16.1–16.3 are likely to occur in the future.

In common with antisense and siRNA therapeutics, as well as gene therapy, the overarching challenge for aptamers is for much more effective delivery of these nucleic acid-based therapeutics – emphatically referred to as 'delivery, delivery, delivery' by Greco *et al.*[101] However, aptamers have a uniquely important advantage over these other classes of nucleic acid-based therapeutics in that, as discussed in the introduction, aptamers can be used as drug entities *per se*, drug targeting agents, chimeric targeting or drug entities, as well as imaging agents, and as biosensors[102] for theranostics.[103] This remarkable pharmacological and analytical versatility increases the probability of successful clinical development of aptamers, including simple delivery through intact human skin, according to findings in a recent study[104] said to challenge 'orthodoxy.' In addition, and importantly, there are now numerous companies[105] offering selection of aptamers as a custom service, which allows biomedical researchers to simply purchase aptamers for targets of interest. Such service companies greatly facilitate progress in basic and preclinical research and development. Taken together, the aforementioned factors indicate that prospects for the future 25 years of aptamer clinical development look promising.

References

1. C. Tuerk and L. Gold, *Science*, 1990, **249**, 505.
2. A. D. Ellington and J. W. Szostak, *Nature*, 1990, **346**, 818.
3. S. Poolsup and C. Y. Kim, *Curr. Opin. Biotechnol.*, 2017, **48**, 180.
4. T. Tse, R. J. Williams and D. A. Zarin, *Chest*, 2009, **136**, 295.
5. K. T. Urak, S. Shore, W. M. Rockey, S. J. Chen, A. P. McCaffrey and P. H. Giangrande, *Methods*, 2016, **103**, 167.
6. E. Ouellet, J. H. Foley, E. M. Conway and C. Haynes, *Biotechnol. Bioeng.*, 2015, **112**, 1506.
7. Y. Yu, C. Liang, Q. Lv, D. Li, X. Xu, B. Liu, A. Lu and G. Zhang, *Int. J. Mol. Sci.*, 2016, **17**, 358.
8. S. Kraemer, J. D. Vaught, C. Bock, L. Gold, E. Katilius, T. R. Keeney, N. Kim, N. A. Saccomano, S. K. Wilcox, D. Zichi and G. M. Sanders, *PLoS One*, 2011, **6**, e26332.
9. B. Hall, S. Arshad, K. Seo, C. Bowman, M. Corley, S. D. Jhaveri and A. D. Ellington, *Curr Protoc Mol Biol*, 2009, ch. 24, Unit 24 3.
10. L. Gold, D. Ayers, J. Bertino, C. Bock, A. Bock, E. N. Brody, J. Carter, A. B. Dalby, B. E. Eaton, T. Fitzwater, D. Flather, A. Forbes, T. Foreman, C. Fowler, B. Gawande, M. Goss, M. Gunn, S. Gupta, D. Halladay, J. Heil, J. Heilig, B. Hicke, G. Husar, N. Janjic, T. Jarvis, S. Jennings, E. Katilius, T. R. Keeney, N. Kim, T. H. Koch, S. Kraemer, L. Kroiss, N. Le, D. Levine, W. Lindsey, B. Lollo, W. Mayfield, M. Mehan, R. Mehler, S. K. Nelson, M. Nelson, D. Nieuwlandt, M. Nikrad, U. Ochsner, R. M. Ostroff, M. Otis, T. Parker, S. Pietrasiewicz, D. I. Resnicow, J. Rohloff, G. Sanders, S. Sattin, D. Schneider, B. Singer, M. Stanton, A. Sterkel, A. Stewart, S. Stratford, J. D. Vaught, M. Vrkljan, J. J. Walker, M. Watrobka, S. Waugh, A. Weiss, S. K. Wilcox, A. Wolfson, S. K. Wolk, C. Zhang and D. Zichi, *PLoS One*, 2010, **5**, e15004.
11. B. N. Gawande, J. C. Rohloff, J. D. Carter, I. von Carlowitz, C. Zhang, D. J. Schneider and N. Janjic, *Proc. Natl. Acad. Sci. U. S. A.*, 2017, **114**, 2898.
12. S. Gupta, M. Hirota, S. M. Waugh, I. Murakami, T. Suzuki, M. Muraguchi, M. Shibamori, Y. Ishikawa, T. C. Jarvis, J. D. Carter, C. Zhang, B. Gawande, M. Vrkljan, N. Janjic and D. J. Schneider, *J. Biol. Chem.*, 2014, **289**, 8706.

13. M. Hirota, I. Murakami, Y. Ishikawa, T. Suzuki, S. Sumida, S. Ibaragi, H. Kasai, N. Horai, D. W. Drolet, S. Gupta, N. Janjic and D. J. Schneider, *Nucleic Acid Ther.*, 2016, **26**, 10.
14. M. Abifadel, M. Varret, J. P. Rabes, D. Allard, K. Ouguerram, M. Devillers, C. Cruaud, S. Benjannet, L. Wickham, D. Erlich, A. Derre, L. Villeger, M. Farnier, I. Beucler, E. Bruckert, J. Chambaz, B. Chanu, J. M. Lecerf, G. Luc, P. Moulin, J. Weissenbach, A. Prat, M. Krempf, C. Junien, N. G. Seidah and C. Boileau, *Nat. Genet.*, 2003, **34**, 154.
15. W. He, M. A. Elizondo-Riojas, X. Li, G. L. Lokesh, A. Somasunderam, V. Thiviyanathan, D. E. Volk, R. H. Durland, J. Englehardt, C. N. Cavasotto and D. G. Gorenstein, *Biochemistry*, 2012, **51**, 8321.
16. G. L. Lokesh, H. Wang, C. H. Lam, V. Thiviyanathan, N. Ward, D. G. Gorenstein and D. E. Volk, *Methods Mol. Biol.*, 2017, **1632**, 151.
17. H. Wang, C. H. Lam, X. Li, D. L. West and X. Yang, *Biochimie*, 2018, **145**, 125.
18. V. Thiviyanathan and D. G. Gorenstein, *Proteomics: Clin. Appl.*, 2012, **6**, 563.
19. A. P. Mann, A. Somasunderam, R. Nieves-Alicea, X. Li, A. Hu, A. K. Sood, M. Ferrari, D. G. Gorenstein and T. Tanaka, *PLoS One*, 2010, **5**, e13050.
20. D. E. Volk and G. L. R. Lokesh, *Biomedicines*, 2017, **5**, 41.
21. S. Y. Wu, X. Yang, K. M. Gharpure, H. Hatakeyama, M. Egli, M. H. McGuire, A. S. Nagaraja, T. M. Miyake, R. Rupaimoole, C. V. Pecot, M. Taylor, S. Pradeep, M. Sierant, C. Rodriguez-Aguayo, H. J. Choi, R. A. Previs, G. N. Armaiz-Pena, L. Huang, C. Martinez, T. Hassell, C. Ivan, V. Sehgal, R. Singhania, H. D. Han, C. Su, J. H. Kim, H. J. Dalton, C. Kovvali, K. Keyomarsi, N. A. McMillan, W. W. Overwijk, J. Liu, J. S. Lee, K. A. Baggerly, G. Lopez-Berestein, P. T. Ram, B. Nawrot and A. K. Sood, *Nat. Commun.*, 2014, **5**, 3459.
22. Y. Morita, M. Kamal, S. A. Kang, R. Zhang, G. L. Lokesh, V. Thiviyanathan, N. Hasan, S. Woo, D. Zhao, M. Leslie, S. Suh, W. Razaq, H. Rui, D. G. Gorenstein, D. E. Volk and T. Tanaka, *Mol. Ther.--Nucleic Acids*, 2016, **5**, e399.
23. Y. Hu, J. Duan, B. Cao, L. Zhang, X. Lu, F. Wang, F. Yao, Z. Zhu, W. Yuan, C. Wang and X. D. Yang, *Clin. Transl. Oncol.*, 2015, **17**, 647.
24. Q. Liu and A. Deiters, *Acc. Chem. Res.*, 2014, **47**, 45.
25. A. Heckel and G. Mayer, *J. Am. Chem. Soc.*, 2005, **127**, 822.
26. R. S. Lovely, M. Moaddel and D. H. Farrell, *J. Thromb. Haemostasis*, 2003, **1**, 124.
27. A. Heckel, M. C. Buff, M. S. Raddatz, J. Muller, B. Potzsch and G. Mayer, *Angew. Chem., Int. Ed. Engl.*, 2006, **45**, 6748.
28. L. Li, R. Tong, H. Chu, W. Wang, R. Langer and D. S. Kohane, *Proc. Natl. Acad. Sci. U. S. A.*, 2014, **111**, 17099.
29. C. P. Rusconi, J. D. Roberts, G. A. Pitoc, S. M. Nimjee, R. R. White, G. Quick Jr, E. Scardino, W. P. Fay and B. A. Sullenger, *Nat. Biotechnol.*, 2004, **22**, 1423.
30. J. C. Gilbert, T. DeFeo-Fraulini, R. M. Hutabarat, C. J. Horvath, P. G. Merlino, H. N. Marsh, J. M. Healy, S. Boufakhreddine, T. V. Holohan and R. G. Schaub, *Circulation*, 2007, **116**, 2678.

31. W. H. Thiel, C. L. Esposito, D. D. Dickey, J. P. Dassie, M. E. Long, J. Adam, J. Streeter, B. Schickling, M. Takapoo, K. S. Flenker, J. Klesney-Tait, V. Franciscis, F. J. Miller Jr and P. H. Giangrande, *Mol. Ther.*, 2016, **24**, 779.
32. W. H. Thiel, T. Bair, A. S. Peek, X. Liu, J. Dassie, K. R. Stockdale, M. A. Behlke, F. J. Miller Jr and P. H. Giangrande, *PLoS One*, 2012, 7, e43836.
33. A. Varshney, J. Bala, B. Santosh, A. Bhaskar, S. Kumar and P. K. Yadava, *Mol. Cell. Biochem.*, 2017, **427**, 157.
34. Alzheimer's.Net, *Alzheimer's Will be a Global Epidemic by 2050*, https://www.alzheimers.net/2013-12-11/alzheimers-global-epidemic-by-2050/.
35. H. Liang, Y. Shi, Z. Kou, Y. Peng, W. Chen, X. Li, S. Li, Y. Wang, F. Wang and X. Zhang, *PLoS One*, 2015, **10**, e0140733.
36. F. Shabanpoor, S. M. Hammond, F. Abendroth, G. Hazell, M. J. A. Wood and M. J. Gait, *Nucleic Acid Ther.*, 2017, **27**, 130.
37. S. M. Hammond, G. Hazell, F. Shabanpoor, A. F. Saleh, M. Bowerman, J. N. Sleigh, K. E. Meijboom, H. Zhou, F. Muntoni, K. Talbot, M. J. Gait and M. J. Wood, *Proc. Natl. Acad. Sci. U. S. A.*, 2016, **113**, 10962.
38. J. Zhou, S. Satheesan, H. Li, M. S. Weinberg, K. V. Morris, J. C. Burnett and J. J. Rossi, *Chem. Biol.*, 2015, **22**, 379.
39. Q. Zhu, T. Shibata, T. Kabashima and M. Kai, *Eur. J. Med. Chem.*, 2012, **56**, 396.
40. I. Alves Ferreira-Bravo, C. Cozens, P. Holliger and J. J. DeStefano, *Nucleic Acids Res.*, 2015, **43**, 9587.
41. Centers for Disease Control and Prevention, *Rift Valley Fever (RVF)*, https://www.cdc.gov/vhf/rvf/index.html.
42. T. J. Green, R. Cox, J. Tsao, M. Rowse, S. Qiu and M. Luo, *J. Virol.*, 2014, **88**, 3766.
43. M. Ellenbecker, L. Sears, P. Li, J. M. Lanchy and J. S. Lodmell, *Antiviral Res.*, 2012, **93**, 330.
44. Scholar Works University of Montana, *Characterization of RNA aptamer binding to Rift Valley fever virus nucleocapsid protein*, https://scholarworks.umt.edu/umcur/2016/pmposters/7/.
45. Centers for Disease Control and Prevention, *Japanese Encephalitis*, https://www.cdc.gov/japaneseencephalitis/.
46. S. R. Han and S. W. Lee, *Biochem. Biophys. Res. Commun.*, 2017, **483**, 687.
47. J. I. Jung, S. R. Han and S. W. Lee, *Biotechnol. Lett.*, 2018, **40**, 315.
48. Centers for Disease Control and Prevention, *Dengue*, https://www.cdc.gov/dengue/index.html.
49. J. Vivekananda, C. Salgado and N. J. Millenbaugh, *Biochem. Biophys. Res. Commun.*, 2014, **444**, 433.
50. O. S. Kolovskaya, A. G. Savitskaya, T. N. Zamay, I. T. Reshetneva, G. S. Zamay, E. N. Erkaev, X. Wang, M. Wehbe, A. B. Salmina, O. V. Perianova, O. A. Zubkova, E. A. Spivak, V. S. Mezko, Y. E. Glazyrin, N. M. Titova, M. V. Berezovski and A. S. Zamay, *J. Med. Chem.*, 2013, **56**, 1564.
51. J. B. Robbins, G. H. McCracken Jr, E. C. Gotschlich, F. Orskov, I. Orskov and L. A. Hanson, *N. Engl. J. Med.*, 1974, **290**, 1216.

52. S. Marton, F. Cleto, M. A. Krieger and J. Cardoso, *PLoS One*, 2016, **11**, e0153637.
53. Y. Lyu, G. Chen, D. Shangguan, L. Zhang, S. Wan, Y. Wu, H. Zhang, L. Duan, C. Liu, M. You, J. Wang and W. Tan, *Theranostics*, 2016, **6**, 1440.
54. N. Duan, X. Ding, S. Wu, Y. Xia, X. Ma, Z. Wang and J. Chen, *J. Microbiol. Methods*, 2013, **94**, 170.
55. S. H. Suh and L. A. Jaykus, *J. Biotechnol.*, 2013, **167**, 454.
56. P. W. Gething, A. P. Patil, D. L. Smith, C. A. Guerra, I. R. Elyazar, G. L. Johnston, A. J. Tatem and S. I. Hay, *Malar. J.*, 2011, **10**, 378.
57. N. A. A. Nik Kamarudin, N. A. Mohammed and K. M. F. Mustaffa, *Biomedicines*, 2017, **5**.
58. A. Barfod, T. Persson and J. Lindh, *Parasitol. Res.*, 2009, **105**, 1557.
59. J. C. Niles, J. L. Derisi and M. A. Marletta, *Proc. Natl. Acad. Sci. U. S. A.*, 2009, **106**, 13266.
60. M. Keighobadi, A. K. Tarsi, M. Fakhar and S. Emami, *Trop. Parasitol.*, 2017, 7, 49.
61. S. Shigdar, *Ther. Delivery*, 2017, **8**, 53.
62. C. Jin, J. Zheng, C. Li, L. Qiu, X. Zhang and W. Tan, *J. Mol. Evol.*, 2015, **81**, 162.
63. H. AlShamaileh, T. Wang, D. Xiang, W. Yin, P. H. Tran, R. A. Barrero, P. Z. Zhang, Y. Li, L. Kong, K. Liu, S. F. Zhou, Y. Hou, S. Shigdar and W. Duan, *Sci. Rep.*, 2017, **7**, 5898.
64. X. She, L. Chen, Z. Yi, C. Li, C. He, C. Feng, T. Wang, S. Shigdar, W. Duan and L. Kong, *Mini-Rev. Med. Chem.*, 2018, **18**, 976–989.
65. L. Pascual, C. Cerqueira-Coutinho, A. Garcia-Fernandez, B. de Luis, E. S. Bernardes, M. S. Albernaz, S. Missailidis, R. Martinez-Manez, R. Santos-Oliveira, M. Orzaez and F. Sancenon, *Nanomedicine*, 2017, **13**, 2495.
66. S. Taghavi, M. Ramezani, M. Alibolandi, K. Abnous and S. M. Taghdisi, *Cancer Lett.*, 2017, **400**, 1.
67. S. M. Taghdisi, N. M. Danesh, M. Ramezani, P. Lavaee, S. H. Jalalian, R. Y. Robati and K. Abnous, *Eur. J. Pharm. Biopharm.*, 2016, **102**, 152.
68. M. A. Harris, T. R. Pearce, T. Pengo, H. Kuang, C. Forster and E. Kokkoli, *Nanomedicine*, 2018, **14**, 85.
69. S. Yoon and J. J. Rossi, *Curr. Opin. Oncol.*, 2017, **29**, 366.
70. J. Zhou and J. Rossi, *Nat. Rev. Drug Discovery*, 2017, **16**, 440.
71. M. S. Weinberg, *Mol. Ther.--Nucleic Acids*, 2014, **3**, e194.
72. S. M. Nimjee, R. R. White, R. C. Becker and B. A. Sullenger, *Annu. Rev. Pharmacol. Toxicol.*, 2017, **57**, 61.
73. J. W. Lee, H. J. Kim and K. Heo, *BMB Rep.*, 2015, **48**, 234.
74. United States Library of Medicine, ClinicalTrials.gov, https://clinicaltrials.gov/.
75. P. Sundaram, H. Kurniawan, M. E. Byrne and J. Wower, *Eur. J. Pharm. Sci.*, 2013, **48**, 259.
76. A. Vater and S. Klussmann, *Drug Discovery Today*, 2015, **20**, 147.

77. R. D. Jager, W. F. Mieler and J. W. Miller, *N. Engl. J. Med.*, 2008, **358**, 2606.
78. E. W. Ng, D. T. Shima, P. Calias, E. T. Cunningham Jr, D. R. Guyer and A. P. Adamis, *Nat. Rev. Drug Discovery*, 2006, **5**, 123.
79. G. Biesecker, L. Dihel, K. Enney and R. A. Bendele, *Immunopharmacology*, 1999, **42**, 219.
80. E. N. Dunn, S. M. Hariprasad and V. S. Sheth, *Ophthalmic Surg. Lasers Imaging Retina*, 2017, **48**, 100.
81. K. G. Carrasquillo, A. P. Adamis, J. W. Miller and E. S. Gragoudas, *US Pat.* Appl., 0175708 A1, 2005.
82. K. E. Maier and M. Levy, *Mol. Ther.--Methods Clin. Dev.*, 2016, **5**, 16014.
83. D. L. Dornbos, H. Harris, S. Gnyawali, D. G. Wheeler, A. Huttinger, N. Venetos, S. Talentino, H. Vaccaro, N. Musgrave, C. Jones, P. Amaya, C. Rink, J. L. Zweier, C. K. Sen and S. M. Nimjee, *Stroke*, 2017, **48**, A187.
84. A. M. Lincoff, R. Mehran, T. J. Povsic, S. L. Zelenkofske, Z. Huang, P. W. Armstrong, P. G. Steg, C. Bode, M. G. Cohen, C. Buller, P. Laanmets, M. Valgimigli, T. Marandi, V. Fridrich, W. J. Cantor, B. Merkely, J. Lopez-Sendon, J. H. Cornel, J. D. Kasprzak, M. Aschermann, V. Guetta, J. Morais, P. R. Sinnaeve, K. Huber, R. Stables, M. A. Sellers, M. Borgman, L. Glenn, A. I. Levinson, R. D. Lopes, V. Hasselblad, R. C. Becker, J. H. Alexander and R.-P. Investigators, *Lancet*, 2016, **387**, 349.
85. F. Schwoebel, L. T. van Eijk, D. Zboralski, S. Sell, K. Buchner, C. Maasch, W. G. Purschke, M. Humphrey, S. Zollner, D. Eulberg, F. Morich, P. Pickkers and S. Klussmann, *Blood*, 2013, **121**, 2311.
86. J. Menne, D. Eulberg, D. Beyer, M. Baumann, F. Saudek, Z. Valkusz, A. Wiecek, H. Haller and G. Emapticap Study, *Nephrol., Dial., Transplant.*, 2017, **32**, 307.
87. A. Vater, J. Sahlmann, N. Kroger, S. Zollner, M. Lioznov, C. Maasch, K. Buchner, D. Vossmeyer, F. Schwoebel, W. G. Purschke, S. Vonhoff, A. Kruschinski, K. Hubel, M. Humphrey, S. Klussmann and F. Fliegert, *Clin. Pharmacol. Ther.*, 2013, **94**, 150.
88. J. Hoellenriegel, D. Zboralski, C. Maasch, N. Y. Rosin, W. G. Wierda, M. J. Keating, A. Kruschinski and J. A. Burger, *Blood*, 2014, **123**, 1032.
89. E. K. Waters, R. M. Genga, M. C. Schwartz, J. A. Nelson, R. G. Schaub, K. A. Olson, J. C. Kurz and K. E. McGinness, *Blood*, 2011, **117**, 5514.
90. J. Zhou and J. Rossi, *Nat. Rev. Drug Discovery*, 2017, **16**, 181.
91. R. K. Stuart, K. Stockerl-Goldstein, M. Cooper, M. Devetten, R. Herzig, B. Medeiros, G. Schiller, A. Wei, G. Acton and D. Rizzieri, *J. Clin. Oncol.*, 2009, **27**, 7019.
92. Anonymous, *Genetic Engineering & Biotechnology News*, 2011.
93. D. W. Drolet, L. S. Green, L. Gold and N. Janjic, *Nucleic Acid Ther.*, 2016, **26**, 127.
94. S. T. Crooke, *Curr. Mol. Med.*, 2004, **4**, 465.
95. S. T. Crooke, S. Wang, T. A. Vickers, W. Shen and X. H. Liang, *Nat. Biotechnol.*, 2017, **35**, 230.
96. D. Kong, W. Yeung and R. Hili, *J. Am. Chem. Soc.*, 2017, **139**, 13977.
97. A. M. Kabza and J. T. Sczepanski, *ChemBioChem*, 2017, **18**, 1824.

98. N. Iwamoto, D. C. D. Butler, N. Svrzikapa, S. Mohapatra, I. Zlatev, D. W. Y. Sah, Meena, S. M. Standley, G. Lu, L. H. Apponi, M. Frank-Kamenetsky, J. J. Zhang, C. Vargeese and G. L. Verdine, *Nat. Biotechnol.*, 2017, **35**, 845.
99. X. Xu, D. D. Dickey, S. J. Chen and P. H. Giangrande, *Methods*, 2016, **103**, 175.
100. A. Neveol, R. Islamaj Dogan and Z. Lu, *J. Biomed. Inf.*, 2011, **44**, 310.
101. O. Greco, S. D. Scott, B. Marples and G. U. Dachs, *Front. Biosci.*, 2002, **7**, d1516.
102. L. C. Fetter, J. Jacobs and A. J. Bonham, *FASEB J.*, 2017, **31**, 767.13.
103. X. Meng, Z. Liu, Y. Cao, W. Dai, K. Zhang, H. Dong, X. Feng and X. Zhang, *Adv. Funct. Mater.*, 2017, **27**, 1605592.
104. J. D. Lenn, J. Neil, C. Donahue, K. Demock, C. V. Tibbetts, J. Cote-Sierra, S. H. Smith, D. Rubenstein, J. P. Therrien, P. S. Pendergrast, J. Killough, M. B. Brown and A. C. Williams, *J. Invest. Dermatol.*, 2018, **138**, 282.
105. International Society on Aptamers, *Aptamer Companies*, http://aptamer-society.org/aptamer-companies/.
106. D. S. Wishart and A. Wu, *Curr. Protoc. Bioinf.*, 2016, **54**, 14.4.1.

CHAPTER 17

CRISPR-based Technologies for Genome Engineering: Properties, Current Improvements and Applications in Medicine

SYLVAIN GENY, ELAHEH SADAT HOSSEINI, JEAN-PAUL CONCORDET AND CARINE GIOVANNANGELI*

Museum National d'Histoire Naturelle, INSERM U1154, CNRS UMR 7196, Sorbonne Universités, 43 rue Cuvier, Paris F-75231, France
*E-mail: carine.giovannangeli@mnhn.fr

17.1 Introduction

Clustered regularly interspaced palindromic repeats (CRISPR) technology is among the most important technological developments in biology, with the potential to provide novel treatments for diseases with a genetic component. CRISPR technology enables genome editing, the introduction of a desired modification in a chosen genomic DNA sequence, and very numerous associated applications. It is based on the use of programmable nucleases, which are able to cleave at a specific target sequence in the genome.

Drug Discovery Series No. 68
Advances in Nucleic Acid Therapeutics
Edited by Sudhir Agrawal and Michael J. Gait

Published by the Royal Society of Chemistry, www.rsc.org

Three major platforms of custom nucleases have been developed in the last twenty years:[1] Zinc-finger nucleases (ZFNs), transcription activator-like effector nucleases (TALENs) and most recently CRISPR systems. All of them are able to introduce a double-strand break (DSB) into chromosomal DNA in a site-specific manner and were demonstrated to be efficient genome-editing tools, by triggering endogenous DNA repair machineries that result in targeted genome modification. However, among these nucleases, the CRISPR system is much easier to design and produce compared with the previous nuclease generations. It is an RNA-guided nuclease. CRISPR-associated (Cas) endonuclease protein, that forms a complex with the guide RNA, cleaves a specific DNA sequence following base pairing of the guide RNA and binding of the Cas protein to a protospacer adjacent motif (PAM) sequence which is adjacent to the guide RNA target.[2,3] Designing a nuclease for each genomic site of interest is of great ease and this has allowed rapid development of the CRISPR technology in all fields of life sciences, such as functional genomics, development of new animal models of diseases and human therapeutics.

In this chapter we summarize some recent developments in CRISPR technology and present some applications in medicine.

17.2 Sequence-specific CRISPR Nucleases and Improved Variants

17.2.1 Cas 9 from *Streptococcus pyogenes*, Orthologues and Variants

17.2.1.1 CRISPR Systems from the Bacterial Immune Response

The CRISPR–Cas systems are prokaryotic immune systems.[4,5] The effector ribonucleoprotein complexes (RNP) consist of CRISPR associated protein(s) (Cas proteins) and CRISPR RNA(s) (crRNAs) that have nuclease activity that is responsible for the cleavage and subsequent degradation of foreign nucleic acids. The CRISPR–Cas nucleases cleave target DNA complementary to the crRNA guide, which is flanked by a PAM whose recognition is specific to individual CRISPR systems (Figure 17.1).

The CRISPR–Cas systems are classified into two classes and subdivided into six types.[6] The effector nucleases in class 2 consist of a single Cas protein, which can cleave double-stranded DNA targets. In class 2 are found Cas9 nucleases in the type II system and Cpf1 nucleases in the type V systems[7] (Table 17.1). Several orthologs from different bacteria have been identified and characterized for their cleavage activity,[8] such as Cas9 from *Streptococcus pyogenes* (SpCas9) and *Staphylococcus aureus* (SaCas9)[9] or CRISPR-associated endonuclease in *Prevotella* and *Francisella* 1 (Cpf1) from *Acidaminococcus* species (AsCpf1) and Lachnospiraceae bacteria (LbCpf1).[7] The structures of the corresponding guide RNAs and PAM sequences have been precisely described (see examples in Table 17.1). Cpf1 from the type V system has

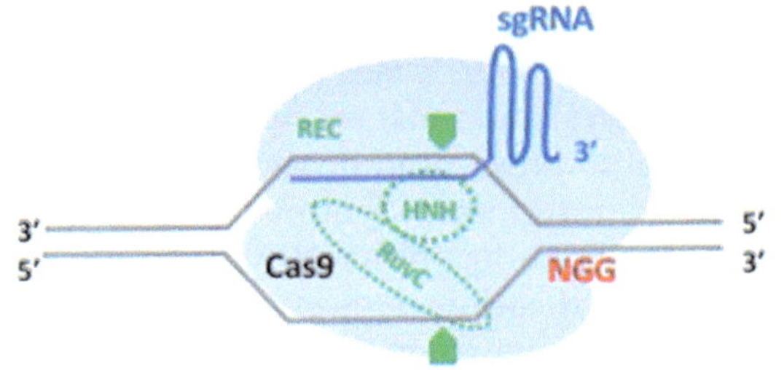

Figure 17.1 Cas9 DNA binding and cleavage. Cas9 complexed with guide RNA scans the genome for PAM sequences and upon PAM recognition, the upstream DNA is directionally unwound to allow RNA guide binding. If the upstream sequence is complementary to the guide RNA, the histidine–asparagine–histidine (HNH) domain undergoes a conformational change and activates resistance to ultraviolet light C (RuvC) and HNH nuclease domains for concerted cleavage. The HNH and RuvC catalytic domains simultaneously cleave the target and non-target strand, respectively. The non-catalytic receiver (REC) domains are essential for HNH reorientation that is necessary for cleavage. Single chimeric guide RNA (sgRNA) is indicated in blue, DNA target in dark grey, PAM sequence on the non-target strand in red, Cas9 protein in light blue with some functional domains indicated in green, and cleavage sites as green arrows.

Table 17.1 Properties of some CRISPR systems and engineered CRISPR nucleases that are used for genome editing.

Cas protein	Bacterial origin	MW (kDa)	AM sequence – Target length[a]	Cleavage site
SpCas9	*Streptococcus pyogenes*	158	(20 nt) NGG	17–18
SaCas9	*Staphylococcus aureus*	124	(21 nt) NNGRRT	18–19
AsCpf1/LbCpf1	*Acidaminococcus* and *Lachnospiraceae*	143	TTTV (23 nt)[b]	19–20/23–24[c]
Variants				
VQR SpCas9	*Streptococcus pyogenes*	158	(20 nt) NGA	17–18
VRER SpCas9	*Streptococcus pyogenes*	158	(20 nt) NGCG	17–18
KKH Sa	*Staphylococcus aureus*	124	(21 nt) NNNRRT	18–19
RVR AsCpf1	*Acidaminococcus*	143	TATV (23 nt)	19–20/23–24[c]
RR AsCpf1	*Acidaminococcus*	143	TYCV (23 nt)	19–20/23–24[c]

[a]The non-target strand is represented (5′–3′).
[b]V = A, G or C.
[c]Cpf1 nucleases generate four-nucleotide 3′ overhangs.

some distinct properties compared with Cas9 from the type II system, especially the nature of crRNAs. Cas9 is directed to its DNA target by two RNAs, a *trans*-activating crRNA (tracrRNA) and a crRNA. In the synthetically reconstituted system, the two short RNAs can be fused into a single chimeric guide RNA (sgRNA). Cpf1 is directed to DNA by a single short crRNA that is produced from a crRNA array through Cpf1 processing. This latter property can be exploited to simultaneously target multiple target genes in gene-editing experiments.[10]

All prokaryotic RNA-guided nucleases mentioned have been used successfully to induce targeted DSBs in mammalian chromosomes by expression of the guide RNA and Cas protein, and these are well established and robust genome-editing tools. The diversity of CRISPR–Cas systems therefore offers a variety of genome-editing approaches. However, not all Cas nucleases tested have been found to be active in eukaryotic cells, possibly due to inefficient binding to nucleosomal DNA.[7] In this chapter we mainly focus on SpCas9, the most commonly used tool in genome-editing applications.

Type VI CRISPR systems target RNA rather than DNA and have also been recently adapted to develop a novel RNA targeting technology. An RNA-guided RNase, Cas13, that mediates precise RNA cleavage (*Leptotrichia wadei* Cas13a, LwaCas13a) has been adapted as a tool for RNA knockdown or RNA base editing in mammalian cells.[11] Unexpectedly, RNA targeting was also recently described for subtypes II-A and II-C Cas9 proteins, for example from *S. aureus* and *Campylobacter jejuni*.[12] Programmable RNA targeting with CRISPR systems is not further described in this chapter, since we have chosen to focus on DNA targeting.

17.2.1.2 Variation Around the SpCas9–sgRNA Complex to Improve Activity

The most commonly used Cas9 protein is the 1368-residue Cas9 from *Streptococcus pyogenes*, SpCas9, and various intrinsic parameters have been shown to be important for its optimal activity. Fine optimization of Cas9 design in specific applications is indeed important to improve gene-editing efficiency. For example, varying the nature, number and position of nuclear localization signals added to the prokaryotic Cas9 protein may improve transfer into the nucleus.[13] The design of guide RNA may also significantly affect efficiency. The single-guide design proposed by Jinek *et al.* fused the crRNA and a shortened tracrRNA into a single RNA molecule and was useful to unravel the basic mechanism of DNA cleavage by Cas9 *in vitro*.[2] However, it was found subsequently to be rather inefficient at supporting DNA cleavage when expressed in cells and this was rapidly replaced by sgRNA designs with longer tracrRNA moieties[14] and a few additional mutations were later included to favor high-level expression from RNA polymerase III promoters and improved Cas9 loading.[15] Cheaper prices and chemical modifications that stabilize guide RNAs in cells have now also made synthetic RNAs one of the preferred ways of obtaining guide RNAs for genome editing with CRISPR–Cas9. Synthetic, modified crRNAs are combined with synthetic tracrRNA to form a dual guide RNA that closely mimics the natural guide RNA design. Several chemical modifications, such as 2′-*O*-methyl groups, are generally introduced in order to favor stability of guide RNAs in cells and have been found to significantly improve efficiency of genome editing.[16]

17.2.1.3 Expanding the Target Repertoire

The simple NGG PAM sequence of SpCas9 occurs on average every 8 bp in the human genome, a frequency that is not very limiting for DSB-based genome editing. However, in some specific situations, for example, when it is necessary to distinguish between a wild-type and mutant allele, or for other genome-editing techniques, such as base editing (see Section 17.4.3), precise targeting of a locus with a single-nucleotide resolution can be essential. In these cases, the PAM requirements can be a limitation. In this context, some studies have significantly expanded the number of target loci for genome editing.

As described above, several natural CRISPR systems have been identified and used in genome-editing experiments. Each CRISPR nuclease can vary in size, PAM sequence and location of the DSB introduced within the target DNA (Table 17.1).[8] The diversity of natural PAM sequences allows one to extend the range of targetable sequences.

Cas9 can also be modified to specify other PAMs. For SpCas9, NGA rather than NGG is possible with the VQR variant, or NGCG with VRER variant,[17] for SaCas9, NNNRRT instead of NNGRRT with the KKH variant,[18] and for Cpf1, TATV and TYCV instead of TTTV (V is A, G, or C) in RVR and RR variants,[19] respectively.

17.2.2 Expanding Targeted Functions with "CRISPR Fusions"

CRISPR systems have been used to target different types of functional domains to specific genomic sites. To do that "CRISPR fusions" have been developed, and CRISPR systems were modified in order to suppress DSB induction while maintaining DNA binding properties and fused to the protein domain(s) of interest. In this way, transcription modulation, modification of epigenetic status (Section 17.4), DNA repair manipulation (Section 17.3), DNA imaging,[15,20] and DNA capture of specific loci[21] have been successfully achieved. Today the most developed approaches are transcription modulation and base editing, which was recently developed, and they will be described in more detail in Section 17.4.

17.2.2.1 Catalytically Dead CRISPR Systems

CRISPR systems incompetent for DNA cleavage can be designed in two main ways. A Cas mutant with undetectable endonuclease activity (deadCas or dCas9) can be used. Indeed Cas9 proteins include two catalytic domains, the resistance to ultraviolet light C (RuvC) and histidine–asparagine–histidine (HNH)-like domains, each capable of cleaving one strand of the DNA target.[2] In Cpf1 a single RuvC-like domain is present and appears to cleave both strands of DNA.[7] For SpCas9, a double point mutant of RuvC and HNH catalytic sites (D10A and H840A or N863A) creates dCas9 and for Cpf1 a single mutation is sufficient (dLbCpf1 with D832A). A modified design of guide RNAs is an alternative. For SpCas9 it was shown that truncated guide RNAs containing only 14–16 nucleotides complementary to the DNA target upstream of the PAM no longer lead to detectable nuclease activity but are still sufficient for strong DNA binding.[22]

17.2.2.2 Multiplexing Strategies

In order to enhance the activity of functional domains targeted to DNA, multiplexing strategies have been developed, allowing the recruitment of multiple domains by a unique guide RNA. A series of structure-guided designs were proposed.[23] Based on the fact that some regions in the guide RNA sequence protrude outside of the RNP complex, RNA aptamers interacting with specific proteins were added at these positions to facilitate the recruitment of effector domains to the Cas9 complex.[24,25] For example, RNA hairpins that bind with very high affinity to the MS2 bacteriophage coat protein were successfully used in combination with the domain of interest fused to MS2 and enable multiple copies of functional domains to be recruited by a single guide RNA.[26] In order to achieve multiplexing, the repeating peptide array, SunTag, has also been fused to Cas9, resulting in recruitment of multiple copies of a single-chain antibody fused to the domain of interest.[27] In the context of DSB-induced genome editing, CRISPR fusions using catalytically active Cas proteins have also started to be used to enhance modification rates and favor precision genome editing (Section 17.3).

Interestingly the two approaches, fusion to Cas protein and fusion to a protein domain interacting with RNA domains in a modified guide RNA, can be combined to reinforce the desired activity or to associate multiple functions. For example, in the context of transcription activation (see Section 17.4), one of the most efficient systems, called synergistic activation mediator (SAM), is based on a dCas9–VP64 fusion in association with a sgRNA including two MS2 RNA aptamers and additional MS2–p65–heat shock factor 1 (HSF1) activation fusion protein.[26] Multiple functions can also be simultaneously achieved with orthogonal CRISPR systems since crRNA structures are specific to a given Cas protein, allowing their simultaneous use.[22,28,29]

17.2.3 Specificity of CRISPR Systems

The specificity of DNA cleavage induced by CRISPR systems has been studied extensively and it has been shown that under some conditions binding and cleavage at off-target sites can be observed. However, the field of Cas9 specificity is moving fast, with improvements in off-target detection methods, in guide RNA selection or in the engineering of high-specificity CRISPR systems (for a review see Tycko *et al.*[30]).

17.2.3.1 Approaches to Determine Specificity

The DNA-targeting specificity of CRISPR-based genome-editing agents has been examined using a variety of approaches. Various methods have been developed to detect off-targets such as Guide-seq,[31] Circle-seq,[32] digested-genome sequencing (Digenome-seq),[33] Site-seq[34] or *in situ* breaks labeling sequencing (BLESS)[35] and breaks labelling *in situ* and sequencing (BLISS).[36] They allow one to identify DSBs induced by nuclease treatment in the whole genome and also to identify off-targets with low mutation frequencies, down to 0.1% depending on the depth of DNA sequencing. The methods showed that off-targets are always related to the guide sequence and in great majority correspond to the target sequence with up to four mismatches. This observation is of practical importance and supports the development of bioinformatics tools for guide RNA selection (see below). Nevertheless, off-targets with more than four mismatches or with bulges in the alignment to the guide RNA sequence can also be detected.[31] The candidate off-target sequences that can be aligned with five or six mismatches or with bulges are just too numerous to be to fully investigated. Off-target mutations are generally extremely low at such sites and they are not a problem for studies in basic research. However, they could be a real concern when considering therapeutic applications and must then be investigated carefully.

Interestingly off-target cleavage sites of Cas9 are different from off-target binding sites of dCas9. It was shown for SpCas9, with a comparison between off-target cleavage sites determined by GUIDE-seq and off-target binding sites determined by chromatin immunoprecipitation (ChIP)-seq. These data are consistent with those from studies showing different structural requirements for binding and cleavage activation.[37]

17.2.3.2 Computational Tools for Guide RNA Selection

Scoring methods have been developed to predict DNA cleavage at off-target sites related to the guide RNA.[38] They are based on systematic quantification of the effects on target DNA cleavage of single or dual point mutations in guide RNAs. The resulting datasets are essential but other parameters also come into play, such as chromatin context. For example, *in vitro* studies have demonstrated that specific positions of target DNA on nucleosomes could inhibit Cas9 binding and preclude efficient cleavage.[39] However, the corresponding information is generally not available and such parameters therefore intrinsically limit the potential of off-target predictions.

Nevertheless off-target detection is also often carried out. The methods exploit the predictive models of highly probable off-target sites and evaluate mutation rates at the top off-target sites by targeted analysis.

However, recent studies have demonstrated that predictive models remain insufficiently accurate.[40] Off-targets identified by experimental methods are always predicted bioinformatically but too many candidates have similar or superior scores and it is not feasible in practice to test them all by polymerase chain reaction (PCR) sequencing (Haeussler M., personal communication). Pending further improvements, the authors have therefore recommended carrying out unbiased detection of off-targets in cultures cells and using high-throughput sequencing in animal models to systematically check for the presence of mutations at the off-targets identified in cultured cells.

17.2.3.3 Approaches to Improve Specificity

Several strategies were developed and significantly improved the specificity of SpCas9.

The first strategy was based on a dimerization approach, requiring two separate Cas9 binding events to take place at the same locus in order to obtain DNA cleavage.[41] A Cas9 mutant, inactivated on one of the catalytic sites is able induce a single-stranded break on DNA with the remaining catalytic domain. Such a Cas9 nickase can be used with a pair of guide sequences in order to provoke a DSB and has been shown to exhibit improved specificity. However, its use is based on the design of two guides with some constraints, PAMs on opposite strands, positioned to generate 5′ overhangs of around 50 nucleotides, and the potential optimizations are thus limited.

Structure-guided engineering of SpCas9 has led to the design of high-fidelity Cas9 proteins (HiFi Cas9).[35,42] The designs were based on the idea that excess DNA-binding energy is a source of non-specific interactions favoring off-target cleavage. Based on the structural data for the ternary complex of Cas9, guide RNA and target DNA, three to four amino acid mutations were introduced into SpCas9 in order to neutralize non-specific electrostatic interactions between the protein and the sugar–phosphate DNA backbone of the target DNA. Such modified Cas9 proteins have strongly increased cleavage specificity and generally maintain on-target activity, only around 10% of guides being less active. Recently a new HiFiCas9 variant, called hyper-accurate Cas9, was designed[43]. It is based on the precise molecular understanding of the balance between DNA recognition and cleavage activation and includes mutations in a non-catalytic domain governing this regulation. Mutations that increase Cas9 specificity could also be transferred successfully to variants with altered PAM sequences.[44]

Some naturally occurring Cas proteins, such as SaCas9 or Cpf1, have longer PAMs and/or guide target sequences and were expected to be more specific since a lower number of off-targets are possible. Such higher specificity was indeed experimentally demonstrated, for example for SaCas9 compared to SpCas9, the former having a longer NGRRT PAM motif.[9]

A second class of approaches to improve specificity did not involve any change to the Cas9 protein. For SpCas9, truncation of the sgRNA, with less than 20 nucleotides complementary to the target DNA, typically 17–18 nucleotides, was demonstrated to improve cleavage specificity under some conditions while maintaining high on-target cleavage.[45] Improved specificity and maintenance of on-target activity probably depend on the guide RNA sequence and need to be checked in each case. Recently, it has also been shown that incorporation of modified bases such as locked nucleic acids (LNA) or bridged nucleic acids (BNA) at specific positions in the guide RNA increases Cas9 cleavage specificity.[46] The modifications appear to destabilize the cleavage-competent conformation of RNP in the presence of mismatches and prevent off-target cleavage.

Another strategy is to decrease Cas9 lifetime in cells to minimize opportunities for off-target cleavage. The direct delivery of Cas9–sgRNA RNPs into cells results in transient Cas9 activity,[47–49] compared with plasmid transfection, which results in long-duration Cas9 and sgRNA expression. Such a short pulse of cleavage was shown to increase the ratio of on-target to off-target activity.[50] Alternatively, several natural CRISPR inhibitors were identified that enable phages to escape immunity and these anti-CRISPR proteins, such as anti-CRISPR IIA4 (AcrIIA4), have started to be used to control Cas9 genome editing in cells.[51] Finally, control of the duration of Cas nuclease activity was also achieved using different types of strategies. Optical control of the duration of Cas9 activity was described, based on the use of split Cas9 fragments without nuclease activity and photoinducible dimerization domains named Magnets.[52] Blue light irradiation induces heterodimerization between the two domains, which enables split Cas9 fragments to associate, thereby recovering nuclease activity. Importantly the activity can be switched off simply by stopping light irradiation. Another approach consists of tagging Cas9 with ubiquitin-proteasomal degradation signals to facilitate the degradation of Cas9.[53]

Together, these advances in improving the specificity of CRISPR-based genome-editing agents are major developments that increase their promise as future therapeutics.

17.3 Genome Editing Mechanisms and Current Improvements

17.3.1 Principles: Genome Editing Takes Place During Repair of DSB Breaks Induced by Cas9

Genome editing occurs through the stimulation of the cellular repair machinery *via* a DNA double-strand break (DSB) induced by the system CRISPR–Cas9. The most common applications for genome editing with CRISPR–Cas9 aim to create mutations to inactivate a specific gene (knockout) or to insert a desired sequence (knock-in) at a specific locus in the genome. These two outcomes are generally achieved through distinct DNA repair pathways. Gene knockouts (also called gene disruptions) are predominantly created by end-joining pathways such as classical non-homologous-end-joining (cNHEJ), whereas gene knock-in usually depends on homology-directed repair (HDR). End joining pathways are based on ligation with little or no homology between the two extremities of a DSB. In contrast, HDR pathways rely on the presence of donor single-stranded (ssDNA) or double-stranded DNA (dsDNA) fragments containing sequences homologous to the sequences flanking the DSB to generate precise nucleotides substitution or DNA sequence insertions.

17.3.1.1 End-joining Pathways

cNHEJ is the main end joining pathway in the natural context, which is involved in numerous biological processes that include programmed DNA rearrangements. NHEJ is capable of precise DSB rejoining but may be error-prone in the context of genome editing because even when the DSB is repaired perfectly, the DNA target is susceptible to another round of cutting.[54] In addition, *in vitro* single-molecule experiments indicated that Cas9 remains bound to the DNA ends after cleavage.[55] In cells, such stable binding might impair the repair process and favor erroneous repair during end joining.[54,56] The direct ligation of the two DNA ends can change the DNA sequence by insertion or deletion (indel) of one to several nucleotides. Systematic analysis of mutations induced by programmable nucleases indicated that the mutation pattern is not random.[57] In particular, deletions are frequently flanked by short stretches of identical sequence that indicate that repair did not take place by cNHEJ but by an alternative end joining pathway called micro-homology-mediated end joining (MMEJ).[58,59] The cNHEJ pathway is well characterized at the molecular level. The first step of cNHEJ involves binding of the Ku70–Ku80 heterodimer to DNA ends, within seconds after DSB.[60] Ku70–Ku80 subsequently recruits the main NHEJ factors, including DNA ligase IV, the catalytic complex DNA-dependent protein kinase, catalytic subunit (DNA-PKcs),[61] the scaffold proteins X-ray cross complementing protein 4 (XRCC4), XRCC4-like factor (XLF)[62] and Aprataxin-and-polynucleotide kinase-like factor (APLF). In contrast, MMEJ is not as well characterized and its importance in the physiological context has been controversial. However, one key protein in the pathway is the Polθ polymerase encoded by the POLQ gene. In zebrafish, it has been demonstrated that inactivating the POLQ gene severely impairs the efficiency of DNA repair of DSB breaks induced by Cas9 and the mutation pattern was changed,[63] providing strong evidence for a crucial role of MMEJ in generating mutations during genome editing with Cas9.

17.3.1.2 Homology-directed Repair

HDR is mechanistically more complex than end joining, involves a larger number of partners, but results in a more accurate repair and preserves genome integrity. HDR-mediated DNA repair relies on the use of a homologous DNA template. Homology-directed repair (HDR) comprises several distinct mechanisms and can be divided into homologous recombination (HR), single-strand annealing (SSA), synthesis-dependent strand annealing (SDSA) and break-induced replication (BIR).[64] The interplay and the relative contributions of the different HDR repair pathways are not completely understood. They start with the same initial resection from the ends of the DSB. Resection is the process of converting double-stranded DNA into single-stranded DNA by the exonucleolytic digestion of one strand. The resection step is crucial because it commits the cell into HDR-mediated repair[65] and generates a ssDNA to search for a homologous DNA template for accurate repair of the DSB.

17.3.1.3 Regulation of DNA Repair Pathways During the Cell Cycle

Eukaryotic cells have highly regulated mechanisms to choose between the numerous repair pathways.[66] The balance between HDR and cNHEJ is tightly synchronized with the cell cycle and also depends on the type of DSB. During the cell cycle, HDR is active during G2 and S phases, when a sister chromatid is available as a homology template. Conversely, cNHEJ does not require a homologous template and is therefore not restricted to a particular cell cycle phase. At the molecular level, control mechanisms restrict resection to initiate only in S and G2 phases and not in G1.[67] During the S and G2 phases, c-NHEJ and HDR compete for DNA repair.[68] However, both pathways are downregulated during M phase.

cNHEJ is a comparatively faster and more efficient repair pathway than HDR. By use of reporter plasmids transfected in cells, cNHEJ of compatible ends or incompatible ends can be completed in approximately 30 min, while HDR is much slower and takes 7 h or longer to complete.[69] In actively cycling cells, cNHEJ of compatible ends is twice as efficient as cNHEJ of incompatible ends that is still three times more efficient than HDR. On the basis of these results, HDR might be considered as a repair system operating when cNHEJ is not active nor successful.[70] As a result, in genome editing the frequency of precise sequence modification based on HDR is generally low and methods to increase its efficiency are highly needed.

17.3.2 Improvement of HDR-based CRISPR Strategies for Programmable Genome Modification

17.3.2.1 Strategies to Repress the NHEJ-mediated Repair

NHEJ can be subdivided into classical NHEJ (c-NHEJ) and alternative NHEJ (alt-NHEJ including MMEJ). NHEJ and MMEJ-based methods are also capable of integrating long exogenous DNA respectively with no homology[71,72] or with microhomologies.[73] These approaches can be very efficient but the random direction of DNA integration and indel mutations at the DSB extremities are drawbacks, and careful screening of genome editing events is needed.

Several approaches have been developed to increase HDR genome editing efficiency through inhibition of cNHEJ (Figure 17.2). DNA ligase IV is needed to ligate the DNA ends in cNHEJ. A straightforward strategy is to inhibit ligase IV through the binding of small molecules[74] or proteins that trigger its ubiquitin-dependent degradation.[75–77] Inhibition of cNHEJ was also attempted by targeting other partners such as DNA-PK.[78] Silencing by RNA interference of Ku70, Ku80 or ligase IV also repressed NHEJ whilst increasing HDR.[75]

Recently, Canny *et al.* engineered a protein inhibitor of 53BP1 that increases the efficiency of HDR-dependent genome editing. 53BP1 is also a target of interest because it regulates commitment of the cell to cNHEJ. 53BP1 inhibits breast cancer associated 1 (BRCA1) recruitment to DSBs in

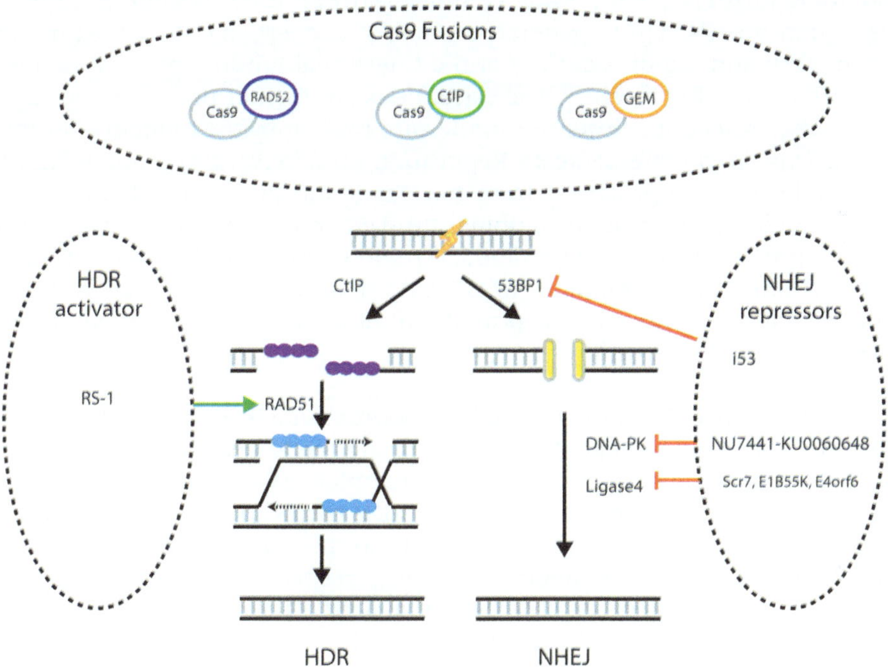

Figure 17.2 The main strategies for enhancing CRISPR–Cas9 HDR-mediated genome editing. After CtIP-dependent resection and binding of the single-stranded DNA binding protein replication protein A (RPA; in purple), the HDR activator RAD51-stimulatory compound 1 (RS1) promotes RecA-like NTPase (RAD51)-assisted (in blue) strand invasion into the DNA homology template, therefore increasing the HDR efficiency. After binding of Ku70–Ku80 (yellow) to the DNA ends, NHEJ inhibitors repress NHEJ and thus indirectly favor HDR by targeting the main partners involved in the NHEJ pathway [tumor suppressor p53-binding protein 1 (53BP1), DNA-PK, DNA ligase IV]. The fusion of Cas9 with a protein involved in HDR C-terminal binding protein interacting protein (CtIP, RAD52) or mediating degradation, during the G1 phase, during which HDR is inactive (GEM, geminin), is another promising strategy to favor HDR-mediated genome editing.

G1 phase and therefore prevents resection.[79,80] This inhibitor was isolated from a library of ubiquitin variants that is able to compete with histone H4 dimethylated as lysine 20 (H4K20me2) for 53BP1 binding *in vitro*. When overexpressed in cells, it prevents 53BP1 recruitment to DSBs in response to gamma irradiation and increased HDR can be detected in genome editing experiments.[81]

17.3.2.2 Strategies to Enhance HDR

17.3.2.2.1 Main Pathways in HDR-mediated DSB Repair. In HDR-mediated repair, the DSB ends are resected to expose 3′ ssDNA tails. Resection starts with a short-range resection initiated by meiotic recombination 11 homolog (Mre11) {of the MRN complex [MRE11–RAD50–Nijmegen breakage syndrome (NBS)]} to generate a short 3′ ss overhang at each extremity of the DSB and requires activity of the C-terminal binding protein interacting protein (CtIP) protein.[82] The extremities are further processed during a long-range resection, either by exonuclease Exo1 or by the tandem of exonuclease Dna2 and helicase RecQ. Replication protein A (RPA) coats the newly resected ssDNA to protect it from nucleases and potential self-annealing. RPA is exchanged for the recombinase RAD51 in response to activity of the BRCA2 protein. RAD51 next promotes strand invasion to the adjacent intact sister chromatid as a template for HDR. After strand invasion and synthesis, the second DSB end can be captured and repair can proceed by gap-filling DNA synthesis and ligation.[83]

17.3.2.2.2 DNA Homology Template Donors for HDR-mediated Genome Editing. In HDR-mediated genome editing, cells accept different types of DNA donor substrate as homology templates: single-strand (ss), double-strand (ds), linear or plasmid DNA. HDR pathways differ with different types of donor DNA.[84] Despite numerous studies, no specific type of DNA donor was found to be superior and their respective efficiencies depend on many factors. However, studies that analyzed the various types of DNA donors resulted in new rules for the design of the homology templates.

Plasmid donors are used typically for the efficient introduction of long DNA fragments (1–5 kb) flanked by large homology regions to ensure efficient HDR integration. In some specific cases, linearization of the plasmid, including *in situ*, increases the frequency of HDR-mediated genome integration.[85,86] Linear dsDNA donors from commercially synthetic dsDNA, PCR, or enzyme-digested plasmids are efficient despite decreasing the length of the homology regions.[87]

Single-stranded oligonucleotide (ssON) donors are a promising alternative for genome editing. They are less prone to random integration compared with dsDNA and require short homology regions (~35 nt) for efficient and precise integration in the genome.[87] They are also commercially available and readily modified with phosphorothioate linkages or LNA, which improves the efficiency of genome editing by increasing stability in cells and thereby the effective concentration of ssDNA during DNA repair.[88] Recently, it was shown that after DNA cleavage, Cas9 releases the PAM distal non-target strand before dissociation, thereby allowing the binding of a ssDNA donor. This observation led to the design of an optimized ssON donor with asymmetric homology regions around the DSB, that can result, in some cases, in a higher HDR rate.[89] Limited, due to synthesis constraints, to 200 nt long ONs, ssONs contain typically a short DNA integration cassette (<100 bp). However, to integrate a longer stretch into the genome, several groups made recent improvements in the synthesis of long ssDNA. The template dsDNA is transcribed into ssRNA and then reverse transcribed into ssDNA. Heteroduplex RNA–DNA is then treated with RNAse to yield a long ssDNA that was an efficient donor for HDR in mice.[90] Another method for the production of long ssDNA relies on the double nicking of a plasmid to release ssDNA. Such ssDNA was used successfully in rats. The mechanism of HDR with ssON is not entirely clear and differs from the integration of a dsDNA donor.[87,88] It has recently been suggested to involve specifically partners of the Fanconi repair pathway.[91]

17.3.2.2.3 Fusion of Cas9 to HDR Partners. Several approaches to enhance HDR include the fusion to Cas9 of another protein thereby combined to obtain a more efficient genome-editing process (Figure 17.2). This important strategy aims to design a Cas9 fusion protein that favors recruitment of proteins that could enhance HDR. The targeting of the protein at the DSB site allows one to avoid global effects on cellular DNA repair. The fusion of Cas9 with a minimal N-terminal fragment of CtIP was sufficient to stimulate HDR.[92] This fragment contains the cell-cycle-dependent phosphorylation sites of CtIP that represent one of the main control points of HDR. CtIP promotes Mre11 activity before it is degraded by the proteasome during G1 phase. A similar fusion approach was applied to RecA, a bacterial analogue of Rad51 and resulted in a higher HDR-mediated integration of ssDNA but a decreased HDR efficiency.[93] Rad52 is associated with SSA and catalyzes the annealing of complementary strands of DNA wrapped in RPA.[94] Overexpression of RAD52 or direct fusion of RAD52 were shown to increase HDR using ss and ds donor DNAs.[95] A variant approach aimed to take advantage of the cell cycle control of HDR using a Cas9 fused with GEMININ, which is therefore degraded in G1 when NHEJ is prominent and HDR inactive, the fusion reducing NHEJ-mediated indels and enhancing, in some cases, HR-based genome editing.[96,97]

17.3.2.2.4 Modifications of the Cell Cycle During Gene Editing and Other Approaches. Orthwein *et al.* showed that the replacement of partner and localizer of BRCA2(PALB2) by a mutant that does not undergo phosphorylation-induced degradation during G1 phase in combination with end resection favored HR in cells by allowing it to take place in G1 and thus enhancing HR-based CRISPR–Cas9 genome editing.[98] The main function of BRCA2 is to deliver its cargo of RAD51 molecules (6–8) to the ssDNA. BRCA2 enables

RAD51 to displace RPA from ssDNA. PALB2 contains two DNA-binding regions but also interacts with BRCA1 and BRCA2. The inhibition of homologous recombination in G1 minimally consists of the suppression of DNA-end resection coupled with the prevention of the assembly between BRCA1 and the PALB2–BRCA2 complex necessary for the formation of the Rad51 nucleofilament.

Another approach based on cell cycle control of HDR includes the synchronization of cells and the delivery of Cas9 in G2 which increases HDR efficiency whilst reducing NHEJ repair.[99] The combined use of CCND1, a cyclin that functions in G1–S transition, and nocodazole, a G2–M phase synchronizer significantly increased HDR efficiency in genome editing.[99]

Additionally, a method of "co-selection" was recently developed based on the multiplex ability of CRISPR to target a locus of interest in combination with the gene for the sodium–potassium pump ATPase Na^+/K^+ transporting subunit alpha 1 (ATP1A1). After HDR-mediated editing of the ATP1A1 gene, cells were resistant to ouabain, a specific inhibitor interacting with a specific extracellular loop of ATP1A1. After selection with ouabain, a strong enrichment of cells with HDR-based editing at the locus of interest is observed, the process therefore being called co-selection.[100]

17.4 Epigenome Editing and Base Editing with the CRISPR Systems

17.4.1 Modulation of Transcription, CRISPRa and CRISPRi Systems

The CRISPR system was also used to edit the epigenome in order to alter the regulation of a target gene by acting directly on transcription or on chromatin environment (see Section 17.4.2), rather than to modify its sequence directly[101,102] (Figure 17.3).

For direct transcriptional modulation of a specific gene, activation or repression, the principle is to use the CRISPR system to target a transcriptional activator or repressor and design it to bind in the vicinity of the transcriptional start site of a gene of interest using CRISPR fusions [CRISPR activation (CRISPRa) and CRISPR inhibition (CRISPRi), respectively]. As described previously (Section 17.2), various designs are possible and were used to direct fusion to Cas9 of transcriptional activators or repressors and alternately, multiple copies can be recruited.

A variety of transcriptional activation domains have been described, which increase the expression of a large variety of genes. The first generation of CRISPRa makes use of direct fusion of VP64 to dCas9.[103] A series of second-generation CRISPRa have been evaluated[104] and among them, the most efficient ones seem to be the tripartite activator VPR[105] (VP64–p65–Rta, where Rta is the transcriptional activation domain of the Epstein–Barr virus), the Suntag system, using VP64 activator, and the SAM system (see Section 17.2).

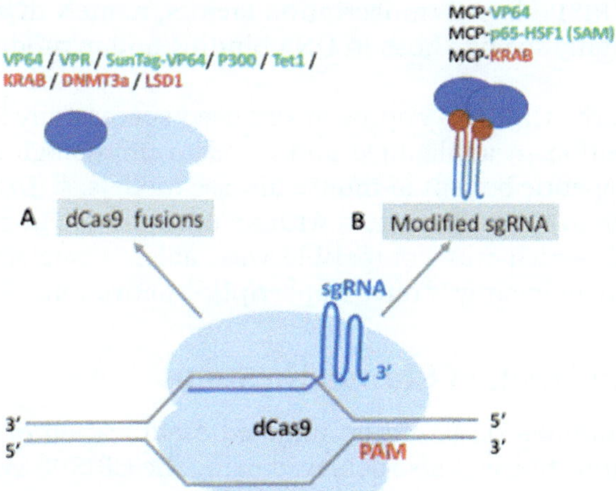

Figure 17.3 CRISPR-based epigenome editing. (A) Transcriptional activators can be generated by direct fusion of dCas9 with the transcriptional activator [VP64, VPR for VP64-p65-Rta, where Rta is the transcriptional activation domain of the Epstein–Barr virus (VP64–p65–Rta) fusion, or the histone acetyltransferase enzyme p300 or the DNA demethylase ten–eleven translocation 1 (Tet1)]. Transcriptional repressors can be generated by fusion of dCas9 to the transcriptional repressor domain Kruppel-associated box (KRAB) or the DNA methyltransferase enzyme DNMT3a or to the lysine-specific histone demethylase (LSD). (B) Alternatively, transcriptional activators or repressors can be obtained by use of a modified sgRNA including MS2 stem loops that recruit (MCP-activator/repressor) fusions. In the drawing, MS2 stem loops and MCP (MS2 bacteriophage coat protein) are represented in red and activator/repressor domains in dark blue.

Conversely, for transcription repression, dCas9 was fused to the Kruppel-associated box (KRAB) transcriptional repressor[106–108] and used successfully to decrease expression of target genes. dCas9 was also shown to be a transcriptional repressor by itself, on the basis of its high affinity for target DNA and ability to block the transcriptional machinery.[109]

A public tool is available to facilitate the design of appropriate sgRNAs to enable efficient CRISPRi and CRISPRa (http://CRISPR-ERA.stanford.edu). CRISPR modulation of transcription is robust enough to be used in high-throughput CRISPRi and CRISPRa screens to identify genes that result in phenotypes of interest when upregulated or downregulated.[26,110] Compared with knockout (KO) screens using wild-type (wt)Cas9,[111,112] which can generate heterogeneous KO among alleles, CRISPRi screens allow knockdown that consists of homogenous loss of function (for a review see Doench[113]).

Genome-wide effects of CRISPR-based transcriptional activators in human cells have been determined using ChIP-seq to evaluate DNA-binding specificity and RNA-seq to measure the specificity of transcriptome modulation. CRISPR-based transcription factors, namely dCas9-VP64, were shown to be highly specific both in DNA binding and in modulation of gene expression.[114]

Importantly, the CRISPRa approach was used successfully for *in vivo* transcription activation in adult mice and it led to physiological phenotypes, including therapeutic benefit in mouse disease models.[115] To reach this goal an optimized SAM CRISPRa system with an optimized RNA guide structure was developed, which was compatible with adeno-associated virus (AAV) delivery and can efficiently induce transcription activation.

17.4.2 Modulation of Chromatin Status

Chromatin manipulation based on targeted modification of histones or of DNA methylation state was also achieved using the CRISPR system.

Histone-modifying CRISPR-based tools have been developed. For example, the catalytic domain of human acetyltransferase p300 was fused to dCas9.[108,116] This fusion catalyzes histone H3 lysine 27 (H3K27) acetylation at loci up to thousands of base pairs from the sgRNA target locus. This results in transcriptional activation of genes, thus considered as robust CRISPRa. Alternately, the lysine-specific histone demethylase LSD1 has been fused to dCas9,[117] inducing demethylation of dimethylated histone H3 lysine 4 (H3K4me2) at sites a few hundred bp from the sgRNA binding site.

The fusion of the DNA methyltransferase enzyme DNMT3A to dCas9 functions as an epigenetic modifier that methylates CpG islands within around 100 bp of the sgRNA genomic target.[118] In contrast, fusion of the catalytic domain of the DNA demethylase ten–eleven translocation 1(Tet1) to dCas9 has been used to induce targeted DNA demethylation. Such fusions can induce quite complete demethylation of CpG islands within 200 bp of the target site.

Recently, a very interesting study showed the possibility of reversing gene inactivation by epigenome editing in a pathological situation, providing a proof of concept for the use of such approaches as therapeutic strategies for disorders that involve epigenetic silencing.[119] It concerns fragile X syndrome (FXS), the most common genetic disorder of intellectual deficiency in males, associated with hypermethylation of CGG repeats at the fragile X mental retardation 1 (FMR1) locus. Fusion of a dCas9 with DNA methylation modification enzymes, DNA methyltransferase (DNMT) or Tet (dCas9–DNMT or dCas9–Tet) were used to reverse the hypermethylation of CGG repeats at the FMR1 locus in FXS patient-derived induced pluripotent stem cells (iPSCs). A single guide RNA in association with the dCas9–Tet1 enabled efficient demethylation at the CGG repeats in the hypermethylated pathological FMR1 locus. Complete demethylation of the CGG expansion increased specific histone modifications at the FMR1 promoter, and suppressed the epigenetic silencing of the FMR1 gene, restoring expression in FXS iPSCs and neurons.

Concerning epigenetic modifications, the on-target methylation efficiency of dCas9-methyltransferases has been reported in different studies. However a paper reported their global activity only recently, with measurements of seeding, spreading, and maintenance.[120] This study was possible thanks to the development of appropriate cell lines engineered to be depleted in global methylation levels but still active for methylation maintenance, and allowing inducible expression of dCas9–methyltransferase, *i.e.* dCas9 fused to the catalytic domain of DNMT3a. Such an experimental setting was essential since in the presence of the endogenous *de novo* DNA methyltransferases the precise evaluation of dCas9–methyltransferase activity might be very tricky. A widespread off-target activity is observed, which occurs even in the absence of sgRNAs. These results contrast with the specificity studies done with the nuclease-active Cas9 that rarely cuts at off-target sites, or with the CRISPR-based transcription repressors or activators. It indicates the need to investigate further and consider this ubiquitous activity of dCas9-fused epigenetic effector proteins, especially those including an enzymatic activity different from the nuclease type that is naturally allosterically controlled in the CRISPR–Cas system.[37]

17.4.3 Base Editors

Base editing is a new strategy to introduce point mutations with the CRISPR system that does not require HDR or the introduction of a double-stranded DNA break.

The first described base editors enable the irreversible conversion of a C–G base pair to a T–A base pair.[121] Different generations of such CRISPR-based base editors were developed and today the most commonly used, called BE3, consists of a single protein including a three-part fusion (Figure 17.4A).

A cytidine deaminase enzyme, apolipoprotein B mRNA editing enzyme, catalytic polypeptide-like (APOBEC) in BE3, is used to hydrolytically deaminate cytidines to uracils, resulting in G–U intermediates. Indeed a small (around five nucleotides) window of ssDNA is formed following DNA binding of the sgRNA–Cas9 complex. Any cytidines within this window are used as substrates. A uracil glycosylase inhibitor (UGI) allows the ability to protect the edited G–U intermediate from excision in cells by uracil DNA glycosylase (UNG), one component of the base excision repair (BER) pathway. Finally, a nickase variant of Cas9 is used to nick the DNA strand opposite to the newly formed uracil, to push the DNA repair mechanism into the conversion of the G–U intermediate into the desired A–U and A–T products.

At some loci, such base editors give rise to undesired byproducts in which the target C–G base pair is converted into a G–C or A–T base pair, rather than to the desired T–A product. Based on the fact that UNG activity is required for the formation of undesired byproducts, base editors were slightly modified, with the use of extended linkers between the different parts of the fusion and of a second copy of UGI.[122] This novel generation, called BE4, carried

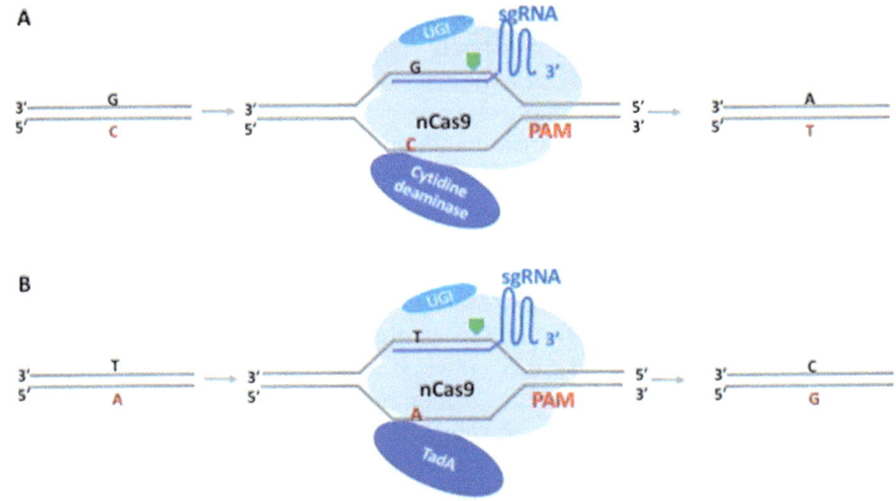

Figure 17.4 CRISPR-based base-editing. Base editors are capable of converting G–C into A–T (A) and T–A into C–G (B) with high efficiency and very low indel rates. The base editor is a three-part fusion of cytidine deaminase or tRNA adenosine deaminase (TadA), Cas9 nickases, and uracil glycosylase inhibitor (UGI).

out base editing with improved product purity. Finally, fusion with Gamma (Gam) protein, a bacteriophage Mu double-stranded DNA end-binding protein, in BE4–Gam, minimizes the formation of undesired indels, likely to be due to Cas9 nickase activity.

These base editors could achieve C–G to T–A conversion, *i.e.* only two kinds of chemical conversions: C to T and G to A. Indeed, there is no known enzyme that can convert A into G or T into C in DNA. David Liu's lab developed one by evolution of a tRNA adenosine deaminase, TadA, an *Escherichia coli* enzyme[123] (Figure 17.4B). The new enzyme converts A into inosine (I) which is read as a G by polymerases. As described above, it was used in a three-part fusion with Cas9 nickase. This base editor, called ABE, works in the other direction, converting T into C and A into G.

One of the current limitations of base editors is that a PAM sequence needs to be located at a specific distance from the targeted C, and therefore there is an inability to distinguish among multiple Cs present within the editing window. However, these base editors have been already used for a wide variety of applications, including mammalian[124] and plant genome editing,[125] for targeted mutagenesis, and knockout studies.[126] Importantly, many human diseases are caused by the mutation of a single base, and base editors are promising candidates to correct the corresponding defective gene. In some systems CRISPR-based nucleases have difficulty correcting such point mutations efficiently and precisely, mainly due to a poorly efficient HDR pathway that is necessary in this process, so base editing could provide a more effective approach.[50,127] For example, CRISPR base editing was used in zebrafish to induce precise base conversion with high efficiency[127] in contrast to other HDR-based genome-editing approaches that are still poorly efficient in this organism. Interestingly the VQR Cas9 variant with NGA PAM instead of NGG was also used successfully in this study, expanding the repertoire of base editors available to generate zebrafish models of human diseases involving point mutations.

The specificity of base editors was assessed by a modified Digested-genome sequencing approach (Digenome-seq), initially developed for identification of DSB by Cas9 nucleases.[128] Base editors were found to be highly specific and, notably, off-target sites of the base editors are often different from those of Cas9 alone, supporting the need to independently evaluate their genome-wide specificities.

17.5 Applications in Medicine and Challenges

17.5.1 Applications in Biomedical Research

Genome editing has tremendous potential for both research and medicine, both in disease modeling for better understanding of pathologies and testing drugs, and in therapy. Here we focus on applications in medicine and present challenges.

In clinics, different genome editing options are available: direct correction of pathological mutations, inactivation of a disease-responsible gene, deletion of disease-responsible sequences, including induction of skipping of an in-frame exon or targeted insertion of therapeutic genes. For example, a pathogenic gene may be disrupted by introduction of loss-of-function mutations in treated cells. Correction of a deleterious mutation in a gene or introduction of a corrective gene in a safe harbor position of the genome can also be considered. Precise gene editing, based on HDR approaches, has higher potential in gene therapy due to the broader spectrum of diseases that can be tackled. The ability to modify precisely the genomes of cells and tissues in animals offers the promise to treat a series of diseases. Several groups have already investigated the potential application of CRISPR–Cas9 to treat genetic diseases such as Duchenne muscular dystrophy,[129] hemophilia,[130] Fanconi anemia,[131] and beta thalassemia.[132]

Genome editing approaches appear highly flexible and precise compared with common gene therapy, that allows gene addition at non-controlled genomic sites.[133,134] However, multiple factors may influence clinical outcomes of therapeutic genome editing, namely off-target effects (see Section 17.2.3), HDR improvement (see Section 17.3.2) and delivery (see Section 17.5.2).

Efficient implementation of genome editing in a large variety of animal models and in iPSCs strongly facilitates and extends approaches for disease modeling, preclinical studies, and drug evaluation. To cite the example of myopathies, which are muscular diseases associated with deficient dystrophin function and for which efficient treatments are still lacking, various dystrophin-deficient animal models have been generated but the mouse model displays only a few features of the human pathology. Genome editing enables the generation of another small animal model for DMD in rats and, interestingly, in this case the disease phenotype is much better recapitulated compared with mice.[135]

The development of robust editing technologies in human pluripotent stem cells (hPSCs) allows one to functionally test the data generated by genome-wide association studies (GWAS).[136] Successful strategies for increasing editing efficiencies in hPSCs have been described. Today the implementation of genome editing in hPSCs enables disease modeling, using isogenic cell lines that have corrected a mutation observed in GWAS studies in iPSCs or generated a mutant allele in wild-type hPSCs.

17.5.2 The Delivery Challenge

In the CRISPR system, Cas9 protein can be delivered in the format of DNA, RNA, or protein, and the guide RNA in the format of DNA or RNA. For HDR-based gene editing, it is also necessary to provide a donor DNA template as a single-stranded or double-stranded DNA. All components of the genome-editing system must access the genome of the target cells in a sufficient amount to ensure expected activity. Multiple methods have been developed to enable the delivery of DNA, including antisense oligonucleotides, RNA, and proteins. They can be generally categorized as mechanical, such as electroporation or injection, chemical, and viral methods. Viral delivery utilizes a viral vector (for example, AAVs or lentiviruses) to encapsulate a gene, in RNA or DNA form.

For research applications, many combinations (reagents, format, and delivery method) have been used successfully in cells, in zygotes, and in adult animals. Development of safe, efficient and specific methods of CRISPR–Cas9 delivery to target tissues and cells is crucial for its success in clinical settings. So far, physical delivery and viral vectors including lentivirus (LV), adenovirus (AV) and AAV are the most applied methods.[137] Physical methods, such as electroporation and microinjection, have been used successfully. However, cell viability and difficulty in application *in vivo* limit them to *in vitro* and *ex vivo* experiments. Concerns regarding immunogenicity, carcinogenesis and mutations associated with viral vectors restrict them as well. Recently, chemical methods and nanocomplexes, such as lipid-based nanoparticles, polyethylenimine (PEI), FuGENE6 transfection reagents, cell-penetrating peptides, cationic arginine gold nanoparticles, core–shell artificial viruses and nanoclews have been exploited, which promise the ability to address many of these limitations.[138]

Generally, viral and non-viral vectors developed for gene therapy can be used directly for introducing DNA-based CRISPR–Cas9 expression systems without further modifications.[139] For example, the delivery systems developed for small interfering RNA (siRNA) could be applied for the Cas9 mRNA and sgRNA with only minor changes.[140] In contrast, delivery of Cas9 protein–sgRNA RNP complex is completely different from existing delivery systems for nucleic acids because of the positive charge of the Cas9 protein.[47]

Zuris *et al.* demonstrated that common cationic lipids can be used to deliver Cas9 protein fused to an engineered super-negatively charged green fluorescent protein (−30GFP) and delivery of the RNP including the (−30GFP)–Cas9 fusion resulted in 80% genome editing in cultured human cells and 20% *in vivo*.[141] Several bio-reducible lipids are designed to promote endosomal escape of the nanoparticles in the cytosol to release the encapsulated Cas9–RNPs. By this approach, the efficiency of gene editing in human HEK cells is higher than 70%.[142] Covalently conjugated non-arginine-based cell-penetrating peptide (CPP) to Cas9 protein is another possibility for Cas9–RNP delivery. In this case, sgRNA could also be transfected into cells in a complex formed by electrostatic interactions with the CPP. The rate of gene editing by Cas9–CPP conjugate and sgRNA–CPP complex ranged from 2.3% to 16% in various cell lines.[143]

Using rolling-circle replication (RCR) technology, a yarn-like DNA nanoclew has been developed for Cas9–RNP delivery. To increase cellular uptake, PEI was added to the complex to supply positive charges and to stimulate endosomal escape. Around 25% gene inactivation was obtained *in vivo via* an intra-tumoral injection of the complex, using U2OS.EGFP tumor bearing mice as models.[144] The emerging field of nanotechnology provides more options, which may have diverse potential for delivery of CRISPR–Cas9.

Recently, a direct and efficient delivery of Cas9–RNP has been described that utilizes gold nanoparticles (GNP) that were co-assembled with Cas9 protein (which was fused to a glutamate peptide tag) and sgRNA. The efficiency of nanoparticle delivery and gene editing in various cell types were 90% and 30%, respectively.[145] Another GNP-based Cas9–RNP delivery vehicle is CRISPR-Gold, which provides a new therapeutic strategy for Duchenne muscular dystrophy (DMD) treatment in mice through local administration.[146] This complex can directly deliver Cas9–RNP and donor DNA *in vivo*, resulting in induction of HDR. It consists of GNP conjugated with DNA and complexed with donor DNA, Cas9–RNP and endosomal disruptive polymer PA_{sp}(DET). This polymer results in cell internalization and endosomal disruption, while in the cytoplasm glutathione releases the DNA from the gold core. CRISPR-Gold and more broadly CRISPR-nanoparticles open a new way for safer and controlled delivery of gene editing tools.

In clinical trials, genome-editing systems can be delivered *ex vivo* or *in vivo* (for a review, see ref. 129). *Ex vivo* delivery is based on extraction of primary cells from patients, genome editing in these cells and then transplantation of the modified cells back into the patient. One potential target tissue is the hematopoietic system, including hematopoietic stem cells (HSCs)[132] and primary lymphocytes. Different challenges are faced by *in vivo versus ex vivo* delivery,[134,147] *in vivo* delivery being the most difficult, and the optimal procedures will differ in each case, depending on the targeted tissue and the modification efficiency necessary to achieve a therapeutic effect. However, therapeutic genome editing of most tissues cannot be carried out *ex vivo*, and effective *in vivo* delivery will be required to enable the broad application of genome-editing therapies.

In addition to the efficiency of delivery, different parameters must be considered such as the duration of nuclease expression that should be limited to decrease potential off-target effects (see Section 17.2.3) or the immune response to the delivery vehicle or the genome-editing reagents themselves[148] or the corrected gene.[149]

17.5.3 The Genome Editing Precision Challenge

Diseases requiring HDR-based genome editing may prove to be more difficult to treat than those benefiting from gene inactivation. As described in Section 17.3, for any HDR application, the competition from highly efficient end-joining repair can strongly affect treatment efficiency. Although NHEJ-mediated genome editing is useful for gene knockout, it is highly undesirable during HDR because it decreases the efficiency of HDR-mediated repair. Improving the efficiency and the specificity of precise genome editing *via* HDR while reducing NHEJ is a major challenge for the development of CRISPR-based therapies. The relative importance of HDR and NHEJ pathways depends on the locus in the genome, the cell type,[150] and the transfection method.[151] Multiple approaches will therefore be needed to address this challenge. One possibility is to increase the HDR : NHEJ ratio either directly by favoring HDR or indirectly by inhibiting NHEJ as described in detail in Section 17.3, and the most promising strategies that are developed and validated in model systems, will probably be evaluated in a clinical setting. Recent studies show that CRISPR–Cas9 genome editing induces a p53-mediated damage response that strongly decreases HDR-based activity, highlighting that p53 function must be carefully monitored in therapeutic applications.[152,153]

Finally, optimization of the conditions of donor delivery, for example, timing of providing the donor template, may facilitate enhanced HDR. In this context, further research to improve HDR efficiency will be crucial to address a broad range of diseases.

17.5.4 Clinical Trials Based on Genome Editing

Many preclinical studies in animals have described the potential of genome editing therapies. Several ongoing clinical trials are evaluating ZFNs, TALENs, and CRISPR to target monogenic diseases, infectious diseases, or cancer.

The first and most advanced trial concerns the modification of the cell-surface receptor CCR5 for the treatment of HIV using ZFNs *ex vivo*.[154] Sangamo BioSciences is currently evaluating the safety of ZFN-mediated *CCR5*-modified T cells in HIV-infected patients in a phase II trial. Companies such as

CRISPR Therapeutics, Intellia Therapeutics, and Editas Medicine have been working on the development of CRISPR–Cas9 technologies as human therapeutics and many clinical trials are underway. A total of 14 such trials are listed in *clinicaltrials.gov*. The first trials concerned *ex vivo* treatments and knockout genome editing using the nuclease alone, without donor DNA. *Ex vivo* treatments of HSCs and chimeric antigen receptor (CAR) T cells, have been evaluated in hemoglobinopathies, immunodeficiencies and cancer immunotherapies. *In vivo* CRISPR treatments are under investigation for diseases of the liver, eye, or lung, organs that are today targetable.

Recently, in November 2017, Sangamo launched the first *in vivo* genome editing trial. In addition, this is the first time that an HDR approach using both nucleases, here ZFNs, and a donor DNA is being utilized. This gene editing therapy is aimed to treat Hunter's syndrome, a lysosomal storage disorder that is caused by mutations in the gene encoding the enzyme iduronate-2-sulfatase. The goal is to insert a corrective gene into a precise location within the highly expressed albumin gene by use of AAV vectors that specifically target the liver. The introduction of the therapeutic gene permanently could enable a patient's liver to produce stably the missing enzyme. Compared with conventional cDNA gene therapy genome editing theoretically allows a therapeutic gene to integrate precisely into the genome and not randomly.

Finally, a promising area of medical applications, even if clinical trials are still not yet underway, is genome-editing-based transplantation. Pigs are appropriate for organ transplant in humans[155] but major problems need to be solved to allow xenotransplantation in humans: immunological rejection, physiological incompatibility, and the risk of transmission of porcine viruses in the human recipient. Genome editing is a way to increase the safety of xenotransplantation by overcoming these roadblocks. Genetically modified pigs have been generated successfully in such a way to prevent rejection by inactivation of specific genes[156] and to prevent transmission of endogenous virus (PERV) by inactivation of all of them.[157,158]

17.6 Conclusions

In summary, advances in the field of genome editing have been remarkable and are still continuing, both in the attempts to improve delivery and in understanding and improving the properties of nucleases, including their specificities, and their genome-editing activity, especially in HDR approaches. Applications in basic research and in applied biology, medicine and also agronomy are very numerous. Finally it is tempting to speculate that the next generations of genome editors with potential for improved activity might be proposed from the new bacterial anti-phage defense systems recently described,[159] once their molecular mechanisms of action are deciphered and adapted into useful molecular tools.

CRISPR technology is very simple and efficient and is now becoming very popular outside the lab, which perhaps helps to democratize science, but is also tempting for "Do It Yourself" scientists, rendering important the regulations of such uses. The development of CRISPR technology for curing genetic diseases in adults is extensively under investigation and seems to be safe in principle. However, germ-line editing, DNA modifications that children will inherit, is today banned, and alternative methods, such as pre-implantation diagnosis, are available. In this context it is important to think about how the technology should be regulated and in what cases its use should be allowed or banned, and many groups have already started such analyses.[160]

Nevertheless, the impact of CRISPR technology on biology is already obvious and it will undoubtedly be included in scientific history as a tremendous discovery. Thus we can be optimistic that the near future might include treatment where genome editing provides new therapeutic approaches for currently untreatable diseases.

References

1. H. Kim and J.-S. Kim, *Nat. Rev. Genet.*, 2014, **15**, 321–334.
2. M. Jinek, K. Chylinski, I. Fonfara, M. Hauer, J. A. Doudna and E. Charpentier, *Science*, 2012, **337**, 816–821.
3. G. Gasiunas, R. Barrangou, P. Horvath and V. Siksnys, *Proc. Natl. Acad. Sci. U. S. A.*, 2012, **109**, E2579–E2586.
4. D. Bhaya, M. Davison and R. Barrangou, *Annu. Rev. Genet.*, 2011, **45**, 273–297.
5. M. P. Terns and R. M. Terns, *Curr. Opin. Microbiol.*, 2011, **14**, 321–327.
6. K. S. Makarova, Y. I. Wolf, O. S. Alkhnbashi, F. Costa, S. A. Shah, S. J. Saunders, R. Barrangou, S. J. J. Brouns, E. Charpentier, D. H. Haft, P. Horvath, S. Moineau, F. J. M. Mojica, R. M. Terns, M. P. Terns, M. F. White, A. F. Yakunin, R. A. Garrett, J. van der Oost, R. Backofen and E. V. Koonin, *Nat. Rev. Microbiol.*, 2015, **13**, 722–736.
7. B. Zetsche, J. S. Gootenberg, O. O. Abudayyeh, I. M. Slaymaker, K. S. Makarova, P. Essletzbichler, S. E. Volz, J. Joung, J. van der Oost, A. Regev, E. V. Koonin and F. Zhang, *Cell*, 2015, **163**, 759–771.
8. I. Fonfara, A. Le Rhun, K. Chylinski, K. S. Makarova, A.-L. Lécrivain, J. Bzdrenga, E. V. Koonin and E. Charpentier, *Nucleic Acids Res.*, 2014, **42**, 2577–2590.
9. F. A. Ran, L. Cong, W. X. Yan, D. A. Scott, J. S. Gootenberg, A. J. Kriz, B. Zetsche, O. Shalem, X. Wu, K. S. Makarova, E. V. Koonin, P. A. Sharp and F. Zhang, *Nature*, 2015, **520**, 186–191.
10. I. Fonfara, H. Richter, M. Bratovič, A. Le Rhun and E. Charpentier, *Nature*, 2016, **532**, 517–521.
11. D. B. T. Cox, J. S. Gootenberg, O. O. Abudayyeh, B. Franklin, M. J. Kellner, J. Joung and F. Zhang, *Science*, 2017, **358**, 1019–1027.
12. S. C. Strutt, R. M. Torrez, E. Kaya, O. A. Negrete and J. A. Doudna, *eLife*, 2018, 7, e32724.

13. K. Suzuki, Y. Tsunekawa, R. Hernandez-Benitez, J. Wu, J. Zhu, E. J. Kim, F. Hatanaka, M. Yamamoto, T. Araoka, Z. Li, M. Kurita, T. Hishida, M. Li, E. Aizawa, S. Guo, S. Chen, A. Goebl, R. D. Soligalla, J. Qu, T. Jiang, X. Fu, M. Jafari, C. R. Esteban, W. T. Berggren, J. Lajara, E. Nuñez-Delicado, P. Guillen, J. M. Campistol, F. Matsuzaki, G.-H. Liu, P. Magistretti, K. Zhang, E. M. Callaway, K. Zhang and J. C. I. Belmonte, *Nature*, 2016, **540**, 144–149.
14. M. Haeussler and J.-P. Concordet, *J. Genet. Genomics*, 2016, **43**, 239–250.
15. B. Chen, L. A. Gilbert, B. A. Cimini, J. Schnitzbauer, W. Zhang, G.-W. Li, J. Park, E. H. Blackburn, J. S. Weissman, L. S. Qi and B. Huang, *Cell*, 2013, **155**, 1479–1491.
16. A. Hendel, R. O. Bak, J. T. Clark, A. B. Kennedy, D. E. Ryan, S. Roy, I. Steinfeld, B. D. Lunstad, R. J. Kaiser, A. B. Wilkens, R. Bacchetta, A. Tsalenko, D. Dellinger, L. Bruhn and M. H. Porteus, *Nat. Biotechnol.*, 2015, **33**, 985–989.
17. B. P. Kleinstiver, M. S. Prew, S. Q. Tsai, V. V. Topkar, N. T. Nguyen, Z. Zheng, A. P. W. Gonzales, Z. Li, R. T. Peterson, J.-R. J. Yeh, M. J. Aryee and J. K. Joung, *Nature*, 2015, **523**, 481–485.
18. B. P. Kleinstiver, M. S. Prew, S. Q. Tsai, N. T. Nguyen, V. V. Topkar, Z. Zheng and J. K. Joung, *Nat. Biotechnol.*, 2015, **33**, 1293–1298.
19. H. Nishimasu, T. Yamano, L. Gao, F. Zhang, R. Ishitani and O. Nureki, *Mol. Cell*, 2017, **67**, 139–147.e2.
20. B. Gu, T. Swigut, A. Spencley, M. R. Bauer, M. Chung, T. Meyer and J. Wysocka, *Science*, 2018, **359**, 1050–1055.
21. X. Liu, Y. Zhang, Y. Chen, M. Li, F. Zhou, K. Li, H. Cao, M. Ni, Y. Liu, Z. Gu, K. E. Dickerson, S. Xie, G. C. Hon, Z. Xuan, M. Q. Zhang, Z. Shao and J. Xu, *Cell*, 2017, **170**, 1028–1043.e19.
22. J. E. Dahlman, O. O. Abudayyeh, J. Joung, J. S. Gootenberg, F. Zhang and S. Konermann, *Nat. Biotechnol.*, 2015, **33**, 1159–1161.
23. H. Nishimasu, F. A. Ran, P. D. Hsu, S. Konermann, S. I. Shehata, N. Dohmae, R. Ishitani, F. Zhang and O. Nureki, *Cell*, 2014, **156**, 935–949.
24. J. G. Zalatan, M. E. Lee, R. Almeida, L. A. Gilbert, E. H. Whitehead, M. La Russa, J. C. Tsai, J. S. Weissman, J. E. Dueber, L. S. Qi and W. A. Lim, *Cell*, 2015, **160**, 339–350.
25. O. Shalem, N. E. Sanjana and F. Zhang, *Nat. Rev. Genet.*, 2015, **16**, 299–311.
26. S. Konermann, M. D. Brigham, A. E. Trevino, J. Joung, O. O. Abudayyeh, C. Barcena, P. D. Hsu, N. Habib, J. S. Gootenberg, H. Nishimasu, O. Nureki and F. Zhang, *Nature*, 2014, **517**, 583–588.
27. M. E. Tanenbaum, L. A. Gilbert, L. S. Qi, J. S. Weissman and R. D. Vale, *Cell*, 2014, **159**, 635–646.
28. W. Deng, X. Shi, R. Tjian, T. Lionnet and R. H. Singer, *Proc. Natl. Acad. Sci. U. S. A.*, 2015, **112**, 11870–11875.
29. F. J. Najm, C. Strand, K. F. Donovan, M. Hegde, K. R. Sanson, E. W. Vaimberg, M. E. Sullender, E. Hartenian, Z. Kalani, N. Fusi, J. Listgarten, S. T. Younger, B. E. Bernstein, D. E. Root and J. G. Doench, *Nat. Biotechnol.*, 2017, **36**, 179–189.

30. J. Tycko, V. E. Myer and P. D. Hsu, *Mol. Cell*, 2016, **63**, 355–370.
31. S. Q. Tsai, Z. Zheng, N. T. Nguyen, M. Liebers, V. V. Topkar, V. Thapar, N. Wyvekens, C. Khayter, A. J. Iafrate, L. P. Le, M. J. Aryee and J. K. Joung, *Nat. Biotechnol.*, 2015, **33**, 187–197.
32. S. Q. Tsai, N. T. Nguyen, J. Malagon-Lopez, V. V. Topkar, M. J. Aryee and J. K. Joung, *Nat. Methods*, 2017, **14**, 607–614.
33. D. Kim, S. Bae, J. Park, E. Kim, S. Kim, H. R. Yu, J. Hwang, J.-I. Kim and J.-S. Kim, *Nat. Methods*, 2015, **12**, 237–243.
34. P. Cameron, C. K. Fuller, P. D. Donohoue, B. N. Jones, M. S. Thompson, M. M. Carter, S. Gradia, B. Vidal, E. Garner, E. M. Slorach, E. Lau, L. M. Banh, A. M. Lied, L. S. Edwards, A. H. Settle, D. Capurso, V. Llaca, S. Deschamps, M. Cigan, J. K. Young and A. P. May, *Nat. Methods*, 2017, **14**, 600–606.
35. I. M. Slaymaker, L. Gao, B. Zetsche, D. A. Scott, W. X. Yan and F. Zhang, *Science*, 2016, **351**, 84–88.
36. W. X. Yan, R. Mirzazadeh, S. Garnerone, D. Scott, M. W. Schneider, T. Kallas, J. Custodio, E. Wernersson, Y. Li, L. Gao, Y. Federova, B. Zetsche, F. Zhang, M. Bienko and N. Crosetto, *Nat. Commun.*, 2017, **8**, 15058.
37. J. S. Chen, Y. S. Dagdas, B. P. Kleinstiver, M. M. Welch, A. A. Sousa, L. B. Harrington, S. H. Sternberg, J. K. Joung, A. Yildiz and J. A. Doudna, *Nature*, 2017, **550**, 407–410.
38. J. G. Doench, N. Fusi, M. Sullender, M. Hegde, E. W. Vaimberg, K. F. Donovan, I. Smith, Z. Tothova, C. Wilen, R. Orchard, H. W. Virgin, J. Listgarten and D. E. Root, *Nat. Biotechnol.*, 2016, **34**, 184–191.
39. J. M. Hinz, M. F. Laughery and J. J. Wyrick, *Biochemistry*, 2015, **54**, 7063–7066.
40. M. Haeussler, K. Schönig, H. Eckert, A. Eschstruth, J. Mianné, J.-B. Renaud, S. Schneider-Maunoury, A. Shkumatava, L. Teboul, J. Kent, J.-S. Joly and J.-P. Concordet, *Genome Biol.*, 2016, **17**, 148.
41. F. A. Ran, P. D. Hsu, C.-Y. Lin, J. S. Gootenberg, S. Konermann, A. E. Trevino, D. A. Scott, A. Inoue, S. Matoba, Y. Zhang and F. Zhang, *Cell*, 2013, **154**, 1380–1389.
42. B. P. Kleinstiver, V. Pattanayak, M. S. Prew, S. Q. Tsai, N. T. Nguyen, Z. Zheng and J. K. Joung, *Nature*, 2016, **529**, 490–495.
43. A. Casini, M. Olivieri, G. Petris, C. Montagna, G. Reginato, G. Maule, F. Lorenzin, D. Prandi, A. Romanel, F. Demichelis, A. Inga and A. Cereseto, *Nat. Biotechnol.*, 2018, **36**, 265–271.
44. J. H. Hu, S. M. Miller, M. H. Geurts, W. Tang, L. Chen, N. Sun, C. M. Zeina, X. Gao, H. A. Rees, Z. Lin and D. R. Liu, *Nature*, 2018, **556**, 57–63.
45. Y. Fu, J. D. Sander, D. Reyon, V. M. Cascio and J. K. Joung, *Nat. Biotechnol.*, 2014, **32**, 279–284.
46. C. R. Cromwell, K. Sung, J. Park, A. R. Krysler, J. Jovel, S. K. Kim and B. P. Hubbard, *Nat. Commun.*, 2018, **9**, 1448.

47. Y. H. Sung, J. M. Kim, H.-T. Kim, J. Lee, J. Jeon, Y. Jin, J.-H. Choi, Y. H. Ban, S.-J. Ha, C.-H. Kim, H.-W. Lee and J.-S. Kim, *Genome Res.*, 2014, **24**, 125–131.
48. S. W. Cho, J. Lee, D. Carroll, J.-S. Kim and J. Lee, *Genetics*, 2013, **195**, 1177–1180.
49. J. A. Zuris, D. B. Thompson, Y. Shu, J. P. Guilinger, J. L. Bessen, J. H. Hu, M. L. Maeder, J. K. Joung, Z.-Y. Chen and D. R. Liu, *Nat. Biotechnol.*, 2015, **33**, 73–80.
50. H. A. Rees, A. C. Komor, W.-H. Yeh, J. Caetano-Lopes, M. Warman, A. S. B. Edge and D. R. Liu, *Nat. Commun.*, 2017, **8**, 15790.
51. J. Shin, F. Jiang, J.-J. Liu, N. L. Bray, B. J. Rauch, S. H. Baik, E. Nogales, J. Bondy-Denomy, J. E. Corn and J. A. Doudna, *Sci. Adv.*, 2017, **3**, e1701620.
52. Y. Nihongaki, F. Kawano, T. Nakajima and M. Sato, *Nat. Biotechnol.*, 2015, **33**, 755–760.
53. Z. Tu, W. Yang, S. Yan, A. Yin, J. Gao, X. Liu, Y. Zheng, J. Zheng, Z. Li, S. Yang, S. Li, X. Guo and X.-J. Li, *Sci. Rep.*, 2017, **7**, 42081.
54. E. K. Brinkman, T. Chen, M. de Haas, H. A. Holland, W. Akhtar and B. van Steensel, *Mol. Cell*, 2018, **70**, 801–813.e6.
55. S. H. Sternberg, S. Redding, M. Jinek, E. C. Greene and J. A. Doudna, *Nature*, 2014, **507**, 62–67.
56. E. K. Brinkman, A. N. Kousholt, T. Harmsen, C. Leemans, T. Chen, J. Jonkers and B. van Steensel, *Nucleic Acids Res.*, 2018, **46**, e58.
57. M. van Overbeek, D. Capurso, M. M. Carter, M. S. Thompson, E. Frias, C. Russ, J. S. Reece-Hoyes, C. Nye, S. Gradia, B. Vidal, J. Zheng, G. R. Hoffman, C. K. Fuller and A. P. May, *Mol. Cell*, 2016, **63**, 633–646.
58. S. Bae, J. Kweon, H. S. Kim and J.-S. Kim, *Nat. Methods*, 2014, **11**, 705–706.
59. H. Ata, T. L. Ekstrom, G. Martinez-Galvez, C. M. Mann, A. V. Dvornikov, K. J. Schaefbauer, A. C. Ma, D. Dobbs, K. J. Clark and S. C. Ekker, *PLoS Genet.*, 2018, **14**, e1007652.
60. J. R. Walker, R. A. Corpina and J. Goldberg, *Nature*, 2001, **412**, 607–614.
61. T. M. Gottlieb and S. P. Jackson, *Cell*, 1993, **72**, 131–142.
62. K. Yano, K. Morotomi-Yano and H. Akiyama, *Int. J. Biochem. Cell Biol.*, 2009, **41**, 1237–1240.
63. S. B. Thyme and A. F. Schier, *Cell Rep.*, 2016, **15**, 1611–1613.
64. K. Rodgers and M. McVey, *J. Cell. Physiol.*, 2016, **231**, 15–24.
65. A. Rupnik, N. F. Lowndes and M. Grenon, *Chromosoma*, 2010, **119**, 115–135.
66. R. Ceccaldi, B. Rondinelli and A. D. D'Andrea, *Trends Cell Biol.*, 2016, **26**, 52–64.
67. K. Karanam, R. Kafri, A. Loewer and G. Lahav, *Mol. Cell*, 2012, **47**, 320–329.
68. P. A. Mateos-Gomez, F. Gong, N. Nair, K. M. Miller, E. Lazzerini-Denchi and A. Sfeir, *Nature*, 2015, **518**, 254–257.
69. Z. Mao, M. Bozzella, A. Seluanov and V. Gorbunova, *DNA Repair*, 2008, **7**, 1765–1771.
70. O. D. Shahar, E. V. S. Raghu Ram, E. Shimshoni, S. Hareli, E. Meshorer and M. Goldberg, *Oncogene*, 2012, **31**, 3495–3504.

71. T. O. Auer, K. Duroure, A. De Cian, J.-P. Concordet and F. Del Bene, *Genome Res.*, 2014, **24**, 142–153.
72. K. Suzuki, Y. Tsunekawa, R. Hernandez-Benitez, J. Wu, J. Zhu, E. J. Kim, F. Hatanaka, M. Yamamoto, T. Araoka, Z. Li, M. Kurita, T. Hishida, M. Li, E. Aizawa, S. Guo, S. Chen, A. Goebl, R. D. Soligalla, J. Qu, T. Jiang, X. Fu, M. Jafari, C. R. Esteban, W. T. Berggren, J. Lajara, E. Nuñez-Delicado, P. Guillen, J. M. Campistol, F. Matsuzaki, G.-H. Liu, P. Magistretti, K. Zhang, E. M. Callaway, K. Zhang and J. C. I. Belmonte, *Nature*, 2016, **540**, 144–149.
73. S. Nakade, T. Tsubota, Y. Sakane, S. Kume, N. Sakamoto, M. Obara, T. Daimon, H. Sezutsu, T. Yamamoto, T. Sakuma and K. T. Suzuki, *Nat. Commun.*, 2014, **5**, 5560.
74. M. Srivastava, M. Nambiar, S. Sharma, S. S. Karki, G. Goldsmith, M. Hegde, S. Kumar, M. Pandey, R. K. Singh, P. Ray, R. Natarajan, M. Kelkar, A. De, B. Choudhary and S. C. Raghavan, *Cell*, 2012, **151**, 1474–1487.
75. V. T. Chu, T. Weber, B. Wefers, W. Wurst, S. Sander, K. Rajewsky and R. Kühn, *Nat. Biotechnol.*, 2015, **33**, 543–548.
76. Z. Hu, Z. Shi, X. Guo, B. Jiang, G. Wang, D. Luo, Y. Chen and Y.-S. Zhu, *Cell Biosci.*, 2018, **8**, 12.
77. T. Maruyama, S. K. Dougan, M. C. Truttmann, A. M. Bilate, J. R. Ingram and H. L. Ploegh, *Nat. Biotechnol.*, 2015, **33**, 538–542.
78. F. Robert, M. Barbeau, S. Éthier, J. Dostie and J. Pelletier, *Genome Med.*, 2015, **7**, 93.
79. A. Bothmer, D. F. Robbiani, N. Feldhahn, A. Gazumyan, A. Nussenzweig and M. C. Nussenzweig, *J. Exp. Med.*, 2010, **207**, 855–865.
80. S. F. Bunting, E. Callén, N. Wong, H.-T. Chen, F. Polato, A. Gunn, A. Bothmer, N. Feldhahn, O. Fernandez-Capetillo, L. Cao, X. Xu, C.-X. Deng, T. Finkel, M. Nussenzweig, J. M. Stark and A. Nussenzweig, *Cell*, 2010, **141**, 243–254.
81. M. D. Canny, N. Moatti, L. C. K. Wan, A. Fradet-Turcotte, D. Krasner, P. A. Mateos-Gomez, M. Zimmermann, A. Orthwein, Y.-C. Juang, W. Zhang, S. M. Noordermeer, E. Seclen, M. D. Wilson, A. Vorobyov, M. Munro, A. Ernst, T. F. Ng, T. Cho, P. M. Cannon, S. S. Sidhu, F. Sicheri and D. Durocher, *Nat. Biotechnol.*, 2017, **36**, 95–102.
82. L. S. Symington, *Cold Spring Harbor Perspect. Biol.*, 2014, **6**, a016436.
83. P. Sung and H. Klein, *Nat. Rev. Mol. Cell Biol.*, 2006, **7**, 739–750.
84. C. D. Richardson, K. R. Kazane, S. J. Feng, N. L. Bray, A. J. Schaefer, S. Floor and J. Corn, *bioRxiv*, 136028; DOI: 10.1101/136028.
85. F. Song and K. Stieger, *Mol. Ther. Nucleic Acids*, 2017, **7**, 53–60.
86. W. Zhang, D. Wang, S. Liu, X. Zheng, H. Ji, H. Xia and Q. Mao, *Gene Ther.*, 2014, **21**, 282–288.
87. A. Paix, A. Folkmann, D. H. Goldman, H. Kulaga, M. J. Grzelak, D. Rasoloson, S. Paidemarry, R. Green, R. R. Reed and G. Seydoux, *Proc. Natl. Acad. Sci. U. S. A.*, 2017, **114**, E10745–E10754.

88. J.-B. Renaud, C. Boix, M. Charpentier, A. De Cian, J. Cochennec, E. Duvernois-Berthet, L. Perrouault, L. Tesson, J. Edouard, R. Thinard, Y. Cherifi, S. Menoret, S. Fontanière, N. de Crozé, A. Fraichard, F. Sohm, I. Anegon, J.-P. Concordet and C. Giovannangeli, *Cell Rep.*, 2016, **14**, 2263–2272.
89. C. D. Richardson, G. J. Ray, M. A. DeWitt, G. L. Curie and J. E. Corn, *Nat. Biotechnol.*, 2016, **34**, 339–344.
90. R. M. Quadros, H. Miura, D. W. Harms, H. Akatsuka, T. Sato, T. Aida, R. Redder, G. P. Richardson, Y. Inagaki, D. Sakai, S. M. Buckley, P. Seshacharyulu, S. K. Batra, M. A. Behlke, S. A. Zeiner, A. M. Jacobi, Y. Izu, W. B. Thoreson, L. D. Urness, S. L. Mansour, M. Ohtsuka and C. B. Gurumurthy, *Genome Biol.*, 2017, **18**, 92.
91. CRISPR-Cas9 Genome Editing In Human Cells Works Via The Fanconi Anemia Pathway, ***bioRxiv***, https://www.biorxiv.org/content/early/2017/05/09/136028, (accessed March 16, 2018).
92. M. Charpentier, A. H. Y. Khedher, S. Menoret, A. Brion, K. Lamribet, E. Dardillac, C. Boix, L. Perrouault, L. Tesson, S. Geny, A. De Cian, J. M. Itier, I. Anegon, B. Lopez, C. Giovannangeli and J. P. Concordet, *Nat. Commun.*, 2018, **9**, 1133.
93. L. Lin, T. S. Petersen, K. T. Jensen, L. Bolund, R. Kühn and Y. Luo, *J. Biotechnol.*, 2017, **247**, 42–49.
94. J. M. Grimme, M. Honda, R. Wright, Y. Okuno, E. Rothenberg, A. V. Mazin, T. Ha and M. Spies, *Nucleic Acids Res.*, 2010, **38**, 2917–2930.
95. S. Shao, C. Ren, Z. Liu, Y. Bai, Z. Chen, Z. Wei, X. Wang, Z. Zhang and K. Xu, *Int. J. Biochem. Cell Biol.*, 2017, **92**, 43–52.
96. T. Gutschner, M. Haemmerle, G. Genovese, G. F. Draetta and L. Chin, *Cell Rep.*, 2016, **14**, 1555–1566.
97. S. E. Howden, B. McColl, A. Glaser, J. Vadolas, S. Petrou, M. H. Little, A. G. Elefanty and E. G. Stanley, *Stem Cell Rep.*, 2016, **7**, 508–517.
98. A. Orthwein, S. M. Noordermeer, M. D. Wilson, S. Landry, R. I. Enchev, A. Sherker, M. Munro, J. Pinder, J. Salsman, G. Dellaire, B. Xia, M. Peter and D. Durocher, *Nature*, 2015, **528**, 422–426.
99. J.-P. Zhang, X.-L. Li, G.-H. Li, W. Chen, C. Arakaki, G. D. Botimer, D. Baylink, L. Zhang, W. Wen, Y.-W. Fu, J. Xu, N. Chun, W. Yuan, T. Cheng and X.-B. Zhang, *Genome Biol.*, 2017, **18**, 35.
100. D. Agudelo, A. Duringer, L. Bozoyan, C. C. Huard, S. Carter, J. Loehr, D. Synodinou, M. Drouin, J. Salsman, G. Dellaire, J. Laganière and Y. Doyon, *Nat. Methods*, 2017, **14**, 615–620.
101. P. I. Thakore, J. B. Black, I. B. Hilton and C. A. Gersbach, *Nat. Methods*, 2016, **13**, 127–137.
102. A. C. Komor, A. H. Badran and D. R. Liu, *Cell*, 2017, **168**, 20–36.
103. P. Perez-Pinera, D. D. Kocak, C. M. Vockley, A. F. Adler, A. M. Kabadi, L. R. Polstein, P. I. Thakore, K. A. Glass, D. G. Ousterout, K. W. Leong, F. Guilak, G. E. Crawford, T. E. Reddy and C. A. Gersbach, *Nat. Methods*, 2013, **10**, 973–976.

104. A. Chavez, M. Tuttle, B. W. Pruitt, B. Ewen-Campen, R. Chari, D. Ter-Ovanesyan, S. J. Haque, R. J. Cecchi, E. J. K. Kowal, J. Buchthal, B. E. Housden, N. Perrimon, J. J. Collins and G. Church, *Nat. Methods*, 2016, **13**, 563–567.
105. S. Kiani, A. Chavez, M. Tuttle, R. N. Hall, R. Chari, D. Ter-Ovanesyan, J. Qian, B. W. Pruitt, J. Beal, S. Vora, J. Buchthal, E. J. K. Kowal, M. R. Ebrahimkhani, J. J. Collins, R. Weiss and G. Church, *Nat. Methods*, 2015, **12**, 1051–1054.
106. P. I. Thakore, A. M. D'Ippolito, L. Song, A. Safi, N. K. Shivakumar, A. M. Kabadi, T. E. Reddy, G. E. Crawford and C. A. Gersbach, *Nat. Methods*, 2015, **12**, 1143–1149.
107. D. Du, A. Roguev, D. E. Gordon, M. Chen, S.-H. Chen, M. Shales, J. P. Shen, T. Ideker, P. Mali, L. S. Qi and N. J. Krogan, *Nat. Methods*, 2017, **14**, 577–580.
108. T. S. Klann, J. B. Black, M. Chellappan, A. Safi, L. Song, I. B. Hilton, G. E. Crawford, T. E. Reddy and C. A. Gersbach, *Nat. Biotechnol.*, 2017, **35**, 561–568.
109. L. S. Qi, M. H. Larson, L. A. Gilbert, J. A. Doudna, J. S. Weissman, A. P. Arkin and W. A. Lim, *Cell*, 2013, **152**, 1173–1183.
110. L. A. Gilbert, M. A. Horlbeck, B. Adamson, J. E. Villalta, Y. Chen, E. H. Whitehead, C. Guimaraes, B. Panning, H. L. Ploegh, M. C. Bassik, L. S. Qi, M. Kampmann and J. S. Weissman, *Cell*, 2014, **159**, 647–661.
111. T. Wang, K. Birsoy, N. W. Hughes, K. M. Krupczak, Y. Post, J. J. Wei, E. S. Lander and D. M. Sabatini, *Science*, 2015, **350**, 1096–1101.
112. T. Wang, J. J. Wei, D. M. Sabatini and E. S. Lander, *Science*, 2014, **343**, 80–84.
113. J. G. Doench, *Nat. Rev. Genet.*, 2017, **19**, 67–80.
114. L. R. Polstein, P. Perez-Pinera, D. D. Kocak, C. M. Vockley, P. Bledsoe, L. Song, A. Safi, G. E. Crawford, T. E. Reddy and C. A. Gersbach, *Genome Res.*, 2015, **25**, 1158–1169.
115. H.-K. Liao, F. Hatanaka, T. Araoka, P. Reddy, M.-Z. Wu, Y. Sui, T. Yamauchi, M. Sakurai, D. D. O'Keefe, E. Núñez-Delicado, P. Guillen, J. M. Campistol, C.-J. Wu, L.-F. Lu, C. R. Esteban and J. C. Izpisua Belmonte, *Cell*, 2017, **171**, 1495–1507.e15.
116. I. B. Hilton, A. M. D'Ippolito, C. M. Vockley, P. I. Thakore, G. E. Crawford, T. E. Reddy and C. A. Gersbach, *Nat. Biotechnol.*, 2015, **33**, 510–517.
117. N. A. Kearns, H. Pham, B. Tabak, R. M. Genga, N. J. Silverstein, M. Garber and R. Maehr, *Nat. Methods*, 2015, **12**, 401–403.
118. X. S. Liu, H. Wu, X. Ji, Y. Stelzer, X. Wu, S. Czauderna, J. Shu, D. Dadon, R. A. Young and R. Jaenisch, *Cell*, 2016, **167**, 233–247.e17.
119. X. S. Liu, H. Wu, M. Krzisch, X. Wu, J. Graef, J. Muffat, D. Hnisz, C. H. Li, B. Yuan, C. Xu, Y. Li, D. Vershkov, A. Cacace, R. A. Young and R. Jaenisch, *Cell*, 2018, **172**, 979–992.e6.
120. C. Galonska, J. Charlton, A. L. Mattei, J. Donaghey, K. Clement, H. Gu, A. W. Mohammad, E. K. Stamenova, D. Cacchiarelli, S. Klages, B. Timmermann, T. Cantz, H. R. Schöler, A. Gnirke, M. J. Ziller and A. Meissner, *Nat. Commun.*, 2018, **9**, 597.

121. A. C. Komor, Y. B. Kim, M. S. Packer, J. A. Zuris and D. R. Liu, *Nature*, 2016, **533**, 420–424.
122. A. C. Komor, K. T. Zhao, M. S. Packer, N. M. Gaudelli, A. L. Waterbury, L. W. Koblan, Y. B. Kim, A. H. Badran and D. R. Liu, *Sci. Adv.*, 2017, **3**, eaao4774.
123. N. M. Gaudelli, A. C. Komor, H. A. Rees, M. S. Packer, A. H. Badran, D. I. Bryson and D. R. Liu, *Nature*, 2017, **551**, 464–471.
124. K. Kim, S.-M. Ryu, S.-T. Kim, G. Baek, D. Kim, K. Lim, E. Chung, S. Kim and J.-S. Kim, *Nat. Biotechnol.*, 2017, **35**, 435–437.
125. B. Ren, F. Yan, Y. Kuang, N. Li, D. Zhang, X. Zhou, H. Lin and H. Zhou, *Mol. Plant*, 2018, **11**, 623–626.
126. C. Kuscu, M. Parlak, T. Tufan, J. Yang, K. Szlachta, X. Wei, R. Mammadov and M. Adli, *Nat. Methods*, 2017, **14**, 710–712.
127. Y. Zhang, W. Qin, X. Lu, J. Xu, H. Huang, H. Bai, S. Li and S. Lin, *Nat. Commun.*, 2017, **8**, 118.
128. D. Kim, K. Lim, S.-T. Kim, S. Yoon, K. Kim, S.-M. Ryu and J.-S. Kim, *Nat. Biotechnol.*, 2017, **35**, 475–480.
129. M. L. Maeder and C. A. Gersbach, *Mol. Ther.*, 2016, **24**, 430–446.
130. H. Li, V. Haurigot, Y. Doyon, T. Li, S. Y. Wong, A. S. Bhagwat, N. Malani, X. M. Anguela, R. Sharma, L. Ivanciu, S. L. Murphy, J. D. Finn, F. R. Khazi, S. Zhou, D. E. Paschon, E. J. Rebar, F. D. Bushman, P. D. Gregory, M. C. Holmes and K. A. High, *Nature*, 2011, **475**, 217–221.
131. M. J. Osborn, R. Gabriel, B. R. Webber, A. P. DeFeo, A. N. McElroy, J. Jarjour, C. G. Starker, J. E. Wagner, J. K. Joung, D. F. Voytas, C. von Kalle, M. Schmidt, B. R. Blazar and J. Tolar, *Hum. Gene Ther.*, 2015, **26**, 114–126.
132. D. P. Dever, R. O. Bak, A. Reinisch, J. Camarena, G. Washington, C. E. Nicolas, M. Pavel-Dinu, N. Saxena, A. B. Wilkens, S. Mantri, N. Uchida, A. Hendel, A. Narla, R. Majeti, K. I. Weinberg and M. H. Porteus, *Nature*, 2016, **539**, 384–389.
133. M. Cavazzana-Calvo, S. Hacein-Bey, G. de Saint Basile, F. Gross, E. Yvon, P. Nusbaum, F. Selz, C. Hue, S. Certain, J. L. Casanova, P. Bousso, F. L. Deist and A. Fischer, *Science*, 2000, **288**, 669–672.
134. L. Naldini, *Nature*, 2015, **526**, 351–360.
135. T. Larcher, A. Lafoux, L. Tesson, S. Remy, V. Thepenier, V. François, C. Le Guiner, H. Goubin, M. Dutilleul, L. Guigand, G. Toumaniantz, A. De Cian, C. Boix, J.-B. Renaud, Y. Cherel, C. Giovannangeli, J.-P. Concordet, I. Anegon and C. Huchet, *PLoS One*, 2014, **9**, e110371.
136. D. Hockemeyer and R. Jaenisch, *Cell Stem Cell*, 2016, **18**, 573–586.
137. L. Li, Z.-Y. He, X.-W. Wei, G.-P. Gao and Y.-Q. Wei, *Hum. Gene Ther.*, 2015, **26**, 452–462.
138. J. L. Gori, P. D. Hsu, M. L. Maeder, S. Shen, G. G. Welstead and D. Bumcrot, *Hum. Gene Ther.*, 2015, **26**, 443–451.
139. F. A. Ran, P. D. Hsu, J. Wright, V. Agarwala, D. A. Scott and F. Zhang, *Nat. Protoc.*, 2013, **8**, 2281–2308.
140. B. Shen, W. Zhang, J. Zhang, J. Zhou, J. Wang, L. Chen, L. Wang, A. Hodgkins, V. Iyer, X. Huang and W. C. Skarnes, *Nat. Methods*, 2014, **11**, 399–402.

141. J. A. Zuris, D. B. Thompson, Y. Shu, J. P. Guilinger, J. L. Bessen, J. H. Hu, M. L. Maeder, J. K. Joung, Z.-Y. Chen and D. R. Liu, *Nat. Biotechnol.*, 2015, **33**, 73–80.
142. M. Wang, J. A. Zuris, F. Meng, H. Rees, S. Sun, P. Deng, Y. Han, X. Gao, D. Pouli, Q. Wu, I. Georgakoudi, D. R. Liu and Q. Xu, *Proc. Natl. Acad. Sci. U. S. A.*, 2016, **113**, 2868–2873.
143. S. Ramakrishna, A.-B. Kwaku Dad, J. Beloor, R. Gopalappa, S.-K. Lee and H. Kim, *Genome Res.*, 2014, **24**, 1020–1027.
144. W. Sun, W. Ji, J. M. Hall, Q. Hu, C. Wang, C. L. Beisel and Z. Gu, *Angew. Chem., Int. Ed. Engl.*, 2015, **54**, 12029–12033.
145. R. Mout, M. Ray, G. Yesilbag Tonga, Y.-W. Lee, T. Tay, K. Sasaki and V. M. Rotello, *ACS Nano*, 2017, **11**, 2452–2458.
146. K. Lee, M. Conboy, H. M. Park, F. Jiang, H. J. Kim, M. A. Dewitt, V. A. Mackley, K. Chang, A. Rao, C. Skinner, T. Shobha, M. Mehdipour, H. Liu, W. Huang, F. Lan, N. L. Bray, S. Li, J. E. Corn, K. Kataoka, J. A. Doudna, I. Conboy and N. Murthy, *Nat. Biomed. Eng.*, 2017, **1**, 889–901.
147. H. Yin, C.-Q. Song, S. Suresh, Q. Wu, S. Walsh, L. H. Rhym, E. Mintzer, M. F. Bolukbasi, L. J. Zhu, K. Kauffman, H. Mou, A. Oberholzer, J. Ding, S.-Y. Kwan, R. L. Bogorad, T. Zatsepin, V. Koteliansky, S. A. Wolfe, W. Xue, R. Langer and D. G. Anderson, *Nat. Biotechnol.*, 2017, **35**, 1179–1187.
148. C. T. Charlesworth, P. S. Deshpande, D. P. Dever, B. Dejene, N. Gomez-Ospina, S. Mantri, M. Pavel-Dinu, J. Camarena, K. I. Weinberg and M. H. Porteus, *bioRxiv*, 243345; DOI: 10.1101/243345.
149. K. H. High, A. Nathwani, T. Spencer and D. Lillicrap, *Haemophilia*, 2014, **20**, 43–49.
150. Y. Miyaoka, J. R. Berman, S. B. Cooper, S. J. Mayerl, A. H. Chan, B. Zhang, G. A. Karlin-Neumann and B. R. Conklin, *Sci. Rep.*, 2016, **6**, 23549.
151. M. Zaboikin, T. Zaboikina, C. Freter and N. Srinivasakumar, *PLoS One*, 2017, **12**, e0169931.
152. R. J. Ihry, K. A. Worringer, M. R. Salick, E. Frias, D. Ho, K. Theriault, S. Kommineni, J. Chen, M. Sondey, C. Ye, R. Randhawa, T. Kulkarni, Z. Yang, G. McAllister, C. Russ, J. Reece-Hoyes, W. Forrester, G. R. Hoffman, R. Dolmetsch and A. Kaykas, *Nat. Med.*, 2018, **24**, 939–946.
153. E. Haapaniemi, S. Botla, J. Persson, B. Schmierer and J. Taipale, *Nat. Med.*, 2018, **24**, 927–930.
154. P. Tebas, D. Stein, W. W. Tang, I. Frank, S. Q. Wang, G. Lee, S. K. Spratt, R. T. Surosky, M. A. Giedlin, G. Nichol, M. C. Holmes, P. D. Gregory, D. G. Ando, M. Kalos, R. G. Collman, G. Binder-Scholl, G. Plesa, W.-T. Hwang, B. L. Levine and C. H. June, *N. Engl. J. Med.*, 2014, **370**, 901–910.
155. J. Y. C. Yang and M. M. Sarwal, *Nat. Rev. Genet.*, 2017, **18**, 309–326.
156. J. R. Butler, L. L. Paris, R. L. Blankenship, R. A. Sidner, G. R. Martens, J. M. Ladowski, P. Li, J. L. Estrada, M. Tector and A. J. Tector, *Transplantation*, 2016, **100**, 571–576.
157. L. Yang, M. Guell, D. Niu, H. George, E. Lesha, D. Grishin, J. Aach, E. Shrock, W. Xu, J. Poci, R. Cortazio, R. A. Wilkinson, J. A. Fishman and G. Church, *Science*, 2015, **350**, 1101–1104.

158. D. Niu, H.-J. Wei, L. Lin, H. George, T. Wang, I.-H. Lee, H.-Y. Zhao, Y. Wang, Y. Kan, E. Shrock, E. Lesha, G. Wang, Y. Luo, Y. Qing, D. Jiao, H. Zhao, X. Zhou, S. Wang, H. Wei, M. Güell, G. M. Church and L. Yang, *Science*, 2017, **357**, 1303–1307.
159. S. Doron, S. Melamed, G. Ofir, A. Leavitt, A. Lopatina, M. Keren, G. Amitai and R. Sorek, *Science*, 2018, **359**, eaar4120.
160. H. Chneiweiss, F. Hirsch, L. Montoliu, A. M. Müller, S. Fenet, M. Abecassis, J. Merchant, B. Baertschi, M. Botbol-Baum, J. A. Houghton, M. Kritikos, J. Mifsud, E. Bartnik, J. Rath, C. Druml, B. Friedrich, A. S. Carvalho, D. Lanzerath and A. Saint-Raymond, *Transgenic Res.*, 2017, **26**, 709–713.

CHAPTER 18

Therapeutic Potential of Ribozymes

DARKO BALKE AND SABINE MÜLLER*

University of Greifswald, Institute of Biochemistry, Felix-Hausdorff-Str. 4, 17487 Greifswald, Germany
*E-mail: smueller@uni-greifswald.de

18.1 Introduction

The importance of ribonucleic acids has been substantially underestimated for a long time, and the three major known classes of RNA [messenger RNA (mRNA), ribosomal RNA (rRNA) and transfer RNA (tRNA)] were merely considered as auxiliary molecules for gene expression. However, the abundance of progressing discoveries that showed the far-reaching functional and regulatory role of RNA in cellular processes led to a paradigm shift in molecular biology. Apart from protein synthesis, RNA is also involved in regulation of transcription or translation, mRNA maturation, tRNA processing and regulation of epigenetic processes that control, for example, differentiation and development.[1]

One important type of RNA, which notably changed our understanding of RNA from a simple intermediate to a dynamic and versatile molecule, is referred to as a ribozyme. Discovered in the 1980s, ribozymes, and thus RNA, were shown for the first time to catalyse chemical reactions,[2,3] a feature that until then had been assigned only to proteins. For that reason, Sidney Altman and Thomas R. Cech, the discoverers of catalytically active

Drug Discovery Series No. 68
Advances in Nucleic Acid Therapeutics
Edited by Sudhir Agrawal and Michael J. Gait

Published by the Royal Society of Chemistry, www.rsc.org

RNAs, were honoured with the Nobel Prize in 1989. Meanwhile, it is known that ribozymes occur in all domains of life. Compared with protein enzymes the reaction spectrum of naturally occurring ribozymes is relatively small and comprises, besides peptide bond formation, mainly hydrolysis and transesterification of RNA. However, the adaptation of the powerful selection strategy for high-affinity RNA binders (aptamers) called SELEX (systematic evolution of ligands by exponential enrichment) led to the selection of ribozymes and made it not only possible to extend the repertoire of novel ribozymes, but also to broaden the reaction spectrum of these catalytically active RNAs.

While the beginning of ribozyme research was dominated by the discovery and characterisation of ribozymes that occur in nature, nowadays the research focus has been changed significantly.[4] The investigations of known ribozymes have led to an abundance of information about their structure and functionality, allowing targeted engineering to equip these ribozyme motifs with new properties or functionalities. Ribozymes and other well-characterized functional RNAs, such as aptamers[5,6] or riboswitches,[7–9] serve as building blocks in the research field referred to as RNA engineering, and such building blocks can be combined or specifically designed for a particular application to form a new molecular tool.

Impressive examples of RNA engineering include the development of biosensors that allow the detection of target molecules for medical or environmental issues[10] as well as the application of molecular switches that enable the control of gene expression.[11,12] One subarea of RNA engineering, namely ribozyme engineering, deals with the development of ribozyme-based tools, which are being used, for instance, in biotechnology for protein engineering or in synthetic biology as allosteric regulatory switches for gene expression.[11] Additionally, they can be used in chemical biology for modification of long-mer RNAs,[13] or for therapeutic applications, by switching off undesired gene expression or by repairing defective genetic information at the mRNA level.[14,15]

In order to generate new ribozymes, two methods are applied: *in vitro* selection and rational design.[16] *In vitro* selection is based on selection of an RNA with the desired characteristic properties out of a large pool of random sequences, whereas rational design is carried out by introduction of a desired modification into known functional RNA motifs. Thus, rational design requires sufficient information about the sequence requirements (conserved nucleotides), secondary and tertiary structure (catalytically active conformation) and functionality (reaction mechanism) of the ribozyme to be adapted.

18.2 *Trans*-cleaving Ribozymes

As already mentioned, ribozymes are promising tools for therapeutic application, since they allow the inhibition of gene expression by sequence-specific cleavage of mRNAs. The most prominent ribozymes for this purpose are the hammerhead (HH), hairpin (HP) and hepatitis delta virus (HDV) ribozyme, which have been thoroughly studied and characterised.[17] Due to their simple structure, the low number of nucleotides and the similar mechanisms of catalysis, these ribozymes are often referred to as small nucleolytic ribozymes. In nature, they catalyse the reversible intramolecular cleavage of RNA.

The hammerhead (HH) ribozyme (Figure 18.1A) was discovered in the satellite RNA of the tobacco ringspot virus,[18,19] whereas the hairpin (HP) ribozyme (Figure 18.1B) was first found in the "minus" strand of the satellite RNA of tobacco ringspot virus.[20] The HDV ribozyme (Figure 18.1C) was identified in genomic and antigenomic RNAs of HDV.[21] For many plant viruses or human pathogens, it has been shown that replication of their circular genomes takes place by a rolling-circle mechanism. The single-stranded, circular RNA genome serves as a template while the polymerase generates a consecutive, repetitive strand. Cleavage of these concatamers at specific positions leads to formation of monomers, *i.e.* single copies of the entire genome. In the case of the satellite RNA of the tobacco ringspot virus, replication of the "positive" strand takes place by generation of a repetitive "minus" strand using the host RNA polymerase. At defined recurring positions catalytically active hairpin ribozyme structures are formed. After self-cleavage, RNA monomers are obtained that can be circularised to the "minus" genome by ligation of the monomer, which then can act as template itself for generation of the "positive" strand RNA. The concatamers of the "positive" strand are processed by the hammerhead ribozyme and after ligation of the monomers the "positive" RNA genome is obtained.

Since their discovery in the late 1980's small nucleolytic ribozymes have been characterised and thanks to their determined crystal structures the precise reaction mechanisms could be revealed.[22–27] All three ribozymes mentioned above share a common enzymatic mechanism. Sequence-specific cleavage takes place by a transesterification reaction. The 2′-hydroxyl

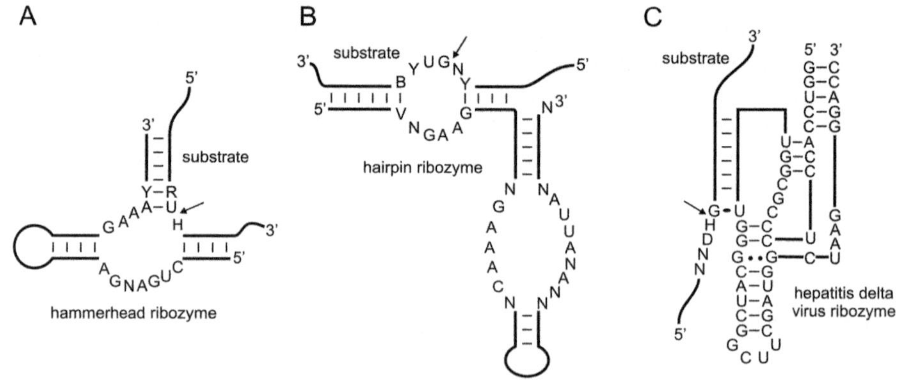

Figure 18.1 Secondary structure of the *trans*-cleaving (A) hammerhead, (B) hairpin and (C) hepatitis delta virus ribozymes. Arrows indicate the cleavage sites.

group attacks the adjacent phosphodiester bond as a nucleophile, and the reaction leads to the formation of 2′,3′-cyclic phosphate and 5′-hydroxyl termini as cleavage products. The hammerhead or hairpin ribozyme-catalysed transesterification reaction is fully reversible. The reverse reaction proceeds by ligation of the cleavage fragments *via* the same intermediate state. Metal ions play a significant role in terms of ribozyme activity. For the HDV ribozyme divalent metal ions are involved in catalysis as co-factors. In the case of the HH or HP ribozyme, metal ions are not directly involved in catalysis, but they are required for formation of, and particularly for stabilisation of, the active ribozyme conformation as well as for neutralisation of negative charges.[28] Transition from naturally occurring, *cis*-cleaving ribozyme motifs into *trans*-cleaving ribozymes enabled an enormous widening of the application spectrum. *Trans*-cleaving ribozymes allow targeted and sequence-specific cleavage of RNAs and can therefore be used like small interfering RNAs (siRNAs) for functional characterisation of genes or in gene therapy for inhibition of undesired gene expression of viral genes or oncogenes.[29] *Trans*-cleaving ribozymes have the advantageous property that they can be easily adapted to the target substrate. However, one has to consider the consensus sequence of the ribozyme, which defines the ribozyme's adaptability to a particular target RNA and thus represents the most crucial aspect for sequence design (Figure 18.1).

As has been extensively reviewed in the past,[14,29–33] ribozymes have been used to target oncogenes and viruses [human immunodeficiency virus 1 (HIV-1), hepatitis C virus (HCV)] and have helped in understanding the role of genes in developmental processes. Major challenges for the application of ribozymes as therapeutics are their efficient delivery into cells, RNA stability and accessibility, as well as their correct binding to the target RNA. Both the availability of divalent cations and the pH in cells also need to be considered, since the intracellular Mg^{2+} concentration (~0.25–1 mmol l^{-1}) is known to be much lower than commonly used for *in vitro* reactions.[34] However, it has been shown that all of the small nucleolytic ribozymes are active in living cells, which makes these ribozymes attractive agents for inhibition of gene expression due to their sequence specificity. Numerous examples can be found in the literature, where ribozymes have been utilised against oncogenes and viruses or have enabled the elucidation of the role of genes in developmental processes.[29–33] Delivery of ribozymes can be achieved either exogenously, where the ribozymes are directly taken up by the target cells, or endogenously, where plasmids encoding the ribozyme are introduced into the cell nucleus. Transcribed ribozymes are then transported into the cytoplasm.[35,36] Since unmodified RNA is susceptible to degradation by cellular ribonucleases, several strategies have been developed to overcome that hurdle. In order to stabilise ribozymes for therapeutic application, various modifications of the phosphodiester backbone or the 2′-OH group have been employed, such as the use of phosphorothioate linkages as a backbone modification and 2′-*O*-methyl, 2′-fluoro or 2′-*O*-methoxyethyl groups as sugar modifications. More recently, 2′-*O*-4′-*C*-methylene linked bicyclic nucleotides called locked nucleic acids (LNA) have entered the field and were also used successfully.[37–40] Furthermore, an interesting strategy has been reported for stabilisation of ribozymes against cleavage by ribonucleases that involves use of L-RNA, the enantiomer of natural RNA.[41] These mirror-image ribozymes, which are able to cleave D-RNA targets, are known as spiegelzymes (in analogy to spiegelmers) and these do not need any further chemical modifications to be nuclease-resistant.

Gene therapy trials have been reported where patients with either cancer or viral infectious diseases were treated with ribozymes, provided by either endogenous transcription from a suitable vector or by direct injections (exogenous delivery). For clinical trials many attempts have been focused on the use of *trans*-cleaving ribozymes (mainly hammerhead or hairpin ribozymes) against HIV-1.[42] In clinical tests against HIV, a gene therapy-based approach has been applied using $CD4^+$ T cells or $CD34^+$ haematopoietic stem cells, which may differentiate into $CD4^+$ T cells.[43–46] Since HIV-1 preferentially infects $CD4^+$ T cells, therapeutically modified $CD4^+$ T cells that produce the *trans*-cleaving ribozyme would prevent production of the functional HIV-1 virus. To this end, the ribozyme was delivered *ex vivo* into the cells that were harvested either from the infected patient (autologous cell therapy) or from an uninfected identical twin (syngeneic cell therapy). Retroviral or lentiviral gene vectors encoding the ribozyme were used to integrate the therapeutic genes into the host genome to ensure long-term gene expression.[43,47,48] However, no clear survival advantage for cells delivered with the ribozyme-encoding vector over cells treated with an empty vector (negative control) could, as yet, be demonstrated. Nevertheless, the safety and the feasibility of this gene therapy approach were clearly shown. A possible interpretation is that transduced cells were poorly engrafted and that ribozyme efficacy was reduced by inappropriate cellular conditions. Alternatively, the suboptimal ribozyme activity may have been caused by chromatin silencing of the ribozyme-coding region integrated into the genome.[49]

Other clinical trials with ribozymes were conducted that target an activated oncogene,[50] a cellular receptor involved in angiogenesis in cancer,[51] or hepatitis B and C virus infections.[52,53] A synthetically modified anti-HER2 ribozyme, called Herzyme, was used to target the human epidermal growth factor receptor type 2 (HER-2) mRNA to inhibit growth of breast, pancreatic and ovarian cancer cells in culture and/or in mouse xenograft models.[54–56] After initially encouraging results in a Phase I clinical trial, showing stabilisation of disease, though without any partial or complete responses, use of Herzyme was discontinued in 2008.

In another clinical study a synthetic hammerhead ribozyme that contained a 5′-phosphorothioate linkage, 2′-*O*-Me- and 2′-deoxy-2′-*C*-allyluridine modifications and terminal inverted 3′–3′ deoxy abasic nucleotides was used to target the mRNA of the vascular endothelial growth factor receptor-1 (VEGFR-1).[57] Since almost all tumours produce VEGF as ligand that can bind to VEGFR-1 in order to induce angiogenesis,[58] targeting of VEGFR-1 may prevent formation of new blood vessels and thus block tumour growth. The ribozyme is called Angiozyme and is currently licensed by Merck-Sirna. The results of a Phase I clinical study have been encouraging. The evaluation of patients with advanced refractory solid tumours has demonstrated that daily subcutaneous injection of Angiozyme was well tolerated without significant side effects.[59] Later, Angiozyme entered Phase II clinical trials as monotherapy or in combination with standard chemotherapy for treatment of advanced breast or colorectal cancers. Administered as a monotherapy, Angiozyme was successful in lowering serum levels of soluble VEGFR-1, but without significant clinical response. After encouraging results when administered in combination with standard chemotherapy regimens in patients with metastatic renal cancer, Angiozyme use was discontinued, until in 2012 a new study of Angiozyme in the treatment of metastatic breast cancer was reported.[60] Since then no further details have been published.

Another synthetic ribozyme, also currently licensed by Merck-Sirna, named Heptazyme, is directed against hepatitis C virus by targeting the 5′-untranslated region of the viral RNA. In a pre-clinical study using cultured cells, application of a Heptazyme led to reduced viral replication by more than 85%, and results of the subsequent Phase I/II clinical trials were also encouraging: Heptazyme was well tolerated.[61] However, after testing this drug in non-human primates, one animal lost its eyesight, and so further trials were terminated.[61]

Likewise, HerBzyme, targeting several hepatitis B virus (HBV) transcripts as well as the HBV pregenomic RNA, was shown to inhibit viral replication at several stages of the HBV life cycle.[53] After testing HerBzyme in Phase I/II clinical trials,[62] once again use of this ribozyme was discontinued in 2003.

Thus, taken together, a rather large number of clinical trials have been carried out to demonstrate the potential of ribozymes to inhibit the expression of pathogenic genes. Whereas results at the cell culture level and in early clinical trials looked promising in many cases, only a very small number of ribozymes have been capable of inducing significant clinical effects or indeed of being developed into a drug.

18.3 Ribozyme-mediated Genetic Repair

Just like siRNAs (see Chapters 10–12), *trans*-cleaving ribozymes as described above are functional RNAs that have been used to inhibit gene expression of therapeutically relevant targets,[15,63,64] and are thus potentially useful for knockdown of undesired gene products associated with a specific disease. However, if the disease is caused by expression of a non-functional protein, as is the case for hereditary diseases, such as cystic fibrosis or phenylketonuria, genetic knockdown is a less suitable strategy. Instead, correction of the genetic information would be a most desirable method of therapeutic intervention. In order to restore the function of proteins being expressed from mutated genes, several powerful strategies have been developed recently. Traditionally, diseases that are caused by defective proteins have been treated by the delivery of DNA containing the correct gene to the cell and by integration of the gene into the genome. However, since the DNA is randomly integrated into the genome, this can cause unwanted events and thus become very problematic and rather dangerous.[65]

In recent years, novel gene-editing-based techniques have been developed to overcome the problem of random integration. Instead of adding a correct gene version, programmable nucleases, such as zinc finger nucleases (ZFNs) or transcription factor-like effector nucleases (TALENs), have been used to correct genetic mutations associated with diseases site-specifically.[66–68] A more recent method for targeted gene editing makes use of the clustered regularly interspaced short palindromic repeat (CRISPR)–CRISPR associated protein (Cas) system, which has been discovered to function as bacterial adaptive immune defence by protecting bacteria against plasmids or phages.[69–71] Adaptation of this programmable RNA-guided DNA endonuclease system to human cells has paved the way for CRISPR–Cas9 to enable genome editing (see Chapter 17).[72–76] Gene editing by programmable nuclease for therapeutic applications seems to be the most promising method of the future, since it bears the potential not only to alleviate symptoms, but to actually eliminate the cause of a particular disease. However, currently a number of challenges have been reported that still need to be addressed, including target site selection and guide RNA design, optimal methods for delivery, nuclease specificity and efficiency, off-target effects and frequency of homology-directed recombination.[68,77] Since introduced changes within the genome are permanent, these genome-editing techniques involve the risk of worsening the situation by undesired editing events at off-target sites. As long as these methods have not been proven to be sufficiently safe, alternative methods allowing the restoration of correct genetic information of disordered genes by repair at the level of RNA possess high significance. In order to address this challenging task, several strategies focusing on mRNA repair, such as spliceosome-mediated RNA *trans*-splicing (SMaRT)[78–80] or *trans*-splicing group I intron ribozymes[81–83] have been developed. Also twin ribozymes, although still being rather in their infancies, offer a promising way for controlled site-specific RNA sequence alteration.[84]

18.3.1 Spliceosome-mediated RNA *Trans*-splicing (SMaRT)

In eukaryotes transcription involves the formation of a primary transcript (pre-mRNA) containing non-coding (introns) and coding regions (exons) that is further processed to form the mature mRNA. Post-transcriptional processes include 5′-capping, 3′-polyadenylation and *cis*-splicing reactions. Pre-mRNA splicing, *i.e.* intron excision and ligation of the remaining exons, takes place in the cell nucleus and is catalysed by the spliceosome, a large molecular machine composed of small nuclear ribonucleoprotein complexes.[85] *Cis*-splicing of the pre-mRNA proceeds *via* two consecutive *trans*-esterification reactions. First, a specific adenosine within the intron, referred to as the branch point, attacks with its 2′-OH function the phosphate that connects the 5′-exon and the intron. During this step a branched structure within the intron is formed and the 5′-exon with free 3′-OH group is released as a leaving group. In the second step, the 3′-OH group of the 5′-exon attacks the phosphate connecting the intron and the 3′-exon. This final *trans*-esterification step leads to the release of the intron as a lariat while connecting both exons. Besides *cis*-splicing, another splicing mechanism has been discovered in eukaryotes.[86–90] Although this is a rarely occurring process, it has been shown that the spliceosome can also mediate *trans*-splicing, *i.e.* a 5′-exon of one pre-mRNA is connected with a 3′-exon of another pre-mRNA. This is possible if 5′- and 3′- splice sites are spatially available on two individual mRNAs. *Trans*-splicing could be observed for transcripts of the same gene (homotypic) as well as for transcripts of different genes (intergenic).[91] The mechanism of *trans*-splicing is similar to that of *cis*-splicing. Again the branch point adenosine of one molecule attacks with its 2′-OH group the 5′-splice site of the second pre-mRNA. Then the 3′-OH group of the 5′-exon attacks the 3′-splice site of the first pre-mRNA by forming the connected exons and a Y-shaped intron as leaving group.

Trans-splicing events can be very useful for therapeutic approaches, since they permit the possibility of restoring defective genetic information by replacing the entire mutation-containing exon with a wild-type sequence exon. Thus, translation of the formed hybrid mRNA leads to production of functional protein. In 1999, Puttaraju *et al.* showed, for the first time, that SMaRT can be used for mRNA repair (Figure 18.2A).[78] Through addition of pre-*trans*-splicing RNA molecules (PTMs) to tumour cells, it was possible to reprogram the cancer-associated human chorionic gonadotropin β polypeptide (CGB) mRNA. The site-specific exon replacement was achieved both in cell culture and in an animal model. Since then, SMaRT has been further optimised and applied in many pre-clinical studies as a tool for RNA repair, allowing the treatment of genetic disorders, such as cystic fibrosis, haemophilia A or X-linked immunodeficiency.[92–95] Additionally, it has been used as suicide-gene therapy to defeat cancer.[96,97] Although repair efficiencies are relatively low, they are high enough for phenotypic correction, and hence not the major hurdle of the strategy.[93,98,99] The main challenges still include target specificity, cell delivery of the PTMs and finally the limited suitability of this method for intron-containing targets. Repair of intronless gene transcripts or non-coding RNAs cannot be addressed this way.

18.3.2 Group I Intron Ribozyme-mediated *Trans*-splicing

One of the first catalytic RNAs to be discovered in 1982 was the group I intron. This ribozyme was identified in an intron from an rRNA transcript of *Tetrahymena thermophila* where it catalyses the self-splice reaction of primary transcripts without any assistance from proteins.[2] Group I introns belong to a class of introns that are categorised by their splicing mechanism. The

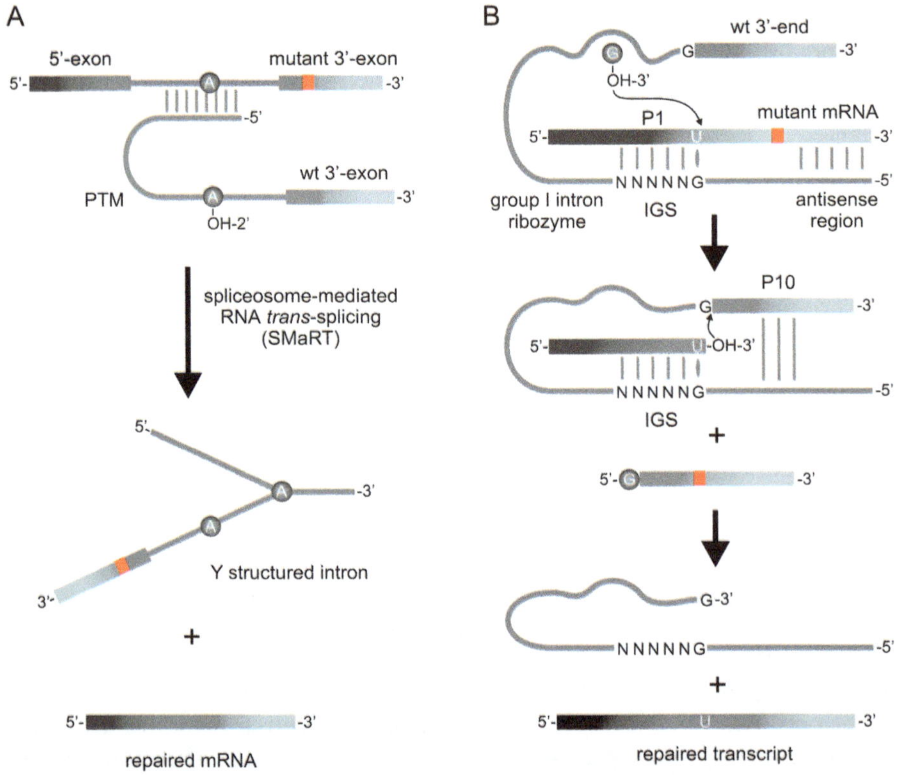

Figure 18.2 RNA repair by (A) spliceosome-mediated RNA *trans*-splicing (SMaRT) or (B) group I intron ribozyme-mediated *trans*-splicing.

autocatalytic splicing of group I introns proceeds *via* a two-step *trans*-esterification reaction involving a guanosine as co-factor.[100] In the first step of the splicing reaction an exogenous guanosine (exoG) is bound to the guanosine-binding site in the catalytic core of the ribozyme. Subsequently, the 3′-OH function of exoG attacks the 5′-splice site. The 5′-splice site is defined by a G–U wobble base pair and additional 4–6 nucleotides that form the P1 helix by base pairing between the 3′-end of the 5′-exon and the internal guide sequence (IGS) located in the intron. After addition of exoG to the 5′-end of the intron, the ribozyme undergoes a conformational rearrangement, so that the 3′-splice site located in the P10 domain is positioned for nucleophilic attack by the 3′-OH group of the 5′-exon. During this final step both exons are joined and the intron is released.

Group I introns can be reprogrammed to allow *trans*-splicing reactions (Figure 18.2B).[81] For that purpose a group I intron ribozyme needs a binding site in its 5′-terminal region, to interact specifically with the target RNA to be *trans*-spliced. The only sequence requirement for that binding site is a uridine residue that allows the formation of a conserved G–U wobble base pair between the IGS of the ribozyme and the 5′-exon of the target

RNA. Since only a uridine needs to be present at the 5′-splice site, group I introns can be engineered to target virtually any RNA for *trans*-splicing. Hence, group I intron-mediated *trans*-splicing may be a useful strategy for replacement of defective genetic information. In initial studies, successful repair of a truncated β-galactosidase (*lacZ*) mRNA was demonstrated in bacteria,[81] as well as in mammalian cells[101] with a repair yield of up to 49%.[102] Following these promising results group I intron ribozyme-mediated *trans*-splicing has been employed in clinically relevant studies, including the repair of a mutated mRNA associated with myotonic dystrophy,[103] and repair of the β-globin mRNA associated with sickle cell anemia.[104] Since then the strategy has been optimised further for RNA-based gene therapy in anti-viral or anti-cancer approaches, discussed in more detail in recent reviews.[82,83]

A disadvantage of *trans*-splicing by the spliceosome or group I intron ribozymes is the fact that it requires a rather large processing ribozyme fused to the "repair exon" to replace the mutated sequence. This can become a very long sequence patch, since the "repair exon" must comprise the entire mRNA sequence downstream of the mutation. For example, *trans*-splicing of the *lacZ* transcript mentioned above required a 416-nucleotide ribozyme followed by 200 nucleotides of the *lacZ* sequence.[35] Most transcripts of functional genes are even longer than that and, depending on the location of the mutation within the entire mRNA, rather large reconstituting sequence patches may be required for *trans*-splicing. Thus, a general method for manipulation at will of any chosen patch of any given mRNA sequence would be a valuable strategy to limit the size of the RNA-processing ribozyme.

18.3.3 Twin Ribozyme-mediated RNA Repair

A general fragment exchange strategy as mentioned above requires an agent that promotes, in a strictly controlled fashion, two chain cleavage events and two ligations. On the basis of its unique cleavage–ligation properties, the hairpin ribozyme is a superior starting point for design to produce the desired activity of supporting such fragment exchange. In general, the hairpin ribozyme is a versatile ribozyme motif (Figure 18.1B) that is eminently suitable for RNA engineering. As mentioned above it can be used as a *trans*-cleaving ribozyme to inhibit gene expression. However, when extending the minimal hairpin ribozyme to a three- or four-way junction motif, substrate binding is stabilised and ligation is strongly preferred. Thus, the specific structure of the ribozyme–substrate complex determines the extent of cleavage *versus* ligation. RNA fragments that are tightly bound to the ribozyme, such that dissociation is thermodynamically and kinetically disfavoured, are preferentially ligated. In contrast, fragments that are less stably bound (but stable enough to allow folding into a catalytically competent structure) favour cleavage.[105–107] This makes the hairpin ribozyme a suitable catalyst for RNA recombination.[108] When

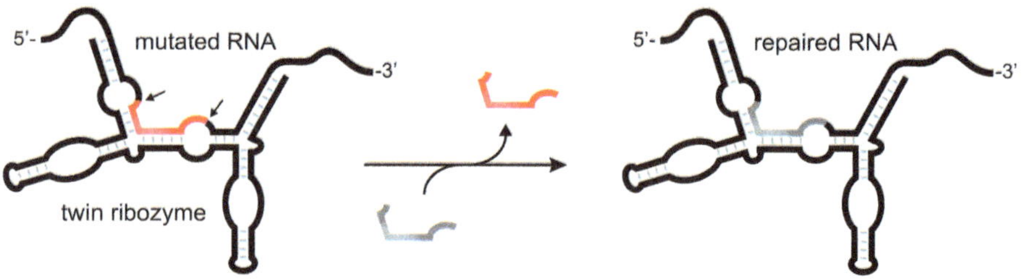

Figure 18.3 Twin ribozyme-mediated RNA repair. Arrows indicate the cleavage and ligation sites.

two three-way junction hairpin ribozymes are combined into one molecule by tandem duplication, the resulting twin ribozyme, because of having two catalytically active motifs, is able to cleave and ligate a substrate RNA at two defined positions in a strictly controlled fashion. This activity can be used to cleave out a short patch of RNA and replace it with an externally added oligonucleotide (Figure 18.3).[84] The key issue for sequence exchange is the design of the twin ribozyme with regard to the formed ribozyme–substrate and ribosome–product complexes. To promote dissociation of the cleavage fragment and binding of the external repair oligonucleotide, the twin ribozyme needs to be designed in such a way that binding of the substrate leads to formation of destabilising structural elements (*e.g.* mismatches or bulges) within the sequence patch to be cut out. While this would promote dissociation of the cleavage fragment, binding of the external repair oligonucleotide should be favoured by formation of a more stable structure, such as a contiguous duplex with the ribozyme. Depending on the specific ribozyme design (mismatches or a destabilizing bulge in the ribozyme or substrate strand), it is possible to exchange fragments of the same length,[109] as well as short fragments against longer versions,[84,110,111] and *vice versa*.[112] These properties qualify twin ribozymes for the repair of short deletion, insertion or base replacement mutations at the RNA level, making them a tool for potential therapeutic applications to treat genetic disorders.

A number of successful repair reactions could be demonstrated using short model substrates, in one case even mimicking a therapeutically relevant target.[110] The oncogenic catenin beta 1 serine 45 deletion (*CTNNB1*-ΔS45) mRNA, encoding an unregulated form of the transcription factor β-catenin, which plays a key role in the wingless–integrated (Wnt) signalling pathway, was a first target to be investigated. Up to 30% of a model substrate could be converted into repaired RNA.[110] Twin-ribozyme-mediated repair of a full-length transcript has been shown for the first time recently. A mutated transcript of an enhanced green fluorescent protein (*EGFP*) mRNA containing a four-base deletion was restored to the wild-type sequence at 32% yield.[111] *In vitro* translation of the repaired *EGFP* mRNA resulted in formation of a functional protein, as detected by the restored fluorescence. Reliable demonstration of twin ribozyme activity in cells has not yet been achieved, although initial results look promising.

18.3.4 Correction of Genetic Disorders by Retro-homing Group II Introns

Group II introns are self-splicing ribozymes, which follow the same reaction mechanism as the spliceosome.[113] In addition, group II introns can have retro-homing activity, *i.e.* self-insertion into specific intronless genes. The group II intron of the long terminal repeat retrotransposon of *Lactococcus lactis* (*Ll.LtrB*) is such a mobile RNA genetic element.[114] After transcription of the LtrB precursor RNA, the intron-encoded multifunctional protein LtrA is translated. It facilitates *cis*-splicing of the catalytic intron, and when the intron has been excised, it becomes bound to LtrA to form a ribonucleoprotein complex with endonuclease and integrase activity.[115] For retro-homing, the target is recognized by base pairing of the exon-binding site (EBS) in the intron to the intron-binding site (IBS) in the intronless gene, assisted by LtrA.[116] The catalytic intron reverse splices into the target gene and the DNA strand opposite the insertion site is cleaved by LtrA to enable reverse transcription of the intron. Second-strand synthesis and ligation of the nicked DNA is completed by the cellular machinery.[117] Based on the fact that the target is recognized by the above mentioned EBS–IBS interaction, group II intron ribozymes may be targeted to specific sites in a desired DNA sequence context by alteration of the sequence of the ribozyme internal EBS.[113,118,119] Indeed, group II introns have been shown to be retargeted to insert efficiently into a target DNA in human cells, thus retaining their activity (Figure 18.4).[118]

Thus, retro-homing introns have been engineered to disrupt pathogenic genes in bacteria and human cells with about 60% efficiency.[118] With regard to correcting genetic disorders, mutant *lacZ* and β-globin genes were shown to be repaired with high fidelity in model bacterial systems. Upon co-transfection of target and intron plasmids into *Escherichia coli*, 60% *lacZ* gene repair was achieved with high specificity, whereas only 2% repair of the β-globin gene was observed.[120]

18.4 Conclusions

A large number of studies in the past have shown that ribozymes have the potential to be applied as tools for the treatment and elimination of viral diseases and cancer and, moreover, to correct genetic disorders at the level of RNA and DNA. *In vitro* and cell culture experiments have been carried out with promising results. Ribozymes have entered clinical trials and in some cases performed well. Nevertheless, ribozyme-mediated therapies are virtually non-existent, and studies regarding therapeutic application of ribozymes have largely stopped.

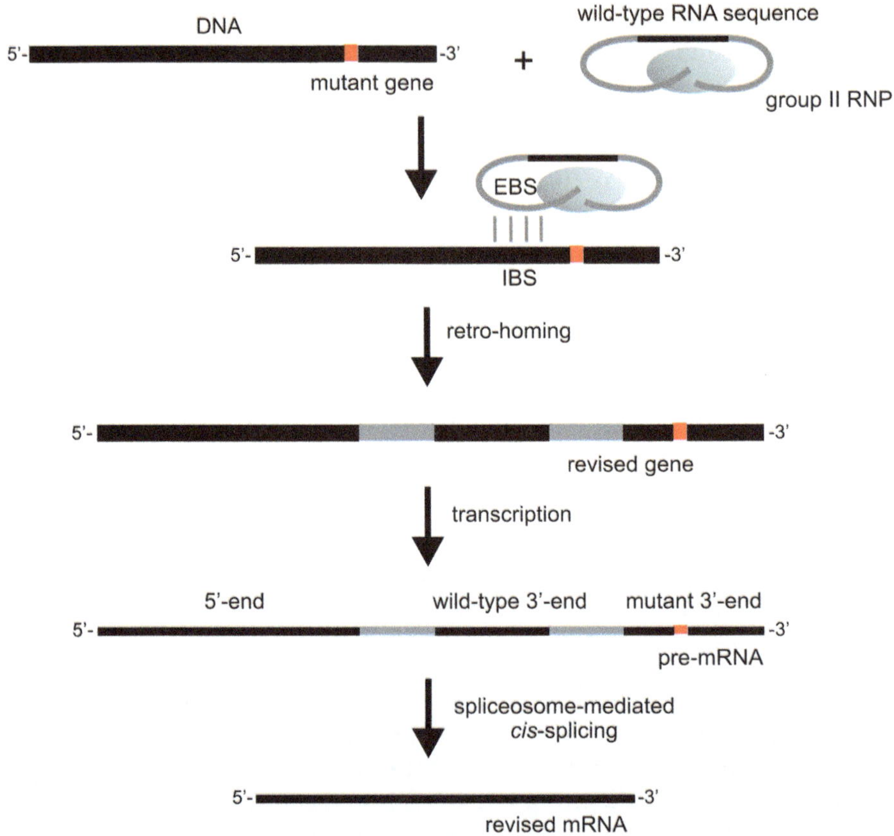

Figure 18.4 Group II intron-mediated gene repair. EBS: exon-binding site; IBS: intron-binding site.

One reason for this is certainly due to the fact that other strategies have outperformed ribozyme technologies. Beyond doubt, the standard technology for knockdown of unwanted genes nowadays is RNA silencing by endogenously or exogenously delivered interfering RNAs (RNAi).[121,122] RNAi can be utilized to silence gene expression in mammalian cells, which gives tremendous hope for new classes of therapeutics to treat cancer, viral infections and other diseases.[15,123] Likewise, a most promising strategy for genome editing is the CRISPR–Cas technology,[124] which has been used in a number of studies aiming at correcting genetic maladies, involving treatment at individual level and also at the embryonic stage in animal models.[125] Here, a synthetic guide RNA needs to be delivered in combination with the Cas protein to cleave the DNA specifically and to remove the mutated sequence, along with a DNA oligonucleotide encoding the correct version, which then is incorporated into the genome by the cellular machinery. In recent years, CRISPR has proved to be a versatile technology for gene therapy, although several drawbacks, such as off-target effects, have to be addressed before it can be utilised for therapy.[68,77]

A second reason is the number of developmental hurdles, which apply also to other RNA-based therapeutic strategies that have not yet been cleared. The mobility of RNA inside cells, cellular factors that may impede ribozyme action, and of course the cell biology involving trafficking and intracellular localisation of RNA, appear to be particularly difficult to overcome in ribozyme technologies and to hamper efficient ribozyme action. A major issue is the development of effective delivery systems capable of targeting specific cell types and to give stable enough expression of ribozymes inside cells to allow them to act on the target RNA for extended periods of time. Beyond efficient entry into cells, ribozyme stability and precise targeting to the substrate RNA sequences is another issue. Unprotected RNA molecules are susceptible to degradation by 3′-exonucleases and pyrimidine-specific endonucleases in human serum. Stabilization of the ribozyme can be achieved by modifications of the phosphodiester linkage and/or the 2′-OH group of the ribose, as well as by conjugation to molecular components that provide protection against exonucleolytic degradation. However, stabilization by modification is limited to synthetic ribozymes that need to be produced and delivered exogenously. Last, and by no means least, are the intracellular conditions under which ribozymes must work, in particular the availability of divalent cations and pH are not ideally suited to ribozyme action, therefore requiring tedious optimization and adaptation of ribozymes to cellular conditions prior to use.

In summary, ribozyme-based therapeutics have not achieved as yet what one may have anticipated after the discovery of ribozymes and their engineering into useful tools. Even though the potential of ribozymes for a number of applications is unquestionably high, the problems associated with intracellular ribozyme action have been a significant factor preventing the development of ribozymes into universal therapeutic molecules. Other methods such as RNAi or CRISPR–Cas hold greater promise in the field, due to their easier design, smaller size of the involved RNAs and higher efficiencies. Overall, the concept of intracellular ribozyme-mediated mRNA cleavage and direct RNA revision is intriguing. Perhaps we will see a renaissance of ribozyme strategies useful for these purposes, based on sophisticated approaches for delivery and performance, and driven by the enthusiasm of researchers in the field. The potential of ribozymes in this area has been demonstrated beyond doubt.

References

1. K. V. Morris and J. S. Mattick, *Nat. Rev. Genet.*, 2014, **15**, 423–437.
2. K. Kruger, P. J. Grabowski, A. J. Zaug, J. Sands, D. E. Gottschling and T. R. Cech, *Cell*, 1982, **31**, 147–157.
3. C. Guerrier-Takada, K. Gardiner, T. Marsh, N. Pace and S. Altman, *Cell*, 1983, **35**, 849–857.

4. S. Müller, B. Appel, D. Balke, R. Hieronymus and C. Nübel, *F1000-Research*, 2016, **5**, 1511.
5. T. Mairal, V. C. Özalp, P. Lozano Sánchez, M. Mir, I. Katakis and C. K. O'Sullivan, *Anal. Bioanal. Chem.*, 2008, **390**, 989–1007.
6. Y. X. Wu and Y. J. Kwon, *Methods*, 2016, **106**, 21–28.
7. A. Wittmann and B. Suess, *FEBS Lett.*, 2012, **586**, 2076–2083.
8. C. Berens and B. Suess, *Curr. Opin. Biotechnol.*, 2015, **31**, 10–15.
9. P. Machtel, K. Bakowska-Zywicka and M. Zywicki, *J. Appl. Genet.*, 2016, **57**, 531–541.
10. J. Liu, Z. Cao and Y. Lu, *Chem. Rev.*, 2009, **109**, 1948–1998.
11. D. Balke, C. Wichert, B. Appel and S. Müller, *Appl. Microbiol. Biotechnol.*, 2014, **98**, 3389–3399.
12. J. Frommer, B. Appel and S. Müller, *Curr. Opin. Biotechnol.*, 2015, **31**, 35–41.
13. D. Balke, J. Frommer, N. Rublack, D. Springstubbe, B. Appel and S. Müller, in *Chemical Biology of Nucleic Acids: Fundamentals and Clinical Applications*, ed. V. A. Erdmann, W. T. Markiewicz and J. Barciszewski, Springer, Berlin Heidelberg, 2014, pp. 477–496.
14. I. Drude, V. Dombos, S. Vauléon and S. Müller, *Mini-Rev. Med. Chem.*, 2007, **7**, 912–931.
15. J. C. Burnett and J. J. Rossi, *Chem. Biol.*, 2012, **19**, 60–71.
16. D. Balke, R. Hieronymus and S. Müller, Challenges and perspectives in nucleic acid enzyme engineering, in *Adv. Biochem. Eng./Biotechnol.*, Springer, Berlin, Heidelberg, 2017, pp. 1–16.
17. A. R. Ferré-D'Amaré and W. G. Scott, *Cold Spring Harbor Perspect. Biol.*, 2010, **2**, a003574.
18. G. A. Prody, J. T. Bakos, J. M. Buzayan, I. R. Schneider and G. Bruening, *Science*, 1986, **231**, 1577–1580.
19. A. C. Forster and R. H. Symons, *Cell*, 1987, **49**, 211–220.
20. A. Hampel and R. Tritz, *Biochemistry*, 1989, **28**, 4929–4933.
21. M. M. Lai, *Annu. Rev. Biochem.*, 1995, **64**, 259–286.
22. H. W. Pley, K. M. Flaherty and D. B. McKay, *Nature*, 1994, **372**, 68–74.
23. W. G. Scott, J. T. Finch and A. Klug, *Cell*, 1995, **81**, 991–1002.
24. A. R. Ferré-D'Amaré, K. Zhou and J. A. Doudna, *Nature*, 1998, **395**, 567–574.
25. P. B. Rupert and A. R. Ferré-D'Amaré, *Nature*, 2001, **410**, 780–786.
26. P. B. Rupert, A. P. Massey, S. T. Sigurdsson and A. R. Ferré-D'Amaré, *Science*, 2002, **298**, 1421–1424.
27. M. Martick and W. G. Scott, *Cell*, 2006, **126**, 309–320.
28. J. B. Murray, A. A. Seyhan, N. G. Walter, J. M. Burke and W. G. Scott, *Chem. Biol.*, 1998, **5**, 587–595.
29. S. Bagheri and M. Kashani-Sabet, *Curr. Mol. Med.*, 2004, **4**, 489–506.
30. S. Schubert and J. Kurreck, *Curr. Drug Targets*, 2004, **5**, 667–681.
31. M. Sioud and P. O. Iversen, *Curr. Drug Targets*, 2005, **6**, 647–653.
32. L. Citti and G. Rainaldi, *Curr. Gene Ther.*, 2005, **5**, 11–24.

33. A. U. Khan, *Clin. Chim. Acta*, 2006, **367**, 20–27.
34. R. D. Grubbs, *BioMetals*, 2002, **15**, 251–259.
35. A. Michienzi and J. J. Rossi, *Methods Enzymol.*, 2001, **341**, 581–596.
36. O. Seksek and J. Bolard, *Methods Mol. Biol.*, 2004, **252**, 545–568.
37. N. Usman and L. M. Blatt, *J. Clin. Invest.*, 2000, **106**, 1197–1202.
38. C. F. Bennett and E. E. Swayze, *Annu. Rev. Pharmacol. Toxicol.*, 2010, **50**, 259–293.
39. S. Shukla, C. S. Sumaria and P. I. Pradeepkumar, *ChemMedChem*, 2010, **5**, 328–349.
40. Y. S. Sanghvi, *Curr. Protoc. Nucleic Acid Chem.*, 2011, Chapter 4, Unit 4.1.1-4.1.22.
41. E. Wyszko, F. Mueller, M. Gabryelska, A. Bondzio, M. Popenda, J. Barciszewski and V. A. Erdmann, *PLoS One*, 2014, **9**, e86673.
42. M. H. Nielsen, F. S. Pedersen and J. Kjems, *Retrovirology*, 2005, **2**, 10.
43. F. Wong-Staal, E. M. Poeschla and D. J. Looney, *Hum. Gene Ther.*, 1998, **9**, 2407–2425.
44. A. Michienzi, D. Castanotto, N. Lee, S. Li, J. A. Zaia and J. J. Rossi, *Ann. N. Y. Acad. Sci.*, 2003, **1002**, 63–71.
45. R. G. Amado, R. T. Mitsuyasu, J. D. Rosenblatt, F. K. Ngok, A. Bakker, S. Cole, N. Chorn, L. S. Lin, G. Bristol, M. P. Boyd, J. L. MacPherson, G. C. Fanning, A. V. Todd, J. A. Ely, J. A. Zack and G. P. Symonds, *Hum. Gene Ther.*, 2004, **15**, 251–262.
46. J. L. Macpherson, M. P. Boyd, A. J. Arndt, A. V. Todd, G. C. Fanning, J. A. Ely, F. Elliott, A. Knop, M. Raponi, J. Murray, W. Gerlach, L. Q. Sun, R. Penny, G. P. Symonds, A. Carr and D. A. Cooper, *J. Gene Med.*, 2005, **7**, 552–564.
47. G. Bauer, P. Valdez, K. Kearns, I. Bahner, S. F. Wen, J. A. Zaia and D. B. Kohn, *Blood*, 1997, **89**, 2259–2267.
48. R. G. Amado, R. T. Mitsuyasu, G. Symonds, J. D. Rosenblatt, J. Zack, L. Q. Sun, M. Miller, J. Ely and W. Gerlach, *Hum. Gene Ther.*, 1999, **10**, 2255–2270.
49. J. C. Burnett and J. J. Rossi, *Gene Ther.*, 2009, **16**, 1178–1179.
50. T. J. Parry, C. Cushman, A. M. Gallegos, A. B. Agrawal, M. Richardson, L. E. Andrews, L. Maloney, V. R. Mokler, F. E. Wincott and P. A. Pavco, *Nucleic Acids Res.*, 1999, **27**, 2569–2577.
51. P. A. Pavco, K. S. Bouhana, A. M. Gallegos, A. Agrawal, K. S. Blanchard, S. L. Grimm, K. L. Jensen, L. E. Andrews, F. E. Wincott, P. A. Pitot, R. J. Tressler, C. Cushman, M. A. Reynolds and T. J. Parry, *Clin. Cancer Res.*, 2000, **6**, 2094–2103.
52. P. A. Lee, L. M. Blatt, K. S. Blanchard, K. S. Bouhana, P. A. Pavco, L. Bellon and J. A. Sandberg, *Hepatology*, 2000, **32**, 640–646.
53. D. V. Morrissey, P. A. Lee, D. A. Johnson, S. L. Overly, J. A. McSwiggen, L. Beigelman, V. R. Mokler, L. Maloney, C. Vargeese, K. Bowman, J. T. O'Brien, C. S. Shaffer, A. Conrad, P. Schmid, J. D. Morrey, D. G. Macejak, P. A. Pavco and L. M. Blatt, *J. Viral Hepatitis*, 2002, **9**, 411–418.

54. T. Suzuki, B. Anderegg, T. Ohkawa, A. Irie, O. Engebraaten, M. Halks-Miller, P. S. Holm, D. T. Curiel, M. Kashani-Sabet and K. J. Scanlon, *Gene Ther.*, 2000, **7**, 241–248.
55. V. W. Lui, Y. He and L. Huang, *Mol. Ther.*, 2001, **3**, 169–177.
56. A. Thybusch-Bernhardt, A. Aigner, S. Beckmann, F. Czubayko and H. Juhl, *Eur. J. Cancer*, 2001, **37**, 1688–1694.
57. H. Kobayashi, S. G. Eckhardt, J. A. Lockridge, M. L. Rothenberg, A. B. Sandler, C. L. O'Bryant, W. Cooper, S. N. Holden, R. D. Aitchison, N. Usman, M. Wolin and M. L. Basche, *Cancer Chemother. Pharmacol.*, 2005, **56**, 329–336.
58. J. A. Sandberg, V. P. Parker, K. S. Blanchard, D. Sweedler, J. A. Powell, A. Kachensky, L. Bellon, N. Usman, T. Rossing, E. Borden and L. M. Blatt, *J. Clin. Pharmacol.*, 2000, **40**, 1462–1469.
59. D. E. Weng, P. A. Masci, S. F. Radka, T. E. Jackson, P. A. Weiss, R. Ganapathi, P. J. Elson, W. B. Capra, V. P. Parker, J. A. Lockridge, J. W. Cowens, N. Usman and E. C. Borden, *Mol. Cancer Ther.*, 2005, **4**, 948–955.
60. P. K. Morrow, R. K. Murthy, J. D. Ensor, G. S. Gordon, K. A. Margolin, A. D. Elias, W. J. Urba, D. E. Weng, H. S. Rugo and G. N. Hortobagyi, *Cancer*, 2012, **118**, 4098–4104.
61. G. R. Foster, *Semin. Liver Dis.*, 2004, **24**(Suppl 2), 97–104.
62. P. Karayiannis, *J. Antimicrob. Chemother.*, 2003, **51**, 761–785.
63. S. Fu and L. Q. Sun, *Future Med. Chem.*, 2015, **7**, 1701–1707.
64. A. Wittrup and J. Lieberman, *Nat. Rev. Genet.*, 2015, **16**, 543–552.
65. S. Hacein-Bey-Abina, C. von Kalle, M. Schmidt, F. Le Deist, N. Wulffraat, E. McIntyre, I. Radford, J. L. Villeval, C. C. Fraser, M. Cavazzana-Calvo and A. Fischer, *N. Engl. J. Med.*, 2003, **348**, 255–256.
66. M. Bibikova, M. Golic, K. G. Golic and D. Carroll, *Genetics*, 2002, **161**, 1169–1175.
67. M. J. Moscou and A. J. Bogdanove, *Science*, 2009, **326**, 1501.
68. T. Gaj, C. A. Gersbach and C. F. Barbas 3rd, *Trends Biotechnol.*, 2013, **31**, 397–405.
69. A. Bolotin, B. Quinquis, A. Sorokin and S. D. Ehrlich, *Microbiology*, 2005, **151**, 2551–2561.
70. F. J. Mojica, C. Diez-Villasenor, J. Garcia-Martinez and E. Soria, *J. Mol. Evol.*, 2005, **60**, 174–182.
71. R. Barrangou, C. Fremaux, H. Deveau, M. Richards, P. Boyaval, S. Moineau, D. A. Romero and P. Horvath, *Science*, 2007, **315**, 1709–1712.
72. S. W. Cho, S. Kim, J. M. Kim and J. S. Kim, *Nat. Biotechnol.*, 2013, **31**, 230–232.
73. L. Cong, F. A. Ran, D. Cox, S. Lin, R. Barretto, N. Habib, P. D. Hsu, X. Wu, W. Jiang, L. A. Marraffini and F. Zhang, *Science*, 2013, **339**, 819–823.
74. M. Jinek, A. East, A. Cheng, S. Lin, E. Ma and J. Doudna, *eLife*, 2013, **2**, e00471.
75. P. Mali, L. Yang, K. M. Esvelt, J. Aach, M. Guell, J. E. DiCarlo, J. E. Norville and G. M. Church, *Science*, 2013, **339**, 823–826.

76. P. Liang, Y. Xu, X. Zhang, C. Ding, R. Huang, Z. Zhang, J. Lv, X. Xie, Y. Chen, Y. Li, Y. Sun, Y. Bai, Z. Songyang, W. Ma, C. Zhou and J. Huang, *Protein Cell*, 2015, **6**, 363–372.
77. R. Peng, G. Lin and J. Li, *FEBS J.*, 2016, **283**, 1218–1231.
78. M. Puttaraju, S. F. Jamison, S. G. Mansfield, M. A. Garcia-Blanco and L. G. Mitchell, *Nat. Biotechnol.*, 1999, **17**, 246–252.
79. L. G. Mitchell and G. J. McGarrity, *Gene Ther.*, 2005, **12**, 1477–1485.
80. V. Wally, E. M. Murauer and J. W. Bauer, *J. Invest. Dermatol.*, 2012, **132**, 1959–1966.
81. B. A. Sullenger and T. R. Cech, *Nature*, 1994, **371**, 619–622.
82. T. Fiskaa and A. B. Birgisdottir, *New Biotechnol.*, 2010, **27**, 194–203.
83. C. H. Lee, S. R. Han and S.-W. Lee, *WIREs RNA*, 2018, **9**, e1466.
84. R. Welz, K. Bossmann, C. Klug, C. Schmidt, H. J. Fritz and S. Müller, *Angew. Chem., Int. Ed.*, 2003, **42**, 2424–2427.
85. N. A. Faustino and T. A. Cooper, *Genes Dev.*, 2003, **17**, 419–437.
86. W. J. Murphy, K. P. Watkins and N. Agabian, *Cell*, 1986, **47**, 517–525.
87. R. E. Sutton and J. C. Boothroyd, *Cell*, 1986, **47**, 527–535.
88. J. Eul, M. Graessmann and A. Graessmann, *EMBO J.*, 1995, **14**, 3226–3235.
89. C. Finta and P. G. Zaphiropoulos, *J. Biol. Chem.*, 2002, **277**, 5882–5890.
90. G. Flouriot, H. Brand, B. Seraphin and F. Gannon, *J. Biol. Chem.*, 2002, **277**, 26244–26251.
91. T. Takahara, B. Tasic, T. Maniatis, H. Akanuma and S. Yanagisawa, *Mol. Cell*, 2005, **18**, 245–251.
92. X. Liu, Q. Jiang, S. G. Mansfield, M. Puttaraju, Y. Zhang, W. Zhou, J. A. Cohn, M. A. Garcia-Blanco, L. G. Mitchell and J. F. Engelhardt, *Nat. Biotechnol.*, 2002, **20**, 47–52.
93. H. Chao, S. G. Mansfield, R. C. Bartel, S. Hiriyanna, L. G. Mitchell, M. A. Garcia-Blanco and C. E. Walsh, *Nat. Med.*, 2003, **9**, 1015–1019.
94. M. Tahara, R. G. Pergolizzi, H. Kobayashi, A. Krause, K. Luettich, M. L. Lesser and R. G. Crystal, *Nat. Med.*, 2004, **10**, 835–841.
95. X. Liu, M. Luo, L. N. Zhang, Z. Yan, R. Zak, W. Ding, S. G. Mansfield, L. G. Mitchell and J. F. Engelhardt, *Hum. Gene Ther.*, 2005, **16**, 1116–1123.
96. R. G. Pergolizzi, A. E. Ropper, R. Dragos, A. C. Reid, K. Nakayama, Y. Tan, J. R. Ehteshami, S. H. Coleman, R. B. Silver, N. R. Hackett, A. Menez and R. G. Crystal, *Mol. Ther.*, 2003, **8**, 999–1008.
97. K. Nakayama, R. G. Pergolizzi and R. G. Crystal, *Cancer Res.*, 2005, **65**, 254–263.
98. M. Puttaraju, J. DiPasquale, C. C. Baker, L. G. Mitchell and M. A. Garcia-Blanco, *Mol. Ther.*, 2001, **4**, 105–114.
99. Z. N. Adelman, I. Sanchez-Vargas, E. A. Travanty, J. O. Carlson, B. J. Beaty, C. D. Blair and K. E. Olson, *J. Virol.*, 2002, **76**, 12925–12933.
100. T. R. Cech, *Annu. Rev. Biochem.*, 1990, **59**, 543–568.
101. J. T. Jones, S. W. Lee and B. A. Sullenger, *Nat. Med.*, 1996, **2**, 643–648.
102. J. T. Jones and B. A. Sullenger, *Nat. Biotechnol.*, 1997, **15**, 902–905.
103. L. A. Phylactou, C. Darrah and M. J. Wood, *Nat. Genet.*, 1998, **18**, 378–381.

104. N. Lan, R. P. Howrey, S. W. Lee, C. A. Smith and B. A. Sullenger, *Science*, 1998, **280**, 1593–1596.
105. S. M. Nesbitt, H. A. Erlacher and M. J. Fedor, *J. Mol. Biol.*, 1999, **286**, 1009–1024.
106. M. J. Fedor, *Biochemistry*, 1999, **38**, 11040–11050.
107. C. P. Donahue, R. S. Yadava, S. M. Nesbitt and M. J. Fedor, *J. Mol. Biol.*, 2000, **295**, 693–707.
108. R. Hieronymus, S. P. Godehard, D. Balke and S. Müller, *Chem. Commun.*, 2016, **52**, 4365–4368.
109. S. Vauléon, S. A. Ivanov, S. Gwiazda and S. Müller, *ChemBioChem*, 2005, **6**, 2158–2162.
110. D. Balke, I. Zieten, A. Strahl, O. Müller and S. Müller, *ChemMedChem*, 2014, **9**, 2128–2137.
111. D. Balke, A. Becker and S. Müller, *Org. Biomol. Chem.*, 2016, **14**, 6729–6737.
112. I. Drude, S. Vauléon and S. Müller, *Biochem. Biophys. Res. Commun.*, 2007, **363**, 24–29.
113. M. B. Long, J. P. Jones 3rd, B. A. Sullenger and J. Byun, *J. Clin. Invest.*, 2003, **112**, 312–318.
114. M. J. Curcio and M. Belfort, *Cell*, 1996, **84**, 9–12.
115. M. Matsuura, R. Saldanha, H. Ma, H. Wank, J. Yang, G. Mohr, S. Cavanagh, G. M. Dunny, M. Belfort and A. M. Lambowitz, *Genes Dev.*, 1997, **11**, 2910–2924.
116. G. Mohr, D. Smith, M. Belfort and A. M. Lambowitz, *Genes Dev.*, 2000, **14**, 559–573.
117. B. Cousineau, D. Smith, S. Lawrence-Cavanagh, J. E. Mueller, J. Yang, D. Mills, D. Manias, G. Dunny, A. M. Lambowitz and M. Belfort, *Cell*, 1998, **94**, 451–462.
118. H. Guo, M. Karberg, M. Long, J. P. Jones 3rd, B. Sullenger and A. M. Lambowitz, *Science*, 2000, **289**, 452–457.
119. A. K. Johnson, J. Sinha and S. M. Testa, *Biochemistry*, 2005, **44**, 10702–10710.
120. J. P. Jones 3rd, M. N. Kierlin, R. G. Coon, J. Perutka, A. M. Lambowitz and B. A. Sullenger, *Mol. Ther.*, 2005, **11**, 687–694.
121. S. E. Fischer, *Curr. Protoc. Mol. Biol.*, 2015, **112**, 26.1.1-26.1.5.
122. C. Tam, J. H. Wong, R. C. F. Cheung, T. Zuo and T. B. Ng, *Appl. Microbiol. Biotechnol.*, 2017, **101**, 7091–7111.
123. M. L. Bobbin and J. J. Rossi, *Annu. Rev. Pharmacol. Toxicol.*, 2016, **56**, 103–122.
124. J. D. Sander and J. K. Joung, *Nat. Biotechnol.*, 2014, **32**, 347–355.
125. S. M. Riordan, D. P. Heruth, L. Q. Zhang and S. Q. Ye, *Cell Biosci.*, 2015, **5**, 33.

CHAPTER 19

Large-scale Automated Synthesis of Therapeutic Oligonucleotides: A Status Update

YOGESH S. SANGHVI*

Rasayan Inc., 2802 Crystal Ridge Road, Encinitas, CA 92024-6615, USA
*E-mail: ysanghvi@rasayan.us

19.1 Introduction

The use of synthetic oligonucleotides for therapeutic applications has proven to be a fast track path for rational drug design and discovery. In recent years, we have witnessed enormous progress where more than 150 products are undergoing various stages of human clinical trials for the treatment of viral infections, cancers, inflammatory disorders, neurological diseases, cardiometabolic and renal diseases and multiple rare orphan diseases. These advances have led to the FDA approval of eight oligonucleotide-based products: Vitravene (fomivirsen), Macugen (pegaptanib), Mipomersen (kynamro), Spinraza (nusinersen), Exondys 51 (eteplirsen), Heplisav B (rHBsAg-1018 ISS), Onpattro (patisiran) and Tegsedi (inotersen). The overall clinical and commercial success of these drugs has triggered demand for a large number of chemically modified synthetic oligonucleotides. It is projected that these oligonucleotides will need to be

Drug Discovery Series No. 68
Advances in Nucleic Acid Therapeutics
Edited by Sudhir Agrawal and Michael J. Gait

Published by the Royal Society of Chemistry, www.rsc.org

manufactured collectively in metric tonne quantities per year. A real effort is under way to meet the surge in demand by both private and commercial companies with an objective to develop environmentally safe, robust, reliable, scalable and cost-efficient protocols for the manufacturing of revolutionary medicines in the 21st century. This chapter summarizes the current state-of-the-art in manufacturing technologies and offers a brief update on the various advances that may be game-changers in how oligonucleotides are assembled and purified on commercial-scale in the near future.

19.2 Chemical Modifications in Clinical and Commercial Products

Therapeutic oligonucleotides are endowed with drug-like properties due to the chemical modifications strategically placed in their structure. Excellent review articles are available describing various base, sugar and backbone modifications accessible for design of oligonucleotide-based drugs. Readers are encouraged to consult the literature for a comprehensive review (see also Chapter 2).[1] The purpose of this section is to capture a snapshot of the six chemical modifications (Figure 19.1) used most often to improve the pharmacokinetic (PK) and pharmacodynamics (PD) properties of oligonucleotides.

19.2.1 First-generation Backbone Modifications – the Phosphorothioate (PS) Internucleotide Linkage

The replacement of a single non-bridging oxygen atom of the natural phosphate internucleotide linkage with a sulfur atom is the easiest and most rewarding chemical modification known today for the design of therapeutic oligonucleotides. Gratifyingly, the incorporation of a PS linkage is easily accommodated using standard phosphoramidite (amidite) chemistry on an automated synthesizer currently practiced on a very large scale. PS modification is often referred to as first-generation chemistry because it provided the first easy access to chemically modified oligonucleotide with drug-like properties. Today, PS chemistry remains at the front and center in the majority of therapeutic oligonucleotides as the first choice for chemical modification. The automated synthesis of PS oligonucleotides using standard phosphoramidite chemistry results in the formation of a mixture of two diastereoisomers (*R*p and *S*p) in almost equal ratio. Therefore, most of the PS oligonucleotides currently on the market and in clinical trials contain a mixture of diastereoisomers unless stated otherwise. It is noteworthy that recently Wada and co-workers[2] have enabled the synthesis of stereo-controlled PS oligonucleotides further improved by researchers from Wave Life Sciences.[3] This effort at Wave has led to two products in clinical trials containing stereo-pure PS oligonucleotides and more are expected in the near future.

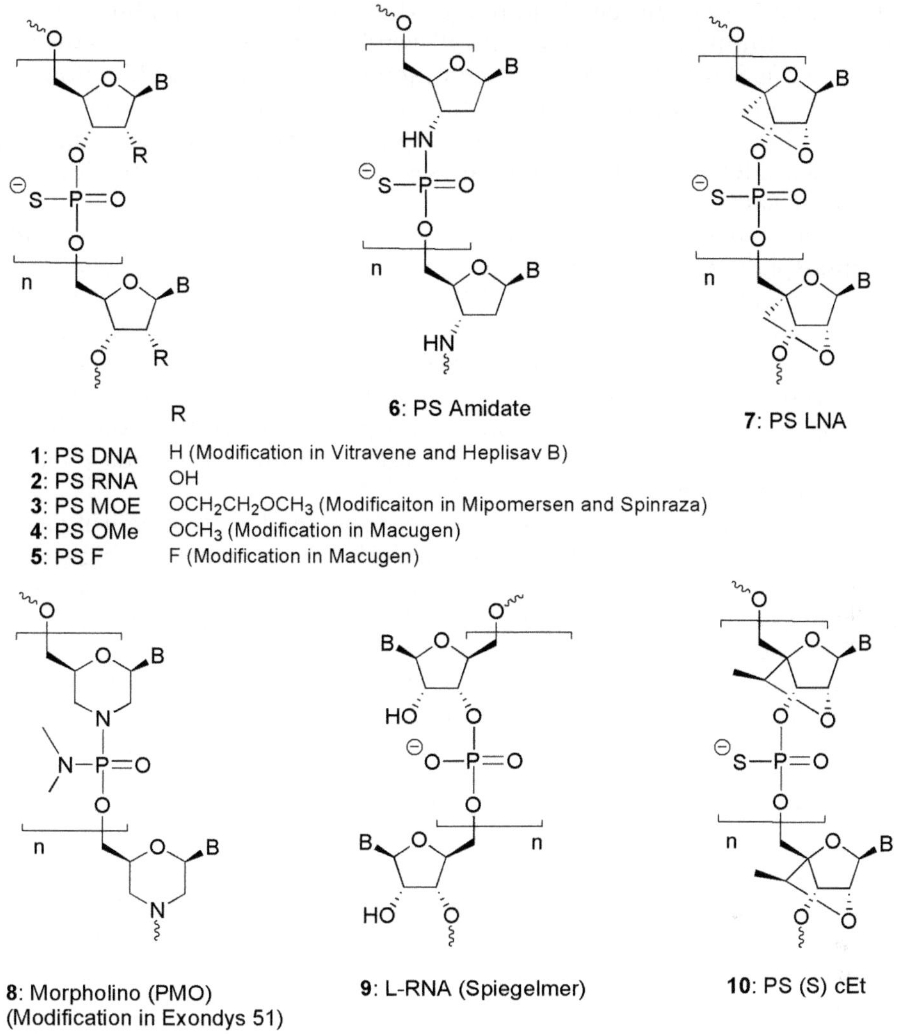

Figure 19.1 Common chemical modifications present in commercial and clinical oligonucleotides.

19.2.2 Second-generation Sugar Modifications

Although PS modification remains high on the list of preferred chemical modifications, the first generation all PS 2′-deoxyoligonucleotides have run into dose-dependent toxicities. PS modification is now used in combination with second-generation sugar modifications to improve PK and PD properties. Among the various 2′-modifications, the 2′-*O*-methoxyethyl (MOE) has proven to be the sugar modification of choice and has made it all the way to three commercial products and a large number of active ongoing clinical trials (Figure 19.1).[4] The rapid success with

MOE-chemistry could be attributed to the commercial availability of the amidite building-blocks from multiple vendors. Additionally, 2′-*O*-Me and 2′-F modifications are two other commonly used sugar modifications in the design of therapeutic oligonucleotides. Delightfully, a 2′-sugar modified amidite maintains its excellent coupling efficiency during automated synthesis allowing fast assembly of second-generation oligonucleotides using the same platform.

Imetelstat (GRN163L) is a 13-mer oligonucleotide containing a N3′→P5′ thiophosphoramidate backbone currently in Phase II/III clinical trials for the treatment of various malignancies.[5,6] Here, the 3′-hydroxyl group of the sugar is replaced by a 3′-amino group offering excellent affinity for the target RNA and extended nuclease resistance. Once again, kilo-quantities of Imetelstat have been manufactured that employ solid-phase amidite chemistry, demonstrating the broad utility of the single process. The commercial accessibility of amidite building blocks may permit further evaluation of this valuable sugar modification.

Locked nucleic acid (LNA) is yet another sugar modification offering ideal sugar pucker with a pre-organized conformation for exceptionally high affinity for targeting RNA whilst maintaining resistance to nuclease-mediated degradation. Despite the multi-step synthesis of LNA amidites, multiple vendors have scaled up the synthesis and offer LNA on a commercial scale. Given the steric crowding in the molecule, LNA amidites are slow to couple compared with MOE and require some modification of the coupling cycle during automated synthesis. Nevertheless, high-quality LNA oligonucleotides have been manufactured on a kilo-scale by use of identical synthesis hardware. Several LNA-containing oligonucleotides are progressing through various stages of clinical trials.[7]

The talented team of chemists at Ionis discovered (*S*)-constrained ethyl (cEt) as the latest arsenal of sugar modifications and named it generation 2.5. As expected, cEt chemistry instills the high affinity for target RNA while increasing (approximately tenfold) the potency of the therapeutic oligonucleotide compared with an analogous MOE sequence. Currently two cEt oligonucleotides are undergoing Phase II clinical trials in patients with malignancies.[1] Like LNA, the assembly of oligonucleotides containing cEt monomers was accomplished *via* conventional DNA synthesis equipment.

Exondys 51 is the first FDA approved drug containing the phosphorodiamidate morpholino-oligonucleotide (PMO) chemistry (see Chapter 1). Sarepta is engaged in the discovery and development of the next-generation of PMO modified oligonucleotide-based drugs. Both solid-phase and solution-phase synthesis of PMO oligonucleotide has been accomplished and will be discussed later in this chapter. The growing number of publications containing PMO chemistry is a further endorsement of its utility as a desirable modification to explore therapeutic modalities.[8]

NOXXON is developing therapeutic aptamers using unnatural L-RNA building blocks – also known as Spiegelmers.[9] The extraordinary biostability of L-RNA aptamers allowed their use as potential therapeutics and two L-RNAs are currently in active clinical trials. Because the assembly of Spiegelmers employs L-RNA building blocks, the prior art of D-RNA chemistry is fully transferrable during synthesis. Therefore, manufacturing of Spiegelmers mirrors well established small interfering RNA (siRNA) synthesis using solid-phase amidite chemistry.

The six chemical modifications summarized above have permitted tremendous strides in delivering the promise of new medicines for a broad spectrum of diseases (Figure 19.1). Additionally, conjugation of oligonucleotides with molecules such as carbohydrates [*e.g.* *N*-acetyl galactosamine (GalNAc)], peptides, lipids and polyethylene glycol (PEG) have clearly demonstrated that further improvements in target delivery are possible.[10–13] Most importantly, the clinical and commercial success all oligonucleotides have enjoyed is primarily due to the ability to assemble a variety of sequences utilizing a single solid-support automated synthesis platform employing robust phosphoramidite chemistry. The following discussion summarizes the key steps during solid-support synthesis applicable to wide range of the chemically modified therapeutic oligonucleotides.

19.3 The Oligonucleotide Manufacturing Process

An overview and historical perspective of oligonucleotide synthesis is well described in Chapter 1 of this book. Therefore, this update descries the current protocol for large-scale synthesis, purification, isolation and analysis of single-stranded oligonucleotides in use for clinical and commercial applications. Before synthesis is initiated, appropriate starting materials, reagents and solvents are purchased and tested to ensure that specifications are met. The next step is to program the automated computer-controlled synthesizer to execute the assembly of the desired sequence using a solid support of choice. Upon completion of the automated synthesis, the oligonucleotide sequence undergoes a two-step base deprotection, releasing the crude product attached to the solid support. The crude product then undergoes a number of purification steps followed by lyophilization to obtain the purified drug substance. A general description of each step is presented below.

19.3.1 Use of an Automated Synthesizer

For the past two decades, GE Healthcare (GE) synthesizers have been pivotal in the rapid synthesis and provision of therapeutic oligonucleotides offering seamless scale-up from millimolar to molar on three types of scalable equipment, namely, the ÄKTA Oligopilot for gram-scale, the OligoPilot 400 for hundred-gram scale and the OligoProcess for kilogram-scale production of therapeutic oligonucleotides under Current Good Manufacturing Practice (cGMP). The synthesis on GE equipment is carried out by use of a Code of Federal Regulations Title 21 part 11, compatible UNICORN software. The Asahi Kasei Bioprocess offers an alternative suite of solid-phase flow-through Asahi Oligosynthesizer options in a similar operating range (millimolar to molar) along with matching synthesis columns. Both GE and Asahi synthesis equipment have proven to be very efficient for the assembly of numerous chemically modified oligonucleotides at gram to kilogram scales.[14]

Advances in the design and automation of these synthesizers have permitted excellent coupling efficiencies (>99%) under optimal conditions. For example, conventional DNA amidites are required in 1.5 equivalents excess and RNA amidites in 2.0 equivalents excess to accomplish almost quantitative coupling of monomers during automated synthesis. The equipment is programmed to deliver all reagents and solvents with great precision and with very high flow-rates to execute the synthesis of a 20-mer oligonucleotide in single day. The key design feature is to pump all reagents through a fixed-bed pressurized stainless-steel reactor packed with solid support. The packing and bed height of the solid support in the column is critical to the successful outcome of the synthesis. The objective is to pump a precise flow of reagents in a uniform manner, creating a compact zone for the reactions and washing of the excess reagents at the same time. This innovation requires significantly less solvent and reagent usage with a packed bed of solid support compared with the conventional batch-type stirred reactor used in a manufacturing plant. The upshot of the automated process is in the reduction of total cycle time compared with a solution-phase synthesis of the same sequence. The entire synthesis is executed without human intervention, thus reducing the chance of human exposure to hazardous chemicals.

19.3.2 Starting Materials

Under fully optimized conditions, the current solid-phase technology offers crude oligonucleotide of remarkably high quality and yield. The correct choice and high quality of the starting materials is mandatory for achieving the best outcome at the end of automated synthesis. A starting material is defined as a significant structural fragment that is present in the structure of the drug substance. Therefore, a starting material should be a substance of defined chemical structure and properties confirmed by appropriate analytical methods.

19.3.2.1 Nucleoside-loaded Solid Support

In the past, the solid supports were loaded with a nucleoside that represented the 3′-terminal unit of the oligonucleotide sequence (Figure 19.2). Therefore, the solid support was considered as the starting material. An in-depth review describing linker chemistry and loading protocols for a variety of solid supports is available in the literature.[15–17] Although controlled-pore glass (CPG) was the solid support of choice during the early days of oligonucleotide synthesis, today polymeric solid support is the first choice for kilo-scale production of oligonucleotide-based drugs. Among the various solid supports available, NittoPhase and Primer Support are the two leaders in the market place. NittoPhase-HL solid support is a cross-linked polystyrene bead with a loading of 200–400 mmol g^{-1}.[15] Primer Support 5G is the latest generation of solid support from GE with the loading capacity reaching 350 mmol g^{-1} for conventional synthesis. Both solid supports perform well when packed in a fixed-bed column reactor offering high crude purity both for DNA and RNA synthesis on commercial scale. As an example, an antisense MOE-gapmer (5–10–5) oligonucleotide was synthesized on 900 mM scale using NittoPhase-HL furnishing >83% crude full-length product under optimized conditions.[15]

During the last decade, there has been a conscious effort to move away from the use of nucleoside-loaded solid support and to replace it by universal supports – solid supports without an attached nucleoside residue. Today, UnyLinker-loaded solid support is perhaps the most commonly used support in the industry. Both NittoPhase and Primer Support are available as a UnyLinker-loaded support in commercial quantities. The key benefit in use of the UnyLinker-loaded support is the ability to buy, store and use a single support to synthesize any sequence at short notice. Because UnyLinker-loaded support does not contribute structural atoms to the drug substance, it is no longer a starting material. Recently, the UnyLinker-chemistry has been further optimized to avoid some of the limitations of the first-generation linker. In a nutshell, UnyLinker-loaded support offers the speed, quality and performance needed for the commercial production of therapeutic oligonucleotides on large-scale (Figure 19.2).[18–20]

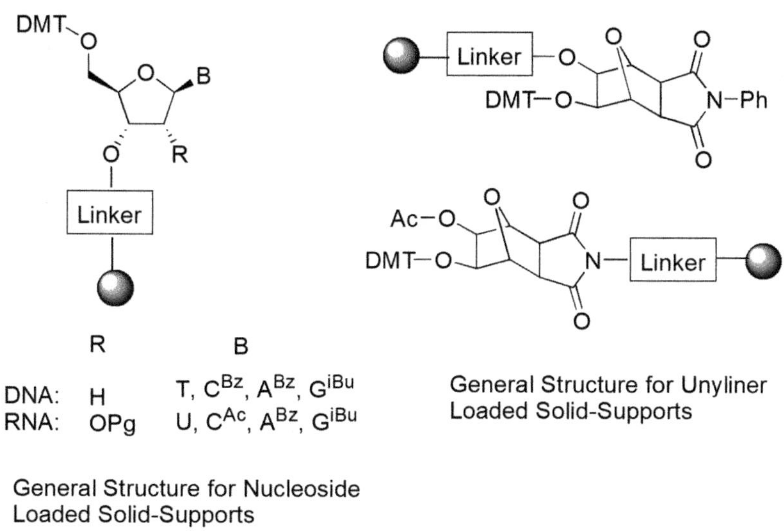

Figure 19.2 General structures of commercial solid supports.

19.3.2.2 Nucleoside Phosphoramidites

Clearly, nucleoside phosphoramidites contribute the lions' share of molecular weight to an oligonucleotide drug substance, making them the most important starting material. A detailed account of the synthesis, quality and stability of various phosphoramidites is available in the literature.[21] Therefore, a concise summary and some recent findings are summarized below. A general structure of a nucleoside phosphoramidite is shown in Figure 19.3. The P(III) reactive building blocks are decorated with multiple orthogonal protecting groups to avoid side reactions during solid-supported synthesis. The most commonly used protecting groups are: (i) an acid-labile 5′-hydroxyl group [*e.g.* dimethoxytrityl (DMT)]; (ii) a base-labile acyl exocyclic amino-protecting group (*e.g.* acetyl, benzoyl and isobutyryl); (iii) another base-labile protecting group (*e.g.* cyanoethyl) for the P(III) functionality and (iv) a 2′-hydroxyl protecting group [*e.g. tert*-butyldimethylsilyl (TBDMS)] in ribonucleoside amidites. The synthesis of stereo-pure antisense oligonucleotides[3] has recently been recently accomplished by use of a modified oxazaphospholidine building block (Figure 19.3).

Since nucleoside phosphoramidites are the key components during assembly of oligonucleotides, a thorough analysis of their purity and quality is essential. An impurity present in an amidite may affect the overall quality of the drug substance. A good number of impurities originating from protecting groups, nucleosides and reagents have been identified, analyzed and their fate and tolerance have been discussed in great detail and published in the literature.[15] Broadly, these impurities are classified into three categories. First, non-reactive and non-critical are often impurities that are due to the conversion of a reactive P(III) functionality to more stable P(V) products as shown in Figure 19.4. These impurities, if present, do not pose serious concern and are easily purged during automated synthesis because they do not couple. Second, reactive and non-critical impurities are generated

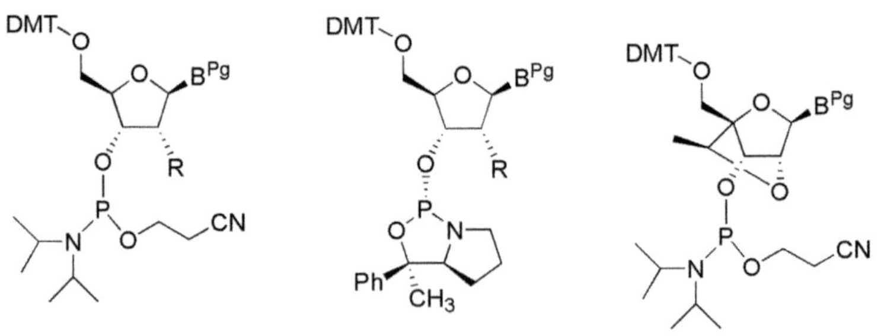

11a: DNA Amidite
R = H
11b: RNA amidite
R = *O*-Protecting group
11c: MOE Amidite
R = $O(CH_2)_2OCH_3$

12: Stereopure Amidite (Reference 3)

13: (S) cEt Amidite (Reference 1)

Figure 19.3 Common structures of amidites as starting material.

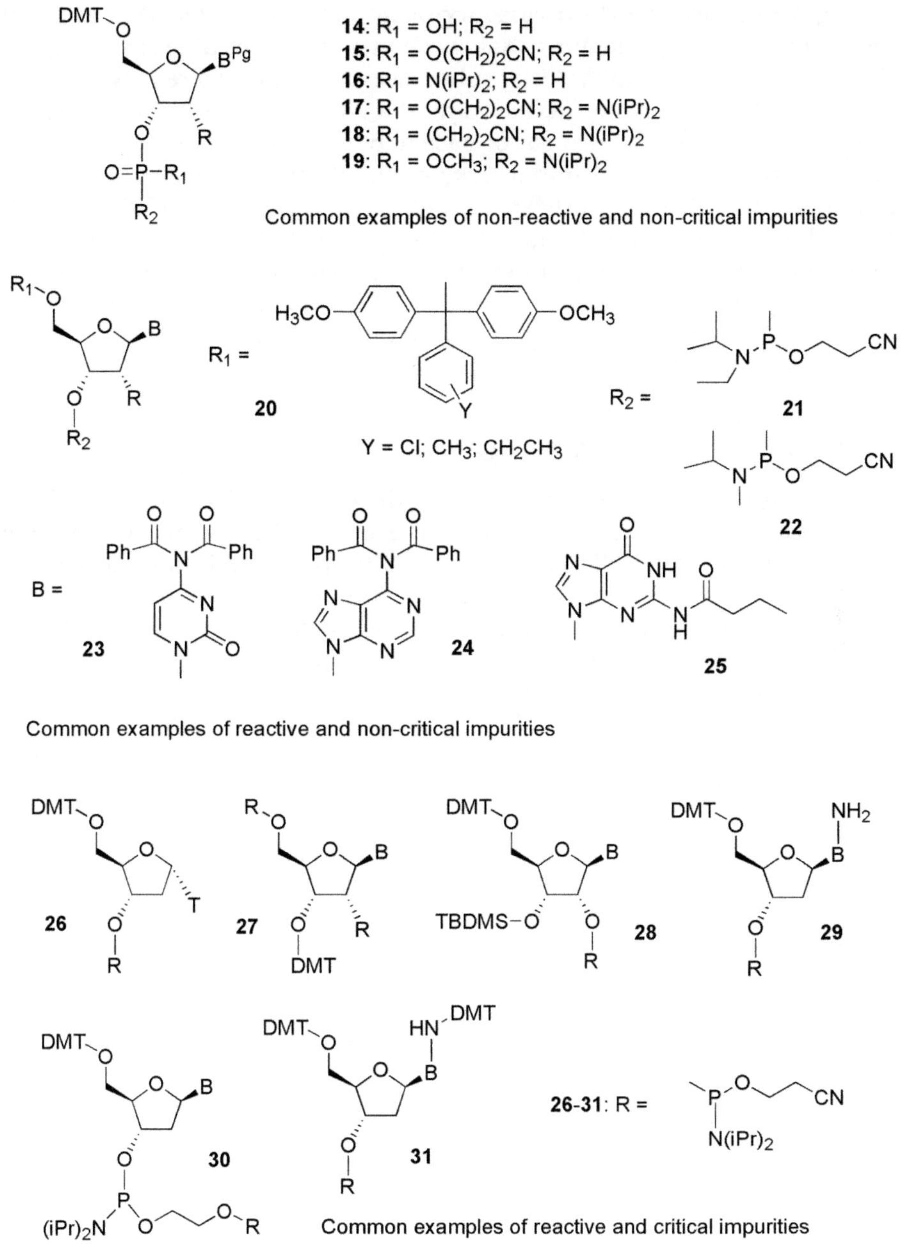

Figure 19.4 Selected examples of impurities in amidites as starting material.

due to impurities present in the protecting groups, which are carried forward into the corresponding amidites. A select group of these impurities are shown in Figure 19.4. Although these impurities are reactive, they do not cause concern because the protecting groups are eventually removed during synthesis. The third category of impurities is those that are both reactive and critical, meaning that these may incorporate into an oligonucleotide and show up in the drug substance. The potential reactivity of critical impurities may result in insertion of a chemical modification altering the phosphate backbone, sugar or base depending on its origin. A selected group of reactive critical impurities are shown in Figure 19.4. Identification and quantification of these impurities has been accomplished by use of UV, high-performance liquid chromatography (HPLC; relative retention time) and liquid chromatography–mass spectrometry (LC–MS) protocols and tightly controlled as a part of amidite specifications. An outstanding account of the various impurities is available in a recently published white paper review article.[22]

19.3.2.3 Sulfur-transfer Reagents

During automated PS oligonucleotide synthesis, the sulfur atom in the internucleotide linkage is introduced by means of a sulfur-transfer reagent (Figure 19.5). In this step, the P(III) linkage is transformed into a more stable P(V) phosphorotriester linkage by means of an appropriate oxidizing agent. Excellent (>99%) sulfurization efficiency is accomplished with appropriately "aged" phenylacetyl disulfide (PADS) reagent for the synthesis of PS oligonucleotides.[23] Today, the PADS reagent is a favorite of industry due to its low-cost, crystalline state, high solubility in acetonitrile, good stability and, most importantly, high reactivity for conversion of P(III) into P(V) within minutes. The mystery of PADS "aging" was solved by recently published intriguing mechanistic work demonstrating that "polysulfides" generated *in situ* are responsible for the efficient sulfurization.[24] Because the use of PADS is patented and there is a foul odor of the polysulfides solution, other reagents have been developed over the years. These include, *N*, *N*-dimethyl-*N*′-(3-thioxo-3*H*-1,2,4-dithazol-5-yl)methanimidamide (DDTT)[25] and 3-phenyl 1,2,4-dithiazoline-5-one (POS) as popular alternative sulfur-transfer reagents.[26] Both reagents have been successfully utilized for the large-scale synthesis of PS oligonucleotides on solid support. Dimethylthiarum disulfide (DTD) is another reagent that is worthy of consideration because it offers both a saving of solvent consumption (~20%) and a reduction of the cycle-time (~25%) during synthesis.[27]

19.3.3 Reagent-related Impurities

Oligonucleotide synthesis requires a repetitive four-step cycle to assemble the desired length of the product. The first step in this process is deprotection or removal of the acid labile DMT-group from the 5′-hydroxyl group of the terminal nucleoside. A 10% solution of dichloroacetic acid [DCA; acid dissociation constant (p*K*a) 1.5] in toluene is the preferred deprotection reagent during large-scale synthesis. Interestingly commercial DCA may

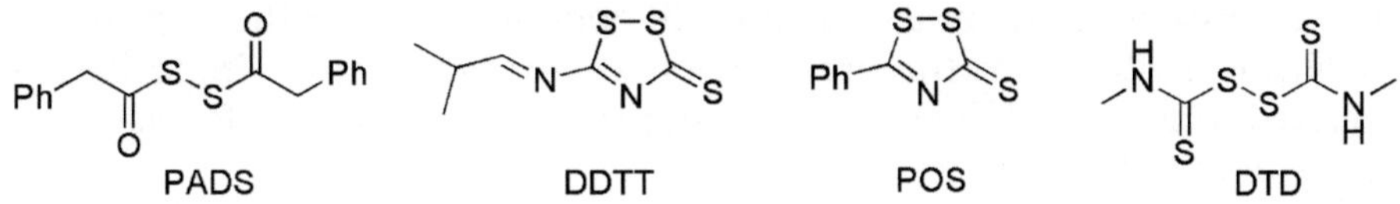

Figure 19.5 List of common sulfur transfer reagents utilized on a large scale.

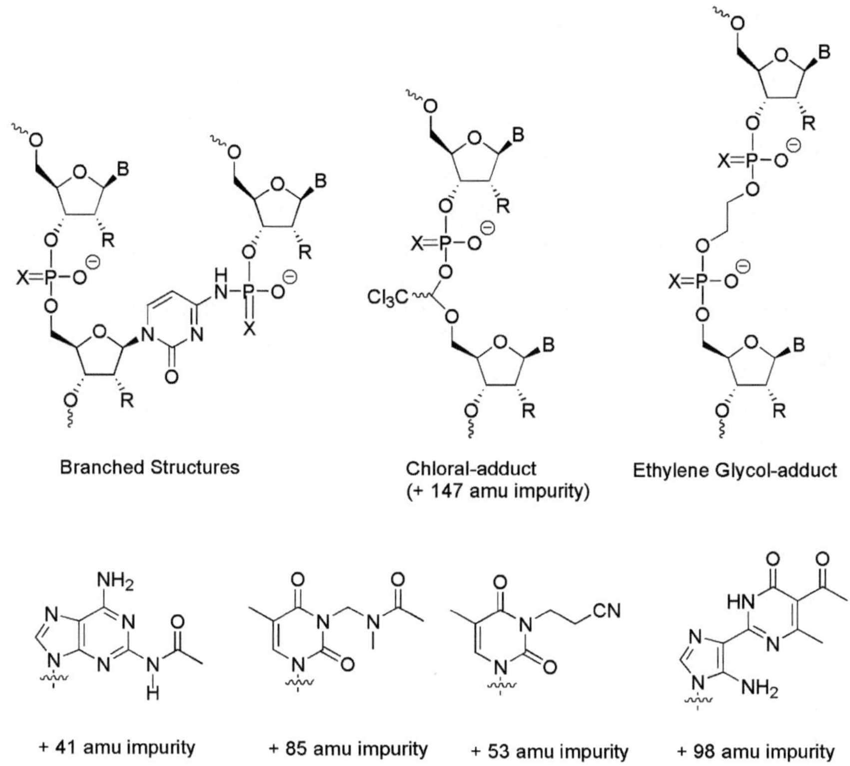

Figure 19.6 Partial list of additional impurities.

contain low levels of trichloroacetaldehyde (chloral), which react during the deprotection step to generate an impurity (+147 amu; Figure 19.6) identified as the chloral adduct.[28] For example, 100 ppm of chloral in DCA may introduce 0.8% of impurity during synthesis. Therefore, it is recommended that the DCA solution must be free of chloral to avoid the formation of an undesired product. The last step in each synthesis cycle is capping with a mixture of two reagents to carry out the acylation of the uncoupled 5′-hydroxyl group. *N*-methylimidazole (NMI) is often used as the Cap B solution and may contain a trace amount of trimethylhexahydrotriazine (TMHHT), which is reported to furnish a thymine-modified impurity (+85 amu).[22] In summary, all synthesis reagents must be carefully monitored using the appropriate analytical methods for the presence of reactive and critical impurities. If detected, one should consider rejecting the lot to avoid the formation of undesired products.

19.3.4 The Four-step Synthesis Cycle

For an introductory review of basic steps during solid-support oligonucleotide synthesis see Chapter 1. Although the software on the synthesizer may offer a set of programs for the assembly of oligonucleotides, it is highly recommended that for best results each step of large-scale synthesis (deprotection, coupling, sulfurization/oxidation and capping; Figure 19.7) must be fully optimized. For example, one must consider carrying out design of experiment (DoE) to establish the correct exposure of acid during the deprotection cycle to minimize depurination – a major concern during the synthesis of DNA oligonucleotides. It is noteworthy that the rates of de-dimethoxytritylation for the four nucleosides are not identical (dA > dG > dC > T) and may require a customized deprotection cycle for each residue. Similarly, each modified amidite may have a different rate of coupling depending on its chemical structure. As an example, a cEt amidite is slower in coupling compared with a 2′-deoxynucleoside amidite due to its increased steric hindrance. Therefore, a higher equivalent of certain amidites may be required to drive the coupling efficiencies to >99% for each cycle. The current practice is to utilize a 1.5–2.0 equivalent excess of each amidite during the coupling cycle. The preferred nucleophilic activator for coupling is 4,5-dicyanoimidazole (DCI; p*K*a 5.2) for the synthesis of antisense PS oligonucleotides and 5-ethylthiotetrazole for synthesis of siRNA. The sulfurization or oxidation step is carried out after coupling, where a reactive phosphite triester is converted into more stable phosphorothioate triester. The sulfurization step is fast and very efficient under optimal conditions. The fourth step in each

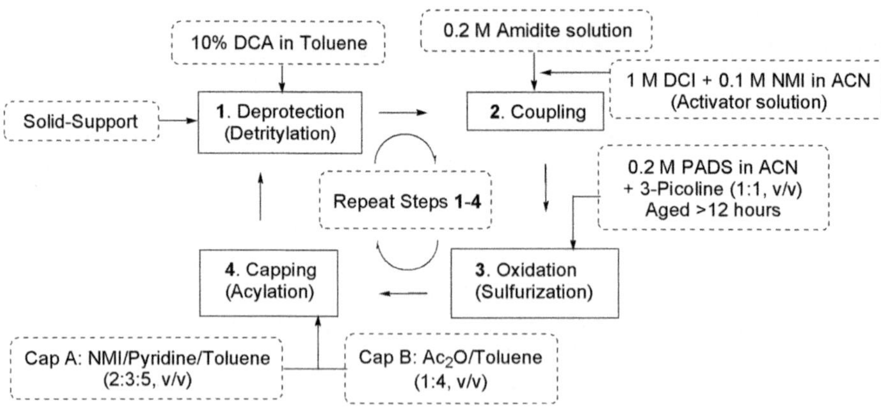

Figure 19.7 An overview of automated four-step solid-support synthesis.

synthesis cycle is the capping reaction to minimize the n-1 deletion-mer impurities by acylating the terminal nucleoside that fail to couple. Recently, the capping reaction has been responsible for the formation of an impurity where an adenine residue was modified with an additional mass of 98 amu (Figure 19.6).[29] Yet another impurity formation was attributed to a capping reaction where a protected guanine residue was transformed into N^2-acetyl-2,6-diaminopurine (+41 amu; Figure 19.6). This study recommended the use of half the recommended volume (8.2 equivalents of acetic anhydride over 45 seconds) of capping reagents to minimize the formation of the +41 amu impurity.[30] Currently efforts are in progress to fully optimize the reagent excess, contact time and flow rates for the first three synthesis cycles where the highest efficiencies are accomplished therefore eliminating the need for capping altogether.

19.3.5 Cleavage and Deprotection

Upon completion of the automated synthesis cycles, the crude oligonucleotide remains attached to the support with the protecting groups intact. Next, the support-bound oligonucleotide is processed through a two-step base treatment protocol in a batch mode. First, treatment of the solid support with a solution of triethylamine in acetonitrile (1:1 v/v) removes the cyanoethyl group from the phosphate linkage, generating acrylonitrile that is easily washed away circumventing the formation of undesired Michael N^3-thymine adduct (CNET; Figure 19.6). Additionally, acrylonitrile is postulated to add to the PS linkage making it susceptible toward attack by a hydroxide ion during aqueous ammonia treatment, increasing the formation of a PO linkage. Second, after purging the acrylonitrile away from the solid support, it is treated with aqueous ammonia to deblock the acyl protecting groups and release the crude oligonucleotide into solution. Next, the solid support is removed *via* filtration and the product concentrated under vacuum in a stripping tank. Today, the two-step cleavage and deprotection protocol allows the synthesis of high purity crude oligonucleotide minimizing the formation of impurities. For example, a crude purity of >75% full-length antisense gapmer is routinely accomplished *via* this procedure (Figure 19.8).

For manufacturing of RNA oligonucleotides, an additional step is required to deprotect the 2′-*O*-TBDMS group – the most commonly used group today. Use of a fluoride reagent such as triethylamine trihydrofluoride (TEA.3HF) is now well established and often carried out at slightly elevated temperature (40–60 °C) in the presence of DMSO to increase the solubility of the TBDMS-protected crude RNA sequence. Often for RNA-amidites more labile base protecting groups are employed to minimize the long exposure to basic pH, which is not well suited for the stability of RNA strands. Also, the 2′-fluoro modification is well known to undergo elimination during extended base treatment to generate an anhydro nucleosidic impurity.[15]

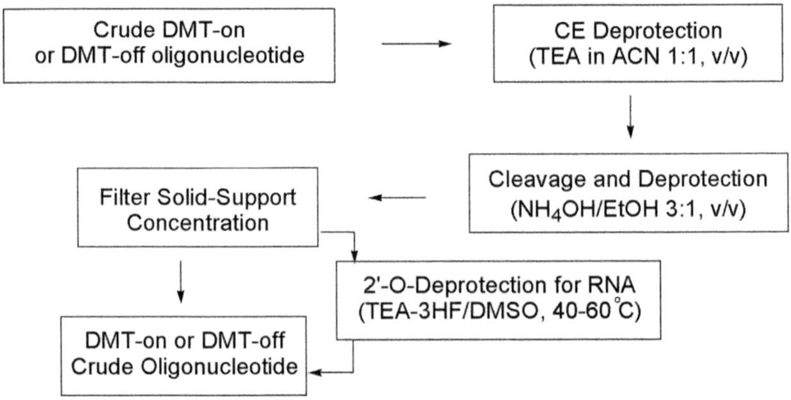

Figure 19.8 An overview of post-synthesis chemical deprotection steps.

19.3.6 The Purification Process

Comprehensive review articles on large-scale oligonucleotide purification are available in the literature.[31] This section provides an overview of current technologies and a summary of recent advances made in this arena. Although a well optimized, solid-phase automated synthesis results in >75% crude purity for oligonucleotides, this must undergo a substantial increase in purity because of patient safety during clinical trials. Each therapeutic oligonucleotide sequence is unique and may contain sequence- and chemistry-related impurities that have not been observed before with other analogous sequences. Therefore, purification of oligonucleotides is a constantly evolving process where significant efforts must be dedicated to generating a high-purity drug substance. Oligonucleotides are negatively charged, highly polar molecules that contain hydrophobic elements. Oligonucleotides with certain sequences (*e.g.* G-rich) often aggregate into a secondary structure, complicating the purification protocol.

Today, anion-exchange (AX) and reversed-phase (RP) are the two HPLC strategies used in practice for the purification of oligonucleotides. AX chromatography is a widely used technique for the large-scale purification of oligonucleotides because of the following reasons (Figure 19.9).[32] First, AX chromatography is the method of choice for the purification of biomolecules offering easy access to low-pressure equipment and media. Second, the system requires aqueous buffers, avoiding the organic solvents needed for RP chromatography. Third, compared with RP, the loading during AX chromatography is significantly higher and permits effective separation of PO impurities. Lastly, the relative purity of the drug substance obtained from AX is often better than the purity of product obtained from an RP protocol. Currently, Q Sepharose and TSK Super-Q are the two resins that have been used in large-scale purification of various oligonucleotides. The maximum loading for crude RNA product on two resins is 20–30 mg ml^{-1}.

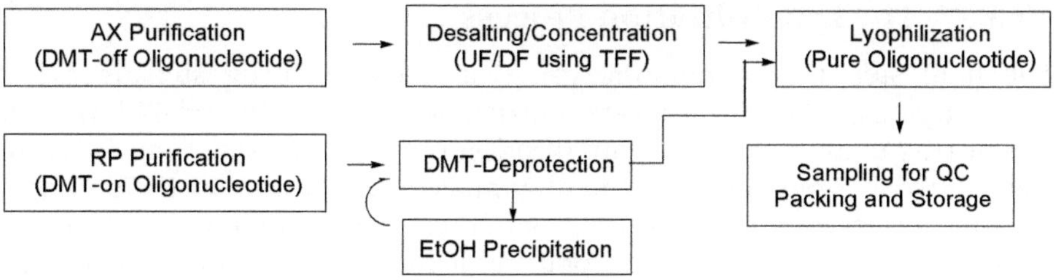

Figure 19.9 An exemplary downstream process flow of drug substance.

The optimal pH for good selectivity for a DNA oligonucleotide is between 8 and 9 and for RNA between 6 and 7 (because of the lability of the ribose sugar under alkaline conditions). Single-step purification often resulted in >95% product purity but this could be further improved by using an optimized step-gradient elution. The pure fractions post AX chromatography contain a high concentration of salts that are easily removed using tangential flow filtration (TFF). This protocol permits the use of appropriately sized filter membranes, maximizing the product recovery (>98%). TFF has become a mainstay for desalting and concentration after AX chromatography due to the availability of commercial-scale equipment from PALL and EMD.

The RP purification takes advantage of the hydrophobic DMT group as a separation handle where all DMT-off short sequences elute early on and the desired DMT-on product is eluted later. Therefore, crude DMT-on oligonucleotide is loaded onto the polymeric C18 type of stationary phase with low organic mobile phase, such as methanol or acetonitrile in sodium or ammonium acetate buffer. Typically for a PS oligonucleotide, the DMT-on full-length product elutes as two peaks due to the chiral backbone. The column is eluted with an increasing amount of organic content to remove the DMT-off failure sequences and products generated from the protecting groups. Next, the amount of organic content is further increased, eluting the DMT-on product. Compared with AX chromatography, the RP offers higher recovery of the full-length product after purification. The purified DMT-on product is then precipitated from ethanol to remove the buffer salt and any excess organic solvent. This simple and elegant process of room temperature precipitation has been successfully implemented and is currently in use for kilo-scale production of antisense oligonucleotides.[33] The precipitated DMT-on product is then treated with sodium acetate solution (10 mM, pH 3.0) under precise control of pH, concentration and time to drive the detritylation to completion, furnishing DMT-off product in excellent yield with minimum depurination. More recently, Biogen researchers presented a combination of hydrophobic interaction chromatography and AX as an alternative offering higher quality (>94%) of the isolated drug substance.

19.3.7 The Lyophilization Process

The final step in the manufacturing of kilo-scale drug substance is a freeze-drying or lyophilization step carried out using commercial VirTis Genesis or Hull equipment. The lyophilization process consists of the following steps. First, is preparation of the oligonucleotide solution at the appropriate concentration in high-purity low-endotoxin water, followed by filtration through a 0.2-micron filter and pouring it into freeze-drying trays or containers (*e.g.* Gore LyoGuard). Second, there is freezing or solidification of the content of the trays followed by a primary drying step (ice sublimation) when the condenser temperature is lowered to −70 °C and connected to high vacuum for 2–3 days. The purpose of this step is to remove the bulk of the water that is unbound to the oligonucleotide. Upon completion of the primary drying step, the internal temperature is gradually increased to 25–30 °C under higher vacuum than used for primary drying. When the secondary drying cycle (moisture desorption) is complete, the water content of the product is around 2–5% w/w. Next the product trays are transferred to a low-humidity chamber. The product cakes are then transferred to a polyethylene bag and the cakes are manually manipulated to form free-flowing solids. Homogenization of the solids results from tumbling the contents of the bag. The drug substance is then transferred to clean high-density polyethylene (HDPE) bottles for storage at −20 °C in a freezer.

Clearly, the overall success during oligonucleotide manufacturing is dependent on several factors such as selection of good quality reagents, solvents, amidites, solid support, resin for purification, choice of the right size equipment, proper understanding of the equipment loading and performance, void volumes in transfer lines, heat transfer during the lyophilization cycle and, most importantly, having a good understanding of synthesis steps because "chemistry matters".

19.4 Analytical Protocols

Manufacturing of oligonucleotides requires not only efficient and scalable synthesis and purification protocols, but also an extensive set of analytical methods to monitor the quality of in-process products, drug substance and drug product. An excellent collection of analytical methods for the complete characterization of oligonucleotides have been published in a book.[34] These methods offer utilization of modern day HPLC or ultra-performance liquid chromatography (UPLC) in a variety of modes such as RP, ion-pair–reversed-phase (IP–RP), ion-exchange (IE), hydrophilic interaction liquid chromatography (HILIC) and size-exclusion chromatography (SEC) teasing-out all impurities in a well-defined manner. Furthermore, a complementary mass spectral method such as HPLC–mass spectrometry (HPLC–MS), tandem mass spectrometry (MS–MS) and matrix assisted laser desorption/ionization–time of flight (MALDI–TOF) provides additional information for the structural determination of oligonucleotides.[35] Additionally, ultra-performance liquid chromatography–tandem mass spectrometry (UPLC–MS–MS) analysis of siRNA oligonucleotides is used for the identity confirmation of the target sequence-related impurities in a facile manner with a run time of less than 15 min. Fragmentation patterns of modified oligonucleotides have been investigated using RNA 2′-*O*-Me, PS and LNA modifications. Typically an oligonucleotide drug substance is tested by multiple protocols summarized in Table 19.1.

A recent marketing application for Spinraza included thorough assessment of clearance of genotoxic and small-molecule impurities following

Table 19.1 List of conventional analytical test protocols, equipment used and acceptance criteria.

Name of the test performed	Acceptance criterion	Method and equipment
Appearance	Free flowing white powder	Visual inspection
Molecular sequence for identity check	Consistent with expected oligonucleotide sequence	Electrospray ionization (ESI)–MS–MS or HPLC–TOF–MS
Molecular weight for identity check	Consistent with expected molecular weight	ESI–MS or IP–HPLC–UV–MS
HPLC for purity check	≥ 90% for antisense oligonucleotides ≥ 85% for single-strand RNA used for siRNA duplex	RP– and IE–HPLC RP– and IE–HPLC
HPLC or LC–MS for impurities	Report all ≥1% by area	RP– and IP–HPLC and UV–MS
Duplex purity for siRNA	≥ 85% area	SEC–HPLC
Purity and content analysis	≥ 90% for most oligonucleotides	Capillary gel electrophoresis
^{31}P nuclear magnetic resonance (NMR) for impurities	≤ 2–3% PO in antisense PS oligonucleotides	High-field NMR
Moisture/water content	≤ 10% w/w	Karl Fisher
Melting temperature™ for identity	Consistent with expected T_m ± 2–3 °C	UV spectrometer
Residual solvents	Reported percentage of solvents used in the process: acetonitrile, toluene, pyridine *etc.*	Headspace GC
Bacterial endotoxins	≤ 3–5 endotoxin units mg^{-1}	*Limulus* amebocyte lysate US Pharmacopeia (USP) <85>
Bioburden	≤ 100 colony forming units g^{-1}	Plate count method: USP <61>
pH of 1% w/v aqueous solution	Reported observed pH (6–8)	USP <791>
Sodium content	Consistent with expected value	Flame atomic absorption spectrometry (Flame–AA)
Heavy metals	Reported Ag, As, Bi, Cu, Hg, Mo, Pb, Sb, Sn *etc.* (≤1–2 ppm)	Inductively coupled plasma–mass spectrometry (ICP–MS)

the International Conference on Harmonization (ICH) M7 guidelines. For example, the drug substance was tested for the presence of acrylonitrile [permitted daily exposure (PDE) 6 mg per day], benzamide (PDE 1.5 mg per day), isobutyramide, acetamide, succinamide, Uny-CTP (a fragment of Uny-linker of undisclosed chemical structure), aniline, sodium acetate and DMT-OH and confirmed to be well below PDEs where applicable.

19.5 Synthesis Yield and Product Purity

Over the years, the oligonucleotide synthesis protocol has been streamlined to reduce the cost of goods while improving the overall yield and the quality. For example, the average purity of oligonucleotides entering clinical trial is >90% full-length product with less than 5% water content. A typical 500 mM solid-support campaign for a 20-mer antisense gapmer oligonucleotide would furnish 400–450 grams of drug substance. Furthermore, an automated synthesizer is now routinely used on 500–1000 mM scale for kilo-scale production of therapeutic oligonucleotides. The synthesis of siRNA and other modified oligonucleotides has reached a similar scale in a single campaign. Today, there are multiple commercial manufacturers of synthetic oligonucleotides with a capacity to produce metric-tonne quantities of a single sequence per year to meet the projected demand.

19.6 Conclusions and Future Outlook

Clearly, solid-supported synthesis is the method of choice for oligonucleotide drugs because of its automated protocols and remarkably short synthesis cycle times. For the current demand for both clinical and marketed products there is sufficient capacity to manufacture hundreds of oligonucleotides in a fast and cost-effective manner annually. However, should there be a need for a single product required in very large amounts, solid-supported synthesis may have a few limitations. For example, the heterogeneous nature of the solid support process leads to a high mass-transfer resistance between the reagents and the support surface, introducing steric hindrance and demanding an excess usage of amidites to drive the reaction to completion. Also, the loading capacity of the solid support is limited, particularly for the synthesis of longer sequences. Lastly, the ratio of the mass of waste per mass of product (E-factor) during solid-supported synthesis is unacceptable for the process to be environmentally friendly.

Gratifyingly both researchers from academia and industry have been exploring alternative options that offer hope for an improved process for metric-tonne production of a single oligonucleotide in the near future. AJIPHASE (Ajinomoto) is one such protocol that has been recently practiced for large-scale synthesis in solution-phase.[36] The protocol utilized an anchor molecule with long-chain branched alkyl linkers highly soluble in organic solvents. Standard phosphoramidite building blocks were employed and each synthesis cycle was carried out in a single conventional reaction vessel. The excess reagents were washed away *via* a precipitation protocol between each step. Upon completion of the synthesis, the anchor was cleaved off to generate crude product. As a demonstration, 21-mer PS oligonucleotide was assembled with crude purity of over 80% on a large scale. Recently, researchers from Imperial College have demonstrated that liquid-phase oligonucleotide synthesis using organic solvent nanofiltration (OSN) could offer yet another protocol with high production capacity. This work utilized an aromatic three-armed homostar anchor, which allowed good rejection by a membrane and facilitated in-process analysis by UPLC–MS. The chain elongation was carried out using standard phosphoramidites. Upon completion of the reaction, the solution was passed through a polybenzimidazole (PBI) membrane module (spiral wound) to remove unreacted reagents and the product pumped back into the reaction chamber for the next step. This protocol was used to synthesize 2′-*O*-Me 20-mer oligonucleotide on a gram-scale.[37] The key finding in these two liquid-phase synthesis protocols is the ability to design an anchor that remains soluble during the synthesis cycle, providing excellent coupling efficiency and utilization of a conventional amidite chemistry platform. Further advances in these two technologies may set the stage for alternative options that may be amenable to scale-up.

Since amidite chemistry remains as the most powerful tool during the assembly of oligonucleotides, more efforts to improve the coupling efficiency with a reduced excess of building block would be worthy of further investigation. Next, the ability to recover or recycle solvents, reagents and amidites, easy manipulation of automated steps and increased in-process control and simplified product isolation would all be very attractive objectives to pursue. The application of real-time mid-infrared and Raman spectroscopic monitoring combined with modeling is worthy of consideration to minimize the risk of batch failure during automated synthesis of oligonucleotides.[38] With the advances summarized herein, process chemists will continue to have options in cheaper raw materials, new reagents and processes to choose from in their endeavor to synthesize oligonucleotides at large scales. Further oligonucleotide manufacturing processes are likely to integrate a variety of these options, ultimately delivering safe and cost-effective methods for large-scale production of oligonucleotides.

References

1. (a) S. T. Crooke, *et al.*, *Cell Metab.*, 2018, **27**, 714; (b) A. Khvorova and J. K. Watts, *Nat. Biotechnol.*, 2017, **35**, 238.
2. Y. Nukaga, N. Oka and T. Wada, *J. Org. Chem.*, 2016, **81**, 2753.
3. N. Iwamoto and D. C. Butler, *et al.*, *Nat. Biotechnol.*, 2017, **35**, 845.

4. X. Shen and D. R. Corey, *Nucleic Acid Res.*, 2018, **46**, 1584.
5. G. Mosoyan, *et al.*, *Leukemia*, 2017, **31**, 2458.
6. A. Tefferi, *et al.*, *Blood Cancer J.*, 2016, **6**, e405.
7. M. H. Ree, *et al.*, *Aliment. Pharmacol. Ther.*, 2016, **43**, 102.
8. S. Miyatake, *et al.*, *Methods Mol. Biol.*, 2018, **1687**, 123, and reference cited therein.
9. A. Vater and S. Klussmann, *Drug Discovery Today*, 2015, **20**, 147.
10. H. Ludwig, *et al.*, *Leukemia*, 2017, **31**, 997.
11. K. Schmidt, *et al.*, *Nucleic Acids Res.*, 2017, **45**, 2194.
12. Y. Huang, *Mol. Ther. Nucleic Acids*, 2017, **6**, 116.
13. R. L. Juliano, *Nucleic Acids Res.*, 2016, **44**, 6518.
14. Y. S. Sanghvi, in *Current Protocols in Nucleic Acid Chemistry*, ed. M. Egli, P. Herdewijn, A. Matsuda and Y. S. Sanghvi, John Wiley & Sons, 2011, vol. 46, pp. 4.1–4.1.22.
15. E. Pardes, *et al.*, in *Comprehensive Medicinal Chemistry*, Elsevier, 3rd edn, 2018, vol. 6, pp. 233–279.
16. A. P. Guzaev, in *Current Protocols in Nucleic Acid Chemistry*, ed. M. Egli, P. Herdewijn, A. Matsuda and Y. S. Sanghvi, John Wiley & Sons, 2013, vol. 53, pp. 3.1.1–3.1.60.
17. A. P. Guzaev and R. T. Pon, in *Current Protocols in Nucleic Acid Chemistry*, ed. M. Egli, P. Herdewijn, A. Matsuda and Y. S. Sanghvi, John Wiley & Sons, 2013, vol. 52, pp. 3.2.1–3.2.23.
18. I. E. Cedillo, *et al.*, *US Pat.* 9701708, 2017.
19. R. K. Kumar, *et al.*, *Tetrahedron*, 2006, **62**, 4528.
20. V. T. Ravikumar, *et al.*, *Org. Process Res. Dev.*, 2008, **12**, 399.
21. Y. S. Sanghvi, *Chim. Oggi – Chem. Today*, 2014, **32**, 10.
22. D. Capaldi, *et al.*, *Nucleic Acid Ther.*, 2017, **27**, 309.
23. A. H. Krotz, *et al.*, *Org. Process Res. Dev.*, 2004, **8**, 852.
24. J. L. Scotson, *Org. Biomol. Chem.*, 2016, **14**, 8301.
25. A. P. Guzaev, *Tetrahedron Lett.*, 2011, **52**, 434.
26. S. K. Roy and J.-Y. Tang, *US Pat.* 6500944 B2, 2002.
27. Z. Wang, Q. Song and Y. S. Sanghvi, Oligonucleotide synthesis: methods and applications, in *Methods in Molecular Biology*, ed. P. Herdewijn, Humana Press, Totowa, NJ, 2005, vol. 288. pp. 51–63.
28. H. Gaus, *et al.*, *Bioorg. Med. Chem. Lett.*, 2005, **15**, 4118.
29. A. A. Rodriguez, *et al.*, *Bioorg. Med. Chem. Lett.*, 2016, **26**, 3468.
30. A. A. Rodriguez, *et al.*, *Bioorg. Med. Chem. Lett.*, 2014, **24**, 3243.
31. N. D. Sinha and K. E. Jung, in *Current Protocols in Nucleic Acid Chemistry*, ed. M. Egli, P. Herdewijn, A. Matsuda and Y. S. Sanghvi, John Wiley & Sons, 2015, vol. 61, pp. 10.5.1–10.5.39.
32. Y. S. Sanghvi and M. Schulte, *Curr. Opin. Drug Discovery Dev.*, 2004, **7**, 765.
33. D. C. Capaldi and A. N. Scozzari, in *Antisense Drug Technology: Principles, Strategies and Applications*, ed. S. T. Crooke, CRC Press, Boca Raton, FL, 2008, pp. 401–434.

34. J. V. Bonilla and G. S. Srivatsa, *Handbook of Analysis of Oligonucleotides and Related Products*, CRC Press, Boca Raton, FL, 2011.
35. S. G. Roussis, *et al.*, *Rapid Commun. Mass Spectrom.*, 2018, **32**, 1099.
36. D. Takahashi, *et al.*, *US Pat.* 20170218361 A1, 2017.
37. J. F. Kim, *et al.*, *Org. Process Res. Dev.*, 2016, **20**, 1439.
38. J. W. Rydzak, *et al.*, *Org. Process Res. Dev.*, 2015, **19**, 203.

CHAPTER 20

Preclinical and Clinical Drug-metabolism, Pharmacokinetics and Safety of Therapeutic Oligonucleotides

PATRIK ANDERSSON*[a] AND CATHALINE DEN BESTEN[b]

[a]Advanced Medicines Safety, Drug Safety and Metabolism, IMED Biotech Unit, AstraZeneca, Pepparedsleden 1, 431 83 Mölndal, Gothenburg, Sweden; [b]ProQR Therapeutics BV, Zernikedreef 9, 2333 CK Leiden, The Netherlands
*E-mail: Patrik.J.Andersson@astrazeneca.com

20.1 Introduction

Synthetic oligonucleotide drugs (ONDs) continue to make steady progress on their journey to a mature and clinically validated therapeutic platform.[1] The genomic revolution has rapidly increased our understanding of the genetic basis of health and disease and fueled the innovation of RNA-modulating therapies. A key advantage is the rational sequence design of ONDs to their cognate RNA target, driven by highly selective complementary Watson–Crick base pairing. This avoids the need to identify compounds with complex ligand–protein binding sites, interactions that are critical to small-molecule drugs. Another key advantage is the high degree of selectivity without being restricted to extracellular epitopes, as is the case for most antibody drugs.

Drug Discovery Series No. 68
Advances in Nucleic Acid Therapeutics
Edited by Sudhir Agrawal and Michael J. Gait

Published by the Royal Society of Chemistry, www.rsc.org

Another feature supporting a rational drug discovery approach is that oligonucleotides of a given chemistry and design demonstrate a consistent and predictive 'class' behavior in terms of drug-metabolism and pharmacokinetics (DMPK) and target organs for toxicity.

The DMPK and toxicological properties of single-stranded phosphorothioate (PS) backbone ONDs and 2′-*O*-Methoxyethyl (2′MOE) modified gapmers (see Chapter 2) were comprehensively summarized[2–5] about a decade ago, and these reviews shaped the concept of oligonucleotide class effects. Since then, the field has progressed considerably, with an expanding toolbox of chemical modifications and therapeutic OND approaches. We have increased our understanding of the molecular mechanisms critical to efficacy, distribution and toxicity and how these are affected by OND sequence, chemical modifications and design. Moreover, preclinical and clinical safety databases covering subchronic and chronic treatments are now becoming large enough for meaningful analyses across multiple targets and sequences,[6–9] providing a body of data to confirm or disprove clinical translation of preclinical observations.

With a focus on new knowledge obtained in the last decade, the current overview summarizes key DMPK and toxicological aspects of OND therapeutics and how these properties are influenced by design and chemistry. In scope are OND classes that utilize hybridization-dependent mechanisms, such as single-stranded antisense ONDs (ASOs) acting *via* RNase H degradation, splice interference or anti-miRNA mechanisms and double-stranded ONDs directed at RNA interference [microRNA (miRNA) and small interfering RNA (siRNA)]. Out of scope are other promising oligonucleotide modalities acting *via* other mechanisms than Watson–Crick hybridization, *e.g.* protein-coding mRNA[10] and aptamers[11,12] designed to act at the protein level rather than RNA. However, some of the principles described for hybridization dependent ONDs may also apply to the non-hybridization ONDs.

The chapter ends with a brief description of regulatory considerations of bringing ONDs into clinical testing. There are to date no formal regulatory guidelines dedicated to ONDs, but examples of best practice and recommendations covering several areas have been published by the Oligo Safety Working group (OSWG).[13]

20.2 Oligonucleotide Chemistries and Mode of Action (MOA)

20.2.1 Chemistry and Design Considerations of Therapeutic ONDs

The DMPK and safety profile of OND therapeutics are highly affected by design and chemical modification of the oligonucleotide. Significant progress has been made to optimize these parameters for a balanced improvement of affinity, nuclease resistance, distribution, pharmacokinetics (PK), cellular uptake and safety. The design and chemistry requirements of the various oligonucleotide classes have been described in previous chapters of this book and in excellent recent review articles.[14–16] Key features are summarized below to facilitate the understanding of the complex interplay between chemical design and RNA-modulating mechanism, on the one hand, and DMPK and safety properties, on the other hand.

20.2.1.1 OND Chemistry

Chemical optimization strategies have been implemented since the late 1990s to improve the poor drug-like properties of naturally occurring oligonucleotides. Incorporation of PS linkages in the OND backbone provided resistance against nuclease degradation, improved tissue uptake and drug retention and increased plasma protein binding, resulting in reduced renal excretion.[17–19] As alternatives to negatively charged PS backbone ASOs, phosphorodiamidate morpholinos (PMOs) and peptide nucleic acids (PNA) have neutral OND backbones with excellent nuclease resistance, but they have high renal clearance and relatively low cellular uptake due to their low degree of protein binding.[16] To improve delivery, tissue-targeting approaches are increasingly being explored to advance this chemistry to clinical meaningful potential.[20]

A variety of modifications at the 2′ ribose position increased hybridization affinity.[21,22] Commonly used 2′ ribose modifications include 2′-fluoro (2′F), 2′-*O*-methyl (2′OMe) and 2′MOE.[14,16,23] The newer 2′-*O*-4′-*C*-bridged ribose modifications locked nucleic acid (LNA) and constrained ethyl (cEt) have been implemented in single-stranded ASOs to achieve even higher binding affinity to the RNA targets. The most common chemistries used for double-stranded RNA ONDs are 2′F and 2′OMe as well as some PS linkages for stability.[14,16,23]

20.2.1.2 OND Design

ONDs acting *via* hybridization-dependent mechanisms are single- or double-stranded stretches in the range of 12–30 nucleotides, depending on the intended mechanism. The shorter ONDs (12–16 nucleotides) generally carry higher affinity 2′-ribose modifications, such as LNA or cEt. The clinically most advanced OND classes are ASOs acting *via* RNase H[24,25] or splice modifying[26–28] mechanisms and siRNAs acting *via* mechanism(s) dependent on the RNA induced silencing complex (RISC).[29,30] Although nucleobase sequence is absolutely key for hybridization specificity, the chemistry and design determine most other properties. The degrees of freedom of OND chemistry and design are dictated by the intended mechanism of action, with some RNA-modulating approaches being more tolerant of variations than others. Thus, for a given RNA modulating mechanism, only certain designs and combinations of chemical modifications are accepted. For example, RNase H-mediated RNA knockdown tolerates a fully modified PS backbone, but does not tolerate neutral PMO or PNA backbones.[16] The RNase H mechanism also requires a central stretch of DNA, allowing ribose modification only in the wings; a so-called gapmer design.[31,32] Splice-correcting and anti-miR ASOs act *via* a steric blocking mechanism and can carry PS and ribose modifications throughout the full sequence of the OND.[33] These mechanisms allow neutral backbones, such as PMO and PNA.[34–36] For double-stranded ONDs acting *via* RISC, full PS-modification is not tolerated and the higher affinity LNA and cEt ribose modifications negatively affect separation of the antisense and sense strands of a duplex.[37]

The nomenclature that will be used, together with the most prominent chemistry and design features of the OND classes covered in this chapter are summarized in Table 20.1.

Table 20.1 Terminology and most common key features of hybridization-dependent OND classes discussed in this chapter.

Term	Mechanism	Design	Approximate length (nucleotides)	Backbone	2′ ribose modification[a]
PS ASO	RNase H degradation, steric blocking	Single-stranded	18–25	Full PS	None
ASO gapmer	RNase H-mediated RNA degradation	Single-stranded, central DNA gap with ribose-modified wings	12–20	Full PS, mixed PS–PO	MOE, LNA, cEt
ASO non-gapmer	Steric blocking, *e.g.* splice modulating or anti-miR	Single-stranded, fully ribose-modified	15–30	Full PS, mixed PS–PO	F, OMe, MOE, LNA, cEt
		Single-stranded	15–30	PMO, PNA	NA
siRNA	RISC-mediated RNA degradation	Double-stranded, 5′-phosphate, partly or fully 2′-ribose-modified	15–25	Mixed PO–PS backbone	H, OH, F, OMe
miR mimic	RISC-dependent generation of miRNA	Double stranded, 5′-phosphate, partly ribose-modified	15–25	Mixed PO–PS backbone	OMe, F,
Conjugate	Receptor-mediated facilitated uptake	OND conjugated to ligand *via* linker	12–30	Any of the above	Any of the above

[a]Single-stranded ASOs generally contain one type of ribose modification (*e.g.* 2′MOE or LNA); double stranded ONDs may include multiple 2′-ribose modifications (*e.g.* 2′OMe and 2′F). NA, not applicable.

Within each of these OND classes, the chemistry and design variations are kept relatively constant, with only the nucleobase sequence being varied to match the target transcript. Some chemistry and design variations are more dominant than others, resulting in certain common properties across the OND classes. For example, the prominent polyanionic nature of PS-modified ASOs results in a relatively consistent protein binding and PK profile regardless of mechanism, although sequence-specific exceptions have been described.[38,39] *N*-acetyl galactosamine (GalNAc) conjugation (Section 20.3.2.3) results in enhanced hepatocyte uptake of both single- and double-stranded ONDs, regardless of whether they have a full PS backbone or not.[40,41] On the other hand, other seemingly subtle changes such as various 2′ ribose modifications and their position in the oligonucleotide can have significant effects on a given property. This also includes the chirality of the PS linkages, which is known to affect affinity, RNase H activity and nuclease resistance.[5,42–45] In recent years, more evidence has been generated on PS backbone stereochemistry also affecting efficacy and safety.[38] Given such variability, extrapolation of DMPK and safety properties between different oligonucleotide classes and chemistries should therefore be carried out with caution, supported by robust experimental data.

Safety findings referred to as "oligonucleotide class effects" are mainly based on findings with single-stranded PS backbone ASOs with or without 2′MOE and 2′OMe modifications.[2,3] Although many of these safety effects are still observed at higher doses with newer ASO classes, a better mechanistic understanding, leading to improved OND chemistry, design and screening cascades, has resulted in successful mitigation of many of these "oligonucleotide class effects" for more recent clinical candidates.

20.2.2 Delivery Approaches

Complex formulation strategies, such as lipid nanoparticles (LNP), need to be applied for non-stabilized siRNA and miR mimetics to avoid rapid degradation and excretion.[46–49] Due to the uncharged backbone of PMO- and PNA-based ONDs, protein binding is low, resulting in low tissue distribution and rapid excretion. To improve uptake and distribution to target tissues, ONDs can be conjugated to different targeting moieties, such as lipids, peptides, carbohydrates and even antibodies, see Section 20.3.2.3. The best characterized targeting approach to date is GalNAc conjugation of both single- and double-stranded ONDs, which significantly increases uptake into hepatocytes.[40,41]

When applying conjugation strategies, the chemistry and design of the OND part is generally unchanged and much of the understanding of the class profile with regard to intrinsic potency and kinetic properties such as tissue half-life and metabolism also translates to these conjugated molecules.

20.3 Distribution, Metabolism and Pharmacokinetics (DMPK)

The DMPK properties of ONDs are largely sequence-independent and driven by their physicochemical characteristics as determined by the chemical design. Most of the knowledge on PK and toxicity of ONDs is based on a relatively limited number of chemical modifications. Overall, the concept of a consistent and predictive DMPK behavior driven by the plasma protein-binding properties of PS backbone ASOs as described 10 years ago still stands,[5] but we now know that multiple additional protein interactions have a significant influence on cellular uptake and intracellular trafficking, and thereby efficacy.[50–52] Compared with ASOs with an anionic PS-ribose backbone, ASOs with a neutral morpholino backbone have different PK and cellular uptake properties that follow from their different physicochemical characteristics. This includes primarily their weak binding affinity for proteins and as a result their rapid excretion from plasma by glomerular filtration.[53]

The preclinical and clinical DMPK properties of PS, 2′MOE and 2′OMe ASOs have been comprehensively described elsewhere.[5,54] Most of this understanding is applicable also for PS ASOs with newer, high affinity modifications. The following section therefore focuses on new knowledge, such as productive uptake and targeted delivery, and only briefly summarizes the established DMPK understanding to support the subsequent discussions of OND toxicity.

20.3.1 DMPK Properties of ONDs

20.3.1.1 Plasma and Tissue Pharmacokinetics

One of the main challenges to the OND platforms has been to achieve sufficient productive uptake into the target tissue.[51,55] In contrast to most small-molecule drugs, ONDs are poorly absorbed from the gastrointestinal tract.[56] More similar to antibodies and therapeutic peptides and proteins, ONDs are therefore administered *via* subcutaneous (SC) or intravenous (IV) injection to obtain clinically meaningful systemic exposures. Alternatively, they can be administered *via* local delivery to *e.g.* the central nervous system (CNS) or the eyes (see Section 20.3.2.1). Bioavailability after SC injection is generally complete, although low bioavailability has occasionally been observed for isolated sequences (Den Besten, personal communication). The typical PK profile of PS backbone ASOs is characterized by a multiphasic plasma concentration *versus* time profile, with an initial distribution phase where the ASO is cleared from the systemic circulation by tissue uptake within hours.[5] The plasma half-life (and as a consequence, dose-normalized systemic exposure) may increase with higher doses due to saturation of tissue uptake.[5,57] The distribution phase is followed by a much slower elimination phase, when plasma concentrations are in equilibrium with tissue concentrations.[54] At clinically relevant doses, plasma concentrations in the distribution phase are generally in the μg ml^{-1} range, with low ng ml^{-1} or pg ml^{-1} concentrations in the post-distribution elimination phase.[5,58] The terminal half-life in plasma reflects the clearance from tissues, mostly through nuclease digestion to shorter fragments,[58,59] and may range from a few days for PS ASOs to one to many months for gapmer and non-gapmer PS ASOs, depending on their chemical modification pattern and the tissue. A consistent liver-to-plasma ratio of approximately 5000 was reported for 2′MOE gapmers, independent of sequence, which indicates that steady-state plasma exposure can be used to estimate tissue exposure.[5,58,60,61] Even though the absolute ratio may differ, initial studies in non-human primates (NHPs) with GalNAc-conjugated ASO gapmers showed a liver-to-plasma ratio during the elimination phase that can also be used to approximate tissue concentrations.[62]

The main determinant of ASO PK properties is the high degree of protein binding. PS backbone ASOs bind to hydrophilic sites on plasma proteins, such as albumin and alpha-2-macroglobulin. These binding sites are distinct from the binding sites to which small molecules bind.[5,63] Hence, ONDs demonstrate negligible risk for drug interactions through protein binding displacement. Plasma protein binding capacity is generally lower in mice than in other species. As a consequence, saturation of plasma protein binding is observed at lower drug concentrations in mice, which results in a higher free fraction with increased renal excretion. This explains why the mouse is somewhat of an outlier regarding PK.[5,54] The plasma protein binding can be affected by 2′-ribose modifications. For example, 2′MOE tends to reduce, whereas other modifications, such as 2′OMe, 2′F and LNA, tend to increase plasma protein binding (generally >99%).[5]

Compared with double-stranded ONDs, such as siRNA, which have more defined helical structures, single-stranded ASOs are conformationally more flexible and can therefore interact with proteins in more diverse ways.[64] As a consequence, siRNA shows less plasma protein binding than single-stranded PS ASOs, resulting in higher renal clearance and lower tissue exposure.[47] Non-formulated, chemically unmodified siRNA is cleared within 1–2 minutes from plasma of mice dosed intravenously.[47,49] In addition, unmodified siRNA was rapidly degraded in human plasma[48] or calf serum,[46] indicating that the rapid clearance of unmodified siRNA may be the result of both kidney clearance and nuclease degradation. For this reason, siRNA needs to be either formulated (*e.g.* LNP, Section 20.3.2.2) or conjugated to *e.g.* GalNAc (Section 20.3.2.3) and further optimized to increase plasma half-life[47] and achieve significant tissue uptake and pharmacology. Chemical modifications like 2′F, 2′OMe, some PS linkages and cholesterol conjugation increase plasma stability and tissue exposure of non-formulated siRNA.[47–49,65–67]

Pharmacokinetics of uncharged PMOs reflect their different physicochemical properties, resulting in low protein binding and rapid clearance from plasma. Their plasma concentration *versus* time profile is therefore primarily dependent on glomerular filtration, rather than on tissue distribution and may be more variable with sequence than that of a PS ASO.[53]

20.3.1.1.1 Anti-drug Antibodies. Increasing experience with longer-term treatment duration of PS gapmer and non-gapmer ASOs in animal studies and clinical practice revealed that low-avidity anti-drug antibodies (ADA) can be induced, which tend to prolong plasma residence time of the ASO. This results in increased plasma trough levels and overall systemic exposure, with little effect on peak plasma levels.[68] These ADAs appear to be non-specific, binding to the PS backbone rather than specific OND sequences, and are not known to affect efficacy or side effects.

20.3.1.2 Tissue Distribution

Due to their molecular size and charge, formulated and non-formulated ONDs are primarily taken up by phagocytically active cells or tissues with fenestrated endothelium.[51] The PS ASOs have the broadest tissue distribution profile because of the high degree of binding to plasma proteins, which keeps them in circulation and binding to cell surface proteins, and this facilitates cellular uptake.[5] The highest concentrations are found in the kidney and liver.[5,54,69] To a lesser extent, ASOs are also taken up in other organs, such as spleen, lymph nodes, adipocytes, bone marrow and vascular endothelial cells.[54] At high doses, uptake is observed in multiple other tissues.[54,70] There is an overall low uptake in circulating blood cells but ASOs readily accumulate in tissue macrophages, and ASO-containing macrophages can often be detected in many tissues, particularly in rodents. There is minimal uptake of systemically administered PS ASOs into organs such as CNS and across the placenta,[54,70–73] indicating that innovative approaches are needed for therapeutic intervention in those tissues. On the other hand, the much lower exposures in most tissues at least partly explain the relative lack of toxicity observed in organs other than liver and kidney.

Despite being highly vascularized, the relatively low uptake in lung, heart and kidney glomerulus indicates that high blood flow has limited relevance for uptake of PS backbone ONDs. Degree of plasma protein binding and length of the OND seem to be more important factors for tissue uptake, where ONDs with a longer PS backbone tend to reach higher concentrations in liver than in the kidney and shorter ONDs show the opposite pattern following increased glomerular filtration due to less protein binding.[5,63,74] Perhaps somewhat unexpected with a given design and chemistry, the OND sequence can have a significant effect on uptake. For example, PS backbone 3–10–3 LNA gapmers against the same target differing only in sequence ranged fivefold in liver concentration and up to 25-fold in liver-to-kidney ratio.[38]

The uptake of PS backbone ONDs within a given organ is strikingly non-homogenous. Although the hepatocyte parenchymal cells show good uptake and pharmacology in the liver, sinusoidal endothelial and Kupffer cells show the highest concentration of non-conjugated ASOs.[69,70] Targeting strategies are now successfully applied to increase hepatocellular uptake (see Section 20.3.2.3). In the kidney, the proximal tubular epithelial cells of the convoluted tubule show the highest uptake, whereas uptake in tubular cells in kidney medulla is much lower.[69,70] Similar cell-type differences in uptake are observed in other organs, and despite significant research, the molecular basis for this remarkable cell specific difference in uptake is still incompletely understood.

Chemically stabilized but unformulated and unconjugated ^{32}P-labelled siRNA was found in several tissues 30 minutes after intravenous administration. The distribution pattern differed between type of modification but included lung, heart, kidney, spleen and liver. Apart from a weak signal from lung, no intact siRNA was found after 24 hours after administration.[47] The distribution pattern for LNP-formulated siRNA is quite similar to that observed for non-formulated ASOs. Autoradiographic detection of ^{3}H-labelled guide strand and matrix-assisted laser desorption/ionization–mass spectrometry (MALDI–MS) imaging of the cationic lipid in the LNP showed a similar distribution pattern after IV administration to mice. Highest concentrations were observed in the spleen, followed by liver, kidney and adrenals.[75,76]

Uncharged ONDs (such as PMOs) accumulate to a much lower extent in tissue, due to their low degree of protein binding and rapid glomerular filtration. This results in the kidney having by far the highest PMO concentration.[77] To achieve clinically meaningful doses in other tissues, frequent intravenous administration of high doses is required, as illustrated with eteplirsen, an FDA approved therapy in Duchenne muscular dystrophy patients at a weekly IV infusion of 30 mg kg^{-1}. Notably, once PMOs are taken up in cells, they have a long tissue residence time due to their nuclease-resistant backbone.[78]

The non-homogenous tissue-uptake properties for ONDs have important implications for both pharmacodynamics and safety. To have the desired pharmacodynamic effect, the drug needs to have access to the target cell type. For safety assessment, one needs to be aware of the cells in tissues that possess the highest concentrations of drug.

20.3.1.3 Productive Uptake

For a given OND and cell type, there is a positive correlation between tissue concentration and pharmacological activity. However, this is less evident when comparing across different cell types. This can be illustrated by observations from *in vitro* experiments where most cells readily take up PS-backbone ASOs, whereas activity is seen only in a subset of cell types.[50,52,79,80] *In vivo* studies comparing bolus and slow infusion as well as competition with non-active ASO gapmers demonstrated discordance between liver concentration of the active ASO gapmer and level of target gene knockdown.[81] Uptake and endosomal escape of gold-labelled siRNA delivered *in vitro* in LNP formulation estimated that only 1–2% of the siRNA was released from the endosomes and thus available for activity.[82] Such observations have led to the use of the expression of "productive *vs.* non-productive uptake", *i.e.* referring to uptake that does or does not lead to a pharmacodynamic effect, such as transcript knockdown or splice-correction. Besides being a key consideration for efficacy, understanding productive *vs.* non-productive uptake also affects the likelihood of seeing both on- and off-target exaggerated pharmacology (Sections 20.5.1 and 20.5.2) in a given cell type.

To understand productive uptake, it is important not only to measure tissue concentrations of the OND, but also to analyze the pharmacodynamic effects. This is ideally done at cellular resolution. Hung and colleagues made a detailed characterization of productive uptake across multiple tissues in monkeys and mice dosed with 2′MOE- and cEt-modified PS backbone gapmers targeting the long non-coding RNA metastasis associated lung adenocarcinoma transcript 1 (MALAT-1). This transcript is ubiquitously expressed at high levels, making it an ideal target for studies of productive uptake. RNA knockdown was analyzed by reverse transcription polymerase chain reaction (RT-PCR) and *in situ* hybridization (ISH) and ASO uptake was assessed at the cellular level by use of immunohistochemistry (IHC) with antibodies against the PS ASO backbone.[70] All ASOs were administered at high doses, reaching saturated knockdown (*i.e.* productive uptake) in liver and kidney in both species. Despite a similar distribution pattern on the level of whole tissue, the shorter cEt ASOs showed a significantly higher degree of knockdown and thus productive uptake in a wider range of tissues. Semi-quantitative IHC analysis showed a good correlation between ASO uptake and degree of knockdown for a given cell type. Except for lung, adrenal gland and gall bladder, the mice showed a significantly higher degree of knockdown than monkeys,[70] indicating a possible species difference in productive uptake of this class of ASOs.

The molecular mechanisms by which non-conjugated ONDs are taken up into cells or "escape" from the endosome or lysosome compartments to interact with its target RNA are not fully understood. This is recognized as one of the main barriers for the OND platforms to overcome.[51,55] Functional uptake of PS ASOs is an endosomal process that has been studied in the hepatocyte-derived cell line MHT, which shows good productive uptake.[79] Pretreatment using siRNA against adaptor related protein complex 2 subunit mu 1 (AP2M1) significantly decreased antisense activity of a 2′MOE gapmer both *in vitro* and *in vivo*, indicating an important role for this adaptor protein in productive uptake. Surprisingly, siRNA against clathrin and caveolin receptors did not influence target gene knockdown, indicating a limited role of these proteins in productive hepatocyte uptake of non-conjugated ASOs.[79] Some receptors have been identified in the liver that participate in cellular uptake of PS backbone ASOs.[52] These include Stabilin 1 and 2 receptors that are expressed on sinusoidal cells in many tissues, including liver and spleen, where they participate in productive uptake of PS backbone ASOs.[19,83] These receptors are also involved in vesicle trafficking from endosomes to the trans-Golgi network, indicating a possible pathway of endosomal escape.[83] Other molecular components involved in endosomal release of PS ASOs have also been described.[80,84] As with any receptor-mediated process, saturation of uptake can occur. Saturation of non-specific uptake (or non-productive intracellular localization) in liver by pre- or co-administration of an inactive sequence enhanced the efficacy of an active PS gapmer ASO.[85] Interestingly, emerging *in vitro*[52] and *in vivo*[64] data indicate that the asialoglycoprotein receptor (ASGR) is not only instrumental in the uptake of GalNAc-conjugated ONDs (see Section 20.3.2.3), but also plays a role in basal uptake of non-conjugated PS backbone ASOs.

To exert pharmacological activity, the OND needs to reach the right subcellular compartment. RNase H1 is located in both the nucleus and cytoplasm and ASO gapmers are active in both these compartments.[86] Splice-correcting ASOs act in the nucleus and the RISC-mediating siRNA and miRNA OND activities are found in the cytosol. Crooke and colleagues have made a series of *in vitro* studies analyzing kinetics and impact of ASO chemistry and delivery method on subcellular localization and intracellular protein interactions.[50] More than 50 intracellular proteins interact with PS ASOs. Binding to some of these proteins [*e.g.* Ku70–80, 54 kDa Nuclear RNA- and DNA-binding protein (p54nrb), heterogeneous nuclear ribonucleoprotein K (hnRNPK)] increase RNase H activity, whereas binding to others [*e.g.* heat shock protein 90 (HSP90), T-complex protein 1 (TCP1), Annexin A2] decrease RNase H1 activity. Some, but not all, of these effects are associated with alterations in subcellular ASO localization. Interestingly, both ASO sequence and type of ribose modification influence binding affinity to a given protein.[39,50] The time required to observe significant target gene knockdown differs between delivery approaches. The time to target gene knockdown is much longer after free uptake than after lipid transfection, correlating with different kinetics for the ASOs to distribute to various subcellular compartments.[50]

The understanding of productive and non-productive OND uptake has dramatically increased over the last decade, revealing a complex picture of critical protein-binding interactions that may be tissue- and cell-specific. Further elucidation of the key mechanisms that contribute to productive uptake and intracellular trafficking is essential to optimize chemistry and design, aiming to increase the potency and reduce toxicity of ONDs as well as to successfully expand OND therapies to a wider range of cell types and tissues.

20.3.1.4 Metabolism

ONDs are degraded by endonucleases and exonucleases that are ubiquitously present in plasma and tissues. Both single-stranded and double-stranded OND therapeutics are chemically modified to reduce degradation. Replacing PO with PS linkages in the ASO backbone significantly reduces nuclease activity, and most 2′ ribose modifications confer additional nuclease-resistance. ASOs acting *via* a RNase H mechanism require a design with a central DNA "gap" and are therefore only 2′ ribose-modified in the wings (Chapter 2 and Section 20.2.1.2). Such chemical modifications to the 5′ and 3′ ends result in blunted exonuclease metabolism, but allow cleavage in the deoxy gap region by endonucleases. This is followed by exonuclease degradation of the unprotected ends.[87]

In contrast, fully 2′ribose modified non-gapmer ASOs are degraded exclusively by exonucleases, mostly (but not exclusively) from the 3′-end.[88] ASOs with neutral PMO backbone resist nuclease-mediated degradation and are excreted unchanged.[53] The ultimate fate of an internalized PMO has not been established.

The stereochemistry of PS backbone linkages has a major effect on nuclease resistance, with the Sp configuration being more stable than the Rp configuration.[5,42,44] However, the Sp isomer has lower hybridization affinity, leading to reduced RNase H activity,[43,45] and a stereo-random mixture has been described to provide the best balance of these properties for ASO gapmers.[45]

Unformulated and unmodified siRNA is rapidly degraded in plasma and tissues,[46–49] and many of the siRNAs currently in clinical trials have some degree of modification at the ends to limit degradation by nucleases. For example, introduction of PS linkages at the 5′-end of GalNAc-conjugated siRNA[40,66] or an increase in the proportion of the more nuclease-resistant ribose modification 2′OMe over 2′F[65] resulted in significantly improved resistance to exonuclease degradation plasma and tissues.[66]

ONDs are not substrates for phase I (Cytochrome P450) or phase II metabolizing enzymes (glucuronyl and sulfotransferases) or drug transporters [such as P-glycoprotein (PgP)], nor do they inhibit or induce these enzymes that are critical for small-molecule metabolism, uptake and excretion.[58,62,89] Hence, drug–drug interaction potential of ONDs is negligible as confirmed by clinical experience to date.[90–93]

20.3.1.5 Excretion

ONDs are primarily excreted *via* urine. Mass-balance studies indicate that for a typical 2′MOE gapmer, approximately 80% of the total dose is excreted into the urine after 90 days, whereas only a few percent is found in feces and the remaining fraction can be found in the carcass.[5] The renal excretion rate is inversely proportional to plasma protein binding, with lower protein binding resulting in higher degree of glomerular filtration and excretion.[5,88] Plasma protein binding can be saturated at high doses in toxicity studies (particularly in mice) resulting in an increased excretion of the parent ASO in urine.[59]

As previously described, ONDs with lower protein binding, *e.g.* neutral backbone PMOs or double stranded siRNA, show higher renal clearance and reduced cellular uptake and tissue distribution. As a consequence, these ONDs require relatively high doses or delivery vehicles, such as LNP formulations, or conjugation to targeting ligands, such as GalNAc, in order to reach therapeutic efficacious concentrations in the clinic.[16]

20.3.2 Delivery Strategies

Most therapeutic ONDs are administered *via* parenteral administration to reach adequate systemic exposures. However, reaching effective local concentrations in a broader range of tissues and cell types has been an ongoing challenge from the early days of these platforms.[14,16,51,55,94,95] Moreover, unstable ONDs, such as siRNA, need protection against nuclease degradation to facilitate cellular uptake. Successful strategies to overcome these hurdles have included the use of local administration, advanced formulations (LNP) or conjugation for targeted delivery (*e.g.* GalNAc). Specific considerations for DMPK and safety are briefly discussed below.

20.3.2.1 Local Delivery

To overcome the poor uptake of oligonucleotides in some tissues, direct administration to target organs has been proven successful to reach clinically relevant tissue levels. Local application, for example intravitreal,[96–99] inhalation[100,101] or intrathecal/intracerebroventricular,[102–104] resulted in extensive local distribution and accumulation. The ease of rapidly achieving therapeutically relevant concentrations after local administration combined with slow clearance provides the advantage of a low dosing frequency. The low doses used in local applications combined with slow clearance to the circulation minimizes systemic exposure for many local administration routes.[98,100,105] Side effects, if any, are therefore limited to locally administered tissues, with low risk of systemic toxicity. Notably, systemic exposure following direct CNS application is more pronounced than with other local application routes, due to the turnover of CSF resulting in more OND leaking out of the CNS into the systemic circulation.

Clinical validation of local administration has been achieved for ocular injection (fomivirsen) and, more recently, for CNS administration (nusinersen). Fueled by the latter success, the exploration of therapeutic opportunities of ONDs in other CNS indications is currently an active area of research and development.[102]

20.3.2.2 Advanced Formulation Approaches

OND designs with nuclease-resistant chemical modifications normally do not require protective formulations. However, compared with single-stranded ASOs that are highly protein bound, effective delivery of siRNA requires either receptor-mediated transport into the cell or a lipid formulation. The most common approaches in clinical practice are conjugation to GalNAc (Section 20.3.2.3) or encapsulation in lipid nanoparticles (LNP), often containing cationic lipids.[51] LNP formulations improve potency of both siRNA and ASO gapmers,[106] but need to be administered intravenously together with steroids and anti-histamines to suppress hypersensitivity immune responses elicited by the LNP particles[107] and are therefore normally avoided unless required. An OSWG subcommittee has published a white paper with recommendations and points to consider for formulated ONDs.[76]

20.3.2.3 Targeted Delivery

Successful productive uptake into target cells is recognized as a key hurdle for full clinical optimization and maturation of ONDs. One way to overcome this is to conjugate the OND or OND carrier system to various ligands.[51] Early examples included conjugation to lipids, such as cholesterol and alpha-tocopherol, with improved activity in animal studies. Rather than binding to a specific receptor on the cell surface, the improved pharmacodynamics (PD) were attributed to the altered physicochemical properties of the conjugate, with enhanced hydrophobicity and association with serum lipoproteins facilitating delivery of the polar ASO across the plasma membrane.[16,51]

A more specific approach is conjugation to a ligand that specifically binds to an internalizing receptor, allowing the OND to "hitchhike" into the cell. The most successful example is the conjugation of siRNA and ASO gapmers to GalNAc. This carbohydrate binds to the ASGR, which is a high-capacity, rapidly recycling cell-surface receptor that is almost exclusively expressed at high density on hepatocytes.[51,108] The efficient liver targeting of GalNAc conjugates has enabled siRNA delivery to move away from complex LNP formulations in favor of simple SC injections in saline.[40] This approach has also significantly reduced doses for single-stranded ASOs,[41] illustrated in a clinical study demonstrating a 30-fold potency improvement of GalNAc-conjugated compared with non-conjugated gapmers of the same sequence.[109] The ASGR-mediated uptake, despite its high capacity, can become saturated. This was shown by the less than dose-proportional liver uptake at higher doses of GalNAc-conjugated ASO gapmers in NHPs.[62] Additional support comes from lower liver exposure, despite higher plasma C_{max}, with IV- compared to SC-administered GalNAc-conjugated siRNA.[66]

The chemistry and design of the OND part is generally unchanged in conjugates. After tissue uptake, most of the understanding of what affects, for example, intrinsic potency, tissue half-life and metabolism of the oligonucleotide moiety, also translates to conjugated molecules. The overall liver exposure is only slightly increased with GalNAc conjugates. Instead, the biggest improvement is the larger fraction of total liver exposure that is taken up by hepatocytes *versus* non-parenchymal cells.[41] It is important to note that the main benefit of these targeted delivery examples is to increase delivery to the target cell, leading to significantly lower dose levels required for efficacy. This contrasts with many other targeting approaches, such as antibody–drug conjugates (ADCs), which are frequently used in oncology and mainly aim to avoid exposure of the (often cytotoxic) cargo in other tissues than the tumor.[110] In fact, uptake of ASO gapmers is observed in other tissues, such as the kidney, also when conjugated to GalNAc,[62,111] most probably *via* the PS backbone-dependent mechanism(s). However, since the doses required for hepatocyte pharmacology are 10-fold to 30-fold lower for GalNAc conjugates, the net uptake and exposure in cells other than hepatocytes becomes proportionally much lower.

Interestingly, many of the early experiments that evaluate different conjugates utilized cargo that was only partially modified. When comparing fully modified siRNA conjugated to cholesterol, GalNAc or docosahexanoic acid (DHA), Hassler and colleagues achieved significantly improved uptake and target gene knockdown in several tissues compared with partially modified siRNA cargo of the same sequence.[112] It has also been demonstrated that GalNAc binding affinity to the receptor ASGR was significantly affected by the chemistry of the conjugated ASO.[64] Hence, not only the choice of targeting ligand, but also the chemistry of the cargo affects the overall properties of conjugated ONDs.

Targeted delivery is directed at changing the relative tissue distribution to favor productive uptake in target cells, and this could be associated with a novel pattern of toxicity when cell types are exposed to concentrations and activity at levels much higher than it has previously been possible to achieve. Also, distribution, expression levels, capacity and even function of the internalization receptor could differ between species and between states of health and disease. These factors may introduce more variability in uptake than previously observed for un-conjugated ONDs, so careful characterization of pharmacodynamics, DMPK and toxicity behavior is warranted for each new targeting approach.

The success of the GalNAc targeting approach to the liver has stimulated the search for other "GalNAc-like" ligand–receptor pairs in therapeutic tissues of interest beyond the liver. Many other approaches have been described,[51] but their clinical utility is yet to be demonstrated.

20.4 Class Profile of Toxicity

Based on early experience with PS and 2′MOE ASOs, a common profile of 'class toxicities' was identified, which highly correlated with the PK properties of these large, hydrophilic poly-anionic molecules.[2,3] Toxicities were largely considered to be independent of antisense hybridization and more related to tissue accumulation, pro-inflammatory mechanisms or aptameric effects.[2,113] Chemical modifications modulated these toxicities to some extent, probably *via* changing the degree of protein binding.[2,4,114]

In the last decade, the rapidly expanding experience beyond PS and 2′MOE ASOs has provided more nuance to the concept of class toxicity. The significant efforts using sophisticated bioinformatic analytics on large safety data sets of defined sequences and chemical design, has improved our mechanistic understanding.[115–117] We now see a more composite toxicity profile where the class fingerprint of potential target organs has not changed from earlier days, but the manifestations and underlying mechanisms causing toxicity might be different.

Apart from sporadic cases of thrombocytopenia recently being reported[118] (See Section 20.6.2.2), no other "new" toxicities or target organs of clinical relevance have been added to the list of potential liabilities that can be observed with ONDs. Thus, even with today's increased knowledge, we can still recognize a consistent pattern of potential safety liabilities related to ONDs as a modality. However, some of the class toxicities are primarily observed for a given chemistry (*e.g.* PS backbone, 2′OMe), a specific OND mechanism (*e.g.* ASO gapmer or siRNA RNA degradation) or sequence (*e.g.* unmethylated CpG motifs). The main features of these OND toxicities are summarized in Figure 20.1 and Table 20.2 and are discussed in the following sections.

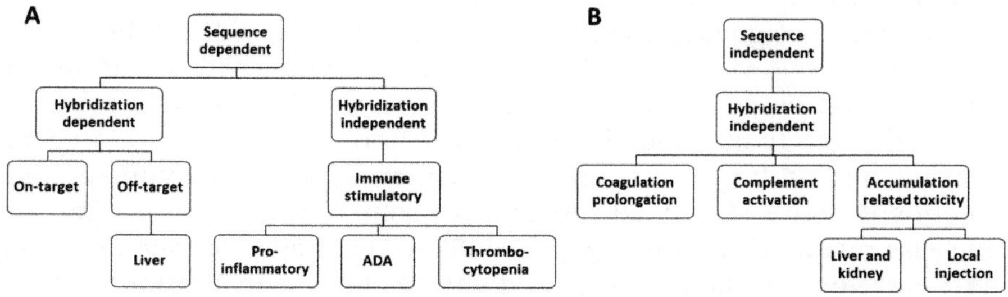

Figure 20.1 Schematic overview of the most common OND-associated toxicities and the effects of sequence and Watson–Crick hybridization. Some of these toxicities have only been confirmed in animal studies. (A) Toxicities mainly dependent on OND sequence. (B) Toxicities mainly independent of OND sequence. ADA: Anti-drug antibody.

Table 20.2 Summary and main features of the most common toxicities observed with ONDs.

Watson–Crick hybridization	Sequence-dependent	Toxicity	Main organ or cell type affected	OND class	Discussed in section
Dependent	Yes	Exaggerated pharmacology	Many possible	Gapmer ASO, siRNA, miR mimics, anti-miR	20.5.1
	Yes	Off-target	Liver, possibly kidney	Gapmer ASO, siRNA, miR mimics, anti-miR	20.5.2, 20.6.3.1, 20.6.3.2
Independent	Yes	Pro-inflammatory	Immune system, injection site	PS-backbone ASOs, siRNA	20.6.2.1
	Yes	Thrombocytopenia	Platelets	PS-backbone ASOs	20.6.2.2
	No	Coagulation prolongation	Blood	PS-backbone ASOs	20.6.1.1
	No	Complement activation	Cardiovascular, kidneys	PS-backbone ASOs	20.6.1.2
	No	Accumulation-related	Liver, kidney	PS-backbone ASOs	20.6.3

20.5 Hybridization-dependent Toxicities

As with any pharmaceutical modality, concerns with on-target safety (exaggerated pharmacology) and off-target interactions need to be considered. For ONDs, this translates to Watson–Crick hybridization to the target RNA and to non-target RNA species. By virtue of their RNA-modulating mechanism, ASO gapmers and siRNAs are more liable to hybridization-dependent toxicities than fully 2′-ribose modified splice-modulating ASOs.

If carefully designed and evaluated, ASOs usually provide excellent selectivity for a specific sequence within the intended target RNA, minimizing the risk of off-target hybridization-dependent toxicities. Preclinical activities carried out to understand and manage potential hybridization-dependent on- and off-target risks should consider both parent compound and potentially active shortened metabolites sometimes observed after exonuclease metabolism of non-gapmer ASOs. These analyses should ideally start early, during the discovery and design stage, with continued awareness during the development process to de-risk for hybridization dependent toxicities.

20.5.1 On-target Safety and Exaggerated Pharmacology

As with any drug platform, exaggerated on-target pharmacology of ONDs, by excessive or prolonged activity in the intended or non-intended organ, could lead to toxicity. De-risking on-target toxicities for any drug modality starts with a thorough understanding of the biology of the pharmacological target and the patient population to identify and assess likelihood and potential effects of on-target toxicities. Valuable information for this assessment includes expression levels and tissue distribution of the target RNA, phenotypes associated with known human mutations and polymorphisms and genetically modified animal models, as well as data from other compounds modulating the same target. The next steps are to confirm experimentally or refute the risks of concern, establish safety margins and, if needed, identify biomarkers to monitor for the specific toxicity in preclinical and clinical studies.

Published examples of OND-mediated on-target safety findings include mice treated with high SC doses of the antithrombin siRNA, fitusiran, which showed findings suggestive of thrombosis and disseminated intravascular coagulation, consistent with the intended pro-coagulant effect.[119] Other examples include the adverse kidney effects that are proposed to be caused by knockdown of the renal target sodium–glucose co-transporter-2 (SGLT2)[74] and the liver toxicity seen in patients treated with mipomersen due to exaggerated knockdown of the primary target apolipoprotein B100 (ApoB100).[120]

Compared with small-molecule and antibody drugs, some specific considerations need to be taken into account when assessing potential on-target safety for OND therapeutics.[121] Although certain cells in the liver and kidney should always be considered for potential on-target safety effects, there are normally only a limited number of other tissues and cell types that will experience sufficient productive uptake for significant on-target toxicity to occur (see Sections 20.3.1.2 and 20.3.1.3). Another aspect is that target sequences and consequences of target RNA modulation may not be conserved across species. The likelihood for this depends on the OND mechanism and design. Sequences for siRNA and ASO gapmers are often identified by screening the full transcript to find the most potent site for hybridization. For ASO gapmers and non-gapmers, such sites can often be intronic sequences in the pre-mRNA.[122] Introns are generally less conserved between species, often leading to limited activity of the human candidate in animal species, including NHPs. In contrast, the critical seed sequences in miRNAs are highly conserved between species, so binding of the human candidate is often observed also in animal species.

In cases of insufficient cross-species potency of the clinical candidate, a surrogate OND that is active in one of the preclinical species (normally a rodent) may be used to support on-target risk assessment. For monoclonal antibodies, species-specific surrogates may be less relevant, due to difficulties in reproducing sufficiently similar epitope binding to the human candidate. In contrast, species-specific surrogate ONDs of the same chemistry and design are relevant, since they have the same degree of productive uptake, distribution and pharmacological kinetics as the human candidate. Importantly, any surrogate OND to be used to assess potential exaggerated on-target (or off-target) pharmacology, should be selected with stringency to the human candidate to minimize the variability in 'class effect'-related changes that could complicate the final interpretation.

For splice-correcting ONDs, the intended pharmacology is often patient (mutation)-specific and addressing potential consequences of exaggerated pharmacology in preclinical species would not be meaningful. Hence, this obviates the need for inclusion of a surrogate OND in the preclinical development program. For certain targets, the management of potential on-target toxicity in clinical settings may be challenging, for example due to the long tissue half-life or the long duration of action inherent to the mechanism (*e.g.* siRNA). In those cases, development of antidotes to reverse the pharmacological action of ASO gapmers[123] and siRNA[124] may be appropriate.

Risk assessment of exaggerated pharmacology of miRNA modulation is challenging. In contrast to ASO gapmers and siRNA that are intended to degrade a single transcript, miRNAs have subtle effects on the expression of several targets in multiple signaling pathways, affecting expression of different sets of genes depending on species and cell type.[125] Broad-omics analyses with refined statistical analyses should be used to provide insight into the downstream consequences of exaggerated pharmacology of anti-miRs and miR mimics to obtain information on potential on-target toxicities.

20.5.2 Off-target Pharmacology

In earlier years of ASO development, the potential for off-target effects was of little concern for optimized clinical candidates. In fact, the authors are not aware of any reports on clinically meaningful off-target toxicities with optimized gapmers or splice-switching ASOs of 20–25 nucleobases in length. Off-target hybridization effects were regarded to be well understood in terms of the rational *in silico* design process. Moreover, and in contrast to small-molecule therapeutics, the relatively few cell types being amenable to sufficient productive uptake also limit the number of organs and cell types where hybridization dependent off-target effects can occur.

However, with increased mechanistic understanding of siRNA liver toxicity and use of high-affinity chemistries (LNA and cEt) resulting in potent ASOs of shorter lengths (12–16-mers), the appreciation of off-target hybridization and associated toxicities has significantly increased.[116] In fact, exaggerated RNase H-dependent off-target activity has been proposed as a potential mechanism behind the sequence-dependent liver and possibly kidney toxicity of high-affinity ASO gapmers[115,126,127] (see also Section 20.6.3).

siRNA-induced liver toxicity caused by RISC-dependent off-target activity has been described in multiple preclinical publications.[128–132] Using different approaches, Janas and colleagues elegantly demonstrated that RISC loading and the sequence and affinity of the seed region (*i.e.* nucleotides 2–8) were key determinants of off-target activity causing liver toxicity.[132] This could be reduced by introduction of lower-affinity chemistry in the seed region. Other chemical modifications such as abasic spacers in the so-called pivot region[133] or placing unnatural triazolyl nucleotide at position 1 [134] have been reported to reduce off-target effects for siRNA.

Splice-modulating ASOs do not act *via* degradation of RNA transcripts, and are therefore devoid of the risks associated with the (unintended) degradation of RNA transcripts. Splice modulating ASOs need to bind one of the few critical sequences in the target pre-mRNA to produce the desired effect (*i.e.* steric blocking of splicing factors). Any splice modulating ASOs non-specifically bound to unintended (pre-)mRNAs have been shown to be removed by helicase activity of ribosomes,[135] thereby minimizing off-target pharmacodynamic or toxicological risks.

Importantly, unintended reduction of non-target transcripts is not unique to RNA-targeting therapeutics. Analysis of publicly available transcriptomics data from 315 FDA approved small molecules and 25 ASO gapmers revealed that the median number of affected transcripts was around 60 for both modalities.[116]

In 2012, the OSWG subcommittee on off-target analysis published recommendations on how to assess and manage potential hybridization-dependent off-target activities.[136] Although progress in the field has led to increased insights and understanding, with an OSWG update in preparation, the basic steps in the risk assessment of hybridization-dependent off-target effects described in the 2012 recommendations remain valid.[136]

Several different approaches can be used for the *in silico* analysis. Character-based searches like BLAST and FASTA are relatively rapid but are less precise than more demanding search algorithms based on hybridization energy.[116] For ONDs such as ASO gapmers or splice switching ASOs that (mainly) exert their activity in the nucleus, off-target sequence analysis needs to cover also introns of the unspliced pre-mRNA.[122] Defining the stringency in search criteria regarding tolerated mismatches and gaps is also critical and should ideally be informed by experience from the given OND design and chemical modifications used.[116] There is currently no *in silico* model that can reliably predict whether a given gapmer or siRNA sequence, with or without mismatches, will lead to RNase H- or RISC-mediated degradation or not, so experimental confirmation is important. In fact, many gapmers with perfect sequence match can be totally inactive, possibly due to inaccessibility of the target site due to protein binding or secondary structures of the transcript.[116] Moreover, with the high-affinity LNA or cEt modifications, minor changes in gapmer design and number or positions of modified nucleotides in the wings can have a dramatic effect on knockdown potency of both on- and off-target transcripts.[38]

Experimental verification of predicted off-targets for clinical candidates and comparison to on-target potency must be done in human cell lines that express the suspected transcripts. Off-targets may be verified by quantitative RT-PCR that is sensitive enough to detect changes in less abundant transcripts. Different transcriptomics approaches can also be used, but with the caveat that potential reductions in transcripts with low levels of expression may go undetected due to lower sensitivity. When using transcriptomics, at least two different gapmer sequences with similar on-target potency but hitting different sites of the target transcript should be assessed in parallel to avoid classifying changes secondary to intended knockdown of the primary target transcript as off-target activities.[116]

Any confirmed off-targets with insufficient margins should be followed up by theoretical and experimental quantitative risk assessment, following the same principles as assessing risks of exaggerated on-target effects (see Section 20.5.1). Testing the clinical candidate in human cells can be valuable if the appropriate readouts can be assessed *in vitro*. However, since potent rodent-specific ONDs can be designed in a rational way, studies informing off-target hybridization risk assessment are preferably carried out *in vivo* using rodent-active surrogates for the off-target.[116]

The examples above describe some of the challenges for off-target analyses and risk assessment for RNase H-dependent gapmers and siRNA aimed at targeting one single transcript. For oligonucleotide approaches aimed at simultaneously affecting multiple transcripts, such as anti-miRNA and miR mimics, the off-target detection and subsequent risk assessment becomes even more complex. Interestingly, when downstream effects of two anti-miRs targeting miR-122 of 8 or 15 nucleobases in length were compared using transcriptomics, there were surprisingly few differences in the number of affected transcripts, despite the anticipated lower specificity of the short "tinymer".[137]

20.6 Hybridization-independent Toxicities

The majority of the OND-induced toxicities are not caused by Watson–Crick base pairing to RNA, but are instead due to accumulation, pro-inflammatory potential and/or protein binding. Although sequence can have an effect, these toxicities are largely driven by the OND chemistry and design.

Hybridization-independent effects make up a highly consistent and predictive class profile of toxicities, which has been summarized comprehensively by others.[2,3,113] Some of the common observations are well understood in terms of mechanisms and species translatability. This includes coagulation prolongation (Section 20.6.1.1) and complement activation effects (Section 20.6.1.2) related to binding to specific proteins. The specific molecular mechanisms of other class effects, however, are less well understood, such as the pro-inflammatory effects that are characterized by overt species differences, or severe thrombocytopenia. It is today clear that within each of the class toxicities more than one mechanism appears to be involved. This is well illustrated for liver toxicity of gapmer ASOs, acting mainly *via* high tissue accumulation for 2′MOE gapmers, but *via* off-target mechanisms for those with higher affinity modifications such as LNA and cEt (Section 20.6.3.1). This underscores the need to provide more nuance to the OND class profile concept for novel OND classes with different properties.

20.6.1 Effects Related to Transient Protein Binding

20.6.1.1 Inhibition of the Coagulation Cascade

Prolongation of the clotting time is a well-characterized effect of PS ASOs[138–140] directly correlated to the PS content and extrapolating well across all species and different sequences.[2,138,141–143] The effect results from transient binding and inhibition of the intrinsic tenase complex, a key activator of the intrinsic coagulation cascade, resulting in prolongation of the coagulation time of the intrinsic pathway.[139,144] Prolongation of coagulation is a highly predictable, plasma peak serum concentration (C_{max})-related class toxicity,[4] but has little clinical significance as it is usually not associated with increased bleeding risk or other symptoms of coagulation disorders at clinically relevant exposure levels. It can be well managed by dosing regimens avoiding high plasma levels (*e.g.* SC administration or longer IV infusions instead of bolus IV), or by reducing the degree of protein binding through shorter oligonucleotide length, chemical ribose and/or backbone modifications. This effect is clearly linked to the overall PS backbone content and is therefore not observed with siRNA or ONDs with neutral backbones, *e.g.* morpholinos or PNA.

20.6.1.2 Complement System Activation

Another well-characterized protein-binding feature of PS ASOs is activation of the alternative complement pathway,[2,145,146] a complex system of proteolytic cascades contributing to innate immunity. NHPs are particularly sensitive to this effect,[147] but at high bolus IV doses it may also be seen in other species[142] including man when given as prolonged infusion.[141,148,149] Aggressive IV dose regimens in NHPs may lead to pronounced complement-related clinical symptoms and/or mortality,[114] but lower-dose IV infusions or SC administration are usually without acute symptoms. Complement activation is generally not seen in clinical settings with SC injection at doses up to 1000 mg kg^{-1}, even when circulating drug levels were similar to or exceeded those associated with complement activation in monkeys.[6,150] Thus, although complement activation is a dose-dependent toxicity in NHPs, it appears to have little clinical relevance.[151]

The mechanism of complement activation for PS backbone ASOs is well established. They inhibit complement factor H (CFH),[152,153] an endogenous inhibitor of the alternative pathway of complement activation, resulting in a transient increase in complement split factors (Bb, C3a) at the cost of complement C3 protein. The particular sensitivity of NHPs does not seem to lie in a higher binding affinity to CFH, but rather in the lower intrinsic inhibitory activity of monkey CFH compared with human CFH.[147,150] Whether binding to other complement factors, such as identified for C3,[147] also plays a role in OND-induced complement activation and potential species differences awaits further investigations.

Decreased circulating levels of complement factors may have detrimental effects on the overall function of the complement system and its role in innate immune surveillance, including clearance of immune complexes. Chronic complement activation following repeat dosing in NHPs led to increased basal levels of split products and sustained complement C3 depletion, with secondary increased inflammation, particularly in the vascular beds of organs like heart, liver and kidneys.[154] As a consequence, vasculitis is a commonly encountered complement-related pathology in NHP studies.[154] However, the clinical experience to date has not shown evidence of sustained complement activation or vascular inflammation in humans,[6] confirming that complement activation and downstream pathologies is a NHP-specific concern with limited relevance to humans.

In vitro assays using whole blood or serum obtained from monkeys and humans that measure specific split products of the alternative pathway [*e.g.* activated complement factor B (Bb)] are limited in predicting *in vivo* concentration–effect relationships, but can be used to screen-out obvious bad actors and contribute to the understanding of structure–activity relationships. The overall number of PS linkages (and thus PS OND length[2]) and the influence on protein-binding by the 2′-ribose modifications significantly affect complement activation. Although sequence seems less critical,[2,147,150] there are examples of (unexpectedly) potent complement activation within a sequence series of ONDs (Den Besten, personal communication).

Complement activation is generally not observed with neutral backbone PMO or unformulated double-stranded ONDs such as siRNA and miRNA-mimics containing only a limited number of backbone PS linkages. However, lipid-formulated ONDs can induce complement activation in several species, including humans, although the pattern and probably also the mechanism is different from that observed for unformulated PS-backbone ASOs.[76]

20.6.2 Immune-mediated Effects

20.6.2.1 *Pro-inflammatory Effects*

Immune-stimulatory effects have long been a prominent feature of oligonucleotides, where responses may vary widely between species and depend on oligonucleotide design and sequence, as well as chemical modifications of the base, ribose or backbone. The immunomodulatory potential can be used deliberately to design nucleotide-based immunotherapies and vaccine adjuvants, often harboring so-called CpG motifs,[101,155–157] but in most other cases these effects are unwanted.

20.6.2.1.1 Manifestations of Immune Stimulation. The manifestations of immune stimulation by DNA- and RNA-based ONDs have been described in detail elsewhere[2,113,158–168] and are summarized in Figure 20.2. Rodents are particularly sensitive to the immunostimulatory effects of ONDs, with a typical T-helper (Th)-1 like response illustrated by a dose-dependent lymphoid hyperplasia and enlargement of spleen and lymph nodes. Lymphohistiocytic cell infiltration is often seen in multiple tissues, and when it occurs in liver it is sometimes associated with mild to moderate, non-progressive elevations in liver enzymes (See also Section 20.6.3.1). Variable increases can be detected in cytokines [*e.g.* interleukin (IL)-1, IL-6, IL-10, IL-12, IP-10, interferon gamma (IFNγ)] and/or chemokines [particularly monocyte chemoattractant protein-1 (MCP-1)], as well as increased serum immunoglobulin levels related to polyclonal B cell stimulation.[113,166] Upon chronic exposure (generally more

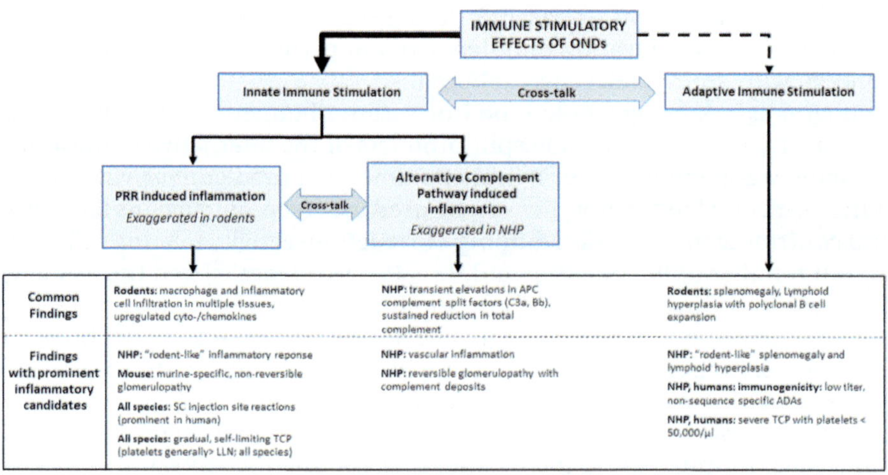

Figure 20.2 Immune stimulatory effects of ONDs – primary and secondary effects and pathologies related to immune stimulation (PRR, pattern recognition receptor; NHP, non-human primate, ADA, anti-drug antibodies; TCP, thrombocytopenia; APC, alternative pathway of complement; LLN, lower limit of normal).

than 3 months), unresolved inflammatory processes associated with those ASOs with stronger inflammatory effects may become associated with cell degeneration and ultimately fibrosis in affected tissues. A rather extreme example is the irreversible glomerulonephritis in mice, reported for some ONDs[113] (see also Section 20.6.3.2). However, for 2′MOE ASOs designed to minimize the inflammatory effects, there is typically no progression of the inflammation to degenerative or fibrotic changes.[169] Compared with rodents, inflammatory responses in NHPs,[2,170,171] minipig[142] and man are usually milder and less pronounced. However, prominent inflammatory responses, including vasculitis, have occasionally be seen in NHPs,[113,154,172] particularly with sequences that also show pronounced complement activation (see Section 20.6.1.2). The underlying mechanisms for the rodent sensitivity are not fully understood, but are thought to include differences in expression patterns of pattern recognition receptors (PRR) of the innate immune system, ligand specificity and signaling pathways.[173,174]

Although PS DNA ONDs and, to a lesser extent, 2′MOE gapmers or non-gapmer ASOs at high doses, have been associated with injection-site reactions and systemic flu-like symptoms of fever, chills and fatigue,[148,175] other pro-inflammatory effects observed in rodents are not prominent in man. Tolerability was improved when administration regimens were adapted to avoid high circulating drug levels[148] or by lower clinical doses of more potent ONDs. Stricter screening criteria to minimize the inflammatory effects have also contributed to improving the overall tolerability.[176] Local injection site reactions following SC administration have been described as erythema, induration, itching, pain and sometimes ulceration or necrosis.[148,175,177] These local responses were dose-limiting with early candidates (PS backbone DNA), but better tolerated with the 2′MOE gapmers. However, more recent reports on long-term clinical experience with a 2′OMe-modified ASO (drisapersen) revealed how the initial mild responses of erythema and hyperpigmentation developed with chronic dosing into a severe presentation including sclerosis and lipo-atrophy, with the appearance of sclerodermic scars or morphea-like skin changes that were poorly reversible.[178,179] The pathophysiology underlying these local skin reactions remains unclear, but probably involves the persistent high local concentration of the pro-inflammatory OND.[148,172] Inflammatory SC injection site reactions in animal studies appear less severe and are mostly reversible, indicating that SC tolerability is lower in man, possibly due to species differences in SC skin composition, such as a higher adipocyte content in humans than in laboratory animal species.

ONDs administered in lipid formulations have been shown to induce inflammatory responses. Humans seem to be more sensitive to these effects than both rodents and NHPs.[76] TKM-080301 is an LNP-formulated siRNA that triggered flu-like infusion-related reactions which were associated with cytokine increases and dyspnea at the top dose.[180] Despite pretreatment with anti-inflammatory and anti-histamine, intravenous administration of the LNP-formulated siRNAs patisiran and ALN-VSP triggered some cases of infusion-related reactions in clinical studies, including erythema, tachycardia and breathing difficulties and activation of complement. This was associated with cytokine increases.[181,182]

20.6.2.1.2 Pro-inflammatory Mechanisms. The pro-inflammatory properties of non-formulated ONDs stem from the activation of the innate immune system. A large number of pattern-recognition receptors (PRR) have evolved to recognize viral and bacterial RNA and DNA. The best characterized PRRs involved in the recognition of synthetic ONDs are the membrane-bound Toll-like receptors (TLRs) (see Chapter 14), particularly the endosomal TLR3, TLR7 and TLR8, recognizing different RNA variants,[183,184] and TLR9, that responds to unmethylated CpG motifs in DNA.[155,185] Other PRRs residing in the cytoplasm that also respond to synthetic RNAs, include retinoic acid-inducible gene I (RIG-I),[186,187] melanoma differentiation-associated protein 5 (MDA-5),[188,189] Protein kinase RNA-activated (PKR)[190,191] and a growing number of other receptors.[167,192,193] Binding to these various PRRs and activation of downstream signaling pathways discriminate between oligonucleotide design, chemistry and sequence, so may not be relevant for all ONDs. Table 20.3 summarizes the main PRRs known to date that respond to the different ONDs.

A key parameter for proinflammatory properties of ONDs is nucleotide sequence. A well-known example is DNA CpG motifs, which strongly activate

Table 20.3 Most common pattern recognition receptors (PRRs) involved in the immune stimulation by ONDs. Modified from Whitehead *et al.* (2011),[194] Meng and Lu (2017),[184] Lässig and Hopfner (2017)[195] and Chan and Gack (2016).[196]

PRR	Subcellular location	Ligand	Stimulatory motif	Dampening modification	Synthetic ligand
TLR3	Cell surface Endosome	dsRNA siRNA	Not defined	Not defined	Poly(I:C) Poly U
TLR7	Endosome	ssRNA	U-repeats	2′ribose modifications	Imidazoquinolines and guanosine analogues
TLR8	Endosome	siRNA	5′UGUGU-3′ 5′GUCCU-UGAA-3′ GU or AU rich		
TLR9	Endosome	ssDNA	XX–CpG–YY[a]	5-Methylated cytosine	CpG ONDs
RIG-I	Cytoplasm	dsRNA (ssRNA)	5′di- and tri-phosphates Duplex structures	2′*O*-methylation of first nucleotide	Poly(I:C)
MDA5	Cytoplasm	Long dsRNA >2000 bp	Not defined	Not defined	Poly(I:C)
PKR	Cytoplasm	dsRNA	Not defined	Not defined	Not defined

[a]Optimal CpG flanking sequences for activating TLR9 vary among species.[165]

TLR9,[185,197] particularly in the presence of optimal flanking sequences that are different between rodents and NHPs or man.[165,198] This has spurred significant efforts to identify potent, immune-stimulatory sequences for development in therapeutic indications where immune stimulation is beneficial, such as cancer, infectious diseases and asthma or allergy.[155] Examples of other proinflammatory sequences include RNA rich in GU or AU motifs, which trigger TLR 7 and TLR8 responses[199,200] and U-rich sequences in siRNA.[159,161,201–203] The cytoplasmic PRRs, like RIG-I and PKR, are not activated by a specific nucleobase sequence but are instead triggered by structural features typical for viral RNA, such as the presence of uncapped 5′ triphosphate in ssRNA or dsRNA, or blunt-ends in dsRNA.[187]

For ASO gapmers and siRNA, multiple alternative sequences can be identified along the pre-mRNA and mRNA to allow selection of a candidate with a good balance of potency and low toxicity, including minimal proinflammatory effects. For OND classes that are much more restricted in the choice of sequence, such as splice-switching or anti-miR ASOs, avoiding proinflammatory sequence motifs may be more challenging.

20.6.2.1.3 Mitigating Proinflammatory Effects. Chemical modifications can modify the immune-stimulatory potential of ONDs of a given sequence. PS modification of the backbone has long been known to increase the immune-stimulatory properties of ONDs,[204–206] but some phosphodiester (PO) backbone ONDs have been reported to be more inflammatory than the same sequence on a PS backbone.[204,207,208] Interestingly, the PS backbone chirality has also been reported to influence the inflammatory response, with a stronger impact of the *R* stereoisomer.[155] The neutral backbone in PMOs does not evoke an immune response, which contributes to their good tolerability.[53] Methylation and other post-transcriptional modifications are more abundant in mammalian DNA and RNA than in pathogenic nucleic acids and help the immune system to distinguish between self and non-self.[185,209] 5′-Methylation of cytosine is frequently used to suppress the immune stimulatory effect of CpG DNA sequences.[165,210] 2′OMe modification of ssRNA or siRNA sequences inhibit immune stimulation *via* suppressing recognition by TLR7 or TLR8[211,212] or RIG-I,[187] whereby modifications even at a single nucleotide can significantly reduce the cytokine upregulation of inflammatory siRNA motifs.[161,208] Other 2′-ribose modifications (2′F, 2′H, 2′MOE, LNA) have also been described as reducing proinflammatory effects.[203,210]

For most disease indications, unintended immunostimulation of OND sequence or delivery vehicle remains a liability that warrants a careful design strategy and screening process. The increased awareness and understanding of immune-stimulatory sequence motifs and immunosuppressive chemical modifications has helped the design of safer ASO and siRNA sequences. Moreover, the increased pharmacological potency, leading to lower doses and injection volumes, has significantly reduced the proinflammatory risk. Nevertheless, dedicated screening remains warranted, and in early drug development this is often done *in vitro* using human-derived cells or whole blood.[213] Due to the complexity of the mammalian immune system, *in vitro* to *in vivo* translation is challenging, and mitigation of the pro-inflammatory risk often requires testing of promising clinical candidates in animal studies of four to six weeks duration, before progressing to investigational new drug (IND) enabling studies.

20.6.2.1.4 Immunogenicity – Anti-drug Antibodies (ADAs). PS backbone ONDs were long considered to be devoid of antigenic properties based on the available literature that DNA and RNA are poor antigens and the difficulty of generating an acute antibody response without using the stimulatory carrier protein KLH (keyhole limpet hemocyanin). However, the growing clinical experience with 2′-ribose-modified ONDs has revealed that antidrug-antibodies (ADAs) can be detected in NHPs and patients upon longer-term dosing. Primary publications in this area are still pending, but unpublished data show that ADAs emerge with a relatively late onset, between three and six months of repeated administration.[214] ADA's have been reported in 65% of the patients receiving mipomersen[215] and 30% of patients receiving drisapersen, generally with low median titers.[172] The occurrence of ADA's correlated with significantly higher drug trough levels in plasma, indicating an ADA-induced increase in plasma half-life of the ASOs. However, there is no indication of adversely affected pharmacokinetic [C_{max}, area under the curve (AUC) or tissue concentration], efficacy or safety outcomes in ADA-positive patients. These antibodies are generally of low avidity and directed against the chemically modified portions, such as PS backbone or 2′-modifications and do not bind to native PO DNA backbone[68] and generally do not react with dsDNA, alleviating concerns about an increased autoimmune potential. Since animal studies are poorly predictive for the potential to develop ADAs in humans, this risk should be characterized in patients when ONDs are progressing to larger-scale clinical testing. The mechanism for the ADA response is not clear, but is probably associated with the proinflammatory properties described above. With the trend towards lower (systemic) doses and exposures and infrequent administration regimens, the risk of inducing ADAs is expected to be reduced.

Many lipid formulations contain conjugates of polyethylene glycol (PEG). Development of anti-PEG antibodies is fairly common but has not been reported to constitute a safety concern in the context of lipid-formulated ONDs.[76]

20.6.2.2 Thrombocytopenia

Thrombocytopenia (TCP) is a common side effect of many pharmaceuticals[216,217] and has also been observed with PS oligonucleotides. Dose-dependent reductions in platelet counts have been reported in mice and NHPs following repeated dosing of PS ASOs, but it is not clear whether the mechanism is the same between the species. In rodents this effect is often accompanied by splenomegaly and seems to be associated with the proinflammatory effect of ASOs commonly observed in rodents.

Based on the broad nonclinical NHP and clinical human experience with PS gapmer and non-gapmer ASOs, two distinct phenotypic presentations of platelet reductions are now recognized, each with a different clinical implication. The more common phenotype presents as a mild to moderate, gradual and dose-dependent decline in platelet count, which is non-progressive upon continued treatment.[9] The platelet count in this reproducible profile of platelet reduction normally does not fall below the lower limit of normal (150 000 platelets μl^{-1} in humans). Recent evaluations using large proprietary safety databases on 2′MOE gapmers showed that 40% of the 102 independent sequences induced mild to moderate platelet reductions (*i.e.* >100 000 platelets μl^{-1}) in NHPs[9] *vs.* <20% of 16 independent sequences in clinical trials,[8] indicating sequence-dependency of the effect. Since platelet counts generally remain within the normal range and do not further decline with longer treatment duration and are not associated with any evidence of clotting impairment or increased bleeding risk,[8,178] this mild platelet reduction is considered to be of limited safety concern.

The second phenotype is a low incidence of severe TCP. In the database of 2′MOE gapmer studies in NHPs, sporadic decreases (2–4%) to <50 000 μl^{-1} were observed above a threshold dose of ≥5 mg kg^{-1} per week.[9] These events increase in frequency with higher doses and longer treatment duration and are not seen with all sequences or even between studies using the same sequence and dose levels. Several observations, including recurring platelet count reductions after re-dosing of affected monkeys, indicate an immune-mediated mechanism. Severe TCP in NHPs has also been reported for the splice-switching ASO, drisapersen.[172] The sensitivity to this severe TCP was dependent on the origin of the NHPs. Cynomolgus macaques from Mauritius with a lower genetic diversity,[218] are more sensitive to ASO-induced TCP than monkeys from Southeast Asia. This was clearly demonstrated by ≥50% of Mauritian cynomolgus macaques showing platelet counts <50 000 $\mu 1^{-1}$ in all drisapersen groups dosed >8 mg kg^{-1}, whereas none of 12 cynomolgus macaques of Chinese origin dosed at 12 mg kg^{-1} experienced platelet counts <150 000 (Table 20.4).

Although not observed in the majority of the clinical trials of 2′-MOE ASOs,[8] this type of sporadic and severe TCP has been reported in a few studies. For example, there are cases of TCP reported for early PS DNA ASOs in oncology trials where the doses were relatively high (2 to 6 mg kg^{-1} per day) and patients had prior exposure to multiple chemotherapeutics.[141,219–222] In addition, recent experience in longer duration clinical programs has revealed incidental Grade 3 or 4 TCP with three unrelated ASOs - the 2′MOE gapmers, volanesorsen and inotersen, and the 2′OMe PS ASO drisapersen, which are SC administered at high doses (≥4.3 mg kg^{-1} per week). [118,172,178,223]

Subjects treated chronically with drisapersen had platelet counts generally within the normal range, but after treatment for at least 9 months, some experienced a rapid decline in platelet count over one to two weeks (summarized in Table 20.4). For the 2′MOE gapmers, the time to onset of severe TCP was faster, with severe TCP reported after three to nine months of treatment.[223] The decline in platelet counts to <50 000 μl^{-1} is relatively rapid, generally over the course of some weeks. When treatment is withdrawn, platelet counts recover

Table 20.4 Incidence and time to onset of severe TCP in NHPs treated for 26 weeks and Duchenne muscular dystrophy (DMD) boys treated for up to 188 weeks with drisapersen. Both NHPs and humans were treated *via* SC injection twice weekly in the first two weeks and once weekly thereafter. Mau, Mauritian; Chi, Chinese origin of NHPs. Unpublished NHP data are courtesy of Biomarin.[e]

Species	Dose SC (mg kg^{-1})	Treatment duration in weeks (NHP origin)	Incidence of TCP (%)	Platelet count nadir (×1000 µl^{-1})[b,c]	Time to TCP onset (weeks)[b]	Total (cumulative) dose at first onset (mg kg^{-1})	C_{max} (µg ml^{-1})	AUC (µg h^{-1} ml^{-1})
NHP[a]	0	26 (Mau)	0/8 (0)	—	—	—	—	—
	4	26 (Mau)	0/6 (0)	—	—	—	12	78.9
	8	26 (Mau)	3/6 (50)	5–27	11–18	104	26	158
	12	26 (Mau)	4/9 (44)	6–16	7–15	108	31	219
	24	26 (Mau)	5/9 (56)	3–59	7–17	216	60	534
	12	39 (Chi)	0/12	>150	—	—	24	492
Human	6	Up to 188[d]	8[e]/285 (3)	5–35	43–112	270	5.8	58.4

[a]NHP were of Mauritian or Chinese origin.
[b]Minimum–maximum.
[c]Platelets were monitored once every two weeks in NHP and humans.
[d]A total of 285 patients were treated up to 188 weeks, corresponding to a total of 433 patient-years at 6 mg kg^{-1} and 490 patient-years (all doses, *i.e.* 3–9 mg kg^{-1}).
[e]Grade 3 TCP (<50 000 µl^{-1}) was noted in two out of eight patients and Grade 4 TCP in six out of eight were grade 4 (<25 000 µl^{-1}), with five out of eight reporting spontaneous bleedings and five out of eight positive for antiplatelet antibodies.

to normal values over a similar time course (one to several weeks), although sometimes requiring supportive treatment. Clinical symptoms of clotting impairment, such as ecchymosis, bleeding gums, bruises and petechia were apparent when platelet counts dropped very low (generally <10 000 µl^{-1}), and a single fatality due to intracranial bleeding has been reported in the inotersen program. It should be noted that the cases of severe TCP occurred in patients that received relatively high doses of ASOs (200 mg per week SC drisapersen or 300 mg per week SC volanesorsen and inotersen) and where the disease background may have increased the sensitivity to platelet clearance. In particular, the FCS (Familial Chylomicronemia Syndrome) patient population treated with volanesorsen are reported to have low baseline levels of platelets and a high degree of fluctuation in platelet count.[224] Notably, although TCP has occasionally been reported as a dose-limiting side effect in phase 1/2 clinical trials with formulated siRNA,[182] severe TCP has not appeared to be a clinical risk for this class of ONDs based on the available experience to date.

Without mechanistic understanding, there is a lack of predictive *in vitro* models and it is uncertain if NHP toxicity studies can predict whether a given ASO candidate has the potential to induce severe TCP in humans or not. Until this is clarified and factors involved in patient susceptibility are better understood so patients at risk can be identified earlier, risk for severe TCP should be managed by frequent monitoring of platelet count until sufficient clinical experience is available for the specific candidate compound. The collective experience of TCP in NHPs shows a clear dose-dependency and an apparent threshold dose above which severe TCP may develop.[9] Given the lower doses achieved by increased OND potency and conjugate approaches, future clinical candidates should have reduced risk of developing severe TCP.

The fact that these ASOs showing severe TCP in clinical settings cover two different mechanisms of action (RNase H1 and splice switching) and three different targets indicate that the TCP is not caused by exaggerated pharmacology. ASO-induced TCP is not associated with general bone marrow toxicity, impaired platelet production or presence of antidrug antibodies.[225,226] Several different mechanisms have been proposed for TCP induced by other drug classes, both immune-mediated and non-immune-mediated, and it is possible that more than one mechanism is involved in ASO-induced TCP. Recent investigations have started to provide some understanding of the interaction between ASOs and platelets. Heparin-induced TCP (HIT) is caused by induction of IgG antibodies formed to a complex of heparin binding to platelet factor 4 (PF4), with subsequent clearance of platelets. Binding to PF4 has been demonstrated for unmodified DNA and RNA structures[227] and PS ASOs without ribose modifications.[228] However, this interaction was blunted by introduction of ribose modification.[228] Other findings contradicting a HIT-like mechanism for ASO-induced TCP include non-HIT-like *in vivo* manifestations of severe TCP in patients and monkeys, such as an apparent dose-dependency, lack of thrombosis and the lack of very low platelet nadir counts. Moreover, anti-PF4 immunoglobulin G (IgG) antibodies were not detected in a monkey with severe TCP,[229] but this monkey did have increased levels of anti-platelet IgG antibodies that were independent of the presence of drug.[226] Similarly, the presence of antiplatelet IgG antibodies was also confirmed in five out of the eight patients that developed severe TCP on drisapersen.[172] The collective evidence to date indicates an immune-mediated, PF4-independent mechanism for severe TCP that involves anti-platelet IgG antibodies.

Other studies addressed the direct and indirect effect of PS backbone ONDs on platelet functionality, using an immunostimulatory tool sequence containing a CpG motif,[230] or a series of 5–11 AC-dinucleotide repeats.[228] Both studies demonstrated that PS-modified ONDs bound to the glycoprotein VI receptor (GPVI) on the platelet surface and (directly) activated platelets in a PS (length)- and dose-dependent manner. Similarly to the PF4 binding, this effect was attenuated when ribose modifications were introduced in a gapmer design,[228] indicating that other mechanisms than direct (*i.e.* agonist-independent) platelet activation are involved in the platelet decline with ribose-modified ASOs. Recent *in vitro* data indicate that both PS ODNs and 2′MOE gapmers lower the threshold of platelet activation of known platelet activators like ADP in a sequence-dependent manner.[231] Similar experiments using the fully modified 2′OMe PS ASO, drisapersen, showed little effect on platelet activation in the presence of ADP, but a reproducible sensitization was observed for 2′ ribose modified ASOs in the presence of collagen-related peptide (CRP), see Figure 20.3. CRP is a different PLT activator that signals through the glycoprotein (GP)-VI receptor, to which PS ASOs have been shown to bind.[230] Thus, depending on sequence

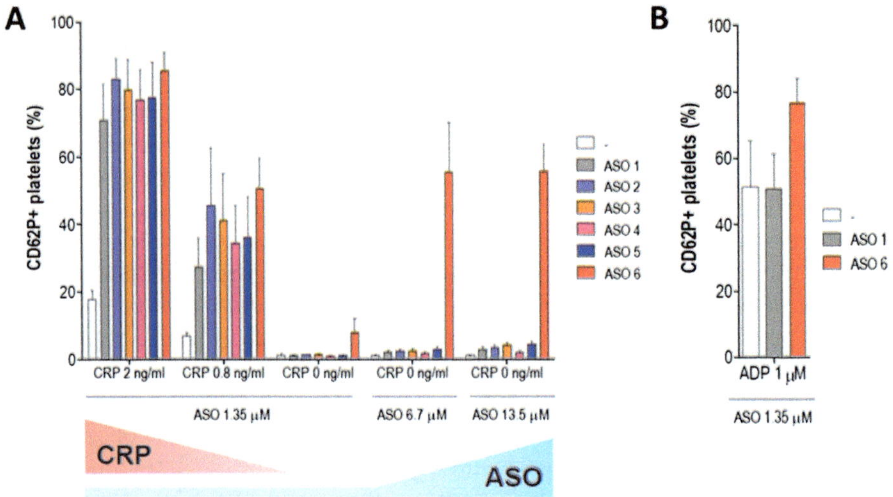

Figure 20.3 (A) ASOs can sensitize platelets for cross-linked collagen-related peptide (CRP-XL)-induced platelet activation *in vitro*. Citrated whole blood was diluted fourfold in HEPES [4-(2-hydroxyethyl)-1-piperazineethanesulfonic acid] buffer and incubated with cross-linked collagen-related peptide (CRP-XL) (0–2 ng ml^{-1}) and the indicated compounds (0–13.5 μM) for 30 min at room temperature. ASO 1 is drisapersen. ASO 1 and ASO 2 are fully modified 2′-OMe PS compounds of 20 and 25 nt respectively. ASO 3, 4 and 5 are 2′MOE PS gapmers of 20 nt and ASO 6 is CpG class C DNA of 22 nt. (B) ASO 1 (drisapersen) did not sensitize platelets in the presence of ADP. Samples were stained with a *R*-phycoerythrin (PE) anti-human CD62P Ab and fixed in 1% paraformaldehyde (PFA) before flow cytometric analysis. $n = 3$; mean + SD are depicted. Unpublished data, courtesy of BioMarin.

and chemical modification, ASOs may differentially influence platelet activation, but it remains to be established if and to what extent this contributes to platelet declines seen in clinical settings.

It cannot be excluded that other, more general mechanisms contribute to the platelet reductions seen in animal species and man. A chronic, low level of inflammatory cytokines and macrophage activation could potentially contribute to the observed platelet declines through several mechanisms, including increased rate of platelet clearance through phagocytosis. Moreover, recent data provide support for the involvement of an innate immune-mediated polyclonal IgM antibody response in the mild to moderate platelet reductions, whereas drug-independent anti-platelet IgG antibodies may contribute to the severe TCP.[118,226,229] Altered pre-study baseline levels of inflammation-related cytokines like B-cell activating factor (BAFF) and IL-23 were observed in the FCS and ATTR (transthyretin amyloidosis) patient populations treated with the TCP-inducing ASOs volanesorsen and inotersen, respectively,[225,232] indicating a possible disease-related predisposition. Lastly, although no clear correlation has been observed for 2′MOE gapmers,[225] it is not known whether complement activation in any way contributes to the observed platelet effects, particularly in monkeys, considering their high sensitivity towards ASO-induced complement activation.

In conclusion, grade 3 or 4 TCP is now recognized as a rare, but serious clinical risk of single-stranded ASOs, which requires a well-defined, frequent monitoring strategy. Improved mechanistic insights will contribute to the optimization and selection of safer ASO candidates and could ultimately facilitate early identification of patients at risk. Given that the few clinical cases of severe TCP known to date were observed at relatively high doses, the improved potency of next-generation ASO candidates should translate to a reduced risk for this effect.

20.6.3 Toxicity in High-exposure Organs

As discussed in Section 20.3.1.2, ONDs within a class show consistent tissue distribution and PK characteristics across sequences with highest OND concentrations in liver and kidney. Due to their first-order kinetics, once steady-state is reached, there is no further increase in tissue concentration resulting from subsequent administrations. This understanding is important when considering potential accumulation-related toxicities.

Highly concentrated ASOs can be observed as basophilic granules in tissue sections stained with hematoxylin and eosin (HE), primarily in cells with high uptake, like the Kupffer cells of the liver and in proximal tubular cells of the kidney. Single-cell necrosis and Kupffer cell hypertrophy, as well as basophilic granules in other organs are occasionally seen at high doses in animals.[2,113,233] Since these granules disappear with dosing cessation and do not lead to degenerative changes *per se*, they are regarded as non-adverse, adaptive changes. On formalin fixation of kidney sections from animals that had received high doses, vacuoles can be seen occasionally in the renal proximal tubular cells. This is a fixation artefact that should not be interpreted as a degenerative change.[234]

Tissue macrophages, often referred to as histiocytes, are frequently observed with a foamy appearance in many tissues in rodents administered higher doses of PS backbone ASOs. In contrast to the basophilic granules containing ASOs, the granular cytoplasm in macrophages is regarded to represent stored cytokines in response to an activated state. These are normally not seen in monkeys, reflecting the higher rodent sensitivity to proinflammatory effects of PS-backbone ASO.[113]

20.6.3.1 Liver Toxicity

Although the liver can achieve high local concentrations of ONDs, hepatotoxicity observed in animals is more dependent on sequence and the sensitivity of various species than absolute concentration. Through careful screening and selection of sequences,[176] hepatoxicity has not been a general feature of most PS and 2′OMe and 2′MOE ASOs in clinical settings,[2,148] with the exception of toxicity secondary to exaggerated pharmacological inhibition of apoB-100 as described for mipomersen.[120] Another clinical case is the GalNAc-conjugated siRNA ALN-AAT, which was terminated after liver enzymes increased in a Phase 1 study,[235] described to be caused by miRNA-like off-target activity.

The overall lack of hepatic toxicity in clinic settings corresponds with absent or mild liver findings in monkeys, which are usually confined to hypertrophy and basophilic granules in Kupffer cells, regarded as non-adverse.[2] Liver toxicity in rodents has been more common, partly due to their particular sensitivity to the proinflammatory effects, with increased lymphohistiocytic cell infiltrates as a key pathological feature. This is usually associated with slow onset, mild to moderate liver enzyme elevations (less than fivefold) that do not progress with chronic treatment, and single-cell necrosis, hepatocellular hypertrophy and, occasionally, increased mitoses and karyomegaly, particularly in studies of longer duration.[2] Overall, NHPs are considered to better predict liability to clinically relevant hepatotoxic related to tissue accumulation.

With the introduction of ASO gapmers with high-affinity 2′-ribose modifications, such as LNA and cEt, a distinct and more severe type of liver toxicity has emerged that is not caused by ASO accumulation or proinflammatory effects *per se*. This type of hepatotoxicity can be observed as early as after a single or a few doses in mice.[115,117,236–238] Mild pathologies at low doses or early time-points may be similar to those described above. However, hepatic inflammation is not common, and, if present, it is more secondary to degenerative changes.[236] In more severe cases, changes are more profound with larger areas of overt necrosis,[117,238] reflected by pronounced liver enzyme elevations, morbidity and mortality. Rodent tolerability studies have proven to be an effective approach for filtering out these hepatotoxic candidates before progressing to clinical use, indicated by only one known clinical example of LNA-gapmer induced liver toxicity to date,[220] despite the increasing clinical use of LNA- and cET-modified gapmer ASOs.

20.6.3.1.1 GalNAc-conjugation. At a given dose, GalNAc conjugation significantly increases hepatocyte uptake of both single-stranded and double-stranded ONDs[40,41] (Section 20.3.2.3). As a consequence, effective concentrations in hepatocytes are reached at more than ten times lower doses of the conjugate compared with the unconjugated OND. Hence, for a given degree of hepatocyte pharmacology much lower doses are required, with correspondingly reduced exposure in non-hepatocyte cells. There is no evidence to date that GalNAc conjugates induce liver toxicity *per se*. Mice lacking the GalNAc receptor ASGR1 are phenotypically normal,[239] and no adverse effects have been observed with multiple GalNAc-conjugated siRNAs and ASO gapmers in clinical settings[109,240–242] or in preclinical toxicity studies, even after administration of high doses of GalNAc oligonucleotides.[243,244]

Taken together with the low clinical doses of GalNAc-conjugated ONDs and the high capacity of this receptor system,[108] it is unlikely that the transient utilization of the receptor for delivery of GalNAc-oligonucleotide drugs will affect the physiological function(s) of ASGR. As indicated above, the greatly reduced dose of GalNAc conjugates required to achieve desired hepatocyte pharmacology leads to a significantly improved therapeutic index. However, GalNAc conjugation of hepatotoxic OND sequences will reach critical hepatocyte threshold concentrations and induce liver toxicity at lower administered doses compared with unconjugated versions of the same sequences. In a recent publication, candidate GalNAc-conjugated siRNAs were reported to have an attrition rate of about 40%, due to liver toxicity in repeat dose screening studies in rats,[132] further described below. Although most GalNAc-conjugated ONDs in clinical trials have been reported as safe, there have been a few isolated clinical examples of liver safety findings with GalNAc-conjugated siRNA[235,245] and an anti-miR ASO.[246]

20.6.3.1.2 Toxic Sequence Motifs. Important efforts have been made to better understand the sequence component and define particular sequence motifs that are associated with hepatotoxic liabilities. Applying machine learning to *in vivo* data from a large number of LNA gapmers, two groups identified motifs comprising two or three nucleotides that correlated fairly well with the toxic potential of sequences.[117,236] The position of the toxic sequence in the gapmer (gap *vs.* wings), the position of the LNA's in the wings and the size of the gap all affected the toxicity profile.[117,236,238]

20.6.3.1.3 Mechanistic Hypotheses. The understanding of the mechanism involved in the high-affinity gapmer hepatotoxicity has greatly evolved in the last decade, facilitating today's screening efforts to advance safer candidates. We know that hepatotoxicity is seen in different strains of mice[117,236,238] and in rats,[238] but there is a clear knowledge gap as to the translation to NHPs or man, because these toxic candidates are usually not taken further in development. Toxicity is not caused by metabolites or impurities[238] and appears unrelated to local concentration or knockdown of the primary target mRNA.[115,117,238] This was recently confirmed using a series of 3–10–3 LNA gapmers designed against the same mRNA target. Profound alanine aminotransferase (ALT) increases were seen for sequences with low liver concentrations and limited target mRNA knockdown, whereas no ALT increases were observed for sequences with much higher liver concentrations and a stronger target mRNA knockdown.[38] A similar lack of correlation between liver toxicity and activity on primary target or liver concentration has been recently reported also for GalNAc-conjugated siRNA.[132] Hence, although being dose-dependent for a given sequence, this type of liver toxicity is not caused by liver accumulation *per se*.

Current evidence indicates a critical contribution of off-target RNA knockdown in gapmer-induced hepatotoxicity. Although sequences are carefully designed to minimize hybridization to RNA other than the target transcript, transcriptomic analysis revealed that hepatotoxic sequences are often associated with downregulation of a significantly higher number of non-target transcripts than non-hepatotoxic sequences. For example, Burel and colleagues analyzed transcriptional profile in mouse livers of several 3–10–3 LNA gapmers with varying degrees of liver toxicity.[115] For the most toxic sequences, ALT levels reached over 10 000 IU ml^{-1} within 72 hours after a single dose of 33 mg kg^{-1}, whereas ALT levels for non-toxic sequences were not elevated despite being dosed up to 300 mg kg^{-1}. Although no ALT increase was observed at 24 hours post dose for any of these sequences, transcriptomics analysis showed downregulation of up to 1300 unique transcripts, whereas the number of downregulated transcripts for safe sequences administered at higher doses did not exceed 121.[115] Similar effects have been observed with cEt-modified gapmers. Importantly, knocking down RNase H1 dramatically reduces the toxicity of many liver toxic ASOs,[115,126,247] clearly indicating that this severe hepatotoxicity is RNase H-dependent. Minor positional shifts on the target mRNA sequence could reduce liver toxicity while maintaining primary target potency.[236] Moreover, addition of only one LNA nucleotide to a 13-mer gapmer abrogated liver toxicity,[117] possibly *via* increasing hybridization specificity to the target transcript. Taken together, these data indicate that sequences producing severe liver toxicity act *via* RNase H-dependent off-target activity, a mechanism that is more evident with high-affinity ribose modifications, like LNA and cEt.

A very similar pattern of liver toxicity caused by primary-mechanism-dependent off-target activity has been described also for GalNAc-conjugated siRNA.[132] Transcriptomics analysis of rat hepatocytes and rat livers demonstrated an enrichment of perfect seed region matches in the genes that are downregulated by liver toxic sequences. Analogous to the use of RNase H1 knockdown for ASO gapmers, Janas and colleagues used multiple approaches to show the importance of RISC activity for causing the liver toxicity. Keeping sequence and chemical modifications the same, capping of the siRNA to prevent RISC-loading or blocking downstream RISC activity by administrating REVERSIR™ antagonists 24 hours before or after the siRNA abrogated liver toxicity. Further support for an off-target mechanism was demonstrated by the reduced liver toxicity when replacing the seed region of a toxic RNA with seed region from a safe siRNA.[132]

A liver toxicity mechanism that is dependent on reduction of single critical transcripts seems unlikely, given the common toxicity pattern across multiple different high-affinity gapmer sequences and siRNAs, each with different off-target knockdown patterns. More likely is the hypothesis that the concentration of cleaved RNAs can reach a level that is toxic to the cells.[38] This was supported by *in vitro* studies by Dieckmann and colleagues, where lipid transfection of LNA gapmers into mouse 3T3 fibroblasts induced apoptosis with a rank order that correlated with the *in vivo* hepatotoxicity in mice, target affinity [calculated melting temperature (T_m)] and number of non-target transcripts affected. Reducing RNase H1 activity by either pretreating with a siRNA knocking down RNase H or introducing 2′OMe ribose modifications in the DNA gap region significantly reduced apoptosis as well as the potency on the primary transcript and the number of off-targets,[126] recapitulating *in vitro* the findings observed *in vivo*. Interestingly, similar results were also seen in human cancer cells of different cell-type origin (HeLa, HepG2 and A549),[126] indicating that these mechanisms are cell- and species-independent.

It cannot be excluded that the chemistry- and sequence-dependent hepatotoxicity is partly due to certain sequences binding with high affinity to important intracellular proteins, resulting in perturbation of critical pathways.[236,248] More than 50 intracellular proteins have been identified to which gapmer ASOs bind, where some interactions show clear sequence dependence.[39,50] For a given gapmer sequence, protein-binding affinity is dependent on the 2′-ribose modification with the rank order 2′F > 2′cEt > 2′MOE.[50] In a recent study using 2′F-substituted 5–10–5 gapmers in mice, hepatotoxicity was linked to binding to and degradation of *Drosophila* behavior/human splicing (DBHS) proteins, rather than RNase H1 activity. Gapmers with the same sequence but with 2′MOE modifications instead of 2′F were non-toxic, whereas LNA and cEt modifications resulted in an intermediate degree of toxicity.[249]

Taken together, the available evidence indicates that the high-affinity (LNA, cEt) gapmer liver toxicity observed with some sequences is caused by accumulation of RNase H-cleaved transcripts and/or protein interactions above a critical threshold level. This threshold level is reached at lower doses for GalNAc-conjugated gapmers, due to more efficient productive uptake into hepatocytes. Identification and removal of liver-toxic sequences from progression into clinical use has so far relied on rodent studies.[2,250] Although further refinement may be needed before replacing *in vivo* studies, the emerging examples of reasonably predictive *in silico*[117,236] and *in vitro* models[126,250] indicate that alternative methods for filtering out liver-toxic ASO gapmers can be established, thereby reducing the use of animals for screening for liver toxicity.

20.6.3.2 Kidney Toxicity

High local concentrations of oligonucleotide make the kidney one of the common organs for toxicity of all OND modalities, including the charge-neutral-backbone PMOs.[251–253] Changes at the proximal tubule are more common than in other tubular sections or in the glomeruli. Renal effects of PS backbone ASOs have been summarized in multiple reviews,[2,113,234] and a brief overview of current knowledge is presented below.

20.6.3.2.1 Tubular Effects. The morphological and functional tubular changes in animal studies follow a predictable course, whereby initial microscopic changes of cytoplasmic basophilic granules at low doses become associated with tubular atrophy, regeneration and degeneration at high doses and tissue concentrations.[2,233,254] Functional tubular changes in NHPs are usually only seen at doses that are high multiples of the clinically relevant dose, as reported for 2′MOE gapmers.[2,254] NHPs appear to predict well for the tubular toxicity risk in man, since there was no evidence of clinically significant renal dysfunction following evaluation of a large safety database on over 2400 subjects involving up to 52 weeks of treatment data from multiple clinical trials of different 2′MOE gapmer sequences.[7] Proteinuria was reported in most Duchenne muscular dystrophy (DMD) patients upon chronic treatment with the splice-switching PS 2′OMe ASO drisapersen.[178] This proteinuria was mostly mild in severity, primarily affecting low-molecular-weight proteins, reversible with temporary treatment interruptions and not progressive over time or associated with other signs of renal dysfunction.[178,255] A similar, reversible low-molecular-weight proteinuria was observed in NHPs treated with drisapersen at and above clinically relevant doses,[256] confirming good monkey-to-man translation of this effect. The mechanism for this mild, reversible tubular dysfunction is not entirely clear, but could be related to transient and competitive interference with proximal tubular reabsorption of solutes, including proteins, from the glomerular filtrate, since both ASOs and proteins are known to be re-absorbed from the glomerular filtrate *via* scavenging-receptor-mediated endocytic processes (such as *via* the megalin–cubulin receptor). This would result in increased excretion of low-molecular-weight proteins and, to a lesser extent, albumin and is readily reversed when treatment is interrupted.[178,234] Notably, acute tubular dysfunction, as illustrated by a dose-dependent increase in serum creatinine and increased urinary excretion of beta-2 microglobulin and kidney injury molecule-1 (KIM1), was reported in human volunteers treated with an unusually short (12-mer) 2′MOE gapmer designed to accumulate in the kidney and target the renal SGLT2 transporter. Although a direct tubular effect could not be excluded,[257] the intended pharmacology is likely to be the main contributor,[74] since serum creatinin levels normalized upon continued treatment are indicative of tubular adaptation.

The tubular effects described above are regarded as being little affected by sequence, other than for those effects directly related to tissue accumulation. However, for ASOs with higher-affinity modifications, a different pattern of tubular toxicity is emerging,[258] where sequence appears to be more important than tubular accumulation above a critical concentration threshold. This can be exemplified by overt kidney toxicity in mice after only two weeks of dosing of 3–8–3 LNA gapmers that were also toxic to the liver.[259] A clinical example of a highly sequence-dependent renal toxicity is the acute tubular necrosis after three administrations (5 mg kg^{-1}) of the LNA gapmer SPC5001.[260] There were no indications of renal changes in the pivotal mouse and monkey toxicity studies, but a retrospective high-dose study in rats confirmed the tubular toxicity for this sequence.[261] LNA gapmers that cause renal toxicity *in vivo* were shown to impair EGF receptor signaling in primary and immortalized proximal tubules *in vitro*, providing a robust EGF-based assay that contributes to the *in vitro* safety profiling of this class of ONDs.[261] Thus, in analogy to the sequence-dependent liver toxicity of high-affinity gapmers, similar mechanisms of excessive accumulation of RNase H dependent off-target transcripts[115,126] and/or aberrant protein binding[249] may be involved in the sequence-dependent renal toxicity of ASOs with high affinity modifications.

20.6.3.2.2 Glomerular Effects. Glomerular lesions are not commonly observed with PS ONDs and 2′MOE gapmers,[234] but glomerulopathies were a reproducible observation in mouse and monkey toxicity studies of 3 months or more duration using 2′OMe ASOs in development for the treatment of DMD.[113,262] These glomerular lesions were generally seen at or below the clinical doses and exposures and were not associated with glomerular accumulation. They were characterized by profound species differences where local inflammatory activity and mouse-specific conditions appeared to be important determinants in the progressive and irreversible murine pathology.[262] The reversible monkey lesions indicated an underlying chronic complement activation and ASO-induced inflammatory effects,[262] which is known to be caused by ASOs when administered to monkeys above a certain dose threshold[147,151,154] (see Section 20.6.1.2). The lower susceptibility of humans to ASO-induced complement activation and inflammatory mechanisms indicate that animal toxicity studies over-predict the clinical risk of glomerular lesions. This has been confirmed by clinical practice, where only isolated examples of confirmed ASO-related glomerulonephritis have been observed.[255]

20.7 Getting to the Clinic

20.7.1 Regulatory Considerations for Preclinical Development

The existing regulatory guidance does not directly address OND therapeutics but general principles of drug safety assessment also apply to this therapeutic platform. Being chemically synthesized, ONDs are formally regulated as new chemical entities (NCE) by most regulatory authorities, and their nonclinical development should therefore follow small-molecule guideline recommendations from the International Conference on Harmonisation (ICH M3[R2]). This includes assessment of repeat-dose toxicity in two species for up to 6 months in one rodent and 9 months in one non-rodent species, as well as assessment of safety pharmacology, genotoxicity, carcinogenicity and reproductive toxicity, normally *via* the intended route of administration. However, some OND characteristics, such as their larger size, species-specificity, predictive degradation pathways and predictive tissue distribution, are more like those of biological therapeutics, and therefore principles from ICH S6 should also be considered when designing a non-clinical testing strategy for ONDs. Strict application of ICH S6 is not appropriate however. For example, rodent active surrogates can be applied much more easily in the case of OND therapeutics for the purpose of assessing on-target safety (see Section 20.5.1), and there are generally no issues with regard to species-specific drug-neutralizing antibodies. This said, although rodents are generally more useful for OND safety assessment than stated in the ICH S6 guidance, NHPs are considered to be the most predictive species. Other considerations, such as route of administration and size of the intended patient population, may influence the design of the nonclinical data package. Such hybrid ICH M3/S6 approaches that are tailored to the characteristics of the specific OND approach fulfill the regulatory expectations of a scientifically and clinically meaningful risk assessment.

The lack of formal, oligonucleotide-specific guidelines should be viewed as a blessing in disguise rather than a hurdle, because it provides companies with the opportunity to closely liaise with regulators and design programs that are more tailored to the specific challenges associated with the use of a specific OND approach. Informal, best practice recommendations are provided by a series of white papers authored by members of OSWG.[76,121,136,146,263–267] This work is part of an ongoing process, with new subgroups being formed to either cover new topics or to revise existing recommendations when needed. The regulatory considerations for documenting pharmacodynamics, DMPK and safety that apply to single-stranded ASOs are summarized in Table 20.5.

Table 20.5 Considerations regarding nonclinical safety evaluation strategies[a] for OND therapeutics of well characterized chemistries.

End-point (main applicable guidelines[b])	Strategy and comments	GLP compliance required	Key references[c]
Pharmacodynamics (PD)			
On-target toxicity, hybridization-dependent	Characterize by assessment of potential on-target safety and exaggerated pharmacology concerns, and by repeat-dose toxicity studies with pharmacological activity in at least one species, or consider testing of species-specific surrogate OND	No	121,175
Off-target activity, hybridization-dependent	Characterize by *in silico* evaluation and on a case by case basis, experimental verification *in vitro* using human cells and *in vivo* using (rodent) active surrogates	No	116,136
Safety pharmacology			
Cardiovascular, human ether-a-go-go-related gene (hERG) assay (*in vitro*) (ICH S7B)	Generally not required. Physico-chemical properties of ONDs (*e.g.* size and charge) preclude interaction with hERG channel. Large dataset available demonstrating lack of OND effect on hERG channel.	Yes	265,268,269
Cardiovascular, (*in vivo*), respiratory, CNS (ICH S7A)	Stand-alone studies generally not required. Incorporate relevant endpoints in repeat-dose toxicity studies if meaningful systemic or local (heart, lung, CNS) exposure is expected. Systemically administered ASOs do not pass across the blood–brain barrier	Yes	265,268
DMPK characterization			
Absorption, kinetics (ICH S3A)	Characterize PK parameters (*e.g.*, C_{max}, AUC, target tissue half-life)	No	5,54,88

Table 20.5 (*continued*)

End-point (main applicable guidelines[b])	Strategy and comments	GLP compliance required	Key references[c]
Distribution (ICH S3B, S6[R1])	Characterize in context of established class profile for known chemistries. Routine mass balance studies and quantitative whole-body autoradiography (QWBA) are generally not required until late development stages and may be waived under certain circumstances, *e.g.* local route of administration , or availability of sufficient predictive class profile data. Quantitative assessment of OND concentrations in target PD tissue to assess PK–PD relationships, and in peripheral tissues known to accumulate ASOs (liver, kidney)	No	5,54,88
Metabolism [ICH S6(R1), metabolites in safety testing (MIST[270])]	Characterize in context of established class profile for known chemistries. Predictive metabolism pathways *via* nuclease-degradation that is qualitatively similar between species. Commonly-used modified nucleotide metabolites are not regarded as reactive or toxic	No	5,54,88
Excretion	Characterize in context of established class profile for known chemistries. Hydrophilic ONDs are largely excreted *via* urine, with minimal fecal excretion	No	5,54,88
Drug–drug interaction (DDI)[271]	Generally not required. ONDs do not interact with small-molecule drug-metabolizing enzymes or transporters	No	
Safety characterization			
Acute toxicity	Dedicated studies not required; assessment included in repeat-dose toxicity studies. ONDs have low risk of acute toxicity.	Yes	
Repeat dose toxicity (ICH M3[R2], S4, S6[R1])	Characterize in context of established class profile of known chemistries. By default, assessment in two species, one rodent and one non-rodent, but one species may be acceptable if justified (ICH S6). Long tissue half-life may affect study duration of main and recovery period.	Yes	2,175,273
Genetic toxicity (ICH S2[R1], EMA reflection paper[272])	Not required in EU and some divisions of FDA, unless OND has novel chemistry or structures. Avoid OND motifs which promote unwanted triplex DNA formation	Yes	264,272

(*continued*)

Table 20.5 (*continued*)

End-point (main applicable guidelines[b])	Strategy and comments	GLP compliance required	Key references[c]
Reproductive toxicity (ICH S5[R2], S6[R1])	Characterize in context of established class profile of known chemistries and intended patient population. ONDs have a low risk for embryofetal developmental (EFD) or male-fertility toxicity, because they do not penetrate placental or testes barriers to a significant degree. The need for these studies can be discussed case by case. For EFD studies, daily dosing is recommended to cover for daily peak exposure levels during organogenesis.	Yes	267
Juvenile toxicity (ICH S11, FDA 2006[274])	Characterize in case of inclusion of pediatric patients. Rare diseases often include treatment of pediatric patients early in the program.	Yes	
Carcinogenicity (ICH S1, S6[R1])	Characterize, based on a case-by case-approach. Design and species selection may be challenging due to frequent lack of PD in rodents, or due to rodent high sensitivity to the pro-inflammatory effects of some ONDs. Usually post marketing commitment in rare disease applications.	Yes	
Immunotoxicity (ICH S8, S6[R1])	Characterize, with focus on immunostimulatory potential rather than immunosuppression. Immunogenicity potential should be assessed in clinical settings, where relevant, as animal data on immunogenicity are not predictive for humans.	Yes	175
Local tolerability (ICH S6[R1], EMA 2015[275])	Characterize tolerability at site of administration as part of repeat-dose toxicity studies, rather than dedicated local tolerability studies	Yes	263
Photosafety (ICH S10)	Not required. ASOs do not have the characteristics to elicit a phototoxic response.	Yes	

[a]This table is not comprehensive in content and references, but is meant to illustrate that the unique chemical characteristics of OND, their PK profile and the intended therapeutic indication and patient population require a science-based nonclinical testing approach.
[b]ICH Guidelines can be found at http://www.ich.org/products/guidelines.html.
[c]Key references include, for example, OSWG white papers and reviews on OND class characteristics.

20.7.2 Precision Medicine and Opportunity for Accelerated Timelines

In the field of innovative medicines, the road from an initial pharmacological concept to clinical demonstration of therapeutic benefit is a tedious one, paved with many challenges. Regardless of drug modality, the overall approach is the same, covering target selection, hit identification and lead optimization, which is followed by regulatory PK and safety characterization before progressing a candidate into clinical testing. For small molecules, this process can easily take three to five years. The platform of RNA-modulating therapeutics provides advantages at each of these stages, with dramatically shorter timelines. In contrast to most other modalities where large libraries of diverse compounds are screened to identify hits requiring further optimization of potency, DMPK and safety parameters, today's vast and detailed genomic information allows a very rational and fast design of hybridization-dependent ONDs. Coupled with the efficient synthesis of ONDs, this allows rapid generation of OND candidates with a high likelihood of showing promising potency. This cuts timelines for generating a set of compounds for evaluation as clinical candidates but it also allows rapid generation of tool compounds for target validation in human *in vitro* systems and in animal disease models. The well-understood and largely predictive behavior regarding DMPK and potential toxicities allows implementation of a robust screening cascade that can be applied for any project and target within a given OND class. In principle, a clinical OND candidate with promising PD, PK and early safety characteristics could be selected within one to two years from target selection and be ready for clinical testing. This is a significantly shorter time compared with the timelines for most other established drug modalities.

The Watson–Crick base pairing of DNA and RNA is highly selective and by utilizing this paradigm, ONDs have an intrinsic high degree of selectivity that is difficult to achieve with small molecules. Compared with antibodies, that are also highly selective, ONDs have the big advantage that they can modulate expression and activity of both extracellular and intracellular proteins. When the options for targeted intracellular delivery increase, this will expand the number of tissues and cell types amenable to treatment. Together with the rapid and rational discovery and development process, these aspects make ONDs well suited to target otherwise very challenging or undruggable targets. This is particularly true for diseases caused by monoallelic mutations, where the small number of patients normally does not warrant the timelines and investments required for most other drug modalities. However, to fully unlock the potential of this platform, a discussion about modifying the regulatory framework for rare genetic diseases is warranted.

20.8 Conclusions

During the last ten years, the area of therapeutic ONDs has undergone a remarkable evolution and maturation that has resulted in several regulatory approvals and a large number of ongoing clinical trials. Meanwhile, our understanding of how chemistry and sequence affect DMPK and safety properties has greatly evolved, opening up new therapeutic opportunities. GalNAc-conjugation of siRNA and ASOs is a promising showcase of a major increase in productive uptake into target cells, demonstrating that the big hurdle of delivering ONDs to target cells can be overcome. Higher-affinity ribose modifications for ASOs and enhanced designs for siRNA have led to a great reduction in the doses needed for desired PD effects in the liver and enabled siRNA to be delivered *via* SC injection in saline without the need for complex formulations. Use of more robust screening cascades has resulted in the ability to filter out unsafe candidates, leading, for example, to a significantly reduced incidence and severity of SC injection-site reactions and other pro-inflammatory responses in clinical trials. Neutral backbone PMO's appear to have an advantage over PS-backbone ONDs in terms of tolerability, but the true assessment of safety of this class needs confirmation once next-generation PMOs achieve higher tissue concentrations.

Compared with earlier OND generations, today's high-affinity OND designs with significantly increased potency are also more prone to sequence-dependent liver and kidney toxicity, due to increased RNase H and RISC-dependent degradation of off-target transcripts. This may become increasingly relevant also for other organs, when higher uptake in other tissues is achieved. The broader clinical experience with higher patient numbers and longer duration of treatment for multiple clinical candidates has revealed a sporadic, serious side effect of severe TCP for three PS ASOs to date, but the improved potency of next-generation ASO candidates is expected to reduce this risk in future programs.

In the coming decade, further elucidation of OND-delivery mechanisms will broaden the therapeutic opportunities for ONDs. The success of GalNAc conjugation has paved the way for development of novel targeting approaches that will increase productive uptake in cell types other than hepatocytes, expanding the range of tissues amenable to therapeutic OND approaches. This will require new PK–PD relationships to be established and may reveal novel safety issues in tissues previously not in focus due to low uptake. The coming decade will also witness a much deeper understanding of the mechanisms behind the sequence-dependent OND toxicities with a significant improvement of predictive *in silico* and *in vitro* models to filter out toxic sequences at an early stage. The growing databases containing both clinical and preclinical safety information will provide critical information about the translational value between preclinical and clinical observations. Taken together, these will further improve the efficient and rational design process to identify potent and safe OND clinical candidate compounds. This will be a necessary step for ONDs to be the first platform that is truly adapted for the exciting era of personalized medicine.

Acknowledgements

We would like to thank Scott Henry (Ionis Pharmaceuticals) and Art Levin (Avidity Biosciences) for critical review and valuable suggestions.

References

1. S. T. Crooke, J. L. Witztum, C. F. Bennett and B. F. Baker, *Cell Metab.*, 2018, **27**, 714–739.
2. S. P. Henry, T.-W. Kim, K. Kramer-Stickland, T. A. Zanardi, R. A. Fey and A. A. Levin, in *Antisense Drug Technology: Principles, Strategies and Applications*, ed. S. T. Crooke, CRC Press, Taylor and Francis group, Boca Raton, FL, USA, 2nd edn, 2008, pp. 327–363.
3. A. A. Levin, *Biochim. Biophys. Acta*, 1999, **1489**, 69–84.
4. A. A. Levin, S. P. Henry, D. Monteith and V. T. Templin, in *Antisense Drug Technology – Principles, Strategies, and Applications*, ed. S. T. Crooke, Marcel Dekker, Inc., New York, USA, 2001, pp. 201–267.
5. A. A. Levin, R. Z. Yu and R. S. Geary, in *Antisense Drug Technology – Principles, Strategies, and Applications*, ed. S. T. Crooke, CRC Press, Taylor and Francis Group, Boca Raton, FL, USA, 2nd edn, 2008, pp. 183–215.
6. S. T. Crooke, B. F. Baker, T. J. Kwoh, W. Cheng, D. J. Schulz, S. Xia, N. Salgado, H. H. Bui, C. E. Hart, S. A. Burel, H. S. Younis, R. S. Geary, S. P. Henry and S. Bhanot, *Mol. Ther.*, 2016, **24**, 1771–1782.
7. S. T. Crooke, B. F. Baker, N. C. Pham, S. G. Hughes, T. J. Kwoh, D. Cai, S. Tsimikas, R. S. Geary and S. Bhanot, *Nucleic Acid Ther.*, 2018, **28**, 10–22.
8. S. T. Crooke, B. F. Baker, J. L. Witztum, T. J. Kwoh, N. C. Pham, N. Salgado, B. W. McEvoy, W. Cheng, S. G. Hughes, S. Bhanot and R. S. Geary, *Nucleic Acid Ther.*, 2017, **27**, 121–129.
9. S. P. Henry, P. Narayanan, L. Shen, S. Bhanot, H. S. Younis and S. A. Burel, *Nucleic Acid Ther.*, 2017, **27**, 197–208.
10. U. Sahin, K. Kariko and O. Tureci, *Nat. Rev. Drug Discovery*, 2014, **13**, 759–780.
11. S. M. Nimjee, R. R. White, R. C. Becker and B. A. Sullenger, *Annu. Rev. Pharmacol. Toxicol.*, 2017, **57**, 61–79.
12. J. Zhou and J. Rossi, *Nat. Rev. Drug Discovery*, 2017, **16**, 181–202.
13. D. Schubert, A. A. Levin, D. Kornbrust, C. L. Berman, J. Cavagnaro, S. Henry, R. Seguin, N. Ferrari and S. B. Shrewsbury, *Nucleic Acid Ther.*, 2012, **22**, 211–212.
14. A. Khvorova and J. K. Watts, *Nat. Biotechnol.*, 2017, **35**, 238–248.
15. P. P. Seth and E. E. Swayze, in *Natural Products in Medicinal Chemistry*, ed. S. Hanessian, Wiley-VCH Verlag, 1st edn, 2014, pp. 403–439.
16. W. B. Wan and P. P. Seth, *J. Med. Chem.*, 2016, **59**, 9645–9667.
17. F. Eckstein, *Antisense Nucleic Acid Drug Dev.*, 2000, **10**, 117–121.
18. S. M. Freier and K. H. Altmann, *Nucleic Acids Res.*, 1997, **25**, 4429–4443.
19. C. M. Miller, A. J. Donner, E. E. Blank, A. W. Egger, B. M. Kellar, M. E. Ostergaard, P. P. Seth and E. N. Harris, *Nucleic Acids Res.*, 2016, **44**, 2782–2794.
20. S. M. Hammond, G. Hazell, F. Shabanpoor, A. F. Saleh, M. Bowerman, J. N. Sleigh, K. E. Meijboom, H. Zhou, F. Muntoni, K. Talbot, M. J. Gait and M. J. Wood, *Proc. Natl. Acad. Sci. U. S. A.*, 2016, **113**, 10962–10967.
21. M. Egli, G. Minasov, V. Tereshko, P. S. Pallan, M. Teplova, G. B. Inamati, E. A. Lesnik, S. R. Owens, B. S. Ross, T. P. Prakash and M. Manoharan, *Biochemistry*, 2005, **44**, 9045–9057.

22. M. Manoharan, *Biochim. Biophys. Acta*, 1999, **1489**, 117–130.
23. E. E. Swayze and B. Bhat, in *Antisense Drug Technology – Principles, Strategies, and Applications*, ed. S. T. Crooke, CRC Press, Taylor & Francis, Boca Raton, FL, USA, 2nd edn, 2008, pp. 143–182.
24. C. F. Bennett, B. F. Baker, N. Pham, E. Swayze and R. S. Geary, *Annu. Rev. Pharmacol. Toxicol.*, 2017, **57**, 81–105.
25. S. T. Crooke, *Nucleic Acid Ther.*, 2017, **27**, 70–77.
26. P. Disterer, A. Kryczka, Y. Liu, Y. E. Badi, J. J. Wong, J. S. Owen and B. Khoo, *Hum. Gene Ther.*, 2014, **25**, 587–598.
27. M. A. Havens and M. L. Hastings, *Nucleic Acids Res.*, 2016, **44**, 6549–6563.
28. F. Rigo, P. P. Seth and C. F. Bennett, *Adv. Exp. Med. Biol.*, 2014, **825**, 303–352.
29. R. Titze-de-Almeida, C. David and S. S. Titze-de-Almeida, *Pharm. Res.*, 2017, **34**, 1339–1363.
30. M. L. Bobbin and J. J. Rossi, *Annu. Rev. Pharmacol. Toxicol.*, 2016, **56**, 103–122.
31. S. Agrawal, S. H. Mayrand, P. C. Zamecnik and T. Pederson, *Proc. Natl. Acad. Sci. U. S. A.*, 1990, **87**, 1401–1405.
32. B. P. Monia, E. A. Lesnik, C. Gonzalez, W. F. Lima, D. McGee, C. J. Guinosso, A. M. Kawasaki, P. D. Cook and S. M. Freier, *J. Biol. Chem.*, 1993, **268**, 14514–14522.
33. P. Sazani, M. A. Graziewicz and R. Kole, in *Antisense Drug Technology – Principles, Strategies, and Applications*, ed. S. T. Crooke, CRC Press, Taylor and Francis Group, Boca Raton, FL, USA, 2nd edn, 2008, pp. 89–114.
34. J. Alter, F. Lou, A. Rabinowitz, H. Yin, J. Rosenfeld, S. D. Wilton, T. A. Partridge and Q. L. Lu, *Nat. Med.*, 2006, **12**, 175–177.
35. S. Cirak, V. Arechavala-Gomeza, M. Guglieri, L. Feng, S. Torelli, K. Anthony, S. Abbs, M. E. Garralda, J. Bourke, D. J. Wells, G. Dickson, M. J. Wood, S. D. Wilton, V. Straub, R. Kole, S. B. Shrewsbury, C. Sewry, J. E. Morgan, K. Bushby and F. Muntoni, *Lancet*, 2011, **378**, 595–605.
36. P. Sazani, F. Gemignani, S. H. Kang, M. A. Maier, M. Manoharan, M. Persmark, D. Bortner and R. Kole, *Nat. Biotechnol.*, 2002, **20**, 1228–1233.
37. M. Manoharan and K. G. Rajeev, in *Antisense Drug Technology – Principles, Strategies, and Applications*, ed. S. T. Crooke, CRC Press, Taylor and Francis Group, Boca Raton, FL, USA, 2nd edn, 2008, pp. 437–464.
38. P. H. Hagedorn, R. Persson, E. D. Funder, N. Albaek, S. L. Diemer, D. J. Hansen, M. R. Moller, N. Papargyri, H. Christiansen, B. R. Hansen, H. F. Hansen, M. A. Jensen and T. Koch, *Drug Discovery Today*, 2018, **23**, 101–114.
39. T. A. Vickers and S. T. Crooke, *PLoS One*, 2016, **11**, e0161930.
40. J. K. Nair, J. L. Willoughby, A. Chan, K. Charisse, M. R. Alam, Q. Wang, M. Hoekstra, P. Kandasamy, A. V. Kel'in, S. Milstein, N. Taneja, J. O'Shea, S. Shaikh, L. Zhang, R. J. van der Sluis, M. E. Jung, A. Akinc, R. Hutabarat, S. Kuchimanchi, K. Fitzgerald, T. Zimmermann, T. J. van Berkel, M. A. Maier, K. G. Rajeev and M. Manoharan, *J. Am. Chem. Soc.*, 2014, **136**, 16958–16961.

41. T. P. Prakash, M. J. Graham, J. Yu, R. Carty, A. Low, A. Chappell, K. Schmidt, C. Zhao, M. Aghajan, H. F. Murray, S. Riney, S. L. Booten, S. F. Murray, H. Gaus, J. Crosby, W. F. Lima, S. Guo, B. P. Monia, E. E. Swayze and P. P. Seth, *Nucleic Acids Res.*, 2014, **42**, 8796–8807.
42. F. Eckstein, *Nucleic Acid Ther.*, 2014, **24**, 374–387.
43. M. Koziolkiewicz, A. Krakowiak, M. Kwinkowski, M. Boczkowska and W. J. Stec, *Nucleic Acids Res.*, 1995, **23**, 5000–5005.
44. M. Koziolkiewicz, M. Wojcik, A. Kobylanska, B. Karwowski, B. Rebowska, P. Guga and W. J. Stec, *Antisense Nucleic Acid Drug Dev.*, 1997, **7**, 43–48.
45. W. B. Wan, M. T. Migawa, G. Vasquez, H. M. Murray, J. G. Nichols, H. Gaus, A. Berdeja, S. Lee, C. E. Hart, W. F. Lima, E. E. Swayze and P. P. Seth, *Nucleic Acids Res.*, 2014, **42**, 13456–13468.
46. F. Czauderna, M. Fechtner, S. Dames, H. Aygun, A. Klippel, G. J. Pronk, K. Giese and J. Kaufmann, *Nucleic Acids Res.*, 2003, **31**, 2705–2716.
47. S. Gao, F. Dagnaes-Hansen, E. J. Nielsen, J. Wengel, F. Besenbacher, K. A. Howard and J. Kjems, *Mol. Ther.*, 2009, **17**, 1225–1233.
48. J. M. Layzer, A. P. McCaffrey, A. K. Tanner, Z. Huang, M. A. Kay and B. A. Sullenger, *RNA*, 2004, **10**, 766–771.
49. D. V. Morrissey, J. A. Lockridge, L. Shaw, K. Blanchard, K. Jensen, W. Breen, K. Hartsough, L. Machemer, S. Radka, V. Jadhav, N. Vaish, S. Zinnen, C. Vargeese, K. Bowman, C. S. Shaffer, L. B. Jeffs, A. Judge, I. MacLachlan and B. Polisky, *Nat. Biotechnol.*, 2005, **23**, 1002–1007.
50. S. T. Crooke, S. Wang, T. A. Vickers, W. Shen and X. H. Liang, *Nat. Biotechnol.*, 2017, **35**, 230–237.
51. R. L. Juliano, *Nucleic Acids Res.*, 2016, **44**, 6518–6548.
52. C. M. Miller, M. Tanowitz, A. J. Donner, T. P. Prakash, E. E. Swayze, E. N. Harris and P. P. Seth, *Nucleic Acid Ther.*, 2018, **28**, 119–127.
53. P. L. Iversen, in *Antisense Drug Technology: Principles, Strategies and Applications*, ed. S. T. Crooke, CRC Press, Taylor and Francis group, Boca Raton, FL, USA, 2nd edn, 2008, pp. 565–582.
54. R. S. Geary, *Expert Opin. Drug Metab. Toxicol.*, 2009, **5**, 381–391.
55. S. F. Dowdy, *Nat. Biotechnol.*, 2017, **35**, 222–229.
56. L. G. Tillman, R. S. Geary and G. E. Hardee, *J. Pharm. Sci.*, 2008, **97**, 225–236.
57. R. Z. Yu, J. S. Grundy, S. P. Henry, T. W. Kim, D. A. Norris, J. Burkey, Y. Wang, A. Vick and R. S. Geary, *Mol. Ther. Nucleic Acids*, 2015, **4**, e218.
58. R. Z. Yu, K. M. Lemonidis, M. J. Graham, J. E. Matson, R. M. Crooke, D. L. Tribble, M. K. Wedel, A. A. Levin and R. S. Geary, *Biochem. Pharmacol.*, 2009, **77**, 910–919.
59. R. Z. Yu, T. W. Kim, A. Hong, T. A. Watanabe, H. J. Gaus and R. S. Geary, *Drug Metab. Dispos.*, 2007, **35**, 460–468.
60. R. S. Geary, D. Norris, R. Yu and C. F. Bennett, *Adv. Drug Delivery Rev.*, 2015, **87**, 46–51.
61. R. S. Geary, R. Z. Yu, A. Siwkowski and A. A. Levin, in *Antisense Drug Technology: Principles, Strategies and Applications*, ed. S. T. Crooke, CRC Press, Taylor and Francis group, Boca Raton, FL, USA, 2nd edn, 2008, pp. 305–326.

62. R. Z. Yu, R. Gunawan, N. Post, T. Zanardi, S. Hall, J. Burkey, T. W. Kim, M. J. Graham, T. P. Prakash, P. P. Seth, E. E. Swayze, R. S. Geary, S. P. Henry and Y. Wang, *Nucleic Acid Ther.*, 2016, **26**, 372–380.
63. T. A. Watanabe, R. S. Geary and A. A. Levin, *Oligonucleotides*, 2006, **16**, 169–180.
64. K. Schmidt, T. P. Prakash, A. J. Donner, G. A. Kinberger, H. J. Gaus, A. Low, M. E. Ostergaard, M. Bell, E. E. Swayze and P. P. Seth, *Nucleic Acids Res.*, 2017, **45**, 2294–2306.
65. D. J. Foster, C. R. Brown, S. Shaikh, C. Trapp, M. K. Schlegel, K. Qian, A. Sehgal, K. G. Rajeev, V. Jadhav, M. Manoharan, S. Kuchimanchi, M. A. Maier and S. Milstein, *Mol. Ther.*, 2018, **26**, 708–717.
66. J. K. Nair, H. Attarwala, A. Sehgal, Q. Wang, K. Aluri, X. Zhang, M. Gao, J. Liu, R. Indrakanti, S. Schofield, P. Kretschmer, C. R. Brown, S. Gupta, J. L. S. Willoughby, J. A. Boshar, V. Jadhav, K. Charisse, T. Zimmermann, K. Fitzgerald, M. Manoharan, K. G. Rajeev, A. Akinc, R. Hutabarat and M. A. Maier, *Nucleic Acids Res.*, 2017, **45**, 10969–10977.
67. J. D. Thompson, D. J. Kornbrust, J. W. Foy, E. C. Solano, D. J. Schneider, E. Feinstein, B. A. Molitoris and S. Erlich, *Nucleic Acid Ther.*, 2012, **22**, 255–264.
68. S. P. Henry, unpublished data.
69. M. Butler, K. Stecker and C. F. Bennett, *Lab. Invest.*, 1997, **77**, 379–388.
70. G. Hung, X. Xiao, R. Peralta, G. Bhattacharjee, S. Murray, D. Norris, S. Guo and B. P. Monia, *Nucleic Acid Ther.*, 2013, **23**, 369–378.
71. S. P. Henry, K. H. Denny, M. V. Templin, R. Z. Yu and A. A. Levin, *Birth Defects Res., Part B*, 2004, **71**, 359–367.
72. S. P. Henry, K. H. Denny, M. V. Templin, R. Z. Yu and A. A. Levin, *Birth Defects Res., Part B*, 2004, **71**, 368–373.
73. N. V. Soucy, J. P. Riley, M. V. Templin, R. Geary, A. de Peyster and A. A. Levin, *Birth Defects Res., Part B*, 2006, **77**, 22–28.
74. T. A. Zanardi, S. C. Han, E. J. Jeong, S. Rime, R. Z. Yu, K. Chakravarty and S. P. Henry, *J. Pharmacol. Exp. Ther.*, 2012, **343**, 489–496.
75. J. Christensen, K. Litherland, T. Faller, E. van de Kerkhof, F. Natt, J. Hunziker, J. Boos, I. Beuvink, K. Bowman, J. Baryza, M. Beverly, C. Vargeese, O. Heudi, M. Stoeckli, J. Krauser and P. Swart, *Drug Metab. Dispos.*, 2014, **42**, 431–440.
76. J. L. Marlowe, V. Akopian, P. Karmali, D. Kornbrust, J. Lockridge and S. Semple, *Nucleic Acid Ther.*, 2017, **27**, 183–196.
77. FDA, https://www.accessdata.fda.gov/drugsatfda_docs/nda/2016/206488Orig1s000PharmR.pdf, Accessed March 2018.
78. R. M. Hudziak, E. Barofsky, D. F. Barofsky, D. L. Weller, S. B. Huang and D. D. Weller, *Antisense Nucleic Acid Drug Dev.*, 1996, **6**, 267–272.
79. E. Koller, T. M. Vincent, A. Chappell, S. De, M. Manoharan and C. F. Bennett, *Nucleic Acids Res.*, 2011, **39**, 4795–4807.
80. S. Wang, H. Sun, M. Tanowitz, X. H. Liang and S. T. Crooke, *Nucleic Acids Res.*, 2017, **45**, 5309–5322.
81. R. S. Geary, E. Wancewicz, J. Matson, M. Pearce, A. Siwkowski, E. Swayze and F. Bennett, *Biochem. Pharmacol.*, 2009, **78**, 284–291.

82. J. Gilleron, W. Querbes, A. Zeigerer, A. Borodovsky, G. Marsico, U. Schubert, K. Manygoats, S. Seifert, C. Andree, M. Stoter, H. Epstein-Barash, L. Zhang, V. Koteliansky, K. Fitzgerald, E. Fava, M. Bickle, Y. Kalaidzidis, A. Akinc, M. Maier and M. Zerial, *Nat. Biotechnol.*, 2013, **31**, 638–646.
83. C. M. Miller, W. B. Wan, P. P. Seth and E. N. Harris, *Nucleic Acid Ther.*, 2018, **28**, 86–96.
84. S. Wang, H. Sun, M. Tanowitz, X. H. Liang and S. T. Crooke, *Nucleic Acids Res.*, 2016, **44**, 7314–7330.
85. A. J. Donner, E. V. Wancewicz, H. M. Murray, S. Greenlee, N. Post, M. Bell, W. F. Lima, E. E. Swayze and P. P. Seth, *Nucleic Acid Ther.*, 2017, **27**, 209–220.
86. X. H. Liang, H. Sun, J. G. Nichols and S. T. Crooke, *Mol. Ther.*, 2017, **25**, 2075–2092.
87. M. S. Baek, R. Z. Yu, H. Gaus, J. S. Grundy and R. S. Geary, *Oligonucleotides*, 2010, **20**, 309–316.
88. M. Dirin and J. Winkler, *Expert Opin. Biol. Ther.*, 2013, **13**, 875–888.
89. C. S. Shemesh, R. Z. Yu, M. S. Warren, M. Liu, M. Jahic, B. Nichols, N. Post, S. Lin, D. A. Norris, E. Hurh, J. Huang, T. Watanabe, S. P. Henry and Y. Wang, *Mol. Ther. Nucleic Acids*, 2017, **9**, 34–47.
90. R. S. Geary, J. D. Bradley, T. Watanabe, Y. Kwon, M. Wedel, J. J. van Lier and A. A. VanVliet, *Clin. Pharmacokinet.*, 2006, **45**, 789–801.
91. R. S. Geary, S. P. Henry and L. R. Grillone, *Clin. Pharmacokinet.*, 2002, **41**, 255–260.
92. J. J. Laskin, G. Nicholas, C. Lee, B. Gitlitz, M. Vincent, Y. Cormier, J. Stephenson, Y. Ung, R. Sanborn, B. Pressnail, F. Nugent, J. Nemunaitis, M. E. Gleave, N. Murray and D. Hao, *J. Thorac. Oncol.*, 2012, **7**, 579–586.
93. R. Z. Yu, R. S. Geary, J. D. Flaim, G. C. Riley, D. L. Tribble, A. A. vanVliet and M. K. Wedel, *Clin. Pharmacokinet.*, 2009, **48**, 39–50.
94. R. Juliano, M. R. Alam, V. Dixit and H. Kang, *Nucleic Acids Res.*, 2008, **36**, 4158–4171.
95. R. L. Juliano, S. Alahari, H. Yoo, R. Kole and M. Cho, *Pharm. Res.*, 1999, **16**, 494–502.
96. S. P. Henry, R. C. Miner, W. L. Drew, J. Fitchett, C. York-Defalco, L. M. Rapp and A. A. Levin, *Invest. Ophthalmol. Visual Sci.*, 2001, **42**, 2646–2651.
97. R. P. Danis, S. P. Henry and T. A. Ciulla, *Expert Opin. Pharmacother.*, 2001, **2**, 277–291.
98. J. M. Leeds, S. P. Henry, S. Bistner, S. Scherrill, K. Williams and A. A. Levin, *Drug Metab. Dispos.*, 1998, **26**, 670–675.
99. E. C. Solano, D. J. Kornbrust, A. Beaudry, J. W. Foy, D. J. Schneider and J. D. Thompson, *Nucleic Acid Ther.*, 2014, **24**, 258–266.
100. R. A. Fey, M. V. Templin, J. D. McDonald, R. Z. Yu, J. A. Hutt, A. P. Gigliotti, S. P. Henry and M. D. Reed, *Inhalation Toxicol.*, 2014, **26**, 452–463.
101. S. Jackson, A. F. Candia, S. Delaney, S. Floettmann, C. Wong, J. D. Campbell, S. Kell, J. Lum, E. M. Hessel, P. Traquina, M. McHale, I. Robinson, J. Bell, R. Fuhr, D. Keeling and R. L. Coffman, *Clin. Pharmacol. Ther.*, 2018, **104**, 335–345.

102. O. Khorkova and C. Wahlestedt, *Nat. Biotechnol.*, 2017, **35**, 249–263.
103. W. Querbes, P. Ge, W. Zhang, Y. Fan, J. Costigan, K. Charisse, M. Maier, L. Nechev, M. Manoharan, V. Kotelianski and D. W. Sah, *Oligonucleotides*, 2009, **19**, 23–29.
104. F. Rigo, S. J. Chun, D. A. Norris, G. Hung, S. Lee, J. Matson, R. A. Fey, H. Gaus, Y. Hua, J. S. Grundy, A. R. Krainer, S. P. Henry and C. F. Bennett, *J. Pharmacol. Exp. Ther.*, 2014, **350**, 46–55.
105. J. M. Leeds, S. P. Henry, L. Truong, A. Zutshi, A. A. Levin and D. Kornbrust, *Drug Metab. Dispos.*, 1997, **25**, 921–926.
106. T. P. Prakash, W. F. Lima, H. M. Murray, S. Elbashir, W. Cantley, D. Foster, M. Jayaraman, A. E. Chappell, M. Manoharan, E. E. Swayze and S. T. Crooke, *ACS Chem. Biol.*, 2013, **8**, 1402–1406.
107. T. Coelho, D. Adams, A. Silva, P. Lozeron, P. N. Hawkins, T. Mant, J. Perez, J. Chiesa, S. Warrington, E. Tranter, M. Munisamy, R. Falzone, J. Harrop, J. Cehelsky, B. R. Bettencourt, M. Geissler, J. S. Butler, A. Sehgal, R. E. Meyers, Q. Chen, T. Borland, R. M. Hutabarat, V. A. Clausen, R. Alvarez, K. Fitzgerald, C. Gamba-Vitalo, S. V. Nochur, A. K. Vaishnaw, D. W. Sah, J. A. Gollob and O. B. Suhr, *N. Engl. J. Med.*, 2013, **369**, 819–829.
108. A. L. Schwartz, D. Rup and H. F. Lodish, *J. Biol. Chem.*, 1980, **255**, 9033–9036.
109. N. J. Viney, J. C. van Capelleveen, R. S. Geary, S. Xia, J. A. Tami, R. Z. Yu, S. M. Marcovina, S. G. Hughes, M. J. Graham, R. M. Crooke, S. T. Crooke, J. L. Witztum, E. S. Stroes and S. Tsimikas, *Lancet*, 2016, **388**, 2239–2253.
110. M. J. Hinrichs and R. Dixit, *AAPS J.*, 2015, **17**, 1055–1064.
111. R. Z. Yu, M. J. Graham, N. Post, S. Riney, T. Zanardi, S. Hall, J. Burkey, C. S. Shemesh, T. P. Prakash, P. P. Seth, E. E. Swayze, R. S. Geary, Y. Wang and S. Henry, *Mol. Ther. Nucleic Acids*, 2016, **5**, e317.
112. M. R. Hassler, A. A. Turanov, J. F. Alterman, R. A. Haraszti, A. H. Coles, M. F. Osborn, D. Echeverria, M. Nikan, W. E. Salomon, L. Roux, B. Godinho, S. M. Davis, D. V. Morrissey, P. D. Zamore, S. A. Karumanchi, M. J. Moore, N. Aronin and A. Khvorova, *Nucleic Acids Res.*, 2018, **46**, 2185–2196.
113. K. S. Frazier, *Toxicol. Pathol.*, 2015, **43**, 78–89.
114. S. P. Henry, G. Beattie, G. Yeh, A. Chappel, P. Giclas, A. Mortari, M. A. Jagels, D. J. Kornbrust and A. A. Levin, *Int. Immunopharmacol.*, 2002, **2**, 1657–1666.
115. S. A. Burel, C. E. Hart, P. Cauntay, J. Hsiao, T. Machemer, M. Katz, A. Watt, H. H. Bui, H. Younis, M. Sabripour, S. M. Freier, G. Hung, A. Dan, T. P. Prakash, P. P. Seth, E. E. Swayze, C. F. Bennett, S. T. Crooke and S. P. Henry, *Nucleic Acids Res.*, 2016, **44**, 2093–2109.
116. P. H. Hagedorn, B. R. Hansen, T. Koch and M. Lindow, *Nucleic Acids Res.*, 2017, **45**, 2262–2282.
117. P. H. Hagedorn, V. Yakimov, S. Ottosen, S. Kammler, N. F. Nielsen, A. M. Hog, M. Hedtjarn, M. Meldgaard, M. R. Moller, H. Orum, T. Koch and M. Lindow, *Nucleic Acid Ther.*, 2013, **23**, 302–310.

118. X. Chi, P. Gatti and T. Papoian, *Drug Discovery Today*, 2017, **22**, 823–833.
119. A. Sehgal, S. Barros, L. Ivanciu, B. Cooley, J. Qin, T. Racie, J. Hettinger, M. Carioto, Y. Jiang, J. Brodsky, H. Prabhala, X. Zhang, H. Attarwala, R. Hutabarat, D. Foster, S. Milstein, K. Charisse, S. Kuchimanchi, M. A. Maier, L. Nechev, P. Kandasamy, A. V. Kel'in, J. K. Nair, K. G. Rajeev, M. Manoharan, R. Meyers, B. Sorensen, A. R. Simon, Y. Dargaud, C. Negrier, R. M. Camire and A. Akinc, *Nat. Med.*, 2015, **21**, 492–497.
120. N. Hashemi, R. D. Odze, M. P. McGowan, R. D. Santos, E. S. Stroes and D. E. Cohen, *J. Clin. Lipidol.*, 2014, **8**, 606–611.
121. D. Kornbrust, J. Cavagnaro, A. Levin, J. Foy, P. Pavco, C. Gamba-Vitalo and A. Guimond, *Nucleic Acid Ther.*, 2013, **23**, 21–28.
122. P. J. Kamola, J. D. Kitson, G. Turner, K. Maratou, S. Eriksson, A. Panjwani, L. C. Warnock, G. A. Douillard Guilloux, K. Moores, E. L. Koppe, W. E. Wixted, P. A. Wilson, N. J. Gooderham, T. W. Gant, K. L. Clark, S. A. Hughes, M. R. Edbrooke and J. D. Parry, *Nucleic Acids Res.*, 2015, **43**, 8638–8650.
123. J. R. Crosby, C. Zhao, H. Zhang, A. R. MacLeod, S. Guo and B. P. Monia, *Nucleic Acid Ther.*, 2015, **25**, 297–305.
124. I. Zlatev, A. Castoreno, C. R. Brown, J. Qin, S. Waldron, M. K. Schlegel, R. Degaonkar, S. Shulga-Morskaya, H. Xu, S. Gupta, S. Matsuda, A. Akinc, K. G. Rajeev, M. Manoharan, M. A. Maier and V. Jadhav, *Nat. Biotechnol.*, 2018, **36**, 509–511.
125. A. L. Jackson and A. A. Levin, *Nucleic Acid Ther.*, 2012, **22**, 213–225.
126. A. Dieckmann, P. H. Hagedorn, Y. Burki, C. Brugmann, M. Berrera, M. Ebeling, T. Singer and F. Schuler, *Mol. Ther. Nucleic Acids*, 2018, **10**, 45–54.
127. T. Kasuya, S. Hori, A. Watanabe, M. Nakajima, Y. Gahara, M. Rokushima, T. Yanagimoto and A. Kugimiya, *Sci. Rep.*, 2016, **6**, 30377.
128. A. Birmingham, E. M. Anderson, A. Reynolds, D. Ilsley-Tyree, D. Leake, Y. Fedorov, S. Baskerville, E. Maksimova, K. Robinson, J. Karpilow, W. S. Marshall and A. Khvorova, *Nat. Methods*, 2006, **3**, 199–204.
129. Y. Fedorov, E. M. Anderson, A. Birmingham, A. Reynolds, J. Karpilow, K. Robinson, D. Leake, W. S. Marshall and A. Khvorova, *RNA*, 2006, **12**, 1188–1196.
130. A. L. Jackson, S. R. Bartz, J. Schelter, S. V. Kobayashi, J. Burchard, M. Mao, B. Li, G. Cavet and P. S. Linsley, *Nat. Biotechnol.*, 2003, **21**, 635–637.
131. A. L. Jackson, J. Burchard, J. Schelter, B. N. Chau, M. Cleary, L. Lim and P. S. Linsley, *RNA*, 2006, **12**, 1179–1187.
132. M. M. Janas, M. K. Schlegel, C. E. Harbison, V. O. Yilmaz, Y. Jiang, R. Parmar, I. Zlatev, A. Castoreno, H. Xu, S. Shulga-Morskaya, K. G. Rajeev, M. Manoharan, N. D. Keirstead, M. A. Maier and V. Jadhav, *Nat. Commun.*, 2018, **9**, 723.
133. H. S. Lee, H. Seok, D. H. Lee, J. Ham, W. Lee, E. M. Youm, J. S. Yoo, Y. S. Lee, E. S. Jang and S. W. Chi, *Nat. Commun.*, 2015, **6**, 10154.
134. S. R. Suter, J. Sheu-Gruttadauria, N. T. Schirle, R. Valenzuela, A. A. Ball-Jones, K. Onizuka, I. J. MacRae and P. A. Beal, *J. Am. Chem. Soc.*, 2016, **138**, 8667–8669.

135. S. Takyar, R. P. Hickerson and H. F. Noller, *Cell*, 2005, **120**, 49–58.
136. M. Lindow, H. P. Vornlocher, D. Riley, D. J. Kornbrust, J. Burchard, L. O. Whiteley, J. Kamens, J. D. Thompson, S. Nochur, H. Younis, S. Bartz, J. Parry, N. Ferrari, S. P. Henry and A. A. Levin, *Nat. Biotechnol.*, 2012, **30**, 920–923.
137. S. Obad, C. O. dos Santos, A. Petri, M. Heidenblad, O. Broom, C. Ruse, C. Fu, M. Lindow, J. Stenvang, E. M. Straarup, H. F. Hansen, T. Koch, D. Pappin, G. J. Hannon and S. Kauppinen, *Nat. Genet.*, 2011, **43**, 371–378.
138. S. P. Henry, W. Novotny, J. Leeds, C. Auletta and D. J. Kornbrust, *Antisense Nucleic Acid Drug Dev.*, 1997, **7**, 503–510.
139. J. P. Sheehan and H. C. Lan, *Blood*, 1998, **92**, 1617–1625.
140. T. L. Wallace, S. A. Bazemore, D. J. Kornbrust and P. A. Cossum, *J. Pharmacol. Exp. Ther.*, 1996, **278**, 1313–1317.
141. R. Advani, B. L. Lum, G. A. Fisher, J. Halsey, R. S. Geary, J. T. Holmlund, T. J. Kwoh, F. A. Dorr and B. I. Sikic, *Invest. New Drugs*, 2005, **23**, 467–477.
142. A. Braendli-Baiocco, M. Festag, K. Dumong Erichsen, R. Persson, M. J. Mihatsch, N. Fisker, J. Funk, S. Mohr, R. Constien, C. Ploix, K. Brady, M. Berrera, B. Altmann, B. Lenz, M. Albassam, G. Schmitt, T. Weiser, F. Schuler, T. Singer and Y. Tessier, *Toxicol. Sci.*, 2017, **157**, 112–128.
143. H. X. Chen, J. L. Marshall, E. Ness, R. R. Martin, B. Dvorchik, N. Rizvi, J. Marquis, M. McKinlay, W. Dahut and M. J. Hawkins, *Clin. Cancer Res.*, 2000, **6**, 1259–1266.
144. J. P. Sheehan and T. M. Phan, *Biochemistry*, 2001, **40**, 4980–4989.
145. S. P. Henry, P. C. Giclas, J. Leeds, M. Pangburn, C. Auletta, A. A. Levin and D. J. Kornbrust, *J. Pharmacol. Exp. Ther.*, 1997, **281**, 810–816.
146. S. P. Henry, R. Seguin, J. Cavagnaro, C. Berman, J. Tepper and D. Kornbrust, *Nucleic Acid Ther.*, 2016, **26**, 210–215.
147. S. P. Henry, M. A. Jagels, T. E. Hugli, S. Manalili, R. S. Geary, P. C. Giclas and A. A. Levin, *Nucleic Acid Ther.*, 2014, **24**, 326–335.
148. T. J. Kwoh, in *Antisense Drug Technology: Principles, Strategies and Applications*, ed. S. T. Crooke, CRC Press, Taylor and Francis group, Boca Raton, FL, USA, 2nd edn, 2008, pp. 365–399.
149. C. M. Rudin, J. Holmlund, G. F. Fleming, S. Mani, W. M. Stadler, P. Schumm, B. P. Monia, J. F. Johnston, R. Geary, R. Z. Yu, T. J. Kwoh, F. A. Dorr and M. J. Ratain, *Clin. Cancer Res.*, 2001, **7**, 1214–1220.
150. L. Shen, A. Frazer-Abel, P. R. Reynolds, P. C. Giclas, A. Chappell, M. K. Pangburn, H. Younis and S. P. Henry, *J. Pharmacol. Exp. Ther.*, 2014, **351**, 709–717.
151. L. Shen, J. A. Engelhardt, G. Hung, J. Yee, R. Kikkawa, J. Matson, B. Tayefeh, T. Machemer, P. C. Giclas and S. P. Henry, *Nucleic Acid Ther.*, 2016, **26**, 236–249.
152. S. Henry, P. Giclas, J. Leeds, M. Pangburn, C. Auletta, A. Levin and D. Kornbrust, *J. Pharmacol. Exp. Ther.*, May 1997, **281**(2), 810–816.
153. S. Henry, M. Jagels, T. Hugli, S. Manalili, R. Geary, P. Giclas and A. Levin, *Nucleic Acid Ther.*, October 2014, **24**(5), 326–335.

154. J. A. Engelhardt, P. Fant, S. Guionaud, S. P. Henry, M. W. Leach, C. Louden, M. S. Scicchitano, J. L. Weaver, T. S. Zabka, K. S. Frazier and Society of Toxicologic Pathology Vascular Injury Working Group, *Toxicol. Pathol.*, 2015, **43**, 935–944.
155. A. M. Krieg, *Nat. Rev. Drug Discovery*, 2006, **5**, 471–484.
156. U. Holtick, M. E. Scheulen, M. S. von Bergwelt-Baildon and M. R. Weihrauch, *Expert Opin. Invest. Drugs*, 2011, **20**, 361–372.
157. J. Vollmer and A. M. Krieg, *Adv. Drug Delivery Rev.*, 2009, **61**, 195–204.
158. A. J. Bridge, S. Pebernard, A. Ducraux, A. L. Nicoulaz and R. Iggo, *Nat. Genet.*, 2003, **34**, 263–264.
159. V. Hornung, M. Guenthner-Biller, C. Bourquin, A. Ablasser, M. Schlee, S. Uematsu, A. Noronha, M. Manoharan, S. Akira, A. de Fougerolles, S. Endres and G. Hartmann, *Nat. Med.*, 2005, **11**, 263–270.
160. A. D. Judge, G. Bola, A. C. Lee and I. MacLachlan, *Mol. Ther.*, 2006, **13**, 494–505.
161. A. D. Judge, V. Sood, J. R. Shaw, D. Fang, K. McClintock and I. MacLachlan, *Nat. Biotechnol.*, 2005, **23**, 457–462.
162. K. Kariko, P. Bhuyan, J. Capodici and D. Weissman, *J. Immunol.*, 2004, **172**, 6545–6549.
163. K. Kariko, H. Ni, J. Capodici, M. Lamphier and D. Weissman, *J. Biol. Chem.*, 2004, **279**, 12542–12550.
164. A. M. Krieg, *Curr. Opin. Drug Discovery Dev.*, 2000, **3**, 214–221.
165. A. M. Krieg, *Annu. Rev. Immunol.*, 2002, **20**, 709–760.
166. D. K. Monteith and A. A. Levin, *Toxicol. Pathol.*, 1999, **27**, 8–13.
167. S. Paz, J. Hsiao, P. Cauntay, A. Soriano, L. Bai, T. Machemer, X. Xiao, S. Guo, G. Hung, H. Younis, C. F. Bennett, S. Henry, T. J. Yun and S. Burel, *Nucleic Acid Ther.*, 2017, **27**, 272–284.
168. M. Robbins, A. Judge and I. MacLachlan, *Oligonucleotides*, 2009, **19**, 89–102.
169. T. A. Zanardi, T. W. Kim, L. Shen, D. Serota, C. Papagiannis, S. Y. Park, Y. Kim and S. P. Henry, *Nucleic Acid Ther.*, 2018, **28**, 233–241.
170. S. P. Henry, H. Bolte, C. Auletta and D. J. Kornbrust, *Toxicology*, 1997, **120**, 145–155.
171. D. K. Monteith, R. S. Geary, J. M. Leeds, J. Johnston, B. P. Monia and A. A. Levin, *Toxicol. Sci.*, 1998, **46**, 365–375.
172. EMA, http://www.ema.europa.eu/docs/en_GB/document_library/Application_withdrawal_assessment_report/2016/09/WC500212619.pdf, Accessed March 2018.
173. W. Barchet, V. Wimmenauer, M. Schlee and G. Hartmann, *Curr. Opin. Immunol.*, 2008, **20**, 389–395.
174. M. Jurk, A. Kritzler, B. Schulte, S. Tluk, C. Schetter, A. M. Krieg and J. Vollmer, *Eur. J. Immunol.*, 2006, **36**, 1815–1826.
175. E. K. Mustonen, T. Palomaki and M. Pasanen, *Regul. Toxicol. Pharmacol.*, 2017, **90**, 328–341.
176. B. P. Monia, R. S. Yu, W. Lima and A. Siwkowski, in *Antisense Drug Technology – Principles, Strategies, and Applications*, ed. S. T. Crooke, CRC Press, Taylor and Francis Group, Boca Raton, FL, USA, 2nd edn, 2008, pp. 487–506.

177. L. van Meer, M. Moerland, J. Gallagher, M. B. van Doorn, E. P. Prens, A. F. Cohen, R. Rissmann and J. Burggraaf, *Br. J. Clin. Pharmacol.*, 2016, **82**, 340–351.
178. N. M. Goemans, M. Tulinius, M. van den Hauwe, A. K. Kroksmark, G. Buyse, R. J. Wilson, J. C. van Deutekom, S. J. de Kimpe, A. Lourbakos and G. Campion, *PLoS One*, 2016, **11**, e0161955.
179. A. L. Nguyen, E. H. Niks, E. J. van Zuuren and R. van Doorn, *Ned. Tijdschr. Dermatol. Venerol.*, 2016, **26**, 670–673.
180. R. K. Ramanathan, S. I. Hamburg, M. J. Borad, M. Seetharam, M. N. Kundranda, P. Lee, P. Fredlund, M. Gilbert, C. Mast, S. C. Semple, A. D. Judge, B. Crowell, L. Vocila, I. MacLachlan and D. W. Northfelt, *Cancer Res.*, 2014, **73**, LB-289.
181. O. B. Suhr, T. Coelho, J. Buades, J. Pouget, I. Conceicao, J. Berk, H. Schmidt, M. Waddington-Cruz, J. M. Campistol, B. R. Bettencourt, A. Vaishnaw, J. Gollob and D. Adams, *Orphanet J. Rare Dis.*, 2015, **10**, 109.
182. J. Tabernero, G. I. Shapiro, P. M. LoRusso, A. Cervantes, G. K. Schwartz, G. J. Weiss, L. Paz-Ares, D. C. Cho, J. R. Infante, M. Alsina, M. M. Gounder, R. Falzone, J. Harrop, A. C. White, I. Toudjarska, D. Bumcrot, R. E. Meyers, G. Hinkle, N. Svrzikapa, R. M. Hutabarat, V. A. Clausen, J. Cehelsky, S. V. Nochur, C. Gamba-Vitalo, A. K. Vaishnaw, D. W. Sah, J. A. Gollob and H. A. Burris, 3rd, *Cancer Discovery*, 2013, **3**, 406–417.
183. S. Akira and K. Takeda, *Nat. Rev. Immunol.*, 2004, **4**, 499–511.
184. Z. Meng and M. Lu, *Front. Immunol.*, 2017, **8**, 331.
185. A. M. Krieg, A. K. Yi, S. Matson, T. J. Waldschmidt, G. A. Bishop, R. Teasdale, G. A. Koretzky and D. M. Klinman, *Nature*, 1995, **374**, 546–549.
186. A. Ablasser, F. Bauernfeind, G. Hartmann, E. Latz, K. A. Fitzgerald and V. Hornung, *Nat. Immunol.*, 2009, **10**, 1065–1072.
187. V. Hornung, J. Ellegast, S. Kim, K. Brzozka, A. Jung, H. Kato, H. Poeck, S. Akira, K. K. Conzelmann, M. Schlee, S. Endres and G. Hartmann, *Science*, 2006, **314**, 994–997.
188. S. A. Burel, T. Machemer, F. L. Ragone, H. Kato, P. Cauntay, S. Greenlee, A. Salim, W. A. Gaarde, G. Hung, R. Peralta, S. M. Freier and S. P. Henry, *J. Pharmacol. Exp. Ther.*, 2012, **342**, 150–162.
189. L. Gitlin, W. Barchet, S. Gilfillan, M. Cella, B. Beutler, R. A. Flavell, M. S. Diamond and M. Colonna, *Proc. Natl. Acad. Sci. U. S. A.*, 2006, **103**, 8459–8464.
190. A. J. Sadler and B. R. Williams, *Curr. Top. Microbiol. Immunol.*, 2007, **316**, 253–292.
191. C. A. Sledz, M. Holko, M. J. de Veer, R. H. Silverman and B. R. Williams, *Nat. Cell Biol.*, 2003, **5**, 834–839.
192. P. Broz and D. M. Monack, *Nat. Rev. Immunol.*, 2013, **13**, 551–565.
193. A. Iwasaki and R. Medzhitov, *Nat. Immunol.*, 2004, **5**, 987–995.
194. K. A. Whitehead, J. E. Dahlman, R. S. Langer and D. G. Anderson, *Annu. Rev. Chem. Biomol. Eng.*, 2011, **2**, 77–96.
195. C. Lassig and K. P. Hopfner, *J. Biol. Chem.*, 2017, **292**, 9000–9009.
196. Y. K. Chan and M. U. Gack, *Nat. Rev. Microbiol.*, 2016, **14**, 360–373.

197. S. Bauer, C. J. Kirschning, H. Hacker, V. Redecke, S. Hausmann, S. Akira, H. Wagner and G. B. Lipford, *Proc. Natl. Acad. Sci. U. S. A.*, 2001, **98**, 9237–9242.
198. G. Hartmann, R. D. Weeratna, Z. K. Ballas, P. Payette, S. Blackwell, I. Suparto, W. L. Rasmussen, M. Waldschmidt, D. Sajuthi, R. H. Purcell, H. L. Davis and A. M. Krieg, *J. Immunol.*, 2000, **164**, 1617–1624.
199. A. Forsbach, J. G. Nemorin, C. Montino, C. Muller, U. Samulowitz, A. P. Vicari, M. Jurk, G. K. Mutwiri, A. M. Krieg, G. B. Lipford and J. Vollmer, *J. Immunol.*, 2008, **180**, 3729–3738.
200. F. Heil, H. Hemmi, H. Hochrein, F. Ampenberger, C. Kirschning, S. Akira, G. Lipford, H. Wagner and S. Bauer, *Science*, 2004, **303**, 1526–1529.
201. S. S. Diebold, C. Massacrier, S. Akira, C. Paturel, Y. Morel and C. Reis e Sousa, *Eur. J. Immunol.*, 2006, **36**, 3256–3267.
202. M. Sioud, *J. Mol. Biol.*, 2005, **348**, 1079–1090.
203. M. Sioud, *Eur. J. Immunol.*, 2006, **36**, 1222–1230.
204. S. Jeske, R. Pries and B. Wollenberg, *Nucleic Acid Ther.*, 2013, **23**, 118–124.
205. H. Liang, Y. Nishioka, C. F. Reich, D. S. Pisetsky and P. E. Lipsky, *J. Clin. Invest.*, 1996, **98**, 1119–1129.
206. Q. Zhao, J. Temsamani, P. L. Iadarola, Z. Jiang and S. Agrawal, *Biochem. Pharmacol.*, 1996, **51**, 173–182.
207. M. Jurk, G. Chikh, B. Schulte, A. Kritzler, D. Richardt-Pargmann, C. Lampron, R. Luu, A. M. Krieg, A. P. Vicari and J. Vollmer, *Nucleic Acid Ther.*, 2011, **21**, 201–214.
208. S. Tluk, M. Jurk, A. Forsbach, R. Weeratna, U. Samulowitz, A. M. Krieg, S. Bauer and J. Vollmer, *Int. Immunol.*, 2009, **21**, 607–619.
209. K. Kariko, M. Buckstein, H. Ni and D. Weissman, *Immunity*, 2005, **23**, 165–175.
210. S. Henry, K. Stecker, D. Brooks, D. Monteith, B. Conklin and C. F. Bennett, *J. Pharmacol. Exp. Ther.*, 2000, **292**, 468–479.
211. S. Hamm, E. Latz, D. Hangel, T. Muller, P. Yu, D. Golenbock, T. Sparwasser, H. Wagner and S. Bauer, *Immunobiology*, 2010, **215**, 559–569.
212. M. Robbins, A. Judge, L. Liang, K. McClintock, E. Yaworski and I. MacLachlan, *Mol. Ther.*, 2007, **15**, 1663–1669.
213. C. Coch, C. Luck, A. Schwickart, B. Putschli, M. Renn, T. Holler, W. Barchet, G. Hartmann and M. Schlee, *PLoS One*, 2013, **8**, e71057.
214. Ionis Pharmaceuticals and Biomarin, personal communication, 2018.
215. EMA, http://www.ema.europa.eu/docs/en_GB/document_library/EPAR_-_Public_assessment_report/human/002429/WC500144511.pdf, Accessed March 2018.
216. R. H. Aster and D. W. Bougie, *N. Engl. J. Med.*, 2007, **357**, 580–587.
217. G. P. Visentin and C. Y. Liu, *Hematol. Oncol. Clin. North Am.*, 2007, **21**, 685–696, vi.
218. N. Osada, N. Hettiarachchi, I. Adeyemi Babarinde, N. Saitou and A. Blancher, *Genome Biol. Evol.*, 2015, **7**, 821–830.
219. A. A. Adjei, G. K. Dy, C. Erlichman, J. M. Reid, J. A. Sloan, H. C. Pitot, S. R. Alberts, R. M. Goldberg, L. J. Hanson, P. J. Atherton, T. Watanabe, R. S. Geary, J. Holmlund and F. A. Dorr, *Clin. Cancer Res.*, 2003, **9**, 115–123.

220. A. W. Tolcher, A. Patnaik, K. P. Papadopoulos, J. Agnew, F. M. Lokiec, K. Rezai, S. Kalambakas and A. Buchbinder, *Mol. Cancer Ther.*, 2011, **10**, A216.
221. J. S. Waters, A. Webb, D. Cunningham, P. A. Clarke, F. Raynaud, F. di Stefano and F. E. Cotter, *J. Clin. Oncol.*, 2000, **18**, 1812–1823.
222. A. R. Yuen and B. I. Sikic, *Front. Biosci.*, 2000, **5**, D588–D593.
223. FDA, in https://www.fda.gov/downloads/AdvisoryCommittees/CommitteesMeetingMaterials/Drugs/EndocrinologicandMetabolicDrugsAdvisoryCommittee/UCM606857.pdf, Accessed May 2018.
224. Akcea, https://www.fda.gov/downloads/AdvisoryCommittees/CommitteesMeetingMaterials/Drugs/EndocrinologicandMetabolicDrugsAdvisoryCommittee/UCM606860.pdf, Accessed May 2018.
225. R. S. Geary, presented in part at the DIA Oligonucleotide Conference, North Bethesda, MD, October 2017.
226. P. Narayanan, L. Shen, B. R. Curtis, M. A. Bourdon, J. P. Nolan, S. Gupta, C. Hoffmaster, F. Zhou, B. Christian, J. L. Schaubhut, S. Greenlee, S. A. Burel, J. L. Witztum, J. A. Engelhardt and S. P. Henry, *Toxicol Sci.*, 2018, **164**, 613–626.
227. M. E. Jaax, K. Krauel, T. Marschall, S. Brandt, J. Gansler, B. Furll, B. Appel, S. Fischer, S. Block, C. A. Helm, S. Muller, K. T. Preissner and A. Greinacher, *Blood*, 2013, **122**, 272–281.
228. S. Sewing, A. B. Roth, M. Winter, A. Dieckmann, C. Bertinetti-Lapatki, Y. Tessier, C. McGinnis, S. Huber, E. Koller, C. Ploix, J. C. Reed, T. Singer and A. Rothfuss, *PLoS One*, 2017, **12**, e0187574.
229. S. P. Henry, presented in part at the DIA Oligonucleotide Conference, North Bethesda, MD, October 2017.
230. U. Flierl, T. L. Nero, B. Lim, J. F. Arthur, Y. Yao, S. M. Jung, E. Gitz, A. Y. Pollitt, M. T. Zaldivia, M. Jandrot-Perrus, A. Schafer, B. Nieswandt, R. K. Andrews, M. W. Parker, E. E. Gardiner and K. Peter, *J. Exp. Med.*, 2015, **212**, 129–137.
231. J. E. Italiano, presented in part at the DIA Oligonucleotide Conference, North Bethesda, MD, October 2017.
232. Ionis Pharmaceuticals, manuscript in preparation.
233. D. K. Monteith, M. J. Horner, N. A. Gillett, M. Butler, R. Geary, T. Burckin, T. Ushiro-Watanabe and A. A. Levin, *Toxicol. Pathol.*, 1999, **27**, 307–317.
234. J. A. Engelhardt, *Nucleic Acid Ther.*, 2016, **26**, 199–209.
235. Alnylam Press Release, in http://www.alnylam.com/2016/09/28/aln-aat-data-and-rnai-platform-optimization/, Accessed March 2018.
236. A. D. Burdick, S. Sciabola, S. R. Mantena, B. D. Hollingshead, R. Stanton, J. A. Warneke, M. Zeng, E. Martsen, A. Medvedev, S. S. Makarov, L. A. Reed, J. W. Davis, 2nd and L. O. Whiteley, *Nucleic Acids Res.*, 2014, **42**, 4882–4891.
237. P. P. Seth, A. Jazayeri, J. Yu, C. R. Allerson, B. Bhat and E. E. Swayze, *Mol. Ther. Nucleic Acids*, 2012, **1**, e47.
238. E. E. Swayze, A. M. Siwkowski, E. V. Wancewicz, M. T. Migawa, T. K. Wyrzykiewicz, G. Hung, B. P. Monia and C. F. Bennett, *Nucleic Acids Res.*, 2007, **35**, 687–700.
239. R. Tozawa, S. Ishibashi, J. Osuga, K. Yamamoto, H. Yagyu, K. Ohashi, Y. Tamura, N. Yahagi, Y. Iizuka, H. Okazaki, K. Harada, T. Gotoda, H.

Shimano, S. Kimura, R. Nagai and N. Yamada, *J. Biol. Chem.*, 2001, **276**, 12624–12628.

240. K. Fitzgerald, S. White, A. Borodovsky, B. R. Bettencourt, A. Strahs, V. Clausen, P. Wijngaard, J. D. Horton, J. Taubel, A. Brooks, C. Fernando, R. S. Kauffman, D. Kallend, A. Vaishnaw and A. Simon, *N. Engl. J. Med.*, 2017, **376**, 41–51.
241. M. J. Graham, R. G. Lee, T. A. Brandt, L. J. Tai, W. Fu, R. Peralta, R. Yu, E. Hurh, E. Paz, B. W. McEvoy, B. F. Baker, N. C. Pham, A. Digenio, S. G. Hughes, R. S. Geary, J. L. Witztum, R. M. Crooke and S. Tsimikas, *N. Engl. J. Med.*, 2017, **377**, 222–232.
242. K. K. Ray, U. Landmesser, L. A. Leiter, D. Kallend, R. Dufour, M. Karakas, T. Hall, R. P. Troquay, T. Turner, F. L. Visseren, P. Wijngaard, R. S. Wright and J. J. Kastelein, *N. Engl. J. Med.*, 2017, **376**, 1430–1440.
243. N. D. Keirstead, DIA/FDA Oligonucleotide-based Therapeutic Conference, 2015.
244. T. P. Prakash, J. Yu, M. T. Migawa, G. A. Kinberger, W. B. Wan, M. E. Ostergaard, R. L. Carty, G. Vasquez, A. Low, A. Chappell, K. Schmidt, M. Aghajan, J. Crosby, H. M. Murray, S. L. Booten, J. Hsiao, A. Soriano, T. Machemer, P. Cauntay, S. A. Burel, S. F. Murray, H. Gaus, M. J. Graham, E. E. Swayze and P. P. Seth, *J. Med. Chem.*, 2016, **59**, 2718–2733.
245. K. Garber, *Nat. Biotechnol.*, 2016, **34**, 1213–1214.
246. Regulus Press Release, in http://ir.regulusrx.com/news-releases/news-release-details/regulus-announces-pipeline-updates-and-advancements, Accessed March 2018.
247. S. Kakiuchi-Kiyota, P. H. Koza-Taylor, S. R. Mantena, L. F. Nelms, A. E. Enayetallah, B. D. Hollingshead, A. D. Burdick, L. A. Reed, J. A. Warneke, L. O. Whiteley, A. M. Ryan and N. Mathialagan, *Toxicol. Sci.*, 2014, **138**, 234–248.
248. S. Kakiuchi-Kiyota, L. O. Whiteley, A. M. Ryan and N. Mathialagan, *Nucleic Acid Ther.*, 2016, **26**, 93–101.
249. W. Shen, C. L. De Hoyos, H. Sun, T. A. Vickers, X. H. Liang and S. T. Crooke, *Nucleic Acids Res.*, 2018, **46**, 2204–2217.
250. S. Sewing, F. Boess, A. Moisan, C. Bertinetti-Lapatki, T. Minz, M. Hedtjaern, Y. Tessier, F. Schuler, T. Singer and A. B. Roth, *PLoS One*, 2016, **11**, e0159431.
251. M. P. Carver, J. S. Charleston, C. Shanks, J. Zhang, M. Mense, A. K. Sharma, H. Kaur and P. Sazani, *J. Neuromuscular Dis.*, 2016, **3**, 381–393.
252. P. Sazani, K. P. Ness, D. L. Weller, D. Poage, K. Nelson and A. S. Shrewsbury, *Int. J. Toxicol.*, 2011, **30**, 322–333.
253. P. Sazani, K. P. Ness, D. L. Weller, D. W. Poage, K. Palyada and S. B. Shrewsbury, *Int. J. Toxicol.*, 2011, **30**, 313–321.
254. S. P. Henry, M. Johnson, T. A. Zanardi, R. Fey, D. Auyeung, P. B. Lappin and A. A. Levin, *Toxicology*, 2012, **301**, 13–20.
255. N. Goemans, E. Mercuri, E. Belousova, H. Komaki, A. Dubrovsky, C. M. McDonald, J. E. Kraus, A. Lourbakos, Z. Lin, G. Campion, S. X. Wang, C. Campbell and DEMAND III Study Group, *Neuromuscular Disord.*, 2018, **28**, 4–15.

256. C. Den Besten, unpublished data.
257. L. van Meer, M. van Dongen, M. Moerland, M. de Kam, A. Cohen and J. Burggraaf, *Pharmacol. Res. Perspect.*, 2017, **5**, e00292.
258. K. S. Frazier, J. A. Engelhardt, P. Fant, S. Guionaud, S. P. Henry, M. W. Leach, C. Louden, M. S. Scicchitano, J. L. Weaver, T. S. Zabka and Society of Toxicologic Pathology Vascular Injury Working Group, *Toxicol. Pathol.*, 2015, **43**, 915–934.
259. R. Stanton, S. Sciabola, C. Salatto, Y. Weng, D. Moshinsky, J. Little, E. Walters, J. Kreeger, D. DiMattia, T. Chen, T. Clark, M. Liu, J. Qian, M. Roy and R. Dullea, *Nucleic Acid Ther.*, 2012, **22**, 344–359.
260. E. P. van Poelgeest, R. M. Swart, M. G. Betjes, M. Moerland, J. J. Weening, Y. Tessier, M. R. Hodges, A. A. Levin and J. Burggraaf, *Am. J. Kidney Dis.*, 2013, **62**, 796–800.
261. A. Moisan, M. Gubler, J. D. Zhang, Y. Tessier, K. Dumong Erichsen, S. Sewing, R. Gerard, B. Avignon, S. Huber, F. Benmansour, X. Chen, R. Villasenor, A. Braendli-Baiocco, M. Festag, A. Maunz, T. Singer, F. Schuler and A. B. Roth, *Mol. Ther. Nucleic Acids*, 2017, **6**, 89–105.
262. K. S. Frazier, C. Sobry, V. Derr, M. J. Adams, C. D. Besten, S. De Kimpe, I. Francis, T. L. Gales, R. Haworth, S. R. Maguire, R. C. Mirabile, D. Mullins, B. Palate, Y. P. Doorten, J. E. Ridings, M. S. Scicchitano, J. Silvano and J. Woodfine, *Toxicol. Pathol.*, 2014, **42**, 923–935.
263. E. W. Alton, H. A. Boushey, H. Garn, F. H. Green, M. Hodges, R. J. Martin, R. D. Murdoch, H. Renz, S. B. Shrewsbury, R. Seguin, G. Johnson, J. D. Parry, J. Tepper, P. Renzi, J. Cavagnaro and N. Ferrari, *Nucleic Acid Ther.*, 2012, **22**, 246–254.
264. C. L. Berman, S. A. Barros, S. M. Galloway, P. Kasper, F. B. Oleson, C. C. Priestley, K. S. Sweder, M. J. Schlosser and Z. Sobol, *Nucleic Acid Ther.*, 2016, **26**, 73–85.
265. C. L. Berman, K. Cannon, Y. Cui, D. J. Kornbrust, A. Lagrutta, S. Z. Sun, J. Tepper, G. Waldron and H. S. Younis, *Nucleic Acid Ther.*, 2014, **24**, 291–301.
266. D. Capaldi, A. Teasdale, S. Henry, N. Akhtar, C. den Besten, S. Gao-Sheridan, M. Kretschmer, N. Sharpe, B. Andrews, B. Burm and J. Foy, *Nucleic Acid Ther.*, 2017, **27**, 309–322.
267. J. Cavagnaro, C. Berman, D. Kornbrust, T. White, S. Campion and S. Henry, *Nucleic Acid Ther.*, 2014, **24**, 313–325.
268. T. W. Kim, K. S. Kim, J. W. Seo, S. Y. Park and S. P. Henry, *J. Pharmacol. Toxicol. Methods*, 2014, **69**, 49–60.
269. R. Z. Yu, R. Gunawan, R. S. Geary, S. G. Hughes, S. P. Henry and Y. Wang, *Nucleic Acid Ther.*, 2017, **27**, 285–294.
270. FDA, https://www.fda.gov/downloads/Drugs/GuidanceCompliance-RegulatoryInformation/Guidances/UCM079266.pdf, Accessed March 2018.
271. FDA, https://www.fda.gov/downloads/Drugs/GuidanceComplianceRegulatoryInformation/Guidances/UCM581965.pdf, Accessed March 2018.

272. EMA, http://www.ema.europa.eu/docs/en_GB/document_library/Scientific_guideline/2009/09/WC500003149.pdf, Accessed in March 2018.
273. H. S. Younis, M. Templin, L. O. Whitely, D. Kornbrust, T.-W. Kim and S. P. Henry, in *A Comprehensive Guide to Toxicology in Preclinical Drug Development*, Academic Press, 2013, pp. 647–664.
274. FDA, https://www.fda.gov/downloads/Drugs/GuidanceComplianceRegulatoryInformation/Guidances/ucm079247.pdf, Accessed March 2018.
275. EMA, http://www.ema.europa.eu/docs/en_GB/document_library/Scientific_guideline/2015/11/WC500197321.pdf, Accessed March 2018.

Subject Index

absent in melanoma 2 (AIM2)-like receptors (ALRs), 307, 308
agonists
preclinical studies, 314–315
of TLR3, 310–311
TLR 7 and TLR 8, 311–315
of TLR9, 315–322
alternative splicing, 106–108
Alzheimer's disease (AD), 196–197
Angelman's syndrome (AS), 195
anti-HIV-1 RNAi therapeutics
clinical development of, 282–283
combinatorial RNAi, 275–276
effectors, *in vivo* delivery, 276–281
HIV-1 viral genome, targeting, 268–270
host factors, targeting, 270–272
potential targets for, 268–272
rational design and chemical modifications, 274–275
in vivo RNAi efficacy, barriers, 272–274
anti-HIV-1 shRNAs, *ex vivo* delivery, 281–282
antimiR medicinal chemistry development, 154–162
antagomirs, 157
constrained ethyl derivatives, 159–161
2′-fluoro-modified antimiRs, 158–159
locked nucleic acids (LNAs), 159–161
2′-O-methoxyethyl antimiRs, 158
2′-O-methyl-modified antimiRs, 155–157
morpholino oligonucleotides, 162
peptide nucleic acids (PNA), 161–162
antimiRs
antimiR-17, ADPKD, 170
in clinical trials, 167–170
miR-21, Alport Syndrome, 169
miR-155, 170
oligonucleotides, 154–162
targeting miR-122, HCV treatment, 167–169
antisense conjugates, 13–14
antisense oligodeoxynucleotide modifications, 3–12
2′-O-alkyl sugars, 10–11
backbone modifications, 3–9
bridged nucleic acids (BNA), 11–12
heterocyclic bases, 9
non-anionic internucleotide linkages, 7–9
oligoribonucleotides, 10
phosphorothioates (PS), 4–7
sugar modifications, 10–12

antisense oligonucleotides (ASOs)
frataxin activation, case study, 27–29
gene expression control, 29
mechanisms of, 22–29
RNase H, 23–25, 32–52
sense strand of RNA, target, 23
splicing regulation, 25–27
steric blocking, 26
antisense strategies, 2, 138–141
ASOs directed against DNA, 138
ASOs directed against RNA, 139–140
RNA interference, 140–141
antisense technology, 62–73
apolipoproteins (apo), 63
aptamer-based targeting, drugs, 381–384
aptamers, 339–342
active studies, 391–392
clinical applications, 367–394
completed clinical studies, 385–390
DNA *vs.* RNA, 342–343
drugs registered in clinicaltrials.gov, 384–392
non-ocular diseases, 376–381
photo-regulated aptamers, 374–376
preclinical studies of, 369–381
SOMAmers and X-aptamers, 369–373
structures, 369–376
terminated or withdrawn studies, 390–391
thioaptamers, 373–374
in vitro cell-binding, 384
in vivo biodistribution studies, 384
aptamer–target interactions, 340
artificial *trans*-splicing, 141
Asahi Kasei bioprocess, 458
ataxia telangiectasia (AT), 110–111
atherosclerosis, 91–92
autoimmune diseases, 81
autosomal dominant polycystic kidney disease (ADPKD), 170

Becker muscular dystrophy (BMD), 111, 112
blood–brain barrier (BBB), 48, 218, 378

C46 peptide, 283
cell-penetrating peptides (CPP), 216
cell-specific delivery, 49–50
cell-surface proteins
cell-specific delivery, gapmers, 49–50
interaction with gapmer ASOs, 48–49
chiral PS DNA, 39–40
cholesterol-conjugated siRNAs (Chol–siRNAs), 213
cisplatin, 358
class effect, 6
clinical candidates, RNAi therapeutics, 240–257
fitusiran, 256
givosiran, 255
inclisiran, 254–255
Onpattro™ (Patisiran), 240, 254
revusiran, 256–257
TTRsc02, 256
clustered regularly interspaced palindromic repeats (CRISPR) technology, 400–424
CNS-targeting ligands, 218
cobomarsen, 170
cognate binding site, 70
conjugated modalities, 213–218

conjugated siRNAs
broad functional delivery of, 213–216
clearance kinetics, reducing, 223–224
local delivery of, 219–221
PK properties, enhancing, 223–224
systemic delivery of, 221–222
targeted delivery of, 216–218
target tissue uptake, 223–224
constitutive exons, 106
constitutive splicing, 106
constrained ethyl 2′-4′-bridged nucleic acids (cEt), 42–43
CRISPR–Cas9 protein system, 141–142, 423
CRISPR nucleases, 401–424
cryptic splicing, 108, 109–111
ataxia telangiectasia (AT), 110–111
restoring, 109–111
β-thalassemia, 109–110
cyclic GMP–AMP synthase (cGAS), 309–309
cytomegalovirus (CMV), 5

dengue, 380
disease molecular mechanisms, 128–130
gain-of-function, 130
loss-of-function, 129
distribution, metabolism and pharmacokinetics (DMPK), 479–488
advanced formulation approaches, 486
anti-drug antibodies, 481
delivery strategies, 485–488
excretion, 485
local delivery, 486
metabolism, 484–485
plasma and tissue pharmacokinetics, 479–481
productive uptake, 482–484
properties of ONDs, 479–485
targeted delivery, 487–488
tissue distribution, 481–482
DNA, 22–24, 28, 32, 33, 35, 36
DNA–RNA heteroduplex, 35, 39
docetaxel (DTX), 358
double-stranded DNA breaks (DSB), 198, 199
Dravet's syndrome (DS), 194–195
duchenne muscular dystrophy (DMD), 90, 111–114
Duchenne muscular dystrophy (DMD), 27
dystrophia myotonica protein kinase, 71
dystrophic epidermolysis bullosa, 115

enhancer RNA (eRNA), 187
epigenome editing
base editors, 417–419
chromatin status modulation, 416–417
CRISPR system, 414–416
transcription modulation, 414–416
epithelial to mesenchymal transition (EMT), 200
eteplirsen, 27, 114
exondys 51, 456
exon inclusion, 116–118
glycogen storage disease type II or Pompe disease, 118
spinal muscular atrophy (SMA), 116–118
exon skipping, 112–114
expanded repeats, 130–134
antisense strategies, 138–141
imperfect hairpin, 131–133
quadruplexes, 133
R-loops, 131
small molecules targeting, 136–137
structures formed by, 131–133

therapeutic strategies, 135–136
therapeutic targeting, structural implications, 134
triple helix, 131

fitusiran, 256
2′-fluoro-arabino nucleic acid (FANA), 40
5-fluorouracil (5-FU), 382
fragile X mental retardation 1 (FMR1), 197
Friedreich's ataxia (FA), 27

gapmer ASOs
cEt gapmers, 47
design of, 45–47
duplexes, 47
interaction with cell-surface proteins, 48–49
interaction with plasma proteins, 48
LNA gapmers, 46
MOE, 45–46
gapmers, 12–13, 25, 32–33
gap modifications, structure–activity relationships, 36–41
2′-fluoro-arabino nucleic acid (FANA), 40
boranophosphonate DNA, 40
chiral PS DNA, 39–40
DNA-like modifications, 41
methyl phosphonates, 40
phosphoramidates, 40
phosphorothioate (PS) DNA, 38–39
phosphotriesters, 40
GEM91, 6
Gen 2.0 ASOs, 70
generation 2.5 ASO chemistry, 70
gene silencing effects, 224
genome editing
biomedical research applications, 419–420
Cas9 fusion, 413
cell cycle modifications, 413–414
clinical trials, 422–423
delivery challenge, 420–422
DNA homology template donors, 412–413
DNA repair pathways regulation, 410
double-strand break (DSB) breaks, Cas9, 408–410
end-joining pathways, 409
HDR-based CRISPR strategies, 410–414
HDR-mediated DSB repair, 412
homology-directed repair, 409–410
NHEJ-mediated repair, 410–412
precision challenge, 422
genome engineering, 400–424
givosiran, 255

hepatitis C virus (HCV), 167–169
high-exposure organs, 505–511
kidney toxicity, 509–511
liver toxicity, 505–509
HIV-1, 92–93
human RNase H1, 33–36
biochemistry of, 33–34
structural biology of, 34–36
huntingtin *(HTT)* gene, 119
huntingtin transcripts, 41
Huntington disease (HD), 119–120
Huntington's disease (HD), 220
Hutchinson–Gilford Progeria syndrome, 118–119
hybridization-dependent toxicities, 490–493
off-target pharmacology, 491–493
on-target safety and exaggerated pharmacology, 490–491

hybridization-independent toxicities, 493–511
high-exposure organs, 505–511
immune-mediated effects, 496–505
transient protein binding, 494–495
hydroxychloroquine (HQ), 81, 82

imetelstat, 456
imiquimod, 312
immune-mediated effects, 496–505
pro-inflammatory effects, 496–500
thrombocytopenia (TCP), 500–505
immune-stimulatory toxicities, 50–51
IMO-3100, 84
IMO-8400, 84
IMO-9200, 84
inclisiran, 254–255
inflammatory bowel disease (IBD), 92, 202
innate immune receptors, 14–15
inotersen, 46
interleukin 4 (IL4), 72
intracerebroventricular (ICV) injections, 220
ionizable lipid nanoparticles (iLNPs), 238–239

Japanese Encephalitis (JE) Virus (JEV), 379

kidney toxicity, 509–511
glomerular effects, 511
tubular effects, 509–510

lead TLR antagonist candidates
clinical development of, 93–95
moderate-to-severe plaque psoriasis, 93–94
Waldenström's Macroglobulinemia (WM), 94–95
lipid-conjugated oligonucleotides, 213
liver-targeted RNAi therapeutics, 233–257
chemistry, role of, 234–237
clinical candidates, 240–257
liver targeting, 64–67
beginning, 64–66
early liver target programs, 65
innovations in, 67–70
pipeline, broadening, 66–67
liver toxicity, 505–509
GalNAc conjugation, 506–507
mechanistic hypotheses, 507–509
toxic sequence motifs, 507
locked nucleic acid (LNA), 11, 42–43, 456
long non-coding RNAs (lncRNAs)
allelic expression regulation, 186
antisense oligonucleotides, 192
biology and functions of, 185–190
cardiology, 201–202
classification of, 190–191
CRISPR–Cas9 system, 193
functional classification, 191
gastroenterology, 202
genomic classification, 191
history of, 182–185
neurology and psychiatry, 194–198
oligonucleotide targeting of, 181–202
oncology, 198–201
perspectives, 202
post-transcriptional processing regulators, 187–189
siRNAs, 192–193
targeting by oligonucleotides, 192–193

therapeutic applications, 194–202
transcription regulators, 186–187
translation regulators, 189–190
low-density lipoprotein–cholesterol (LDL–C), 63

mesoporous silica nanoparticles (MSNs), 382
2′-*O*-methoxyethyl RNA (MOE), 42
methylphosphonate oligodeoxynucleotides (MP-ODNs), 7
microRNA (miRNAs)
antimiR-mediated tailing and trimming of, 163
antimiR-mediated unloading, RISC, 164–165
biogenesis, 151–153
oligonucleotides targeting, 151–171
targeting, chemistries and approaches, 165–167
targeting, RISC, 163–164
targets for drugs, 154
moderate-to-severe plaque psoriasis, 93–94
modified oligodeoxy/oligoribonucleotides, 12–13
multisystemic repeat diseases, 142–143
MyD88 L265P-positive B cell lymphoma, 90–91

N-ethylmaleimide (NEM), 34
neurodegenerative disease, intrathecal administration, 72
NLRP3 inflammasome, 308–309
NOD-like receptors, 308–309
non-anionic internucleotide linkages, 7–9
peptide nucleic acids (PNA), 9
phosphorodiamidate morpholino oligonucleotides (PMO), 8–9
non-ocular diseases, 376–381
Alzheimer's Disease (AD), 377–378
antibacterial applications, 380–381
antiparasitic applications, 381
antiviral applications, 378–380
cardiovascular and cerebrovascular diseases, 376–377
nucleic acid innate immune receptors, 292–300
toll-like receptors, 293–298
nucleic acids sensing
in cytoplasm, 298–299
cytosolic DNA sensors, dsDNA, 298–299
immune responses, MAVS activation, 299
RIG-I and MDA5, dsDNA, 299
nusinersen, 27

oligonucleotide based miRNA targeting, 162–165
oligonucleotide-based TLR antagonists, 82–87
structure–activity relationship studies, 83
oligonucleotide drugs (ONDs), 474, 475–476, 516
OND chemistry, 476
OND design, 476–478
oligonucleotide manufacturing process, 457–468
automated synthesizer, 457–458
cleavage and deprotection, 465–466
four-step synthesis cycle, 464–465
lyophilization process, 468
nucleoside-loaded solid support, 458–459
nucleoside phosphoramidites, 460–462

oligonucleotide manufacturing process (*continued*)
purification process, 466–467
reagent-related impurities, 462–464
starting materials, 458–462
sulfur-transfer reagents, 462
oligonucleotide synthesis, developments, 2–3
one gene–one protein concept, 183
Onpattro™ (Patisiran), 240, 254
organic solvent nanofiltration (OSN), 471

pegaptanib, 392
pegnivacogin, 360
peptide nucleic acids (PNA), 9
peripheral blood mononuclear cells (PBMCs), 85–87
phosphorodiamidate morpholino oligomers (PMOs), 27
phosphorodiamidate morpholino oligonucleotides, 8–9
phosphorothioate (PS) DNA, 38–39
post-transcriptional processing regulators, 187–189
competing endogenous RNAs (ceRNAs), 188
endogenous siRNA, 189
interactions with miRs, 189
miRNA generation, 189
mRNA stability, 188
polyadenylation, 188
splicing, 188
pro-inflammatory effects, 496–500
immune stimulation, manifestations, 496–498
immunogenicity – anti-drug antibodies, 500
mitigating, 499–500
pro-inflammatory mechanisms, 498–499
prostate cancer imaging, 339–360
prostate-specific membrane antigen (PSMA) aptamers, 343–348
applications, 348–359
bispecific PSMA aptamer conjugates, 356
computed tomography (CT) applications, 349
DNA, 347–348
functionalized nanoparticles, 357–359
magnetic resonance imaging (MRI) applications, 349–350
optical imaging applications, 350–351
protein toxin conjugates, 356
PSMA aptamer–CRISPR conjugates, 355–356
radio-imaging applications, 348–349
RNA, 345–347
RNAi conjugates, 351–355
small-molecule conjugates, 357
targeted delivery applications, 351–359
therapeutic inhibitors, 351
protein binding control, 47–51
cell-specific delivery, 49–50
gapmer-ASOs, cell-surface proteins interaction, 48–49
gapmer ASOs interaction, 48
intracellular distribution, optimizing, 51
TLR receptors, avoiding, 50–51
protein effectors, 141–142
protein frataxin (FXN), 27
psoriasis, 87–88
psychiatric diseases, 197–198

repeat-expanded transcripts, structures, 133–134
repetitive sequences, expanded
disease molecular mechanisms, 128–130
human disease and, 127–130
repeat instability, 127–128
restenosis, 91–92
retinoic acid-inducible gene 1-like (RIG-I-like) receptors, 307–308
revusiran, 256–257
rheumatoid arthritis (RA), 89–90
ribozyme-mediated genetic repair, 439–445
group I intron ribozyme-mediated *trans*-splicing, 441–443
retro-homing group II introns, genetic disorders, 445
spliceosome-mediated RNA *trans*-splicing (SMaRT), 440–441
twin ribozyme-mediated RNA repair, 443–445
ribozymes
ribozyme-mediated genetic repair, 439–445
spliceosome-mediated RNA *trans*-splicing (SMaRT), 440–441
therapeutic potential, 434–447
trans-cleaving, 435–439
RNA, 29, 35, 36
RNA–DNA hybrids, 23
RNAi-based anti-HIV therapeutics, 266–285
RNAi-based therapeutics, conjugation chemistry, 224–226
RNA-induced silencing complex (RISC), 207, 212, 267
RNA interference (RNAi) conjugates, 351–355
RNase H, 23–25, 29
RNase H-activating antisense oligonucleotides, 32–52
gapmer ASOs, 45–47
gap modifications, structure-activity relationships, 36–41
protein binding control, 47–51
wing modifications, SAR, 41–44
RNA splicing, 103–106
sequence motifs, 104–106
splicing machinery, 104
Rous sarcoma virus (RSV) replication, 1, 2

serinol nucleic acids (SNAs), 165
small interfering RNAs (siRNAs), 189, 192–193
clearance effects, 207–212
conjugate-mediated delivery of, 207–212
distribution effects, 207–212
GalNAc conjugates, discovery, 240
ionizable lipid nanoparticles (iLNPs), 238–239
lipid-conjugated, RNAi therapeutics, 239
liver-specific delivery of, 237–240
sustained gene silencing, 207–212
Soibam, 187
spinal muscular atrophy (SMA), 27, 195–196
splice-modulating ASOs, 120–122
ASO delivery, 121
future perspective, 121–122
personalized medicine development, 120–121
spliceosome-mediated RNA *trans*-splicing (SMaRT), 440–441
splicing modulation, 103–122

stimulator of IFN genes (STING) pathway, 309
structure–activity relationship (SAR)
gap modifications, 36–41
of wing modifications, 41–44
survival motor neuron 2-(SMN2), 71
sustained gene silencing, 207–212
systematic evolution of ligands by exponential enrichment (SELEX) process, 339–342, 359, 367
systemic lupus erythematosus (SLE), 88–89

targeted *versus* broad functional delivery, 213–218
β-thalassemia, 109–110
therapeutic exon skipping options
ASOs and chemical modifications, 108–109
cryptic splicing, restoring, 109–111
exon inclusion, 116–118
frame restoration, reading, 111–115
less protein or non-toxic protein, 118–120
therapeutic oligonucleotides
accelerated timelines, opportunity, 515
analytical protocols, 468–470
biodistribution, modulating, 213–218
chemical modifications, 454–457
chemistry and design considerations, 475–478
delivery approaches, 478
distribution, metabolism and pharmacokinetics (DMPK), 479–488
endosomal barrier, overcoming, 218–219
first-generation backbone modifications, 454
hybridization-dependent toxicities, 490–493
hybridization-independent toxicities, 493–511
large-scale automated synthesis, 453–471
oligonucleotide chemistries and mode of action (MOA), 475–478
oligonucleotide manufacturing process, 457–468
OND chemistry, 476
OND design, 476–478
phosphorothioate (PS) internucleotide linkage, 454
precision medicine, 515
productive delivery of, 218–219
regulatory considerations, preclinical development, 511–514
second-generation sugar modifications, 455–457
synthesis yield and product purity, 470
toxicity, class profile, 488–489
thrombocytopenia (TCP), 500–505
TLR9 agonists, 315–322
asthma and allergies, treatment, 324–325
cancer, clinical trials, 326–327
cancer, preclinical studies, 322–324
clinical development of, 325–327
hepatitis C patients, clinical trials, 325–326
role of 5′-end, CpG motif, 317–319
secondary structures, 320–322
synthetic immune-stimulatory motifs, 319–320

therapeutic applications of, 322–325
as vaccine adjuvants, 325
TLR antagonists, 83–84
clinical candidates, inhibitory activity, 85–87
in disease models, 87–93
duchenne muscular dystrophy (DMD), 90
HIV-1, 92–93
inflammatory bowel disease (IBD), 92
inhibitory activity of, 85
MyD88 L265P-positive B cell lymphoma, 90–91
psoriasis, 87–88
restenosis and atherosclerosis, 91–92
rheumatoid arthritis (RA), 89–90
systemic lupus erythematosus (SLE), 88–89
toll-like receptors (TLRs), 80, 208, 293–298, 309–310
chaperones, NA-sensing TLRs, 295–296
NA-sensing TLRs, proteolytic cleavage, 296
ssRNA and guanosine or uridine, 293–294
synthetic agonists of, 306–328
TLR3 recognizes dsRNA, 293
TLR7 and TLR8, 293–294
TLR7 and TLR9, trafficking, 297–298
TLR7 and TLR9 responses, 296
TLR9 recognizes CpG-DNA, 294–295
type I interferon, pDCs, 297–298
Unc93B1, 296
toxic repeats, targeting, 126–143
transcription activator-like effector nucleases (TALENs), 141–142
transient protein binding, 494–495
coagulation cascade inhibition, 494
complement system activation, 494–495
TTRsc02, 256
twin ribozyme-mediated RNA repair, 443–445

ubiquitin-protein ligase E3A (UBE3A), 195

volanosorsen, 46

Waldenström's Macroglobulinemia (WM), 94–95
wing modifications, SAR, 41–44
α-L-LNA and analogs, 43
constrained ethyl 2′-4′-bridged nucleic acids (cEt), 42–43
F-CeNA, 43–44
F-HNA, 43–44
locked nucleic acids (LNA), 42–43
2′-*O*-methoxyethyl RNA (MOE), 42
phosphorodiamidate linked morpholinos (PMOs), 44
tricyclo DNA (tcDNA), 44

zinc finger nucleases (ZFNs), 141–142